PROCEEDINGS OF THE THIRD U.S. NATIONAL CONFERENCE ON EARTHQUAKE ENGINEERING

August 24–28, 1986
Charleston, South Carolina

PROCEEDINGS OF THE THIRD U.S. NATIONAL CONFERENCE ON EARTHQUAKE ENGINEERING

August 24–28, 1986
Charleston, South Carolina

VOLUME II

5. Response of Structures, Seismic Analysis, and Spectra
6. Tests on Structures and Components, Experimental Methods
7. Nonstructural Systems and Building Components

The Earthquake Engineering Research Institute assumes no responsibility for the statements made in the papers of these Proceedings. Any opinions expressed are those of the individual authors. Interested readers should contact the individual authors for necessary clarification.

The material contained herein reflects reproduction and reduction from original materials submitted by the individual authors. The variable quality of reproduction reflects differences in the copy provided.

©1986 by Earthquake Engineering Research Institute

All rights reserved. No part of this book may be reproduced, in any form or by any means, without permission in writing from:

Earthquake Engineering Research Institute
6431 Fairmount Avenue, Suite 7
El Cerrito, CA 94530-3624
(415) 525-3668

Printed in the United States of America
1 2 3 4 5 6 7 8 9 10
ISBN# 0-943198-07-0

ORGANIZATION OF PROCEEDINGS VOLUMES

Preface

VOLUME I

1. Seismic Hazard and Risk	1
2. Ground Motion and Seismicity	331
3. Geotechnical, Soil Stability, Soil-Structure Interaction, and Foundations	475
4. Special Structures, Critical Facilities, and Dams	693

VOLUME II

5. Response of Structures, Seismic Analysis, and Spectra	819
6. Tests on Structures and Components, Experimental Methods	1187
7. Nonstructural Systems and Building Components	1597

VOLUME III

8. Seismic Structural Design—Seismic Codes and Standards	1635
9. Repair, Strengthening, Isolation, and Retrofit	1895
10. Lifelines—Utilities, Transportation Systems, Telecommunications, and Other Facilities	2045
11. Urban Design, Socioeconomic Issues, Public Policy and Preparedness	2329

Index of Proceedings Authors

Typical damage to many of the stores and store fronts along East Bay Street.

SPONSORING ORGANIZATION

Earthquake Engineering Research Institute

HOSTING ORGANIZATIONS

South Carolina Seismic Safety Consortium
Oak Ridge National Laboratory
in cooperation with:

American Concrete Institute
American Institute of Architects
American Institute of Steel Construction
American Nuclear Society
American Society of Civil Engineers, South Carolina Section
American Society of Mechanical Engineers
Applied Technology Council
Association of Engineering Geologists
Baptist College at Charleston
Building Seismic Safety Council
Duke Power
Electric Power Research Institute
Federal Emergency Management Agency
Geological Society of America
National Bureau of Standards
National Science Foundation
National Society of Professional Engineers
Nuclear Regulatory Commission
Seismological Society of America
Southeastern U.S. Seismic Safety Consortium
Southern Eastern Electric Exchange
Technology Transfer and Development Council
Tennessee Valley Authority
The Citadel
United States Army Corps of Engineers
United States Department of Energy
United States Geological Survey
Veterans Administration

EARTHQUAKE ENGINEERING RESEARCH INSTITUTE
6431 Fairmount Avenue, Suite 7
El Cerrito, California 94530-3624

President
    Robert V. Whitman
President-Elect
    Frank E. McClure
Vice-President
    Robert A. Olson
Secretary/Treasurer
    Henry J. Lagorio
Board of Directors
    Clarence R. Allen
    Neil M. Hawkins
    Henry J. Lagorio
    Frank E. McClure
    Robert A. Olson
    Anne E. Stevens
    Delbert B. Ward
    Robert V. Whitman
    Loring A. Wyllie, Jr.
Association Director
    Susan B. Newman
Technical Director
    Roger E. Scholl
Publications Manager
    Gail H. Shea

The Earthquake Engineering Research Institute is a U.S. national professional society devoted to finding better ways to protect life and property from the effects of earthquakes. Its objectives are: (1) advancement of the science and practice of earthquake engineering; and (2) the solution of national earthquake engineering problems.

ORGANIZING COMMITTEES

PROGRAM COMMITTEE

CONFERENCE CHAIRMAN

James E. Beavers   Martin Marietta Energy Systems Inc., Oak Ridge, Tennessee

LOCAL ARRANGEMENTS CHAIRMAN

Charles Lindbergh   The Citadel, Charleston, South Carolina

ADVISORY COMMITTEE

M. S. Agbabian	University of Southern California, Los Angeles, California
J. B. Bagwell	Baptist College at Charleston, Charleston, South Carolina
L. L. Beratan	Nuclear Regulatory Commission, Washington, D.C.
B. A. Bolt	University of California at Berkeley, Berkeley, California
H. J. Degenkolb	H. J. Degenkolb Associates, San Francisco, California
R. G. Domer	Tennessee Valley Authority, Knoxville, Tennessee
A. J. Eggenberger	National Science Foundation, Washington, D.C.
W. J. Hall	University of Illinois, Urbana, Illinois
R. D. Hanson	University of Michigan, Ann Arbor, Michigan
W. W. Hays	United States Geological Survey, Reston, Virginia
G. W. Houser	California Institute of Technology, Pasadena, California
R. P. Kennedy	Structural Mechanics Associates, Newport Beach, California
R. W. Krimm	Federal Emergency Management Agency, Washington, D.C.
H. J. Lagorio	University of California at Berkeley, Berkeley, California
E. V. Leyendecker	National Bureau of Standards, Gaithersburg, Maryland
R. D. McConnell	Veterans Administration, Washington, D.C.
O. W. Nuttli	St. Louis University, St. Louis, Missouri
R. A. Parmelee	Alfred Benesch & Co., Chicago, Illinois
H. C. Shah	Stanford University, Stanford, California
R. L. Sharpe	Engineering Decision Analysis Co., San Jose, California
J. C. Stepp	Electric Power Research Institute, Palo Alto, California
C. C. Thiel	Stanford University, Stanford, California

TECHNICAL PROGRAM COMMITTEE

CHAIRMAN   W. J. Hall
            University of Illinois, Urbana, Illinois

CO-CHAIRMEN D. A. Foutch and K. D. Hjelmstad
            University of Illinois, Urbana, Illinois

COMMITTEE

R. D. Borcherdt	United States Geological Survey, Menlo Park, California
M. A. Cassaro	University of Louisville, Louisville, Kentucky
R. E. Elling	Clemson University, Clemson, South Carolina
M. P. Gaus	National Science Foundation, Washington, D.C.
A. K. Gupta	North Carolina State University, Raleigh, North Carolina
R. B. Herrmann	St. Louis University, St. Louis, Missouri
J. R. Hill	United States Department of Energy, Washington, D.C.
R. J. Hunt	Tennessee Valley Authority, Knoxville, Tennessee
A. C. Johnston	Tennessee Earthquake Information Center, Memphis, Tennessee
E. Jones	Central U.S. Earthquake Consortium, Marion, Illinois
L. F. Kahn	Georgia Institute of Technology, Atlanta, Georgia
J. Kariotis	Kariotis & Associates, South Pasadena, California
G. C. Lee	State University of New York at Buffalo, Buffalo, New York
J. Loss	University of Maryland, College Park, Maryland
W. F. Marcuson, III	USAE Waterways Experiment Station, Vicksburg, Mississippi
J. P. Moehle	University of California at Berkeley, Berkeley, California
A. J. Murphy	Nuclear Regulatory Commission, Washington, D.C.
J. M. Nau	North Carolina State University, Raleigh, North Carolina
J. M. Nigg	Arizona State University, Tempe, Arizona
R. I. Palm	University of Colorado, Boulder, Colorado
J. B. Radziminski	University of South Carolina, Columbia, South Carolina
A. J. Schiff	Stanford University, Stanford, California
L. T. Youd	Brigham Young University, Provo, Utah

ACTIVITY COMMITTEES

CONFERENCE COORDINATOR	Norma F. Cardwell
EXHIBITS	Kenneth E. Fricke
	Gus A. Aramayo
ASSISTANT TO THE CHAIRMAN	Barbara A. Luttrell
TECHNICAL PROGRAM ASSISTANTS	Susan A. Warsaw
	Truth A. Callison
	Mary K. Pearson
SPOUSE/GUEST PROGRAM	Russell L. Lindbergh
HOSPITALITY	William B. Wright
CONFERENCE BUDGET	David H. Williamson
SOCIAL FUNCTIONS	Charles T. McLoughlin
PUBLICATIONS/PRINTING	Vivian A. Jacobs
	W. Dale Jones
PUBLIC RELATIONS	W. Alex Dallis, Jr.
	Joanne T. Dennett
REGISTRATION	E. H. (Chip) Stehmeyer, Jr.
MUSEUM & BOAT TOURS	I. D. Smith

PUBLIC AWARENESS ACTIVITIES

TOWN MEETING	Maurice R. Harlan
4-8 GRADE EARTHQUAKE POSTER/ART CONTEST	Joyce B. Bagwell
MIDDLE AND HIGH SCHOOL EARTHQUAKE ESSAY CONTEST	W. Alex Dallis, Jr.
MIDDLE AND HIGH SCHOOL EARTHQUAKE TOOTHPICK BUILDING CONTEST	Michael Woo
COLLEGE EARTHQUAKE PAPER CONTEST	Rudolph E. Elling
COLLEGE EARTHQUAKE BALSA WOOD BUILDING CONTEST	Russell H. Stout, Jr.
SPEAKERS BUREAU & ETV PRODUCTIONS	James B. Radziminski
REGIONAL STATE CHAIRMEN FOR LOCAL ACTIVITIES	Alabama—David E. Elton
	Florida—Winfred O. Carter
	Georgia—Lawrence F. Kahn
	North Carolina—Clay Sams
	Tennessee—Neil E. Johnson
	South Carolina—Charles Lindbergh
	Virginia—Gerald W. Clough

PURPOSE

The purpose of the Conference is to:
- Discuss state-of-the-art knowledge in earthquake hazards and mitigation
- Address planning and implementation needs to mitigate future seismic risks

PAST U.S. NATIONAL CONFERENCES

First—June 18–20, 1975
University of Michigan

Second—August 22–24, 1979
Stanford University

FINANCIAL SPONSORS

EERI is greatly appreciative of the financial support provided by government, industry, and consultants. To ensure recognition of each financial contributor, a complete list will be provided in the *Post-Conference Proceedings*.

TABLE OF CONTENTS

VOLUME I

Organization of Proceedings Volumes v
Sponsoring Organization vii
Hosting Organizations vii
Organizing Committees ix
Preface xxxiii

1. SEISMIC HAZARD AND RISK

Evidence for Three Moderate to Large Prehistoric
Holocene Earthquakes Near Charleston, S.C. 3
    R. E. Weems, S. F. Obermeier, M. J. Pavich,
    G. S. Gohn, M. Rubin, R. L. Phipps, and
    R. B. Jacobson

Seismotectonics of the Charleston Region 15
    P. Talwani

The Charleston Earthquake Hypotheses—A Classification
by Fundamental Tectonic Processes 25
    L. T. Long, R. E. White, and J. Dwyer

An Evaluation of the Quality of the 1886 Charleston, South Carolina
Intensity Data Using Indicator Functions 33
    J. R. Carr and C. E. Glass

Strike-Slip on Reactivated Triassic(?) Basin Boundary
Fault Zones as Sources of Earthquakes Near Charleston, S.C. 43
   J. C. Behrendt and A. Yuan

Earthquake Recurrence Rates and Probability Estimates for the Occurrence
of Significant Seismic Activity in the Charleston Area: The Next 100 Years 55
   D. Amick and P. Talwani

Comparison of Parameters for Charleston and
Other Seismic Zones in Eastern North America 65
   H. Acharya

NRC Participation in Studies of the Charleston Seismicity 77
   E. G. Zurflueh

Magnetotelluric Soundings in the Charleston, South Carolina Area 83
   C. T. Young, M. R. Kitchen, J. C. Rogers,
   and J. C. Mareschal

Overview and Summary of Results for the Project: Seismic Hazard
Characterization for the Eastern United States 93
   J. B. Savy, D. L. Bernreuter, and
   R. W. Mensing

Three-Dimensional Modeling of Crustal Stress
and the Seismicity of the East Coast 105
   J. C. Mareschal, J. Kuang, and L. T. Long

Source Scaling Relations of Large Eastern North American
Earthquakes and Implications for Ground Motions 117
   P. Somerville

A Feature of the 3 March 1985 Chile Earthquake—
Possible Terrain Amplification 125
   M. Celebi

On Nonstationary Stochastic Models for Earthquakes 137
   E. Safak and D. M. Boore

Time-Dependent Seismic Hazard Estimates From the New Madrid Fault Zone 149
   A. S. Kiremidjian and S. Suzuki

On Use of Fuzzy Sets To Characterize Earthquake
Effects Based on Building Damage Records 161
   C. H. Juang and D. J. Elton

A Comparison of Seismotectonics and Seismic Hazard
to Either Side of the North Atlantic 173
   R. M. Wood and G. Woo

An Evaluation of Seismic Hazard in California 185
   S. G. Wesnousky

Holocene and Late Pleistocene(?) Earthquake-Induced
Sand Blows in Coastal South Carolina 197
                  S. F. Obermeier, R. B. Jacobson,
                  D. S. Powars, R. E. Weems, D. C. Hallbick,
                  G. S. Gohn, and H. W. Markewich

Seismicity and Seismic Hazards in the New York City Area 209
                  G. J. P. Nordenson and C. T. Statton

Evaluation of Seismic Hazard With Fuzzy Algorithm 221
                  M. Lamarre and W. Dong

Assessment of Earthquake Hazards in New Mexico 233
                  G. D. Johnpeer, D. W. Love,
                  and M. Hemingway

A Methodology to Correct for the Effect of the
Local Site's Characteristics in Seismic Hazard Analyses 245
                  D. L. Bernreuter, J. C. Chen, and J. B. Savy

Seismic Hazard Analysis in the North-Central United States 257
                  D. J. Becker and J. E. Topi

Seismic Hazard Analysis Including Source Directivity Effect 269
                  R. Araya and A. Der Kiureghian

Modeling Spatial Dependence in Earthquake Occurrence 281
                  T. Anagnos

Sensitivity of Earthquake Hazard in the Central
and Eastern U.S. to Alternative Input Interpretations 293
                  R. K. McGuire and J. C. Stepp

Capturing Uncertainty in Probabilistic Seismic Hazard
Assessments Within Intraplate Tectonic Environments 301
                  K. J. Coppersmith and R. R. Youngs

Probabilistic Seismic Hazard Studies for a Site in Northern Switzerland 313
                  E. Berger and V. Langer

2. GROUND MOTION AND SEISMICITY

Characteristics of the Ground Motions of the
Charleston, South Carolina Earthquake 327
                  E. A. Marciano and D. J. Elton

Ground Motion Amplification Studies for Sites in the Charleston Area 333
                  P. C. Rizzo, P. F. O'Hara, and E. G. Zullo

The Engineering Characteristics of the 1982 Miramichi Earthquake Records 345
    A. C. Heidebrecht and N. Naumoski

Site Amplification of Earthquake Ground Motion 357
    W. W. Hays

Variation of Earthquake Ground Motion With Depth 369
    M. S. Power, C. Y. Chang, and I. M. Idriss

Estimation of Spatial Correlation of Seismic Waves
Using the SMART-1 Strong Motion Array Data 381
    C. H. Loh

Effects of Recorder Nonsynchronization on Interpretation of
Strong Motion Records at the Meloland Road Overpass 393
    S. D. Werner and J. L. Beck

Response Characteristics of Soil Deposits 405
    *C. P. Aubeny, J. L. Von Thun,
    and N. Y. Chang*

Scattering of SH Waves by an Alluvial Valley
of Arbitrary Shape: A Boundary Integral Approach 417
    M. Dravinski, H. Eshraghi, and F. J. Sabina

Earthquake Ground Motions for a NPP in Eastern Canada 427
    T. S. Aziz and M. M. Elgohary

EQGEN—A User-Friendly Artificial Earthquake Simulation Program 439
    *N. Y. Chang, M. J. Huang, B. H. Lien,
    and F. K. Chang*

A US-Japan Comparison on Earthquake Disasters
by 1900–1979 Damage Data Statistics 451
    H. Oshashi and Y. V. Ohta

Statistical Analysis Based on the Newly
Completed Seismic Intensity Map in Japan 463
    *K. Muto, T. Usami, O. Hamamatsu,
    T. Sugano, M. Miyamura, and M. Kamata*

3. GEOTECHNICAL, SOIL STABILITY, SOIL-STRUCTURE INTERACTION, AND FOUNDATIONS

Seismic Stability of Dredged Slopes in the Charleston Area 477
    J. L. Withiam, P. F. O'Hara, and M. J. Biasko

Soil-Structure Interaction at the Waterfront 489
        J. M. Ferritto

Site Period Study for Charleston, S.C. 497
        D. J. Elton and J. R. Martin

Nonlinear Dynamic Analysis of a Centrifuge Model Embankment 505
        W. H. Roth, R. F. Scott, and P. A. Cundall

Evaluation of Seismic Lateral Pile Capacity,
Mark Clark Expressway, Charleston, South Carolina 517
        G. N. Richardson and R. C. Chaney

Soil-Structure Interaction Effects on the
Reliability Evaluation of Reactor Containments 529
        J. Pires, H. Hwang, and M. Reich

Effects of Soil-Structure Interaction on the
Liquefaction Characteristics of a Site 541
        M. M. Zaman and S. M. Mamoon

Dynamic Structure-to-Structure Interaction for Closely Spaced Buildings 553
        H. L. Wong and J. E. Luco

Seismically Induced Settlements: Two Models for New England 565
        C. Soydemir

Risk of Structural Damage in Liquefaction 575
        A. Haldar and S. M. Luettich

Experimental Investigation of Relative Foundation and Free-Field Response 587
        M. R. Somerville, B. B. Redpath,
        R. B. Whorton, and D. Williams

The Nonlinear Seismic Response of Retaining Walls 599
        T. J. Siller, J. A. Coronato, J. Bielak,
        and P. P. Christiano

Seismic Response Analysis of Soil Deposits
by Computer On-Line Testing System—A Hybrid System 611
        Z. Wang, C. K. Shen, X. S. Li, and H. W. Yang

Effects of Site Stratigraphy on Soil-Structure
Interaction Impedance Functions 623
        P. J. Gallagher, N. R. Vaidya, and P. C. Rizzo

Nonstable Rotational Stiffness of a Pile Group 635
        G. M. Norris

Time Domain Response Analysis of Nonlinear
Deep Foundations Subjected to Axial Dynamic Load 647
        T. Nogami, K. Konagai, and J. Otani

Portable Mini-Cone System for Field Liquefaction Studies 659
        B. Sweeney and G. W. Clough

Effect of Soil-Structure Interaction on Damping
and Frequencies of Base-Isolated Structures 671
        M. C. Constantinou and M. C. Kneifati

On the Validity of Modal Superposition
Technique in Soil Structure Interaction Analysis 683
        *N. R. Vaidya, E. Bazan-Zurita,
        and P. C. Rizzo*

4. SPECIAL STRUCTURES, CRITICAL FACILITIES, AND DAMS

Earthquake Evaluation of Existing Earth Dams With Earth Foundations 695
        S. D. Stone and E. L. Spearman

Seismic Stability and Stress Analysis for Bluestone Dam 703
        *J. A. Hribar, J. D. Mozer, P. M. Wimberly,
        and R. E. Yost*

Vibration Study of the Richard B. Russell Concrete Gravity Dam 715
        V. P. Chiarito

Earthquake Analysis and Design of Concrete Gravity Dams 717
        L. G. Guthrie

Analysis of Strong-Motion Earthquake Records
From a Well-Instrumented Earth Dam 729
        J. J. Fedock

Earthquake Analysis of Concrete Dams 741
        A. K. Chopra

Mode Identification of an Arch Dam by a Dynamic Air-Gun Test 753
        H. P. Liu, J. J. Fedock, and J. B. Fletcher

Nonstationary Hydrodynamic Forces
on Arch Dams From Random Earthquakes 765
        M. Debessay and C. Y. Yang

An Introduction to Protecting Museum Artifacts from Earthquake Damage 777
        A. J. Schiff

New Considerations for Offshore Seismic Response Studies 789
        G. Bureau

Reliability of Offshore Platform Pile Design—A Case Study 797
        D. J. Wisch and T. Hadj-Hamou

Index of Proceedings Authors xxxv

VOLUME II

Organization of Proceedings Volumes v
Sponsoring Organization vii
Hosting Organizations vii
Organizing Committees ix
Preface xxxiii

5. RESPONSE OF STRUCTURES, SEISMIC ANALYSIS, AND SPECTRA

Coupled Lateral-Rocking-Torsional Response of
Structures With Embedded Foundations Due to SH-Waves 811
    J. Bielak, H. Sudarbo, and D. V. Morse
Reversing Cyclic Elasto-Plastic Demands on
Structures During Strong Motion Earthquake Excitation 823
    V. Perez, A. G. Brady, and E. Safak
Consideration of Floor Flexibility
in Dynamic Analysis of High-Rise Buildings 835
    M. A. M. Torkamani and J. T. Huang
Analysis of the Seismic Performance
of the Imperial County Services Building 847
    C. A. Zeris, S. A. Mahin, and V. V. Bertero
Vulnerability of Tall Buildings in Atlanta, Chicago, Kansas City,
and Dallas to a Major Earthquake on the New Madrid Fault 859
    L. E. Malik and O. Nuttli
Seismic Performance of Existing Buildings 871
    P. Gergely, R. White, and G. R. Fuller
The Role of Cladding in Seismic Response
of Lowrise Buildings in the Southeastern U.S. 883
    B. J. Goodno and J. P. Pinelli
Partial Uplift of Inverted-Pendulum Tower on Flexible Ground 895
    B. Pacheco, Y. Fujino, and M. Ito

Response of Structures to a Spatially Random Ground Motion 907
    A. Mita and J. E. Luco

Seismic Risk Analysis For Codified Structural Design 919
    C. J. Turkstra, A. G. Tallin, and M. Brahimi

Regional Risk Assessment of Existing Buildings:
An Update of Current Research at Stanford University 931
    H. M. Thurston, W. Dong,
    A. C. Boissonnade, and H. C. Shah

An Interpretation of Failure Probabilities 943
    A. H. Hadjian and J. Goodman

Estimation of Response Spectra of Relatively Long-Period Ground Motions
From Sloshing Heights of Liquid Storage Tanks and Seismograms 955
    Y. Yamada, H. Iemura, S. Noda,
    and S. Shimada

On the Modeling of a Class of Deteriorating
Structures Subjected to Severe Earthquake Loading 967
    A. O. Cifuentes and W. D. Iwan

Seismic Energy and Structural Damage 979
    J. M. Tembulkar and J. M. Nau

Earthquake Responses of Multistory Buildings
Under Stochastic Biaxial Ground Motions 991
    Y. J. Park and A. M. Reinhorn

Simplified Procedures for Earthquake Analysis of Buildings 1003
    A. K. Chopra and E. F. Cruz

Inelastic Analysis of RC Sections 1015
    A. E. Aktan and G. E. Nelson

A Hysteresis Model for Biaxial Bending of Reinforced Concrete Columns 1027
    G. E. Ghusn and M. Saiidi

Consensus-Opinion Earthquake Damage Probability
Matrices and Loss-of-Function Estimates for Facilities in California 1039
    C. Rojahn, R. L. Sharpe, R. E. Scholl,
    A. S. Kiremidjian, R. V. Nutt, and R. R. Wilson

Analysis of Multistory Structure-Foundation
Systems Including P-Delta Effects 1051
    K. S. Sivakumaran

Identification of Time-Dependent Dynamic Characteristics
of Structures Responding to Strong Ground Motion 1063
    V. Sotoudeh and H. C. Shah

Coupled Lateral-Torsional Response of
MDOF Systems to Nonstationary Random Excitation 1075
    W. J. Sun and A. Kareem

Prediction of the Largest Peaks in Single-Degree-Of-Freedom
Oscillator Response to Strong Ground Motion 1087
    J. M. Pauschke and J. Chatterjee
Seismic Behaviour of Friction Damped Braced Frames 1099
    P. Baktash and C. Marsh
Modification of Earthquake Response
Spectra With Respect to Damping Ratio 1107
    K. Kawashima and K. Aizawa
Observations on Spectra and Design 1117
    W. J. Hall and S. L. McCabe
Supplemental Mechanical Damping for
Improved Seismic Response of Buildings 1129
    R. D. Hanson, D. M. Bergman,
    and S. A. Ashour
Determining Design Response Spectra for a Site
of Low Seismicity on the Basis of Historical Records 1141
    D. D. Hunt, J. T. Christian, T. Y. H. Chang,
    and P. A. Cadena
Response Spectra for Building Design 1153
    N. C. Donovan and A. M. Becker
Higher Modes Contribution to Total Seismic Response 1165
    A. H. Hadjian and S. T. Lin

6. TESTS ON STRUCTURES AND COMPONENTS, EXPERIMENTAL METHODS

Behavior and Design of Reinforced Concrete Joints Using Special Materials 1179
    R. J. Craig
T-Beam Effect in Structures Subjected to Lateral Loading 1191
    C. W. French and A. Boroojerdi
Behavior of a Strengthened Reinforced Concrete Frame 1203
    T. D. Bush, C. E. Roach, E. A. Jones,
    and J. O. Jirsa
Design and Behavior of a Strengthened Reinforced Concrete Frame 1215
    L. A. Wyllie, Jr., C. D. Poland, J. O. Malley,
    and M. T. Wagner

Influence of Transverse Reinforcement on Seismic Performance of Columns 1227
     L. S. Johal, D. W. Musser, and W. G. Corley

Seismic Behavior of Precast Walls 1239
     P. Mueller

Results of Full-Scale Tests of Steel-Deck-Reinforced
Concrete Floor Diaphragms 1251
     M. L. Porter and W. S. Easterling

Dynamic Response of R/C Frames With Irregular Profiles 1263
     S. L. Wood

Strong-Motion Instrumentation of Structures
in Charleston, South Carolina and Elsewhere 1273
     M. Celebi and R. Maley

Low-Cycle Fatigue of Semi-Rigid Steel Beam-To-Column Connections 1285
     J. B. Radziminski and A. Azizinamini

Earthquake Resistance and Behavior of Wood-Framed Building Partitions 1297
     S. S. Rihal

Pseudo-Dynamic Testing and Model Identification 1311
     J. L. Beck and P. Jayakumar

Seismic Strengthening of Structural Masonry Walls With External Coatings 1323
     S. P. Prawel, A. M. Reinhorn,
     and S. K. Kunnath

Shear Strength of Reinforced Masonry Walls Under Earthquake Excitation 1335
     P. Hidalgo and C. Luders

Seismic Behavior of Precast Concrete Shear Walls—
Correlation of Experimental and Analytical Results 1347
     V. Caccese and H. G. Harris

Stresses in Composite Masonry Shear Walls Due to Earthquake Loads 1359
     S. C. Anand and M. A. Rahman

Measured Hysteresis in a Masonry Building System 1371
     D. P. Abrams

Earthquake Simulated Performance of an Unreinforced and a
Reinforced Masonry Single-Story House Model 1383
     G. C. Manos and R. W. Clough

Tests of Squat Shear Wall Under Lateral Load Reversals 1395
     S. Wiradinata and M. Saatcioglu

Measured Dynamic Response of Hyperbolic Shell 1407
     A. N. Lin, H. Mozaffarian,
     and M. Helpingstone

Confinement Requirements for Prestressed
Concrete Columns Subjected to Seismic Loading 1419
     A. J. Durrani and H. E. Elias

Analytical Prediction of the Biaxial Response
of a Reinforced Concrete Shake Table Model 1427
     C. A. Zeris and S. A. Mahin

Mechanisms of Stiffness and Strength
Deterioration in Biaxially Loaded R.C. Beam-Column Joints 1439
     R. T. Leon

Strain Rate Effect on the Stress Strain Behavior of
Structural Steel Under Seismic Loading Conditions 1451
     K. C. Chang and G. C. Lee

High Strength Concrete Frames Subjected to Inelastic Cyclic Loading 1463
     M. R. Ehsani, A. E. Moussa,
     and C. R. Vallenilla

Behavior of Reinforced Concrete Columns
Subjected to Lateral and Axial Load Reversals 1475
     M. E. Kreger and L. Linbeck

Low-Cycle Fatigue Based Seismic Design of Steel Structures 1487
     B. Lashkari

Earthquake Damage Prediction for Buildings Using Component Test Data 1493
     O. Kustu

Seismic Response of Slab-Column Frames 1505
     J. P. Moehle

Earthquake Response of Reinforced Concrete Frames With Yielding Columns 1517
     A. E. Schultz

Effect of Cyclic Loading Rate on Response of
Model Beam-Column Joints and Anchorage Bond 1529
     S. P. Shah and L. Chung

Evaluation and Improvement of the Pseudodynamic Test Method 1541
     P. B. Shing

Interstory Displacement Recording 1553
     S. E. Pauly

Advanced On-Line Computer Control
Methods for Seismic Performance Testing 1563
     C. R. Thewalt, S. A. Mahin,
     and S. N. Dermitzakis

Roof-Top Ambient Vibration Measurements 1575
     J. G. Diehl

7. NONSTRUCTURAL SYSTEMS AND BUILDING COMPONENTS

Special Issues in the Secondary System
Response by the Floor Response Spectrum Method 1589
    J. W. Jaw and A. K. Gupta
Dynamic Response of Nonstructural Components 1601
    A. G. Hernried and H. Jeng
Seismic Qualification of Interior Mechanical Systems 1613
    A. H. Chowdhury

Index of Proceedings Authors xxxv

VOLUME III

Organization of Proceedings Volumes v
Sponsoring Organization vii
Hosting Organizations vii
Organizing Committees ix
Preface xxxiii

8. SEISMIC STRUCTURAL DESIGN— SEISMIC CODES AND STANDARDS

ATC-14—A Methodology for the Seismic Evaluation of Existing Buildings 1627
    C. D. Poland and J. O. Malley

Torsional Response of Three Instrumented
Buildings During the 1984 Morgan Hill Earthquake 1639
    A. F. Shakal and M. J. Huang

Some Lessons From the Study of a Collapse 1651
    E. Del Valle

Evaluation of Response Reduction Factors
Recommended by ATC and SEAOC. 1663
    V. V. Bertero

Simulation Modeling of Fire Following Earthquake 1675
    C. Scawthorn

The Adequacy of Existing Building Codes for Charleston,
South Carolina in Light of Recent Discoveries of Seismic Hazard 1687
    W. J. Johnson and E. Bazan-Zurita

A Description of Proposed Changes in California Seismic Code Requirements 1699
    T. Zsutty

Probabilistic Limit States Design of
Moment-Resistant Frames Under Seismic Loading 1709
    M. A. Austin, K. S. Pister, and S. A. Mahin

Damage-Limiting Aseismic Design of Buildings. 1721
    Y. J. Park, A. H. S. Ang, and Y. K. Wen

Assessment of ATC-3-06 Based on Structural Optimization 1731
    K. Z. Truman, D. S. Juang, and F. Y. Cheng

Seismic Design of Tall Steel Buildings. 1743
    M. M. Ali

Occupant Behavior Related to Seismic
Performance in a High-Rise Office Building 1755
    C. Arnold

Seismic Design Guidelines for Tilt-Up-Wall Buildings
Based on Experimental and Analytical Models 1767
    S. A. Adham and M. S. Agbabian

Moving Beam Plastic Hinging Zones for
Earthquake Resistant Design of R/C Buildings. 1779
    J. Wight, M. Al-Haddad, and B. A. Fattah

Semi-Rigid Composite Frames for Seismic Loading 1791
    R. T. Leon and D. J. Ammerman

Seismic Resistance of Reinforced Concrete
Structural Systems of Limited Deformability. 1803
    S. K. Ghosh

Optimum Seismic Design of Low-Rise Brittle Structures—Design Problems
and Optimality Criteria for Maximum Crack Resistance 1815
    R. Razani

Seismic Behavior of Unbraced Steel
Building Frames With Semi-Rigid Connections 1827
    M. S. L. Roufaiel and F. F. Monasa

Seismic Risk Analysis of a Multi-Site Portfolio of Buildings 1839
    T. A. Sabol and G. C. Hart

Seismic Vulnerability Studies of Buildings at
Military Facilities in the Southeastern United States 1851
    S. Tandowsky, C. Hanson, and C. Beauvoir

The U.S. Navy's Earthquake Safety Program 1863
    J. V. Tyrrell and B. Curry

Cost Considerations in Seismic Structural System Selection 1873
    C. C. Thiel, Jr.

9. REPAIR, STRENGTHENING, ISOLATION, AND RETROFIT

Elastomeric Bearings Seismically Retrofit Bridge.............................. 1887
    *J. B. Hoerner, G. M. Snyder,
    and R. C. Van Orden*

Control of Structures by Coupling Rubber Isolators and Active Pulses 1899
    *A. M. Reinhorn, G. D. Manolis,
    and C. Y. Wen*

Quality Assurance and Control of Fabrication
for a High-Damping-Rubber Base Isolation System 1911
    *A. G. Tarics, J. Kelly, D. Way,
    and R. Holland*

Design of Ductile Energy Absorbing Devices
for Enhancement of Earthquake Resistance 1923
    S. F. Stiemer

Two Types of Adaptable Bearings for
Isolating Buildings Against Strong Seismic Motions 1935
    N. Dimitrov and A. Pocanschi

Economic Aspects Involved in Retrofitting a Commercial Building 1947
     R. Shepherd and C. M. Platt

Seismic Strengthening for the Veterans Administration
Medical Center, Charleston, South Carolina 1959
     J. A. Baldelli

Active Control of Buildings Structures During Earthquakes 1971
     A. M. Reinhorn, T. T. Soong,
     and L. L. Chung

Validity of Approximate Analysis Techniques for
Base-Isolated Structures With Nonlinear Isolation Elements 1981
     A. K. Karamchandani and J. W. Reed

Response of Structures Supported on Resilient-Friction Base Isolator (R-FBI) 1993
     N. Mostaghel, M. Hejazi,
     and M. Khodaverdian

Structural Dampers: An Alternative Procedure for Retrofitting Buildings 2005
     R. E. Scholl

Seismic Strengthening—A Structural Engineer's Perspective 2017
     R. L. Sharpe

Relative Benefits of Alternative Strengthening
Methods for Low Strength Masonry Buildings 2023
     C. Scawthorn and A. Becker

10. LIFELINES—UTILITIES, TRANSPORTATION SYSTEMS, TELECOMMUNICATIONS, AND OTHER FACILITIES

Implementation of the Recently Developed
ATC Procedures for Seismic Design and Retrofitting of Bridges 2037
     R. A. Imbsen and R. V. Nutt

Responses of a Long Span Arch Bridge to Differential Abutment Motions 2049
     R. A. Dusseau and R. K. Wen

Design Ground Motions for Cooper River Bridge,
Mark Clark Expressway, Charleston, South Carolina 2061
     G. N. Richardson and W. B. Wright

Design and Damage Prediction for Bridges Under Seismic Loads 2073
    D. A. Foutch and M. E. Barenberg

Earthquake Resistant Analysis of Cable-Stayed
Bridges in Eastern and Central United States 2085
    A. M. Abdel-Ghaffar and A. S. Nazmy

The Effect of Wall Piers on Seismic Response of Bridges 2097
    C. A. Issa and R. M. Barker

Precast Concrete Bridges in Seismic Regions 2107
    F. Sauter

Earthquake Hazard Reduction Techniques at Petroleum Facilities in Japan 2119
    G. Selvaduray

Dynamic Response of a Broad Storage Tank Model
Under a Variety of Simulated Earthquake Motions 2131
    G. C. Manos

Behaviour of Elevated Storage Tanks During Earthquakes 2143
    M. R. Resheidat and H. Sunna

A Preuplift Method for "Anchoring" Fluid Storage Tanks 2155
    R. Peek, P. C. Jennings, and C. D. Babcock

A Case Study of Seismic Hazards and
Pipeline System Response for San Francisco 2167
    T. D. O'Rourke and P. L. Lane

Buckling of Pipelines in Seismic Environments 2179
    H. D. Yun and S. Kyriakides

Ground Deformation Spectra 2191
    T. Harada and M. Shinozuka

Earthquake Hazard Mitigation of Utility Lifeline Systems 2203
    R. T. Eguchi, S. D. Werner, and J. P. Masek

Early Post-Earthquake Damage Detection for Lifeline Systems 2215
    R. T. Eguchi and J. D. Chrostowski

Fundamental Modal Behavior of an Earthquake-Excited Bridge 2225
    A. G. Brady and M. Celebi

Parametric Studies on a Simple Model
for the Rigid Body Rotation of Skew Bridges 2237
    E. A. Maragakis

Nonlinear Stiffness Analysis of Buried Jointed Pipelines 2249
    M. S. Zuroff and A. C. Singhal

Evolution and Application of the
Seismic Qualification Program on the GCEP Project 2261
    *R. M. Drake, E. A. Smietana, P. J. Richter,
    and J. E. Beavers*

Ductility of Bridges in Yellowstone National Park 2271
        B. A. Suprenant and J. R. Kattell

Quantification of the Seismic Hazard
Due to Process Piping in Industrial Facilities 2283
        *V. N. Vagliente, B. Lashkari,
        and C. A. Kircher*

Development of Lifeline Design Criteria for Incoherent Ground Motions 2293
        L. R. L. Wang

The Mexico Earthquake of September 19, 1985—
Performance of Power and Industrial Facilities 2305
        S. J. Eder and S. W. Swan

11. URBAN DESIGN, SOCIOECONOMIC ISSUES, PUBLIC POLICY AND PREPAREDNESS

Earthquake Prediction Response Planning in California 2319
        R. K. Reitherman

Earthquake Recovery: A Triage Approach
to the Physical Reconstruction of Housing 2329
        D. Evans and C. Arnold

Long-Term Earthquake Preparedness in Charleston, South Carolina 2341
        P. L. Gori and M. R. Greene

An Analysis of State Government Orientation
Toward Earthquake Mitigation: Comparing Six States 2353
        E. Mittler

Earthquake Recovery Planning for Business and Industry 2365
        J. D. Lichterman

The Role of the Volunteer Engineer in Post-Earthquake Disaster Mitigation 2377
        SEAONC (DES)

Earthquake Engineering and Public Policy: Key Strategies for Seismic Safety 2389
        S. Scott

Dilemmas in Post-Earthquake Recovery:
The Case of the 1983 Coalinga, California Earthquake 2401
        K. J. Tierney

Development of Earthquake Hazard Reduction Equipment in Japan 2413
        G. Selvaduray

Recent Experience With the Implementation
of Nonstructural Hazard Reduction Programs 2425
        R. K. Reitherman

Political, Social, Economic Aspects of Life-Safety
of Hazardous Structures in Earthquake Zones 2435
        S. J. K. Rao, B. Gulkarov, and W. J. Petak

Existing Hazardous Structures: Techno-Economic/Social
Systems Analysis With Case Studies of Seismic Upgrading 2447
        S. J. K. Rao, B. Gulkarov, and J. Day

Earthquake Loss Studies in the Central U.S.:
Public Policy Implications and Research Needs................................. 2459
        L. E. Malik and R. E. Scholl

Vulnerability of Energy Distribution Systems
in the Eastern United States to Earthquakes 2471
        J. E. Beavers, R. G. Domer, and R. J. Hunt

        Index of Proceedings Authors xxxv

PREFACE

During the past quarter century the North American continent has experienced a number of damaging earthquakes, among which were the 1964 Alaska earthquake, the 1971 San Fernando, California, earthquake, and most recently the 1985 Mexico City earthquake. A large number of smaller earthquakes have occurred during this period, all of which, along with large earthquakes that have occurred in other parts of the world, serve to remind us that the earthquake hazard is real and constantly with us. In view of potential loss of life and the economic losses that could result from large earthquakes, it is important that the United States continue its vigorous efforts towards mitigating the hazards of earthquakes—including developing and implementing safe and economic methods of earthquake-resistant design and construction.

In the light of the foregoing observations it is fitting that this Third U.S. National Conference on Earthquake Engineering be held in 1986 at Charleston, South Carolina, on the one-hundred-year anniversary of the 1886 Charleston earthquake. Although intended primarily for participation by U.S. practitioners and researchers, participants from many other parts of the world will also be present. From the more than 300 papers offered for publication and presentation, over 200 papers will finally be published in the three volumes of *Proceedings* and the single volume of *Post-Conference Proceedings*.

We wish to take this opportunity to thank the many individuals who worked so hard in preparing these *Proceedings* and in developing the program; these include: Norma F. Callaham, Conference Coordinator; Charles Lindbergh, Local Arrangements Chairman; the Advisory Committee and the Technical Program Committee Members; Gail Shea, EERI Publications Manager; Douglas A. Foutch and Keith D. Hjelmstad who served as Co-chairmen of the Technical Program Committee; and last, but by no means least, Barbara Luttrell of Martin Marietta Energy Systems, Inc. and Susan Warsaw, Truth Callison, and Mary Pearson of the University of Illinois who literally performed miracles in coordinating the technical program and assembling the *Proceedings*.

Finally, we wish to thank the cooperative and financial sponsors of the Third U.S. National Conference on Earthquake Engineering, without whose help the Conference would not be the success we envision.

        W. J. Hall, Chairman
        Technical Program Committee

        J. E. Beavers
        Conference Chairman

5. RESPONSE OF STRUCTURES, SEISMIC ANALYSIS, AND SPECTRA

COUPLED LATERAL-ROCKING-TORSIONAL RESPONSE OF STRUCTURES WITH EMBEDDED FOUNDATIONS DUE TO SH-WAVES

J. Bielak[I], H. Sudarbo[II] and D. V. Morse[III]

ABSTRACT

A one-story structure with a rigid cylindrical foundation embedded into an elastic halfspace is used to examine the effects of combined geometric eccentricity and nonuniform base excitation on system response. The excitation consists of horizontally or vertically incident harmonic SH-waves. Numerical results indicate that for tall structures the base shears and base moments (except for the torsional moment) are not sensitive to the direction of propagation (vertical or horizontal) of the incident waves. Squatty structures, by contrast, are more strongly influenced by this effect, especially for embedded foundations and soft soils. The results also show that for structures supported on flexible soils the coupled lateral-torsional response due to geometric eccentricity follows the same general trends as for fixed-base structures with some important differences. In particular, coupling effects decrease with decreasing soil stiffness, especially for squatty structures.

INTRODUCTION

The 19 September 1985 Mexico earthquake provides added evidence that earthquakes can induce lateral and rotational motion in buildings, which can, in turn, give rise to large torsional forces. Torsional response in buildings occurs if the centers of mass and resistance are not coincident or if the horizontal ground motion is not uniform throughout the base, even if the building is symmetric. The problem of coupled lateral-torsional response of buildings due to geometric eccentricity between center of mass and center of resistance has been studied extensively under the assumption of a uniform base excitation. (See, e.g. [1,2] and references therein.) The effects of a flexible soil have been investigated in [3] and [4] under the same assumption.

The problem of torsion in symmetric buildings due to traveling incident waves was apparently first examined in [5] and has been investigated further by several authors, e.g. [6], [7]. A recent paper [8] includes effects both due to geometric eccentricity and due to accidental eccentricity arising from traveling SH-waves in an elastic halfspace, but it ignores the rocking motion in the system caused by the base overturning moment. If the foundation is embedded, a traveling SH-wave will produce rocking, in addition to translation and rotation about a vertical axis. On the other hand, embedment has the effect of reducing the translational motion of the footing [9].

[I] Professor, Department of Civil Engineering, Carnegie-Mellon University, Pittsburgh, PA 15213

[II] Graduate Student, Department of Civil Engineering, Carnegie-Mellon University, Pittsburgh, PA 15213; also Teaching Staff, Department of Civil Engineering, University of Indonesia, Jakarta.

[III] Advisory Industry Specialist, IBM, Southfield, MI; formerly, Graduate Student, Department of Civil Engineering, Carnegie-Mellon University, Pittsburgh, PA 15213

Even though the dynamic response of structures with embedded foundations has been studied previously [10] under uniform base excitation, it is not yet clear how different the structural response will be if the excitation is not uniform throughout the base. This paper has two related objectives. It seeks to throw light on the question of how the direction of travel of the incident wave affects the response of structures supported on surface or embedded foundations. It also seeks to examine torsional effects in buildings due to combined nonuniform excitation and geometric eccentricity. To these ends we consider a simple one-story structure with a rigid foundation which is embedded into an elastic halfspace with damping, and is subjected to horizontally or vertically incident SH-waves. In analyzing this system we make use of the impedance functions and effective input motion presented in [11].

ANALYSIS OF THE SYSTEM

The Structure

The idealized one-story structure consists of a rigid deck supported on massless, axially inextensible linearly elastic columns and walls, as shown on Fig. 1. These elements are attached to a rigid right-circular cylindrical foundation which is embedded into an elastic halfspace. The system is subjected to incident SH-waves that produce horizontal translation, rotation about the vertical axis and rocking of the base. Thus the system has eight degrees of freedom, three corresponding to the deck and five to the base, namely, lateral displacements of the centers of mass of the deck and of the base along the principal axes of resistance, x and y, of the structure; torsional displacements of the deck and the base, and rocking of the base about the x and y axes.

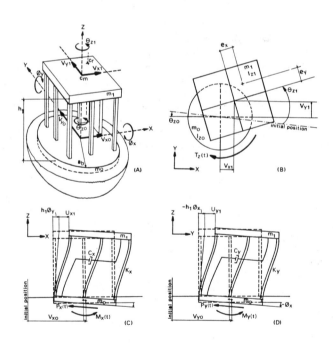

Figure 1:   Model of Structure-Foundation System

The center of resistance of the deck is located at distances e_x and e_y from the coordinate axes. Expressions for e_x and e_y, as well as for the lateral structural and torsional stiffnesses of the fixed-base structure, in terms of the lateral stiffnesses of the individual columns and wall elements are given in [12]. We denote by ω_x, ω_y and ω_θ the natural uncoupled frequencies for fixed-base response in the x, y and torsional direction, respectively, that the system would exhibit if it were torsionally uncoupled ($e_x = 0 = e_y$).

The Foundation

The motion of the foundation is related to the incoming wave motion and to the forces P_b transmitted by the structure to the halfspace by an expression of the form

$$P_b = \hat{K}_{bb} (v_b - v_b^*) \tag{1}$$

in which $P_b = (P_x, M_y, P_y, M_x, T_z)^T$ acts at the center of mass of the foundation; $v_b = (v_{xo}, \phi_y, v_{yo}, \phi_x, \theta_{zo})^T$ is the vector of total base displacements; $\hat{K}_{bb}$ is the impedance matrix of the foundation and v_b^* represents the displacements that a massless rigid base subjected to a prescribed incoming wave would experience in the absence of any external forces.

In writing Eq. 1 it has been assumed that the incident motion, as well as the response, is harmonic, and thus the common factor $\exp(i\omega t)$, where ω = frequency of excitation has been suppressed. In this work the incident wave motion consists of SH-waves that can propagate either vertically or horizontally. Thus, four different cases will be considered, depending on the direction of polarization of the particle motion. Two additional cases, corresponding to uniform free-field surface motion, will be included for comparison. The six cases are summarized in Table 1.

Table 1: Types of incident SH-wave motion considered

Case No.	Incidence	Direction of Particle Motion	Type of Line Used in Results Section
1	Horizontal	x	—·—·—
2	Horizontal	y	—··—··—
3	Vertical	x	— — —
4	Vertical	y	—·—·—
5	Vertical (free field)	x	————
6	Vertical (free field)	y	————

The entries for the matrix $\hat{K}_{bb}$ and the effective displacement v_b^* are given in graphical form in [11] for an elastic halfspace, as functions of the normalized frequency of excitation $a_0 = \omega a_b/\beta$, where a_b = radius of cylindrical base and $\beta = \sqrt{\mu/\rho}$ = shear wave velocity of elastic medium, μ = shear modulus and ρ = density, for a Poisson's ratio ν = 0.25. In order to obtain the corresponding results for a dissipative medium with material damping, use is made of the principle of correspondence in viscoelasticity by replacing μ by $\mu(1 + 2Di)$ in the original results. D measures the amount of structural material damping in the halfspace.

Equations Of Motion

The equations of motion for the steady-state harmonic response of the building-foundation system shown on Fig. 1 can be written symbolically as:

$$\left(-\omega^2 \begin{bmatrix} M_{ss} & M_{sb} \\ M_{bs} & M_{bb} \end{bmatrix} + i\omega \begin{bmatrix} C_{ss} & 0 \\ 0 & 0 \end{bmatrix} + \begin{bmatrix} K_{ss} & 0 \\ 0 & 0 \end{bmatrix} \right) \begin{Bmatrix} u_s \\ v_b \end{Bmatrix} + \begin{Bmatrix} 0 \\ P_b \end{Bmatrix} = \begin{Bmatrix} 0 \\ 0 \end{Bmatrix} \qquad (2)$$

where $u_s = (u_{x1}, u_{y1}, \theta_{z1} - \theta_{z0})^T$ is vector of relative generalized displacements of the deck with respect to the base, excluding rocking. The mass matrices M_{ss}, $M_{sb} = M_{ss}^T$, M_{bb} depend on the masses m_1 and m_0 of the deck and base mass, the total moments of inertia $I_x = I_{x1} + I_{x0}$, $I_y = I_{y1} + I_{y0}$ about the centers of mass, and the height h_1 of the deck. The matrix K_{ss} is the structural stiffness matrix given in [12]. C_{ss} is a similar damping matrix.

Expressing $\hat{K}_{bb}$ as:

$$\hat{K}_{bb} = K_{bb} + i\omega C_{bb} \qquad (3)$$

where K_{bb} and C_{bb} are real, and substituting Eq. 1 into Eq. 2 there results:

$$\left(-\omega^2 \begin{bmatrix} M_{ss} & M_{sb} \\ M_{bs} & M_{bb} \end{bmatrix} + i\omega \begin{bmatrix} C_{ss} & 0 \\ 0 & C_{bb} \end{bmatrix} + \begin{bmatrix} K_{ss} & 0 \\ 0 & K_{bb} \end{bmatrix} \right) \begin{Bmatrix} u_s \\ v_b \end{Bmatrix} = \begin{Bmatrix} 0 \\ P_b \end{Bmatrix} \qquad (4)$$

in which:

$$P_b = \hat{K}_{bb} v_b^* \qquad (5)$$

is the effective generalized input force applied at the center of mass of the bass mass.

NUMERICAL RESULTS

In this section we examine the response of the building-foundation system shown on Fig. 1 for representative values of the system parameters. For displacements we will use the total quantities v_{x1}, $h_1\phi_y$, v_{y1}, $h_1\phi_y$, $a_b\theta_{z1}$, v_{x0}, v_{y0} and $a_b\theta_{z0}$, all normalized with respect to the amplitude of the free-field displacement, u_0. Generalized forces will be expressed as follows: base shear along the x and y axes by $P_i/m_1 \ddot{u}_0$, $i = x, y$; base overturning moment about the x and y axes by $M_i/m_1 h_1 \ddot{u}_0$, $i = x, y$, and torsional moment, referred to the center of resistance, by $T_r/m_1 a_b \ddot{u}_0$, where $\ddot{u}_0$ is the amplitude of the free-field acceleration.

Previous studies of lateral-torsional coupling in structures due to geometric eccentricity show that coupling effects are greatest for structures whose uncoupled lateral and torsional frequencies coincide. Thus, throughout this study ω_θ/ω_x will be taken as unity. Other parameters that will remain fixed are: ratio of the uncoupled lateral frequencies $\omega_y/\omega_x = 1.2$; the fixed-base system is viscously damped in its three uncoupled modes, with two percent critical damping in each; deck to soil mass ratio $m_1/\pi \rho a_b^3 = 0.03$; ratio of the deck's mass moments of inertia $I_{x1}/I_{y1} = 1.2$; ratio of the radius of gyration r of the deck ($m_1 r^2 = I_{x1} + I_{y1}$) to the radius of the base, $r^2/a_b^2 = 2.5$.

The relative stiffness between the structure and the halfspace is measured by $\bar{\beta} = \beta T_x/h_1$, where $T_x = 2\pi/\omega_x$ is the natural period of the structure in the x direction if $e_x =$

$e_y = 0$. Notice that if T_x is proportional to the building height h_1 then $\bar{\beta}$ becomes a pure measure of the soil stiffness. Even if this is not the case $\bar{\beta}$ will depend, in general, only weakly on the structural properties. Two values of $\bar{\beta}$ will be considered, $\bar{\beta} = 6, 3$, to represent medium stiff and soft soils, respectively; soil material damping $D = 0.025$ and 0.04, respectively, to account for the increase in damping for the softer material. Embedment of the foundation is measured by the ratio of the height of the base mass to a_b; in this study $d/a_b = 0, 1$, in order to model both a surface foundation and one for which the embedment is half the base diameter. Tall structures are represented by $h_1/a_b = 5$, with $m_0/m_1 = 0.5$ and squatty structures by $h_1/a_b = 2$, with $m_0/m_1 = 1$. Two cases of geometric eccentricity will be considered: $e_x = 0$, $e_y = 0.4$ and $e_x = e_y = 0.4$; symmetric buildings will be included for comparison, and to examine the effects of nonuniform base excitation, by itself.

In Figs. 2 and 3 are shown the frequency response spectra for the amplitudes of the total displacements of the building-foundation system under study for various combinations of the system parameters. Six curves are shown for each response quantity corresponding to the six possible forms of excitation described in Table 1. The following trends are worthy of note in these figures and similar ones not shown here for lack of space.

- For tall structures on a medium stiff soil the total lateral translations of the deck are only ten percent less than the corresponding values for the fixed-base structure. For the deformable soil, however, the total motion includes interstory deformation as well as displacement due to base rocking. The rocking component for the structure on a surface foundation is about three times greater than for the embedded foundation, independently of geometric eccentricity. The amount of rocking is smaller if the system has an embedded foundation. Also, short structures experience less rocking than tall structures.

- Response of systems with geometric eccentricities in both principal directions exhibit lateral-rocking-torsional coupling in their three principal "modes". As a consequence of this coupling, the amplitude of lateral motion is significantly less than that for structures with torsional coupling in one direction only.

- Torsional rotation due to nonuniform excitation in symmetric structures is significantly less than that due to geometric eccentricity if the soil is stiff, but they become comparable for soft soils.

- Peak torsional rotation for symmetric structures is greater for squatty than for tall structures.

- The deck response of tall structures on medium stiff or soft soils is practically the same for all cases of incident wave motion considered.

- The base motion of tall structures with surface foundations is essentially equal to the surface free-field motion. Some reduction occurs if the base is embedded on a soft soil. By contrast, the base response of squatty structures depends strongly on the direction of the incoming wave motion, especially for embedded foundations and soft soils. The deck response is also affected, although to a lesser extent.

In Figures 4, 5 and 6 response spectra are shown for the amplitudes of the normalized shear forces, overturning moments and torque at the center of mass of the base of the system depicted in Fig. 1. For comparison, Table 2 shows the corresponding peak values for a fixed-base structure and Table 3 those for a symmetric structure on a flexible halfspace. From Table 2 it is seen that base shears and overturning moments for fixed-base structures decrease as a consequence of torsional-lateral coupling. This reduction, however, is accompanied by a (possibly large) base torque. Table 3 shows that for symmetric buildings the base shear and overturning moment decrease as the soil gets softer. Base torque also occurs in this case for horizontal incident waves. A comparison between the results of Tables 2 and 3 and Fig. 4 shows that for a tall structure with geometrical eccentricity along one axis and supported on a medium stiff soil base shear and base moment are not significantly different from the values corresponding to a rigid soil. They are also similar to those of the corresponding symmetric building on medium stiff soil. The response decreases, however, for short buildings and for structures with geometric eccentricity along the two principal axes. For soft soils, the effects due to geometric eccentricity become less important. Notice also that whereas the response for tall structures is largely independent of the direction of propagation of the incident wave, for squatty structures the response varies with the angle of propagation, especially for soft soils. The response in all cases is largest for the free-field motion.

Table 2: Peak amplitude of normalized base forces for fixed-base structure

e_x	e_y	$\bar{P}_x$	$\bar{M}_y$	$\bar{P}_y$	$\bar{M}_x$	$\bar{T}_r$
0	0	25.3	25.0	25.3	25.3	0
0	0.4	13.6	14.3	25.0	25.0	42.6
0.4	0.4	16.9	15.9	11.8	11.6	42.0

Table 3: Peak amplitude of normalized base forces for symmetric structure on flexible halfspace

$\beta T_x/h_1$	h_1/a_b	d/a_b	$\bar{P}_x$	$\bar{M}_y$	$\bar{P}_y$	$\bar{M}_x$	$\bar{T}_r$
6	5	0	24.4	24.4	23.7	23.7	6.6
		1	22.8	22.9	22.5	21.9	7.1
	2	0	20.4	21.0	17.9	18.4	17.5
		1	20.2	20.4	18.2	18.1	16.2
3	5	0	15.1	15.1	13.9	13.3	13.5
		1	14.6	14.0	12.5	12.5	14.4
	2	0	8.6	9.1	7.2	7.4	21.2
		1	9.8	9.5	7.4	7.2	18.3

CONCLUSIONS

The following conclusions may be drawn from the present study. These conclusions, however, can be only tentative in view of the limited range of values of the system parameters.

- Effects due to geometric eccentricity on dynamic response of structures supported on flexible soils are similar to those for fixed-base buildings, although there are important differences. In particular, such effects tend to decrease as the soil becomes softer, especially for squatty structures.

- Base torque in symmetric buildings is greater for squatty than for tall buildings.

- The generalized base forces are affected strongly by the direction of propagation (vertical or horizontal) of the incident wave motion only for the case of squatty structures with embedded foundations supported on soft soils. The translation of the base, however, is very susceptible to the particular wave incidence. Therefore, except for tall buildings resting on stiff soils, the motion recorded at the base of buildings may be significantly different from the free-field surface motion.

REFERENCES

[1]   Kan, C.L. and A.K. Chopra, "Torsional Coupling and Earthquake Response of Simple Elastic and Inelastic Systems," Journal of the Structural Division, ASCE, Vol. 107(8), 1569-1588, 1981.

[2]   Bozorgnia, Y. and W.K. Tso, "Inelastic Earthquake Response of Asymmetric Structures," Journal of Structural Engineering, Vol. 112(2), 383-400, 1986.

[3]   Balendra, T., C.W. Tat and S-L. Lee, "Vibration of Asymmetrical Building-Foundation Systems," Journal of Engineering Mechanics, Vol. 109(2), 430-449, 1983.

[4]   Tsicnias, T.G. and G.L. Hutchinson, "Soil-Structure Interaction Effects on the Steady-State Response of Torsionally Coupled Buildings," Earthquake Engineering and Structural Dynamics, Vol. 12, 237-262, 1984.

[5]   Newmark, N.M., "Torsion in Symmetric Buildings, "Proceedings 4th World Conference Earthquake Engineering, Chile, Vol. I, A319, 1969.

[6]   Luco, J.E., "Torsional Response of Structures for Obliquely Incident Seismic SH-waves," Earthquake Engineering and Structural Dynamics, Vol. 4, 107-219, 1976.

[7]   Scanlan, R.H., "Seismic Wave Effects on Soil-Structure Interaction," Earthquake Engineering and Structural Dynamics, Vol. 4, 379-388, 1976.

[8]   Wu, S.T. and E.V. Leyendecker, "Dynamic Eccentricity of Structures Subjected to SH-Waves," Earthquake Engineering and Structural Dynamics, Vol. 12, 619-628, 1984.

[9]   Luco, J.E., H.L. Wong and M.D. Trifunac, "A Note on the Dynamic Response of Rigid Embedded Foundations," Earthquake Engineering and Structural Dynamics, Vol. 4, 119-127, 1975.

[10] Bielak, J., "Dynamic Behavior of Structures with Embedded Foundations," Earthquake Engineering and Structural Dynamics, Vol. 3, 259-274, 1974.

[11] Day, S.M., "Finite Element Analysis of Seismic Scattering Problems," Ph.D. Thesis University of California, San Diego, 1977.

[12] Kan, C.L. and A.K. Chopra, "Effects of Torsional Coupling on Earthquake Forces in Buildings," Journal of the Structural Division, ASCE, Vol. 103(4), 805-819, 1977.

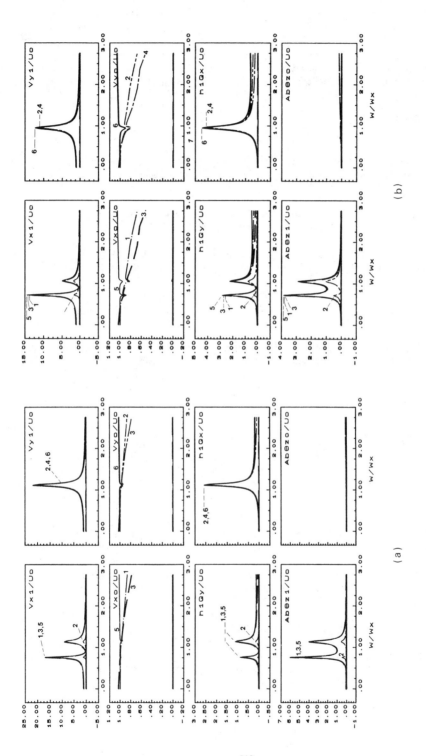

Figure 2: Frequency Response Spectra for System Total Displacements; $d/a_b = 1$, $h_1/a_b = 5$, $e_y/a_b = 0.4$, $e_x/a_b = 0$;
(a) $\beta = 6$, (b) $\beta = 3$

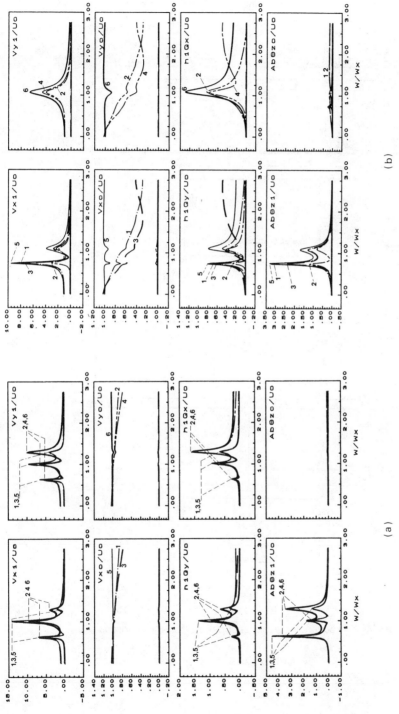

Figure 3: Frequency Response Spectra for System Total Displacements; $d/a_b = 1$, $h_1/a_b = 2$, $e_y/a_b = 0.4$;
(a) $\bar{\beta} = 6$, $e_x/a_b = 0.4$, (b) $\bar{\beta} = 3$, $e_x/a_b = 0$

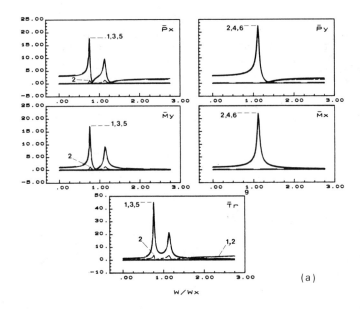

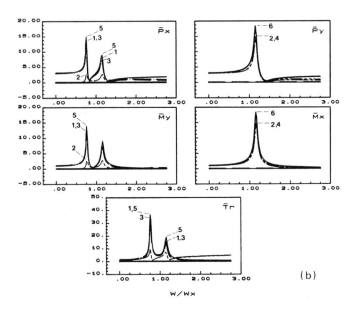

Figure 4: Frequency Response Spectra for Base Forces; $d/a_b = 1$, $\bar{\beta} = 6$, $e_y/a_b = 0.4$, $e_x/a_b = 0$; (a) $h_1/a_b = 5$, (b) $h_1/a_b = 2$

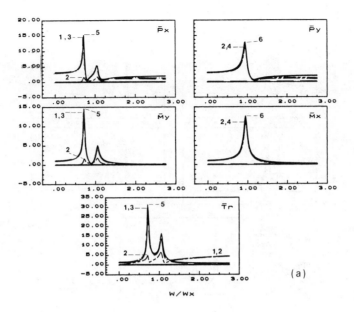

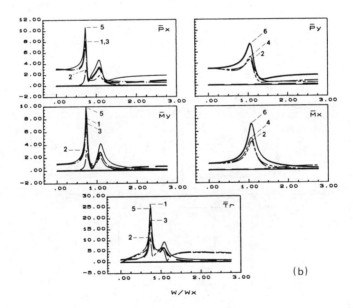

Figure 5: Frequency Response Spectra for Base Forces; $d/a_b = 1$, $\bar{\beta} = 3$, $e_y/a_b = 0.4$, $e_x/a_b = 0$; (a) $h_1/a_b = 5$, (b) $h_1/a_b = 2$

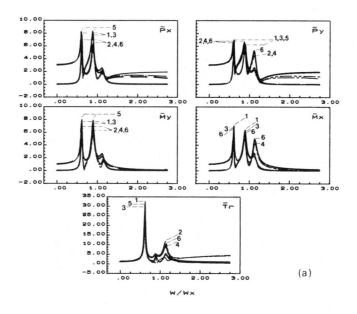

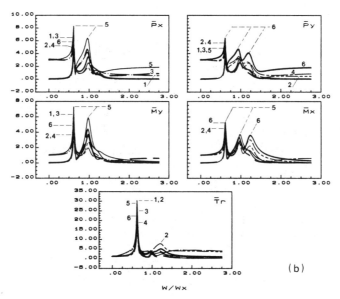

Figure 6: Frequency Response Spectra for Base Forces;
$d/a_b = 1$, $\bar{\beta} = 3$, $e_y/a_b = 0.4$, $e_x/a_b = 0.4$;
(a) $h_1/a_b = 5$, (b) $h_1/a_b = 2$

REVERSING CYCLIC ELASTO-PLASTIC DEMANDS ON STRUCTURES DURING STRONG MOTION EARTHQUAKE EXCITATION

V. Perez[I], A. G. Brady[II], E. Safak[III]

ABSTRACT

Using the horizontal components from El Centro 1940, Taft 1952, and 4 accelerograms from the San Fernando earthquake of 2/9/71, the time history of the elasto-plastic displacement response was calculated for oscillators having periods within the range of 1 to 6 s and ductility factors within the range of 3 to 6. The Nth largest peak of the elasto-plastic response (N=2,4,8,16), when expressed as a percentage of maximum response (that is, N=1), is fairly independent of period within our period range. When considering only plastic peaks occurring, sometimes in a one-directional group of peaks, in the reverse direction from the preceeding plastic peak, the amplitude of the Nth reversing plastic peak is similar to the Nth elastic peak, regardless of the ductility factor.

INTRODUCTION

The response spectrum has been an important tool in the design of earthquake resistant structures [1,2]. But, by focusing on maximum response, the response spectrum ignores all other peaks found in the time history of the response. For structures becoming plastic, the accumulation of damage (and failure) is directly proportional to the number of displacement peaks that exceed the yield level. The purpose of this study is to investigate a) the most severe peaks of the elasto-plastic displacement response and the relation to its maximum value, b) the most severe plastic peaks and their relation to maximum elasto-plastic deformation and c) the relationship between the elasto-plastic response peaks and elastic response peaks. The study considers the response of simple oscillators with periods 1 to 6 s and ductility factors 3 to 6.

The period range and ductilities selected represent values for flexible structures (i.e. buildings with 10 stories and higher), whose response is controlled by the maximum ground displacement both in elastic and elasto-plastic ranges. As will be shown here various measures of response peaks are similar for elastic and elasto-plastic oscillators in this period range. Response peaks for shorter period oscillator are currently being studied, and will be the subject of another paper. This work expands previous studies by Perez and Brady [3] that considered only elastic response.

I.   Mathematician, U.S. Geological Survey, Menlo Park, California
II   Physical Scientist, U.S. Geological Survey, Menlo Park, California
III. Research Structural Engineer, U.S. Geological Survey, Menlo Park, California

The difference between elastic and elasto-plastic oscillator stiffness is that in the latter the restoring force of the oscillator remains constant for any displacement exceeding the elastic limit of the oscillator relative to its current neutral position. That is, the elasto-plastic response is sinusoidal-like in character, but the oscillation occurs about a neutral position that alters with each plastic excursion.

THE AMPLITUDE SUSTAINED FOR M CYCLES, SDAC(M,T), AND ITS APPROXIMATION TO THE NTH (N=2M) DISPLACEMENT RESPONSE PEAK X(N,T)

The cyclical properties of the response are important in the development of time-dependent spectral analysis [3,4,5]. The study of the response as a function of time entails the detailed examination of the displacement amplitudes as the local peaks build up and decay on either side of the maximum peak. Due to the cyclical characteristic of response, (see Fig. 1 for the elastic response and Fig. 2 for the elasto-plastic response,) a valid measurement from the structural engineering point of view is the specific displacement response amplitude that is sustained for a duration equal to that of an integral number of cycles. This measurement consists of first constructing an envelope around the response of an oscillator with specific damping and period; secondly, lowering a horizontal line of a given duration (of 1, 2, 4, 8, 16 or 32 cycles) through the envelope of local peaks until its cumulative length just fits within the envelope; thirdly, measuring the amplitude of the line's location; and fourthly, defining this value as the amplitude sustained for a given number of cycles M [5] (see Fig. 1 for this measurement for the elastic response and Fig. 4 for the elasto-plastic response.) This amplitude is designated SDAC(M,T) ("SD" normally stands for relative displacement response, "A" stands for amplitude sustained, and "C" refers to cycles). The value obtained for SDAC(M,T) is a function of the number of cycles M and the elastic period T. The envelope for the elastic response is obtained relative to its zero position; the envelope for the elasto-plastic response is obtained relative to its neutral position. In the latter case, the neutral position changes over time during plastic yielding and it is equal to the cumulative plastic deformation as shown in Fig. 3. The SDAC(M,T) nomenclature is used for either elastic or elasto-plastic response,

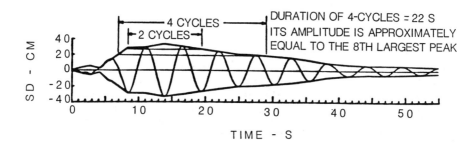

Fig. 1   The envelope of the elastic displacement response for 250 E First Street, bsmnt, Los Angeles, California, February 9, 1971, N54W, with period of 5.5 s and 5% damping. The amplitude sustained for M (M=2,4) cycles (SDAC(M,T)) and its approximation of the Nth displacement peak (N=2M) X(N,T)

depending on the characteristics of the oscillator used. Whereas response spectra use amplitude vs period for a family of dampings, new spectra can be generated by using either the elastic or the elasto-plastic SDAC(M,T) amplitude vs period for a family of cycles of response and one specific damping. Here, the damping used is 5 percent of critical, which is typical of many structures.

There exists a relationship between SDAC(M,T) obtained from a particular oscillator and the peaks of the displacement response of the same oscillator listed in order of decreasing amplitudes. For example, the amplitude of the 8th largest peak is approximately equal to the amplitude sustained for 4 cycles, see Fig. 1. Amplitudes sustained for 1, 2, 4, 8, 16, and 32 cycles are approximately the same as the amplitudes of the 2nd, 4th, 8th, 16th, 32nd and 64th largest peak respectively. This Nth peak is designated X(N,T), where N=2M. Fig. 4 shows that the same approximation can be made for the elasto-plastic response. In this case the Nth peak of the elasto-plastic peak is designated $X_{ep}(N,T)$.

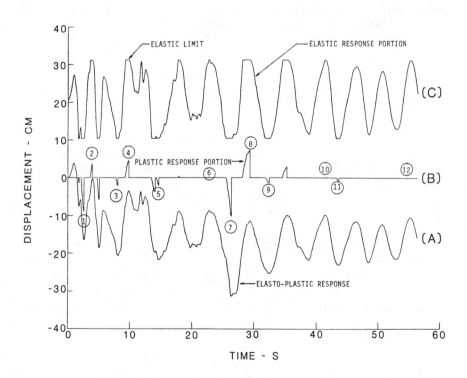

Fig. 2 Elasto-plastic displacement response (A), the plastic portion of the response (B) and the elastic portion of the response (C) - 1.5 scale - for El Centro, May 18, 1940, S00E component. The ductility factor is 5, the damping is 5% of critical and the period of the oscillator is 5 s.

THE NTH REVERSING PLASTIC PEAK: $X_{rp}(N,T)$

Fig. 3 shows the various characteristics of the elasto-plastic response as a function of time. The response behaves elastically until the oscillator reaches its yield point at which time the deformation becomes plastic and stays plastic until it reverses in direction (see vertical lines numbered 1 and 2 in Fig. 3). When it reverses, it becomes elastic again, but now the oscillator has a new neutral position about which it oscillates. This neutral position remains constant until the response becomes plastic again. The neutral position represents the cumulative plastic deformation experienced by the elasto-plastic spring of the oscillator. Consecutive plastic excursions in the same direction sometimes are only momentary interruptions experienced by the oscillator during a swing in a particular direction, see Fig. 2 during the time immediately prior to the encircled numbers 1, 3, 5, 6, and 10. Fig. 2 shows 17 plastic excursions, of which 12 are the last of a series of consecutive plastic deformations in the same direction, and are indicated by the encircled numbers 1 to 12. These are each followed by a reversal. The amplitude of the last plastic peak before a reversal is defined as a reversing plastic peak and the Nth such peak is designated $X_{rp}(N,T)$. The amplitude of each reversing plastic peak can be calculated by adding the cumulative plastic deformation that has occurred up to that time plus the elastic limit of the oscillator. The set of the highest N peaks of an elasto-plastic system's response contains both elastic and plastic peaks, whereas the set of $X_{rp}(N,T)$ peaks contains only the reversing plastic peaks. The maximum plastic deformation is equal to the maximum elasto-plastic response minus the elastic limit. This follows directly from the definition of ductility factor, the ratio of maximum displacement to the elastic limit [6]. In all calculated elasto-plastic time histories in this study, where specific ductility factors are targeted, several iterative choices of the elastic limit are sometimes necessary before the maximum displacement provides this targeted ration within 1 percent. The final plastic deformation is defined as the final amplitude of the neutral position.

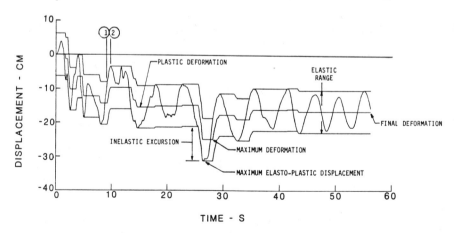

Fig. 3 Elasto-plastic displacement response showing the cumulative plastic deformation and the elastic range as a function of time for El Centro, May 18, 1940, S00E component.

The total number of local peaks experienced during an earthquake by an elastic system can be approximated by dividing the earthquake record length in seconds by half the period of the oscillator. Fig. 2 shows the number of plastic excursions and the number of reversing plastic peaks (the latter are indicated by the encircled numbers). The number of plastic excursions can be much less than the expected number of local elastic peaks.

NORMALIZED AMPLITUDES SUSTAINED FOR M CYCLES $SDAC_n(M,T)$ AND ITS CORRESPONDING NORMALIZED NTH PEAK (N=2M) $X_n(N,T)$

A useful concept is to express the value of $SDAC(M,T)$ as a percentage of the maximum displacement response SD, leaving the ratio as a dimensionless value. The normalized value of $SDAC(M,T)$ is designated $SDAC_n(M,T)$. The Nth elastic peak amplitude is normalized by expressing it as a percentage of the maximum elastic displacement. This normalized value of $X(N,T)$ is designated $X_n(N,T)$. Similarly the normalized values of $X_{ep}(N,T)$ and $X_{rp}(N,T)$ are expressed as a percentages of maximum elasto-plastic displacement and are designated $X_{nep}(N,T)$ and $X_{nrp}(N,T)$ respectively. By representing any peak value of the displacement response as a percentage of maximum response one can retain an effective description of that peak as well as gaining the ability to compare that particular response peak with any other response peak obtained from that earthquake or any other earthquakes.

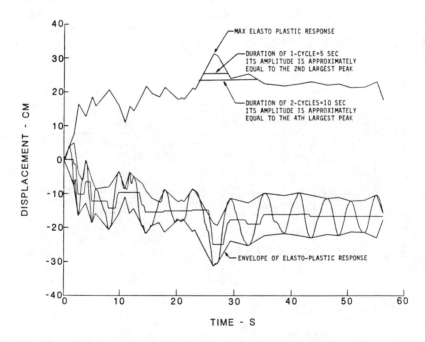

Fig. 4 The envelope of the elasto-plastic displacement response for El Centro, May 18, 1940, S00E component. The amplitude sustained for M (M=1,2) cycles ($SDAC(M,T)$) and its approximation of the Nth displacement peak (N=2M) $X_{ep}(N,T)$

For long periods, generally greater than 2 s, the maximum elasto-plastic and maximum elastic displacement response spectrum values are approximately the same, although they may not occur at the same time.

AVERAGING THE NTH DISPLACEMENT PEAK $X_n(N,T)$ OVER ALL COMPONENTS AT EACH INDIVIDUAL PERIOD T: $X_{mn}(N,T)$

The elastic $X_n(N,T)$ values were calculated for 12 horizontal components from the El Centro, May 18, 1940, the Taft, July 21, 1952 earthquake, and the San Fernando earthquake of February 9, 1971 (see Table 1). Perez and Brady [3] also calculated the elastic $X_n(N,T)$ values for 44 horizontal components from the Imperial Valley earthquake of October 15, 1979. The trends of $X_n(N,T)$ values for the 1979 earthquake were similar to the trends of $X_n(N,T)$ values obtained in the other 3 earthquakes. The mean value of $X_n(N,T)$ for various number of cycles was calculated for the 12 components listed in Table 1. These mean values are designated $X_{mn}(N,T)$ and are shown in Fig. 5A. Note that there are 28 periods ranging in value from 1 to 6 sec. The period range chosen covers the range for flexible structures, i.e. buildings with 10 stories and above.

Averaging the elastic $X_n(N,T)$ for the 12 components for each period in the 1 to 6 sec range produces relatively flat period independent curves with the exception that the $X_{mn}(N,T)$ curves for 8, 16 and 32 cycles tend to zero as the period increases, because the record is not long enough to allow completion of these cycles, see Fig. 5A. Averaging the elasto-plastic $X_{nep}(N,T)$ for the 12 components for the same period range again produced relatively flat period independent curves, with slightly higher values for 4, 8, 16 and 32 cycles, see Fig. 6A. When the reversing plastic peaks

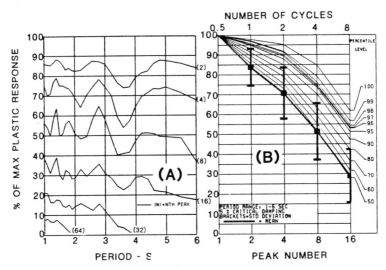

Fig. 5 Averaging the Nth normalized elastic displacement peak $X_n(N,T)$ over 12 components (see Table 1) for each individual period: $X_{mn}(N,T)$ (A). Averaging the Nth elastic displacement peak $X_n(N,T)$ over 12 components and over the period range 1 to 6 s: $X_{2mn}(N)$ (B).

TABLE 1 - Ground motion data

Station Name Date and time of earthquake	Distance to epicenter or fault(km)	Magnitude	Maximum[1] acceleration (g)
5-18-40, 2037 PST El Centro, Calif.	6.4 to fault	7.1	0.36
7-21-52, 0453 PDT Taft, Calif.	40	7.7	0.20
2-9-71, 0600 PST Pacoima Dam, Calif.	8.0	6.6	1.25
8244 Orion Blvd., L.A., Calif. 1st floor	19.3		0.26
445 Figueroa St., L.A., Calif. Sub-basement	40		0.15
250 E. First St., L.A., Calif.	29		0.13

[1]Peak horizontal motion on one of two orthogonal horizontal components.

$X_{nrp}(N,T)$ are averaged for the 12 components and plotted against period (Fig. 7A), they produce relatively flat curves similar to those for elastic and elasto-plastic $X_{mn}(N,T)$ and $X_{mnep}(N,T)$, respectively.

The number of reversing plastic peaks experienced during an earthquake are much less than the number of peaks from the elastic or elasto-plastic response. Fig. 8 shows the average number of reversing plastic peaks experienced by the 12 components in Table 1 for ductility factors 3, 4, 5 and 6.

The scatter of the individual data points from which Figs. 5A, 6A and 7A is plotted is not shown, but will be addressed later.

AVERAGING THE NTH DISPLACEMENT PEAK $X_n(N,T)$ OVER ALL COMPONENTS AND OVER THE PERIOD RANGE 1 TO 6 SECONDS: $X_{2mn}(N)$

Figs. 5A, 6A and 7A indicate that the normalized mean Nth peak $X_{mn}(N,T)$, $X_{mnep}(N,T)$ and $X_{mnrp}(N,T)$ when plotted vs period is fairly flat over the 1 to 6 sec range. Using this characteristic of period independence, these means can be averaged again over this period range. This new mean, designated $X_{2mn}(N)$, $X_{2mnep}(N)$ and $X_{2mnrp}(N)$, is plotted vs N in Figs. 5B, 6B, and 7B. These plots refer to structures in the period range 1 to 6 sec, with 5 percent damping. Perez and Brady [3] found that for the elastic response the $X_{2mn}(N)$ plot for 44 horizontal components for the Imperial Valley earthquake of October 15, 1979 was very similar to the plot shown in Fig. 5B even though a different set of records is represented. We assumed, therefore, that the plots are reasonably representative, and that the trends indicated by Fig. 5B, and subsequently in Figs. 6B and 7B, are meaningful.

Fig. 6B shows that the amplitude attained by the Nth elasto-plastic peak falls off linearly as N increases by factors of 2. In Fig. 6B, the first ordinate, of 100%, represents the maximum response. The standard deviation for the data for 2nd, 4th, 8th, and 16th peak indicates the degree of scatter of points for all components and all periods within the period range, and is represented by the dark bracketed vertical lines. In this investigation, Figs. 5B, 6B and 7B show the results of taking the mean values, $X_{2mn}(N)$, $X_{2mnep}(N)$ and $X_{2mnrp}(N)$ obtained from 12 horizontal components, generated by the three earthquakes listed in Table 1, for the period range 1 to 6 sec. Specifically, this period range has 28 periods and each $X_{2mnep}(N)$ value in Fig. 6B therefore represents the mean obtained from 336 points of $X_{nep}(N,T)$. The value for the 8th largest elasto-plastic peak is 66 percent of maximum response, with a standard deviation of 20 percent. This can be interpreted to mean that when excited by an earthquake representative of this study, a simple structure at any location, with a period between 1 and 6 seconds, would respond in such a way that the 8th largest elasto-plastic peak, occurring at the end of the most severe four cycles of the structure's response, would be 66 percent of the maximum peak, on the average, with a standard deviation of 20 percent. In order to represent graphically other significant high values of the $X_{nep}(N,T)$ data, in addition to the mean and standard deviation, the different percentile levels are shown. The 100 percent curve, plotted for the 2nd, 4th, 8th, and 16th largest peak, indicates the upper bound or the one worst possible case. That is, among the 336 points, in the period range 1-6 sec for all 12 components recorded for these 3 earthquakes, this was the largest $X_n(N,T)$ peak value found. Also shown is the 99, 98, 97, 96, 95, 90, 80, 70, 60, and 50th percentile levels. When these $X_{nep}(N,T)$ percentile levels are presented in the form shown in Fig. 6B the

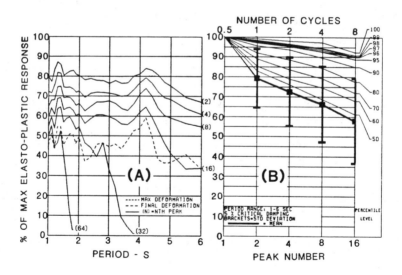

Fig. 6 Averaging the Nth normalized elasto-plastic displacement peak $X_{nep}(N,T)$ over 12 components (see Table 1) for each individual period: $X_{mnep}(N,T)$ (A). Averaging the Nth elasto-plastic displacement peak $X_{nep}(N,T)$ over 12 components and over the period range 1 to 6 s: $X_{2mnep}(N)$ (B).

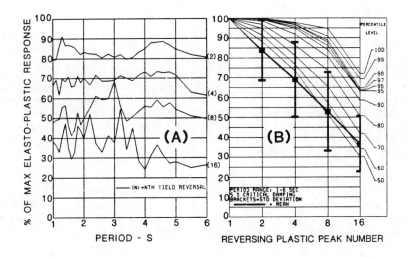

Fig. 7 Averaging the Nth normalized reversing plastic displacement peak $X_{nrp}(N,T)$ over 12 components (see Table 1) for each individual period: $X_{mnrp}(N,T)$ (A). Averaging the Nth reversing plastic displacement peak $X_{nrp}(N,T)$ over 12 components and over the period range 1 to 6 s: $X_{2mnrp}(N)$ (B).

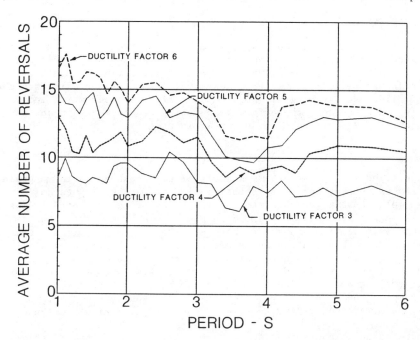

Fig. 8 Curves representing the average number of reversing plastic peaks (ductility factors of 6, 5, 4 and 3 from top to bottom respectively) for the 12 horizontal components listed in Table 1.

emphasis is on those critical upper limits that are of concern in seismic engineering and that must be taken into account when design criteria demand a design for the worst possible case, given by the 100th percentile, or any other percentile level deemed appropiate for a particular structure.

Comparing average values of the Nth elasto-plastic peaks $X_{2mnep}(N)$ and the average values of the Nth elastic peaks $X_{2mn}(N)$, it appears that the elasto-plastic values are in general greater than the elastic values. Although the plot of $X_{2mnep}(N)$ shows only ductility 5, further analysis shows that the greater the ductility factor, the greater the average normalized value of the Nth elasto-plastic peak.

The average normalized value for the reversing plastic peaks shown in Fig. 7B are similar to those shown in Fig. 5B. The standard deviation for the reversing plastic peaks is greater than for the elastic peaks. Nevertheless, there seems to be a correlation between the normalized amplitude of the Nth elastic peak and the normalized amplitude of the Nth reversing plastic peak. It must be noted that in the case of the reversing plastic peaks, their number is much less than the number of local elastic peaks experienced during the time history of the response. Fig. 8 shows that the average number of reversing plastic peaks is around 11 to 17 for elasto-plastic response with a period range of 1 to 6 sec and a ductility factor of 5. As the ductility decreases, so does the number of reversing plastic peaks.

DISCUSSION

In order to use the parameters defined here for prediction and design for future earthquakes they need to be tied to more fundamental ground motion paramenters, such as magnitude, duration and, distance. Although the curves are given in normalized form by their peak value it does not eliminate the effect of all these parameters. An extension of this study which includes normalizing the curves by the effective duration of the record such that each curve is duration-independent is currently in progress.

CONCLUSIONS

The detailed study of the time history of the elasto-plastic response obtained by using a selection of strong-motion earthquake inputs has given rise to a relationship governing the larger peaks of the response. Depending on the Nth peak of the elasto-plastic response being considered, the amplitude of that Nth peak remains a specific percentage of the maximum peak response. It should be noted that, although averaging is broadly applied over all oscillator responses within a certain period range, and over all the earthquake records considered, the trends are consistent. A typical specific result is that structures with periods in the range of 1 to 6 seconds, and with 5% damping and a ductility factor of 5, the 8th largest elasto-plastic peak will have, on the average, a value of 66% of the maximum response.

When compared to a previous study that considered only the elastic response, the Nth elasto-plastic peak (N=2,4,8 and 16) is in general slightly greater in value than the corresponding Nth elastic response peak.

When only reversing plastic peaks are considered, that is, the penultimate plastic peak is measured only if it is in the reverse direction than the present plastic peak, then the amplitude of the Nth reversing plastic peak is similar to the Nth elastic peak, regardles of the ductility factor. The Nth reversing plastic peak is related to the amplitude of the cumulative plastic deformation that has occured up to the time of the measuring of this Nth peak. The number of reversing plastic peaks is much less than the number of elastic peaks.

Further extension of these results would lead to correlations with the degradation of column strength that occurs when this different elasto-plastic peaks are experienced by structures during an earthquake. Experimental results will be needed for this correlation.

REFERENCES

[1] Hudson, D. E., "Response Spectrum Techniques in Engineering Seismology", Proceedings World Conference Earthquake Engineering, University of California, Berkley, 1956.

[2] Merchant, H. E. and D. E. Hudson, "Mode Superposition in Multi-degree of Freedom Systems Using Earthquake Response Spectra", Bulletin of the Seismological Society of America, Vol. 52(2), 405-416, April 1962.

[3] Perez, V and A. G. Brady, "Reversing Cyclic Elastic Demands on Structures During Earthquakes and Applications to Ductility Requirements", Earthquake Spectra, Vol. 1(1), 7-32, November 1984.

[4] Perez, V, "Velocity Response Envelope Spectrum as a Function of Time, for Pacoima Dam, San Fernando Earthquake, February 9, 1971", Bulletin of the Seismological Society of America, Vol. 63(1), 299-313, February 1973.

[5] _____, "Spectra of Amplitudes Sustained for a Given Number of Cycles: An Interpretation of Response Duration for Strong-motion Earthquake Records", Bulletin of the Seismological Society of America, Vol. 70(5), 1943-1954, October 1980.

[6] Newmark, N. M., "Current Trends in the Seismic Analysis and Design of High-rise Structures", in Earthquake Engineering, R. L. Wiegel (ed.) Prentice-Hall, Englewood Cliff, NJ, 403-424, 1970.

CONSIDERATION OF FLOOR FLEXIBILITY
IN DYNAMIC ANALYSIS OF HIGH-RISE BUILDINGS

M. A.M. Torkamani[I] and J. T. Huang[II]

ABSTRACT

Modal and transient analyses of a linearly elastic building subjected to ground accelerations are core and time intensive computations. To save computing time and to solve the problem at a lower core requirement, a unique combination of reduction procedures, with fixed-interface component mode synthesis as the central theme augmented by static condensation and Guyan reduction, is formulated and implemented for the given structure and load case.

A twelve-story three-dimensional building with an L-shape floor plan was analyzed. The results indicate that the combined procedures are advantageous in terms of convergence, the structural characteristics preserved and the percentage of reduction achieved. The results also confirm the importance of floor flexibility in the example studied.

INTRODUCTION

The main theme of the research effort is the application of fixed-interface component mode synthesis, augmented by static condensation and Guyan reduction, in order to evaluate dynamic characteristics and displacement response of a linearly elastic multistory building subjected to ground accelerations.

The date, application of the fixed-interface component mode method to buildings has been limited to a few highly idealized cases. Efforts are made here to formulate and implement the method as applied to seismic analyses of large buildings, and also to test as well as examine its feasibility, advantages and disadvantages. In formulating the modal synthesis, several simplified transformations are derived to upgrade computing efficiency.

The component mode method is a dynamic substructuring technique within the general domain of the finite element approach. By this method, normal mode shapes are extracted from components (substructures) and then used to obtain a reduced master model defined over physical coordinates (boundary coordinates) and generalized coordinates (component normal mode shapes). The master model is then analyzed at lower time and core requirements than that

---

I. Associate Professor, Department of Civil Engineering, University of Pittsburgh, Pittsburgh, PA 15260.

II. Former Graduate Student, Department of Civil Engineering, University of Pittsburgh, PA 15260.

required of an unreduced full scale model. 'The fixed-interface component mode method' is selected for its compatibility at the boundaries and its clarity in implementation.

To take advantage of the high stiffness in a floor plane, static condensation is applied before modal synthesis and Guyan reduction is applied afterward. The Guyan reduction serves to reduce the boundary DOF, which are wholly retained after synthesis. This unique combination of procedures resulted in a substantial reduction in the model size at both component and system levels when applied to solve a twelve-story 3-D building.

The preservation of the dynamic characteristics of components and part of the saving in core and computing time are achieved by the use of a truncated set of component modes. A small truncated set may be used for the given type of problem because of low input energy imparting onto higher modes and low participation by higher modes. Additional savings are achieved by sharing the same allocated core for sequential computations of many components. The penalties partially offsetting the above advantages are the needs to solve component eigenvalue problems and to perform many additional transformations.

MODEL REDUCTION

One way to stretch hardware capacity so that the same amount of available core can be used to solve a larger problem and to save computing cost is to reduce the size of a full scale model. This can be accomplished by using reduction techniques discussed below.

Substructuring and Static Condensation

The key idea of static condensation in reducing the stiffness matrix is to eliminate 'unwanted' interior degrees of freedom (DOF) by expressing them in terms of a set of DOF to be retained. The operation is equivalent to partially executed Gaussian eliminations. Static condensation can be applied to reduce a global model. It can also be applied to substructures before they are assembled.

Many computer codes adopt the static condensation technique. For example, ANSYS provides a "super-element" feature permitting the user to apply static substructuring to reduce model size. Another example is the TAB program family, i.e. ETABS, TABS and TABS'77, which was specifically developed for analysis of large buildings. For a three-dimensional building, the program automatically performs static condensation floor by floor, retaining only three DOF per floor, namely, two horizontal translation DOF and one rotational DOF about the vertical axis passing through the mass centroid of the floor.

When the method is applied to dynamic problems, the drawbacks are: (a) the local mode shapes involving eliminated DOF are lost, and (b) the lumping of masses to the retained DOF is done by judgement. In the case of TAB program family, the reduction scheme implies that, in all vibration mode shapes extracted from the reduced model, every floor collectively acts like a rigid body having only three out of six possible rigid body DOF. This is a good approximation for the type of buildings in which floor systems are very stiff and floor plans are convex shapes with low aspect ratios. In reality, many buildings do not fall in this category. Incidentally, another problem with the TAB programs is that they cannot accommodate bracing members that run in a vertical plane across several floors, a design feature that is incorporated in many high-rise buildings.

Guyan Reduction

To facilitate reduction in dynamic problems, Guyan [1] extended static condensation. In his formulation, the same transformation relating the complete set and the reduced set of coordinates was used to reduce the mass matrix so that the kinetic energy is invariant to coordinate transformation. It is a significant improvement over static condensation in that the mass lumping is based on stiffness relationships rather than judgement. But, again, the local mode shapes involving eliminated DOF are lost.

When local mode shapes reflecting floor flexibility are significant, an appropriate way to economically include them in the system model is the method of component mode synthesis.

Component Mode Synthesis

Since Hurty's [2] first proposal in 1960, the method of component mode synthesis (CMS) has been extensively applied in the aerospace industry. The method was initially slow in spreading, but recently there has been rapid proliferation in application to other fields.

Excellent reviews of the subject were provided by Craig [3], Noor [4], Nelson [5], and Meirovitch [6]. Their reports have served as a guide to this short survey.

The procedures of component mode synthesis are as follows:

1. Form stiffness and mass matrices and solve the eigenvalue problem for all substructures.

2. Perform coordinate transformations to reduce all component matrices. The new set of DOF consists of physical coordinates and a truncated set of normal coordinates.

3. Assemble all respective component matrices to obrain system stiffness and mass matrices.

4. Solve the master model for static or dynamic responses. Provide adjustment at the boundaries if necessary.

The key is the use of a truncated set of component normal modes as generalized coordinates. It is really an extension of the Ritz method. Without truncation, the process would simply be extra exercises. Without the use of normal modes, the convergence will most likely be very poor.

Methods of component mode synthesis differ in the way compatibility at the boundaries (components interfaces) is enforced. The first method is Hurty's 'fixed-interface normal mode' method [2]. His method requires that all boundary DOF are retained and that for the purpose of calculating component normal modes the component boundaries are fixed. The consequences are these: (a) Compatibility at the boundary DOF is not impaired. After component matrices are assembled, it is not necessary to adjust the boundaries to account for component interactions. (b) The reduction is carried out in the interior regions only. The total number of boundary DOF remains the same.

The second approach was proposed by Gladwell [7]. A component with a fixed interface is attached to another component which is free at the same interface. The modes of a substructure are calculated with all other connected substructures assumed to be rigid. This approach is called the 'branch component' method. It is suitable for chain-like structures. The third method, proposed by Goldman and Hou [8], is called the 'free-interface normal mode' method. There are hybrid versions of these three methods by MacNeal and Klosterman [9]. Details of these methods can be found in the literature cited; however, the main focus here is the fixed-interface method.

FINITE ELEMENT MODEL, REDUCTION AND SOLUTION

Model Reduction

As stated previously, by the fixed-interface component mode method, only interior DOF are reduced. All boundary DOF must be retained. This works out nicely for small plane frames. For larger building structures, the model size after synthesis is still large. The Guyan reduction used here serves to reduce DOF at the boundaries. The application of both static condensation and Guyan reduction therefore enhances the merit of the component mode method when applied to building structures. The combined procedures are appropriate because of favorable structure realities.

The TAB program family retains only three out of six possible rigid body DOF of a floor. As discussed previously, it is a good approximation when the floor plan is a convex shape with a low aspect ratio. For buildings in which the floor flexibility is a significant factor, failure to account for it can lead to detrimental errors in assessing design adequacy. The reduction procedures employed here provide a good compromise between an unreduced model and oversimplified ones.

During the combined reduction processes, the stiffness and mass matrices are defined over a total of six coordinate systems. They are:

1. Coordinates before static condensation at the component level. With respect to the references, component matrices and vectors are formed.

2. Coordinates after static condensation at the component level. With respect to the references, the reduced component matrices and vectors are defined.

3. Mixed coordinates for components. With respect to the references,

further reduced component matrices and vectors are defined. The reduction is the outcome of discarding higher component modes.

4. Mixed coordinates for the system after synthesis. The component matrices and vectors are transformed and assembled.

5. Mixed coordinates for the system after Guyan reduction. Based on the new references, the reduced system matrices and vector are defined.

6. Normal coordinates of the system after decoupling. System matrices and vector are redefined. A truncated set of system normal modes is then taken.

The combined reduction in model size is substantial, but the resulting increase in programming efforts for transformations and book-keeping is enormous.

EQUATION OF MOTION

For a component, the unreduced equation of motion subjected to ground acceleration is

$$[M]\{\ddot{u}^t(t)\} + [C]\{\dot{u}(t)\} + [K]\{u(t)\} = \{F\} \qquad (1)$$

where
[M] = component mass matrix
[K] = component stiffness matrix
[C] = component damping matrix
$\{\ddot{u}^t(t)\}$ = absolute or total accelerations
$\{u(t)\}$ = displacement relative to the ground
$\{F\}$ = interaction forces at the common boundaries

In these terms, a substript 'i' indicating the component number is implied, although not explicitly printed. These variables are defined over global coordinates (X,Y,Z). A component mass matrix is formed by directly lumping masses to the boundary DOF and to the interior DOF that are to be retained. A component stiffness matrix is formed by assembling element stiffness matrices in global coordinates.

The total acceleration may be expressed as

$$\{\ddot{u}^t(t)\} = \{\ddot{u}(t)\} + \{a\}\ddot{u}_g(t) \qquad (2)$$

in which the scalar time series $\ddot{u}_g(t)$ are ground accelerations, and $\{a\}$ is a vector indicating the scale factors. It is constructed as follows: assign value '0' to all rotational DOF and assign values a_x, a_y, and a_z to translational DOF parallel to global axes X, Y and Z, respectively. The horizontal direction of the earthquake is indicated by the vecotr (a_x, a_y). Equation 1 can now be rewritten as

$$[M]\{\ddot{u}(t)\} + [C]\{\dot{u}(t)\} + [K]\{u(t)\}$$
$$= -[M]\{a\}\ddot{u}_g(t) + \{F\} \qquad (3)$$

The seismic load vector is based on an unreduced diagonal mass matrix. The scalar time function is factored out for convenience in programming.

The initial finite element models of the components are subsequently reduced through static condensation and component mode synthesis at the component level, and through Guyan reduction and modal decoupling at the system level before solution for responses. Each of these operations results in a new set of stiffness and mass matrices as well as load vector. After synthesis, the system equation of motion remains the same in form as that of a component shown above; except that at all boundaries the respective sum of component interactions vanishes. They are internal forces of the system, and they must cancel (or be in equilibrium) themselves at every common boundary.

The damping matrix [C] is never formed. Instead, damping ratios are assigned to the uncoupled modes of the synthesized system. This is a matter of choice, because these two methods of assigning damping are directly related. Detailed derivations of the equation of motion is given in Ref. [10].

EXAMPLE

This section presents an example that was done to demonstrate that a fairly large 3-D building can be analyzed by the procedures using a limited amount of core. The results confirm the importance of floor flexibility in the example studied. Assuming inadequate diaphragm design, other cases in which the floor flexibility can be significant factor are: buildings with U, T or H-shape floor plan, buildings having setback or local irregularities, buildings supporting heavy masses on floors. The procedures are suitable for the given structure and load case because of the stiffness characteristics of a building and the predominance of lower component and system modes.

Figure 1 shows a perspective view of the twelve story 3-D building. Fig. 2 shows a typical floor framing plan. The floor plan and lump mass distribution are applicable to all floors. The dead weight is 940 kips (4181 kN) per floor per floor, which is equivalent to 133.2 psf (6.378 kPa). X-bracing members are used in vertical planes 4-6, 7-8 and 13-14. The bracing layout is similar to that of a floor plan. Table 1 shows the size of structure members. In the table, I_c and I_{sp} are component number and section property number, respectively. Although the design features and dimensions are assumed, they are realistic.

The building was represented by four components numbered sequentially upward. There were three common boundaries and an optional roof boundary.

The latter was included to enhance accuracy, the former were needed to maintain compatibility.

For both components No. 2 and 3, the DOF number was reduced from 336 (or 4x14x6) to 252 (or 2x84+2x42) when all rotational DOF of interior nodes were condensed out. (The program permits further reduction of some translational DOF in the interior). For the other two components, the DOF number was similarly reduced. On the roof boundary, all rotational DOF

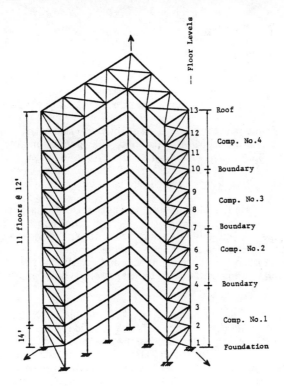

Figure 1. A Perspective View of the Building

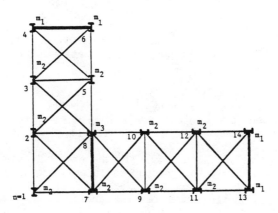

Note:
1. The signs I & H indicate positioning of columns.
2. n = node number
3. Lamp masses:
   $m_1 = 0.101367$ kip-Sec^2/in   (17.752×10^3 kg)
   $m_2 = 0.202733$ (35.504×10^3)
   $m_3 = 0.304097$ (53.256×10^3)
4. Heavy line indicates a vertical plane frame with diagonal bracing.

Figure 2. Typical Floor and Roof Framing Plan

were condensed out at the component level.

There were two interior floors in each of the four components. The dimension of all matrices entering the component eigensolution was 84 or 2X14X3. The number of normal modes in a component was therefore 84.

To simplify presentation, the following modeling parameters are defined: (a) N_B = the number of retained DOF per boundary, (b) N_R = the number of retained DOF on the roof, (c) N' = the number of retained modes per component, and (d) N_C = the number of components taken. By changing these parameters, the following cases were solved.

(a) N_B=24, N_R=42, N'=12, and N_C=4. After assembling, the system model had 342 DOF: 4x12 component modes, 42 DOF on the roof and 3x14x6 DOF on the common boundaries. The Guyan reduction cut the model size down to 162 DOF, which included 24 retained DOF per common boundary as shown in Fig. 1. This is an 84% reduction from the unreduced model having a total of 1008 DOF. The operations required a main array of 57K plus nominal common areas.

(b) N_B=24, N_R=42, N_C=4, and N'=1,4,6,24,36. Table 2 shows the calculated natural frequencies of the first 13 modes. Evidently, when all the other conditions remain the same, the calculated natural frequencies become lower and lower to approach the 'true values' as more and more component modes are included. These trends are in agreement with the known fact that the calculated natural frequencies are upper bounds. It is also evident that it is not necessary to include many component modes in this case. This is not surprising. We know that a cantilever beam modeled by four elements with consistent mass can produce good results. We may similarly expect Guyan reduction to yield good results if there are four components or four boundaries, and if there are enough retained DOF per boundary. When no component mode is taken, the method of CMS is equivalent to Guyan reduction. Therefore it can only do better when some component modes are included.

(c) N_B=9, N_R=9, N_C=4, N'=1,4,6. Figure 3 shows the retention pattern. Table 3 shows the results. This case proves that it is not necessary to retain many DOF per boundary. Note that if N'=6, the size of the reduced system is 42, which is equivalent to 3.5 DOF per floor.

(d) N_B=9, N_R=9, N'=12,21, and N_C=2. In the previous cases, the use of four boundaries helped. Could the procedures do well if there are only two boundaries? This case demonstrates that they can. It is remarkable that, after so much number crunching, the results are so close to that of the previous cases in which the sequences and domains of formations and reductions were quite different. We can attribute the success to the capability of the method of component mode synthesis to preserve structure properties effectively. From the case, it is judged that the model representing a twenty-story building with the same floor plan can be reduced to a system model consisting of (6 to 12)X4+9X4 = 60 to 84 DOF or 3 to 4.2 DOF per floor, and yields comparable answers.

In summary, the solution made of a twelve-story 3-D building using the procedures produced high percentage reductions and yet perserved the most important characteristics of the building.

TABLE 1. SECTION PROPERTIES

I_c	I_{sp}	A	I_y	I_z	J	Remarks
1	1/2	117.0	2170./6000.	6000./2170.	272.	column I/H
	3	24.8	2850.	106.	2.79	floor beam
	4	5.5	0.	0.	0.	floor bracing
	5	7.5	0.	0.	0.	frame bracing
2	6/7	68.5	1150./3010.	3010./1150.	62.6	column I/H
	8	24.7	2370.	94.4	3.72	floor beam
	9	4.865	0.	0.	0.	floor bracing
	10	6.6	0.	0.	0.	frame bracing
3	11/12	46.7	748./1900.	1900./748.	19.5	column I/H
	13	22.4	2100.	82.5	2.70	floor beam
	14	4.22	0.	0.	0.	floor bracing
	15	5.7	0.	0.	0.	frame bracing
4	16/17	42.7	677./1710.	1710./677.	14.2	column I/H
	18	20.0	1830.	70.4	1.86	floor beam
	19	3.555	0.	0.		floor bracing
	20	4.805	0.	0.		frame bracing

SUMMARY AND CONCLUSION

A unique combination of reduction procedures, with fixed-interface component mode synthesis as the central theme augmented by static condensation and Guyan reduction, was therefore presented and implemented for the given structure and load case.

In this work, the applicability and consequences of each method as well as the similarities and differences among them were examined. How they may be justifiably applied in a specified sequence was explained. In essence, static condensation reduces the matrices entering component eigensolution. The method of CMS transforms component matrices to reduced matrices defined over boundary DOF and a truncated set of normal mode shapes extracted from components with fixed boundaries. Guyan reduction eliminates DOF on the boundary after synthesis. In addition, by the choice of the manner in which a few intermediate steps can be treated, several simplifed transformations for carrying out modal synthesis were derived to up grade computing efficiency.

A program package was developed. The matrices former, reducers and solvers as well as the package were validated. The package will direct the computer to read data and form component matrices, accept specifications for retaining interior and boundary DOF that are arbitrarily patterned, and then perform three stage reductions and solve for natural frequencies, mode

TABLE 2. CALCULATED SYSTEM NATURAL FREQUENCIES, CPS

$N_R=42$, $N_B=24$, $N_C=4$

Mode number	N'=1	N'=4	N'=6	N'=12	N'=24	N'=36
1	0.3845	0.3841	0.3841	0.3841	0.3841	0.3841
2	0.5126	0.5111	0.5111	0.5111	0.5111	0.5111
3	0.7660	0.7624	0.7623	0.7621	0.7619	0.7619
4	1.1219	1.1117	1.1117	1.1115	1.1115	1.1115
5	1.4789	1.4582	1.4576	1.4573	1.4572	1.4572
6	2.0486	1.9809	1.9809	1.9765	1.9762	1.9762
7	2.1434	2.0975	2.0959	2.0930	2.0914	2.0913
8	2.2070	2.1568	2.1524	2.1452	2.1407	2.1407
9	2.5917	2.5513	2.5446	2.5413	2.5400	2.5400
10	2.9629	2.9116	2.9115	2.8801	2.8787	2.8787
11	3.3481	3.2149	3.2079	3.1894	3.1803	3.1801
12	3.4132	3.3027	3.2837	3.2755	3.2729	3.2726
13	3.6196	3.5504	3.5012	3.4907	3.4872	3.4872

TABLE 3. CALCULATED SYSTEM NATURAL FREQUENCIES, CPS

$N_R=9$, $N_B=9$

Mode number	$N_C=4$			$N_C=2$	
	N'=1	N'=4	N'=6	N'=12	N'=21
1	0.3854	0.3850	0.3850	0.3854	0.3854
2	0.5134	0.5119	0.5119	0.5121	0.5121
3	0.7664	0.7627	0.7626	0.7626	0.7624
4	1.1393	1.1289	1.1289	1.1374	1.1371
5	1.4952	1.4731	1.4725	1.4786	1.4782
6	2.1230	2.0539	2.0537	2.1005	2.0988
7	2.1483	2.1025	2.1010	2.1182	2.1175
8	2.2573	2.1993	2.1937	2.1796	2.1767
9	2.6466	2.6088	2.6008	2.6398	2.6388
10	3.1806	3.1083	3.1083	2.7813	2.7759
11	3.4051	3.2829	3.2721	3.2667	3.2571
12	3.4348	3.3034	3.2899	3.2796	3.2720
13	3.6972	3.6165	3.5573	3.4204	3.4018

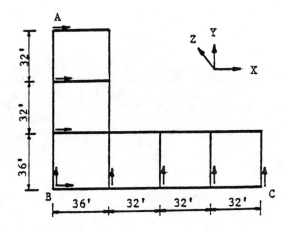

Figure 3   Retained DOF on Boundary Floor-Pattern B

shapes and displacement responses on a much reduced system model. The program uses dynamic core allocation and out-of-core operations so that until reduced forms are obtained, only one major matrix, whether it be stiffness or mass, component or assembled system, will occupy the CPU at a given time. In developing the package, much attention was given to economizing computing and core use. For example, several subroutines were written to replace the subroutine 'NROOT' in the IBM Scientific Subroutines Package (SSP). Roughly one third of the core need is thus saved.

The combined reduction procedures were applied to carry out dynamic analyses of a twelve-story 3-D building. Several solutions were made of reduced models by changing parameters such as the number of components, the number of retained component modes, and the number of retained DOF per boundary. The results demonstrated the importance of the floor flexibility in modes as low as the fourth for the case studied. The results also consistently showed that good convergence was achieved by much-reduced models, a pleasant but not at all surprising finding indeed. A rational is offered in the next paragraph.

Much credit should be attributed to the Ritz or component mode method and to Guyan reduction. But perhaps the characteristics of the given structure and load case deserve some attention. The given structure and load case can be characterized as follows: (a) The floor systems are stiff compared to the whole building laterally. For a typical floor, only its most flexible local modes need to be represented in the reduced model. (b) The energy contents in the high frequency components of ground accelerations are lower than that in the low frequency components. (c) Due to the zigzagging of higher mode shapes, their participation in the total response of a building is lower than that of the lower modes. Therefore, during the three stage reduction process, we have a choice to (a) retain a relatively small number of interior DOF for component eigensolution, (b) retain a relatively small number of component normal modes for transformation and synthesis, (c) retain a relatively small number of boundary DOF in the synthesized matrices, and (d) retain a relatively small number of decoupled normal modes of the reduced system, and still expect to obtain system results without significant loss in accuracy.

Admittedly, the procedures are subjected to the following penalties: (a) Component eigensolutions are required. (b) Many transformations are needed. But the payoffs are large savings in core achieved by substructuring, and huge savings in computing time to be gained by performing eigensolution and transient analyses on a much smaller model. By comparing the alternatives, it is obvious that the gains far exceed the penalties.

ACKNOWLEDGMENTS

This research is partially sponsored by the National Science Foundation (NSF) under Grant PFR-8001506. The authors thank NSF for this support.

REFERENCES

1. Guyan, R. J., "Reduction of Stiffness and Mass Matricies," AIAA J., No. 3(2), Feb 1965, pp. 380.

2. Hurty, W. C., "Vibration of Structural Systems by Component Mode Synthesis," Journal of the Engineering Mechanics Division, ASCE, No. , Aug 1960, pp. 51-69.

3. Craig, R.R., Jr., "Methods of Component Mode Synthesis," The Shock and Vibration Digest, November 1977, pp. 3-10, The Shock and Vibration Information Center, Naval Research Laboratory, Washington, D. C.

4. Noor, A. K., Kamel, H. A. and Fulton, R. E., "Substructuring Techniques - Status and Projections," Computers and Structures, Vol. 8, 1978, pp. 1313-1319.

5. Nelson, F. C., "A Review of Substructuring Analysis of Vibrating Systems," The Shock and Vibration Digest, May 1980, .

6. Meirovitch, L., Computational Methods in Structural Dynamics, Noordhoff, Rockville, Maryland, 1980.

7. Gladwell, G. M. L., "Branch Mode Analysis of Vibrating Systems," J. Sound Vibration, No. 1, Jan 1964, pp. 41-59.

8. Goldman, R. L., "Vibration Analysis by Dynamic Partitioning," AIAA J., Vol. 7, No. 6, June 1969, pp. 1152-1154.

9. Klosterman, A. L., On the Experimental Determination and Use of Modal Representations of Dynamic Characteristics, PhD dissertation, Dept. Mech. Engr., University of Cincinnati, 1971.

10. Torkamani, M. A.M. and Huang, J. T., "Dynamic Analysis of Multistory Buildings by Component Mode Synthesus", Department of Civil Engineering, University of Pittsburgh, SETE-CE-85-008, November, 1984.

ANALYSIS OF THE SEISMIC PERFORMANCE OF THE IMPERIAL COUNTY SERVICES BUILDING

C. A. Zeris [†], S. A. Mahin [††] and V. V. Bertero [‡]

The Imperial County Services Building (ICSB) is one of the few cases available where complete information on ground motion and structural response have been recorded. As such, it provides an excellent opportunity to evaluate the soundness of the seismic provisions used and the adequacy of the structural system and detailing employed. In this paper the recorded response is analyzed and nonlinear analyses of individual subassemblages and of the building as a whole are performed in order to understand the failure sequence of the building and to assess the adequacy of present analytical methods for predicting the observed damage and response.

INTRODUCTION

The ICSB was a six story reinforced concrete (RC) frame, which suffered damages and partial collapse during the 1979 Imperial County earthquake. The structure, located in El Centro, California, was designed in 1968 according to the 1967 Uniform Building Code [1]. Due to the proximity of the structure to an active fault and the form of structural system used, the ICSB was selected for instrumentation under the Strong Motion Instrumentation Program by the California Division of Mines and Geology (CDMG). All the installed accelerometers functioned properly, recording the structural response as well as the free-field and the foundation earthquake motions.

In order to best make use of the available information, a comprehensive program of field, experimental and analytical studies has been pursued. Tests on reinforcement and concrete specimens recovered from the building have been performed. These revealed that the strength of the structural materials at the time of the earthquake were above the minimum design specifications [2]. The design of the building and its compliance to more current seismic provisions has also been reviewed. The review indicated that the building incorporated many features associated with current seismic-resistant design and that most members were conservatively proportioned and detailed [2,3]. Linear elastic analyses of the building have been performed in order to establish the model which most closely simulates the actual state of the building during the earthquake. The results of the above studies have been reported elsewhere [4] and are only briefly reviewed herein.

The recorded response of the building is briefly examined herein. Purpose of this study is twofold; a) to evaluate the soundness of the filtering parameters used for processing the records, hence identifying a set of reliable records for use in analytical correlations and b) to correlate the observed damage in the building with dominant dynamic response patterns evident in the time histories. Descriptive displacement studies have also been performed by other investigators [5,6]. However, these studies were based on the originally processed records from the building [7], rather than the

---

[†] Research Engineer, Department of Civil Engineering, Univ. of California, Berkeley, California 94720.

[††] Associate Professor, Department of Civil Engineering, Univ. of California, Berkeley, California 94720.

[‡] Professor, Department of Civil Engineering, Univ. of California, Berkeley, California 94720.

newer ones used herein.

Results of planar nonlinear analyses of the building are also reported herein. The adopted structural model reflects current modelling practices in the routine analysis of reinforced concrete structures. This is done in order to assess the potential of those assumptions to analytically predict the nonlinear response of systems similar to the ICSB. However, a more refined model is adopted to analyze a corner column subassemblage in more detail. The purpose of this analysis is to identify the effect of bi-directional building response in the observed column damage.

General Information

Structural System The earthquake resisting system of the ICSB consisted of four moment resisting frames in the longitudinal direction and a frame-shear wall system in the transverse direction (Fig.1). In the longitudinal direction, the exterior frames were designed satisfying the requirements for ordinary frames, while the interior ones were detailed in compliance with ductile moment resistant frame specifications. In the transverse direction two three bay wide shear walls were provided at the exterior frame lines along the upper five stories. These walls were discontinued at the second floor. Shears from the upper walls were transfered through a 5 inch (12.7 cm) slab to four one bay wide walls in the ground story. The barbell shaped walls used were asymetrically positioned about the transverse centerline of the building (Fig.1b). The building was founded on pile groups.

From the point of view of structural details it is important to note that deep nonstructural precast panels were used in the exterior longitudinal frames. The geometry, boundary conditions and detailing at the base of all the ground story columns is also of particular importance in the overall distribution of damage observed, forcing the formation of the member critical region just above the ground level. The importance of these factors is discussed more extensively later.

Performance of the ICSB. Structural damages consisted of yielding of the reinforcement in all the ground story columns primarily at the base. Depending on the location of these columns this yielding was accompanied by longitudinal cracking along the corner reinforcement, spalling of the cover, loss of concrete core and even buckling of the longitudinal steel. This damage was consistent with a soft story action in the longitudinal frame direction of the building. Columns at the exterior frame lines were relatively more heavily damaged, particularly at the corners. The absence of a shear wall near the east end of the building permitted a catastrophic failure of all four columns at line G (Fig.1), resulting in a partial collapse of the supported structure. In addition, yielding of the beams in the lower stories in all frames, and separation of the nonstructural precast infills at the exterior facades were evident.

Design Adequacy The structure would fail to comply with current code requirements in that a) perimeter frames are no longer permitted to be non-ductile, and b) columns supporting discontinuous walls are required to be confined along the entire column height, rather than just in the ends [2,3].

The base shear coefficients used in the transverse and longitudinal directions were 0.08 and 0.04, respectively. For the longitudinal direction the above value was estimated assuming a ductile moment resisting frame (hence a structural system reduction coefficient -K- of 0.67 was chosen). The design lateral load was distributed equally among the four frames and subsequently increased for the exterior frames by the ratio of $1/0.67 = 1.5$. The exterior frame columns were detailed to provide limited ductile behavior, with the specification of confining hoops at the column critical regions similar to those of the interior frames, but with the spacing increased by 1.5 times.

Analytical Investigations

A simplified static load to collapse analysis for two typical exterior and interior frames acting independently is performed. The pattern of the applied lateral load is aasumed to be uniform with height. The analysis reveals that the actual capacity of the interior frame was about 25% smaller than the exterior frame (Fig.2). Moreover, the lateral stiffness of the exterior frame was almost twice as high compared to the interior one (with some uncertainty, depending on how the exterior nonstructural precast panels are modelled). Clearly, the difference in the resistance, ductility demand and energy dissipation of the two systems would be great; assuming that the two frames are constrained to have the same roof displacement, initiation of yielding in the exterior frames (excluding all bidirectional effects) will occur when only half the interior frame's resistance has been mobilized, at about 1.80 inches (4.6 cm) roof displacement. In order to mobilize the full strength of the inner frame a roof deformation of 4.0 inches (10.0 cm) is required. Hence, the exterior frame is expected to require at least twice the ductility as the interior frame. Furthermore, the flexural stiffness ratios of the exterior members and the high flexural resistance of the second floor beams due to the second floor deep slab promote the formation of a soft story at the ground level. The exterior ground story columns were not detailed to be as ductile as the interior ones. As a result of the increased deformation demand and reduction in the ductility capacity, the observed severity of damage of the exterior ground columns appears justified. Analyses assuming large member deformation capacities indicate that the expected ultimate lateral resistance of the entire four frame configuration is equal to 24% of the total reactive mass of the structure, close to the value estimated in [8] by interpolating the recorded longitudinal accelerations. This is nearly equal to 4.0 times the design ultimate strength stipulated by the code.

RECORDED HISTORIES AT THE SITE

Ground Motion Records

The instrument at the base of the building recorded a peak ground acceleration of 0.32g, 6.5 seconds after triggering (Fig.3a). However, the damage potential of the ground motion (for the range of periods of concern here) is mostly influenced by an earlier acceleration peak of 0.23g magnitude, associated with a 1.0 second period sinusoid (Fig.3b) that results in an incremental ground velocity of 24 to 40 in/sec. (60 to 100 cm/sec), depending on the filtering parameters used. Analysis of the building has indicated that this velocity pulse motion is responsible for initiating much of the damage in the structure.

Evaluation of the Processing of the Recorded Response

The recorded accelerations were originally processed by the CDMG using standard processing techniques. High and low frequency noise in the records was filtered using an Ormsby filter with 0.1-0.16 Hertz cutoff low frequency limits. Use of the above filter limits results in an awkward long period component that when doubly integrated causes unreasonably large floor displacement estimates during the first two seconds. The validity of this long period component has been questioned by several studies [9,10]. Such a long period component in the absolute floor displacement traces could be attributed to rigid body rocking due to foundation response to the passage of the triggering wave. However, this component is also inconsistently present in different deformation responses within the building. In an effort to improve this situation the records were reprocessed by Rojahn and Mork [10] of the United States Geological Survey (USGS), using a more stringent low frequency cutoff limit, namely 0.2 to 0.5 Hz. A comparative study of the two sets of processed records seems to justify the use of the more stringent filter parameters. Consider for instance the recorded in-plane slab deformation history at the east end of the roof in the transverse direction, given in Fig.4; this is obtained by subtracting the recorded motion (trace 3) from the record obtained by linear extrapolation of the records 1 and 2.

The two histories are markedly different; the CDMG trace estimates twice as high maximum slab deformations compared to the USGS trace, during the early portion of the record. The CDMG history is dominated from triggering time by the same long period component present in the absolute deformation traces indicating an abnormal response, The more stringent filtering does exclude this abnormal behavior while still preserving the significance of the slab in-plane deformations, characteristic of the structural system used. The estimated natural period of oscillation for this mode ranges around 1.0 seconds [11], close to the USGS trace periodicity. The difference in the recorded absolute roof displacement up to 13.5 sec. depicted by the two sets is shown in Fig.5. Eliminating the long period component results in reductions of the displacements by nearly 50% in both principal axes of the building.

Recorded Response of the ICSB

Confining our attention to the more stringently filtered record set it is possible to estimate the yielding and failure sequence at the east end of the structure. In Fig.6a, the in-plane deformation histories of the two ends of the roof slab relative to the middle are compared, obtained by subtracting trace 2 from 1 and trace 3 from 2 respectively. Hence the slab rotates undeformed as a rigid body if the two histories are equal and of the same sign. Because of the high rigidity of the end walls and the slab flexibility in plan, derived in-plane slab deformations are a direct indication of the wall rocking. It can be observed that up to about the 6.7 second, the end walls vibrate more or less in phase, with the interior frames lagging behind the wall motion and the slab bending symmetrically (in plan) in the middle. After 6.7 seconds one more similar cycle occurs, at double the deformation amplitude. In Fig.6b the relative rotation of the roof with respect to second floor slab history at the two ends of the building indicates that this in-plane deformation is in fact concentrated in the east end of the building, implying an overall increase in the east end wall rocking flexibility. This is attributed to initiation of yielding in the ground columns and reduction in their axial stiffness, confirming earlier analytical predictions ([4]). From this point on, a phase change of 180 degrees is evident in the histories of Fig.6a, implying that the recorded roof plan deformations are primarily due to slab rotation with respect to the vertical axis. At the next deformation peak, occuring at 8.5 seconds, the building rotates as a whole in such a way that the east end wall compresses column G4; the relative in-plan rotations between roof and second floor at this instant (Fig.6b) are almost none, implying that the entire east half of the building rocks as a rigid body on the column, resulting in the failure of the member. Upon reversal of the upper part of the building column G1 fails and possibly also B1 during the ninth second. Progressive failure is evident from then on, since the east end histories (both Fig.6a,6b) increase in period. Differences between traces are no longer possible to explain from then on using the flexural deformation assumption for the slabs; a large yield line forms across the building at bays F and G, with the settlement of the structure at the east end.

Prediction of the Linear Response of the ICSB

Linear elastic three-dimensional analyses of the entire building have been performed and are reported in [4]. The most reaslistic model of the initial state of the building should match the actual displacement traces recorded during the earthquake rather than the small amplitude vibration frequencies measured before the earthquake [12]. Indeed, it was observed that by using the uncracked gross section properties of the members one could match very closely the two longitudinal and first transverse mode periods from the ambient tests (used in [13],[14]). This however failed to correlate well with the actual recorded response. Use of the averaged cracked section stiffnesses of the members instead provided a model that matched closely the observed displacement histories during the entire linear part of the seismic response. The periods of the above model were 1.00, 0.36 and 0.44 seconds for the two longitudinal and the first transverse displacement modes, respectively.

ANALYSIS OF THE GROUND LEVEL CORNER COLUMN

A typical ground story corner column is analyzed in uniaxial bending response under 600 and 1200 kips (2670 KN and 5340 KN) axial load. From linear three dimensional analysis, the former is a close estimate of the gravity force on the member while the latter is the estimated axial load at which the ultimate bending capacity of the member was exceeded. It is important to note that a large portion of this load is attributed to the transverse rocking of the supported discontinuous wall rather than just longitudinal frame action. The flexural demands on the column come primarily from the longitudinal frame action. Hence, the behavior of the member is strongly influenced by the bidirectional response of the structure. The detailing of the column is shown in Fig.7. The purpose of the study reported herein is to investigate the importance of the offset in the longitudinal reinforcement and the as built boundary conditions of the member on the post yield behavior and concentration of damage. A more refined finite element model capable of simulating biaxial response of reinforced concrete columns is used.

The element used is a distributed damage fiber model. Equilibrium is imposed internally while flexibilities are linearly interpolated between monitoring slices. Plane sections are assumed to remain plane. Compared to other elements, advantages of the formulation are that: a) it is possible to reliable model crushing; b) coupling between axial and flexural stiffnesses can be taken into account; and c) cyclic characteristic phenomena of RC members can naturally be included in dynamic analysis without any a priori assumptions. A similar formulation for uniaxial bending and the applicability of the scheme has been reported in [15]. The refined element has been included in the general purpose nonlinear analysis program ANSR-I [16].

A sectional elevation of the column is shown in Fig.7. The column is reinforced with ten 1.37 in. (3.5 cm) diameter bars. Section dimensions change locally from 24 in. (59. cm) square to 22.5 in. (55. cm) just above ground level. The reduction in section area causes a reduction in the axial load-moment capacity of the column section shown in Fig.8a, evaluated at 0.3% maximum compressive strain. Lateral confinement consists of 0.5 in. (12.5 mm) diameter ties at 3. in. (7.6 cm) spacing in the critical regions, reducing to 0.37 in. (9.5 mm) diameter bars at 12. in. (30.5 cm.) along the middle part of the column. This reduction in hoop spacing affects the overall ductility capacity of the intermediate portion of the 24 in. square section. Analysis of this section under a range of axial loads demonstrates that the overall section behavior is far from elastic perfectly plastic at maximum flexural resistance (Fig.8b), and that the available curvature ductilities are generally small above an axial load of 600 kips (2670 KN). The column base was assumed for design to be at the the pile cap, 2. feet (0.61 m) below ground level. To accomodate the recess and change in cover the column reinforcement is bent above the section recess by 0.75 in. (2. cm), with a restraining hoop specified at the elbow. Field inspection of exposed reinforcement in the columns damaged at this location indicated that the hoop was missing in several instances. The behavior of a single offset rebar in compression was studied in [4]. The absence of a restraining hoop at the offset results in a reduction of the ultimate axial compressive resistance of the bar by 50%.

The analytical model of the column includes several elements in order to account for the three different types of section size and detailing used along the height of the column. To simulate the presence of the offset, the end section compression reinforcement in the low confined region above the recess is specified at a reduced ultimate resistance as indicated above. Furthermore, the concrete is assumed to spall in the vincinity of the compression reinforcement when this buckles, due to the lateral movement of the buckled offset elbow. Four columns are analyzed, 1) the as-built column with discontinuous lateral confinement but excluding the offset capacity reduction. 2) with discontinuous confinement including the rebar offset at the recessed part. 3) with continuous high confinement along the height including the part where there is a recess. 4) with continuous high confinement excluding the recess. In all four cases the column is analyzed under tip displacement control. The column is fixed at ground level. At the top rotational springs are assigned to model the flexibility of the adjoining members. Lack of fixity at the top moves the point of contraflexure above the midheight of the column. The tip displacement is increased monotonically from 0. to 1.2 in. (3.05 cm), the recorded interstory drift at the first major inelastic excursion of the building. At the same time the axial load is increased from the static value of 600k to 1200 k (2.67 and 5.34 MN).

The influence of the change in cross section at the base of the column is demonstrated in Fig.9a, where the base moment-rotation demands for the fully confined uniform section versus the recessed base section (models 3 and 4) are compared. It can be seen that in the case of a recessed section, the recorded interstory drift causes the column to reach its ultimate capacity of 8600 kin (980 KNm), at the maximum axial load. The non-recessed section has a demand of 10730 kin (1210 KNm) which it apparently is able to resist with half the amount of base rotation, demanded in the previous case. No evidence of softening is observed in the latter case. For comparison, the demand in the base of the non-recessed section at the pile level, had the column been fully unrestrained at the ground level is equal to 9200 kin (1040 KNm), marked in Fig.9a. In this case the axial load bending moment demand at the base would be fully supplied without major damage of the member. Moment-rotation demands are also compared at the level just above the section recess where most of the damage of the member was concentrated, in Fig. 9b. As a result of the change in cross section below the location considered, the demand at this level is now limited by yielding of the base section, while in the uniform section member (no section reduction) the demand increases; since the confinement in all cases is adequate, no excessive damage is predicted. For the base of the member at the pile cap level, yielding at this point is never reached. The design of the column would therefore be adequate to resist the recorded interstory drift and axial loads.

The ideal continuous high confinement case (3) and the as-built low confined column including or not the effects of the offset rebar strength reduction (cases 1 and 2) are compared in Figs. 10a and b. Again, section moment-rotation demands for the prescribed load-deformation history are compared at the offset and fixed ground levels. In amost all cases, the ground section resistance is governed by the ultimate recessed section capacity given by the interaction diagram of Fig.8a. The as-built column however being more flexible above the recess (due to lower confinement along the midheight of the column) has twice as much rotation demand than the continuously confined member at the base. In both cases the critical region forms at the base. The low confinement column starts softening as the load approaches the balanced load. Such a behavior cannot be predicted by simplistic beam-column elements. The case is completely different, however, if the effect of the offset is also included. As Fig.10a shows, for the same interstory drift the fixed rotation demands for the column with the offset are less than 30% of those without the offset at the base. The reason for this is clearly seen in Fig.10b, where the rotations at the offset level are compared: in this case, the column fails catastrophically at the offset level limiting the demands in the fixed column base below. In the absence of the offset, plastic unloading at the base results in an increase of the moment taken by the undamaged member top and therefore in unloading of the demands at the offset level with increasing axial load, as shown in Fig.10b by the dark line. Clearly, this behavior is closest to the observed damage distribution in the member. Considering the axial elongation of the low confined portion of the column (Fig.10c), it is obvious that the as built column fails to support the imposed axial load, at a maximum resistance of 1020 kips (4.5 KN), transfering the axial load to the interior frame columns.

Summarising, the presence of the ground restraint and the low confined region above the recess increases the base column demands forcing the critical region at the offset. Local weakening of the column at the offset reduces considerably the ductility capacity of the member at this location and allows only limited redistribution of moments, while limiting the axial resistance of the member considerably.

PLANAR NONLINEAR ANALYSES OF THE ICSB

Nonlinear dynamic analyses of the ICSB have been performed by other investigators, using the originally released ground motions. The building has been analyzed both in plane [14] and as a three dimensional space frame [13]. In all cases displacement correlations were poor, due to the use of bilinear member models, selection of member structural properties on the basis of the gross section dimensions, ignoring the foundation flexibility, and in one case, allowing for limited nonlinearity in the ground floor east end of the structure only.

The building is analyzed in the longitudinal sense using the proportions and properties of the three dimensional linear model proposed in [4], modified for planar analysis. Previous linear analyses of the entire structure under the longitudinal acceleration input only gave nearly exact displacement correlations until the initiation of yielding. Good correlations were therefore expected in the first 6.5 seconds of the present two-dimensional analysis.

An interior and an exterior frame were included in the model, with horizontal degrees of freedom slaved at each floor. The longitudinal ground motion recorded under the building is specified at the ground level. Only 10.0 seconds of motion are considered. Foundation rotational and vertical springs simulate the effect of the piles under the columns. Lateral masses are considered only, lumped at the floor levels. Damping is mass and initial elastic stiffness proportional, selected to give modal damping ratios of 5% of critical in the first two modes. The initial periods of the model are identical to the corresponding longitudinal periods of the space frame analyzed in [4]. The member capacities were estimated based on the true material characteristics of the structure. Reduction in section size of the ground story columns at the base is taken into account. Exterior beam capacities are computed ignoring the presence of the nonstructural precast elements. Two analyses are performed as follows:

1) The resistance characteristics of the elements are assumed to be bilinear. The ratio of the post yield to initial elastic stiffness used is 10% for all members, but the lower three story columns, where 0.1% is used. Axial load-bending moment interaction is taken into account.

2) The entire structure is modeled using stifness degrading beam elements, The initial and post yield stiffnesses varied as in the previous case. Axial load-bending interaction is ignored, however. The column yield capacity was computed based on the static gravity load.

The roof displacement history for the different models is compared with the recorded trace in Fig.11. In all models yielding in the structure first occurs at 6.7 seconds at the first floor exterior beams. Until then, displacement correlations are excellent. The difference in the peak post-yield displacement estimate between different models is small. Overall, the Takeda model has a greater flexibility reaching a peak displacement upon first reversal (at 6.8 seconds) of 3.5 inches (9 cm); this value compares better with the recorded peak than the bilinear model one. Beyond this peak the planar models fail to correlate with recorded motions. Some period elongation associated with yielding of the ground columns is evident but the amplitudes are incomparable. The elimination of the transverse ground component results in extremely low fluctuation in the axial loads and longitudinal moments of the perimeter columns (Fig. 12). As a result only a small nonlinearity occurs at the column G4 at 6.8 seconds. All inner frame columns yield at a constant axial load. Maximum rotation ductilities in these members based on the Takeda and the bilinear model have been calculated to be around eight.

It can be concluded that the simplified nonlinear planar analysis fails to predict the observed structural damage. Furthermore, the predicted location of nonlinearity does not correspond to the observed concentration of damage at the ground story corner columns, and is therefore qualitatively and quantitatively misleading. Primary causes for such a discrepancy is the inability of the simplistic elements used to adequately model the actual cyclic resistance-deformation characteristics of the ground perimeter columns. Furthermore, structural idealizations such as the one used for this particular system fail to take into account the significant contribution of the bi-directional response of the structure to the decrease in stiffness and strength, and to the increase in the axial load and rotational demands on the columns in the longitudinal sense.

CONCLUSIONS

In summary, the following conclusions can be drawn. The more strictly filtered set of the building records is more realistic for analytic correlations since the presence of the unacceptable long period component is eliminated from the structural deformation patterns. As observed in the transverse deformation histories, failure of the building at the east end occurs due to the rocking of the supported wall between 8.5 to 9.5 seconds. Already, yielding of the second floor beams and most importantly the ground columns has taken place at 6.6 seconds, due to the first energy burst of the record.

Due to the axial load levels resisted and the boundary conditions the behavior of the ground corner columns was far from desirable. This effect was accentuated by local changes in the section size and details at the bottom critical region of these members, located above the design anticipated location. From the design point of view, the need for correct field implementation of the design assumed details is of great significance. Planar nonlinear analyses of the building with simplified elements have been unable to predict not only the global displacements, but also the actual yielding pattern within the structure. It can be concluded that in analyzing similar structural systems, a refined three dimensional analysis is necessary in order to better estimate expected response and damage.

ACKNOWLEDGEMENTS

The authors would like to acknowledge the financial support of the National Science Foundation during the course of this study. We are thankful to Chris Rojahn and Pete Mork for providing us with the refiltered records of the ICSB histories. The opinions and statements expressed in this paper are a responsibility of the authors and not necessarily of the sponsoring organization.

REFERENCES

[1] Uniform Building Code, *Intnl. Conf. of Building Officials*, Whittier, California, 1967.

[2] Allawh N., A Study of the Behavior of the Imperial County Services Building During the 1979 Imperial Valley Earthquake, Master of Eng. Thesis, *Dept. of Civil Engin.*, Univ. of Calif., Berkeley, California, June 1982.

[3] Applied Technology Council, An Evaluation of the Earthquake Response and Associated Damage of the El Centro Imperial County Services Building, Rep. No **ATC-9**, Menlo Park, 1984.

[4] Zeris C. and Altmann R., Implications of the Damages of the Imperial County Services Building to Earthquake Resistant Design, **IV** , *Proc.* , *VIII World Conf. of Earthqu. Eng.*, San Fransisco, 1984.

[5] Hart G.C., Huang S. and Nakaki D., Imperial County Services Building: Displacement Analysis of the 15 October 1979 Strong Motion Earthquake Records, Rep. UCLA ENG 80-83, Univ. of Calif., Los Angeles, California, Nov. 1980.

[6] Pauschke J.M., Oliveira C.S., Shah H.C. and Zsutty T., A Preliminary Investigation of the Dynamic Behavior of the Imperial County Services Building During the October 15, 1979 Imperial Valley Earthquake, Rep. 49, Stanford Univ., Palo Alto, California, Jan. 1981.

[7] Mc Junkin R.D. and Ragsdale J.T., Compilation of Strong Motion Records and Preliminary Data From the Imperial Valley Earthquake, of October 15 1979, Rep. 26, *Calif. Div. of Mines* and *Geol.*, Sacramento, California, 1980.

[8] Kreger M. and Sozen M., A Study of the Causes of Column Failures in the Imperial County Services Building During the 15 October 1979 Imperial Valley Earthquake, Rep. UILU-ENG-83-2013, *Univ. of Illinois at Urbana Champaign* , Illinois, August 1983.

[9] Stephens J. and Yao J.T.P., Data Processing of Earthquake Acceleration Records Rep.No. CE-STR-85-5, Purdue Univ., July 1985.

[10] Rojahn C. and Mork P.N., An Analysis of Strong-Motion Data From a Severely Damaged Structure, the Imperial county Services Building, El Centro, California, Rep. 81-194, *U.S. Geol. Survey*, Menlo Park, California, 1981.

[11] Jain S.K., Analytical Models for the Dynamics of Buildings, Rep. EERL 83-02, *Earth. Eng. Res. Lab.*, Calif. Inst. of Technology, Pasadena, California, May 1983.

[12] Pardoen G.C., Imperial County Services Building Ambient Vibration Test Results, Rep. 79-14, Univ. of Canterburry, Christchurch, New Zealand, Dec. 1979.

[13] Shepperd R. and Plunkett A.W., Damage Analysis of the Imperial County Services Building, *Jnl. of Am. Soc. of Civil Eng.*,**109** , ST7, July 1983, 1711 .

[14] Moss P.J., Carr A.J. and Pardoen G.C., Inelastic Analysis of the Imperial County Services Building, *Bull. of New Zealand Natl. Soc. of Earthqu. Eng.* , **16** No 2, June 1983.

[15] Zeris C. and Mahin S., Significance of Nonlinear Modeling of R/C Columns to Seismic Response, *Am. Soc. of Civil Eng.*, Third Specialty Conf. on Dynamic Response of Structures, Unic. of California, Los Angeles, April 1986.

[16] Mondkar D. and Powell G., ANSR-I General Purpose Program for Analysis of Nonlinear Structural Response, Rep. EERC 75-37, *Earthqu. Eng. Res. Center*, Univ. of California, Berkeley, Dec. 1975.

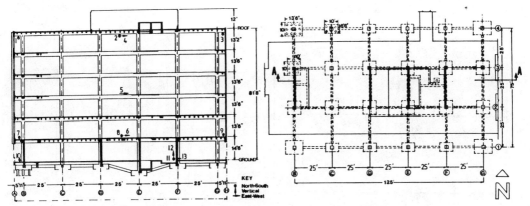

Fig. 1a ICSB ground level and foundation plan

Fig. 1b Elevation and instrument location

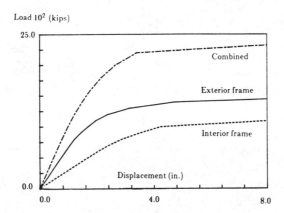

Fig. 2 Uniform lateral load-roof deformation of the ICSB

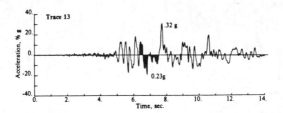

Fig. 3a Recorded ground acceleration under the building, EW.

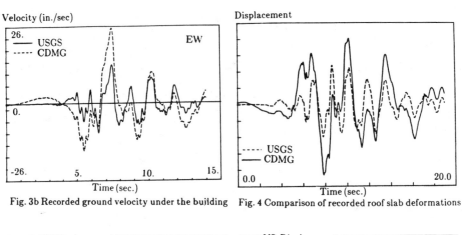

Fig. 3b Recorded ground velocity under the building

Fig. 4 Comparison of recorded roof slab deformations

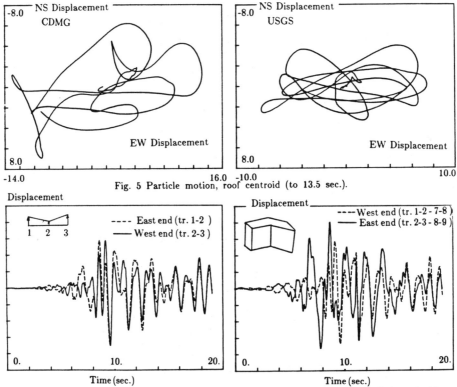

Fig. 5 Particle motion, roof centroid (to 13.5 sec.).

Fig. 6a In-plan slab deformation of the ICSB, roof

Fig. 6b Torsional deformation, roof relative to 2nd floor.

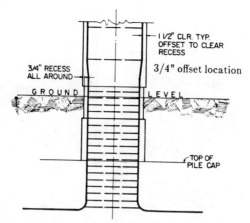

Fig. 7 Typical column reinforcement detail, ground level.

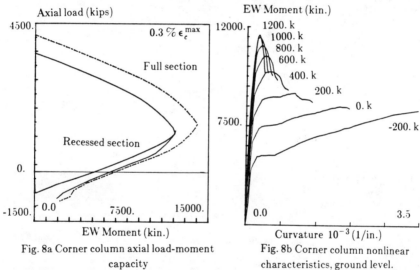

Fig. 8a Corner column axial load-moment capacity

Fig. 8b Corner column nonlinear characteristics, ground level.

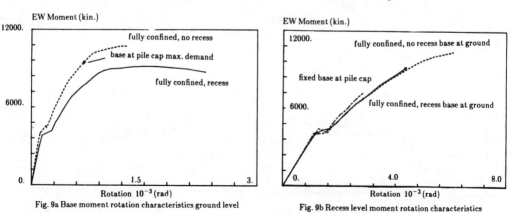

Fig. 9a Base moment rotation characteristics ground level

Fig. 9b Recess level moment rotation characteristics

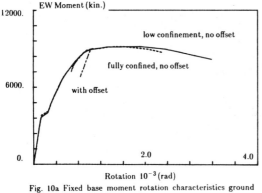

Fig. 10a Fixed base moment rotation characteristics ground

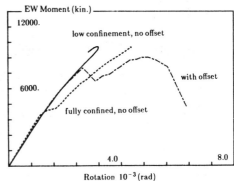

Fig. 10b Recess level moment rotation characteristics

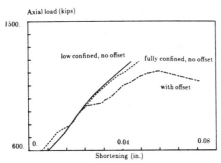

Fig. 10c Axial load-elongation, middle portion of the column

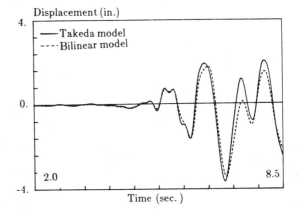

Fig. 11 Correlation, EW recorded-predicted roof rel. displacement

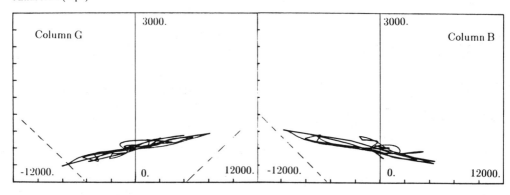

Fig. 12 Predicted axial load-bending demands, corner columns

VULNERABILITY OF TALL BUILDINGS IN ATLANTA, CHICAGO,
KANSAS CITY AND DALLAS TO A MAJOR EARTHQUAKE ON THE NEW MADRID FAULT

Lincoln E. Malik[I] and Otto Nuttli[II]

Abstract

Attenuations east of the Rockies is such that shaking from major earthquakes is felt hundreds of miles from the epicenter. Vulnerability of tall buildings in the four cities to magnitude 7.6 and 8.6 earthquakes on the New Madrid fault is assessed. Dynamic characteristics of shaking and seismic demand on generic 40 story buildings is estimated. Building capacities are assumed to be per wind and drift code requirements. Vulnerability of tall buildings' structural and non-structural elements is evaluated. Recommendations for future studies are proposed.

Introduction

This study is a qualitative evaluation of the performance of tall buildings in the subject cities to a major earthquake in the New Madrid Seismic Zone. The events considered are magnitude (M_s) 7.6 and 8.6 anywhere in the New Madrid Seismic Zone. Tall buildings considered are commercial and/or residential buildings with 20 or more floors.

This effort was not intended to be a comprehensive evaluation as there are very little data available to establish the relevant characteristics of the anticipated ground shaking at such large distances from the New Madrid Seismic Zone. Furthermore, a detailed evaluation would require identification of the dynamic characteristics of the structures which was not performed.

This study attempts to ascertain the general level and type of ground shaking expected in the four cities due to earthquakes in the New Madrid Seismic Zone and to evaluate in a generic sense the behavior of tall buildings to such shaking. The evaluation is then used to arrive at some broad conclusions and recommendations.

Ground Motion

It is difficult to estimate ground motions at long distances from the New Madrid Seismic Zone since there are no instrumentally recorded data at such distances from a major earthquake in this Zone.

[I.] Vice President, URS/John A. Blume & Associates, Engineers, San Francisco, California.

[II.] Professor of Geophysics, Saint Louis University, Saint Louis, Missouri.

Hypothetical regional intensities developed by the United States Geologic Survey (USGS) (1) for an M_S = 8.6 earthquake anywhere along the New Madrid Seismic Zone estimate Modified Mercalli Intensity (MMI) VII for Chicago, Atlanta and Kansas City.

Recognizing the difficulty of establishing rigorous estimates of ground motion at such large distances, estimates of peak ground accelerations, velocities and displacements as a function of wave periods (see Figures 1 through 6) for earthquake magnitudes 7.6 and 8.6 in the New Madrid seismic zone were developed. The estimates are for periods of 1 to 20 seconds, as tall buildings are expected mostly to have fundamental periods of vibration in the range of 2 to 10 seconds. These estimates are only approximate values, and may be two times too large or too small. It should be noted that this study was completed before the Mexican Earthquake of September 19, 1985, or the earthquake in north eastern Ohio on January 31, 1986.

At distances of 600 to 700 km the peak ground motion will be caused by surface waves. They will have a long duration, of the order of 1 to 3 minutes, and will be almost purely sinusoidal. This maximum acceleration is assumed to occur at periods near 1 sec, and their maximum displacement at periods of 8 sec and longer. Because the waves are sinusoidal, $v = 2\pi fd$ and $a = (2\pi f)^2 d$. Peak acceleration, at 1-sec period, is calculated from Nuttli and Herrman (4). Peak dispalcement, at 20-sec period, is calculated from Gutenberg and Richter (5).

It is expected that due to the long source-site distances being considered, the seismic waves will be well separated in accordance with their frequency. Thus, the ground motion at the four cities is expected to last much longer than in sites close to the earthquake as trains of various frequency waves arrive at different time intervals.

Amplifications of the seismic base rock motions due to effects of local soil conditions were not considered in the study. These effects are suspected to have contributed significantly to the damage in Mexico City during the recent earthquake there. Such amplifications should be investigated in future studies.

Response of Tall Buildings to Seismic Shaking (Seismic Demand)

The seismic demand on a structure is a function of the ground shaking and the dynamic characteristics of the building. Inventory data compiled by the Council on Tall Buildings (2) in 1980 is shown in Table 1. This inventory is dated and more tall buildings have since been erected in the four cities. Most of the tall buildings in this inventory are within the range of 20 to 40 stories with concrete and steel construction dominating. For the purposes of establishing benchmark values for comparing seismic demand on tall buildings, studies were performed on a generic control tall building with the following characteristics:

```
No. of stories          =   40
Height per story        =   12.5 ft
Total height            =   500 ft
Width (square shape)    =   143 ft
```

Aspect ratio = 1:3.5
Average floor weight = 170 psf

It is further assumed that the control building is well behaved, without significant architectural eccentricities or significant material or structural discontinuities.

The seismic response of the control tall building described above is expected to be dominated by its fundamental mode of vibration which will most likely produce a generally linear maximum story shear envelope. Given the assumption that the first mode is the dominant mode, the seismic forces acting on the building are a function of the spectral accelerations and the floor mass of the structure.

Envelopes of maximum story shears on the control building described above is given in Figures 7 to 10 for the four cities for a magnitude 7.6 and 8.6 earthquakes in the New Madrid Seismic Zone. These story shears are based on the peak ground accelerations shown in Figures 1 to 6, the assumed mass and geometry of the control building and simplifying assumptions regarding the dynamic building response.

Capacity of Tall Buildings to Resist Earthquake Shaking

Building capacities to resist earthquake shaking are a function of their vertical and horizontal force resisting structural systems. We assume that the buildings in the study area are only designed to resist gravity and wind loads and to satisfy code requirements for drift control and other design provisions.

Inquiries at City Engineer's offices in the four cities indicate that Atlanta enforces wind provisions of the American Standards Institute Inc. (A58.1-1972) while Chicago, Kansas City and Dallas enforce the wind provisions of the Uniform Building Code (UBC). In recent years, Chicago has reduced its wind design requirements to 30 psf of area which is below UBC requirements. Figures 7 through 10 show the maximum wind story shears on the control building required by code in the four cities. It should be noted that these are based on current wind load provisions which are not necessarily those in force when a particular tall building was designed.

A major difference between seismic and wind demand on a building is that maximum story shear forces on a building are proportional to the floor mass in the case of earthquake loads and proportional to the exposed area in the case of wind loads. The wind forces given in Figures 7 through 10 are based on a square floor area.

Figures 7 through 10 also give the envelope of maximum story shears on the control building according to the seismic provisions of the UBC. In these calculations, conservative values of horizontal force factor (K) = 1.0 and site-structure resonance coefficient (S) = 1.5 were used. Chicago, Kansas City and Dallas are identified in the UBC as being in seismic zone 1 and Atlanta in seismic zone 2. Even though its been assumed that the existing tall buildings in these cities were not designed to any seismic provisions, it is informative to establish what the design seismic lateral loads would have been if the UBC was enforced.

The envelopes of maximum wind story shears shown in Figures 7 through 10 are the code forces to be used in design. They give an indication of the capacity of the building to resist lateral forces but are far below the full capacity of the building because it is common that drift requirements, rather than wind loads, control the design of lateral force resisting systems of tall buildings. Furthermore, buildings usually have capacity to resist significantly higher loads than they are designed for because of factors of safety in engineered structures. Some of these factors of safety are required by the code, such as, providing allowable stresses which are significantly below yield stresses (working stress design), increasing the estimated design loads (factored loads in ultimate strength design) and limiting material strengths to minimum specified values. Futhermore, engineers commonly use conservative methods to calculate loads on the building which overestimate the ratio of demand to capacity and often provide higher than required strength and stiffness into building components such as beams and slabs to limit deflections for functional purposes.

The envelopes of maximum shears presented in Figures 7 through 10 show that for the cities in UBC seismic zone 1 (Chicago, Kansas City and Dallas), the wind loads control the design of the 40 story control structure except for a few top stories. UBC seismic loads for Atlanta (zone 2) are slightly higher than wind loads.

Table 2 gives the maximum base shears for the control building for code design wind forces, UBC earthquake forces and estimated forces from magnitude 7.6 and 8.6 earthquakes in the New Madrid Seismic Zone as well as the ratio of earthquake to wind base shears. Figures 7 through 10 and Table 2 show that shears from a magnitude 7.6 earthquake are nominally higher than those due to wind design for Chicago and Dallas and only moderately higher for Kansas City and Atlanta. This pattern is expected as Kansas City and Atlanta are closer to the New Madrid Seismic Zone. Furthermore, Atlanta uses ANSI wind provision which are lower than those by the UBC. The magnitude 8.6 earthquake story shears are significantly higher than design wind forces.

Performance of Tall Buildings to Earthquake Shaking

The above discussions of seismic capacity and demand on tall buildings are only intended to establish general parameters to guide the evaluation of the performance of tall buildings during a major earthquake in the New Madrid Seismic Zone.

Following is a qualitative assessment of the performance of structural and non-structural elements of tall buildings in the four cities. This assessment is for well behaved structures with average or above average design and construction quality. It must be emphasized that significantly higher damage could occur in ill behaved buildings such as:

a. Buildings with largely eccentric geometries
b. Buildings with material or stiffness discontinuities as happens

with buildings constructed of mixed materials or additions to older buildings
c. Buildings with poor diaphragms , poor design details and substandard construction quality
d. Brittle construction such as unreinforced brick or concrete block

<u>Overall Performance of Tall Buildings:</u> It is our considered opinion that well behaved tall buildings in the four cities will maintain structural integrity and have light structural damage due to a magnitude 7.6 earthquake in the New Madrid Seismic Zone.

A magnitude 8.6 earthquake in the New Madrid Seismic Zone is estimated to produce lateral forces significantly larger (2.5 to 3.5 times) than design wind loads. Well behaved structures are expected to maintain structural integrity for this level of shaking and possibly have light structural damage. Moderate localized structural damage is possible in some cases.

This qualitative assessment of damage is based on consideration of factors of safety inherent in structural design, performance of tall buildings in previous earthquakes and judgment. It can be shown that buildings designed to code allowable stresses can resist 2.5 to 3.0 times the design forces with minimal structural damage. It is possible that at these load levels some structural members will undergo limited localized yielding. Tall buildings in Los Angeles were surveyed for damage after the San Fernando Valley earthquake in 1971. Many of these buildings were later analyzed by URS/Blume and others (3). It was found that the buildings sustained little damage despite having to resist estimated earthquake forces several times their design loads. Table 3 gives ratios of estimated earthquake maximum base shears to design base shears and damage data for five such tall buildings.

Surveys of the performance of tall buildings in other earthquakes confirm the observation that tall buildings are capable of resisting significantly higher demands than their design loads with minimal damage.

<u>Performance of Structural Elements.</u> Structural elements are considered to be the principal load resisting components of the building. It is recognized that non-structural elements may in reality carry some loads but it is common engineering practice to neglect such capacity in the evaluation of structures.

Reinforced concrete buildings commonly use shear walls or moment resisting frames composed of floor beams and columns to resist lateral loads. The potential for structural damage to either system from a magnitude 7.6 earthquake is relatively small. A magnitude 8.6 earthquake may produce structural damage as follows:

a) <u>Shear Wall Buildings</u>: Buildings using a shear wall system to resist lateral loads are expected to perform well with possibility of some diagonal cracks in the shear walls which will not impair the structural integrity of the walls.

b) <u>Reinforced Concrete Moment Resisting Frames</u>: Buildings with a moment resisting frame are generally more flexible than shear wall systems. Hence, they are expected to amplify accelerations from the long period waves more than shear wall buildings and thus will produce higher forces and displacements. It is possible that slender columns may develop shear cracks at their center or flexural cracks at the joints. Buildings employing strong-girder-weak-column systems or very tall unbraced slender columns will be more vulnerable. There is also a possibility of moderate localized damage to columns in buildings with significant eccentricities. Nevertheless, this is not expected to impair the structural integrity of the buildings.

c) <u>Braced Steel Frames</u>: Braced steel frame structural systems are expected to perform best during an earthquake. Such systems have sufficient stiffness to limit story drifts as well as sufficient ductility to limit damage. Such buildings are not expected to have damage for a magnitude 7.6 earthquake. A magnitude 8.6 earthquake may produce localized yielding at some connections but such yielding is not expected to be large enough to produce appreciable damage.

d) <u>Steel Moment Resisting Frames</u>: Buildings with a steel moment resisting frame are more flexible than a braced steel frame which leads to higher story drifts in the building. Nevertheless, the inherent ductility in a properly designed steel frame offers sufficient excess capacity to resist loads significantly higher than design loads without damage. Steel frame structures are not expected to exhibit structural damage for a magnitude 7.6 earthquake. It is possible that some elements will sustain localized yielding at the joints for a magnitude 8.6 earthquake. This will be especially true for eccentric buildings.

<u>Non-Structural Elements.</u> Such components as utility lines, parapets, windows, partitions, facade, large fixtures and large pieces of furniture can all be considered as non-structural elements. These building components are especially vulnerable to earthquake shaking because they are not expected to be designed for other than gravity and operating loads.

Non-structural elements are expected to perform adequately during a magnitude 7.6 earthquake but are susceptible to damage during a magnitude 8.6 earthquake as described below:

a) <u>Utility Lines</u>: The expected displacement characteristics of the ground motion, raises some concern for hazards to utility lines. This is true mostly at the connection of these lines to the buildings and less likely for distribution lines within the buildings.

b) <u>Building Attachments</u>: Attachments such as parapets and heavy facade may break and fall. Many of these attachments are especially susceptible to failure due to exaggerated drifts and lateral displacements during a magnitude 8.6 earthquake. Failure of these components may impact some building functions (e.g. failure of antennas) or be a life threatening hazard (e.g. falling debris into the street).

c) <u>Partitions and Windows:</u> It is expected that partitions in tall buildings will have some diagonal cracking. It is also possible that glass in large windows will break which would produce a life threatening hazard from shards of glass falling into streets.

Conclusions and Recommendations

The study reported herein investigated in a qualitative manner the expected performance of well behaved tall buildings in Atlanta, Chicago, Kansas City and Dallas to a magnitude 7.6 and 8.6 earthquake anywhere in the New Madrid Seismic Zone. The conclusions from the study are:

1. The two earthquakes are expected to produce shaking at the four cities in the form of long period waves with amplitudes proportional to distance from the fault. These waves are expected to be separated in time and arriving in trains of essentially sinusoidal waves.

2. Tall buildings in the four cities are mostly less than 40 stories high with concrete and steel construction accounting for more than 2/3 of the buildings. In general, well behaved tall buildings are expected to have the capacity to resist a magnitude 7.6 and 8.6 earthquake in the New Madrid Seismic Zone with some structural damage and light to moderate damage to non-structural elements. The hazard to life safety is mostly from failures in non-structural elements such as falling debris and toppling of bookcases.

3. A magnitude 8.6 earthquake is expected to produce lateral loads significantly larger than design wind loads but clearly within the capacity of well behaved structural systems. Cracking in reinforced concrete buildings, which does not impact the structural integrity of the building, is expected. Buildings with significant stiffness discontinuities, or with large eccentricities will suffer more extensive damage. Structures with mixed structural materials or those constructed of brick masonry could be vulnerable to moderate damage.

4. Non-structural elements are vulnerable to failure from a magnitude 8.6 earthquake. Utility lines, window glass and building attachments such as heavy facade and parapets are especially vulnerable to failure. It is also possible that heavy slender items in the buildings such as bookcases and storage racks may topple.

   Failure of utility lines is the most life threatening hazard as it may initiate fires or affect the functioning of vital systems. Falling debris from failed attachments and broken glass from windows also poses life threatening hazards.

The conclusions from this study indicate that a magnitude 8.6 earthquake may produce more than minimal hazards to certain types of tall buildings and some categories of non-structural elements. On that basis we recommend further studies as follows:

1. A better inventory of tall buildings in the four cities should be compiled. The number of buildings with soft story construction, staggered truss system or with large eccentricities and buildings with substantial brick masonry construction should be identified.

2. A survey, based on sampling procedures, of non-structural elements in the four cities is recommended. Such a survey should identify the general condition of utility line connections to buildings and heavy attachments such as facades, parapets and windows. The survey should identify the degree of hazard from the non-structural elements and offer recommendations whether remedial action is necessary.

3. Further study of the characteristics of the expected ground shaking in these cities due to a magnitude 8.6 earthquake in the New Madrid Seismic Zone will be helpful, especially if the study produces ground response spectra. Data from the Mexican earthquake of September 19, 1985, and the earthquake in north eastern Ohio on January 31, 1986, should be considered in such a study.

4. Amplification of base rock motions in the four cities due to deep alluvial depoists should also be considered in determining the expected ground shaking in the four cities.

References

1. United States Department of the Interior, Geologic Survey, "A workshop on Continuing Actions to Reduce Potential Losses from Future Earthquakes in Arkansas and Nearby States", September 20-22, 1983, North Little Rock, Arkansas.

2. Council on Tall Buildings, Monograph on the Planning and Design of Tall Buildings, Lehigh University Bethlehem, Pennsylvania, 1980.

3. National Oceanic and Atmospheric Administration (NOAA), "San Fernando, California Earthquake of February 9, 1971", U.S. Department of Commerce, Washington, D.C., 1973.

4. Nuttli, O.W. and R. B. Herrmann, "Ground Motion of Mississippi Valley Earthquakes," Jour. Technical Topics in Civil Engineering, ACSE, v. 110-, 54-69, 1984.

5. Gutenberg, B. and C.F. Richter, "On Seismic Waves, 3, "Gerland's Beitraege Geophysik, v. 47, 73-131, 1936.

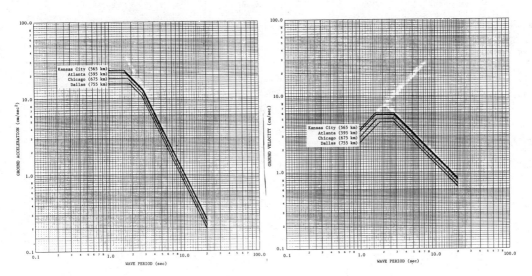

FIGURE 1: TIME-DOMAIN ESTIMATES OF STIFF SOIL HORIZONTAL GROUND ACCELERATION (cm/sec^2) FOR A NEW MADRID FAULT EARTHQUAKE OF M_s = 7.6

FIGURE 2: TIME-DOMAIN ESTIMATES OF STIFF SOIL HORIZONTAL GROUND VELOCITY (cm/sec) FOR A NEW MADRID FAULT EARTHQUAKE OF M_s = 7.6

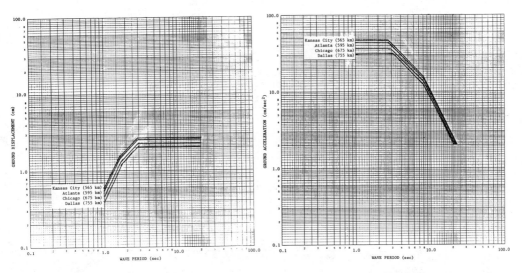

FIGURE 3: TIME-DOMAIN ESTIMATES OF STIFF SOIL HORIZONTAL GROUND DISPLACEMENT (cm) FOR A NEW MADRID FAULT EARTHQUAKE OF M_s = 7.6

FIGURE 4: TIME-DOMAIN ESTIMATES OF STIFF SOIL HORIZONTAL GROUND ACCELERATION (cm/sec^2) FOR A NEW MADRID FAULT EARTHQUAKE OF M_s = 8.6

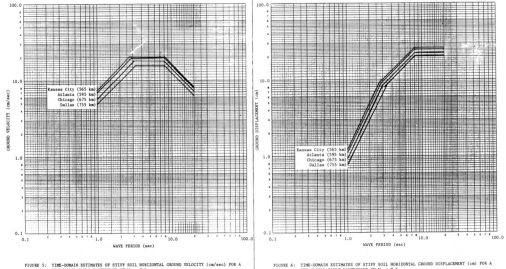

FIGURE 5: TIME-DOMAIN ESTIMATES OF STIFF SOIL HORIZONTAL GROUND VELOCITY (cm/sec) FOR A NEW MADRID FAULT EARTHQUAKE OF $M_s = 8.6$

FIGURE 6: TIME-DOMAIN ESTIMATES OF STIFF SOIL HORIZONTAL GROUND DISPLACEMENT (cm) FOR A NEW MADRID FAULT EARTHQUAKE OF $M_s = 8.6$

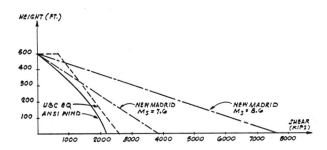

FIGURE 7 ESTIMATED ENVELOPE OF MAXIMUM STORY SHEARS ON 40 STORY CONTROL BUILDING IN ATLANTA

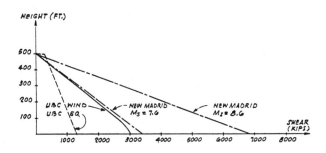

FIGURE 8 ESTIMATED ENVELOPE OF MAXIMUM STORY SHEARS ON 40 STORY CONTROL BUILDING IN CHIGAGO

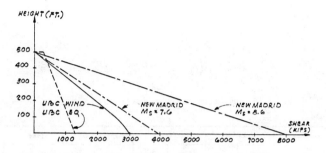

FIGURE 9 ESTIMATED ENVELOPE OF MAXIMUM STORY SHEARS ON 40 STORY CONTROL BUILDING IN KANSAS CITY

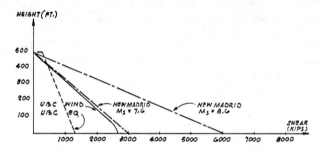

FIGURE 10 ESTIMATED ENVELOPE OF MAXIMUM STORY SHEARS ON 40 STORY CONTROL BUILDING IN DALLAS

TABLE 1

INVENTORY OF TALL BUILDINGS IN ATLANTA, CHICAGO, KANSAS CITY AND DALLAS

No. of Stories	Type of Construction	Atlanta	Chicago	Kansas City	Dallas
20 to 39	Concrete	7	11	5	5
	Steel	13	10	10	10
	Other*	12	7	3	14
40 to 59	Concrete	1	11	1	–
	Steel	1	12	–	–
	Other*	–	9	–	–
60 to 80	Concrete	1	6	–	–
	Steel	–	2	–	–
	Other*	–	–	–	–
>80	Concrete	–	–	–	–
	Steel	–	2	–	–
	Other*	–	–	–	–
Total		35	70	19	34

*Includes buildings whose construction material is not available, buildings constructed of brick and buildings constructed of mixed materials.

TABLE 2

COMPARISON OF MAXIMUM BASE SHEARS*
FOR CONTROL BUILDING

Loading	Atlanta		Chicago		Kansas City		Dallas	
	Base Shear	% Wind	Base Shear	% Wind	Base Shear	% Wind	Base Shear	% Wind
Design Wind	2006	100	3050	100	3050	100	2657	100
UBC EQ.	2606	118	1303	43	1303	43	1303	49
New Madrid EQ. 7.6	3825	173	3400	112	3968	130	2975	112
New Madrid EQ. 8.6	7651	347	6801	223	7935	260	5951	224

*Maximum base shears in kips

TABLE 3

SURVEY OF PERFORMANCE OF TALL BUILDINGS TO
THE SAN FERNANDO EARTHQUAKE, 1971

Building	Construction	No. of Stories	V_e/V_d	Structural Damage
Bank of California 15250 Ventura Blvd.	Concrete Frame	12	2.2	light
Holiday Inn 8244 Orion Ave.	Concrete Frame	7	4 to 5	light
Holiday Inn 1640 Marengo St.	Concrete Frame	7	4 to 5	light
Bunker Hill Tower 800 W. First St.	Steel Frame	32	2.8	none
Sheraton-Universal	Concrete	20	2.1	none

V_e = Estimated base shear during earthquake
V_d = Estimated base shear used for design

SEISMIC PERFORMANCE OF EXISTING BUILDINGS

P. Gergely,[I] R. N. White,[II] and G. R. Fuller[III]

ABSTRACT

The paper summarizes the results of three U.S.-Japan workshops held in 1983, 1984, and 1985 on the evaluation of existing buildings for resistance to earthquakes. The main methods of evaluation discussed are: quick screening, simple rapid evaluation, multilevel procedures, and detailed analysis. Most methods were applied to six benchmark structures.

INTRODUCTION

A series of three workshops were organized under the auspices of Task Committee D, Evaluation of the Performance of Structures, of the U.S.-Japan Cooperative Program of Natural Resources (UJNR). The primary purpose of the workshop series was to compare the various methods of evaluation and to examine the range of applicability of the most common methods used in Japan and in the U.S.

A variety of evaluation methods for buildings were presented during the first workshop, held in Japan in 1983. The topics included practical methods, detailed analysis, computer-aided nonlinear analysis, full scale testing, and post-earthquake evaluation strategies. The second workshop, held in the U.S. in 1984, concentrated primarily on safety evaluation methods, selection and preliminary analysis of benchmark structures, and damage analysis of various types of structures. The papers presented in the first two workshops were published in two volumes [1, 2]. The third workshop was again in Japan in 1985 and dealt with the comparison of the analyses of the benchmark structures by various methods, as well as with a discussion of the recent developments in evaluation methods.

These workshops and several concurrent efforts and publications by others [3] have indicated that safety evaluation of existing structures may be classified into four categories: (1) Screening methods for the evaluation of large groups or classes of buildings. These or similar methods can also

---

I. Professor and Department Chairman, Structural Engineering, Cornell University, Ithaca, New York.

II. Professor, Department of Structural Engineering, Cornell University, Ithaca, New York.

III. Chief, Compliance Branch, Division of Manufactured Housing and Construction Standards, HUD, Washington, DC.

be used for the classification of structures after an earthquake, (2) Rapid evaluation procedures that require little or no computer analysis, although an estimate of the nonlinear load-deflection characteristics of the building may be required, (3) Multilevel procedures that involve several choices of assumptions, depending on the accuracy required and the type of structure, and (4) Detailed dynamic analysis of the structures, resulting in quantities or indices that are used to judge the safety of the building.

The above simple classification proved to be useful in the discussion and comparison of the various methods. However, it became clear during the workshops that there is no agreement among researchers and engineers on what constitutes a simple analysis and what quantities one can estimate easily. In general, our Japanese colleagues require more elaborate calculations in the approximate techniques. Yet, it is evident that all four classes of methods are necessary. Since there are many approximate analysis methods and most have several variations, only a few typical methods are outlined in this paper.

SCREENING METHODS

The purpose of quick classifications is to be able to judge the safety of typical classes of buildings. These methods are useful in surveying large groups of buildings at the municipal or regional level. The calculations typically require about two to four hours per building, and often much less.

The key aspect of these approaches is that various allowable stress levels are assigned to supporting element (columns or walls) depending on their composition, length, and ductility. These stress levels were established from tests and from studies of actual damage. The great advantage of these rapid classification methods is that they can easily be modified to account for new knowledge gained about the behavior of systems. The interaction of various types of structural elements can be accounted for. Poor behavior of certain materials or elements, such as the problems with short (captive) columns are readily included in the evaluation.

These simple classification approaches are not usable or at least are not reliable for unusual structures. For example, highly unsymmetrical structures or poor soil conditions are so variable that only much more rigorous analysis can assess safety in these cases. The rapid evaluation methods are normally quite conservative, and higher level approaches are necessary for screening structures that fail the rapid test.

Three screening procedures are given by Lew [1]. These methods, developed by the National Bureau of Standards, the General Services Administration, and the State of California Seismic Safety Commission, can be used for a large number of buildings to identify hazardous conditions and to supply decision makers with information on the extent of hazard and the feasibility of retrofit measures.

The three methods are based on structural performance indices which are derived from past performance records of particular types of buildings, from engineering judgement, and on seismicity considerations for particular sites. The structural performance indices are related to structural characteristics, structural configurations, and the degree of deterioration of buildings, whereas the seismicity index is obtained by considering the distance from active faults and expected magnitude and frequency of earthquakes.

The preparation for and the management of quick screening program is of utmost importance. People must be trained, evaluation forms designed, and technical data provided. Depending on the reliability desired and the time available, building data may range from visual inspection, design drawings, to as-built drawings or even field testing. In many cases the local soil characteristics are also an important factor. Finally, the results of the evaluation should be tuned to the accepted level of risk in the area.

Screening of buildings provides useful data for assessing seismic risk of cities, neighborhoods, and government buildings. The results of such surveys can help legislators to estimate the cost of hazard reduction. Generally the criteria used for gaging the risk are less severe than those used for the design of new buildings.

Rapid screening methods have also been developed for classifying the safety of buildings immediately after a major earthquake. Trained teams of engineers, architects, and contractors are sent to the damaged area and each team is assigned a neighborhood consisting of about 100 buildings. The time pressures are so great that this classification is much cruder than the screening used for undamaged buildings. After the El Asnam, Algeria earthquake of 1981, all buildings in the damaged area were marked green (safe), red (unsafe for entering and to be demolished), or yellow (must be reevaluated, and can probably be strengthened). This approach proved to be successful and alleviated many problems.

RAPID ANALYSIS METHODS

The common characteristic of these approaches is that the nonlinear load-deflection relationship for the structure, or for each story, is required. Structures behave inelastically when subjected to a major earthquake and it is necessary to know the deflections beyond the elastic limit. This is especially important for buildings composed of several types of lateral resisting elements (frames, stairways, walls) because there is then a second line of defense.

Some engineers question whether we can establish the load-displacement curve for complex structures. In most cases only rough estimates of the elastic limits and the stiffnesses are necessary in these evaluation methods. The second point is that since the structure will deform inelastically, with various post-elastic stiffnesses, a realistic estimate of the safety must rely on an approximate knowledge of the load-displacement relationships.

Three different rapid analysis methods are summarized here:

a. The Substitute Structure Method

In this approach (developed by Sozen and his associates) the inelastic response is estimated by calculating an increased period and an increased damping to account for the energy absorption due to hysteretic response. The changes in period and damping are functions of estimated damage indices which represent a measure of the inelastic deformation level of the structure.

This method is relatively simple because only an elastic modal analysis is required. However, it has been calibrated only for concrete frames and more work is needed to extend it to other types of structures. In essence, the use of inelastic response spectra is replaced by elastic spectra but with changed periods and damping whereas in the use of inelastic spectra the period is not changed. Of course, this approach can be used only on the high-period (descending) side of the spectrum.

b. The Capacity Spectrum Method

In this method the inelastic load-deflection curve is first estimated; often rough calculations will suffice. This curve is best drawn as the roof displacement versus the base shear for a lateral load representing the inertia force vector (Fig. 1). For low buildings a straight line load vector is acceptable. A static linear computer program can be used as follows: the load is increased until a few hinges form and then the rotational stiffnesses are greatly reduced at these locations. The load is increased until another group of hinges form and in this manner a piece-wise linear load-displacement relation is produced. In some cases a rough estimate of the initial stiffness, the elastic limit, and the subsequent stiffness can be done without a computer.

In the next step the roof displacement is related to the corresponding spectral displacement S_d, and the base shear is related to the spectral

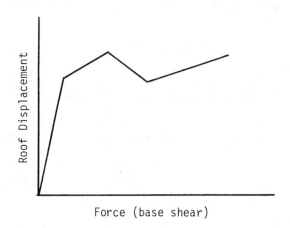

Fig. 1   Roof Displacement vs. Base Shear for Building

acceleration S_a through the well-known modal equations. These transformations are ordinarily based on the first mode only. The periods corresponding to S_d and S_a values are calculated.

The plot of S_a versus the period T is called the capacity spectrum. It is next superimposed on the design spectrum curve (Fig. 2). The relative positions of the capacity and response spectra determines the safety of the structure. An interpolation is necessary because the damping changes as inelasticity develops.

The capacity spectrum method is a relatively simple and attractive method when the first mode dominates, although the contribution of higher modes to the roof deflection can be incorporated in an approximate manner. It is usually satisfactory to estimate the bilinear stiffness of the structure.

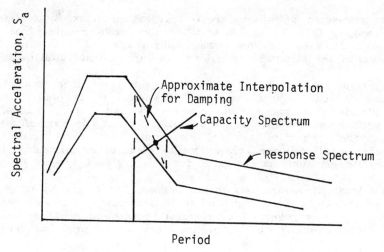

Fig. 2 Capacity Spectrum Method Plot

c. The Reserve Energy Technique

This approach was proposed by John Blume in the early 1960's but it has not been used widely (perhaps because its title is not descriptive). It is also based on the nonlinear load-displacement curve but instead of developing a capacity spectrum, the input kinetic energy is calculated using the velocity spectrum and modal masses $0.5\bar{m}S_v^2$. The area under the load-displacement curve represents the absorbed strain energy and by equating the two energies, one can estimate the deflection and the corresponding damage.

In both the capacity spectrum method and the reserve energy technique, the static load-displacement curve is estimated. Cyclic behavior is not accounted for directly, although it may be possible to modify the curve to do that. As with all approximate methods, it is important for the engineer to have a good feel for the real behavior of structures.

MULTI-LEVEL PROCEDURES

a. Japanese Standard, 1977

The Japanese "Standard for Seismic Capacity Evaluation of Existing Reinforced Concrete Buildings" was compiled in 1977 under the Ministry of Construction. It is described in some detail by Aoyama [1]. It is intended to be applied primarily to low rise buildings up to 6 to 7 stories.

The three-level procedure involves calculation of a Seismic Index

$$I_s = E_o G S_D T \tag{1}$$

where $E_O = \phi(C \times F)$ is the basic seismic index. With the term E_O, C is a strength matrix, F is a ductility matrix, and ϕ is a story index that relates single DOF system response to that of the multi-story system. Thus it is apparent that this method evaluates both strength and ductility capacities and expresses the result E_O as a product of these capacities.

The other necessary definitions are G = geological index, S_D = structural design index ranging from 0.4 to 1.2, and T = time index ranging from 0.4 to 1.0 to cover such items as previous damage, shrinkage and settlement effects, and deterioration of the structure from loadings and environmental factors.

The first level procedure is very easy to apply — it is based on the average stress in the column and wall areas. It is well-suited for wall-dominated structures but is quite conservative for ductile frames. The second level procedure concentrates on potential problems in vertical members by identifying five types of vertical elements: extremely brittle short columns, shear-critical columns, shear walls, flexure-critical columns, and flexural walls. Simple equations are given for treating each type of element.

The third level procedure recognizes flexural hysteresis by considering the response of three additional types of vertical members: columns governed by shear beams, columns governed by flexural beams, and rotating walls with uplift effects from overturning. Equations are again provided for these calculations.

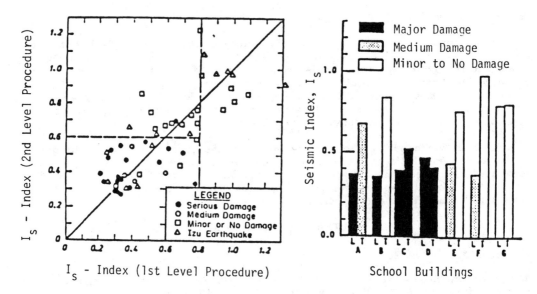

a. Seismic index and damage in 1978 earthquakes in Japan

b. Seismic index and damage of school buildings in 1968 Tokachi-Oki EQ (2nd level procedure)

Fig. 3 Seismic Indices and Damage, Japanese Standard (1977)

Results of the application of these procedures to damaged structures are given in Fig. 3, where the correlation between first and second level procedures applied to buildings from 1978 earthquakes in Japan is given in Fig. 3a. A comparison of damage levels and seismic index I_s values calculated for both the longitudinal (L) and transverse (T) motions of a group of school buildings is shown in Fig. 3b.

Additional examples of damage level and index values are given by Aoyama [1] and in some of the benchmark structure analyses to be described later in this paper.

While seismic capacity is best judged by comparing I_s to an E_t index that is tied to specific ground acceleration levels and other critical factors, general guidelines on performance may be stated. Superior performance is indicated by I_s values of 0.9, 0.7, and 0.6 for levels 1, 2, and 3 evaluation procedures, respectively. I_s values less than 0.4 at the 2nd and 3rd levels indicate questionable performance, and the borderline of damage for level 1 is often taken as 0.8.

b. U.S. Navy Procedure

A methodology is presented by Freeman [1] for the preliminary evaluation of the safety of existing buildings for earthquake effects. The procedure was developed for the U.S. Navy to evaluate its existing structures throughout the naval shore establishment that are located in areas of high seismicity. The scope of the methodology includes a screening process for a large number of buildings, a rapid procedure for approximating building capacity, and a graphical procedure for estimating damage due to postulated earthquake motion and the costs of their corrections. The fundamental ideas in this approach have been described above in the discussion of the capacity spectrum method.

Freeman states that it cannot be expected that the rapid evaluation method will give an accurate prediction of building response, but it is highly useful in giving a general picture of behavior and it is also useful as an aid in determining which buildings will require a full-scale dynamic analysis.

c. The HUD Method

The United States Department of Housing and Urban Development (HUD) published a three-volume manual in November 1978 titled "A Methodology for Seismic Evaluation of Existing Multistory Residential Buildings" to be used to satisfy the HUD requirement that all multistory residential buildings in Seismic Zones 3 and 4 (Uniform Building Code) shall be evaluated for earthquake hazard and seismic resistance. Since that time, HUD has published a series of additional guidelines on building rehabilitation. This HUD methodology for evaluation and structural assessment of building rehabilitation is summarized by Fuller [1].

The methodology has proven to be a valuable tool in analyzing existing buildings that have been proposed for renovation in HUD programs, and it has been used a number of times for specific buildings in the U.S. The paper gives a case study on a building constructed in 1900 and upgraded recently to meet 1979 UBS Seismic Zone 3 forces.

Tabulated cost data related to degree of compliance with the Uniform Building Code have proven to be invaluable to program administrators and building officials who must make decisions on strengthening or demolishing buildings. This data not only helps in economic and engineering planning, but it is also a major factor in determining levels of acceptable risk.

d. Rule-Based Inference Procedures

The rule-based inference procedure for structural damage assessment called SPERIL-I is reviewed by Yao [1]. This approach is a highly structured formal damage assessment method utilizing structural reliability, the theory of fuzzy sets, and expert systems to arrive at consistent results to imperfectly formulated problems. The status of this promising area for the future is captured in the following quote: "The fundamental question regarding the use of inference machine evaluation methods is whether such methods can be relied upon without a careful review by a human expert. It must be recognized that subjective knowledge and experience of an expert imply a higher level of intellectual organization than is possible in data bases containing large sets of information. . . . Nevertheless, the use of appropriate evaluation techniques utilizing inference machine approaches can greatly enhance the effectiveness and reliability of human expert evaluations. Furthermore, it may also contribute to improvements in human expertise through increased communication among experienced engineers." [1].

DETAILED ANALYSIS METHODS

A thorough evaluation of the seismic safety of a building is sometimes desired, regardless of the high cost of such an analysis. Many computer programs are available for the dynamic analysis of structures. In one paper [1] Aoyama lists nearly 70 programs developed mainly in Japan.

There are several difficulties with the use of sophisticated dynamic analysis programs. It is evidently necessary to know the physical characteristics of the structure, especially its stiffness and damping. There are often differences between the designed and built structure and the structure may be so old that design drawings are not available. It is therefore necessary to conduct a detailed field survey of the structure and note all stiffening elements. Even minor cracking and especially prior seismic damage can significantly reduce the stiffness of the structure and increase the damping. Certain key elements could lose stiffness prematurely and affect the response. Finally, special soil conditions can have an overriding effect on the dynamic behavior. All these factors must be ascertained before an analysis can begin.

Even after the above hurdles have been passed, the question remains what to do with the results of nonlinear analyses. The response depends on the properties of the input, which is a statistical quantity. In most cases the main conclusion of such analyses is the identification of weak elements that must be retrofitted. It is claimed that the simpler methods of evaluation often yield the same information with less effort.

APPLICATIONS TO BENCHMARK STRUCTURES

Six benchmark structures were selected for evaluation by the various methods; they are described in Appendix A of Ref. [2]. The buildings are:

1. 7 story reinforced concrete frame tested under the U.S.-Japan Cooperative Research Program
2. Imperial County Services Building, California
3. 9 story ATC benchmark structure, reinforced concrete, 1964
4. 7 story reinforced concrete building (Holiday Inn), instrumented
5. Reinforced concrete building, Japan
6. 6 story steel frame tested under U.S.-Japan Cooperative Research Program

Inasmuch as the evaluation procedures are still being applied to these structures, and critical comparisons of all results are not yet complete, we will provide only selected results in this paper. Final comparisons will be given in the final report on this project to be issued in mid-1986 (and to be reported upon at the 1986 UJNR meeting).

a. 7 Story U.S.-Japan Reinforced Concrete Building

Several papers (Yoshimura and Kabeyasawa [1] provide detailed experimental and analytical results for this structure. This wealth of data is extremely valuable for comparison with both approximate and detailed analytical techniques.

Okamoto [1] evaluated the seven story reinforced concrete building using the Japanese Standard described earlier. The evaluation was made only for the direction in which the seismic forces were applied during the loading in the laboratory, and the weight of each story was taken as the dead weight of the floor level plus the weight of the loading frames. The indices G, S, T were all taken as unity.

Results are quoted directly from Okamoto: "The Basic Structure Performance Index, E_O, was computed as 0.36 (1st Screening), 0.68 (2nd Screening), and 0.59 (3rd Screening). The index given by the first screening is found to be significantly lower than those given the second and third screenings. This low value of index was obtained primarily because of small areas of the columns and shearwalls. In the second and third screenings, on the other hand, E_O was much larger because the structure was estimated to exhibit flexural failure. Nevertheless, if the weight is increased to 1 ton/m^2 by taking live load into account, E_O will decrease significantly to a level at which the structure is considered to be unsafe."

b. Imperial County Services Building

This structure is characterized by its very poor shear transfer properties at one end of the building, where the shear wall stops at the top of the ground floor story.

Force-deformation relations for this building were obtained by Stephens and Yao [2] from structural acceleration records using a structural identification technique. The study showed that these relations are highly useful in studying the structural response both qualitatively and quantitatively; it is possible to extract such information as the range and number of cycles of inelastic response for use in damage expressions. The planned extension of this research includes identification of damage expressions to more effectively use the quantitative information from the force-deformation relations, calculation of energy dissipation and its correlation to structural damage, comparison of results with other investigators, and possible extension of the methodology to include torsional effects and overturning.

Aoyama [2] applied the Japanese Standard procedures (levels 1 and 2) to the structure, using $S_D = 0.95$, $G = 1.0$, and $T = 1.0$, and obtained first level procedure I_S values ranging from 0.11 to 0.65 for the first to sixth story, respectively, in the E-W direction, and from 0.48 to 1.06 in the N-S direction. Second level I_S values were also low. In comparison with accepted Japanese standards, it was found that the Imperial County Services Building fell in a region regarded to be dangerous under earthquake loads.

By Japanese standards, the upper level walls would be regarded as "miscellaneous walls" with only 1/3 the effectiveness of regular shear walls. The average axial stress level in the vertical elements was about twice that of typical Japanese building designs. While there is a question on the validity of using the Japanese approach on the N-S framing, it appears that the E-W direction framing does fit within assumptions.

The analysis by Hart [2] showed that the normal assumption of fixed base shear walls was not valid and that the soil-building system must be modeled. He also showed that the floor diaphragms were flexible, not rigid, and that using a rigid diaphragm analysis is misleading in terms of the distribution of loads at the ground floor column level.

Sozen [2] summarizes the key findings from the PhD dissertation of M. Kreger (U. Illinois, August 1983) who analyzed the Imperial County Services Building with particular emphasis on the causes of the column failures. It is shown that it is difficult to explain the failures by considering forces in only one direction at a time. Sozen provides some very interesting philosophical discussion at the end of his paper:

> "Evaluation of existing buildings is better kept simple. A simple analysis tempered by judgement will suffice in most cases. In the interest of the community, it is a better buy to invest less in reanalysis at the risk of missing a few potential disasters than to invest more with the possibility of identifying some of those few. Vulnerable buildings are in general like bad wine. Even a naive analysis can identify the vinegary taste. It is very unlikely that spending more professional resources on analysis routines will automatically identify the buildings with subtle vulnerabilities. The Imperial County Services Building is a good example. It is very unlikely that calculation alone would have identified the problem and it is very likely that a structural engineer who was informed of similar events would have avoided the conditions at the east end of the structure."

c. 9 Story Reinforced Concrete ATC Benchmark Building

This structure was designed with the 1964 UBC provisions, and has been the subject of re-design in the ATC project. Lew [2] applied the rapid, Level 1 Field Evaluation Method of the NBS approach, and concluded that for an earthquake intensity of 9 MMI, the building rated fair. The unsymmetrical distribution of shear walls in the transverse direction controlled the rating of the structure.

Okamoto and Watanabe [2] applied the Japanese Standard 1st level procedures. In the longitudinal direction, I_s values ranged from 1.37 at the top floor to 0.34 at the base; the fifth floor had I_s = 0.37. In the transverse direction, I_s ranged from 2.30 to 0.67, with 0.68 at the 5th floor. These I_s values are low by Japanese standards.

Freeman [2] used the capacity spectrum method for the ATC building, in a very quick one page calculation; more detailed rapid evaluation will be given in the final report on this project.

d. 7 Story Holiday Inn

This structure was studied by Freeman, Shimizu [2], and Moehle [2]. Shimizu applied the Japanese Standard and obtained E_o values between 0.20 and 0.36 by rapid evaluation, and 0.24 to 0.45 by detailed evaluation. By Japanese standards, E_o should be at least 0.6.

Moehle made a relatively simple analysis plus an assessment of connections and details. He concluded that (a) the structure is fairly flexible and excessive nonstructural damage could be expected in a moderate earthquake, (b) inadequate ductility may be a problem during a more intense earthquake, (c) analysis underpredicts initial stiffness by nearly two to one, and (d) the first floor column details are considered inadequate.

e. 6 Story Steel Frame (U.S.-Japan Program)

Lew [2] rated this building using the NBS rapid FEM, and concluded that the structure rated fair for a MMI 9 earthquake. The semi-rigid nature of the steel deck with concrete topping was a critical factor in reaching this rating.

Hart [2] did a detailed structural analysis of the system, based on a perspective which recognizes the use of site dependent earthquake response spectra. The analysis required about 100 engineering labor hours. It provides a rather detailed prediction of the development of plastic hinges in the structure, the lateral displacement (nonlinear), and the variation of period with base shear. Typical U.S. steel frames have a much flatter curve for change in period with change in base shear.

ACKNOWLEDGEMENTS

The financial support of the National Science Foundation and the personal support and encouragement provided to the series of workshops by Dr. John B. Scalzi, NSF Program Manager, are greatly appreciated. The efforts of Dr. M. Watabe and Dr. S. Okamoto in helping to organize the workshops are also much appreciated, as is the enthusiastic involvement of the many individuals from Japan and the United States who participated in the workshops.

REFERENCES

[1] Proceedings of the First Workshop on Seismic Performance of Existing Buildings, May 1983, edited by Richard N. White and Peter Gergely, Department of Structural Engineering, Cornell University, Ithaca, N.Y. 14853, April 1985, 370 pp.

[2] Proceedings of the Second Workshop on Seismic Performance of Existing Buildings, July 1984, edited by Richard N. White and Peter Gergely, Department of Structural Engineering, Cornell University, Ithaca, N.Y. 14853, April 1985, 516 pp.

[3] Seismic Design for Existing Structures, Seminar Course Manual SCM-14, American Concrete Institute, Detroit, 1986.

THE ROLE OF CLADDING IN SEISMIC RESPONSE OF
LOWRISE BUILDINGS IN THE SOUTHEASTERN U.S.

Barry J. Goodno[I] and Jean-Paul Pinelli[II]

ABSTRACT

An in depth seismic evaluation was performed for a
particular lowrise steel frame structure located in Atlanta,
Georgia. The structure was chosen as typical of the many
small office buildings in the Southeastern region. It was
designed for code level wind forces only but was reevaluated
for seismic forces to assess it's potential vulnerability in the
event of a moderate earthquake in the region. The influence
of nonstructural elements was found to be especially
important for the case study building. The effect of modest
changes in building configuration to reduce torsion and
improve overall seismic performance were explored and the
building was reanalyzed and rechecked for compliance with
seismic codes.

INTRODUCTION

In the past, lowrise buildings have received relatively less research
attention than highrise structures even though, overall, they may
represent a greater seismic hazard than tall buildings because of the much
larger inventory of these structures [1]. Budget limitations affecting
design, construction and inspection, and poor or nonexistent design for
lateral forces, contribute to the potential for disaster for this class of
buildings in the event of even a moderate earthquake in the Southeastern
region. Lowrise buildings (five stories or less) do not receive the same
level of design attention as highrise, are often comprised of an assemblage
of components of diverse materials, and are heavily dependent on
connection performance for their overall behavior. Some research on
lowrise buildings has been carried out by previous investigators [1-4]
motivated in part by reports of failures in past earthquakes in the U. S.
and Japan [5,6]. However, the variability in materials and design and
construction practices makes it difficult to apply research findings to the
entire class of lowrise structures.

The present paper presents a possible seismic investigation
methodology for one class of lowrise buildings and uses it to conduct an in

---

I. Associate Professor, School of Civil Engineering, Georgia Institute of
Technology, Atlanta, Georgia.

II. Structural Engineer, Techint Engineering, Buenos Aires, Argentina.

depth evaluation of a particular lowrise steel frame structure located in Atlanta, Georgia. The structure was chosen as typical of the many small office buildings in the region. It was designed for code level wind forces only but, as part of an NSF sponsored research effort, was reevaluated for seismic forces to assess the potential vulnerability of this class of structures. The building's inherent seismic resistance capability was of particular interest, especially since earthquake forces had not been considered in its design. The influence of nonstructural elements was found to be especially important for the case study building. Cladding analytical models, developed for highrise buildings to include the lateral stiffness effect of exterior cladding, were incorporated in the lowrise building model [7]. The effect of modest changes in building configuration to reduce torsion and improve overall seismic performance were explored and the building was reanalyzed and rechecked for compliance with seismic codes. Key findings from the study are presented below along with a description of the structure and cladding models.

ANALYTICAL MODELS

Structure Description

The case study structure which was the subject of this investigation is a two-story steel frame office building whose planform is shown in Fig. 1. Nine moment resistant frames, with semi-rigid beam-column connections at exterior columns and simple connections at interior column locations, were used to resist gravity load and the design wind load of 15 psf (718 Pa) in the E-W direction; the semi-rigid connections were designed to yield at design wind load. In the N-S direction, two unreinforced masonry infill walls at lines D and E were used as part of the small rectangular enclosure on the north end of the building, and were used to resist wind loads in the N-S direction. The infill masonry wall along frame line 9 in the E-W direction was also assumed to resist wind forces in the E-W direction. The infill walls along D, E and 9 run the full height of the building. N-S beams along lines A and C (see Fig. 1) were placed outside of the column lines and were used to support the precast cladding which enclosed the second story of the building. Column sections are W8 and W12 members, pinned at their bases, and beams are W16 shapes at the roof and W24 shapes at the second floor level. All steel is ASTM A36. The roof consists of metal decking, insulation and built-up roofing; the second floor slab is a 4 inch (102 mm) concrete slab on light gage metal decking. Both roof and floor slabs are supported by open web bar joists.

The first story enclosure walls, of unreinforced brick and concrete masonry construction, were positioned inside the exterior column lines and were attached to the concrete encased columns with dovetail anchors. These walls were assumed nonstructural and were not included as part of the lateral force resisting system. The heavy precast concrete cladding at the second story level was composed of 3000 to 3690 pound (13.3 to 16.4 kN) panels (either with or without window cutouts) which measured 19.25 ft x 5 ft x 1.2 ft (5.9 m x 1.5 m x 0.4 m). There are four panels per bay for each of the six 20 ft (6.1 m) bays in the N-S direction along lines A and C (see Fig. 2), and seven panels per bay for the two 36 ft (11.0 m) bays along column line 1. Four panels per bay were used to form the exterior facade for the two bays along column line 8. Four clip angle attachments were welded to the exterior spandrel beams and to plate inserts in the panel to support each panel at the roof and second floor levels. Each panel was supported by a crane until all four attachments had been welded to the spandrel beam.

Cladding Model

The localized response model presented in reference [7] was used to model the precast panels and supporting framing along column lines 1, 8, A and C (see Fig. 1). In this model, any number of panels, represented as flat rigid bodies connected at four points to the spandrel beams by linear elastic spring elements (two translational, one rotational), may be attached to the underlying framework. The stiffness properties of the connection springs are unknown at present (however, current research is expected to provide performance data for selected connection types [11]), but some information is available on the ultimate strength of panel inserts [7]. On this basis, the assumption was made that the panel connections would fail if the horizontal or vertical components of the connection forces exceeded 10 kips (44.5 kN). Then, the maximum interstory drift permitted by the codes [8,9], 0.5% of the story height (0.85 in, 22 mm), was applied to typical bays of cladding in each structure direction and the connection-spring forces computed. After a number of trial analyses, it was determined that the largest value of the connection spring stiffness for which the connection capacity was not exceeded at code drift levels, was 90 kips/in (15762 kN/m). This value was used in all subsequent analyses provided that the assumed connection capacity (10 kips, 44.5 kN) was not exceeded. If the cladding connection forces were found to be excessive, the cladding stiffness contribution was deleted from the building model.

Structure Models

Three separate models of the structure were assembled to represent its lateral force resistance capability at various stages of its response to an assumed moderate ground motion. The models included the semi-rigid framing, infill masonry wall, and exterior cladding in the E-W direction, and only the exterior cladding and infill masonry walls in the N-S direction. The stiffness and ultimate capacity of each component were evaluated in separate analyses, and then the lateral stiffness of each was combined using a translation of axes procedure to form the 16x16 stiffness matrix for the overall model. The roof was assumed to be a flexible diaphragm but the second floor slab was assumed to be perfectly rigid in its own plane. Three degrees of freedom were sufficient to describe the translations and rotation of the second floor slab, but one degree of freedom was assigned to each of the nine frame lines in the E-W direction at the roof level and four more translational degrees of freedom at lines A to D in the N-S direction. Computed mass-inertia properties were lumped at the three degrees of freedom at the second floor level, and tributary mass was lumped at the thirteen roof level translational degrees of freedom.

The framing and infill masonry walls along column lines 9, D and E were represented by member elements and plane stress finite elements, respectively, to assess their in-plane lateral stiffness values at the roof and second floor levels. Equivalent thickness and modulus of elasticity properties appropriate for masonry wall elements were used in performing the analyses with GTSTRUDL. The ultimate shear capacity of the walls, as well as that of the second floor slab, was evaluated. A separate FORTRAN computer program was used to determine the in-plane stiffness values at the floor and roof levels for the frames along column lines 1 to 9 in the E-W direction. The rotational stiffness and plastic moment capacity of the semi-rigid connections was computed according to Method 2 as presented by Blodgett [10].

The first of the three models of the structure, Model 1, was used to represent the building at the start of the earthquake ground motion before any failure had occurred in any of the lateral load resisting systems. Linear elastic behavior was assumed for Model 1, but infill wall and second floor slab shear levels, moment capacities of exterior beam-column joints and cladding connection forces were checked to determine if element capacities had been exceeded. Both code [8,9] and elastic and inelastic response spectrum analyses were performed. Damping was set at 2% (somewhat conservative) and the N-S component of 1940 El Centro, normalized to 0.1g to represent a moderate earthquake in the Southeast, was used for response spectrum studies; ductility was taken as 2 for inelastic analyses of Model 1. Independent analyses were performed in each of the two orthogonal structure directions.

Model 1 was symmetric about a N-S axis; the frame-infill walls along line D and E each contributed 42% of the lateral stiffness in that direction while the clad frames along line A and C each added the remaining 8%. However, in the E-W direction, an eccentricity of 58 ft (17.7 m) (37% of the N-S dimension of the structure) between centers of mass and rigidity was computed; 78% of the E-W stiffness was provided by the infill wall at frame line 9, 6.5% by the clad frame at line 1 and 4.5% by that on line 8, and each of frames 2 to 7 contributed 1.8%. As a result, torsion was expected to play a significant role in E-W direction response.

Code and response spectrum analyses of Model 1 showed that the masonry infill walls at both story levels along column lines D and E would fail, and that the ultimate shear capacity of the first story infill wall along line 9 would also be exceeded. (Selected results are presented below). On this basis, these elements were removed from Model 1 to form Model 2 with a reduced eccentricity of 9.9 ft (3 m). However, Model 2 was found to be much more flexible than Model 1 and interstory drifts were unacceptably high, so a third model was formed which included economical alterations in configuration and lateral stiffness distribution. Masonry infill walls at column lines D, E and 9 were now assumed to be reinforced and a two story reinforced masonry wall was added to one bay of frame 1. The effect of these changes was to reduce the eccentricity of Model to 3.2 ft (1.0 m).

RESULTS

Models 1, 2 and 3 were analyzed in succession by code and response spectrum approaches and selected results are presented in the tables below. First, the responses of Model 1 were computed and are listed in Tables 1 and 2 for the N-S and E-W directions, respectively. In addition to model dynamic properties, values for the base shear (V_1), second story shear

(V_2), base torsional and overturning moments (M_t, OTM), and wall (V_{wi}) and frame (V_{Fi}) shears at story i (i=1,2) are listed in the tables. For ATC 3-06 [9], both equivalent lateral force (ELFP) and modal analysis (MAP) procedures were used, although results differ very little due to restrictions on the periods which can be used in the analyses. In each table, the failure of the masonry infill walls in shear, along with the computed value of the ultimate shear capacity, are noted in the footnote to the table. In Table 1, the values for the N-S direction base shear force can be seen to vary considerably depending upon which analysis procedure is used. The torsional moments are small (based on code-required accidental eccentricity only) and the walls at column line D and E carry most of the base shear. It should be noted that the story shears were distributed on the basis of relative stiffness of the wall and frame-cladding systems at the second floor level, but at the roof the distribution was based on the mass distribution at that level. The structure is more flexible in the E-W direction (see Table 2), and base shears are lower, at least for UBC and response spectrum analyses. However, the large actual eccentricity value at the second floor level greatly magnifies horizontal torsional moments (M_t) and produces undesirable torsional response in the E-W direction which must be reduced to improve overall seismic performance of the structure. The ATC and response spectrum analyses predict that the ultimate shear capacity of the the infill wall at the first story level of frame 9 will be exceeded.

Next, the failed masonry infill walls were removed from Model 1 (except at the second story level of frame 9) to form Model 2 as described earlier. The structure was considerably more flexible in the N-S direction (the fundamental period increased to 1.58 sec) and computed interstory drifts exceeded code allowables at the first story level. The cladding panels along frame lines A and C acted to brace the second story and cladding connection forces were well below the assumed ultimate capacity of 10 kips (44.5 kN) at all locations. The principal results for the E-W direction of Model 2 are summarized in Table 3. The fundamental period increased only 10% compared to Model 1 so UBC [8] and response spectrum base shears (V_1) are very close to Model 1 values in Table 2. ATC [9] results are still higher than UBC and response spectrum values but the difference in values is measurably reduced. As expected, the torsional moments (M_t) are much smaller as a result of the decreased eccentricity. As a final step, the E-W frames were checked for compliance with the AISC code using the code-check capabilities of GTSTRUDL. The columns of frames 1 and 9 were found to have combined bending-axial stress ratios in excess of allowable.

The conclusion drawn from the analyses of Models 1 and 2 was that the case study building is unable to satisfy seismic code provisions and probably could not resist a moderate earthquake in its present form. As a result, modest and economical changes to the structure were studied to bring it into compliance with code requirements. This produced Model 3 which was described above. The code and response spectrum analyses were repeated for Model 3 and the findings are summarized in Tables 4 and 5. Both code and response spectrum values are now in better agreement, suggesting that the increased ductility value of 3 used for the Model 3 analyses is more appropriate for the altered structure. Finally, it should be noted that, while other changes to the structure are possible (and may be preferable) to improve its seismic performance, the suggested changes do accomplish one of the the principal objectives of the codes: to design ductile structures with regular configuration and uniform distributions of stiffness and mass.

CONCLUSIONS

Earthquake engineers have made important strides in designing highrise building structures to resist the effects of damaging earthquakes. Much additional knowledge is expected to result from a concentrated research effort on the effects of the 19 September 1985 Mexico Earthquake. In contrast, the state of knowledge for lowrise buildings is much less advanced. These structures are usually comprised of a diverse assemblage of materials joined by connections of largely unknown strength and ductility. In many cases, nonstructural elements such as cladding, partitions and masonry infill can have a substantial influence on structure response, at least during the initial stages of the ground motion. If not considered in design, nonstructural elements can magnify torsional response and can concentrate forces and create unanticipated patterns of response. The code and response spectrum analysis results presented for the case study building were intended to show that a relatively simple analysis strategy, employing a series of linear models of the structure at various stages during the earthquake, could be used to guide the designer in making economical changes in the basic configuration. All basic components of the system were accounted for in the model: framing, masonry, floor and roof diaphragms, and cladding. Models of this kind can be used to guide in rehabilitation efforts for existing construction which must be brought up to seismic resistant standards, and to assist in the development of new lowrise construction designed in compliance with modern codes.

REFERENCES

1. Gupta, A.K., editor, "Seismic Performance of Low Rise Buildings: State-of-the-Art and Research Needs," Proceedings, Workshop held at Illinois Institute of Technology, Chicago, Illinois, May 13-14, 1980, (proceedings published by ASCE, 1981).

2. Montgomery, A.M., and Hall, W.J., "Seismic Design of Low-Rise Steel Buildings," Journal of the Structural Division, ASCE, Vol. 105, No. ST10, Proc. Paper 14919, Oct. 1979, pp. 1917-1033.

3. Naman, S. K., and Goodno, B. J., "Seismic Evaluation of a Low Rise Steel Building," Engineering Structures, Vol. 8, No. 1, January 1986, pp. 9-16.

4. Bouwkamp, J. G., and Blohm, J. K., "Dynamic Response of a Two-Story Steel Frame Structure," Bulletin of the Seismological Society of America, Vol. 56, No. 6, December 1966, pp. 1289-1303.

5. "Engineering Features of the San Fernando Earthquake of February 9, 1971," EERL 71-02, Jennings, P. C. (Ed), California Institute of Technology, Pasadena, California, June 1971.

6. "Miyagiken-Oki, Japan, Earthquake, June 12, 1978," <u>Reconnaissance Report</u>, Yanev, P. I. (Ed), Earthquake Engineering Research Institute, Berkeley, California, 1978.

7. Goodno, B. J., Palsson, H., and Pless, D. G., "Localized Cladding Response and Implications for Seismic Design," <u>Proceedings</u>, Eighth World Conference on Earthquake Engineering, San Francisco, California, July 21-28, 1984, Vol. V, pp. 1143-1150.

8. <u>Uniform Building Code</u>, International Conference of Building Officials, Whittier, California, 1982 Edition.

9. <u>Tentative Provisions for the Development of Seismic Regulations for Buildings</u>, Applied Technology Council (ATC), Publication ATC-3-06, Palo Alto, California, June 1978.

10. Blodgett, O. W., <u>Design of Welded Structures</u>, The James F. Lincoln Arc Welding Foundation, Cleveland, Ohio, 1968, Section 5.5.

11. Craig, J. I., Goodno, B. J., Keister, M. J., and Fennell, C. J., "Hysteretic Behavior of Precast Cladding Connections," presented at the Third ASCE Engineering Mechanics Specialty Conference on Dynamic Response of Structures, held at UCLA on March 31-April 2, 1986.

ACKNOWLEDGMENTS

The support of the National Science Foundation through Grant CEE-8213803 is gratefully acknowledged. However, the results and conclusions presented in this paper are those of the authors and do not necessarily represent the views of the National Science Foundation.

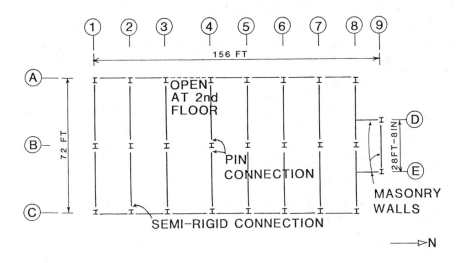

Fig. 1 Floor Framing Plan

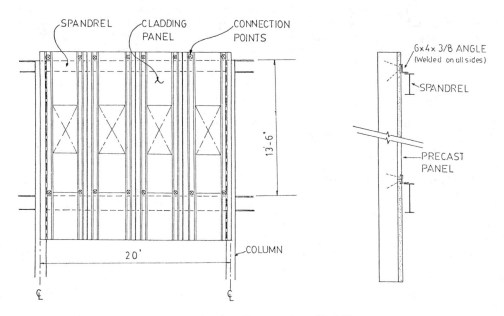

Fig. 2 Typical Bay of Exterior Cladding

Table 1 - Code and Response Spectrum Analysis Results for MODEL 1, N-S Direction

	UBC'82	ATC 3-06 ELFP	ATC 3-06 MAP	Response Spectrum[a]
Period(s) (sec)	0.44	0.19	0.38,0.15	0.44,0.20,0.13
K or R	1(K)	1.5(R)	1.5(R)	---
C_S	0.05	0.167	0.167	0.10
V_1 (kips)*	74.9	238.1	238.1	149.7
V_2 (kips)	32.9	104.8	105.7	69.5
eccentricity (ft)	7.84	3.6	3.6	0
M_t (in-k)	7042	10286	10286	0
OTM (in-k)	17211	43795	49101	33169
Wall Shears at D,E V_{w1} (kips) V_{w2} (kips)	32.0[b] 12.4	100.4[b] 39.3[b]	105.7[b] 39.6[b]	62.9[b] 30.1[b]
Frame Shears at A,C V_{F1} (kips) V_{F2} (kips)	6.3 4.1	19.6 13.1	20.5 13.2	12.1 10.8

* 1 in = 25.4 mm, 1 kip = 4.448 kN
[a] 2% damping, ductility = 2, 1940 N-S El Centro normalized to 0.1g
[b] exceeds shear capacity (26 kips) of masonry wall

Table 2 - Code and Response Spectrum Analysis Results
for MODEL 1, E-W Direction

	UBC'82	ATC 3-06 ELFP	ATC 3-06 MAP	Response Spectrum[a]
Period(s) (sec)	1.25	0.16	0.27,0.14	1.25,0.51,0.41 0.22,0.14,0.13
K or R	1(K)	1.5(R)	1.5(R)	---
C_S	0.03	0.167	0.167	0.044
V_1 (kips)*	47.4	238.1	238.1	63.3
V_2 (kips)	23.7	104.8	98.0	21.8
eccentricity (ft)	58.0	65.5	65.5	58.0
M_t (in-k)	33018	187146	187146	33347
OTM (in-k)	11098	43795	51545	11574
Frame Shears, Line 1				
V_{F1} (kips)	5.5	29.4	31.8	8.8
V_{F2} (kips)	1.7	7.3	6.9	7.6
Frame Shears, Line 2				
V_{F1} (kips)	1.3	7.0	7.6	2.2
V_{F2} (kips)	3.1	13.6	12.7	4.4
Frame Shears, Line 8				
V_{F1} (kips)	3.6	18.8	20.3	5.8
V_{F2} (kips)	2.1	9.4	8.8	8.8
Frame Shears, Line 9				
V_{F1} (kips)	66.7	354.6[b]	382.9[b]	106.5[b]
V_{F2} (kips)	1.6	6.3	5.9	2.7

* 1 in = 25.4 mm, 1 kip = 4.448 kN
[a] 2% damping, ductility = 2, 1940 N-S El Centro normalized to 0.1g
[b] exceeds shear capacity (97 kips) of masonry wall

Table 3 - Code and Response Spectrum Analysis Results
for MODEL 2, E-W Direction

	UBC'82	ATC 3-06 ELFP	ATC 3-06 MAP	Response Spectrum[a]
Period(s) (sec)	1.37	0.16	0.47, 0.19	1.37, 0.41
K or R	1(K)	4.5(R)	4.5(R)	---
C_S	0.032	0.056	0.056	0.043
V_1 (kips)*	45.7	79.9	79.9	61.0
V_2 (kips)	23.0	35.1	31.2	28.4
eccentricity (ft)	9.9	17.7	17.7	9.9
M_t (in-k)	5429	16971	16971	2607
OTM (in-k)	10725	13411	15401	12668

* 1 in = 25.4 mm, 1 kip = 4.448 kN
[a] 2% damping, ductility = 2, 1940 N-S El Centro normalized to 0.1g

Table 4 - Code and Response Spectrum Analysis Results for MODEL 3, N-S Direction

	UBC'82	ATC 3-06 ELFP	Response Spectrum[a]
Period(s) (sec)	0.44	0.19	0.44,0.20,0.15
K or R	1(K)	4.5(R)	---
C_S	0.050	0.056	0.061
V_1 (kips)*	74.9	79.8	87.6
V_2 (kips)	32.9	35.1	49.2
eccentricity (ft)	7.8	3.8	0
M_t (in-k)	7042	3450	0
OTM (in-k)	17211	13771	19355

* 1 in = 25.4 mm, 1 kip = 4.448 kN
[a] 2% damping, ductility = 3, 1940 N-S El Centro normalized to 0.1g

Table 5 - Code and Response Spectrum Analysis Results for MODEL 3, E-W Direction

	UBC'82	ATC 3-06 ELFP	Response Spectrum[a]
Period(s) (sec)	0.52	0.13	0.52,0.22,0.16
K or R	1(K)	4.5(R)	---
C_S	0.052	0.056	0.087
V_1 (kips)*	74.2	79.9	124.1
V_2 (kips)	32.6	35.9	30.6
eccentricity (ft)	7.8	11.0	3.2
M_t (in-k)	6845	10542	7974
OTM (in-k)	16597	13513	19466

* 1 in = 25.4 mm, 1 kip = 4.448 kN
[a] 2% damping, ductility = 3, 1940 N-S El Centro normalized to 0.1g

PARTIAL UPLIFT OF INVERTED-PENDULUM TOWER ON FLEXIBLE GROUND

B. Pacheco [I], Y. Fujino [II], and M. Ito [III]

ABSTRACT

A 5-DOF model is used to analyze the dynamic uplifting of a top-heavy tower with shallow rigid base on viscoelastic foundation, using a concept of variable effective radius. Despite the adverse increases in soil pressure, drift of the top, and base sway, it appears that uplifting and the associated reduction in base shear could generally be expected and tolerated for towers with fixed-base fundamental frequencies in the medium range, say, 0.5-3 Hz.

INTRODUCTION

A structure is said to uplift when part of its base intermittently separates vertically from the ground. While it is true that the displacements involved may be too small to be directly observable, stress- or strain limits are occasionally exceeded at interface or joint elements, as demonstrated by some mathematical models of structure-soil interaction [Refs. 1-3].

The practical questions presently relevant to structural engineers are: under what conditions would uplifting happen; and what benefits or problems would it bring? The present paper attempts to answer these in the case of a top-heavy flexible structure that stands on a shallow rigid base on compliant foundation, in terms of structural stresses, displacements, and soil bearing pressures. Insights from past studies shall serve as introduction.

Uplifting on stiff ground

Meek [4] highlighted uplifting by considering an infinitely stiff floor on top of which a rigid base carrying a shear-type single degree of freedom oscillator is placed unbonded.

The natural choice of uplift criterion in the above model was a force-type criterion: uplift was considered to start whenever the moment at the structural base (" overturning moment ") reaches the product of structural weight and half the base width. To introduce damping as well, every recontact was assumed to be a dull thud that prevented bouncing of the base.

It was concluded that the peak transverse deformation of an uplifting structure, and the shear force at its base, would almost always be less than

---

I. Graduate Student, Dept. of Civil Engineering, University of Tokyo

II. Associate Professor, Dept. of Civil Engineering, University of Tokyo

III. Professor, Dept. of Civil Engineering, University of Tokyo, Japan

the peak response of a corresponding system whose base prevents rotation. It was also pointed out that uplifting may be expected before any yielding in shear, provided that the ratio of base width to height of structural mass is small enough. Ishiyama [5] later made a related conclusion: homogeneous rigid blocks would not slide during uplifting if the ratio of width to height is less than the static coefficient of friction between block and ground.

Meek eventually warned, however, of adverse effects of uplifting on the soil. Chopra and Yim [6] meanwhile developed a similar model to draw analogies for their simplified method of calculating base shear reductions even in the case of structures on non-stiff ground.

Uplifting on flexible ground

When the ground is not infinitely stiff, rocking indeed does not always mean uplifting. The case of a rigid block whose base edges rest on pairs of spring and dashpot (without bond) was studied by Psycharis and Jennings [7], among others.

In this model, a displacement-type uplift criterion was used: uplifting was considered to start whenever the upward displacement of an edge of the base reached the static deflection of the vertical springs due to gravity. The dashpots represented all damping.

It was demonstrated that a solitary horizontal impulse may be large enough to cause uplift in the system and, in the process, drastically prolong the apparent fundamental period and reduce the corresponding damping ratio. The introduction of horizontal springs and dashpots with finite coefficients [8] proved significant particularly in the case of relatively squat block: upthrowing, or a complete separation from all springs, became less likely. Overturning, or an excessive rotation, was shown to be even more remote. A certain scale effect is among the considerations: between two blocks of the same mass, width-to-height ratio, and spring- and dashpot coefficients, the bigger (and less dense) would rotate less and be less likely to overturn.

References 1-14 together illustrate, among other things, that: 1) for large bodies, intermittent uplifting is of more immediate practical interest than either overturning or upthrowing; 2) for the not-so-slender structures, the swaying mode of the the base serves to modify the tendency to uplift or to upthrow; and 3) uplifting often reduces structural stresses but may increase some responses particularly for very high-frequency structures.

Some features of the present model

In this model the base is again considered rigid and circular, with an effective radius that varies with time as partial uplifting starts and stops in a manner that depends on input ground acceleration as well as on properties of structure and ground. It is allowed three rigid-body displacements, namely, swaying (or horizontal translation relative to ground), heaving (or vertical translation), and rocking. The flexibility of the superstructure is accounted for by two additional degrees of freedom.

The soil itself is considered to be linear viscoelastic and infinite in one dimension (if a layer) or two (if a halfspace). Frequency-dependent impedance coefficients for such a foundation system have been obtained for use in substructure approaches to linear interaction. Coefficients available in Refs. 15 and 16 are adapted approximately to the present nonlinear problem.

Since extreme cases of near-overturning are being excluded, the total model does not explicitly account for the overturning moment due to gravity. However, for moderate values of structural deformation and base rocking, the stiffness matrix incorporates geometic effect, due to vertical acceleration as well as gravity. Both vertical and horizontal ground accelerations are simultaneously input directly at the structural base.

THE PRESENT 5-DOF MODEL OF INVERTED-PENDULUM TOWER

A uniform linear elastic beam-column supports the top mass and stands at the center of a rigid basemat whose own mass is nonzero. The base rests by gravity on a linear viscoelastic continuum. Relevant properties and some nondimensionalized parameters for this model are shown in Fig. 1.

Figure 2 illustrates the five degrees of freedom: two associated with tip translation and rotation relative to the base, and three associated with rigid-body sway, rock, and heave.

The tower itself is essentially a Bernoulli-Euler beam with distributed axial force. The two contraint shapes associated with coordinates d and αL are used to reduce its distributed mass, damping, and stiffness properties into discrete values.

Figure 3 schematically shows the two sources of geometric nonlinearity in the present problem: parametric excitation due to vertical ground acceleration as it affects the stiffness of the superstructure; and the last three diagonal elements of both damping and stiffness matrices as they depend on, among other factors, the contact area between base and ground. A step-by-step numerical integration of the nonlinear governing equations may consider an instantaneous linear system at each interval; for intervals where partial uplifting occurs, a reduced effective radius shall be defined.

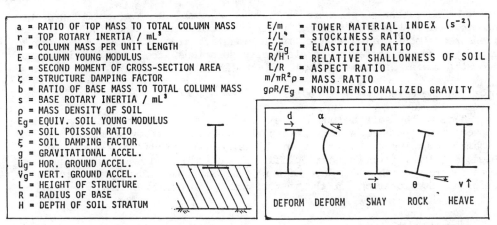

Fig.1. Parameters of the system   Fig.2. Degrees of freedom

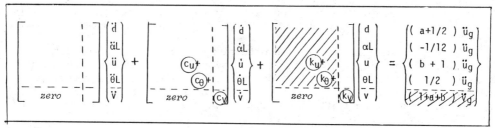

Fig.3. Sources of geometric nonlinearities in the equations

The dynamic nature of foundation impedance, i.e., its dependence on frequency, is also taken into account, even if approximately. Following the approach in many works including Ref. 17, the fundamental natural frequency of the instantaneous (equivalent) linear structure-soil system is taken as the relevant frequency at which constant foundation coefficients for swaying, rocking and heaving are obtained by iteration. Hence, an iteration with respect to frequency is required at each time interval, making the present model "hybrid".

The nondimensionalized frequency, a_o, that is commonly used in presenting dynamic foundation impedance coefficients, and the natural circular frequency of equivalent linear structure-soil system, ω, are related thus:

$$a_0 = \frac{\omega}{\sqrt{E/m}} \sqrt{\frac{2(1+\nu)\ E/E_g}{\pi\ m/\pi R^2 \rho}} \quad . \tag{1}$$

As Fig. 4 suggests, the dynamic stresses are computed at each time step by considering inertia loads proportional to deformations d and αL and also satisfying the geometric end conditions (at top and bottom) of the two constraint shapes. [Ref. 18] Internal forces in the superstructure and bearing pressures under the base are computed after these pseudodynamic loads.

The soil bearing pressure is treated as combination of normal pressure due to base sink and to rocking, under the additional assumption that these pressures almost follow static pressure-distribution patterns. Moreover, to avoid the edge singularity of elasticity solutions, toe presssure is computed at a radial distance of about 98% of effective radius.

Updating the effective base radius

Because the soil medium is flexible, the whole structure has an instantaneous "sink" amounting to v_t, which is the sum of static component v_s and dynamic component v. Depending on this sink and the effective radius of the base, R', uplift or further uplift may be said to impend when the base rotation is such that the base heel is just about to overcome the sink. The critical base rotation for the instantaneous linearized system, θ_{cr}, is:

$$\theta_{cr} = \frac{|v_t|}{R'} \tag{2}$$

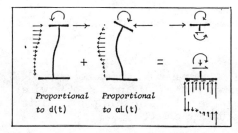
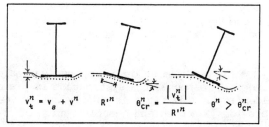

Fig.4. Dynamic stresses   Fig.5. Uplift criterion

on the assumption that $v_t^n < 0$, i.e., the sink is indeed downward. Thus the critical base rotation may be defined at the start of each time step, and be assumed to apply to the whole interval.

At the beginning of any time step n, the effective radius R' and critical base rotation θ_{cr} are defined. At the end of the time step, should the computed base rotation θ exceed the critical, R' will have to be reduced in order to account for uplifting, before proceeding to the next time step. Figure 5 illustrates this displacement-type criterion.

In as much as uplift at the base heel and soil overstress or deterioration at the base toe occur together, it may be sufficient to consider the base to "shrink" without shifting the centroid of the contact area from its initial position beneath the centerline of the superstructure. The reduced effective radius is merely computed such that the associated area is equal to the presumed contact surface under the partially uplifted footing of the system defined at the start of the time step. The new effective radius is to be used at the start of the next interval.

The straightforward equation for reduced radius for interval n+1 is:

$$R'^{n+1} = R'^n \sqrt{\left(\pi - \cos^{-1}\frac{\theta_{cr}^n}{\theta^n} + \frac{1}{2}\sin 2\cos^{-1}\frac{\theta_{cr}^n}{\theta^n}\right)/\pi} . \qquad (3)$$

At the end of any particular time step, it is possible that the computed base rotation θ is less than the corresponding critical value, indicating that uplift is not progressing. Instead, when the current effective radius is less than the full radius R, a recovery of contact must be expressed by increasing R'. The simplest scheme would be to assume that the θ is just equal to the critical base rotation of the next time step:

$$R'^{n+1} = \frac{|v_t^{n+1}|}{\theta_{cr}^{n+1}} \doteq \frac{|v_t^n|}{|\theta^n|} , \qquad R'^{n+1} \leq R . \qquad (4)$$

RESPONSE VARIABLES

While most previous studies on effects of uplifting focused on base shear, e.g., Refs. 4 and 9, or in-structure acceleration, e.g., Refs. 3 and 13, the present paper presents five response variables, namely: a) base shear force; b) soil bearing pressure in the vicinity of the toe; c) drift of the top, being the sum of lateral deformation of the column and rigid-body deflection due to base rocking; d) base sway; and e) effective base radius.

The maximum of each response variable is compared with the corresponding peak response of a system that is prevented from uplifting but otherwise identical. The ratio is simply called "response ratio" in this paper. This provides a convenient quantification of the uplift effect.

It must be remembered, however, that presenting uplift effects in terms of response ratios omits the picture of the qualitative effects, i.e., how uplifting changes the overall appearance of time response histories. These qualitative effects are widely described in the references cited, e.g., Refs. 4, 9, and 11.

To facilitate the interpretation of response ratios, these values are plotted against a logarithmic scale of fixed-base natural frequency, f, i.e., natural frequency neglecting all dynamic interaction with the ground. In each of such graphs, values for all the parameters as defined in Fig.1 are fixed (e.g., Table 1), except the stockiness ratio of the column, I/L^4, and the elasticity ratio, E/E_g.

The stockiness ratio of the column is adjusted to provide a required fixed-base frequency. With this parameter known, the elasticity ratio is subsequently adjusted to provide a prescribed amount of full-contact interaction. This latter value is defined here as the ratio of natural frequencies in the fixed-base condition, f, and in the full-contact interaction, f'. In practice this ratio effectively prescribes also the total damping ratio of the fundamental mode of the full-contact interacting system.

EXAMPLES

Examples in this section will refer to structure-soil systems with: a) either medium interaction (f/f'=1.2) or high interaction (1.4); b) either medium aspect (L/R=5) or high aspect (8); c) either top-heavy (a=1) or uniform (0); and d) either the El Centro 1940 ground acceleration (N-S and

Table 1. Parameters of examples

E/m (s^{-2})	10^6	10.0
r	10^{-3}	20.0 a
b		1.0
s	10^{-3}	25.0
ζ		0.01
R/H		0.0
ξ		0.10
ν		0.33
m/πR^{2}ρ	10^{-3}	10.0
gρR/E$_g$	10^{-3}	5.0

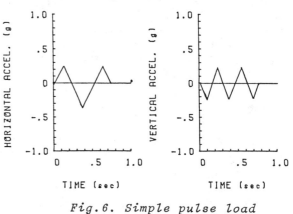

Fig.6. Simple pulse load

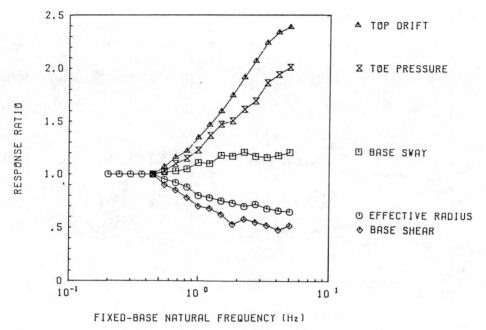

Fig.7. Top-heavy; medium aspect; medium interaction; El Centro

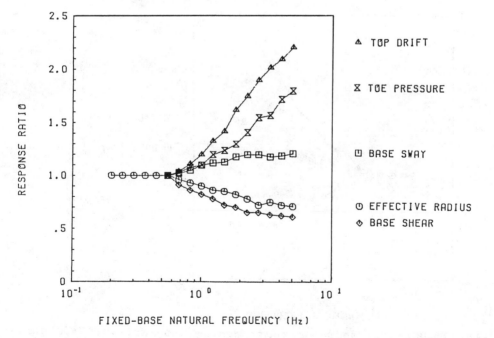

Fig.8. Top-heavy; medium aspect; high interaction; El Centro

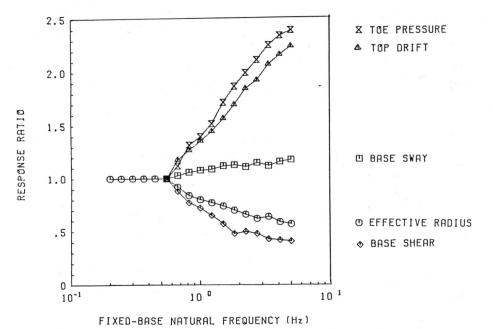

Fig.9. *Top-heavy; high aspect; medium interaction; El Centro*

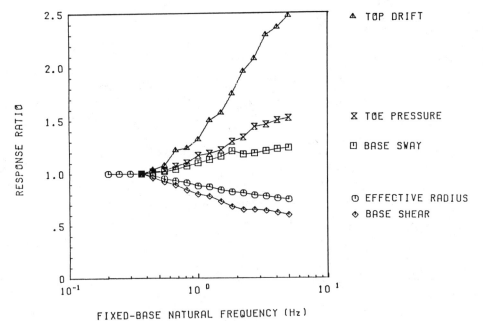

Fig.10. *Uniform; medium aspect; medium interaction; El Centro*

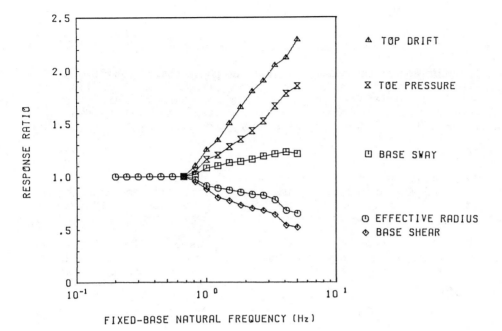

Fig.11. Top-heavy; medium aspect; medium interaction; Pulse

U-D components) or a simplified series of few pulses (see Fig. 6). The simple pulses were adjusted such that the peak horizontal acceleration and the biggest pulse are equal to the corresponding values of the El Centro 1940 N-S component.

Following is a brief discussion of the example response ratios presented in Figs. 7-11. Each graph shall be referred to as a group in as much as it represents a collection of structures that are identical to each other except for their fixed-base fundamental natural frequencies, which in turn are determined by their respective stockiness ratios.

Group 1 (Fig.7)

These structures are top-heavy, of medium aspect ratio, with medium interaction with the ground, and subject to the El Centro ground motion.

It can be seen from Fig.7 that all structures with fixed-base frequencies greater than about 0.5 Hz experience uplift, and the amount of uplifting increases gradually, but monotonically, with frequency. The corresponding reduction in base shear likewise increases as uplifting increases. However, there is an apparent flattening out of this trend at very high frequencies.

Among the "disadvantages" of uplifting, increased base sway seems to be the least problem. The increases in toe pressure and top drift are much bigger.

As may be expected from the fact that toe pressure depends on the square of radius, the increase in toe pressure is greater than the decrease in effective radius. The margin between this pair of "advantage" and "disadvantage" is wider for structures with higher frequencies.

The increase in the amount of drift at the top also rapidly grows as fixed-base frequency increases. There is an apparent slowing of this trend, however, in the high- frequency range. This may be explained by the fact that when such a high-frequency structure uplifts, structural deformation is greatly reduced while rigid-body deflection predominates.

Group 2 (Fig.8)

These structures are identical to Group 1, except that the level of interaction with the ground is higher, i.e., natural frequency reduction and total damping increase are bigger. Generally the damping ratio for the fundamental apparent mode is about double for Group 2, compared to Group 1.

It can be observed from Fig.8 that the threshold frequency, i.e., the minimum fixed-base frequency required before there could be uplifting, is higher for this group. This indicates that fewer structures would uplift.

Moreover, for each uplifting structure, the extent of uplift is less. Correspondingly, the reduction in base shear, the increase in toe pressure, and the increase in drift are also smaller. The increase in base sway is essentially unchanged, however, when compared to group 1.

In the overall sense, it can be seen that higher interaction lessens both the chance and the effect of uplifting. In the present model, this may be explained by the fact that additional damping is assured when rocking and swaying dominate the apparent fundamental mode more and more.

Group 3 (Fig.9)

Except for the higher aspect ratio, this group is also identical to Group 1. As previous studies, e.g., Refs.4 and 9, would predict, the amount of base shear reduction is greater for this group. However, the threshold frequency is higher, implying that fewer structures would uplift.

For an uplifting structure in this group, the bigger uplift also means bigger increase in toe pressure, while the increase in base sway seems unaffected. The increase in drift is somewhat less, compared to group 1.

Group 4 (Fig.10)

This group has less mass at the top, hence the center of mass is lower and the effective aspect ratio is less. Dynamically, the column behaves more and more as the usual cantilever.

The shifting of the threshold frequency to the left of the graph means that more structures would uplift. The amount of uplift is smaller, however.

All the uplift effects are reduced, except the base sway increase and the drift increase. Apparently, the rigid-body deflection during uplifting is so predominant that the reduction in structural deformation becomes rather immaterial as far as the total drift is concerned. (In contrast, for the more slender structures of Group 3, structural deformation remains quite significant relative to the rigid-body deflection.)

Group 5 (Fig.11)

This group is identical to Group 1 except for the loading. The present short-duration loading with large impulse (as large as that of the largest pulse in the El Centro ground motion) causes uplifting in fewer structures. The effects of uplifting are also less, except at the higher frequencies where some response ratios approach those of Group 1. However, the general trends in the various effects of uplifting on response ratios seem to be insensitive to details of the ground motion other than peak acceleration and maximum pulse.

SUMMARY AND CONCLUSIONS

A model that is computationally straightforward has been presented here that allows the use of available dynamic foundation compliance coefficients pertaining to viscoelastic media, even in the nonlinear uplifting problem. The effective-radius approach that was used was rather simple compared to, say, Ref.3.

It has been proposed to assess the effect of uplift on soil presssure and structural displacements as well as base shear. The examples showed that increases in toe pressure and in top drift rapidly grow as the amount of uplifting increases, surpassing the decrease in base shear. The increase in base sway, however, is rather insensitive to many parameters.

The threshold frequency was found to increase when the level of interaction was higher; when the aspect ratio was higher; and when the loading had a short but finite duration. A high threshold only meant that fewer structures were experiencing uplift; meanwhile, for those uplifting structures, the actual amount of uplift again depended on other parameters.

It appears that uplifting would be likely, and rather favorable in an overall sense, for towers with fixed-base fundamental natural frequencies in the medium range of about, say, 0.5-3 Hz. This approximate judgement takes into consideration that the increase in soil pressure should not be so high that the soil behaviour would be very different from the assumption of pseudolinear interaction. For higher natural frequencies, the adverse effects of partial uplift tend to outweigh the favorable reduction in base shear, in the sense that special consideration for the soil may then become necessary.

REFERENCES

[1] Desai,C. and M. Zaman, Influence of Interface Behaviour in Dynamic Soil-Structure Interaction, Proc. 8th World Conf. Earthquake Eng., III, 1049-1056, 1984

[2] Toki, K., T. Sato and F. Miura, Separation and Sliding between Soil and Structure During Strong Ground Motion, Earthquake Eng. Struct. Dyn., 9, 263-277, 1981

[3] Wolf, J., Soil-Structure Interaction with Separation of Base Mat from Soil (Lifting-off), Nuclear Eng. Design, 38, 357-384, 1976

[4] Meek, J., Effects of Foundation Tipping on Dynamic Response, J. Struct. Div., ASCE, 101, ST7, 1297-1311, 1975

[5] Ishiyama, Y., Motions of Rigid Bodies and Criteria for Overturning by Earthquake Excitations, Earthquake Eng. Struct. Dyn., 10, 635-650, 1982

[6] Chopra, A. and S. Yim, Simplified Earthquake Analysis of Structures with Foundation Uplift, J. Struct. Eng., ASCE, 111, 4, 906-930, 1985

[7] Psycharis, I. and P. Jennings, Rocking of Slender Rigid Bodies Allowed to Uplift, Earthquake Eng. Struct. Dyn., 11, 57-76, 1983

[8] Psycharis, I. and P. Jennings, Upthrow of Objects due to Horizontal Impulse Excitation, Bul. Seismological Soc. Am., 75, 2, 543-561, 1985

[9] Yim, S. and A. Chopra, Earthquake Response of Structures with Partial Uplift on Winkler Foundation, Earthquake Eng. Struct. Dyn., 12, 263-281, 1984

[10] Kobori, T., T. Hisatoku and T. Nagase, Nonlinear Uplift Behaviour of Soil-Structure System with Frequency-Dependent Characteristics, Proc. 8th World Conf. Earthquake Eng., III, 897-904, 1984

[11] Psycharis, I., Dynamics of Flexible Systems with Partial Uplift, Earthquake Eng. Struct. Dyn., 11, 501-521, 1983

[12] Fukuzawa, R., O. Chiba, T. Hatori and M. Tohdo, Rocking Vibration of Nuclear Power Plant Considering Uplift and Yield of Supporting Soil, Sixth Int. Conf. Struct. Mech. Reactor Tech., K3/7, 1981

[13] Wolf, J. and P. Skrikerud, Seismic Excitation with Large Overturning Moment: Tensile Capacity, Projecting Base Mat or Lifting-off?, Nuclear Eng. Des., 50, 305-321, 1978

[14] Kennedy, R., S. Short, D. Wesley and T. Lee, Effects on Nonlinear Soil-Structure Interaction due to Base Slab Uplift, Nuclear Eng. Des., 38, 323-355, 1976

[15] Veletsos, A. and B. Verbic, Vibration of Viscoelastic Foundations, Earthquake Eng. Struct. Dyn., 2, 87-102, 1973

[16] Kausel, E. and J. Roesset, Dynamic Stiffness of Circular Foundations, J. Eng. Mech. Div., ASCE, 101, 6, 771-785, 1975

[17] Ghaffar-Zadeh, M. and F. Chapel, Frequency-independent Impedances of Soil-Structure Systems in Horizontal and Rocking Modes, Earthquake Eng. Struct. Dyn., 11, 523-540, 1983

[18] Hurty, W. and M. Rubinstein, Dynamics of Structures, 299-307, 1964

RESPONSE OF STRUCTURES TO A SPATIALLY RANDOM GROUND MOTION

A. Mita[I] and J. E. Luco[II]

ABSTRACT

The seismic response of a flexible three-dimensional structure supported on a rigid circular foundation resting on an elastic half-space when subjected to a spatially random free-field ground motion is studied. Results illustrating the effects of spatial randomness are presented for a 10-story building and for a containment building in a nuclear power plant. The effects of spatial randomness include a reduction of the high-frequency components of the translational response along the axis of the structure and the generation of significant rocking and torsional response components which also affect the motion on the perimeter of the structure.

INTRODUCTION

Strong ground motion records obtained in dense arrays reveal a degree of variability over short distances which suggests the use of a spatially random characterization of the free-field ground motion combined, perhaps, with some deterministic wave passage effects. The spatial variability of the free-field ground motion affects not only the seismic response of structures founded on multiple supports but also the response of structures supported on large mat foundations. In this paper, the effects of the spatial randomness of the free-field ground motion on the response of structures are studied. The particular model under consideration is illustrated in Fig. 1 and it consists of a flexible cylindrical structure of height H supported on a rigid circular foundation of radius a which rests on a uniform elastic half-space characterized by its shear modulus, shear wave velocity β and Poisson's ratio. The excitation corresponds to a spatially random free-field ground motion characterized by a particular spatial coherence function found in studies of wave propagation through random media. The solution incorporates the soil-structure interaction effects including the scattering of the spatially variable free-field ground motion by the rigid foundation. Emphasis is given to comparison of the structural response for a spatially random ground motion with that for spatially coherent ground motion. A second objective is to establish the characteristics of the free-field ground motion that need to be known to completely determine the structural response.

The present study builds upon previous work by Luco and Wong[1] and Luco and Mita[2] who have described procedures to obtain the response of flat massless rigid foundations resting on an elastic half-space and subjected to spatially random ground motion. The general approach[1] is based on the use of an integral representation of the response of the foundation in terms of the free-field ground motion and of the contact tractions between the foundation and the soil when the foundation is subjected to external forces and moments. These tractions are obtained by solution of integral equations

---

I. Graduate Student, Department of Applied Mechanics and Engineering Sciences, University of California, San Diego, La Jolla, California.

II. Professor, Department of Applied Mechanics and Engineering Sciences, University of California, San Diego, La Jolla, California.

representing the contact problem. Extensive numerical results for square[1] and circular[2] foundations have been presented in the studies cited and are used as basis for the structural response calculations reported herein.

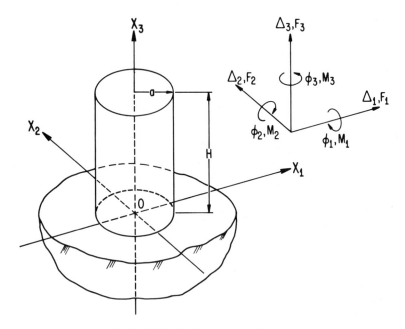

Fig. 1 Model description and coordinate system.

CHARACTERIZATION OF THE FREE-FIELD GROUND MOTION

The free-field ground motion arising from the seismic excitation in absence of the foundation is described on the basis of a Cartesian coordinate system x_1, x_2, x_3 located on the surface of the half-space $(x_3 \leq 0)$. The complex Fourier amplitude of the free-field ground motion vector at a point $\bar{x} = (x_1, x_2, 0)$ on the ground surface is represented by

$$\{U_g(\bar{x},\omega)\} = (U_{g1}(\bar{x},\omega), U_{g2}(\bar{x},\omega), U_{g3}(\bar{x},\omega))^T \tag{1}$$

in which ω is the circular frequency and $U_{gm}(\bar{x},\omega)$ represents the m-th component of the motion at $\bar{x}$. The superscript T denotes transposition.

The components of the free-field ground motion are considered to be random functions of position $\bar{x}$ such that

$$E[U_{gm}(\bar{x},\omega)] = 0 \quad , \quad (m=1,2,3) \tag{2}$$

in which $E[\cdot]$ denotes expected value. The second order properties of the random field are described by the covariance matrix

$$[B(\bar{x},\bar{x}',\omega)] = E[\{U_g(\bar{x},\omega)\}\{\tilde{U}_g(\bar{x}',\omega)\}^T] \tag{3}$$

in which the tilde denotes complex conjugate. The components B_{mn} of the 3×3 matrix $[B]$ are assumed to have the form

$$B_{mn}(\bar{x},\bar{x}',\omega) = D_{gmn}(\omega) f_{mn}(|\bar{x}-\bar{x}'|,\omega) \quad , \quad (m,n=1,2,3) \tag{4}$$

in which $D_{gmn}(\omega)$ defined by

$$D_{gmn}(\omega) = E[U_{gm}(\bar{x},\omega) \tilde{U}_{gn}(\bar{x},\omega)] \quad , \quad (m,n=1,2,3) \tag{5}$$

represents the components of the covariance matrix for the free-field ground motion at any point on the ground surface. In particular, if the free-field ground motion can be represented by a stationary random process with respect to time, then $D_{gmm}(\omega)$ ($m=1,2,3$) can be interpreted as the power spectral density of the m-component of the free-field ground motion while $D_{gmn}(\omega)$ ($m \neq n; m,n=1,2,3$) can be interpreted as the cross power spectral density of the m- and n-components of the free-field ground motion at an arbitrary point on the ground surface. The spatial coherence functions $f_{mn}(|\bar{x}-\bar{x}'|,\omega)$ for points $\bar{x}$ and $\bar{x}'$ on the ground surface are assumed to be decreasing functions of distance $|\bar{x}-\bar{x}'|$ and such that $f_{mn}(0,\omega) = 1$.

In this study, the power spectral density in the x_2-direction is assumed to be the same as that in the x_1-direction, i.e., $D_{g11}(\omega) = D_{g22}(\omega)$. Also, for the purpose of numerical calculations, the coherence functions are selected to be of the form

$$f_{mn}(|\bar{x}-\bar{x}'|,\omega) = \exp[-(\gamma\omega|\bar{x}-\bar{x}'|/\beta)^2] \quad , \quad (m,n=1,2,3) \tag{6}$$

in which γ is a dimensionless spatial incoherence parameter and β is the shear wave velocity of the soil. The motivation for the selection of this form has been presented elsewhere[1].

COVARIANCE FUNCTIONS FOR FOUNDATION INPUT MOTION

As a first step in the solution of the soil-structure interaction problem depicted in Fig. 1 for a spatially random free-field ground motion, the covariance matrix for the foundation input motion must be obtained. The foundation input motion is defined as the response to the seismic excitation (in absence of the superstructure) of a massless rigid foundation with the same shape of the actual foundation. The foundation input motion can be represented by a 6×1 generalized displacement vector

$$\{U_o^*(\omega)\} = (U_{o1}^*, U_{o2}^*, U_{o3}^*, U_{o4}^*, U_{o5}^*, U_{o6}^*)^T \tag{7}$$

in which $(U_{o1}^*, U_{o2}^*, U_{o3}^*) = (\Delta_{o1}^*, \Delta_{o2}^*, \Delta_{o3}^*)$ denotes the translational response at the center of the foundation and $(U_{o4}^*, U_{o5}^*, U_{o6}^*) = (a\phi_{o1}^*, a\phi_{o2}^*, a\phi_{o3}^*)$ corresponds to the normalized rotational response. Luco and Wong[1] have shown that the 6×6 covariance matrix for the foundation input motion, defined by

$$[D_o^*(\omega)] = E[\{U_o^*(\omega)\}\{\tilde{U}_o^*(\omega)\}^T], \tag{8}$$

can be obtained in the form

$$[D_o^*(\omega)] = \iint_S \int_S [\Gamma(\bar{x},\omega)]^T [B(\bar{x},\bar{x}',\omega)][\tilde{\Gamma}(\bar{x}',\omega)] dS(\bar{x}) dS(\bar{x}') \tag{9}$$

in which $[B(\bar{x},\bar{x}',\omega)]$ is the covariance matrix of the free-field ground motion defined by Eq. (3) and $[\Gamma(\bar{x},\omega)]$ is a 3×6 contact traction matrix. Each column of the matrix $[\Gamma(\bar{x},\omega)]$ corresponds to the traction vector at a point $\bar{x}$ on the contact area between the foundation and the soil for unit generalized harmonic forces applied to the rigid foundation at its center in the order ($F_1, F_2, F_3, F_4, F_5, F_6$) where ($F_1, F_2, F_3$) and ($F_4, F_5, F_6$) = ($M_1/a, M_2/a, M_3/a$) represent forces and normalized moments, respectively. The component $\Gamma_{mp}(\bar{x},\omega)$ represents the traction component in the x_m-direction at a point $\bar{x}$ due to a unit generalized force F_p acting on the foundation.

From Eqs. (4) and (9) it is found that the components of the 6×6 covariance matrix $[D_o^*(\omega)]$ can be written in the form

$$D_{opq}^*(\omega) = \sum_{m=1}^{3}\sum_{n=1}^{3} A_{mn}^{pq}(\omega) D_{gmn}(\omega) \quad , \quad (p,q=1,2,..,6) \tag{10}$$

in which the frequency dependent covariance coefficients $A_{mn}^{pq}(\omega)$ are defined by

$$A_{mn}^{pq}(\omega) = \int_S \int_S \Gamma_{mp}(\bar{x},\omega) \tilde{\Gamma}_{nq}(\bar{x}',\omega) f_{mn}(|\bar{x}-\bar{x}'|,\omega) dS(\bar{x}) dS(\bar{x}')$$

$$(p,q=1,2,..,6;\ m,n=1,2,3) \ . \tag{11}$$

Eq. (10) indicates that the relationship between the covariance of the foundation input motion $\{U_o^*\}$ and the covariance of the free-field ground motion $\{U_g\}$ is completely determined by the coefficients $A_{mn}^{pq}(\omega)$. Numerical values for these coefficients for square and circular foundations have been presented by Luco and Wong[1] and Luco and Mita[2], respectively.

In particular, for a circular foundation under relaxed contact conditions(frictionless contact for vertical and rocking vibrations, zero traction components in the vertical direction and in the horizontal direction normal to the applied horizontal force for horizontal vibrations) a number of the coefficients A_{mn}^{pq} are zero. The relation between D_{opp}^* ($p=1,2,...,6$) and D_{gmm} ($m=1,2,3$), in this case, reduces to

$$D_{o11}^*/D_{g11} = D_{o22}^*/D_{g22} = A_{11}^{11} = A_{22}^{22} \tag{12a}$$

$$D_{o33}^*/D_{g33} = A_{33}^{33} \tag{12b}$$

$$D_{o44}^*/D_{g33} = D_{o55}^*/D_{g33} = A_{33}^{44} = A_{33}^{55} \tag{12c}$$

$$D_{o66}^*/(D_{g11}+D_{g22}) = D_{o66}^*/(2D_{g11}) = A_{11}^{66} \tag{12d}$$

which indicates that $(A_{11}^{11})^{1/2}$, $(A_{22}^{22})^{1/2}$ and $(A_{33}^{33})^{1/2}$ can be considered as the amplitude of transfer functions between the p-component($p=1,2,3$) of the foundation input motion and the p-component of the free-field ground motion. Similarly, the terms $(A_{33}^{44})^{1/2}$ and $(A_{33}^{55})^{1/2}$ can be thought of as the amplitudes of transfer functions between the normalized rocking components of the foundation input motion and the vertical component of the free-field ground motion. Finally, $(A_{11}^{66})^{1/2} = (A_{22}^{66})^{1/2}$ can be considered as the amplitude of a transfer function between the normalized torsional component of the foundation input motion and one of the horizontal components of the free-field ground motion.

Numerical values for $(A_{11}^{11})^{1/2} = (A_{22}^{22})^{1/2}$, $(A_{33}^{33})^{1/2}$, $(2A_{11}^{66})^{1/2} = (2A_{22}^{66})^{1/2}$ and $(A_{33}^{44})^{1/2} = (A_{33}^{55})^{1/2}$ as obtained by Luco and Mita[2] for a circular foundation on an elastic half-space(Poisson's ratio $\sigma=1/3$) and for the particular coherence function given by Eq. (6) with $\gamma=0.1$, 0.3 and 0.5 are shown in Fig. 2 versus the dimensionless frequency $a_o=\omega a/\beta$. To facilitate the physical interpretation, the results corresponding to $(A_{11}^{11})^{1/2}$, $(A_{33}^{33})^{1/2}$, $(2A_{11}^{66})^{1/2}$ and $(A_{33}^{55})^{1/2}$ have been labelled $|\Delta_{o1}^*/U_{gH}|$, $|\Delta_{o3}^*/U_{gV}|$, $|a\phi_{o3}^*/U_{gH}|$ and $|a\phi_{o2}^*/U_{gV}|$, respectively. The results in Fig. 2 show a reduction in amplitude of the translational components of the foundation input motion with respect to the corresponding components of the free-field ground motion. The results also reveal the generation of significant rocking and torsional response components. Also shown in Fig. 2 are the corresponding deterministic transfer functions(segmented lines) for nonvertically incident waves with apparent horizontal velocities given by $c=\beta/(2\gamma)$ for the translational and the torsional components and $c=\beta/(\sqrt{2}\gamma)$ for the rocking component. These deterministic transfer functions were calculated by the approach described by Luco and Mita[3]. It appears that for low values of the incoherence parameter γ and for low values of a_o, the effects of spatial randomness can be simulated(in the sense of power spectral densities) by wave passage effects at appropriate effective apparent velocities. A more complete analysis of the relation between the spatial incoherence parameter γ and the equivalent velocity c can be found in reference[1].

To complete the discussion of the covariance matrix $[D_o^*(\omega)]$ it is necessary to refer to the off-diagonal terms D_{opq}^* ($p \neq q$). Considering the relations $D_{opq}^* = \tilde{D}_{oqp}^*$ ($p,q=1,2,..,6$), it is sufficient to list the non-zero off-diagonal terms

$$D_{o12}^*/D_{g12} = A_{12}^{12} = A_{11}^{11} = D_{o11}^*/D_{g11} \tag{13a}$$

$$D_{o13}^*/D_{g13} = A_{13}^{13} \approx \sqrt{A_{11}^{11} A_{33}^{33}} = \sqrt{D_{o11}^* D_{o33}^*}/\sqrt{D_{g11} D_{g33}} \tag{13b}$$

$$D_{o23}^*/D_{g23} = A_{23}^{23} \approx \sqrt{A_{22}^{22} A_{33}^{33}} = \sqrt{D_{o22}^* D_{o33}^*}/\sqrt{D_{g22} D_{g33}} \tag{13c}$$

$$D_{o46}^*/D_{g31} = A_{31}^{46} \approx -\sqrt{A_{33}^{44} A_{11}^{66}} = -\sqrt{D_{o44}^* D_{o66}^*}/\sqrt{2 D_{g33} D_{g11}} \tag{13d}$$

$$D^*_{o56}/D_{g32} = A^{56}_{32} \approx -\sqrt{A^{55}_{33}A^{66}_{22}} = -\sqrt{D^*_{o55}D^*_{o66}}/\sqrt{2D_{g33}D_{g22}} \ . \tag{13e}$$

Eqs. (13) also include some exact and approximate relations with other terms. Eqs. (13a,b,c) indicate that the normalized correlation coefficients $D^*_{opq}/\sqrt{D^*_{opp}D^*_{oqq}}$ ($p,q=1,2,3$) between the translational components of the foundation input motion are equal to or approximately equal to the corresponding coefficients $D_{gmn}/\sqrt{D_{gmm}D_{gnn}}$ ($m,n=1,2,3$) of the free-field ground motion. Eqs. (13d,e) indicate that the correlation between the rocking and torsional components of the foundation input motion are related to the correlation between the horizontal and vertical components of the free-field ground motion. Finally, Eqs. (13a-e) indicate that there is no correlation between the translational and rotational components of the foundation input motion.

RESPONSE OF STRUCTURES TO A SPATIALLY RANDOM GROUND MOTION

Once the foundation input motion has been obtained, the 6×1 generalized total displacement vector $\{U_o(\omega)\} = (\Delta_{o1}, \Delta_{o2}, \Delta_{o3}, a\phi_{o1}, a\phi_{o2}, a\phi_{o3})^T$ at the center of the foundation, including all soil-structure interaction effects, can be readily calculated through the relation (Luco and Wong[4])

$$\{U_o(\omega)\} = [L(\omega)]\{U^*_o(\omega)\} \tag{14}$$

in which

$$[L(\omega)] = ([I] - \omega^2 [C(\omega)]([M_o] + [M_b(\omega)]))^{-1} \tag{15}$$

where $[I]$ is the 6×6 identity matrix, $[C(\omega)]$ is the compliance matrix, $[M_o]$ is the 6×6 mass matrix for the foundation and $[M_b(\omega)]$ is an equivalent frequency-dependent mass matrix for the superstructure. In the case considered here of a superstructure with two vertical planes of symmetry, the only nonzero elements of the matrix $[L(\omega)]$ are L_{11}, L_{22}, L_{33}, L_{44}, L_{55}, L_{66}, L_{15}, L_{51}, L_{24} and L_{42}.

The covariance matrix of the total generalized displacement at the center of the foundation defined by

$$[D_o(\omega)] = E[\{U_o\}\{\tilde{U}_o\}^T] \tag{16}$$

is given by

$$[D_o(\omega)] = [L(\omega)][D^*_o(\omega)][\tilde{L}(\omega)]^T \tag{17}$$

in which $[D^*_o(\omega)]$ is the covariance matrix of the foundation input motion $\{U^*_o\}$. For a circular foundation and a doubly symmetric superstructure, the diagonal elements of $[D_o(\omega)]$, as obtained from Eqs. (17), (10), (12) and (13), are

$$D_{o11} = |L_{11}|^2 A^{11}_{11} D_{g11} + |L_{15}|^2 A^{55}_{33} D_{g33} \tag{18a}$$

$$D_{o22} = |L_{22}|^2 A^{22}_{22} D_{g22} + |L_{24}|^2 A^{44}_{33} D_{g33} \tag{18b}$$

$$D_{o33} = |L_{33}|^2 A^{33}_{33} D_{g33} \tag{18c}$$

$$D_{o44} = |L_{42}|^2 A^{22}_{22} D_{g22} + |L_{44}|^2 A^{44}_{33} D_{g33} \tag{18d}$$

$$D_{o55} = |L_{51}|^2 A^{11}_{11} D_{g11} + |L_{55}|^2 A^{55}_{33} D_{g33} \tag{18e}$$

$$D_{o66} = |L_{66}|^2 (A^{66}_{11} D_{g11} + A^{66}_{22} D_{g22}) = 2|L_{66}|^2 A^{66}_{11} D_{g11} \tag{18f}$$

in which $|L_{mn}|$ denotes the amplitude of the complex transfer function L_{mn}. The last terms in Eqs. (18a,b,d,e) indicate that the vertical components of the free-field ground motion induces horizontal and rocking response components at the center of the foundation.

In spite of the assumed symmetry, a number of off-diagonal elements of the matrix $[D_o(\omega)]$ are nonzero. In particular,

$$D_{o16} = L_{15}\tilde{L}_{66}A^{56}_{32}D_{g32} \tag{19a}$$

$$D_{o26} = L_{24}\tilde{L}_{66}A_{31}^{46}\tilde{D}_{g31} \tag{19b}$$

$$D_{o34} = L_{33}\tilde{L}_{42}\tilde{A}_{23}^{23}\tilde{D}_{g23} \tag{19c}$$

$$D_{o35} = L_{33}\tilde{L}_{51}\tilde{A}_{13}^{13}\tilde{D}_{g13} \tag{19d}$$

$$D_{o45} = L_{42}\tilde{L}_{51}\tilde{A}_{12}^{12}\tilde{D}_{g12} \tag{19e}$$

which indicate that the horizontal and torsional response components and the vertical and rocking response components at the center of the foundation are correlated if the vertical and horizontal components of the free-field ground motion are correlated. The off-diagonal terms of $[D_o(\omega)]$ are required to calculate the variance (or power spectral density) of the motion at points not on the axis of the structure. For a point of coordinates $(x_1, x_2, 0)$ located on the foundation, the following expressions can be obtained

$$E(|U_1(x_1,x_2,0)|^2) = D_{o11} + (x_2/a)^2 D_{o66} - 2(x_2/a)Re(D_{o16}) \tag{20a}$$

$$E(|U_2(x_1,x_2,0)|^2) = D_{o22} + (x_1/a)^2 D_{o66} + 2(x_1/a)Re(D_{o26}) \tag{20b}$$

$$E(|U_3(x_1,x_2,0)|^2) = D_{o33} + (x_2/a)^2 D_{o44} + (x_1/a)^2 D_{o55}$$
$$+ 2Re[(x_2/a)D_{o34} - (x_1/a)D_{o35} + (x_1x_2/a^2)D_{o45}] \tag{20c}$$

in which $Re(\cdot)$ indicates real part. Eqs. (20) and (19) indicate that to calculate the response at a point not on the axis of the structure it is necessary to know not only the variance (or power spectral density) of the three components of the free-field ground motion but also the correlation (or cross power spectral density) between free-field components at a point.

The generalized total displacement vector $\{U(\omega)\} = (\Delta_1, \Delta_2, \Delta_3, a\phi_1, a\phi_2, a\phi_3)^T$ for a point $(0,0,x_3)$ on the axis of the structure and at elevation x_3 can be obtained in the form

$$\{U(\omega)\} = [F(\omega)]\{U_o(\omega)\} \tag{21}$$

in which the 6×6 matrix $[F(\omega)]$ represents the transfer function matrix between the base of structure and a point at elevation x_3. For a structure with two vertical planes of symmetry, the matrix $[F(\omega)]$ has the same form as the matrix $[L(\omega)]$, i.e., the only nozero elements of $[F(\omega)]$ are F_{11}, F_{22}, F_{33}, F_{44}, F_{55}, F_{66}, F_{15}, F_{51}, F_{24} and F_{42}. The covariance matrix of $\{U\}$ defined by

$$[D(\omega)] = E(\{U\}\{\tilde{U}\}^T) \tag{22}$$

can be written in the form

$$[D(\omega)] = [L'(\omega)][D_o^*(\omega)][\tilde{L}'(\omega)]^T \tag{23}$$

in which $[L'(\omega)] = [F(\omega)][L(\omega)]$. The elements D_{mn} of $[D(\omega)]$ can be obtained from Eqs. (18) and (19) after substitution of L_{mn} by $L'_{mn} = \sum_{p=1}^{6} F_{mp}L_{pn}$. Finally, the variance of the components of the total displacement at a point (x_1, x_2, x_3) not on the axis can be obtained from Eq. (20) after substitution of D_{omn} by D_{mn}.

NUMERICAL RESULTS AND CONCLUSIONS

To illustrate the effects of a spatially random ground motion on the response of structures the simplified model shown in Fig. 1 has been selected. The structure is modeled as a uniform cylinder of radius a and height H which is analized as a shear wall for horizontal vibrations and as a uniform elastic bar for vertical and torsional vibrations. The foundation is modeled as a rigid circular disc and the soil as a uniform elastic half-space. Two models corresponding to a 10-story shear wall (Model 1) and a containment structure in a nuclear power plant (Model 2) are considered. The particular characteristics of the two models are listed in Table 1.

Table 1. Properties of the models considered

	Model 1	Model 2
H (m)	40	60
a (m)	10	40
M'_b (kg)	10^7	3×10^7
$S_{b1}/HM'_b = S_{b2}/HM'_b$	0.50	0.50
$I_{b1}/H^2M'_b = I_{b2}/H^2M'_b$	0.33	0.38
$I_{b3}/H^2M'_b$	0.04	0.10
M'_o/M'_b	0.15	0.45
$S_{o1}/HM'_b = S_{o2}/HM'_b$	0.000	0.000
$I_{o1}/H^2M'_b = I_{o2}/H^2M'_b$	0.017	0.017
$I_{o3}/H^2M'_b$	0.033	0.033
Fixed-base natural frequencies (*Hz*):		
horizontal vibrations	2, 6, 10, 14, 18	5, 15, 25
vertical vibrations	3, 9, 15, 21, 27	6, 18, 30
torsional vibrations	3, 9, 15, 21, 27	8, 24, 40
Modal damping coefficients	0.02	0.02
Shear wave velocity in soil (m/sec)	400	600
Shear modulus in soil (N/m^2)	3×10^8	6.75×10^8
Poisson's ratio in soil	0.333	0.333

In Table 1, M'_b represents the mass of the superstructure, and I_{b1}, I_{b2} and I_{b3} the mass moments of inertia about x_1-, x_2- and x_3-axes, respectively. The terms S_{b1} and S_{b2} represent the moments of the masses about x_1- and x_2-axes through the foundation. The corresponding quantities for the foundation are M'_o, I_{o1}, I_{o2}, I_{o3}, S_{o1} and S_{o2}.

Numerical results have been obtained for two excitations. The first case corresponds to horizontal excitation only, *i.e.*, $D_{g11} = D_{g22} = U_{gH}^2$ and $D_{g33} = D_{g32} = D_{g31} = 0$. The second case corresponds to vertical excitation only, *i.e.*, $D_{g11} = D_{g22} = D_{g12} = D_{g13} = D_{g23} = 0$ and $D_{g33} = U_{gV}^2$. The results for horizontal excitation are shown in Figs. 3 and 5 whereas those for vertical excitation are shown in Figs. 4 and 6. The response is presented in the form of transfer functions between the square root of the variance (or power spectral density) of response components at various locations at the top and base of the structure to the square root of the variance (or power spectral density) of the free-field ground motion. Thus, the result labeled $|\Delta_1/U_{gH}|$ in Fig. 3a corresponds to $\sqrt{D_{o11}/D_{g11}}$ while $|a\phi_3/U_{gH}|$ in Fig. 3d represents $\sqrt{D_{o66}/D_{g11}}$. The numerical results have been calculated for four values of the spatial incoherence parameter $\gamma = 0.0$, 0.1, 0.3 and 0.5. The case $\gamma = 0$ corresponds to a motion perfectly coherent with respect to space. It must be noted that the results in frames (e) and (f) of Figs. 3, 4, 5 and 6 are presented in logarithmic scale.

The results obtained show that the spatial incoherence of the free-field ground motion leads to a reduction of the high-frequency components of the translational response on the axis of symmetry of the superstructure (Figs. 3, 4, 5, 6 a, e). These reductions can be observed for frequencies above 5 Hz and increase with the value of the spatial incoherence parameter. The spatial incoherence of the horizontal components of the free-field ground motion leads to marked torsional response components (Fig. 3d and 5d) which also affect the translational response on the perimeter of the structure (Fig. 3b, f, 5b, 5f). The spatial incoherence of the vertical free-field component induces additional rocking components (Figs. 4d and 6d) which are also felt on the perimeter of the structure (Figs. 4b, 4f, 6b and 6f). The amplitude of these induced rotational components increase with γ. In general, the extent of the spatial incoherence effects is highly dependent on the value of the parameter γ. This observation underscores the importance of reliable differential array data. For values of $\gamma < 0.3$, the effects of spatial incoherence are qualitatively and quantitatively similar (in the sense of power spectral density) to those resulting from wave passage.

ACKNOWLEDGEMENTS

The stay of A. Mita at the University of California, San Diego while on leave from the Ohsaki Research Institute has been supported by Shimizu Construction Co., Ltd., Tokyo, Japan. Support from Grant ECE-83 12441 from the National Science Foundation is also acknowledged.

REFERENCES

[1] Luco, J. E. and H. L. Wong, "Response of a Rigid Foundation to a Spatially Random Ground Motion," *Earthquake Engineering and Structural Dynamics*(in press), 1986.
[2] Luco, J. E. and A. Mita, "Response of a Circular Foundation to a Spatially Random Ground Motion," *Journal of Engineering Mechanics*, EMD, ASCE(submitted), 1986.
[3] Luco, J. E. and A. Mita, "Response of a Circular Foundation on a Uniform Half-Space to Elastic Waves," *Earthquake Engineering and Structural Dynamics*(in press), 1986.
[4] Luco, J. E. and H. L. Wong, "Response of Structures to Nonvertically Incident Seismic Waves," *Bull. Seism. Soc. Am.*, 72, 275-302, 1982.

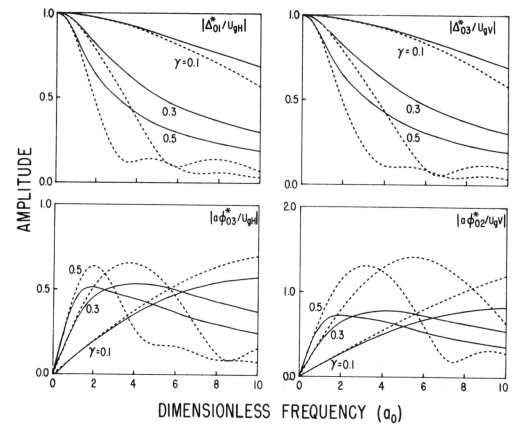

Fig. 2 Transfer functions for the response of the foundation in absence of the superstructure.

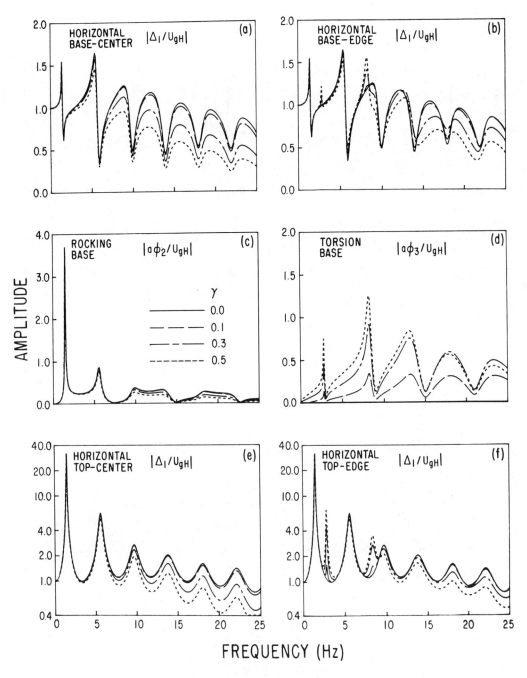

Fig. 3 Transfer functions for horizontal excitation (Model 1).

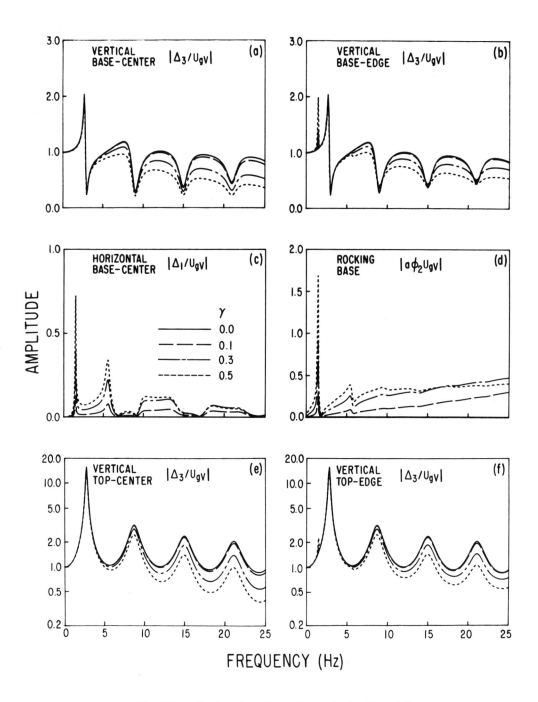

Fig. 4 Transfer functions for vertical excitation (Model 1).

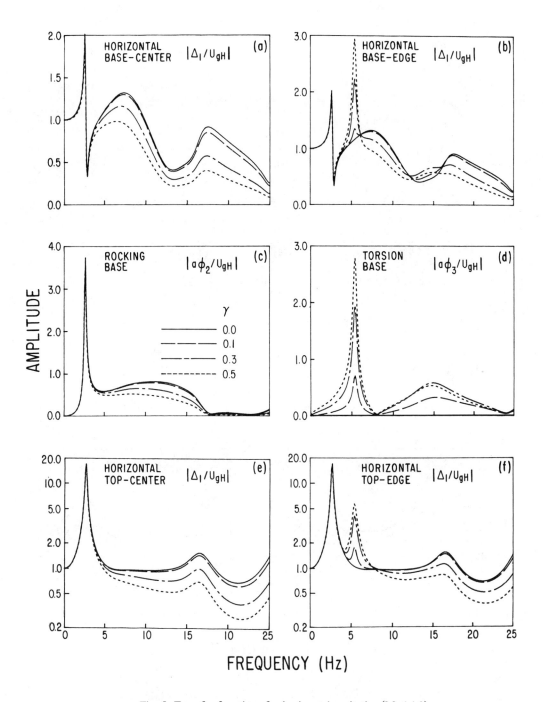

Fig. 5 Transfer functions for horizontal excitation (Model 2).

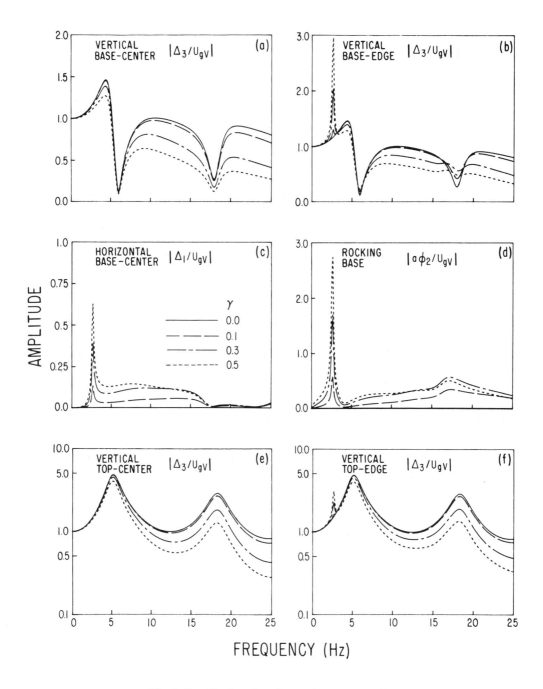

Fig. 6 Transfer functions for vertical excitation (Model 2).

SEISMIC RISK ANALYSIS FOR CODIFIED STRUCTURAL DESIGN

C.J. Turkstra[I], A.G. Tallin[II], and M. Brahimi[III]

ABSTRACT

Published results from advanced safety index analysis suggest that reliability levels in seismic design are significantly less than those for live and wind load design. These results are reviewed briefly and evaluated. An alternative methodology leading to estimates of probability distributions of damage is presented.

INTRODUCTION

Recent developments in structural engineering show a rapid evolution towards limit states design based on safety index analysis. Load and resistance factor design codes for steel (1), concrete (2) and masonry (3) have been developed. Load factors contained in the current ANSI standard (4) are based on a calibration of existing codes by means of an advanced safety index analysis (5) using the Rackwitz-Fiessler algorithm.

An apparent anomaly in the calibration study by Ellingwood et al is a significant and unintentional difference in safety levels implicit in current design practice for different loads. The magnitude of this difference is indicated in Fig. 1 for the earthquake design conditions used as case studies in the following analysis. Also shown are results for reinforced concrete beams designed for wind loads with average design wind to dead load ratio between 0.5 and 3.0 and reinforced concrete beams designed for live load with nominal live to dead load ratio of 0.5.

The objective of this paper is to review the basis for these results, evaluate their significance and suggest an alternative approach to analysis.

I. Professor, Department of Civil Engineering, Polytechnic University, Brooklyn, New York.

II. Assistant Professor, Department of Civil Engineering, Polytechnic University, Brooklyn, New York.

III. Research Assistant, Department of Civil Engineering, Polytechnic University, Brooklyn, New York.

ADVANCED SAFETY INDEX ANALYSIS

Elements of the Approach

A basic case for two moment analysis involves a random resistance R and a random load S with a linear limit state function g = R - S. The function g(R,S) defines a binary partition of the (R,S) space

$$g < 0 \quad : \text{unsafe} \quad (1)$$
$$g \geq 0 \quad : \text{safe}$$

The probability that g < o is the failure probability.

In elementary safety index analysis, safety is defined by the minimum distance from the origin to the line g=o in a reduced space (r,s) where all variables are transformed to have zero mean and unit variance. The safety index β is distribution free and invariant (6). For the basic g = R-S case, the index is given by

$$\beta = \frac{n - 1}{[n^2 V_R^2 + V_S^2]^{0.5}} \quad (2)$$

where n is the ratio of mean resistance ($\bar{R}$) to mean load ($\bar{S}$) and $V_{(.)}$ denotes coefficient of variation. For nonlinear limit state functions, iterative numerical techniques are used to calculate the index.

In advanced analysis, probability distribution functions are introduced. "Equivalent" normal distribution having probability density and cumulative probability equal to the true distribution are established at the point closest to the origin in reduced space (the design point). For non-normal variables, two stages of iteration are required to calculate the index (7).

For the purposes of this analysis, this process involves three major questionable elements.

(1) Changes in the shape of the limit state function resulting from transformations to equivalent normal probability distributions affect the relationship between safety index and failure probability.

(2) Effective linearization of a failure surface at the design point also affects the relationship between reliability and safety index.

(3) Implicitly a safety index similar to Eq. (2) is used in the reduced, linearized space of random variables. As will be shown, Eq. (2) has important limiting properties.

Limiting Properties of the Basic Index

For large central safety factor n, the safety index in Eq. 2 defined in terms of real or "equivalent" means and variances at the design point is readily shown to approach a numerical value of $1/V_R$. Thus, for example, if the coefficient of variation of resistance (real or equivalent) is 0.5, the maximum obtainable safety level is $\beta = 2.0$.

More generally, the relationship between the central safety factor n and the safety index β as a function of V_R are shown in Figs. 2-a,b for two values of V. The dependence of n on V_S for a fixed value of V_R is shown in Fig 2-c.

In practical multi-variate situations with a variety of distribution functions, the dependence of safety on the parameters will generally be more complex. However, since the final result involves a linear limit function with normalized variables, some limiting properties can be expected.

Non-linearization of Failure Surfaces

In the process of advanced safety analysis involving non-normal random variables, transformations to equivalent normal variables is required. Shown in Figs. 3 are examples of the effects on a linear limit function of a transformation of the load variable S from an extreme type I distribution to a normal variate. Similar results when resistance is also transformed from a log-normal distribution are shown in Figs. 4. It is clear that transformations from non-normal distributions can lead to both significant shifts and significant changes in curvature of a limit function.

TABLE 1: CALCULATED FAILURE PROBABILITIES - BASIC CASE

Distributions	Parameters			Safety Index	
	V_S	V_R	n	Advanced Analysis	Monte Carlo
P-Normal	.3	.1	2	2.30	2.27
S-Type I	.3	.9	2	0.58	0.53
	.9	.1	2	1.13	1.13
	.1	.5	6	1.67	1.66
	.1	.5	10	1.80	1.74
R-Lognormal	.3	.1	2	2.29	2.20
S-Type I	.3	.9	2	0.54	0.53
	.9	.1	2	1.12	1.13
	.1	.5	6	3.51	3.54
	.1	.5	10	4.56	--

To assess the importance of these effects on failure probabilities, a series of Monte Carlo simulations was completed. Safety indices computed by advanced analysis and estimated from 5,000 trials with $B = \Phi^{-1}(p_f)$ are compared in Table 1 for a number of basic cases. A comparison of safety index analysis with the exact solution for lognormally distributed load and resistance yielded similar results (7).

In spite of changes in limit state functions due to variable transformation it seems that, at least in basic cases, advanced safety index analysis can yield a good estimate of failure probability. This observation suggests that failure probabilities are dominated by points in the region of the design point. More sophisticated techniques which correct for curvature in limit functions may be used (8) but may not be necessary.

A SEISMIC DESIGN EXAMPLE

To calibrate existing practice in seismic design, Ellingwood, Galambos, MacGregor and Cornell examined a variety of cases (5). As an example, a case with a fixed value of basic live load L_0 to design dead load D_n of $L_0/D_n = 0.5$ with a tributary area $A = 400$ ft^2 has been presented in detail.

A limit state function of the form

$$g = R - L_{apt} - E \qquad (3)$$

is used where R is random resistance, L_{apt} is a random "arbitrary point in time" live load and E is the lifetime maximum earthquake load. The form of Eq. (3) is suggested by an approximate approach to the combination of time dependent loads (9).

Following Ellingwood et al, R, D, and E are assumed normal normal and extreme type II distributed with coefficients of variation of 0.11, 0.10, and 1.38 respectively. For convenience, L_{apt} was assumed lognormally distributed with $V_{apt} = 0.75$. Assumed properties of both loads and resistances have been adjusted to include both statistical variability in physical parameters as well as uncertainties due to modelling errors and analytical simplifications.

For design purposes, nominal beam capacity R_n designed according to the ACI code is governed by the equation

$$0.9 R_n = 0.75 \, [\, 1.4 D_n + 1.7(1.1) E_n + 1.7 \, L_n] \qquad (4)$$

Again, following Ellingwood et al, the ratios of average values to nominal design values are assumed to be $\bar{R}/R_n = 1.05$, $\bar{D}/D_n = 1.05$, $\bar{L}/L_n = 0.35$ and $\bar{E}/E_n = 0.64$ or 1.08 for Boston or Los Angeles respectively.

As mentioned previously, parameter estimates include consideration of modelling errors and simplifications. Although a good deal of judgment has been used and numerical values may be somewhat debatable, the analytical assumptions are within the range of usual values and should be adequate for a comparison of relative safety.

To assess possible effects of non-normal transformations leading to nonlinear limit functions, a series of simulations was performed with 5000 sample points for each case. Resulting safety index levels are compared to those from a Rackwitz-Fiessler analysis in Table 2 for various ratios of design seismic load to dead load.

Table 2 : SEISMIC DESIGN EXAMPLE

Location	$\bar{E}/E_n$	E_n/D_n	Safety Index Simulated	Safety Index Calculated
Boston	0.64	1.0	1.97	2.04
		2.0	1.86	1.91
		5.0	1.79	1.82
Los Angeles	1.08	1.0	1.46	1.51
		2.0	1.34	1.35
		5.0	1.24	1.24

These results suggest that safety index analysis yields a reliable estimate of failure probability under the assumed conditions.

To assess limiting properties of safety levels as design factors are increased, simulation studies were repeated for alternative forms of Eq. 4. Shown in Figs 5 is a comparison of safety indices based on the design equation

$$0.9\, R_n = 0.75\, (2.8\, D_n + 3.4(1.1) E_n + 3.4\, L_n) \qquad (5)$$

It is evident that a substantial increase in load factors does not result in a corresponding increase in safety level.

To further indicate this trend, a design equation of the form

$$0.9\, R_n = \alpha\, (0.75)[1.4\, D_n + 1.7(1.1) E_n + 1.7 L_n] \qquad (6)$$

with a global multiplicative load factor α was investigated. simulation results for the preceeding conditions related to seismic design in Boston are shown in Fig. 6. To raise the safety index to a conventional level of at least 3.0, dead load and live load factors of about 4.2 and 5.1 are required.

One interesting conclusion suggested by this analysis is the following - conventional levels of safety index for seismic design can not be achieved with economically acceptable load factors based on a conventional approach to safety analysis.

An apparent reason for the observed behavior is the type of distributions used for seismic excitation and the very large coefficient of variation involved. In effect, variability in resistance is negligible compared to the uncertainty in loading. Similar characteristics have been suggested in analysis based on fragility curves.

One possible alternative approach is to introduce an upper bound to earthquake magnitude and peak ground acceleration. Although such bounds seem to exist, uncertainty associated with their numerical values is very great and might well lead to similar limits on safety levels.

AN ALTERNATIVE METHODOLOGY

In many ways, seismic design is unlike wind and gravity load design for conventional structures. Firstly, dynamic response is considered either explicitly through modal analysis or implicitly through design horizontal forces. Secondly, irreversible damage is considered through design ductility factors and energy absorption requirements in, for example, detail design.

For these reasons, the concept of a binary safe-unsafe partition of response is naive for seismic design. Instead of a single limit state, there is a continuum of limit states corresponding to different levels of damage. Several approaches to damage definition and prediction have been developed recently (10, 11, 12).

If a set of damage states varying from minor damage to collapse is accepted, the concept of "failure" probability is no longer valid. Instead, a probability can be assigned to each level of damage. Design criteria for different design conditions can then be compared on the basis of expected damage rather than failure risk. Such a measure permits the nature of failure under different load types to be considered.

Theoretically, the probability distribution of damage could be established through Monte Carlo simulation with random excitations and random structural characteristics. Practically, such an approach is not feasible because of the number of random variables involved. These include, for example, earthquake peak ground acceleration, duration and frequency as well as the initial structural frequencies, load-deflection characteristics and the parameters of damage.

One possible approach is a combined simulation and safety index approach. Following the conventional formulation, a damage limit state function can be defined as

$$g(X_j) = d - D(X_j) \qquad (7)$$

where $g(X_j) < 0$ implies that damage level d has been exceeded. To calculate the damage index, all basic variables X_j can be transformed into a set of uncorrelated standard unit normal variables by the Rosenblatt transformation (8).

$$u_1 = \Phi^{-1}[F(X_1)]$$
$$u_2 = \Phi^{-1}[F(X_2/X_1)]$$
$$\cdots$$
$$u_n = \Phi^{-1}[F(X_n/X_1\ldots X_{n-1})] \quad (8)$$

where, for example, $F[X_j/X_K]$ is the conditional distribution of X_j given X_K.

In complex situations, the functions $D(X_j)$ cannot be established in closed form. However, an indirect approach can be used to evaluate the conditional distributions in Eqs.8.

(1) For a given class of earthquakes of specific magnitude at a specific site, a numerical sample of accelerograms can be generated. Several techniques including autoregressive moving average simulation can be used (13,14).

(2) These acceleration records can be used to estimate damage for a deterministic set of structural characteristics (10,11,12). A random sample of damage conditional on the site earthquake class and magnitude as well as structural characteristics can be obtained. Conditional means and variances of damage, $E[D/X_j]$ and $Var[D/X_j]$ can then be calculated.

(3) Marginal distributions of other variables, $F(X_j/X_K)$ related to earthquake class and structural properties can be established in the usual way.

Using the transformation in Eqs (8), with the dependent random variable D yields

$$u_1 = \Phi^{-1}[F(X_1)]$$
$$u_2 = \Phi^{-1}[F(X_2/X_1)]$$
$$\cdots$$
$$u_D = [E(D/X_j) - d]/Var[D/X_j] \quad (9)$$

where the conditional distribution of damage has been assumed normally distributed. If the simulated sample of conditional damage is sufficiently large, the conditional distribution of damage can be estimated and used directly in the transformation.

For a specific value of damage, d, first order second moment methods can be used to calculate a damage index β_d. At each stage of the interation towards the "design point", properties of the conditional distribution of damage given current values of other variables have to be recalculated using simulated accelerograms.

The probability of damage state d would then be given approximately as $p(d) = \Phi(-\beta_d)$. In effect, second moment analysis would be used to provide efficient integration of the half space bounded by a plane tangent to the surface $d - D(X_j) = 0$ at a point of maximum probability density.

As an alternative to repeated simulation of conditional damage at each iteration, a large scale preliminary simulation could be performed to estimate functional relationships between E[D] and Var [D] and the variables X_j.

Implementation of such an approach involves extensive computing but the operations involved are not prohibitively complex. Experience to date using a microcomputer linked to a mainframe for backup storage suggest that the process of earthquake simulation and repeated damage calculations is entirely feasible. In most applications to date, safety index analysis has been found to converge very rapidly.

SUMMARY

Studies of safety levels in current seismic design practice indicate that safety index levels based on conventional methods are relatively very low. Moreover, it seems that seismic safety levels based on such analysis cannot be raised to those common for other design cases by means of economically acceptable load factors. However, it has been pointed out that the concept of a binary limit state is quite unrealistic for seismic design.

As an alternative to conventional safety index analysis, it has been suggested that the probability distribution and expected value of damage to structures should be used as a basis for comparison. A hybrid approach using sets of simulated accelerograms in a dynamic analysis to estimate conditional means and variances of damage can be used. These conditional parameters can be combined with the parameters of the variables held constant during the simulations in a safety index analysis to estimate the probability distributions of damage.

The advantage of such an approach is the ability to introduce a state-of-the-art assessment of excitation and response. In this way, a realistic evaluation of design procedures for alternative loadings can be attempted.

ACKNOWLEDGEMENTS

This study is based upon work supported by the National Science Foundation under Grant No. CEE-8312396 entitled "Safety Levels in Current Codified Structural Design". Any opinions, findings, and conclusions or recommendations are those of the authors and do not necessarily reflect the views of the National Science Foundation.

REFERENCES

1. Galambos, T.V., and Ravindra, M.K., (1973), "Tentative Load and Resistance Factor Design Criteria for Steel Buildings," Rep. No. 18, Dept. of Civil Engrg, Washington University, St. Louis.

2. MacGregor, J.G., et.al.,(1985), "Probabilistic Basis for Design Criteria in Reinforced Concrete," Reinforced Concrete Research Council, Bulletin No. 22, ASCE, New York.

3. Turkstra, C.J.,(1984), "A Safety Index Analysis for Masonry Design," Proc. 4'th, ASCE Specialty Conference on Probabilistic Mechanics and Structural Reliability, Berkeley.

4. ANSI (1982), "Minimum Design Loads for Buildings and Other Structures," A58.1-1982, New York.

5. Ellingwood, B.R., Galambos, T.V., MacGregor, J.G., and Cornell, C.A., (1980), "Development of a Probability Based Load Criterion for American National Standard A58," NBS Special Publication 577, Washington, D.C.

6. Hasofer, A.M., and Lind, N.C., (1974), "An Exact and Invariant First Order Reliability Format," J. Eng. Mech. Div., ASCE, Vol 100, No EM1 pp 111-121.

7. Turkstra, C.J., and Daly, M., (1978), "Two Moment Structural Safety Analysis, " Can. J. Civil Eng., Vol 5, No. 3, pp 414-426.

8. Madsen, H.O., Krenk, S., and Lind, N.C., (1986), "Methods of Structural Safety," Prentice Hall Inc., Englewood Cliffs, pp 403

9. Turkstra, C. J., (1970), "The Theory of Structural Safety, " S.M. Study No. 2, University of Waterloo, Waterloo, pp 124.

10. Lin, J., and Mahin, S.A., (1985), "Effect of Inelastic Behavior on the Analysis and Design of Earthquake Resistant Structures, " Report No. UCB/EERC-85/08, University of California, Berkeley, pp. 138.

11. Park, Y.T., Ang, A.H-S, and Wen, Y.K., (1984), "Seismic Damage Analysis and Damage-Limited Design of R.C. Buildings," St.Res. Series No 516, Dept. of Civil Engineering, University of Illinois, Urbana, pp 163.

12. Zarah, T.F., and Hall, W.J., (1984), "Earthquake Energy Absorbtion in SDOF Structures," Journal of Structural Engineering, ASCE, Vol 110, No. 8, August.

13. Kozin, F., (1977), "Estimation and Modelling of Non-Stationary Time Series, " Proc. Symposium on Applied Computational Methods in Engineering, Univ. of South Calif., Los Angeles, pp 603-612.

14. Shinozuka, M., and Samaras, E., (1984), "ARMA Model Representation of Random Processes, " Proc. 4th ASCE Specialty Conference on Probabilistic Mechanics and Structural Reliability, Berkeley, pp 405-409.

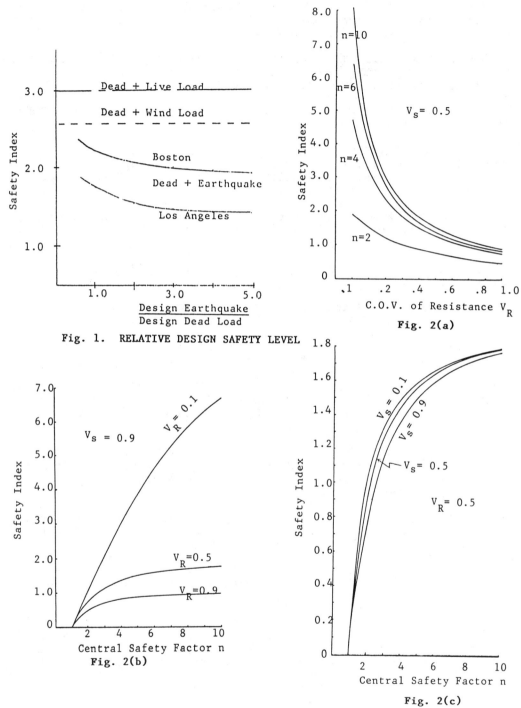

Fig. 1. RELATIVE DESIGN SAFETY LEVEL

Fig. 2 LIMITING PROPERTIES OF BASIC SAFETY INDEX

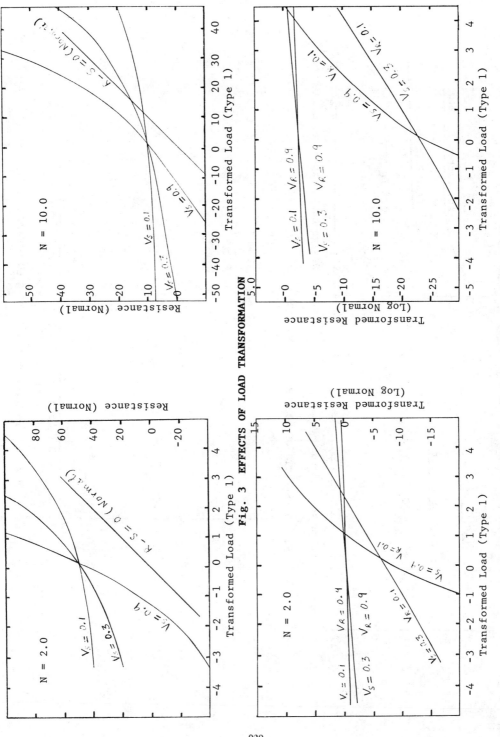

Fig. 3 EFFECTS OF LOAD TRANSFORMATION

Fig. 4 EFFECTS OF LOAD AND RESISTANCE TRANSFORMATIONS

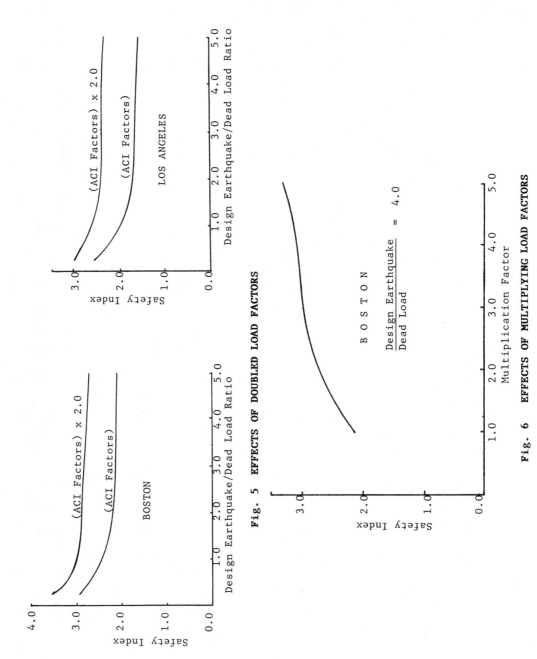

Fig. 5 EFFECTS OF DOUBLED LOAD FACTORS

Fig. 6 EFFECTS OF MULTIPLYING LOAD FACTORS

REGIONAL RISK ASSESSMENT OF EXISTING BUILDINGS:
AN UPDATE OF CURRENT RESEARCH AT STANFORD UNIVERSITY

Howard M. Thurston[I], Weimin Dong[II]
Auguste C. Boissonnade[III] and Haresh C. Shah[IV]

ABSTRACT

This paper summarizes research conducted at Stanford University to identify high risk buildings in seismic regions.

First, high risk subareas are identified in a region. Next, a preliminary list of high risk buildings within these subareas is identified using an expert knowledge base. These buildings, along with any preselected special buildings, are evaluated using a second, more detailed expert knowledge base incorporating data obtained from detailed inspections. The result is a list of high risk buildings for the region. The methodology is being tested on a California city for validation.

INTRODUCTION

There is a general consensus that the greatest source of life and economic loss comes from highly vulnerable, poorly designed, and poorly constructed existing buildings. However, before a decision regarding the strengthening of an existing structure(s) can be made, it is imperative that the following questions be answered:

1. What is the type and level of seismic hazard in the region of interest?

2. For the estimated seismic hazard, what are the risks? In other words, how would the various types of existing structures perform?

3. Which structures pose the greatest life loss, injury, and economic risk?

I. Graduate Student, Stanford University, CA 94305, USA.

II. Visiting Scholar from the People's Republic of China, Stanford University, CA 94305, USA.

III. Post Doctoral Research Scholar, The John A. Blume Earthquake Engineering Center, Stanford University, Stanford, CA 94305, USA.

IV. Chairman, Department of Civil Engineering, Stanford University, Stanford, CA 94305, USA.

4. What is the location of these vulnerable structures relative to critical urban areas? What are the occupants and users of these buildings? Are they considered emergency service facilities and are they critical to the continued functioning of the community

5. What should be the "desirable" levels of seismic resistance for strengthened structures?

6. What are the various economically acceptable and viable alternatives available to strengthen these structures?

The first four questions require an understanding of the level and type of seismic hazard, the vulnerability of the buildings exposed to the seismic hazard, and the overall seismic risk. An answer to the fifth question requires studies on cost-benefit analysis and acceptable risk concepts. The sixth question requires an in-depth study of various methods of upgrading and strengthening existing structures.

A joint project is currently underway between researchers of the People's Republic of China (The Harbin Institute of Engineering Mechanics and the Ministry of Urban and Rural Construction) and the United States (Stanford University and the University of California, Berkeley) to develop a risk-analysis-based methodology for the identification of "hazardous structures," the evaluation of their existing strengths, and the strengthening of these structures to improve their seismic performance. Briefly, the steps are: (1) the identification of high risk structures in a city or a region, (2) the ranking of the screened structures as a function of the available funding and the different levels of strengthening (including replacement), (3) the determination of the seismic retrofit level for a specific structure and, (4) the selection of the optimum retrofit option.

This paper summarizes the work performed by the Stanford researchers on developing a procedure to identify high risk structures in a region (Figure 1). Briefly, this procedure consists of three main steps: (1) the collection of available data for the region, that is the geological hazards (ground shaking, liquefaction potential, ground rupture potential, landslides, etc.), land use and occupancy, construction type, and high priority structures, (2) a first cut screening (Bird's Eye View) to identify high risk subareas from which a preliminary list of high risk structures can be sorted and, (3) a more refined screening using an inspection questionnaire to select a reduced population of hazardous structures.

The innovative aspects of this approach are in the combination of different types of information, the incorporation of expert judgments in the screening procedures, and the use of knowledge-based expert systems to carry out the identification of high risk structures.

DISCUSSION OF METHODOLOGY

High Risk Subarea Identification

The first step in simplifying the assessment of the existing building situation in a region is to reduce the number of buildings which must be evaluated. The reason for this step is obvious. Even in a moderate sized city such as Palo Alto, California (approximate population of 50,000), the number of existing nonresidential buildings is approximately 2000. The manpower and budget available in most cities simply do not allow a detailed inspection of every building. In order to address this problem, Stanford reseachers have developed a procedure known as the "Bird's Eye View" approach. For the region under consideration, the following data are required.

First, the geological hazard is considered. Geological hazards such as liquefaction potential, landslide potential, ground rupture potential, and strong ground shaking must be considered. The methodology developed so far calls for initially defining geological hazard in terms of strong ground shaking. The other hazards are considered later. So, the strong ground shaking is mapped. In Palo Alto, for example, the variation of strong ground shaking is due to either proximity to the San Andreas fault or the presence of so-called Bay Mud, a soil type which amplifies ground motion in certain frequency ranges.

Second, land use and occupancy are mapped from zoning information. This information will indicate areas of commercial buildings and high occupancy.

Third, in very general terms, it is usually possible to identify the distribution of predominant building types throughout the region. For example, the "downtown" area of Palo Alto consists of buildings of which approximately 15% are unreinforced masonry, 60% are reinforced concrete, 15% are steel frame, and 10% are wood frame. Another section of the city consists of industrial buildings of which approximately 80% are pre-1973 tilt-up, 5% are post-1973 tilt-up, 4% are steel frame, and 10% are wood frame. Although this data is collected by various researchers and is admittedly very coarse, it will tend to point out the areas of the city which contain the buildings of most concern based on general performance data of broad building classes. Of course, buildings for which poor seismic performance is usually expected may be located in an area containing generally better buildings. Thus, for the sake of simplicity, the process may tend to overlook some buildings. However, as will be discussed later, the inclusion of "special" buildings in the preliminary list will preclude overlooking important buildings.

Once these three pieces of data are collected, they must be combined in some way. As a simple first try, this information was gathered for Palo Alto and drawn on separate maps. These maps were then overlayed. This process indicated very distinct subareas of concern as follows: (1) the downtown area, where most of the city's unreinforced masonry buildings are located and an area of high density during business hours, (2) two industrial areas consisting of high percentages of pre-1973 tilt-up buildings located in areas of very strong ground motion potential (Bay Mud) and, (3) another industrial area with a high percentage of pre-1973 tilt-up buildings housing industries which are important to Palo Alto's economy. The Stanford team is currently working on a more refined method to combine these data. The methods under study represent theoretical approaches such as "fuzzy" techniques, or more intuitive ones such as a knowledge-based expert system [1].

Expert System Screening

Once the high risk subareas are identified, the proposed methodology will obtain a "first cut" list of buildings selected from all the buildings within those subareas. The research team believes that an expert system approach is the simplest and most consistent way of identifying those buildings for which a detailed evaluation is warranted. During the course of the research, two expert systems were developed, both utilizing available proprietary expert system "shell" programs designed for use on personal microcomputers. The first, "Riskfile," depicted schematically in Figure (2), is designed to identify the "first cut" list of buildings to be examined in detail. "Riskfile" assigns a level of risk to a building based on risk attributes obtained from a few very general pieces of data. The second expert system, described in Reference (3) and shown schematically in Figure (3), takes into account more detailed information and is utilized at a later stage in the methodology. In general, both systems involve creating a knowledge base consisting of rules which indicate the conclusions an expert might reach concerning a building's level of risk given information on geological hazard, importance, vulnerability, and occupancy (risk attributes).

Riskfile: A Building Assessment Knowledge Base

In order to provide a means of assessing the overall risk of an existing building subjected to earthquake loading, it is necessary to identify the attributes of a building which contribute ot its risk. Next, it is necessary to establish a method of assigning a level of risk corresponding to all possible combinations of these attributes. This level of risk is obviously subjective, although it is expected that general agreement by expert structural engineers can be reached in most cases.

In assessing the existing building situation in a particular region, it is necessary to identify a "first cut" list of buildings for detailed evaluation. This initial list will serve to reduce the size of the problem, as well as focus on the buildings which pose the greatest threat to the region. To this end, a knowledge base was created consisting of production (If-Then) attributes and their associated overall building risk level. In this way, consideration is given to the effect of each of four broad attributes on overall building risk. The risk attributes, and their subdivisions, are shown in Figure 2.

Although the selection of the subattributes and their definitions is rather subjective, the following definitions are suggested for the purpose of assessing the existing building situation in a given region.

High Secondary Disaster Potential refers to building damage resulting in the release of toxic or very hazardous substances. Moderate Secondary Disaster Potential refers to building damage resulting in explosion, the release of explosive substances, or the release of somewhat hazardous substances. Low Secondary Disaster Potential applies to all other situations.

Emergency Service Structures are buildings that provide immediate services essential to minimizing life loss and maintaining law and order in the event of a major earthquake. Such buildings include hospitals, fire stations, police stations, and communication centers. Communications centers are buildings that house equipment and personnel necessary to maintain emergency communication between police and fire departments and civil authorities.

Serious Social Impact would result if earthquake damage to buildings seriously affected a region's ability to recover in the short term, or caused serious, long-range economic impact on the region. Such buildings may include banks, food distribution centers and important commercial or industrial buildings upon which a region's economy depends. Such buildings may also include public utility buildings. Moderate Social Impact refers to building damage resulting in significant, though not serious, impact on the region. Such buildings may include schools and other public services. Low Social Impact occurs when damage to buildings has a minimal impact on the region as a whole.

Structure type refers to the general vulnerability of the building under consideration and is based on observed performance data for similar structural systems.

Design characteristics are such things as building symmetry, regularity, structural redundancy, or evidence of some level of aseismic design.

The knowledge base has been structured to be used by Insight^{TM}, an expert system "shell" program. This program provides an interactive environment in which the user's responses, along with a degree of uncertainty in those responses, are used to evaluate the risk of a building, within the context of the knowledge base.

An expert structural engineer may find such an expert system very trivial. It is meant, however, to be a first attempt at providing a quick means of systematically identifying hazardous buildings in a region based on four significant risk attributes.

Special Buildings

Every region has important buildings which should always be considered in a strengthening program. Such buildings might include police and fire stations, hospitals, communication centers, electrical power substations, etc. The definition of these "special" buildings may vary from region to region. The proposed methodology would call upon regional authorities to identify such buildings in their region. These buildings, together with those identified by the expert system as high risk, would be evaluated in detail as described in the next step.

Building Inspection

The Stanford research team has developed a building inspection questionnaire designed to be used by a competent structural engineer to obtain detailed information about a particular building. A detailed inspection will be performed on all "first cut" and "special" buildings described earlier. With the information obtained, more accurate descriptions of the risk attributes can be used by an expert system to obtain a revised assessment of a building's risk. In this way, a final list of high risk buildings may be obtained.

In order to carry out this process, Stanford researchers have created a Building Classification Scheme based on structural system and material. Then, based on available data obtained from various researchers, it was possible to plot Mean Damage Ration (DR) versus MMI for each of the building classifications, where DR is defined as the ratio of repair cost to replacement cost. Since these plots describe mean DR for broad building classes, there may be considerable difference between the predicted DR for a specific building of a particular class and that building's actual DR. However, this informaion can be used to describe the "first cut" vulnerability of a building for use in the expert system. Once again, the actual vulnerability is not as important as the relative vulnerability between the building categories. The information obtained from the inspection questionnaire would be used to assess a specific building's vulnerability. The research team is currently investigating the possibility of using a technique in which an expected DR at a specific intensity level ("equivalent" intensity) can be convolved with the probability of occurrence of that intensity level to determine an expected DR over a prescribed time period. The advantage of this technique is that using the usual convolution equation to obtain expected DR requires evaluating an expected DR for each intensity level. It is more feasible to inspect a building with only one particular intensity level in mind, namely the "equivalent" intensity, then determine the expected DR given the "equivalent" intensity, $E(DR/I^*)$, and the probability of occurrence of the "equivalent" intensity, $P(I^*)$. Convolving these two quantities would then give an expected DR for the particular building under consideration, $E(DR)$. Two problems arise, however. First, how is it possible to obtain $E(DR/I^*)$ for a particular building based on an inspection? Second, assuming the first problem can be solved, how will the expected DR so obtained be used to describe a building's vulnerability? This is an area of active work by the Stanford team.

SRA: Seismic Risk Evaluation System

Once the "first cut" high risk buildings have been identified in each subregion, a more detailed analysis is performed on each structure. During the course of this research, an expert system has been developed utilizing a proprietary expert system "shell" program designed for use on microcomputers [3]. This expert system uses a knowledge base similar to the one shown in Figure 3, consisting of rules which indicate the conclusion an expert might reach concerning the risk level of a structure for a given set of information on seismic hazard, building vulnerability, and importance.

The expert system shell, Deciding Factor^{TM}, has been chosen for application since the risk evaluation fitted the framework of this shell. Deciding Factor uses a decision model formulation, where ideas are organized in a hierarchical tree from the general to the specific. This problem decomposition approach is similar to the one used in the development of "Prospector," a mineral resource evaluation expert system [4]. Deciding Factor contains a backward-chaining inference engine to flexibly connect facts provided by the expert. The system examines all the factors and attributes in the heirarchy, starting from the bottom and asking for the degree of belief (between yes and no inclusive) the user has in the possible occurrence of each of these factors. For example, if the factor is the liquefaction potential, the question is: To what degree do you believe the liquefaction potential is very high? Numerical values ranging from -5.00 (no) and +5.00 (yes) for each response are multiplied by the degree of importance of the factors. The resulting products are then posted as supports to each idea linked above the factors. In other terms, each hypothesis which tends to confirm or reject an idea is examined. The conclusion on the occurrence of a certain idea, based on the knowledge of each supporting factor, is reached using one of the several logical relationships provided by Deciding Factor.

Several buildings have been evaluated by the system and compared to seismic evaluations made by an experienced engineer. This test permitted the calibration of weights assigned to each factor in the hierarchy structure. Another validation took place with the help of city officials in the city of Palo Alto. The system gave similar answers to the ones obtained independently. During consultation, the system asks the user to provide information about the specific building then displays, if necessary, explanations on what type of information is needed, and why the information is needed. Also, it provides explanations on how a particular conclusion has been reached.

Output

The final output of the expert systems developed to date is a list of buildings grouped according to risk level. The next step will be to determine which of these buildings should be strengthened in order to optimize the money spent in strengthening in terms of life safety and other socio-economic considerations. This process may involve the development of another expert system to prioritize the targeted buildings using information on collapse hazard, optimum strengthening level and option, building and market life, and strengthening versus replacement cost.

Two important points should be mentioned. First, the methodology has been designed to identify high risk buildings independently of any strengthening requirements which may be in force in a given city or region. Second, the methodology addresses the issue of buildings which, though not necessarily collapse hazards, may be at high risk for other reasons. (Of course, the methodology also considers buildings which pose a hazard due to collapse).

EXAMPLE BUILDING

To illustrate the use of the methodology, assume that an example building is located in one of the high risk subareas. This building is an electronic instrument production and research facility built in 1960, with subsequent expansions in 1966 and 1977. The superstructure consists of reinforced concrete bents extending in both the longitudinal and transverse directions. Exterior walls are infill, consisting of precast, concrete panels (tilt-up), which do not serve as shear walls. The roof structure is wood planking, supported by structural steel members spanning between bents. The floors are concrete waffle slabs tied into the concrete bents. The building is two stories, and is regular and symmetrical in both elevation and plan. It contains office space and a large open bay for electronic instrument production. There are extensive mechanical and electrical service systems throughout the building. Also, there are interior partitions of unreinforced masonry and an unreinforced masonry freight elevator structure. The day time occupancy is about 200, night time occupancy is 3. There is the possibility of the release of both explosive and hazardous chemicals in the event of earthquake damage.

The above data was input to the expert evaluation systems RISKFILE and SRA. The results of both systems gave this building a ranking of high risk. See Figs. (4) and (5) for input data and results.

CONCLUSION

This paper briefly described the overall methodology used to identify and rank hazardous buildings in a given region. This methodology bears heavily on knowledge obtained from experts and embodied in an expert system program. The methodology is independent of any strengthening requirements and does not determine risk in terms of property and life loss only. The use of knowledge based expert systems is judged appropriate in this case because it identifies the main factors contributing to the overall risk and allows mapping the knowledge in a program, thus ensuring consistency in the rating of the structures. The identified structures can then be evaluated further to determine which should be strengthened in order to optimize the money spent in terms of property and life safety, as well as other socio-economic factors.

ACKNOWLEDGEMENTS

The authors wish to express their appreciation to Professor James Gere, Dr. F. Neghabat, Mr. Fred Hermann, and the members of the advisory board, as well as the graduate students of Stanford's Department of Civil Engineering who have tested the inspection questionnaire for this project. Partial support was obtained from the National Science Foundation, Grant No. CEE-8403516.

REFERENCES

1. Dong. W.M., Shah, H.C., 1986, "Approximate Reasoning for Evaluating the Seismic Rick," Proceedings of the Eight European Conference on Earthquake Engineering, Lisbon, Portugal.

2. Thurston, H.M., Gere, J.M., Shah, H.C., 1985, "Use of Risk Analysis in Evaluating the Seismic Safety of Existing Structures," A progress report presented at the first USA/PRC Workshop, September 25-27, 1985, Beijing, China.

3. Miyasato, G.H., Dong, W.H., Levitt, R.E., Boissonnade, A.C., 1986, "Implementation of a Knowledge Based Seismic Risk Evaluation System on Microcomputers," The International Journal of Applications of Artifical Intelligence in Engineering, Vol. 1.

4. Campbell, A.N., Hollister, V.F., Duda, P.O, and Hart, P.E., 1982, "Recognition of a Hidden Mineral Deposit by an Artificial Intelligence Program," Science, Vol. 217, No. 3.

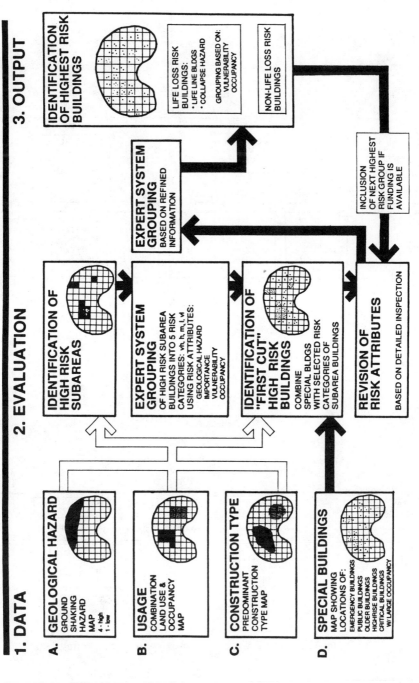

FIG. 1: FLOW CHART FOR IDENTIFYING HIGH RISK BUILDINGS

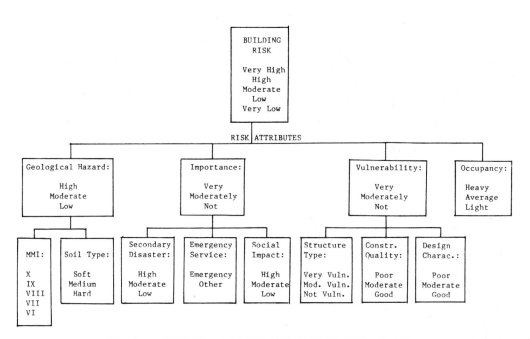

FIG. 2: RISKFILE: A BUILDING RISK ASSESSMENT KNOWLEDGE BASE

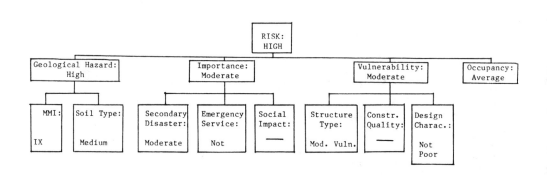

FIG. 4: LIST OF ATTRIBUTES USED IN A SEISMIC RISK EVALUATION

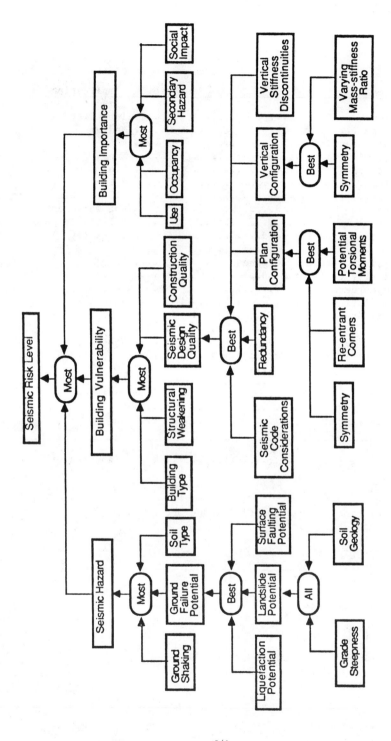

FIG. 3: SIMPLIFIED FLOWCHART OF THE SEISMIC RISK EVALUATION SYSTEM SHOWING THE FACTORS AFFECTING THE SEISMIC RISK LEVEL

Aspects	Response	Implication
Ground motion	2.5	Intensity IX
Soil condition	0.0	Medium
Liquefaction potential	-5	None
Rupture potential	-5	None
Slope of natural grade	-5	Flat
Soil underlain	-3	Slight
Building type	2	Tilt-up with frame
Structure system is weakened	-5	Not weakened
Quality of construction	-3	Relatively good
Seismic consideration included	0	Partially
Redundancy	5	Yes
Discontinuity in vertical stiffness	-5	None
Symmetrical plan configuration	5	Yes
Re-entrance corners	-5	None
Torsional moment	-5	None
Change in diaphragm strength	-5	None
Eccentricity	5	Yes
Change of stiffness in adjacent stories	-5	None
Symmetrical vertical configuration	5	Yes
Essential facility	0	No
Secondary hazard	0	No
High occupancy	1	Day time 100
High rise building	-5	No
Serious social impact	-5	No
Result of evaluation	3.7	High risk

NOTE: Relation between numerial grade and linguistic grade:

very high risk	3.75 -	5.0
high risk	1.25 -	3.75
moderate risk	-1.25 -	1.25
low risk	-3.75 -	-1.25
very low risk	-5 -	-3.75

FIG. 5: DATA AND RESULT FOR SEISMIC RISK EVALUATION BY SRA

AN INTERPRETATION OF FAILURE PROBABILITIES

A. H. Hadjian[I] and J. Goodman[II]

ABSTRACT

The transition from the factor of safety to the probabilistic approach to design is not simple. Failure probabilities in the range of 10^{-3} to 10^{-7} and differing in orders of magnitude are a drastic departure from safety factor concepts used by engineers for decades. An interpretation of failure probabilities in terms of safety factors is necessary. To achieve this objective a parametric study is carried out using three types of distributions (normal, log normal and Johnson), three specific variabilities (coefficient of variations of 0.1, 0.2 and 0.3), three exceedance probabilities for load (0.16, 0.05 and 0.01) and three exceedance probabilities for resistance (0.84, 0.95 and 0.99) to characterize "maximum probable" load and "minimum" allowable stress respectively. The results are discussed and basic relationships established.

INTRODUCTION

The use of failure probabilities to characterize structural safety is evolving at the present time particularly with regards to the issue of seismic margins in nuclear plant facilities. Failure probabilities in the range of 10^{-3} to 10^{-7} and differing in orders of magnitude are a drastic departure from safety factor concepts used by engineers for decades. Factors of safety ranging from one to two have been common in structural engineering and from one to four in geotechnical engineering. Since the factor of safety approach to design has served the profession and the public quite well, such a drastic departure requires an intimate understanding of the relationship between factors of safety and failure probabilities.

There are three basic elements to designing safe and reliable structures: loads, failure stresses and factors of safety. In the working stress design method, for example, the design loads are chosen to be close to the "maximum probable" loads that could occur during the lifetime of the structure. The failure stress is set at a "minimum" value based on laboratory tests and the reciprocal of the factor of safety is that fraction of the "minimum" failure stress that is used to define the allowable stress. In the strength design method, the loads are multiplied by load factors that are equal to or greater than unity and the strength is multiplied by a resistance factor of less than unity. The combination of load factors and resistance

---

I. Principal Engineer, Bechtel Power Corp., Norwalk, CA 90650

II. Senior Engineer, Bechtel Power Corp., Norwalk, CA 90650

factors is intended to produce similar factors of safety as in the working stress method. For clarity of the subsequent discussion the working stress basis to design will be used.

VARIABILITY OF LOAD AND RESISTANCE

There is variability both in the load effects and resistance functions. Uncertainties in the loading environment, load modeling and idealized stress analysis techniques contribute to the variability in the load effects. Material strength variability, laboratory testing and measurements and variations in structural member sizes contribute to the uncertainty in the resistance of structural elements. Fig. 1 depicts both of these variabilities as normal probability density functions with assumed means and standard deviations. R refers to the resistance function and S to the load-induced response function. The Coefficient of Variation, COV, is used as a measure of the variability. Although not specifically determined, the "maximum probable" load induced stress, S_{up}, is usually taken at some multiple of the standard deviation above the mean. In the example of Fig. 1 it is set at mean <u>plus</u> one and one-half the standard deviation. To be of more general use the specification of S_{up} can be set at a given level of exceedance probability. Similarly the "minimum" resistance, R_{low}, is set in Fig. 1 at the mean <u>minus</u> one and one-half the standard deviation. It can also be specified at a comparable exceedance probability. The conventional factor of safety, FS, is then calculated from

$$FS = \frac{R_{low}}{S_{up}} \qquad \left(= \frac{3.4}{2.9} = 1.17 \text{ for this example}\right) \qquad (1)$$

The central factor of safety is the ratio of the means

$$CFS = \frac{R_o}{S_o} \qquad \left(= \frac{4.0}{2.0} = 2.0 \text{ for this example}\right) \qquad (2)$$

An alternative representation can also be used. Define a stress reduction factor, α_R, and a load amplification factor, α_S, as

$$\alpha_R = \frac{R_{low}}{R_o} < 1 \quad \text{and} \quad \alpha_S = \frac{S_{up}}{S_o} > 1 \qquad (3), (4)$$

then the factor of safety, FS, becomes

$$FS = \frac{\alpha_R R_o}{\alpha_S S_o} = \frac{\alpha_R}{\alpha_S} \cdot CFS = \frac{1}{\nu} CFS, \text{ where} \qquad (5)$$

$$\nu = \frac{\alpha_S}{\alpha_R} > 1 \qquad \left(= \frac{1.45}{0.85} = 1.71 \text{ for this example}\right) \qquad (6)$$

PROBABILITY OF FAILURE

Failure is defined to occur when the load-induced stress, S, exceeds the resistance, R. Thus, failure occurs when

$$S \geq R \qquad (7)$$

Using a failure indicator, F, an alternative condition of failure can be defined as
$$F = R - S \leq 0 \quad (8)$$

This alternative definition is shown schematically in Fig. 2. The probability of failure, P_f, is equal to the shaded area. The number of standard deviations, ß, at which failure occurs is referred to as the safety margin, the safety index or the reliability index.

PARAMETRIC STUDY

The emphasis in this paper is on a parametric study. The study establishes the relationship between probability of failure and factors of safety given certain variations of the important parameters. The reader is referred to the Appendix for details of the mathematical derivations. The actual determination of the load and resistance distributions, with their means and variances, the selection of R_{low} and S_{up} for design and the factors of safety used with different types of construction materials are paramount but beyond the scope of this paper.

The parametric study covers the following variables:
- Three distributions: normal, lognormal and Johnson [1]
- Coefficient of variations, for both the load and resistance distributions of 0.1, 0.15, 0.2, 0.25 and 0.3. These values admit of twenty-five combinations.
- Three specifications of S_{up} and R_{low} set at 1%, 5%, 16% and 99%, 95%, 84% exceedance probabilities respectively. These values admit of nine combinations.

The choice of the three distributions is simply due to the fact that the normal distribution has been used extensively in the past, the lognormal is presently one of the favorite distributions and the Johnson, with its bounding limits, may represent engineering reality better than the unbounded normal and partially bounded lognormal distributions. The use of the normal distribution in engineering practice usually leads to untenable conditions when the random variables take on negative values. Negative tails for loads and resistances cannot be justified. This problem has been partially resolved by the use of the lognormal distribution. The lognormal distribution does not admit of negative random variables; however, its lower tail always begins at zero and its upper tail goes to infinity. A more useful distribution, referred to herein as the Johnson distribution, confines the random variable to a finite range, x_{min} to x_{max}. For this condition, it has been shown that, from the viewpoint of the principle of maximum entropy, the best distribution, from among several alternatives, is the Johnson distribution [2].

The limits of the coefficients of variation have been chosen to approximately reflect the reliability of strength characteristics of steel (~0.1), concrete (~0.2) and soil (~0.3) structures. The same range is used to reflect variability of load induced stress: 0.1 possibly for dead load to 0.3 for wind and earthquake. The selection of the exceedance probabilities for S and R covers a range of from one standard deviation, σ, (16/84) to 1.65σ (5/95) to 2.33σ (1/99). And finally the factor of safety ranges from one, obviously a lower bound for engineered structures, to five, a number seldom used except maybe in special situations in geotechnical engineering.

DISCUSSION OF RESULTS

Figs. 3, 4 and 5 present two sets of results from each of the three distributions. The resistance, R_{low} and the load, S_{up} used in the computations are indicated in each of the frames, in terms of the exceedance percentiles. All nine possible COV combinations are also shown. Significant differences exist among the three distribution.

a) Normal Distribution

As shown in Fig. 3 the normal distribution is characterized by two significant trends: the clustering of the curves for varying C_S and the asymtotic trends for each of the clusters. Referring to the Appendix, consider Eq. A12. If FS $\to \infty$, then $\beta_\infty \to 1/C_R$. Thus increasing the factor of safety cannot reduce the probability of failure below same level P_∞, which depends on the resistance variability only. The ramifications from Fig. 3 are very significant. If the normal distribution is an acceptable model for structural safety, then the emphasis in improving safety should concentrate on specifying the resistance functions more accurately; and secondly, that high factors of safety do not buy additional reliability, particularly when the variability of the resistance function increases. Thus a failure probability of 10^{-4} cannot be achieved if the COV for the resistance function, C_R, is larger than 0.27.

b) Lognormal Distribution

The lognormal distribution results shown in Fig. 4 are significantly different from those of the normal distribution. The most important difference is that the failure probabilities for the lognormal are much less than the normal for comparable parameters. In other words, by changing the distribution model from the normal to the lognormal, safety estimates can be improved, particularly for higher resistance variabilities. For example, a factor of safety of three for $C_S = C_R = 0.3$ (Fig. 4a) corresponds to a failure probability of 3.4×10^{-3} for the normal distribution and 3.2×10^{-5} for the lognormal. Thus, a two order of magnitude improvement in reliability can be achieved by changing the distribution model from the normal to the lognormal.

A second difference is the fact that as FS $\to \infty$, $\beta^* \to \infty$ and hence there is no asymtotic behavior and the failure probability tends to zero at infinity factor of safety. Also lacking is the clustering of the curves for a given C_R value. As expected, failure probabilities increase as the load or resistance or both variabilities increase. Additionally, for symmetrical percentiles, when $\lambda_R = \lambda_S$, Eq. A35, the expression for β^*, and therefore, the probability of failure, is symmetric to C_R and C_S. The curves in Fig. 4 are so indicated to reflect this property.

An interesting property of the lognormal distribution is its almost linear behavior on the semi-log plot of Fig. 4. For example the curve for $C_S = C_R = 0.2$, for all practical purposes, can be fitted by
$$\log P_f = -3.08 \, (FS) + 2 \tag{9}$$

c) _Johnson Distribution_
In addition to S_{up}, R_{low}, C_S and C_R, the Johnson distribution requires the specification of the upper and lower limits of the distribution. Fig. 5a is for the case when these limits extend to $\pm 3\sigma$. In Fig. 5b the lower limit is set very close to zero, at 1% of the mean. The purpose for the use of this range is to generate results comparable to the lognormal distribution which has its lower limit at zero.

The same clustering as for the normal distribution is obvious from the results of Fig. 5. Fig. 5a has a special feature: for certain C_S and C_R combinations, $(FS)_{min}$ has been reached and the curves are cut off at these values. Continuing the above example, a factor of safety of three, for $C_X = C_R = 0.3$ corresponds to a failure probability of 2.5×10^{-5} from Fig. 4a and 2.2×10^{-4} from Fig. 4b. There is an order of magnitude difference in these figures indicating the importance of specifying the upper and lower limits of the distribution. Of equal significance is the observation that the lognormal distribution gives a smaller failure probability than the Johnson with its lower limit close to zero (3.4×10^{-5} vs. 2.2×10^{-4}).

Three general observations can be made from Figs 3, 4 and 5. For the case of S_{up} and R_{low} set at one sigma, (Figs, 3a, 4a and 5a), the failure probability is about 10% at FS = 1 for all C_S and C_R combinations. Secondly, as expected, the probabilities from Fig. 3b and 4b are less than their counterpart probabilities in Fig. 3a and 4a. And thirdly, clustering with C_R exists for all distributions, for the normal and Johnson distributions being very pronounced.

The three distributions used in this study have certain behaviors that are schematically shown in Fig. 6. For the normal distribution the FS is asymptotic to $1/C_R$. The lognormal tends to zero probability of failure as FS tends to infinity. And finally, the Johnson distribution is limited to a minimum and a maximum factor of safety. At $(FS)_{max}$ the failure probability is zero.

CONCLUSIONS

Based on rather limited results shown from the study the following conclusions are considered important: a) The choice of distribution functions to characterize load and resistance can lead to significant differences of failure probabilities for given factors of safety; b) The curves tend to cluster with C_R, particularly for the normal and Johnson distributions, indicating that the emphasis should be on the definition of the resistance rather than the load; c) For the normal distribution, increasing the factor of safety cannot reduce the failure probability below a predetermined value; and d) For the Johnson distribution the results are sensitive to the prescription of the upper and lower limits of the distribution.

REFERENCES

[1] Johnson, N. L., "Systems of Frequency Curves Generated by Methods of Translation," _Biometrika_, 36, p. 149, 1949.

[2] Goodman, J., "Structural Fragility and Principle of Maximum Entropy," _Structural Safety_, 3, pp. 37-46, 1985.

APPENDIX - MATHEMATICS OF FAILURE PROBABILITIES

In the following sections the mathematical derivations of the equations used in the paper are given. The presentation follows Figs. 1 and 2 and associated text in the paper.

a) **Normal Distribution**

If the resistance, R, and the load, S, are normally distributed with means R_o and S_o and variances σ_R^2 and σ_S^2 respectively, then the failure indicator, $F = R - S$, is a normal variable with mean and variance

$$F_o = R_o - S_o \quad \text{and} \quad \sigma_F^2 = \sigma_R^2 + \sigma_S^2 \qquad (A1), (A2)$$

The analytical expression for the failure probability, P_f, is
$$P_f = \Phi(-\beta) \qquad (A3)$$

where Φ is a standardized cumulative normal function and

$$\beta = \frac{F_o}{\sigma_F} = \frac{R_o - S_o}{\sqrt{\sigma_R^2 + \sigma_S^2}} \qquad (A4)$$

Usually the level of load and resistance variability is described with variation coefficients C_R and C_S, where

$$C_R = \frac{\sigma_R}{R_o} \quad \text{and} \quad C_S = \frac{\sigma_S}{S_o} \qquad (A5), (A6)$$

Therefore, failure probabilities can be expressed through independent parameters R_o, S_o, C_R, C_S.

The γ-percentile, λ, of the standardized normal distribution is determined from Eq. A7,
$$\Phi(\lambda) = \gamma \qquad (A7)$$

In Fig. 1 assuming that R_{low} is taken at some lower percentile which corresponds to $\lambda = -\lambda_R$ and S_{up} is taken at some upper percentile which corresponds to $\lambda = \lambda_S$, then

$$R_{low} = R_o - \lambda_R \sigma_R = R_o(1 - \lambda_R C_R) \qquad (A8)$$
$$S_{up} = S_o + \lambda_S \sigma_S = S_o(1 + \lambda_S C_S) \qquad (A9)$$

and coefficients α_R and α_S are given by
$$\alpha_R = 1 - \lambda_R C_R \quad \text{and} \quad \alpha_S = 1 + \lambda_S C_S \qquad (A10), (A11)$$

The safety margin β of Fig. 2 can be expressed through the factor of safety, FS, and variation coefficients C_R and C_S, given percentiles λ_R and λ_S:

$$\beta = \frac{\nu \cdot FS - 1}{\sqrt{C_R^2 \nu^2 FS^2 + C_S^2}} \quad , \text{where} \qquad (A12)$$

$$\nu = \frac{\alpha_S}{\alpha_R} = \frac{1 + \lambda_S C_S}{1 - \lambda_R C_R} \tag{A13}$$

b. **Lognormal Distribution**

Assuming that S and R have lognormal distributions

$$f(S) = \frac{1}{\sqrt{2\pi}\, \beta_S S} \exp\left\{-\frac{1}{2}\left(\frac{\ln S - \mu_S}{\beta_S}\right)^2\right\} \tag{A14}$$

$$f(R) = \frac{1}{\sqrt{2\pi}\, \beta_R R} \exp\left\{-\frac{1}{2}\left(\frac{\ln R - \mu_R}{\beta_R}\right)^2\right\} \tag{A15}$$

where medians $\tilde{R}_o$, $\tilde{S}_o$, and variances σ_R^2 and σ_S^2 can be expressed through the following parameters of the lognormal distribution:

$$\tilde{R}_o = e^{\mu_R} \quad \text{and} \quad \tilde{S}_o = e^{\mu_S} \tag{A16}, (A17)$$

$$\sigma_R^2 = R_o^2 \left(e^{\beta_R^2} - 1\right) \quad \text{and} \quad \sigma_S^2 = S_o^2 \left(e^{\beta_S^2} - 1\right) \tag{A18}, (A19)$$

In the above equations R_o and S_o are means

$$R_o = \tilde{R}_o\, e^{\frac{1}{2}\beta_R^2} \quad \text{and} \quad S_o = \tilde{S}_o\, e^{\frac{1}{2}\beta_R^2} \tag{A20}, (A21)$$

and the variation coefficients C_R and C_S are now defined as

$$C_R = \sqrt{e^{\beta_R^2} - 1} \approx \beta_R \quad \text{and} \quad C_S = \sqrt{e^{\beta_S^2} - 1} \approx \beta_S \tag{A22}, (A23)$$

In Eqs. A22 and A23, the approximate expressions are true for $\beta \leq .3$.

The factor of safety, FS, is given by Eq. 1. Instead of the central factor of safety it is more appropriate for the lognormal distribution to use the median factor of safety, MFS, defined as

$$\text{MFS} = \frac{\tilde{R}_o}{\tilde{S}_o} \tag{A24}$$

Taking into account that

$$R_{low} = \tilde{R}_o\, e^{-\lambda_R \beta_R} \quad \text{and} \quad S_{up} = \tilde{S}_o\, e^{\lambda_S \beta_S} \tag{A25}, (A26)$$

the FS and MFS are related by

$$\text{FS} = \frac{1}{\nu_L} \text{MFS}, \quad \text{where} \tag{A27}$$

$$\nu_L = e^{\lambda_R \beta_R + \lambda_S \beta_S} \tag{A28}$$

The failure indicator F in the logarithmic space becomes

$$F = \frac{R}{S} \tag{A29}$$

and failure occurs when $F \leq 1$

The failure indicator F has a lognormal distribution with median $\tilde{F}_o$ and parameter β_F given respectively by

$$\tilde{F}_o = \frac{\tilde{R}_o}{\tilde{S}_o} \quad \text{and} \quad \beta_F^2 = \beta_R^2 + \beta_S^2 \tag{A30, A31}$$

The probability of failure, P_f, is then
$P_f = \Phi(-\beta^*)$, where (A32)

$$\beta^* = \frac{\ln \tilde{F}_o}{\beta_F} = \frac{\ln \frac{\tilde{R}_o}{\tilde{S}_o}}{\sqrt{\beta_R^2 + \beta_S^2}} \tag{A33}$$

The parameter β^* can be expressed through the factor of safety FS as:

$$\beta^* = \frac{\ln FS + \lambda_R C_R + \lambda_S C_S}{\sqrt{C_R^2 + C_S^2}} \tag{A34}$$

where approximations $\beta_R \approx C_R$ and $\beta_S \approx C_S$ are adopted.

c. <u>Johnson Distribution</u>

The probability density function of the Johnson distribution is given by

$$f(x) = \frac{b-x}{\sqrt{2\pi} \, \beta \, (x-a)(b-x)} \exp \left\{ -\frac{1}{2} \left[\frac{\ln\left(\frac{x-a}{b-a}\right) - \mu}{\beta} \right]^2 \right\} \tag{A35}$$

for $a \leq x \leq b$ and where μ and β are parameters of the distribution (do not confuse parameter β with the margin β in Fig. 2).

The median, x_{med}, can be expressed as

$$x_{med} = \frac{be^\mu + a}{1 + e^\mu} \tag{A36}$$

The approximate expressions for mean, m, standard deviation, σ, and variation coefficient, C, are respectively

$$m = \left\{ 1 + (\nu-1)\left(\nu-\frac{1}{2}\right) \beta^2 \right\} x_{med} \tag{A37}$$

$$\sigma = (1-\nu) \beta \, x_{med} \tag{A38}$$

$$C = \frac{1 - \nu}{1 + (\nu-1)(\nu-\frac{1}{2})\beta^2} \cdot \beta \qquad (A39)$$

and $\quad \nu = \dfrac{e^\mu}{1 + e^\mu} \qquad (A40)$

For the symmetrical distribution $\mu = 0$ and $\nu = 1/2$ and the above equations simplify to the following

$$m = x_{med} = \frac{a + b}{2} \qquad (A41)$$

$$\sigma = \frac{1}{2} \beta \, x_{med} \qquad (A42)$$

$$C = \frac{1}{2} \beta \qquad (A43)$$

The load S is determined now (symmetrical case) by three parameters S_{min}, S_{max} and β_S. Similarly, the resistance R is determined by three parameters R_{min}, R_{max}, β_R. Means R_o, S_o, variation coefficients C_R, C_S and R_{low}, S_{up} are thus given by the following expressions:

$$R_o = \frac{R_{max} + R_{min}}{2} \quad \text{and} \quad S_o = \frac{S_{max} + S_{min}}{2} \qquad (A44), (A45)$$

$$C_R = \frac{1}{2} \beta_R \quad \text{and} \quad C_S = \frac{1}{2} \beta_S \qquad (A46), (A47)$$

$$R_{low} = \frac{R_{max} \, e^{-\lambda_R \beta_R} + R_{min}}{1 + e^{-\lambda_R \beta_R}} \quad \text{and} \quad S_{up} = \frac{S_{max} \, e^{\lambda_S \beta_S} + S_{min}}{1 + e^{\lambda_S \beta_S}} \qquad (A48), (A49)$$

For the a symmetric distribution a fourth parameter (μ_R for R and μ_S for S) is required.

In the case of the Johnson distribution the limiting ratio, LR, is introduced:

$$LR = \frac{R_{min}}{S_{max}} \qquad (A50)$$

If the LR $\geq$ 1 then the probability of failure is <u>exactly</u> zero as there is no overlap of the distributions. Therefore, safety can be considered deterministically. The probabilistic approach starts when LR < 1.

For this distribution the FS according to Eq. 1 and the failure indicator F according to Eq. 7 are used.

Using the following notation

$$\rho_R = \frac{R_{max}}{R_{min}} > 1, \quad \rho_S = \frac{S_{max}}{S_{min}} > 1, \quad \text{and} \quad \rho_F = \frac{S_{max}}{R_{min}} > 1 \qquad (A51), (A52), (A54)$$

the factor of safety can be written in the form

$$FS = \frac{1 + \alpha_R^* (\rho_R - 1)}{\rho_F \left(1 - \alpha_S^* \frac{\rho_S - 1}{\rho_S}\right)} \quad , \text{where} \tag{A54}$$

$$\alpha_R^* = \frac{e^{-\lambda_R \beta_R}}{1 + e^{-\lambda_R \beta_R}} \quad \text{and} \quad \alpha_S^* = \frac{1}{1 + e^{\lambda_S \beta_S}} \tag{A55}, (A56)$$

The probability of failure, P_f, can be estimated by using first-order-second-moment method [A1]:

$$P_f = \Phi(-d) \tag{A57}$$

where d is a minimum value of the function d(u):

$$d^2 = \frac{1}{4 c_R^2} \left[\ln\left(\frac{u}{\rho_R - 1 - u}\right)\right]^2 + \frac{1}{4 c_S^2} \left[\ln\left(\frac{u + 1 - \frac{\rho_F}{\rho_S}}{\rho_F - 1 - u}\right)\right]^2 \tag{A58}$$

The independent variable u is limited with the inequality
$$0 < u < \rho_F - 1 \tag{A59}$$

where parameter ρ_F is also limited by
$$1 < \rho_F \leq \rho_{max} \tag{A60}$$

with $\rho_{max} = \min\{\rho_R, \rho_S\}$ \hfill (A61)

Parameters C_R, C_S and ρ_R, ρ_S are characteristics of load and resistance distributions. The parameter ρ_F is the characteristic of the load-resistance overlap. Varying the parameter ρ_F the FS can be calculated according to Eq. A56 and P_f, according to Eq. A59. Comparing FS and P_f we can find P_f as a function of FS.

Note that the factor of safety FS is bounded by
$$(FS)_{min} \leq FS \leq (FS)_{max} \tag{A62}$$

$$(FS)_{min} = \frac{1 + \alpha_R^* (\rho_R - 1)}{\rho_{max} \left[1 - \alpha_s^* \frac{\rho_S - 1}{\rho_S}\right]} \tag{A63}$$

$$(FS)_{max} = \frac{1 + \alpha_R^* (\rho_R - 1)}{1 - \alpha_s^* \frac{\rho_S - 1}{\rho_S}} \tag{A66}$$

It is to be noted that as $FS \to (FS)_{max}$, $P_f \to 0$.

REFERENCES

[A1.] Dolinski, K., "First-Order Second Moment Approximation in Reliability of Structural Systems: Critical Reviews and Alternative Approach," Structural Safety, 1 (1983), pp. 211-241.

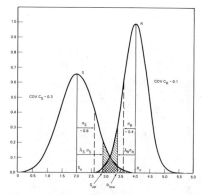

Figure 1 A sample Load Induced Stress, S, and Resistance, R, probability Density Functions

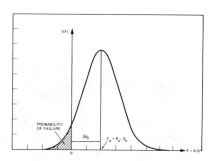

Figure 2 Probability of Failure

Figure 6 Pictorial Comparison of Probability of Failure for Three Distributions

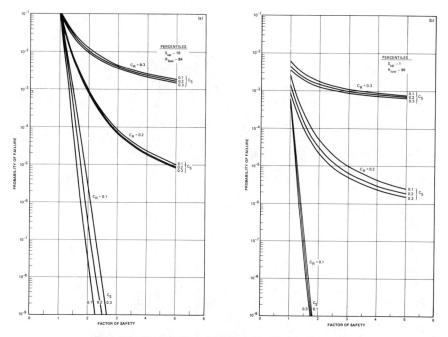

Figure 3 Probability of Failure for Normal Distribution

953

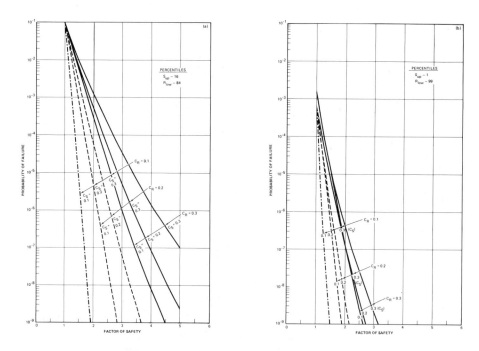

Figure 4 Probability of Failure for Lognormal Distribution

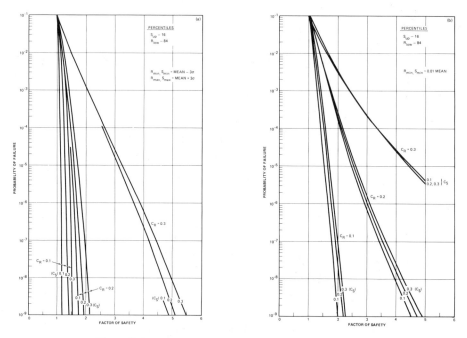

Figure 5 Probability of Failure for Johnson Distribution

ESTIMATION OF RESPONSE SPECTRA OF RELATIVELY LONG-PERIOD GROUND MOTIONS
FROM SLOSHING HEIGHTS OF LIQUID STORAGE TANKS AND SEISMOGRAMS

Yoshikazu Yamada[I], Hirokazu Iemura[II], Shigeru Noda[III] and Saburo Shimada[IV]

ABSTRACT

Response spectra for long-period (5 to 15 sec) earthquake ground motions at several cities in the northern part of Japan during the 1983 Nihonkai-chubu earthquake are estimated, using the observed data of sloshing heights in oil storage tanks and also the recorded data of both the acceleration- and the displacement-type seismographs. Very high sloshings were particularly observed in Niigata city, situated about 270 km away from the epicenter. Mended and corrected displacement- and acceleration-type seismograms were found to give much higher response spectra than the design values for long-period (5 to 10 sec) structures with 0.1-5% damping.

INTRODUCTION

Continuing progress in civil engineering methodology has made possible the design and construction of increasing number of complex large-scale structures with long natural periods of vibration. Inspite of their seeming infallibility against earthquake ground shocks, there poses the danger that structures, such as high-rise buildings, long-span suspension bridges, cable-stayed bridges and large storage tanks, etc., can easily be induced into resonant conditions when subjected to relatively long-period ground motions. Because of the generally very small damping inherent in these types of structures, the duration of vibration may become undesirably prolonged. Due consideration to the overall safety should, therefore, necessitate that an accurate representation of the ground motion characteristics is stipulated in the proper design practices.

However, no clear-cut method has as yet been established for determining the response spectra of relatively long-period ground motions. The usual procedure has been to obtain these by extrapolating from the short-period design spectra, which are rather based on strong motion acceleration records. But spectra inferred in this manner cannot be certified correct; for the reason that the reliability of the acceleration-type strong motion seismographs is not clear at the range peculiar of long-period ground motions. On the other hand, attempts to deduce long-period ground motion characteristics from low-magnification displacement seismographs should likewise be taken with discreetness because of a rather significant offscaled extent of over ± 3 cm In these view of the problems, recently more quantitative work has been made by using recorded data of long-period ground motions and the surface wave theory with a fault dislocation source [1-3].

I., II. and III., Respectively, Professor, Associate Professor and Research Associate, Department of Civil Engineering, Kyoto University, Kyoto, Japan

IV. Senior Staff Engineer, Civil Engineering Design Division, Maeda Construction Co., Ltd., Tokyo, Japan

THE NIHONKAI-CHUBU EARTHQUAKE GROUND MOTIONS

An earthquake registering a magnitude of 7.7 struck the Japan Sea coast of the northern Honshu Island on May 26, 1983 (12:00 JST), with epicenter close to the western coast of Akita and Aomori prefectures (Fig.1). During this 1983 Nihonkai-chubu earthquake, high sloshings in large oil tanks were observed in several far distant cities (Akita, Niigata, Mutsuogawara, Oga and Tomakomai), causing overflows in some tanks. In Akita city, some 113 km from the epicenter, the spillage caused by the sloshing overflow ignited fire in one tank. The seismic intensity in Niigata city, which is located much farther (270 km) from the epicenter, was III in the JMA Intensity Scale, which corresponded to V in the Modified Mercalli Intensity Scale. And yet, high sloshing waves of about 4.5 m high were observed even though the accelerometers were not activated.

Fig.2 shows the low-magnification seismogram recorded at the Niigata Weather Bureau station during the 1983 Nihonkai-chubu earthquake. The unit magnification seismograph has the natural period of 6 sec in the horizontal component and 5 sec in the vertical component. The NS-component shows that ground motion with the period of around 10 sec continued to vibrate for more than 10 minutes after being off-scaled for about 2 minutes. The reason for the amplified ramification was due to the effect of the underground structural composition near the observation site on the directly reaching waves radiated from the epicenter, and more primarily on the surface waves. The surface waves reflect and refract due to the influence of the geographical basin feature; in other words, secondary waves are generated at the irregular boundaries.

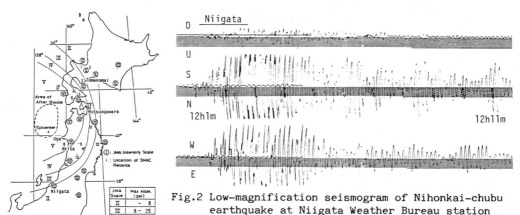

Fig.2 Low-magnification seismogram of Nihonkai-chubu earthquake at Niigata Weather Bureau station

Fig.1 Isoseismal of the Nihonkai-chubu earthquake

Fig.3 shows the distribution of the predominant periods of seismograms as plotted against the maximum amplitudes, the values of which were taken from the Monthly Report of Earthquakes issued by the JMA [4]. Off-scaled data (at Akita, Aomori, Sakata and Niigata) were excluded from the plot. The values of the periods are concentrated in the range of 5 to 15 sec; and an average value of a 10-second period is seem to have been generated, irrespective of the location of the observation sites (as exhibited in Fig.4, predominant period v.s. epicentral distances). This predominant 10-second period occurred at the sites where the maximum displacements of seismograms were more than 1 cm, as substantiated by the maximum aftershock of June 21 (M=7.1). This kind of predominance was not observed at all in the 1968 Tokachi-oki earthquake (M=7.9), the 1964 Niigata earthquake (M=7.5) nor the 1964 Aomoriken-seiho-oki earthquake (M=6.9). Thus, it can generally be said that the predominant periods of these earlier quakes varied with their respective location of the epicenters. In the Tokachi-oki earthquake, it was further observed that the distribution of the frequencies of the predominant period at various districts had two peaks of 5 sec and 15 sec, at both the mainshock and the maximum aftershock (M=7.5).

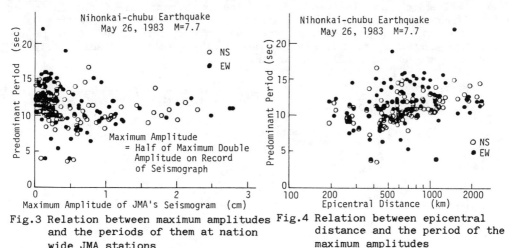

Fig.3 Relation between maximum amplitudes and the periods of them at nation wide JMA stations

Fig.4 Relation between epicentral distance and the period of the maximum amplitudes

RESPONSE SPECTRA ESTIMATED FROM DATA ON SLOSHINGS OF OIL STORAGE TANKS AND COMPARISON WITH SOME CODE-PRESCRIBED SPECTRA

Just after the occurrence of the Nihonkai-chubu earthquake, information and data were obtained from the fire defence headquaters in Tohoku, Hokuriku, and Southern Hokkaido districts. Inquiries were made of the following items: kind of tank (subsurface or on-ground, roof type, shape, tank materials, inner and outer diameters, height of side wall, etc.), support conditions, content and its density, liquid level at the time of earthquake occurrence, ground condition of tank location, sloshing heights and how were they measured, occurrence of overflow, presence of level meter, state of damages, etc. Among the sites considered, only data from those tanks located in Niigata, Akita, Tomakomai, Mutsuogawara and Oga cities were used. In the other sites, no large tanks were situated or no large sloshings were observed.

The fundamental sloshing period can be calculated by adopting an expression derived from the velocity potential theory given in the following form [1]:

$$T_s = 2\pi\sqrt{D/3.682g \cdot \coth(3.682H_l/D)} \qquad (1)$$

where D is the diameter of the tank, H_l is the liquid depth, and g is the gravitational acceleration.

Assuming that the fundamental sloshing response mode of the liquid predominates, maximum sloshing height in a cylindrical tank is given by the following equation derived from the potential theory [1]:

$$\eta_{max} = D/2g \cdot 0.837 S_A$$
$$= D/2g \cdot 0.837 (2\pi/T_s) \cdot S_V \tag{2}$$

where S_A and S_V represent the acceleration and velocity response spectra of an earthquake ground motion, respectively.

Fig.5 and Fig.6 show the values of S_V and S_A obtained by substituting observed sloshing data into Eq.2 above. The interpretation of the solid and broken lines shown in Fig.6 is explained in a subsequent section. This paper presupposes that higher sloshing modes and the dynamic interaction in the liquid-tank-ground system hardly influence the maximum sloshing heights, generalization that can readily be confirmed analytically.

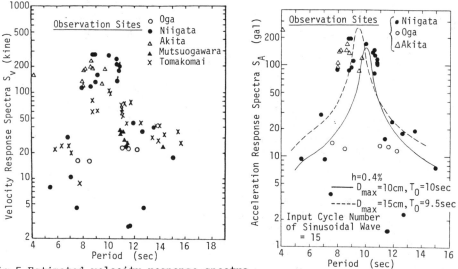

Fig.5 Estimated velocity response spectra from maximum sloshing height

Fig.6 Estimated acceleration response spectra and the calculated response to sinusoidal input waves

From Fig.5, it can be observed that the estimated S_V for period of about 9 sec in Niigata and Akita are larger than 200 kine; and that most of those S_V values are more than twice larger than the values prescribed by the Ministry of Home Affairs discussed later. For instance, estimated S_V for Niigata corresponding to sloshing periods of 8.88, 8.74, 10 and 10.58 sec are 276.3, 272.1, 270.7 and 247.6 kine, respectively; and estimated S_A for the same periods are 195.5, 195.5, 170 and 147 gal, respectively. Maximum of the estimated S_V in Tomakomai is about 100 kine. On the other hand, estimated values of S_V in Mutsuogawara and Oga are very small in comparison with those values obtained for Niigata and Akita. At this point, it is not apt to compare the response spectra estimated from sloshing data directly with the prescribed response spectra of super- and sub-structures such as high-rise buildings and long-span suspension bridges, because the damping constant corresponding to the estimated spectra is as small as 0.1%; whereas those corresponding to the prescribed spectra are practically of damping constant higher than 0.1%. Expectably, the estimated spectra will give a higher response due to the low immanent damping. Reduced response obtained by converting the 0.1% damping into 2-10% damping will be presented in the later part of this paper.

Next, assuming that an earthquake ground motion is represented by a sinusoidal wave with period T_0 and displacement amplitude D_{max}, equivalent ground motions in Niigata and Akita are investigated. Solid line in Fig.6 shows the response spectra with 0.4% damping subjected to 15 cycles of sinusoidal ground motion with D_{max} = 10 cm and T_0 = 10 sec; while that with D_{max} = 15 cm and T_0 = 9.5 sec is represented by the broken line. These periods and amplitudes had been judged from the displacement seismograph recorded at Niigata (Fig.11). The acceleration response amplification to these equivalent earthquake ground motions are about 40, and are found to be very large. The acceleration response spectra corresponding to S_A estimated from the sloshing data can not be obtained, unless the above-mentioned equivalent earthquake ground motions are adopted. The seismic criteria governing the design of liquid storage tanks suggest the use of a 3-cycle sinusoidal wave of the fundamental mode sloshing period as the design earthquake ground motion. Results in Fig.6 show that the number of cycles and the magnitude of the amplitude suggested by the design criteria may not be sufficient.

The design spectra of sloshed tanks and of various structures with long natural periods of vibration ([5]-[10]), and estimated S_V based on sloshing data are shown in Fig.7. Before proceeding further, the standards governing sloshing in oil storage tanks and these prescribed spectra are compared in the same plot.

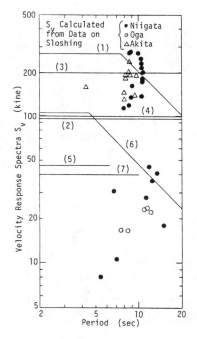

(1) Notification No.515 of the Ministry of International Trade and Industry (1981) ; β_1=1.0, β_2=0.6

(2) Notification No.119 of the Ministry of Home Affairs (1983) ; ν_1=0.85

(3) Design Recommendation for Storage Tanks and Their Supports by AIJ (1984) ; Soil Condition Class 3, T_G=1.28

(4) Ibid.; Soil Condition Class 1, T_G=0.64

(5) Earthquake Resistant Regulation for Tall Buildings by AIJ (1973) ; Upper Limit of C_B

(6) Proposal for Revised Version of API Standard 650 Appendix P (1978) ; Z=1.0, I=1.5, S=1.5

(7) Aseismic Design Specification by Honshu-Shikoku Bridge Authority (1977) ; h=2%

Fig.7 Calculated acceleration response spectra compared with various design specifications

Generalizing on the observations plotted in Fig.7, S_v for the 10-second period estimated from sloshing data exceed the corresponding design spectra prescribed by the design specifications. Especially in Niigata, S_v estimated from sloshing data exceed these design spectra by very large margins, except for spectra prescribed by the "Notification No.515 of the Ministry of International Trade and Industry" [5] and that of the "Design Recommendation for Storage Tanks and Their Supports by the AIJ" [7] for soil condition class 3. Taking into consideration the hazard of high-pressure gases, specification of the Ministry of International Trade and Industry recommends that tanks are closed tightly. Not surprisingly, for stresses to be admissible if caused by liquid sloshing on tank walls, very large design spectra need to be prescribed. In the case of oil storage tanks, while the design spectra of about 100 kine are found to be insufficient, adopting the design spectra used for high-pressure gases may proved to be economically unacceptable on the other extreme.

A state of occurrence of sloshing depends primarily on the ground condition at the tank site. For instance, the thick subsoil and the irregular subsoil structure of Niigata gave rise to large amplification of earthquake ground motions in the rather long-period range [11], which are seemed to have predominated and caused large sloshings in tanks observed during the 1983 Nihonkai-chubu earthquake. Relationship between sloshing phenomena and the input earthquake ground motions in Fig.7 can be used to establish the safety criteria of large oil storage tanks during earthquakes. Based on the above-accounted observations, it should be emphatically stated that the choice of the seismic zone factor may have an important influence on the design spectra.

COMPARISON OF RECORDED DATA OF SMAC ACCELERATION- AND JMA'S DISPLACEMENT-TYPE STRONG MOTION SEISMOGRAPHS

The natural period, the damping constant and the static magnification of the JMA's strong-motion seismograph in the vertical direction are 5 sec, 0.55 and 1, respectively. In Fig.8, calculated response of dynamic model of JMA's

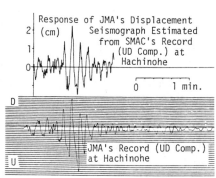

Fig.8 Comparison of recorded seismogram and calculated response of the dynamic model of JMA's seismograph subjected to SMAC's record at Hachinohe

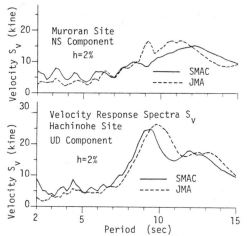

Fig.9 Comparison of velocity response spectra calculated from SMAC and JMA seismographs (h=0.1%)

displacement-meter subjected to SMAC-B_2 accelerogram (upper trace), is compared with the seismogram at Hachinohe site of JMA (lower trace). With respect to wave-form and duration, the reproduced SMAC wave resembles the JMA's recorded seismogram, especially for long-period (about 10 sec) component, whereas the short-period wave calculated from SMAC accelerogram is predominant than the recorded one of JMA. It has been suggested [12,13] that SMAC-B_2 records inevitably contain virtual waves with long periods between about 10 to 15 sec. This may be caused by the irregular performance of the accelerograph and in proper process of the digitization.

In Fig.9, the velocity response spectra with the 2% of critical damping calculated from JMA's seismograms are compared with ones from SMAC's accelerograms, for NS component at Muroran site and UD component at Hachinohe. The velocity response spectrum at Hachinohe calculated from SMAC-B_2 accelerogram have about 25 kine at a period of around 9.4 sec. Although the recorded amplitudes by the SMAC-B_2 type accelerograph were small to be analyzed exactly, the results obtained may have errors somewhat smaller than we expected. However, it is found from Fig.9 that the response spectra calculated from SMAC records are in good agreement with ones from JMA's records, except for periods around 10 sec. The velocity response spectrum ratio of JMA's and SMAC records at Muroran is found to be almost flat over the period range between 5 and 15 sec with approximately value of one. It can be said that relatively long-period ground motions for engineering purposes are approximately inferred from the records obtained by SMAC-B_2 acceleographs.

ESTIMATION OF SATURATED SEISMOGRAM (NS COMPONENT) DURING MAINSHOCK AT JMA STATION IN NIIGATA

Niigata city locates due south of the epicenter of mainshock. Therefore, ground motion from Rayleigh wave radiation is considered to be predominant in NS component (Fig.2). One of the remarkable features of this earthquake is that the sloshing phenomena due to such long-period ground motions were observed. Because the sloshing direction of liquid tanks observed at Niigata is NS component (about N35°W), we tried to mend the saturated seismograph record of the NS component shown in Fig.2.

The record of the JMA's strong motion seismometer was corrected by the following procedures. To mend the longer period waves in the interrupted parts, we must first take into consideration the changes of period and amplitude of the waves recorded before and behind the parts which should be mended, so as to fit the waves lying before and behind the mended part. Next, an enlarged print of the original record was digitized with unequally-spaced data points in time by a digitizer. To obtain the corrected data with an equal time interval, similar procedures outlined in Ref.(2) are adopted.

However, in the case of the record by this seismograph as clearly shown in Fig.2, there is an additional problem to be solved; that is, the time axis varies and is reversed in some stages. Therefore, some care has been taken to estimate the time history wave. As such data cannot be accurately determined in the document now available, we attemped to correct the portion (A-B-C-D) of the original wave by three types of 1) A-B'-C-D, 2) A-B-C'-D and 3) A-B''-C''-D as shown in Fig.10. Herein, with the aid of the analysis for the type 3), the correction was performed.

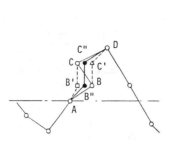

Fig.10 Correction of time history of seismogram

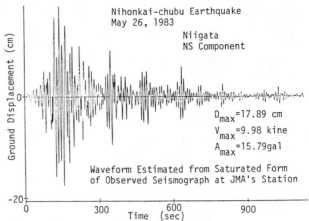

Fig.11 Mended and instrument corrected displacement time history at Niigata

Following the above procedures, the equally spaced data of ground displacement for NS component were obtained as shown in Fig.11. Peak values of ground motions are 17.9 cm, 10 kine and 15.8 gal, respectively. Although the horizontal components of displacement-meter were overloaded and could not be analyzed exactly, the results obtained may be satisfactory.

Fig.12 shows the velocity response spectra with 0.1% of critical damping calculated from JMA's record in Fig.11 for the period range of 5 to 15 sec together with the ones obtained at Seiro-cho in Niigata prefecture (Niigata-Higashi Harbour) by the SMAC-B type accelerograph which were conducted by Kudo and Sakaue [14]. The value of 0.1% corresponds to the sloshing of oil-storage tanks. Two amplitude spectra over the period of 10 sec were presented as an example of not accurate matching, suggesting some error of SMAC record in the range of long periods. The spectra estimated by JMA's record did not differ more than twice from the ones by SMAC record, roughly speakingly. It may be pointed out that the long-period motion was predominant due to the presence of

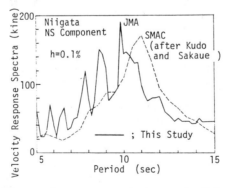

Fig.12 Comparison of response spectra calculated from the mended and corrected JMA record and the SMAC record at Niigata

thick, soft sediments at Niigata. The fact that our result agrees almost with the result of Kudo and Sakaue leads us to the conclusion that the method introduced in this paper is satisfactory, even thogh the SMAC record in the long-period range have errors somewhat greater than we expected.

COMPARISON OF RESPONSE SPECTRA INVERTED FROM SLOSHING HEIGHT IN OIL STORAGE TANKS WITH ONES CALCULATED FROM STRONG MOTION SEISMOGRAMS

Fig.13 shows the seismograms of the mainshock recorded at Akita by the SMAC-type strong motion accelerograph, one of the strongest motions in both time and frequency domain among the records investigated in this study. As a danger of earthquake motions abundant with surface waves is to lengthen the duration of vibration in structures with weak damping, the recorded data with the duration of 3 minutes were used in this analysis, which were newly digitized at the Port and Harbour Research Institute of the Ministry of Transport by our request [15]. In Fig.13, two stage's multiple shock sequence concerning the rupture propagation on fault planes and the occurrence of long-period ground motions is recognized.

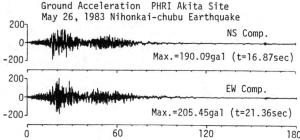

Fig.13 Three minutes time history of SMAC acceleration record at Akita

In Fig.14, velocity response spectra with 0.1% damping of long-period (4 to 15 sec) earthquake ground motions are compared for three cities (Niigata, Akita and Oga), based on sloshing height in the oil-storage tanks, SMAC-B_2 acceleration seismograms at Akita shown in Fig.13, and mended and corrected JMA's displacement type strong motion seismogram at Niigata shown in Fig.11. It is well known that the damping of liquid waves is very small.

The response spectra inverted from sloshing height at Akita agree well with the ones calculated by SMAC records, at portions with periods around 8 and 10 sec, althogh the spectral amplitudes deduced from oil-sloshing for the period of about 9 sec are overestimated as compared with those of the wave composed by two horizontal components of SMAC records. The significant difference in the values estimated by two methods is owing to the difference of locations of oil-sloshing tanks and SMAC accelerograph. The data on sloshing obtained at Niigata roughly match the estimation model (Eqs.1 and 2) for the sloshing of liquid in the cylindrical shell as it is generally understood, but the data that are not confirmed in the model are included. So, clarification is needed on the scattering of the maximum sloshing wave height as revealed as the data, especially at Niigata. Perhaps, in a more detailed discussion, nonlinear sloshing analysis of cylindrical tanks must be taken into account. Work is underway to identify the effect of nonlinear boundary conditions at liquid surface on sloshing height of cylindrical tank under vertical and horizontal ground motions.

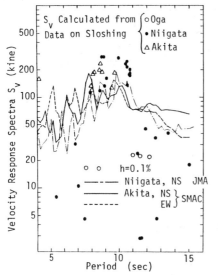

Fig.14 Comparison of velocity response spectra estimated from the sloshing height and calculated from SMAC record (h=0.1%)

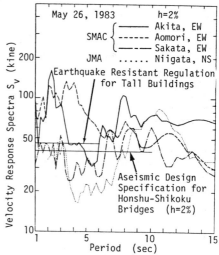

Fig.15 Calculated velocity response spectra and the design specifications for long period bridges and buildings

Fig.15 shows the velocity response spectra with 2% damping of each site for the period range of 1 to 15 sec. The used records are EW component of SMAC accelerograms at Akita, Aomori and Sakata Harbours, and NS component of mended and corrected JMA's seismogram at Niigata shown in Fig.11. The value of 2% corresponds to the steel structures such as suspension bridges and high-rise buildings. For comparison, the current Japanese seismic design specifications for long span suspension bridges [10] and high-rise buildings [9] are also shown in Fig.15.

From Fig.15, the following can be seen; 1) The response spectra at Akita for the periods around 2.2 and 7.8 sec and at Aomori for the period range of 2 to 4.5 sec reach and exceed 100 kine over the design spectra, and 2) Although the maximum ground acceleration at Sakata is nothing but 45 gal, it is worth notice that velocity response spectra in some periods reach 60 kine. Even though the maximum acceleration is small, i.e., short-period motion is not significant, the long-period components included in seismograms enhance the amplitudes and duration of the response of long-period structures. Supposing that large-scale structures such as high-rise buildings and long-span suspension bridges construct in Akita, Niigata and Aomori, the large vibration due to long-period ground motions may be observed under certain circumstances, for example in the case of existence of a thick surface layer of sediment in the crust. These results suggest that those Japanese seismic regulations should be reexamined. Therefore, the determination of the shakeability at the site --- site amplification factor --- as a function of period is keenly required, especially for the purpose of antiseismic design of long-period structures because of its small damping or high resonance at the natural period.

CONCLUSIONS

During the Nihonkai-chubu earthquake (May 26, 1983, M=7.9), oil sloshed from the large tanks in Niigata city, even though the maximum acceleration was less than 15 gal, that is, short-period motion was not significant. As far as we know, it would be the first experience of long-period earthquake motion bringing about the damage. In this study, response spectra of long-period (approximately from 5 to 15 sec) earthquake ground motions were estimated for several cities based on sloshing height in the oil storage tanks, and SMAC-B_2 acceleration- and JMA's displacement-type strong motion seismograms.

Major results of this study may be summarized as follows.

1) By means of questionnaire and interview, field survey has been performed for examing the sloshing phenomena of the large-oil tanks and the generation of surface waves affected by the deep underground structures. From these data, it was found that oil sloshed from the large storage tanks at several cities, i.e., Niigata, Akita and Tomakomai, which are about 270, 113 and 330 km away from the epicenter, respectively. Especially in Niigata city, very high sloshing wave height (about 4.5 m) for the period around 10 sec was observed. The maximum wave heights at Akita and Tomakomai were 3.5 and 2 m, respectively. The oil sloshed out generally for periods of 8 to 11 sec. These results are consistent with the recorded seismograms of JMA.

2) As an inversion problem, we tried to estimate the equivalent response spectra from the maximum sloshing wave height based on the potential theory. The response spectra in Niigata and Akita are predicted to be 200 to 300 kine and 100 to 200 gal for the 10 seconds structures with 0.1% damping. The equivalent velocity response spectrum value at Tomakomai is expected to be about 100 kine.

3) The response spectra estimated from sloshing phenomena is not comparable to ones calculated by response analysis of liquid storage tanks subjected to the equivalent input of 3 cycles of sinusoidal wave. This indicates that a potentially more reliable estimation of seismic response of large tanks can not be achieved by such three wave resonance method. In regulation prescribed by the Ministry of Home Affairs, the design spectra of 110 kine (96 kine especially in Niigata and Akita prefectures) are defined in order to minimize or avoid overflow due to sloshing waves. It can be said that the limit value of 110 kine is reasonable as an average in Japan. However, it should be mentioned that the freeboard provided by the current design practice is not adequate for sloshing waves during the great earthquake when the tank is full, and that it should be increased for typical sites such as Niigata and Akita because of the influence of the local soil conditions to the seismic-waves amplification.

4) Horizontal component records of the JMA's low-magnification displacement seismograph at Niigata were overloaded, so we attempted to mend the interrupted parts of seismogram. Velocity response spectra with 0.1% damping calculated by mended and corrected displacement and acceleration type strong motion seismograms roughly matched ones estimated by the model for the sloshing of liquid in the cylindrical tanks. These seismograms were found to give much higher response spectra than design values with 2% damping for long period structures such as high-rise buildings and long-span suspension

bridges. Especially, the velocity response spectra at Aomori exceed the design spectra for the period range of 1 to 10 sec, and ones at Akita reach 100 kine for the period around 2 and 8 sec.

ACKNOWLEDGEMENTS

Messrs. S. Noda and E. Kurata kindly provided the digital data recorded by the SMAC accelerographs of the Port and Harbour Research Institute of the Ministry of Transport. The original source of the displacement seismograph records reported in this paper is the Japan Meteorological Agency. The fire defence headquaters in several cities of the Japan gave us many valuable data on the sloshing phenomena of large oil-storage tanks. The authors are grateful to all of them. The authors wish to thank Mr. T. Yoshida (Maeda Construction Company) for his assistance in making the computer program of numerical analysis and for helpful discussions during the cource of this work. The writers are also grateful to Mr. T. Shikata (Kyoto University) for the digitization of the displacement seismograph records. Anf finally, the kindness of Dr. N. Shimizu of Chiyoda Chemical Engineering and Construction Co., Ltd., for providing valuable information concerning the aseismic design of oil-storage tanks, is very much appreciated.

REFERENCES

(1) Japan Society of Civil Engineers ; Report on a comprehensive research into the properties of long-period earthquake ground motions for the seismic safety of above-ground storage tanks, December 1982 (in Japanese).
(2) The Public Works Research Institute, Ministry of Construction ; "Studies on the characteristics of long-period strong motions using the JMA's low-magnification displacement seismograms ---- Analysis of the 1968 Tokachi-oki earthquake records ----," Technical Memorandom of PWRI, No.1869, March 1983 (in Japanese).
(3) Yamada,Y. and Noda,S. ; "Sensitivity analysis based on effects of soil condition and source mechanism of fault on relatively long-period ground motions," Journal of Japan Society for Natural Disaster Science, Vol.3, No.2, pp.1 - 28, 1984 (in Japanese).
(4) The Japan Meteorological Agency ; The Seismological Bulletin of the Japan Meteorological Agency for May and June 1983, May and June 1984 (in Japanese).
(5) The Ministry of International Trade and Industry ; "Regulation for earthquake resistant design of equipments and other structures for high pressure gas," An extra of an official gazette No.93, Notification No.515 of the Ministry of International Trade and Industry, pp.3 - 28, October 26, 1981 (in Japanese).
(6) The Ministry of Home Affairs ; "Notification for making a partial amendment of the notification for establishing details of technical standard for regulation of dangerous object," An official gazette No.16870, Notification No.119 of the Ministry of Home Affairs, pp.10 - 12, April 28, 1983 (in Japanese).
(7) Architectural Institute of Japan ; Design recommendation for storage tanks and their supports, 1984 (in Japanese).
(8) American Petroleum Institute (API) ; "Seismic design of storage tanks," in Appendix E of Welded Steel Tanks for Oil Storage, API Standard 650, 6th Edition Revision 3, pp.E1 - E7, October 1979.
(9) Architectural Institute of Japan ; Technical guide for tall buidings ---- An enlarged and revised edition ----, March 1973 (in Japanese).
(10) Honshu-Shikoku Bridge Authority ; Aseismic design specification, pp.1 - 35, March 1977 (in Japanese).
(11) Okada,S. and Kagami,H. ; "A point-by-point evaluation of amplification characteristics in Japan on 1 - 10 sec seismic motions in relation to deep soil deposites," Trans. Architectural Inst. Japan, No.267, pp.29 - 38, May 1978 (in Japanese).
(12) Kuribayashi,E., Toki,K. and Wakabayashi,S. ; "Reliability of ground motions by SMAC-B_2 Accelerographs," Proc. of 4th Japan Earthquake Engineering Symposium, pp.153 - 160, 1975 (in Japanese).
(13) Kawashima,K., Takagi,Y. and Aizawa,K. ; "Accuracy of digitization of strong-motion records obtained by SMAC - accelerograph," Proc. of Japan Society of Civil Engineers, No.323, pp.67 - 75, July, 1982 (in Japanese).
(14) Kudo,K. and Sakaue,M. ; "Oil-sloshing in the huge tanks at Niigata due to the Nihonkai-chubu earthquake of 1983," Bull. Earthq. Res. Inst., Univ. Tokyo, Vol.59, pp.361 - 382, 1984 (in Japanese).
(15) Kurata,E., Fukuhara,T. and Noda,S. ; "Strong-motion earthquake records on the 1983 Nipponkai-chubu earthquake in port areas," Technical Note of the Port and Harbour Research Institute, Ministry of Transport, No.458, September 1983.

ON THE MODELING OF A CLASS OF DETERIORATING STRUCTURES
SUBJECTED TO SEVERE EARTHQUAKE LOADING

Arturo O. Cifuentes[I] and Wilfred D. Iwan[II]

ABSTRACT

This paper presents a numerically efficient system identification algorithm for the analysis of deteriorating structures. This algorithm is to be used in combination with the DDE model recently developed by the authors.

This model was formulated after studying several earthquake records from damaged structures. For small amplitudes of oscillation, the model behaves linearly but for increasing amplitudes of oscillation the model shows progressive stiffness degradation. An example using actual data is discussed.

INTRODUCTION

The problem of modeling the deteriorating behavior of structures subjected to severe ground motion is both complicated and challenging. This is due to the stiffness degradation phenomenon which makes the problem highly non-linear and history dependent. Several models have been proposed to describe this effect [1-9]. Unfortunately, most of the realistic models rely on somewhat complex mathematical representations of the restoring force of the structure. This situation, coupled with a lack of experimental data has limited their applications in analysis and design.

This paper is concerned with the application of a model recently developed by the authors [10,11], for the restoring force behavior of deteriorating structures. This model, although simple, captures the essential features of the stiffness degradation process and therefore can be conveniently used in system identification. The goals of this paper are: 1) to outline a numerically efficient system identification algorithm which is suitable to use with this model, and 2) to demonstrate that this model can accurately predict the response of structures affected by severe ground acceleration by using actual earthquake data.

I Staff Scientist, The MacNeal-Schwendler Corporation, Los Angeles, California.

II Professor, Division of Engineering and Applied Science, California Institute of Technology, Pasadena, California.

THE DETERIORATING DISTRIBUTED-ELEMENT (DDE) MODEL

The DDE model, introduced by the authors [10,11], consists of an ensemble of linear springs and slip dampers arranged as indicated in Fig. 1. This model was formulated after studying a number of earthquake records of reinforced concrete buildings that had suffered significant damage in strong earthquakes. The model consists of three type of elements:

1) A linear spring with stiffness constant equal to K_e.

2) An elasto-plastic element whose linear stiffness is equal to K_{ep} and whose maximum allowable force is equal to $K_{ep} X_{y_{ep}}$, where $X_{y_{ep}}$ is the yielding displacement.

3) A set of N deteriorating elements that have yielding displacement equal to X_{y_i} and a linear stiffness equal to K_i.

Each deteriorating element is similar to the elasto-plastic element except for the fact that it loses all of its strength when the displacement reaches a certain critical value given by βX_{y_i}. At that moment, the deteriorating element no longer contributes to the restoring force.

β is assumed to be 2 for this study, based on a previous investigation [10]. The deteriorating elements have been arranged so that $X_{y_1} < X_{y_2} ... < X_{y_N}$.

It is further assumed that the following relationship exists between the linear stiffness K_i of a typical deteriorating element and its corresponding yielding displacement,

$$K_i = \frac{\gamma}{X_{y_i}^2} \qquad (1)$$

A rationale for this relationship is discussed in [10]. The values X_{y_i} are normally chosen equally distributed within the range of interest. These assumptions leave four parameters to be determined i.e. γ, K_e, K_{ep} and $X_{y_{ep}}$. Once these parameters are identified, the model is completely determined.

SYSTEM IDENTIFICATION

Preliminaries

Assume that records are available for the dynamic response of a multi-story building during an earthquake. At least two records at different levels are needed, say the roof and the basement. These records normally contain the time history of the absolute acceleration, but can be used to generate the time history of the velocity and displacement as well. Assume that the overall response of the building can be adequately simulated using a single structural mode; usually the fundamental mode. The modal equation of motion will have the form

$$\ddot{x} + \bar{f} = -p\ddot{z} \qquad (2)$$

where x is the relative amplitude of the mode at the structure measurement point with respect to the basement; $\ddot{z}$ is the input ground acceleration; $\bar{f}$ is a mass-normalized modal restoring force and p is the participation factor.

Restoring force diagram

Eq. (2) may be written as

$$\bar{f} = -(\ddot{x} + p\ddot{z}) \qquad (3)$$

In this form it is possible to determine $\bar{f}$ for the discrete set of points for which $\ddot{x}$ and $\ddot{z}$ have been recorded [11,12]. Hence, it is possible to determine $\bar{f}$ as a function of x and therefore determine the modal restoring force diagram. A typical restoring force diagram for a single-degree-of-freedom structure that has experienced deterioration is shown in Fig. 2.

Effective Stiffness Diagram

Once the modal restoring force diagram has been computed, it is possible to estimate an equivalent linear stiffness for each cycle of oscillation. The normalized equivalent linear stiffness, K_{eff_i}, can be defined as

$$K_{eff_i} = \frac{\bar{f}_i}{x_i} \qquad (4)$$

where x_i is the maximum amplitude of oscillation for a particular cycle and $\bar{f}_i$ is the corresponding value of the mass-normalized modal restoring force. With this information, one can generate the so called Effective Stiffness Diagram. That is a graph of K_{eff_i} vs. x_i. For a given number of deteriorating elements, this diagram may be used to uniquely determine the value of Y in the DDE model. This is discussed in detail in [11,13].

Determination of K_{ep}, $X_{y_{ep}}$ and K_e.

Let the restoring force diagram of the actual structure be determined by a set of points $(x_j, \bar{f}_j)$ where x_j is the relative modal displacement of the structure at a given time and $\bar{f}_j$ is the corresponding value of the mass-normalized modal restoring force. Similarly, let the set of points $(x_j, \bar{f}_j)$ represent the restoring force diagram given by an arbitrary set of modal parameters of K_e, K_{ep} and $X_{y_{ep}}$. Now, the optimum values of the modal parameters can be determined by minimizing an "error" (measure of agreement) between the two restoring force diagrams. Note that the set of points $(x_j, \bar{f}_j)$ can be computed directly from the time history of response of the structure without requiring the solution of a differential equation. This makes the identification algorithm much more efficient than traditional approaches that are based upon the minimization of an "error" which depends on the time history of the response of the structure.

EXAMPLE

The following example is based on the dynamic response of the Bank of California building (N79W component) during the San Fernando earthquake in 1971. This structure is a twelve story reinforced concrete building. It suffered significant damage during the San Fernando event. It has been shown that it is not possible to model the response of this building using a time invariant linear model [14].

A DDE model having 9 deteriorating elements was employed to model the response of the roof. Table 1 shows the optimal values determined for the modal parameters of the model. A participation factor equal to 1.3 was considered. This value corresponds to the participation factor employed in the design of the building. A viscous damping coefficient equivalent to 1% of critical damping was added to the DDE model. These points are discussed in more depth in [9] and [11].

Fig. 3 compares the effective stiffness diagram of the actual structure and the effective stiffness diagram given by the model. It can be observed that the effective stiffness decreased when the amplitude of oscillation increases. However, when the amplitude of oscillation decays after reaching its maximum value, the effective stiffness does not totally recover its initial value. This indicates that the structure has suffered permanent deterioration.

Fig. 4 shows the restoring force diagram of the actual structure and Fig. 5 shows the restoring force diagram obtained with the DDE model. Both are fairly similar. It is noted, that the loss of stiffness in the structure is apparent from the fact that the slope of the hysteresis loops decreases considerably as the oscillations progress.

Figure 6 depicts the time history of response of the roof as measured and as predicted by the DDE model. It can be seen that there is good agreement between the two results. The model captures the frequency changes of the response as well as the amplitude variation. This result is significant since the effective stiffness reduction of the building during the earthquake was approximately 65%.

CONCLUSIONS

In light of the results reported herein and in [10,13] it is concluded that the DDE model is capable of satisfactorily reproducing the major features of the response of reinforced concrete structures exhibiting deterioration. Once the parameters of the model have been determined for a particular input motion, the model may be used to predict the response of the building to other event. One promising application would be in assessing the likelihood of failure of a damaged building should it be subjected to another earthquake. This subject is treated more extensively in [10,13].

REFERENCES

[1] Takeda, T., Sozen, M.A., and Nielsen, N.N., "Reinforced Concrete Response to Simulated Earthquakes", ASCE, Journal of the Structural Division, Vol. 96, December 1970, pp. 2557-2573.

[2] Toussi, S. and Yao, J., "Identification of Hysteretic Behavior for Existing Structures", Report CE-STR-80-19, School of Civil Engineering, Purdue University, December 1980.

[3] Atalay, B. and Penzien, J., "Inelastic Cyclic Behavior of Reinforced Concrete Flexural Members", Proceedings of the Sixth World Conference on Earthquake Engineering, New Delhi, India, 1977, Vol. III, pp. 3062-3068.

[4] Aoyama, H., "Simple Nonlinear Models for the Seismic Response of Reinforced Concrete Buildings", Proceedings of the Review Meeting U.S.-Japan Cooperative Research Program in Earthquake Engineering, Tokyo, Japan, 1976, pp. 291-309.

[5] Clough, R.W. and Johnston, S.B., "Effect of Stiffness Degradation on Earthquake Ductility Requirements", Proceedings, Japan Earthquake Engineering Symposium, Tokyo, Japan, October 1966, pp. 227-232.

[6] Otani, S., "Nonlinear Dynamic Analysis of 2-D Reinforced Concrete Building Structures", Third Canadian Conference on Earthquake Engineering, Vol. 2, 1979, pp. 1009-1037.

[7] Wen, Y.-K., "Method for Random Vibration of Hysteretic Systems", ASCE, Journal of the Engineering Mechanics Division, Vol. 102, April 1976, pp. 249-263.

[8] Saiidi, M. and Sozen, M., "A Naive Model for Nonlinear Response of Reinforced Concrete Buildings", Proceedings of the Seventh World Conference on Earthquake Engineering, Istanbul, Turkey, 1980, Vol. 7, pp. 8-14.

[9] Muguruma, M., Tominaga, M. and Watanabe, F., "Response Analysis of Reinforced Concrete Structures Under Seismic Forces", Proceedings of the Fifth World Conference on Earthquake Engineering, Rome, Italy, 1974, Vol. 1, pp. 1389-1392.

[10] Cifuentes, A.O., "System Identification of Hysteretic Structures", Ph.D. Thesis, California Institute of Technology, Pasadena, California, September, 1984.

[11] Iwan, W.D. and Cifuentes, A.O., "A Model for System Identification of Degrading Structures", to be published by Earthquake Engineering & Structural Dynamics.

[12] Iemura, H. and Jennings, P.C., "Hysteretic Response of a Nine-Story Reinforced Concrete Building", International Journal of Earthquake Engineering and Structural Dynamics, Vol. 3, 1974, pp. 183-201.

[13] Cifuentes, A.O. and Iwan, W.D., "System Identification of Reinforced Concrete Structures Subjected to Damaging Earthquakes", submitted for publication.

[14] McVerry, G.H., "Frequency Domain Identification of Structural Models from Earthquake Records", Report EERL 79-02, Caltech, Pasadena, California, October 1979.

Table 1. Parameters of the DDE Model

$X_{y_1} = 5\,cm$;	$X_{y_2} = 10\,cm$;	$X_{y_3} = 15\,cm$;
$X_{y_4} = 20\,cm$;	$X_{y_5} = 25\,cm$;	$X_{y_6} = 30\,cm$;
$X_{y_7} = 35\,cm$;	$X_{y_8} = 40\,cm$;	$X_{y_9} = 45\,cm$;

K_e $\frac{1}{sec^2}$	K_{ep} $\frac{1}{sec^2}$	$X_{y_{ep}}$ cm	p	γ $\frac{cm^2}{sec^2}$
2.5	0.99	26.3	1.3	140

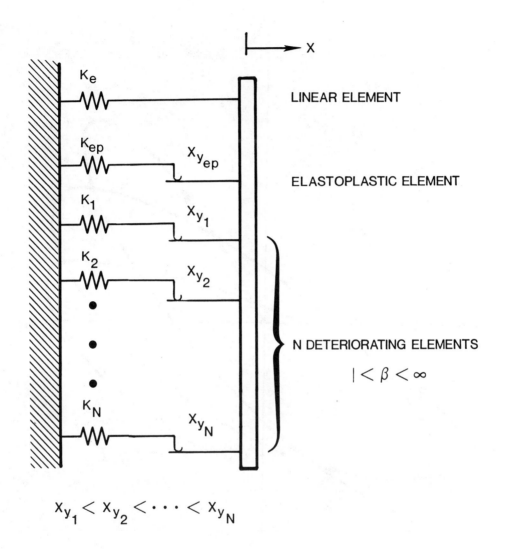

Figure 1

The Deteriorating Distributed Element (DDE) Model.

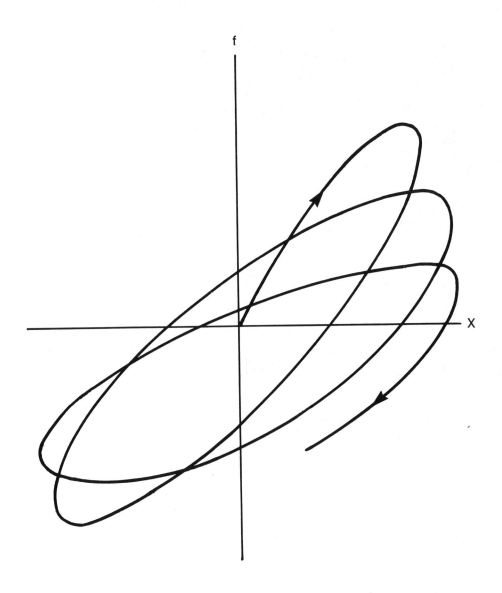

Figure 2

Restoring Force Diagram of a Deteriorating Structure.

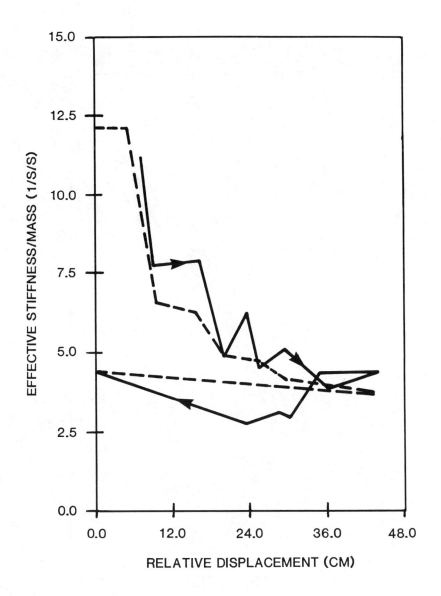

Figure 3

Effective Stiffness Diagram, (Bank of California Building, N79W Component). Comparison between the actual diagram (solid line) and the diagram given by the model (dashed line).

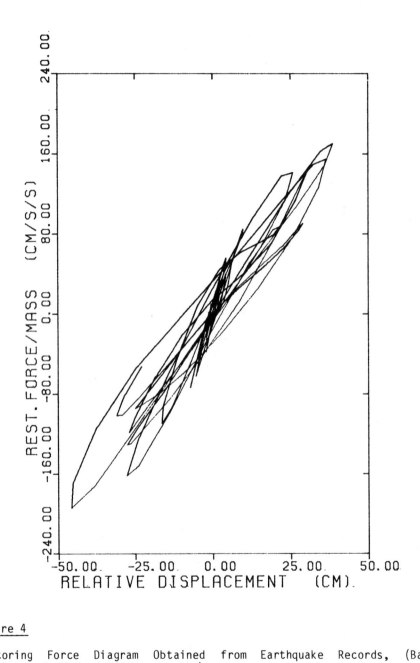

Figure 4

Restoring Force Diagram Obtained from Earthquake Records, (Bank of California Building, N79W component).

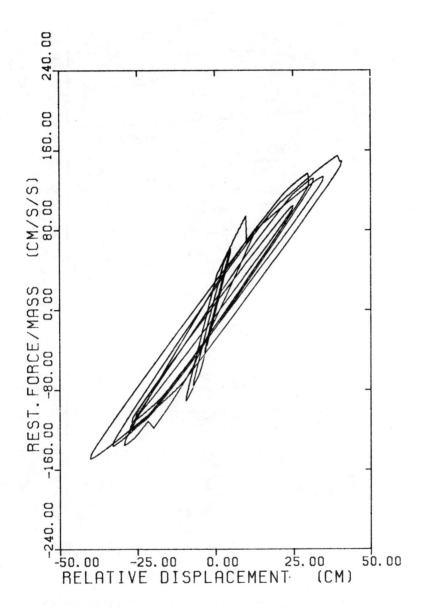

Figure 5

Restoring Force Diagram Given by the DDE Model, (Bank of California Building N79W component).

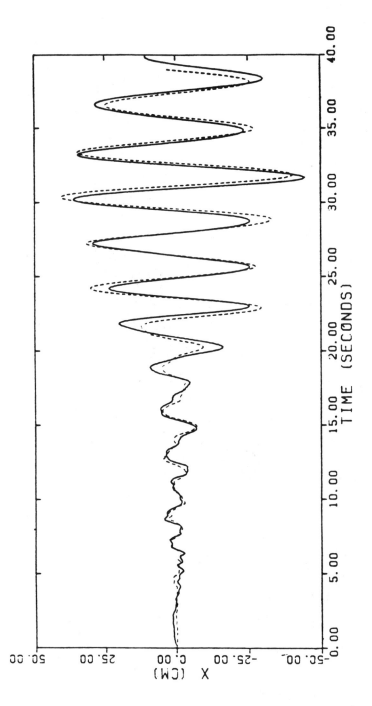

Figure 6

Bank of California Building, N79W component. Comparison between the actual response (solid line) and the response predicted by the model (dashed line).

SEISMIC ENERGY AND STRUCTURAL DAMAGE

J. M. Tembulkar[I] and J. M. Nau[II]

ABSTRACT

Structures subjected to seismic excitation dissipate the imparted energy by viscous damping and inelastic deformations. It is hypothesized that damage incurred depends on the amount of energy dissipated in a structure. The mechanics of the energy dissipation process are studied for simple structures. Energy spectra are derived, and normalizing parameters for the energy spectra are evaluated. A cumulative ductility measure is proposed on the basis of the study of multi-degree-of-freedom shear buildings.

INTRODUCTION

Over the last two decades the response spectrum approach has evolved, in which the earthquake environment is characterized by a smoothed design response spectrum. This smoothed spectrum represents a statistical summary of the response obtained from an ensemble of ground motions. These spectra, both elastic and inelastic, are scaled by peak ground motions. It has been observed that neither the peak ground acceleration, nor the elastic response spectra, are adequate descriptors of damage [1,2]. Mahin and Bertero [3], and Banon et. al. [4], have shown that low cycle fatigue type damage produced by a number of cycles of hysteretic action, can be a crucial element in producing structural damage. Zahrah and Hall [5] have shown that the emphasis on displacement ductility, traditionally used to characterize inelastic response spectra, ignores the cumulative damage that results from numerous inelastic cycles.

Structures subjected to seismic excitation can be modeled to dissipate the imparted energy by viscous damping, and if the excitation is strong enough, by hysteretic action. Since seismic damage refers to the deterioration and possible failure of the elements of a structure, the level of damage should depend on the amount of energy dissipated in, or imparted to, the structural elements. The energy quantities can be obtained from a time

---

I. Graduate Research Assistant, Department of Civil Engineering North Carolina State University, Raleigh, N. C. 27695-7908.

II. Assistant Professor, Department of Civil Engineering, North Carolina State University, Raleigh, N. C. 27695-7908.

history of the structural response, which takes into account all the loading/ unloading cycles, rather than a single response peak.

Housner [6] was one of the first to propose an energy based design approach for seismic design. He estimated the energy per unit mass imparted to a structure as one half times the square of the maximum spectral velocity. Blume [7] introduced the reserve-energy-technique based on empirical factors for estimating energy dissipation demands. Berg [8], Thomaides [9], and Jennings [10] have performed some studies on inelastic deformations and energy dissipation in single-degree-of-freedom (SDOF) systems. McKevitt et. al. [11] have undertaken preliminary studies on correlating energy dissipation in SDOF and multi-degree-of-freedom (MDOF) systems. Kato and Akiyama [12] have proposed a limit strength design procedure for steel frames, based on energy considerations.

ENERGY DISSIPATION IN SDOF SYSTEMS

In order to evaluate the role of energy dissipation, a detailed study of the mechanics of the energy dissipation process was carried out for simple systems. The equation of motion for a SDOF system subjected to seismic excitation is,

$$\ddot{u}(t) + 2\beta\omega\dot{u}(t) + R[u(t)] = -\ddot{u}_g(t) \tag{1}$$

where, u, $\dot{u}$, and $\ddot{u}$ represent the relative displacement, velocity, and acceleration respectively; ω, the circular natural frequency; β, the fraction of critical damping; $R[u(t)]$, the resisting force per unit mass; and $\ddot{u}_g$, the ground acceleration.

The energy imparted to a structure, E_I, the energy dissipated by hysteresis, E_H, and the energy dissipated by damping, E_D, can be formulated per unit mass as,

$$E_I = -\int \ddot{u}_g(t) du \tag{2}$$

$$E_H = \int R[u(t)] du \tag{3}$$

$$E_D = 2\beta\omega \int \dot{u}(t) du \tag{4}$$

where integration is carried out for the duration of the earthquake plus five cycles of free vibration.

Various hysteretic models have been proposed to simulate inelastic force-deformation behavior in analytical studies. The force-deformation models used in this study are the bilinear model, and the stiffness degrading model. The bilinear model is non-deteriorating and represents a broad range of structures which exhibit complete hysteresis loops. The stiffness degrading model characterizes the behavior of reinforced concrete structures, in which there is no significant deterioration due to shear or bond. The complexities of ground motion make it imperative that statistics from an ensemble of earthquakes be examined. The twelve earthquakes used in this study are shown in Table 1.

They represent two major types of ground motion, namely, the short duration impulsive type, and the long duration uniformly severe type.

Step-by-step numerical integration was implemented in the time domain using the Newmark-Beta algorithm. The range of displacement ductilities considered was from 1.5 to 10.0. Viscous damping ratios of two and five percent of critical were used. The frequency range was from 0.035 Hz to 35.0 Hz.

Results of Energy Computations

Some of the more significant results from the study of SDOF systems for the ensemble of earthquakes can be summed up as follows:

i. The mean imparted energy spectra for the ensemble of earthquakes reveal that the influence of displacement ductility is not as great as for response spectra. The distribution of mean imparted energy, E_I, for structures of ductility 2, 5, and 10 is shown in Fig. 1. The imparted energy is greatest in the mid-range of frequencies. This is of particular relevance because most building frequencies are also in this range. The energy imparted is relatively insensitive to the displacement ductility, except for the high frequencies. Traditionally, a higher ductility has implied a higher energy dissipation capacity. This appears to be true only for structures whose frequencies are greater than 2 Hz. Interestingly, in the lower frequency range, energies are usually less for higher displacement ductilities. Thus a higher ductility, while effective in reducing the forces to which the structure is subjected, actually results in less energy imparted to the structure. Similar observations can also be made for the hysteretic energy, E_H.

ii. It has been observed that mean inelastic yield spectra are insensitive to the details of the force-deformation model. Mean imparted energy spectra normalized by peak ground acceleration for the ensemble of earthquakes are shown in Fig. 2. Over the low range of frequencies (less than 2 hz), the two models show good agreement. However, for higher frequencies a greater amount of energy is imparted to the stiffness degrading model. Thus, on the average, energy dissipation estimates for structures of frequencies greater than 2 Hz can be significantly different, depending on the force-deformation model used. The same observations are also valid for the hysteretic energy spectra.

Normalizing Energy Spectra

Various normalizing parameters were considered in order to investigate the appropriate parameter for scaling energy spectra.

The scaling factors, formulated as measures of earthquake strength, are based on ground motion parameters (displacement, velocity, and acceleration). These parameters are shown in Table 2. It is observed from Figs. 3(a) and 3(b) that the integral of squared ground velocity and the integral of squared ground acceleration, give the least dispersion of imparted energy spectra for low and high frequencies respectively. This is of particular interest in light of the developments by Arias [13] and Housner and Jennings [14], of broad band measures of 'the potential destructiveness of ground shaking'. These intensity measures proposed by Housner and Jennings are 'the period ensemble work', W_T, and 'the frequency ensemble work', W_F. It has been shown that the period ensemble work and the frequency ensemble work, are proportional to the integral of the squared ground velocity and the integral of the squared ground acceleration respectively, or,

$$W_T = \int W(T) dT = C_1 \int v^2 dt \qquad (5)$$

$$W_F = \int W(\omega) d\omega = C_2 \int a^2 dt \qquad (6)$$

where, $W(\omega)$, is the energy imparted to an elastic system of circular frequency ω; $v(t)$, the ground velocity; $W(T)$, the work done by a structure of period T; and $a(t)$, the ground acceleration.

The intensity measures formulated by Housner and Jennings are valid only for energy dissipation in elastic systems. It is encouraging to note that these same measures give the least dispersion for inelastic energy spectra. The good correlation between these earthquake intensity measures and energy spectra lends further support to the assumption that seismic energy quantities could be used as damage indices.

It is commonly accepted that instrumental peak ground motions are inadequate descriptors of earthquake strength [15]. Evaluating normalizing parameters for energy spectra gives an estimate of the correlation between energy spectra and various measures of earthquake strength and damage potential. The integrals of squared ground velocity and squared ground acceleration can therefore be used for characterizing 'design' energy spectra.

When hysteretic energy spectra are similarly normalized, the integrals of squared ground velocity and squared ground acceleration again give the least dispersion.

An Alternate Definition of Ductility

Since displacement ductility refers to the ratio of maximum displacement to yield displacement, it neglects the numerous cycles of inelastic deformations and the consequent energy dissipation demands placed on the structure. Various alternative ductility factors have been proposed in the past [3,4,12]. Kato [12] has defined a 'cumulative ductility factor' as the ratio of the hysteretic energy dissipated to the product of the yield force and the yield displacement, or,

$$\eta = E_H / (F_y * \delta_y) \qquad (7)$$

where, η, is the cumulative ductility factor; F_y, the yield force; and δ_y, the yield displacement.

This definition of cumulative ductility incorporates the influence of the energy dissipated in all yield cycles. The value of cumulative ductility is zero when no yielding takes place. Figure 4 shows mean cumulative ductility for the ensemble of earthquakes. Results for displacement ductilities of 2,

5, and 10, for both the bilinear and the stiffness degrading models are shown. It is noted that for large displacement ductilities (e.g. a ductility of 10), the cumulative ductility can be almost an order of magnitude larger. However, for small ductilities (e.g. a ductility of 2) the cumulative ductility is of the same order of magnitude. Thus structures experiencing large displacement ductilities, on the average, can be expected to undergo a larger number of and larger magnitudes of inelastic excursions. It is also seen from Fig. 4 that stiffness-degrading structures experience significantly lower cumulative ductility demands for displacement ductilities of five and greater, over the low and middle range of frequencies.

In some reinforced concrete structures, stiffness degradation is accompanied by strength degradation. It has also been observed that inelastic cyclic loading reduces the shear strength. The influence of these phenomena becomes more critical with the number of inelastic excursions, thereby bringing the structure closer to collapse. Seen in this light, the high cumulative ductility demands corresponding to the higher displacement ductilities have a special significance. This is more so because yet another parameter, namely, the P-Δ effect, can be instrumental in producing unbounded displacement and energy dissipation demands by bringing about strain softening in the inelastic range.

ENERGY DISSIPATION IN MDOF SYSTEMS

Present design and analysis procedures for multistory buildings are based on the response spectrum. The emphasis is on the harnessing of ductility in the design of structures to withstand a certain magnitude of lateral force. Past earthquake damage observations, and preliminary reports from Mexico City (Sept. 1985), suggest that damage is often concentrated in one or more stories. This would imply that the energy dissipation demands on these stories had exceeded the energy dissipation capacity. Since cumulative ductility is based on the energy dissipated by hysteresis, it can therefore be a reasonable focus in the study of energy dissipation in multistory buildings. Thus in order to distribute 'damage' uniformly over the stories of the structure, one may aim for a design which gives equal cumulative ductility demands in all stories. For a given stiffness distribution, the yield levels can be either increased or decreased in order to iterate to equal cumulative ductility in all stories.

Four and seven story one bay shear-building structures have been studied. The mass was assumed to be constant for all stories in a structure, and the ratio of viscous damping was five percent of critical in all modes. To identify the distribution of story stiffness, a factor α (defined as the ratio of the stiffness of the bottom story to that of the top story) was introduced. Intermediate story stiffness was distributed linearly. The values of cumulative ductility iterated to were from 2 to 40. The structural parameters used can be summed up as follows:

        Four story building   Period 0.5 sec.
                                                  5% modal damping
                                                  $\alpha = 1.0$, 1.5, and 2.0.

Seven story building   Period 1.0 sec.
                        5% modal damping
                        α =2.5 and 3.5.

The following observations are based on this phase of the study:

i. Figure 5 shows the distribution of hysteretic energy dissipated per unit mass for SDOF and MDOF systems, when subjected to the San Fernando earthquake. In almost all cases the SDOF estimate of E_H for a given level of cumulative ductility, gives an upper bound on the hysteretic energy dissipated in MDOF systems of the same cumulative ductility. Inelastic behavior of MDOF systems is fairly complex because of the multiplicity of modes of vibration and degrees-of-freedom. In spite of this, the cumulative ductility gives a good estimate of the hysteretic energy dissipated.

ii. The mean displacement ductility (over all the stories) has been plotted against the iterated cumulative ductility for MDOF systems subjected to the San Fernando earthquake in Fig. 6. The scatter in the data is relatively small. On a log-log scale, a linear relationship can be established between the mean displacement ductility and the cumulative ductility. Thus, a design procedure which aims at the uniform distribution of cumulative ductility in all stories, would also give a reasonable estimate of the mean displacement ductility.

iii. Let δ_i and k_i be the yield displacement and stiffness of story i, respectively. We can define a nondimensionalized factor, ε, such that,

$$\varepsilon = (\delta_i k_i)/(\delta_1 k_1) \tag{8}$$

The physical significance of ε is that it reflects the product of the yield level and stiffness of story i, as compared with that of the bottom story. The distribution of ε for the four and seven story buildings, subjected to the San Fernando earthquake, in which the cumulative ductility is constant over the stories is shown in Fig. 7. The ordinate $(i-1)/N$ is the nondimensionalized parameter pertaining to story i of the N story building. A smooth curve is also shown in the figure to fit the data. It is observed that the deviation from this curve is small. Thus a design which aims at the uniform distribution of cumulative ductility over the stories, would have to incorporate stiffness and yield level distributions which agree with the result shown in Fig. 7.

The current approach to seismic design emphasizes the ability of the structure to withstand a required level of lateral force. However, this approach does not address the fact that seismic damage may be concentrated in some parts of the structure, e.g. a 'soft' first story. A design procedure which avoids concentration of 'damage' (represented by cumulative ductility in this instance) shifts the emphasis to the relative distribution of stiffness and yield levels to meet requirements discussed above.

CONCLUSIONS

It has been hypothesized that the energy dissipated by hysteresis should reflect the extent of damage incurred by a structure subjected to strong ground motion. Since displacement ductility refers to a single response peak, it ignores the damage due to numerous cycles of inelastic excursions. A 'cumulative ductility' measure is proposed. This cumulative ductility measure incorporates the contribution from hysteretic energy dissipated in each inelastic excursion. It has been shown that structures exhibiting large displacement ductilities (e.g. $\mu = 10$), would on the average have to meet cumulative ductility demands which are almost an order of magnitude greater. They would therefore be more susceptible to progressive exhaustion of their energy dissipation capacity.

A study of multi-degree-of-freedom shear-building structures has shown that hysteretic energy dissipated can be estimated from single-degree-of-freedom systems of the same viscous damping and cumulative ductility. Mean displacement ductility can be estimated from the cumulative ductility for the multi-degree-of-freedom structure.

Inelastic modal analysis and Equivalent lateral force method, which are the basis of present design procedures for strong ground motion, are empirical, albeit reinforced by observed building performance. However, there have been instances of the failure of structures due to the concentration of damage. An energy based approach incorporating the concept of a cumulative ductility holds promise, and could overcome this shortcoming. An extensive study is called for before a definitive conclusion can be reached as to its veracity and practicality in the design environment.

ACKNOWLEDGEMENTS

This study was carried out in the Department of Civil Engineering at North Carolina State University, under National Science Foundation Grant CEE 84-04744. This support is gratefully acknowledged. Any opinions, findings, and conclusions expressed are those of the authors and do not necessarily reflect the views of the National Science Foundation.

REFERENCES

[1] Housner, G. W., "Earthquake Research Needs for Nuclear Power Plants," *Journal of the Power Division*, ASCE, Vol. 97, P01, pp. 77-91, January 1971.

[2] Kennedy, R. P., "Peak Acceleration as a Measure of Damage," *Proceedings, Fourth International Seminar on Extreme Load Design of Nuclear Power Plant Facilities*, Paris, France, Aug. 1981.

[3] Mahin, S. A. and Bertero, V. V., "An Evaluation of Inelastic Seismic Design Spectra," *Journal of the Structural Division*, ASCE, Vol. 107, No. ST1, pp. 1777-1795, September 1981.

[4] Banon, H., Biggs, J. M., and Irving, H. M., "Seismic Damage in Reinforced Concrete Frames," *Journal of the Structural Division*, ASCE, Vol. 107, No. ST9, pp. 1712-1729, Sept. 1981.

[5] Zahrah, T. F. and Hall, W. J., "Seismic Energy Absorption in Simple Structures," *Structural Research Series No. 501*, Civil Engineering Department, University of Illinois, Urbana, Illinois, 1982.

[6] Housner, G. W., "Limit Design of Structures to Resist Earthquakes," *Proceedings, First World Conference on Earthquake Engineering*, San Francisco, pp. 5-1 to 5-13, 1956.

[7] Blume, J. A., Newmark, N. M., and Corning, L. H., *Design of Multistory Reinforced Concrete Buildings for Earthquake Motions*, Portland Cement Association, Skokie, Illinois, 1961.

[8] Berg, G. V., "Inelastic Deformations in Earthquake Engineering," *Proceedings, 30th Annual Convention, Structural Engineers Association of California*, Yosemite, California, pp. 30-40, Oct. 1963.

[9] Thomaides, S. S., *Effect of Inelastic Action on the Behavior of Structures during Earthquakes*, Ph.D. Dissertation, University of Michigan, Ann Arbor, Michigan, 1961.

[10] Jennings, P. C., "Earthquake Response of Yielding Structures," *Journal of the Engineering Mechanics Division*, ASCE, Vol. 90, No. EM4, pp. 41-68, Aug. 1965.

[11] McKevitt, W. E., Anderson, D. L., Nathan, N. D., and Cherry S., "Towards a Simple Energy Design of Structures," *Proceedings, Second U. S. National Conference on Earthquake Engineering*, EERI, pp. 383-392, 1979.

[12] Kato, B. and Akiyama, H., "Seismic Design of Steel Buildings," *Journal of the Structural Division*, ASCE, Vol. 108, No. ST8, pp. 1709-1921, Aug. 1982.

[13] Arias, A., "A Measure of Earthquake Intensity," in *Seismic Design for Nuclear Power Plants*, R. J. Hansen, editor, Massachusetts Institute of Technology, Cambridge, Massachusetts, pp. 438-483, 1970.

[14] Housner, G. W. and Jennings, P. C., "The Capacity of Extreme Earthquakes to Damage Structures," in *Structural and Geotechnical Mechanics*, W. J. Hall, editor, Prentice-Hall, Inc., Englewood-Cliffs, New Jersey, pp. 102-116, 1977.

[15] Housner, G. W. and Jennings, P. C., *Earthquake Design Criteria*, Earthquake Engineering Research Institute, Berkeley, California, pp. 51-57, 1982.

Table 1 Earthquake Data

Earthquake	Record and Component	Maximum Ground Acceleration (g)	Earthquake	Record and Component	Maximum Ground Acceleration (g)
San Fernando, Calif. Feb. 9, 1971	Pacoima Dam, S16E	1.17	Kern County, Calif. July 21, 1952	Taft-Lincoln School Tunnel, S69E	0.179
Parkfield, Calif. June 27, 1966	Cholame-Shandon No. 2, N65E	0.489	Andreanof Island, Alaska May 1, 1971	Adak, Alaska, U.S. Naval Station, West	0.186
Bear Valley, Calif. Sept. 4, 1972	Melendy Ranch, N29W	0.516	Kilauea, Hawaii April 26, 1973	Hawaii National Park, Namakani Paio Camp, S30W	0.159
Coyote Lake, Calif. Aug. 6, 1979	Gilroy Array No. 6, 230 Deg.	0.417	Managua, Nicaragua Dec. 23, 1972	ESSO Refinery, South	0.324
Imperial Valley, Calif. Oct. 15, 1979	Bonds Corner, 230 Deg.	0.786	Bucarest, Roumania March 4, 1977	Building Research Institute, S-N	0.206
Imperial Valley, Calif. May 18, 1940	El Centro, S00E	0.348	Off Central Chile Coast July 8, 1971	Univ. Of Chile, Santiago Engineering Bldg. N10W	0.159

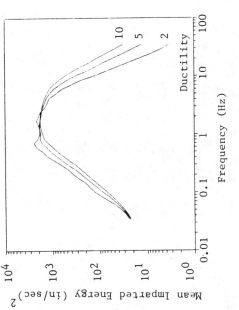

Fig. 1 Mean Imparted Energy per unit mass

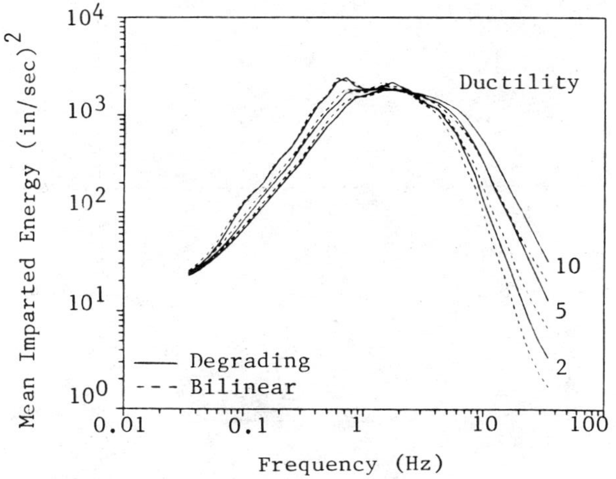

Fig. 2 Influence of Force-Deformation Model on Mean Imparted Energy

Table 2 Normalizing Parameters for Energy Spectra

Peak Ground Motions	—	$x(t)_{maximum}$, Mean Square Ground Motion — $\dfrac{1}{t_{95}-t_{05}} \displaystyle\int_{t_{05}}^{t_{95}} x^2(t)\,dt$
Integral of Squared Ground Motion	—	$\displaystyle\int x^2(t)\,dt$, Root Mean Square Ground Motion — $\sqrt{\dfrac{1}{t_{95}-t_{05}} \displaystyle\int_{t_{05}}^{t_{95}} x^2(t)\,dt}$
Root Square Ground Motion	—	$\sqrt{\displaystyle\int x^2(t)\,dt}$

***Note: $x(t)$=Ground Acceleration/Velocity/Displacement

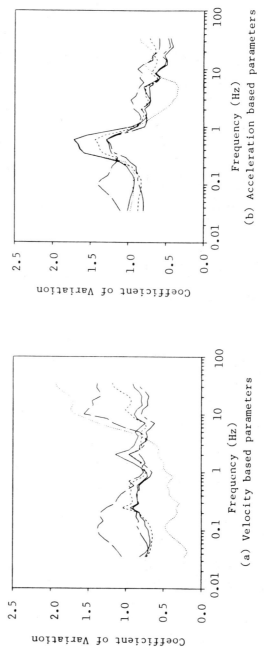

Fig. 3 Coefficients of Variation for Imparted Energy Spectra of Ductility 5, Normalized by Peak (———), Integral of Squared (·······), Root Square (— —), Mean Square (— - —), and Root Mean Square (— - - —) Ground Motions.

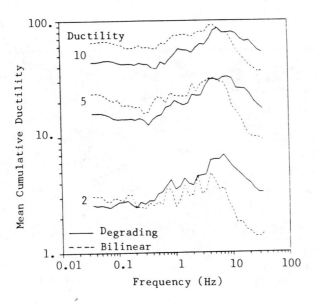

Fig. 4 Mean Cumulative Ductility

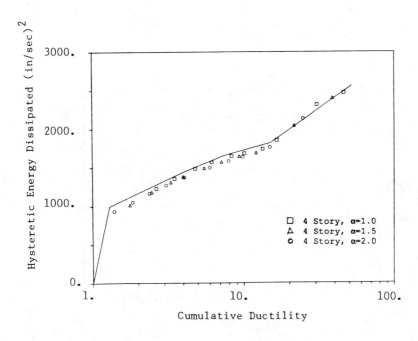

Fig. 5 Hysteretic Energy Dissipated per unit mass for SDOF (solid line) and MDOF systems subjected to the San Fernando Earthquake

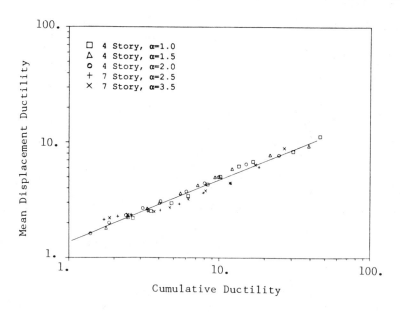

Fig. 6   Mean Displacement Ductility for MDOF systems subjected to the San Fernando Earthquake

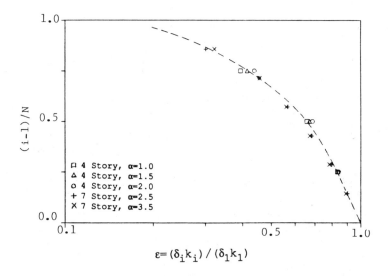

Fig. 7   Distribution of ε for equal Cumulative Ductility

EARTHQUAKE RESPONSES OF MULTISTORY BUILDINGS
UNDER STOCHASTIC BIAXIAL GROUND MOTIONS

Y.J.Park[I] and A.M. Reinhorn[II]

ABSTRACT

A random vibration method is developed for the response analysis of multistory buildings under stochastic bi-directional earthquake motions. The biaxial hysteretic restoring force is modeled by coupled nonlinear differential equations; the reponse statistics are obtained using the equivalent linearizaton method. The complete solution for SDOF systems and the basic formulation for MDOF systems are presented. The validity of the proposed hysteretic model is appraised using available biaxial loading test results of reinforced concrete columns.

INTRODUCTION

Experience from past earthquakes, such as the 1979 Imperial Valley earthquake and the recent 1985 Mexico earthquake, have shown low-rise and medium-rise reinforced concrete structures, designed according to modern seismic code provisions, would sustain sever structural damages, or even collapse [13, 17]. It's been pointed out that the biaxial loading effect and the torsional oscillations due primarily to the eccentric planning of shear walls may contribute largely to such a poor performance of buildings under strong earthquake motions [1,13].

Analytical as well as experimental studies show that the biaxial structural interaction significantly affects the inelastic response and energy absorbing characteristics of reinforced concrete structures [3, 6, 9, 15]. Since large scatter is invaribly associated with biaxial response results, no quantitative conclusion has been obtained regarding the effect of biaxial interaction on the earthquake resistance of structures. This scatter may be attributed largely to the inherent randomness of the ground motion, both in intensity and direction. The problem may require random vibration approaches. Such a method of analysis is presented herein to determine the response statistics and examine the effect of biaxial interaction of structural systems subjected to two-dimensional stochastic earthquake excitations.

I. Visiting Assistant Professor, Department of Civil Engineering, State University of New York, Buffalo, New York.

II. Assistant Professor, Department of Civil Engineering, State University of New York, Buffalo, New York.

BIAXIAL HYSTERETIC MODEL

Following the formulatin of a SDOF system under uniaxial excitiation [16], the equation of motion of a SDOF system subjected to two-dimensional excitations may be written as

$$[m] \begin{Bmatrix} \ddot{u}_x \\ \ddot{u}_y \end{Bmatrix} + [c] \begin{Bmatrix} \dot{u}_x \\ \dot{u}_y \end{Bmatrix} + \begin{Bmatrix} q_x \\ q_y \end{Bmatrix} = \begin{Bmatrix} f_x \\ f_y \end{Bmatrix} \qquad (1)$$

in which [m] and [c] are the mass and viscous damping matrices; u_x and u_y are the displacement, q_x and q_y are the restoring forces, and f_x and f_y are the excitations in the x and y directions, respectively. The restoring froces my be expressed as

$$\begin{Bmatrix} q_x \\ q_x \end{Bmatrix} = \alpha K \begin{Bmatrix} u_x \\ u_y \end{Bmatrix} + (1-\alpha) K \begin{Bmatrix} Z_x \\ Z_y \end{Bmatrix} \qquad (2)$$

in which K is the initial stiffness, α is the post-yielding stiffness ratio, and Z_x and Z_y represent the hysteretic components of restoring forces. For isotropic hysteretic restoring forces, Z_x and Z_y satisfy the following coupled differential equations:

$$Z_x = A\dot{u}_x - \beta|\dot{u}_x Z_x| Z_x - \gamma \dot{u}_x Z_x^2 - \beta|\dot{u}_y Z_y| Z_x - \gamma \dot{u}_y Z_x Z_y \qquad (3)$$

$$Z_y = A\dot{u}_y - \beta|\dot{u}_y Z_y| Z_y - \gamma \dot{u}_y Z_y^2 - \beta|\dot{u}_x Z_x| Z_y - \gamma \dot{u}_x Z_x Z_y \qquad (4)$$

The above formulation is an extension of the one-dimensional hystereitc model [16]. A general two-dimensional force-displacement relationship is illustrated in Fig. 1, where the displacement trajectory is a response result of 2-D white noise-type excitations. The value of the parameters used in Fig. 1 are A=1, α=0.05 and β=γ.

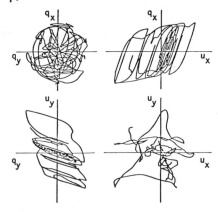

Figure 1 Hysteretic Behavior under White-Noise Excitation

For an orthotropic system, whose stiffness and strength in the two orthogonal directions are different, the following transformation will reduce it to the isotropic case of Eqs. 3 and 4:

$$q_y' = \frac{Q_x}{Q_y} q_y \quad , \quad u_y' = \frac{K_y Q_x}{K_x Q_y} u_y \qquad (5)$$

in which K_x and K_y are the preyield stiffness, and Q_x and Q_y are the yield strength in the x and y directions, respectively. Note that the yield surface (in terms of q_x and q_x) of an orthotropic system may take an elliptic form.

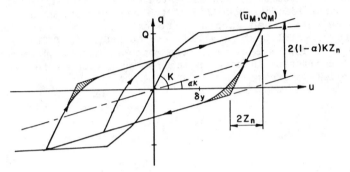

Figure 2 Uni-axial Hysteretic Model for R/C Structures

The biaxial hysteretic behavior of ductile systems, such as steel frame structures, can be modeled by properly selecting the parameters A, K, α, β, and γ. A modification to the foregoing model is necessary for reproducing the deteriorating hysteretic behavior of reinforced concrete structures. Figure 2 illustrates the load-deformation relatonship of the hysteretic model for reinforced concrete structures under uniaxial loading reversals. By specifying the hysteretic parameters A, α, β and γ to be time-dependent functions, the deterioration in both stiffness and strength is obtained as,

$$A = \frac{1}{\mu} \quad , \quad \alpha = \frac{\lambda^2 - \lambda - e\lambda}{\lambda - 1 + e\lambda} \quad , \quad \beta = \gamma = \frac{C_p}{2\mu} \left\{ \frac{1-\alpha}{u_m(\lambda-\alpha)} \right\}^2 \qquad (6)$$

$$C_p = 1 + \frac{\tau}{\delta_y^2} \int_0^t \{E[Z_x \dot{u}_x] + E[Z_y \dot{u}_y]\} dt \qquad (7)$$

where

λ = Q_m/Ku_m, restoring force deteriorating ratio;
e = $E_c(u_m)/(4u_m Q_m)$, energy ratio;
μ = two-dimensional ductility factor (to be determined);
$E_c(\delta)$ = energy dissipation per cycle;
u_m = mean maximum deformation;
Q_m = restoring force amplitude; and
δ_y = yield deformation.

Equation 7 specifies the strength deterioration under repeated loading reversals as a function of dissipated hysteretic energy; the integrand of Eq. 7 gives the mean rate of hysteretic energy and can be obtained from the response covariance matrix. Based on a number of laboratory test data of reinforced concrete components [14], the value of τ may range from 0.1 for brittle components to nearly zero for ductile components.

The above biaxial hysteretic model is compared with experimental data of biaxially loaded reinforced concrete cantilever columns conducted at the University of Tokyo [3]. The test specimen is of rectangular cross-section with 20cm×20cm dimensions, having a shear span ratio of 3.0, reinforced with 4-⌀13 (mm) deformed bars, and subjected to two lateral forces (x and y) and a constant axial load of about 25% of the concrete strength. The parameters e and τ were determined according to the empirical relationship [10] and available uniaxial test data [14], to be $e = 0.06\mu + 0.007$ and $\tau = 0.03$. The comparison of Fig. 4 shows that the proposed model can reproduce the rather complicated interaction between the restoring forces in the two directions under biaxial loadings.

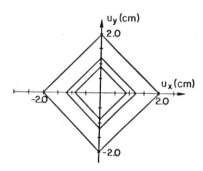

Figure 3 Deformation Path for Experiment

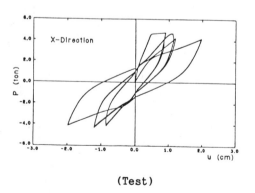

(Test)

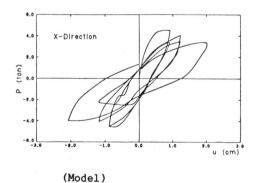

(Model)

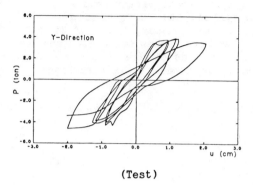 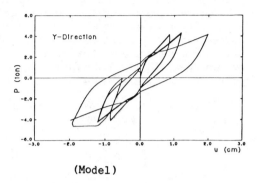

(Test)   (Model)

Figure 4   Biaxial Test of R/C Column

RESPONSE ANALYSIS OF SDOF SYSTEMS

The comprehensive description and modeling of multidimensional earthquake ground motions is available in [8] including the evolution of the frequency content. In this study, each of the two orthogonal components of the horizontal ground motion may be modeled as a filtered shot-noise with a prescribed power spectral density, i.e., the Kanai-Tajimi spectrum, and the Amin-Ang type intensity function [2]. The cross-correlation between the two orthogonal components may be described by the coherence function,

$$\gamma_{xy}(w) = \frac{S_{xy}(w)}{S_{xx}(w)S_{yy}(w)} \qquad (8)$$

in which $S_{xy}(w)$ is the cross spectral density. It may be observed that when the same power density and intensity functions are used in the two orthogonal components, the above coherence function becomes frequency independent and is equal to the correlation coefficient. Such a simplified case is assumed throughout this paper.

Using the equivalent linearization procedure [4], the coupled nonlinear differential equations (Eqs. 3 and 4) can be linearized. Subsequently, the rms response statistics are obtained in terms of the covariance matrix, S; the numerical procedure for this purpose is found in [16].

For evaluating structural performance and safety, the extreme responses of a structure are of primary interest. Under two-dimensional excitation, the problem can be described in terms the probability of the displacement vector, p(t), not outcrossing a safe domain boundary, Γ, over a given duration, t;

$$p(t) \geq 1 - \int_0^t V_D(\Gamma;t)d\tau \qquad (9)$$

in which $V_D(\Gamma;t)$ is the mean outcrossing rate of a random vector from a two-dimensional boundary, Γ. For an istropic system, this safe domain is a circle; whereas for general orthotropic systems, the transformation of Eq. 5 is necessary to obtain a circular boundary. Therefore, under two-dimensional inelastic responses, the ratio of the radius of the circle to the yield displacement in the transformed space (which is the same in all directions) is called the two-dimensional ductility factor.

According to Belyaev [5], the mean outcrossing rate of a random vector process, $\{u\}$, is given by

$$V_D(\Gamma;t) = \int_\Gamma d\Gamma \int_{\dot{u}_n>0} P(\{u_\Gamma\},\{\dot{u}\};t)\dot{u}_n \, d\{\dot{u}\} \qquad (10)$$

in which, $\dot{u}_n = \{n\}\{\dot{u}\}$, $\{n\}$ = unit vector normal to the boundary; $\{u_\Gamma\}$ = the vector $\{u\}$ at the boundary Γ; and $P(\{u\},\{\dot{u}\};t)$ = the joint density function of $\{u\}$ and $\{\dot{u}\}$. So far, the solution to Eq. 10 is available only if the response processes are Gaussian. For implementing the integration of Eq. 10, it is convenient to reduce the original vectors to the following standard form:

$$\{u\} = N(\{0\},[I]) \qquad \{\dot{u}\} = N(\{0\},[V_{xx}]) \qquad (11)$$

in which $N(\{\mu\},[V])$ denotes the Gaussian process with mean vector $\{\mu\}$ and covariance matrix $[V]$. Accordingly, the orginally circular boundary is transformed to an ellipse as shown in Fig. 5, in which $\tan 2\emptyset = 2\sigma u_x u_y/(\sigma u_x^2 - \sigma u_y^2)$ Using a polar coordinate, the mean outcrossing rate can be obtained in an integral form as

$$V_D(r) = \frac{r}{2\pi} \int_0^{2\pi} \frac{f_1(\theta) f_3(\theta)}{f_2(\theta)} \exp\{-\frac{r^2}{2} f_3(\theta)\} [w\bar{\Phi}(w) + \emptyset(w)] d\theta \qquad (12)$$

where:

$f_1(\theta) = \sigma_1^2(\sigma_{\dot{x}}^2 - \sigma_{\dot{x}\dot{y}}^2)\cos^2\theta + \sigma_2^2(\sigma_{\dot{y}}^2 - \sigma_{\dot{x}\dot{y}}^2)\sin^2\theta + 2\sigma_1\sigma_2\sigma_{\dot{x}\dot{y}}\cos\theta\sin\theta$;

$f_2(\theta) = \sigma_1^2\cos^2\theta + \sigma_2\sin^2\theta$; $f_3(\theta) = \cos^2\theta/\sigma_1^2 + \sin^2\theta/\sigma_2^2$; $w = r\, f_1(\theta)$;

$\bar{\Phi}(\cdot)$ = standardized Gaussian distribution function; and

$\emptyset(\cdot)$ = standardized Gaussian density function.

In the above, $\sigma_{\dot{x}}$, $\sigma_{\dot{y}}$ and $\sigma_{\dot{x}\dot{y}}$ are the rms statistics in the transformed space corresponding to $\sigma\dot{u}_x$, $\sigma\dot{u}_y$ and $\sigma\dot{u}_x\dot{u}_y$, respectively. A comparison of the above Eq. 12 with the Monte Carlo simulations (50 samples) for white-noise stationary excitations (uncorrelated) is shown in Fig. 6. In the figures, Q_x and Q_y are the yield strength, and σ_{gx} and σ_{gy} are the rms ground motion intensities in the x and y directions, respectively. The result shows that the analysis gives higher estimates of the crossing rate at low threshold levels; however, the crossing rates at higher threshold levels, which are of more interest in the study of structural performance and safety, are accurately predicted. Based on the above crossing rates, the mean maximum response is also determined empirically [12].

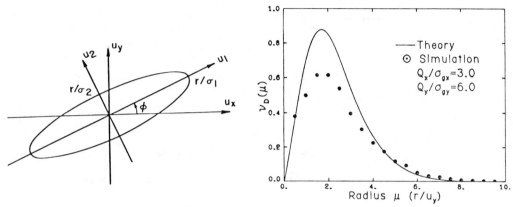

Figure 5   Transformation of Boundary   Figure 6   Mean Outcrossing Rate

MODELING OF MDOF SYSTEMS

The above biaxial hysteretic model can be extended for the analysis of MDOF systems. The problem may be classified into three categories: (A) the biaxial response of symmetric structures subjected to two-directional base motions; (B) the torsional response of asymmetric structures subjected to uni-directional base motions; and (C) the general biaxial response of asymmetric structures.

Using the conventional shear-beam model, the application to the problem - A is strightforward. The restoring forces of each story in two orthogonal directions can be determined through superposition; such a modeling technique is described in detail in [11].

A similar superposition technique can be used in the application to the problem - B. Two degree of freedom, i.e., lateral and rotational interstory deformations, are assigned to each story. A preliminary static analysis is needed to determine the inelastic behavior of rotational spring on each story. A monotonically increasing rotational force may be applied about the center of mass of each story consisting of components having various yield surfaces. The foregoing Eqs. 3 and 4 are interpreted as the coupled equations of the translational and rotational components (after the transformation of Eq. 5). Such a approximate modeling is under progress at the SUNY at Buffalo.

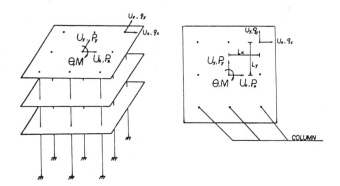

Figure 7 Modeling of MDOF Systems

The formulation for more general cases, i.e., the problem - C, is described herein in some detail.

According to Fig. 7, the restoring forces and displacements, $\{q_x, q_y\}$ and $\{u_x, u_y\}$ are related to those of each story, $\{P_x, P_y, M\}$ and $\{U_x, U_y, \theta\}$, as

$$\begin{Bmatrix} P_x \\ P_y \\ M \end{Bmatrix} = \begin{bmatrix} 1 & 0 \\ 0 & 1 \\ l_y & -l_x \end{bmatrix} \begin{Bmatrix} q_x \\ q_y \end{Bmatrix} \quad \text{and} \quad \begin{Bmatrix} u_x \\ u_y \end{Bmatrix} = \begin{bmatrix} 1 & 0 & l_y \\ 0 & 1 & -l_x \end{bmatrix} \begin{Bmatrix} U_x \\ U_y \\ \theta \end{Bmatrix} \quad (13)$$

Therefore, the restoring forces for each story are expressed as,

$$P_x = (\Sigma a_i Kx_i) U_x + (\Sigma a_i Kx_i l y_i)\theta + \Sigma(1-a_i)Kx_i Zx_i \tag{14}$$

$$P_y = (\Sigma a_i Ky_i) U_y - (\Sigma a_i Ky_i l x_i)\theta + \Sigma \frac{Qy_i}{Qx_i}(1-a_i)kx_i Zy_i \tag{15}$$

$$M = (\Sigma a_i Kx_i l y_i) U_x - (\Sigma a_i Ky_i l x_i) U_x + \{\Sigma a_i (Kx_i l y_i^2 + Ky_i l x_i^2)\}\theta$$
$$+ \Sigma(1-a_i)Kx_i l y_i Zx_i - \Sigma \frac{Qy_i}{Qx_i}(1-a_i)Kx_i l x_i Zy_i \tag{16}$$

in which, Kx_i and Ky_i are the stiffnesses, Zx_i and Zy_i are the strength of the i-the vertical components in x and y directions, respectively. It should be noted that the Z-variables for all the components are still included in the formulation.

A PARAMETRIC STUDY

A parametric study is performed for both nondeteriorating (ductile steel) and deteriorating (reinforced concrete) structures. Only the result of SDOF models is presented herein due to the limitation of space. For the excitations modeled as a filtered shot-noise process, the filter parameters $w_G = 5\pi$ and $\xi_G = 0.5$ are assumed for the Kanai-Tajimi filter, and the strong-phase motion duration is determined to be 15 seconds. Structures are modeled as SDOF systems having a natural period of $T_o = 0.3$ sec. (representing low-rise buildings), and a viscous damping coefficient of $\xi_o = 0.05$. The system shown in Fig. 1 is used to represent a 'nondeteriorating system', whereas for a 'deteriorating system', the foregoing reinforced concrete model shown in Fig. 4 is used.

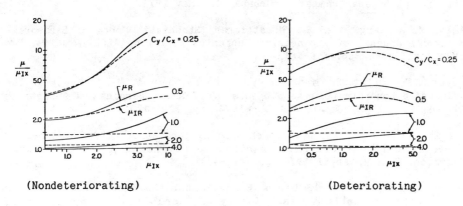

(Nondeteriorating)   (Deteriorating)

Figure 8 Biaxial Response Results

The results are summarized in Fig. 8, in which the two-dimensional ductility factor, μ_R, is expressed as the ratio to the one-dimensional ductility factor under unaxial excitation (x-component), μ_{IX}. In the figure, a comparison is made with the following approximate solutions, μ_{IR}:

$$\mu_{IR} = \sqrt{\mu_{IX}^2 + \mu_{IY}^2} \tag{17}$$

in which μ_{IX} and μ_{IY} are the corresponding uniaxial ductility factors evaluated individually in the X and Y directions. The parameter considered in this study is the relative yield strength ratio of the y-component to that of the x-component C_y/C_x, in which $C_x = Q_x/\alpha_{gx}$ and $C_y = Q_y/\alpha_{gy}$ as used earlier in Fig. 6. The main conclusions of this parameteric study may be summarized as follows:

(i) The effect of biaxial hysteretic interaction is more prominant for deteriorating systems.

(ii) For nondeteriorating systems, the biaxial response may be approximated using Eq. 17 for moderately nonlinear systems (e.g., $\mu_{IX} < 4.0$).

SUMMARY AND CONCLUSIONS

A random vibration method is presented to evaluate the stochastic response statistics of various hysteretic systems subjected to bi-directional earthquake motions. The proposed model is versatile and efficient in numerical treatments, and represent a wide variety of MDOF structural systems under random bi-directional motions.

REFERENCES

[1] Aktan, A.E., Pecknold, D.A.W., and Sozen, M.A., 'Effects of Two-Dimensional Earthquake Motion on a Reinforced Concrete Columns', Civil Engineering Studies, Structural Research Series No. 399, University of Illinois at Urbana-Champaign, Urbana, IL, May 1973.

[2] Amin, M. and Ang, A. H-S., 'Nonsttionary Stochastic Model of Earthquake Motions', Journal of Engineering Mechanics Division, ASCE, Vol. 94, No. EM2, April 1968, pp. 559-583.

[3] Aoyama, H., Fujii, S., Minamino, H., and Yoshimura, M., 'A Study on The Reinforced Concrete Columns Subjected to Biaxial Bending', Transactions AIJ, Extra, Tokyo, Japan, October 1974, pp. 1293-1296.

[4] Atalik, T.S., and Utku, S., 'Stochastic Linearization of Multidegree of Freedom Nonlinear Systems', Earthquake Engineering and Structural Dynamics, Vol. 4, April 1976, pp. 411-420.

[5] Belyaev, Y.K., 'On the Number of Exits Across the Boundary of a Region by a Vector Stochastic Process', Theory of Probabilistic Applications, Vol. 13, 1968, pp. 320-324.

[6] Jirsa, J.O., Maruyama, K., and Ramirez, H., 'The Influence of Load History on the Shear Behavior of Short RC Columns', Proceedings, Sixth World Conference on Earthquake Engineering, New Delhi, India, 1977, pp. 339-346.

[7] Kau, C.L. and Chopra, A.K., 'Effects of Torsional Coupling on Earthquake Forces in Buildings', Journal of the Structural Division, ASCE, Vol. 103, No. ST4, April 1977, pp. 805-819.

[8] Kubo, T. and Penzien, J., 'Time and Frequency Domain Analyses of Three Dimensional Ground Motions, San Fernando Earthquake', Earthquake Engineering Research Center, Report No. EERC 76-6, Univeristy of California, Berkeley, CA, March 1976.

[9] Otani, S., Cheung, V. W-T., and Lai, S.S., 'Reinforced Concrete Columns Subjected to Biaxial Lateral Load Reversals', Proceedings Sixth World Conference on Earthquake Engineering, New Delhi, India, 1977, pp. 525-532.

[10] Park, Y.J. and Ang, A. H-S., 'Mechanistic Seismic Damage Model for Reinforced Concrete', Journal of Structural Engineering, ASCE, Vol. 111, No. 4, April 1985, pp. 722-739.

[11] Park, Y.J., Ang, A. H-S., and Wen, Y.K., 'Seismic Damage Analysis of Reinofrced Concrete Buildings', Journal of Structural Engineering, ASCE, Vol. III, No. 4, April 1985, pp. 740-758.

[12] Park. Y.J., Wen, Y.K., and Ang, A. H-S., 'Random Vibration of Hysteretic Systems under Bi-Directional Ground Motions', Earthquake Engineering and Structural Dynamics, Accepted for Publication (in print), Sept. 1985.

[13] Rosenblueth, E., 'The Mexican Earthquake: A Firsthand Report', Journal of Civil Engineering, Jan. 1986.

[14] Short Column Committee of Japan, 'The Comprehensive Research for the Prevention of Failure in Short Reinforced Concrete Columns', Series 1 to 62, Annual Conference of AIJ, Japan, 1973-77.

[15] Takizawa, H. and Aoyama, H., 'Biaxial Effects in Modeling Earthquake Reponse of R/C Structures', Earthquake Engineering and Structural Dynamics, Vol. 4, 1976, pp. 523-552.

[16] Wen, Y.K., 'Equivalent Linearization for Hysteretic Systems under Random Excitations', Journal of Applied Mechanics, Transactions ASME, Vol. 47, No. 1, March 1980, pp. 150-154.

[17] Wosser, T.D., et al., 'Damage to Engineering Structures in California, The Imperial Valley, California, Earthquake of October 15, 1979' Geological Survey Professional Paper 1254, United States Government Printing Office, Washington, 1982, pp. 273-288.

SIMPLIFIED PROCEDURES FOR EARTHQUAKE ANALYSIS OF BUILDINGS

A. K. Chopra[I] and E. F. Cruz[II]

ABSTRACT

Building code formulas for earthquake forces are evaluated in light of the results of dynamic analysis of buildings. After summarizing a simplified response spectrum analysis procedure for buildings, a hierarchy of four analysis procedures available to the building designer is identified.

INTRODUCTION

While dynamic analysis procedures have been available for many years, the earthquake forces considered in the design of most buildings are computed by the Equivalent Lateral Force Method specified in the governing building code. Formulas for base shear, heightwise distribution of lateral forces, and computation of overturning moments are the key elements of this procedure. Such formulas contained in the three design documents--the Uniform Building Code, ATC-3 design provisions, and Mexico's Federal District Building Code--are evaluated here in light of the results of dynamic analysis of buildings. It is demonstrated that these formulas do not properly recognize the effects of some of the most significant building parameters. An improved code-type analysis procedure which recognizes these parameters is outlined. For those situations where the present or improved code procedures may not be accurate enough for preliminary design, a simplified version of the response spectrum analysis is summarized.

ANALYSIS PROCEDURES

Building Code Analysis

The principal procedure to estimate earthquake forces specified in the Uniform Building Code [1], ATC-3 design recommendations [2], and Mexico's Federal District Code [3] is an Equilvalent Lateral force procedure. Based on an estimate of the fundamental vibration period, formulas are specified in these building "codes" for the base shear and distribution of lateral forces over the height of the building. The design shears and moments for the various stories of the building are determined from static analysis of

---

I. Professor, Department of Civil Engineering, University of California, Berkeley, California.

II. Assistant Professor, Department of Structural Engineering, Universidad Catolica de Chile, Santiago, Chile.

the building subjected to the lateral forces, with some codes permitting reductions in the resulting story moments.

The formula for the design base shear in the above-mentioned building codes and design recommendations can be expressed as $V_0 = CW$ where W is the total weight and the seismic coefficient C depends on the fundamental vibration period T. The seismic coefficients displayed in Fig. 1 are for the Uniform Building Code (UBC), ATC-3 recommendations with $R = 1$, and Mexico's Federal District Code (MFDC) with $\mu = 1$. In order to account for effects of inelastic behavior, the seismic coefficient of Fig. 1 is divided by the reduction factor R or μ' in ATC-3 and MFDC, respectively; μ' is related to the allowable ductility factor μ. Each of the above design codes includes variations in the above formulas to account for soil conditions and importance of the structure, but these factors are not considered in the present evaluation.

Among the three building codes considered, only the MFDC explicity specifies the pseudo-acceleration design spectrum and recognizes that the base shear in buildings with fundamental vibration period larger than T_v, especially in "flexural" structures, exceeds the product of (S_a/g) and the total weight. The ATC-3 recommendations start with a design spectrum and raise its descending branch in the velocity- and displacement-controlled regions to decay at a slower rate with increasing T.

Except for the additional force F_t assigned to the top of the building, the UBC lateral forces are distributed over the height of the building under the assumption of linearly varying floor displacement in the fundamental mode shape. Assignment of an additional force F_t at the top of the building is intended by the code to roughly and implicitly account for the contributions of the higher vibration modes to building response.

The height-wise distribution of ATC-3 lateral forces is based on the assumption that the horizontal accelerations of floor masses are proportional to: (1) the elevation above ground for buildings with $T_1 \leq 0.5$ sec; (2) the square of this elevation for $T_1 \geq 2.5$ sec; and (3) an intermediate power of this elevation for intermediate values of T_1. These force distributions are intended to recognize the changing fundamental mode shape and increasing higher mode contributions to response with increasing T_1.

The base shear in the MFDC is separated into two parts which are distributed over the height, assuming that the accelerations of floor masses are proportional to their elevation above ground, and to the square of this elevation, respectively. The combined acceleration distribution passes smoothly from a straight line when $T = T_v$ to a parabola as T tends to infinity. This variation in acceleration distribution with fundamental period T is intended to recognize the changing fundamental mode shape and increasing higher mode contributions to response with increasing T [3].

The design shears for the various stories of the building are determined from static analysis of the building subjected to the lateral forces computed from the above equations. The story overturning moments determined from the lateral forces by methods of statics are multiplied by a reduction factor in ATC-3 and MFDC.

Response Spectrum Analysis

The maximum planar response of a multistory building associated with a specified earthquake design spectrum can be determined approximately by the standard response spectrum analysis (RSA) procedure ([4]: page 88). In this procedure the maximum response in each natural vibration mode of the building is determined directly from the earthquake design spectrum, and the modal maxima are combined in accordance with the SRSS formula. Because the periods of planar vibration of a building are typically well separated, the SRSS method provides results essentially identical to the CQC method.

Simplified Response Spectrum Analysis

Recognizing that the earthquake response of many buildings can be estimated by considering only the first two modes of vibration in the RSA procedure [5], a simplified response spectrum analysis (SRSA) procedure has been developed [6]. The simplification is achieved mainly in evaluating the natural frequencies and shapes of these two modes of vibration. The fundamental frequency and mode of vibration is determined by the Stodola method. Starting with the code design forces, this method has been presented in a form especially suited for multistory buildings. As is well known, the Stodola method can be modified to include a sweeping matrix to eliminate the first mode contribution in the deflected shape. With this modification introduced in each iteration cycle, the iterative process will converge to the second mode, leading to its vibration properties. However, the contributions of the second vibration mode to building response are relatively small compared to those of the fundamental mode. Thus, it seems unnecessary to compute the vibration properties of the second mode to a high degree of accuracy. Therefore, we avoid the Stodola method with iteration in computing the vibration properties of the second mode. Instead, a simple procedure was developed which directly--without iteration--provides a good approximation of the second vibration mode. In particular, the mode shape is determined as the static deflected shape due to the effective earthquake forces minus their first modal component, and the frequency by Rayleigh's formula.

SYSTEMS AND DESIGN SPECTRUM

Systems Considered

The rectangular plane frames analyzed in this study are idealized as single-bay, moment-resisting plane frames with constant story height = h, and bay width = 2h; the five frames considered are described in [6]. Only flexural deformations are considered in the members which are assumed to be prismatic. The modulus of elasticity E is the same for all members, but the moments of inertia of beams I_b and columns I_c--same for both columns in any story--may vary over the height with the ratio of the two the same in all stories. The mass of the structure is assumed to be concentrated at the floor levels and the rotational inertia is neglected. The damping ratio for all the natural modes of vibration is assumed to be 5 percent.

Each building frame is completely characterized by two additional parameters: the period of the fundamental mode of vibration T_1 and a stiffness ratio ρ [6]. For the one-bay frames considered in this study, this parameter reduces to $\rho = I_b/4I_c$, and it has the same value for all stories. This parameter is a measure of the relative beam-to-column stiffness and hence indicates the degree-of-frame action. The extreme values of ρ, 0 and ∞ represent the following limiting cases of a frame respectively: vertical cantilever with the beams imposing no restraint to joint rotations; and a shear building in which the joint rotations are completely restrained and deformations occur only through double curvature bending of columns. An intermediate value of ρ represents a frame in which beams and columns undergo bending deformations with joint rotation.

Earthquake Design Spectrum

The earthquake excitation is characterized by the smooth design spectrum of Fig. 4 which is constructed by well known procedures [7] for ground motions with maximum acceleration $\bar{a}_g$, velocity $\bar{v}_g$, and displacement $\bar{u}_g$ equal to 1g, 48 in./sec, and 36 in., respectively. The spectrum can thus be divided, as shown in Fig. 2, into acceleration-controlled, velocity-controlled, and displacement-controlled regions. The spectral acceleration is constant in part of the acceleration-controlled region, varies as $1/T$ in the velocity-controlled region, and as $1/T^2$ in the displacement-controlled region. Amplification factors for these regions were selected from [7] for 84.1 percentile response and 5% damping ratio to construct the spectrum.

EVALUATION OF THE SRSA METHOD

The maximum response, computed by the RSA procedure--wherein the contribution of all the natural vibration modes of the frame are included-- and by the SRSA procedure, is plotted against the fundamental vibration period of the frame in the form of response spectra. Such plots are presented in Part II of [8] for a uniform five-story frame for three values of $\rho = 0$, 0.125, and ∞ and six response quantities. For brevity only one of them, the base shear plot, is presented here--in dimensionless form as defined in Fig. 3, where $\bar{a}_g$ is the maximum ground acceleration and W_1^* is the effective weight for the first vibration mode of the building ([3]: page 84). The heightwise variation of story shears and story overturning moments computed by the RSA and SRSA methods were also presented in Part II of [8].

It is obvious from Fig. 3 (and other results presented in [6]) that the responses of the uniform five-story frame (Case 1) computed by the SRSA method are very close to those from the RSA method. The errors in the SRSA results tend to increase with increasing fundamental vibration period T_1, and with decreasing stiffness ratio ρ. This increase in error is closely related to the contributions of the vibration modes higher than the fundamental mode which increase with increase in T_1 and decrease in ρ [5]. The errors in the SRSA results tend to be larger in taller buildings and in buildings with nonuniform variation of mass or stiffness or both over height. The magnitude of errors vary over the height of the frame, being larger in the upper stories where the contributions of the higher vibration modes are shown to be more significant [5]. While the errors in the SRSA results depend on the response quantity and on the heightwise distribution of the mass and stiffness and on the height of the frame, the errors are all below 10 percent for all the frame cases studied over the entire range of ρ, provided the fundamental period T_1 is below the end of the velocity-controlled region of the spectrum.

As shown in [5], for buildings with fundamental vibration period T_1 within the acceleration-controlled region of the earthquake response spectrum, the fundamental mode alone provides essentially the same results as those obtained by including all the modes in the RSA method. Thus, in this range of vibration periods, only the first mode needs to be considered in the SRSA method.

EVALUATION OF BUILDING CODE ANALYSIS PROCEDURES

<u>Base Shear</u>

The maximum response associated with the selected design spectrum, computed by the RSA procedure--wherein the contribution of all the natural vibration modes of the frame are included--is plotted against the fundamental vibration period T_1 in the form of response spectra. Such a plot is presented in Fig. 4 for the base shear in the uniform five-story frame for three values of $\rho = 0$, 0.125, and ∞. The base shear is presented in dimensionless form, having been normalized with respect to the effective weight W_1^* participating in the first vibration mode of the building ([4]: page 84). Also presented is the base shear considering the contribution of only the fundamental mode of vibration, which in the normalized form of Fig. 4 is the same for all ρ values and is identical to the design spectrum of Fig. 2.

It is apparent from Fig. 4 that the normalized base shear for buildings with fundamental vibration period T_1 within the acceleration-controlled region of the spectrum is essentially identical to the contribution of only the fundamental vibration mode. However, for buildings with T_1 in the velocity- or the displacement-controlled regions of the spectrum, the response contributions of the vibration modes higher than the fundamental mode can be significant. They increase with increasing T_1 and decreasing ρ for reasons discussed elsewhere [8].

If the seismic coefficient C in building codes was defined as S_{a1}/g (the pseudoacceleration ordinate at T_1 normalized with respect to the acceleration of gravity), the code formula $V_0 = CW$ would accurately predict the base shear for buildings with T_1 within the acceleration-controlled region of the spectrum, provided the effective weight W_1^* was used instead of the total weight W in computing the code shear. However, the base shear formula in building codes is based on the total weight W, which obviously is always larger than W_1^*, resulting in a larger base shear. This is confirmed by replotting the results of Fig. 4 in the form of a seismic coefficient spectrum as shown in Fig. 5 wherein the base shear is normalized with respect to the total weight. Within the acceleration-controlled region of the spectrum, for buildings with the same total weight, the base shear decreases with decreasing ρ because W_1^* decreases with ρ, and the code value for base shear, $V_0 = CW$ with $C = S_{a1}/g$, exceeds the RSA value for all ρ values. However, in the velocity- or displacement-controlled regions the higher mode contributions can be significant enough for the RSA value of base shear to exceed this code value. The code formula is inadequate for longer period buildings because it does not properly recognize the contributions of higher vibration modes and their dependence on the building parameters T_1 and ρ.

In order to further evaluate the response behavior in this period range, the curve $\alpha T^{-\beta}$ is fitted to the normalized base shear response spectrum of Fig. 4. The parameters α and β, for each of the velocity- and displacement-controlled regions of the spectrum, are evaluated by a least-squared error fit. Comparison of the "exact" response spectra of Fig. 4 with the fitted curves (Fig. 6) indicates that the selected functions provide a reasonable approximation of the computed response.

Recalling that the difference between the normalized pseudoacceleration response spectrum and the normalized base shear represents the contributions of the vibration modes higher than the fundamental mode, it is apparent from Fig. 6 that these contributions can be represented approximately by raising the spectrum curve by changing the exponent $-\beta$ for T. The degree to which the spectrum needs to be raised for the velocity- and displacement-controlled spectral regions depends on the stiffness ratio ρ; the spectrum need be raised very little for shear buildings ($\rho = \infty$), but to an increasing degree with increasing frame action; i.e. decreasing ρ (Fig. 6). The spectral modifications also depend on the number of stories and mass and stiffness distributions of the building, as indicated by the fact that the exponent $-\beta$ varies with the frame height and heightwise variation of mass and stiffness [8].

Presented in Fig. 7 are the fitted curves for normalized base shear from Fig. 6, along with the seismic coefficients specified by the UBC and MFDC codes and ATC-3 design recommendations (Fig. 1). All the curves presented in Fig. 7 have been normalized to a unit maximum value. It is apparent that the seismic coefficient in building codes decreases with increase in period at a rate slower than demonstrated by dynamic analyses, but this does not necessarily imply that the codes are actually conservative. Furthermore, none of the codes recognize that the normalized base shear, as predicted by dynamic analysis, in the long-period range depends significantly on the stiffness ratio ρ. However, the MFDC recognizes that, for long-period buildings, the base shear computed from $V_0 = (S_a/g)W$ should be increased to recognize that, in general, the longer the fundamental period of vibration, the more important will flexural deformations tend to be relative to shear deformations and the more significant will the contributions of higher modes tend to be relative to the fundamental, but even the MFDC does not explicitly recognize that this increase in base shear depends not only on the fundamental period T_1, but also on the stiffness ratio ρ.

In summary, building codes attempt to account for the contributions of the higher modes of vibration to the base shear in a simple, empirical manner by increasing each of the two factors that are multiplied. The total weight W is used instead of the first mode effective weight W_1^*, and for long-period buildings the seismic coefficient used is increased above the design spectrum by raising its descending branch. Both of these concepts lead to the desired result of increasing the design base shear, but the increases are not handled rationally because their dependence on building parameters T_1 and ρ is not recognized.

Story Shears

The distribution of story shears over the height of the uniform five-story frame, computed by the RSA procedure including the contribution of all five modes of vibration, is presented in Fig. 8 for three values of ρ and four values of the fundamental period chosen to be representative of different period regions of the spectrum. The distribution of lateral forces, computed from the story shears of Fig. 8 as the differences between the shears in consecutive stories, is presented in Fig. 9. In a lumped mass system such as the frames considered here, the lateral forces are concentrated at the floor levels and the shear remains constant in each story with discontinuities at each floor. However, such plots of lateral forces and story shears would not be convenient in displaying the differences among various cases, and the alternative presentation of lateral forces and shears varying linearly over story height is used. Also presented in Figs. 8 and 9 is the distribution of lateral forces prescribed by the three building codes and the resulting story shears.

As indicated by this comparison and other results [8] not included here, for buildings within the acceleration-controlled region of the spectrum, the distribution of lateral forces and story shears specified by the three building codes is essentially identical and between the extremes predicted by RSA for $\rho = 0$ and ∞. With increasing fundamental vibration period T_1, the code distributions for lateral forces and story shears increasingly differ from the RSA results, especially for the smaller values of ρ because, under these conditions, the higher mode contributions become more significant [5]. For long-period buildings with T_1 in the velocity- or displacement-controlled regions, the higher mode contributions are pronounced enough to cause reversal of the curvature in the distribution of lateral forces which the code formulas do not recognize.

Overturning Moments

The differences in the overturning moments specified by building codes compared to the values obtained by the RSA procedure are much smaller than for shears because the higher mode contributions to the moments are less significant [8]. It is also shown that some reduction in the story moments relative to the statically computed values is justified in light of the results of dynamic analysis. Thus, no reduction at all as in UBC is inappropriate. However, even the other two codes considered do not recognize that the reduction factor depends significantly on the building parameters T_1 and ρ.

CONCLUDING REMARKS

Recognizing the limitations of present building code formulas identified in this paper, an improved procedure to estimate the earthquake forces for the initial, preliminary design of buildings has been developed ([8]: Part III). Starting with the earthquake design spectrum for elastic or inelastic design and the overall, general description of the proposed building, this procedure provides an indirect approach to estimate the response in the first two vibration modes of the building. The procedure recognizes the important influence of those building properties and parameters that significantly influence its earthquake response without requiring the computations inherent in standard response spectrum analysis.

With the development of the SRSA method a hierarchy of four analysis procedures to determine earthquake forces are available to the building designer. Listed in order of increasing complexity and improving accuracy, these procedures are: code-type procedure; SRSA--simplified response spectrum analysis; RSA--response spectrum analysis; and RHA--response history analysis. These four procedures should be considered in sequence proceeding no farther than the least complex method that leads to sufficiently accurate results. Criteria have been presented [6] to evaluate the accuracy of each procedure and to decide whether it is necessary to improve results by using the next procedure in the hierarchy. In particular, a procedure was developed [6] to evaluate the quality of results from a code-type analysis and, if necessary, to improve results by proceeding to the SRSA method. It was also shown that one such procedure, which was included in the ATC-3 seismic provisions, is conceptually deficient. It is believed that the SRSA method will provide results for earthquake-induced forces and deformations that are sufficiently accurate for the final design of many buildings. In all cases, it will provide the basis for a very good preliminary design. Thus, the SRSA method should be very useful in practical design applications because, although much simpler than the RSA method, it provides very similar estimates of design forces for many buildings.

ACKNOWLEDGEMENTS

The research reported here was supported by Grant Nos. CEE81-05790 and CEE84-02271 from the National Science Foundation, for which the writers are grateful.

REFERENCES

[1] International Conference of Building Officials, Uniform Building Code, 1982.

[2] Applied Technology Council, Tentative Provisions for the Development of Seismic Regulations for Buildings, Report ATC 3-06, NBS Special Publication 510, NSF Publication 78-08, June 1978.

[3] Rosenblueth, E., "Seismic Design Requirements in a Mexican 1976 Code," Earthquake Engineering and Structural Dynamics, Vol. 7, 1979, pp. 49-61.

[4] Chopra, A. K., Dynamics of Structures, A Primer, Earthquake Engineering Research Institute, Berkeley, California, 1982.

[5] Cruz, E. F. and A. K. Chopra, "Elastic Earthquake Response of Building Frames," Journal of Structural Engineering, ASCE, Vol. 112(3), 443-459, March 1986.

[6] Cruz, E. F. and A. K. Chopra, "Simplified Procedures for Earthquake Analysis of Buildings," Journal of Structural Engineering, ASCE, Vol. 112(3), 461-480, March 1986.

[7] Newmark, N. M. and W. J. Hall, "Vibration of Structures Induced by Ground Motion," Chapter 29, Part I, in Shock and Vibration Handbook, 2nd Ed., McGraw-Hill, 1976, pp. 29-1 to 29-19.

[8] Cruz, E. F. and A. K. Chopra, Simplified Methods of Analysis for Earthquake Resistant Design of Buildings, Report No. UCB/EERC-85/01, Earthquake Engineering Research Center, University of California, Berkeley, California, February 1985.

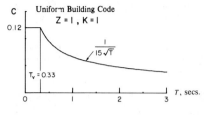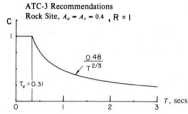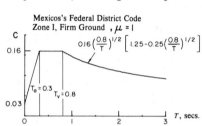

FIGURE 1 Seismic coefficient in building codes.

FIGURE 2 Normalized pseudo-acceleration design spectrum.

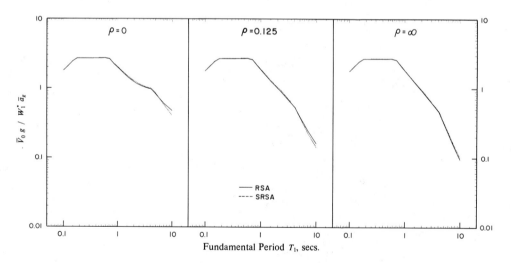

FIGURE 3 Comparison of maximum base shear values computed by RSA and SRSA methods. Results are for Case 1: uniform 5-story frame.

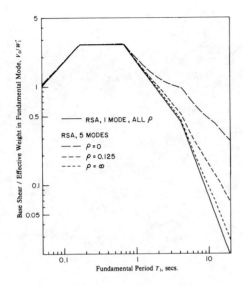

FIGURE 4 Normalized base shear in uniform five-story frame computed by response spectrum analysis (RSA) for three values of ρ.

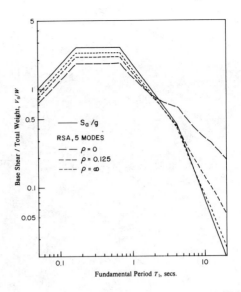

FIGURE 5 Seismic coefficient spectrum for uniform five-story frame computed by RSA for three values of ρ.

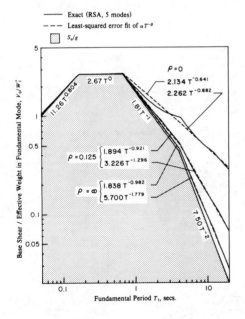

FIGURE 6 Least-squared error fit of functions $\alpha T^{-\beta}$ to the exact base shear response spectrum for long periods. Uniform five-story frame.

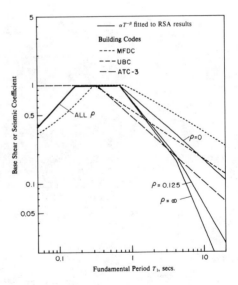

FIGURE 7 Comparison of seismic coefficient spectrum shapes in building codes with base shear spectrum shape computed by RSA for uniform five-story frame.

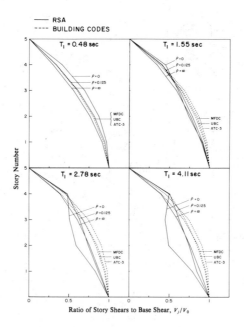

FIGURE 8 Comparison of story shear distributions for uniform five-story frame from building codes and RSA for four values of T_1 and three values of ρ.

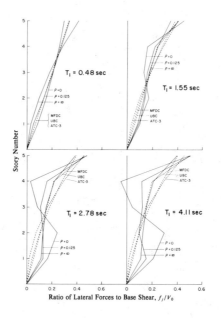

FIGURE 9 Comparison of equivalent lateral forces distributions for uniform five-story frame from building codes and RSA for four values of T_1 and three values of ρ.

INELASTIC ANALYSIS OF RC SECTIONS

A. E. Aktan[I] and G. E. Nelson[II]

ABSTRACT

An assessment of RC section analysis procedures to predict RC element stiffness was carried out. Due to many questionable idealizations and neglect of certain special features of particularly RC wall and beam-slab elements responses, commonly used procedures were observed to lead to significant discrepancies between predicted and measured responses. Modifications were suggested for more reliable and yet practical response prediction of RC sections.

INTRODUCTION

Recent experimental research indicated that the current response prediction methods for RC structures did not lead to results reasonably correlating with their measured counterparts [2,5,8]. Studies indicated that limitations in obtaining reliable estimates of cross sectional stiffness, strength and deformability played an important role in the lack of correlation between observed and predicted structural responses.

Reliable estimates of the supplies of stiffness, strength and energy dissipation of structural elements is considered a prerequisite for reliable seismic design [3,4]. The authors therefore initiated research which currently focuses on the development of reliable and practical tools for RC section analysis.

OBJECTIVES AND SCOPE

The first objective of the paper is to outline the shortcomings in common RC section analysis procedures which were identified as leading to errors in the predicted supplies of stiffness, strength and deformability of RC, and particularly, wall and beam-slab (diaphragm) elements. Some test data will be presented to support and exemplify the authors' viewpoints.

The second objective is to present a sensitivity study, investigating the most critical parameters affecting stiffness, strength and deformability of wall and beam-slab sections. Modifications were made to a recently

---

[I] Associate Professor of Civil Engineering, Louisiana State University, Baton Rouge, Louisiana.

[II] Resident Construction Engineer, Louisiana Department of Transportation and Development, Baton Rouge, Louisiana.

developed microcomputer program for section analysis [11] in an effort to make this program suitable for the analysis of wall and beam-slab sections. These will be described. Conclusions and recommendations for practical implementation will be offered.

EVALUATION OF CURRENT RC SECTION ANALYSIS PROCEDURES

Common Assumptions in Section Analysis

In case the flexural stiffness of an RC element cross section is required for response prediction, relations between its normal stress resultants, i.e., axial force and flexure, versus the corresponding strain resultants, i.e., the average axial strain and curvature, may be based on: (1) a description of geometric characteristics of the element's typical cross section followed by its discretization into fibers or filaments; (2) description of uniaxial stress-strain characteristics of the constituent materials; (3) description of the strain distribution within the cross section; and (4) relating the stress and strain resultants through equilibrium [1,12].

Such formulations based on discrete layer or filament approaches have been suggested in conjunction with uniaxial and biaxial bending, respectively. Although more sophisticated nonlinear analysis procedures based on generalized material constitutive relations and failure theories were subsequently developed for RC [9], the relatively simpler layer or filament models in conjunction with uniaxial (hypo-elastic) material response formulations remained popular due to their practicality.

Certain assumptions common to reported section analysis formulations based on the discrete layer or filament approach were assessed to limit the reliability of the outcome of the analyses, particularly in the case of deep wall and beam-slab sections. Some of these assumptions follow: (1) Only axial-flexural responses are considered, neither shear response nor any interaction between axial-flexure and shear are incorporated. (2) The postulated stress-strain responses of the materials are commonly based on the specified nominal characteristics for these materials. The responses of the actual materials in the "as constructed" structure may be substantially different and will change with time [7]. Even if test results of the actual materials may be available, these are normally based on standard uniaxial tests, and the responses observed from standard tests may not be expected to reflect the responses of the material under the spectrum of stress states, levels, rates and histories which may occur in the critical regions of the structure [7]. (3) Although many uniaxial formulations have been suggested for confined concrete, the applicability of these to all possible types and ranges of confinement is questionable. In the case of wall panels, biaxial formulations are required. Reliable test data incorporating biaxial stress states which typically occur in RC wall panels subjected to high compressive and shear stresses is very limited and has only recently been obtained [13]. (4) Only linear strain profiles are considered along the depth of sections, irrespective of the elements' aspect ratio. (5) Strain distribution along the width of sections are assumed to be constant, irrespective of the proportions of the section. (6) Perfect bond is assumed. (7) Initial stresses, strains, cracking and slip which may occur due to construction or curing effects are commonly neglected.

Due to the above assumptions and idealizations, many errors and uncertainties were introduced in the results of RC wall and beam-slab analyses. Some of these are discussed below.

Test Program to Identify RC Wall Section Stiffnesses

A test program which was carried out at U. C. Berkeley to measure serviceability limit state stiffness characteristics of RC wall elements was particularly helpful in identifying limitations of the RC section analysis procedures [5].

The test specimens were 1/3-scale 4-story RC cantilever walls which were subjected to various combinations of axial force, shear and flexure at the serviceability limit states. The dimensions and reinforcement detailing of the specimen's barbell cross section are given in Fig. 1. The story height to section depth aspect ratio of the wall was 0.5.

The wall specimens were cured while they were exposed to normal environmental conditions for a year before the tests commenced. During this time, reinforcement strains due to shrinkage of concrete and differential shrinkage between panel and edge column concrete were monitored. Strains of the order 0.0006 in./in. were recorded at the end of the curing period. This indicated widespread microcracking and narrow cracking, some of which could even be observed by the naked eye. Several additional "relaxation" cracks were observed after the specimens were exercised by applying and removing compressive axial forces up to 30 percent of the balanced axial force.

The lateral force-first floor level displacement responses measured during a typical test is shown in Fig. 2. The specimen was subjected to a constant axial compression of 30 percent of the balanced force, and a base moment to shear ratio of 4.35 times story height (L), during this test. Two cycles of lateral load was applied, with a maximum level corresponding to the wall base shear due to the seismic "E" force prescribed by 1982 UBC for the 15-story parent structure of the wall [2,5].

The initial loading and unloading segments of the force-displacement responses were not reliable due to limitations in the resolution of instruments. Only the reloading segments were therefore used to extract stiffness. The responses were first corrected for the measured rigid-body rotation and sliding of the foundations. The lateral force versus displacement and rotation at the first floor level, with respect to the foundation, were obtained as illustrated in insert (a) of Fig. 3. The lateral distortion, u, was assumed to consist of contributions from flexural and shear components.

The shear component of the lateral distortion, u_s, was evaluated by measuring the changes in the length of the diagonals of the element, as illustrated in insert (b) of Fig. 3. The remaining component of the lateral distortion was then attributed to the flexural deformation. The shear and flexural components of wall deformation, as well as the contribution of foundation displacements to the total displacement are shown in Fig. 3 for both lateral load directions.

Assuming the typical one-dimensional element flexibility, the total lateral displacement u was related to shear V and moment M as:

$$u = [L^3/(3\ EI) + L/(GA_v)]\ V + [L^2/(2\ EI)]\ M \qquad (1)$$

where L is the story height (Fig. 3). It was thus possible to identify the effective flexural stiffness (EI) and shear stiffness (GA_v) of the section by measuring V, M, u and u_s, and by relating the shear and flexural distortions to their analytical counterparts.

It is important to note, however, that the measured ratio of shear deformation to the total deformation as shown in Fig. 3 indicated that it may be irrational to use the one-dimensional element flexibility to identify section stiffnesses. Expression (1) can be rationally applied only when the term $L/(GA_v)$ is small when compared to the term $L^3/(3\ EI)$, otherwise warping of the section may dominate the strain profile, rendering the one-dimensional idealization in conjunction with "plane sections remain plane" assumption questionable. For the first story of the test specimen, when nominal values of GA_v and EI were used, the ratio of the term $L/(GA_v)$ to $L^3/(3\ EI)$ was over 5. Consequently, modeling each story of the wall as a one-dimensional element was not as suitable as for example a microscopic two-dimensional finite-element discretization.

A significant growth of wall cross sections along their plane, due to "bulging" under compressive stresses, as well as horizontal dilations because of microcracks and narrow cracks, was observed during the tests. This would also support the view that a one-dimensional idealization of the wall may not be rational. Unfortunately, identification of wall stiffnesses in conjunction with more complex models was not attempted. Experiments designed around the concept of "system identification" need to be carried out to identify the parameters of more complex and representative analytical models in order to improve the state-of-the-art in analytical response prediction of RC walls.

Perhaps the most striking observation from the research was the strong interactions between the axial force and stiffness and moment to shear ratio and stiffness, even at the uncracked service stress levels, which could not be simulated by the available elasticity-based section analysis models [5].

Measured versus Estimated Stiffnesses

At an axial compressive force corresponding to 30 percent of the balanced force, the measured section flexural, shear and axial stiffnesses were less than 60 percent of the corresponding stiffnesses based on uncracked sections. Only at axial compression levels approaching the balanced force of the section that the stiffnesses approached the corresponding predicted nominal values [5].

Upon cracking and degradation at serviceability limit state stress levels, the axial, shear and flexural stiffnesses decreased to less than 20 percent of the corresponding nominal values, which approximately corresponded to the stiffness provided only by the reinforcement of the section. The degradation of shear stiffness was quicker and more pronounced upon cracking than the degradation in the flexural stiffness [5].

Experimentally Observed Responses of Beam-Slab Sections

During tests of beam-column-slab subassemblies [10], wall-slab subassemblies [2], and the earthquake-simulator studies of the complete 7-study structure at Berkeley [8], the contribution of slab to the effective flange width of beams were observed to vary with the level of beam rotation [2,8]. Most section analysis procedures do not consider such variations in strain distribution along the width of a section.

CORRELATION BETWEEN MEASURED AND SIMULATED RESPONSES

Microcomputer Program for RC Section Analysis

A microcomputer program recently developed at Berkeley [11] was modified to incorporate general user-defined concrete stress-strain responses. The option for linear or nonlinear user-defined section strain distributions along both the width and/or depth of a section, and which may be updated at any step during analysis, was developed. Hypo-elastic biaxial stress-strain relations for concrete incorporating the interaction between normal and shear stresses which were measured during RC panel tests [13] are being implemented. The code was adapted to the DOS operating environment, and was complemented by screen graphics options. It was used for the following sensitivity study.

Material Characteristics Used in Predicted Responses

Several moment-curvature responses of the wall specimen's section were generated using different concrete responses. The steel responses used in the analyses were obtained from coupon tests. These are not shown in this paper due to space limitations. The concrete responses are depicted in Fig. 4. The response labeled "panel and shell" in Fig. 4 was obtained by testing standard cylinders of the material cured under the same conditions as the specimens. The confined concrete response labeled "core" was obtained by testing RC column specimens with dimensions, lateral and longitudinal reinforcement exactly corresponding to those of the boundary columns of the wall.

The response labeled as "pre-cracked" was obtained as follows. The wall was loaded by increasing axial compressive force while strains in concrete along a cross section were measured. The stress in concrete was found by subtracting the force carried by the reinforcement, and hence an effective uniaxial stress-strain response for the wall concrete was obtained. This test was carried out to 45 percent of concrete strength, and exhibited a lower stiffness than the cylinder tests. This was attributed to the initial microcracks and narrow cracks which existed in the wall, and the response was correspondingly labeled "pre-cracked". As shown in Fig. 4, it was projected to intercept the cylinder response at maximum strength.

The response labeled "modified panel" was postulated to represent the normal component of the biaxial stress-strain response of the panel concrete.

Comparison of Measured and Predicted Wall Stiffnesses

Analytical responses of the section shown in Fig. 5 were generated to assess the effects of the shape of strain profile and other parameters related to concrete response on the service level flexural stiffness characteristics. To generate a reference response, the linear strain profile was assumed, and the concrete responses measured from standard tests of shell and core concrete were incorporated including a measured tensile strength of 500 psi (3447 kPa). This response incorporated gravity axial compression of 250 kips (1112 kN), and is shown by the solid line in Fig. 5.

Section secant stiffness was evaluated at the moment level corresponding to the seismic "E" force specified by 1982 UBC for the parent structure of the specimen, as indicated in Fig. 5. This stiffness coincided with the nominal stiffness based on the uncracked-transformed section. It is worth reiterating that the corresponding measured stiffness prior to cracking the specimen was less than 60 percent of this nominal value [5].

Two other responses were generated using strain profiles which reflected different warping characteristics, as shown in the insert of Fig. 5. The profile labeled "warped", represented the measured strain profile, this type of warping was attributed to the restraining effect of the foundation on the cross section which was instrumented. The other strain profile incorporated the more typical warping due to shear. Both non-planar strain profiles led to reductions in flexural stiffness, as indicated by the corresponding responses in Fig. 5.

Another analysis incorporated the linear strain profile, and the "pre-cracked" concrete response which was discussed in relation to Fig. 4. Commensurate with the assumption that the concrete was "pre-cracked", no tensile strength was incorporated. Section stiffness was affected by the reduced concrete stiffness and absence of tensile strength considerably. The secant flexural stiffness at the "E" force level was only 54 percent of the reference stiffness, close to the value measured prior to cracking of the specimen.

To predict a lower bound of section flexural stiffness, and to simulate the stiffness measured after cracking the specimen, another analysis was carried out. This was based on a reduced axial compression of 150 kips (667 kN) rather than the 250 kips (1112 kN) gravity load, in order to incorporate a pre-existing tensile force of 100 kips (445 kN), attributed to the shrinkage stresses which would exist in concrete. Furthermore, a warped strain profile, and "pre-cracked" concrete stress-strain responses were incorporated. The resulting stiffness was 40 percent of the reference stiffness, still higher than the stiffness which was measured after cracking and degradation of the specimen. The remaining discrepancy may be attributed to bond slip which was not simulated in the analyses.

It is important to note that although the above discussions focused only on the flexural stiffness, similar discrepancies between the nominal reference and measured shear and axial stiffnesses were observed [5]. The existing practical section analysis tools are not suitable for a reliable prediction of wall shear or axial stiffnesses. On the other hand, as observed in Fig. 3, shear deformations were measured to constitute half of the total deformation even when a large moment to shear ratio of 4.25 times the height of the wall element was applied in the tests. Consequently, the implications of such reductions in shear (and axial) stiffnesses would be at least as important as the implications of the discussed reductions in flexural stiffness.

Comparison of Measured and Predicted Wall Strength

The individual wall specimens were coupled and tested to failure [2]. During lateral loading to failure, one of the wall components was subjected to an increase in compression while the other was subjected to a decrease, with respect to their gravity forces. The strength of the coupled wall was controlled by the wall component subjected to an increase in compression, as the concrete in the panel of this wall experienced a semi-brittle splitting-crushing type of failure [2,4]. The axial force, flexure and shear experienced by the walls during the experiments were measured [6]. The history of wall internal forces were therefore known.

Several responses of the wall section were generated to investigate the influence of axial force history on strength, as shown in Fig. 6. The concrete responses labeled "panel" and "shell and core", in Fig. 4, and the linear strain profile were incorporated in these analyses.

One analysis was carried out with a constant axial compression of 750 kips (3336 kN) which was equal to the force measured at the failure of the panel. Another analysis incorporated an initial axial force of 250 kips (1112 kN) equal to the gravity axial compression, which gradually increased to 750 kips (3336 kN), simulating the measured history of axial force. The second analysis indicated a lower initial flexural stiffness as compared to the analysis with a constant axial compression of 750 kips (3336 kN), but a similar strength was attained as indicated in Fig. 6. Both of these analyses overestimated the measured flexural strength of the wall by over 20 percent.

Since recent research indicated significant reductions in the uniaxial stiffness, strength and deformability of concrete due to certain biaxial stress states, typical of the wall panel's stress state at failure [13], an analysis was conducted by assuming a lower stiffness, strength and deformability for the panel concrete, i.e., the "modified panel" concrete response in Fig. 4. This analysis indicated a reduction in the strength and a significant reduction in the deformability of the cross section as observed in Fig. 6. The predicted strength was closer to the measured value. The results indicate that for reliable RC wall analysis, when a combination of high axial compressive and shear stresses are anticipated, it is essential to include appropriate biaxial stress-strain formulations for the panel concrete.

A final analysis was conducted by simulating the axial force history of the wall component which was subjected to a reduction in its compression, as measured during the experiment. The reduction resulted in a net tensile force at the failure of the specimen, due to the strong coupling between the two walls [2]. The corresponding response is shown in Fig. 6, indicating striking reductions in stiffness, strength and deformability of the wall under decreasing axial force as compared to the responses of the wall subjected to an increase in compression. Commensurate with the observed discrepancy in the stiffnesses of the walls subjected to increasing and decreasing compression, respectively, the wall which was under decreasing compression could not effectively contribute to the shear resistance of the coupled wall [2,4]. The analytical model was successful in simulating this phenomenon, which indicated its value as a useful aid in design. In this retrospect, the importance of such sensitivity analyses of critical wall sections of a building in order to assess the advanced limit state responses of the structure cannot be over-emphasized [3].

Simulated Responses in Beam-Slab Sections

Dimensions and reinforcement detailing of a girder cross section, from the prototype structure tested as part of the U.S.-Japan Cooperative Research Program [8,10], is shown in Fig. 7. The distance between parallel girders was 236 in. (6.0 m) and this may be considered as the upper bound of slab contribution to the effective flange width of the girder. As discussed above, tests indicated that the strain distribution in the slab steel varied with the level of girder rotation. At the ultimate limit states, all the reinforcement within the complete slab yielded and contributed to the flexural capacity of the girder [8,10].

This phenomenon was simulated in a cross sectional analysis which was carried out by varying the strain distribution along the flange of the girder. The effective flange width "b_e" was related to the level of curvature "ϕ", as shown in the insert of Fig. 8. Moreover, the strains along the assumed effective flange were not constant, but attenuated towards the extreme ends of the flange as illustrated in the "typical section strain distribution", in Fig. 8.

The response obtained by varying the effective flange width with curvature is compared in Fig. 8 to several responses obtained based on a constant flange width. In all of the analyses, the flange was subjected to tension.

As expected, responses obtained for different effective flange widths indicated significant differences in the effective stiffness, strength and deformability of the section. The section responded as over-reinforced when a flange width of 152 in. (3.86 m) or more was assumed. Based on the assumed nominal material characteristics, concrete strength governed the flexural strength of the section when the flange width exceeded 152 in. (3.86 m). The response attained by varying the effective flange width with curvature, similarly exhibited over-reinforced behavior, once the effective flange width exceeded 152 in. (3.86 m).

Reliable prediction of beam-slab responses would obviously require a correct prediction of the relationship between effective flange width and curvature, which, in turn, would depend on the geometry and deformation kinematics of the complete structural system. The responses in Fig. 8, and the related conclusions of the U.S.-Japan research program [8] indicate the importance of investigating the bounds of beam-slab response while conducting seismic design of RC buildings at regions of high seismic risk.

CONCLUSIONS AND RECOMMENDATIONS

1. The effective stiffnesses of RC elements in the serviceability limit states may be significantly less than the commonly predicted nominal values, even when these may be based on the cracked section assumption. The stiffnesses of wall elements were measured to be strongly influenced by axial

force, to a degree which could not be analytically simulated. By incorporating observed phenomena such as a warped strain profile, initial shrinkage stresses in concrete and effects of initial microcracking and cracking on concrete response, an analytical flexural stiffness close to the measured value could be simulated.

2. The strength and deformability of RC walls, subjected to high axial compressive and shear forces, may be significantly overestimated if predicted by an analysis incorporating only the standard uniaxial test responses of concrete. The biaxial stress states which typically arise due to the shear in the wall panel adversely affect the stiffness, strength and spalling strain of panel concrete. This should be incorporated in the analysis for a reliable prediction of wall response.

3. The stiffness, strength and deformability characteristics of beam-slab sections were significantly affected by the assumed strain distribution along the width of the slab, i.e., assumed effective flange width. Results indicated that this phenomenon should be correctly simulated in analysis to predict advanced limit state stiffness characteristics of RC buildings reliably.

4. Cross sectional analysis of RC may be used as a practical and reliable tool to improve seismic design of RC buildings. The possible bounds of stiffness, strength and deformability of the critical cross sections of a building may be easily obtained in a design office by microcomputer software such as the one described in this paper. Such a study would reveal invaluable information regarding the possible bounds of advanced limit state responses of structural elements.

The authors are currently conducting research to develop practical and reliable building collapse analysis procedures, based on results of section analyses. Such research is advocated as important and necessary in order to advance the state-of-the-art in conceptual seismic design of RC buildings [2,3,4]. The authors would like to emphasize that they are not advocating time-history analyses or sophisticated analyses based on microscopic models. Rather, a simple yet reliable investigation of the possible bounds of stiffness, strength and deformability of the "envelope" responses of the structure is recommended in order to improve seismic design of RC buildings to be constructed in zones of high seismic risk.

ACKNOWLEDGEMENTS

The experimental work which was reported as part of this paper was carried out at U. C. Berkeley, under the sponsorship of the National Science Foundation, Grant CEE 81-07217. Contributions of Professor Bertero, principal investigator of the research, and other members of the research team, Dr. Sakino, Dr. Baleriola and Mr. Ozselcuk, are gratefully acknowledged. The support provided by Louisiana State University in the current research efforts of the authors is deeply appreciated.

REFERENCES

1. Aktan, A. E. and Pecknold, D. A. W., "Response of a Reinforced Concrete

Section to Two-Dimensional Curvature Histories," *Journal* of the American Concrete Institute, May 1974.

2. Aktan, A. E. and Bertero, V. V., "Seismic Responses of R/C Frame-Wall Structures," *Journal* of the Structural Division, American Society of Civil Engineers, August 1984.

3. Aktan, A. E. and Bertero, V. V., "Conceptual Seismic Resistant Design of Frame-Wall Structures," *Journal* of the Structural Division, American Society of Civil Engineers, November 1984.

4. Aktan, A. E. and Bertero, V. V., "R/C Structural Walls: Seismic Design for Shear," *Journal* of the Structural Division, American Society of Civil Engineers, August 1985.

5. Aktan, A. E., Bertero, V. V. and Sakino, K., "Lateral Flexibility Characteristics of R/C Frame-Wall Structures," *Publication SP-86*, "Deflections of Concrete Structures," American Concrete Institute, 1985.

6. Aktan, A. E. and Bertero, V. V., "Measuring Internal Forces of Redundant Structures," *Journal* of Experimental Mechanics, December 1985.

7. Bertero, V. V., Aktan, A. E., Harris, H. G. and Chowdhury, A. A., "Mechanical Characteristics of Materials Used in a 1/5 Scale Model of a 7-Story Reinforced Concrete Test Structure," *Report* No. UCB/EERC-83/21, Earthquake Engineering Research Center, University of California, Berkeley, October, 1983.

8. Bertero, V. V., Aktan, A. E., Charney, F.A. and Sause, R., "Earthquake Simulator Tests and Associated Analytical and Correlation Studies of a 1/5 Scale Replica Model of a Full Scale, 7-Story R/C Frame-Wall Structure," *Publication SP-84*, "Earthquake Effects on Reinforced Concrete Structures, U.S.-Japan Research," American Concrete Institute, 1985.

9. Finite Element Analysis of Reinforced Concrete, American Society of Civil Engineers, 345 East 47th Street, New York, NY 10017, 1982.

10. Joglekar, M., et al., "Full Scale Tests of Beam-Column Joints," *Publication SP-84*, Earthquake Effects of Reinforced Concrete Structures, U.S.-Japan Research, American Concrete Institute, P.O. Box 19150, Redford Station, Detroit, MI 48219.

11. Kaba, S. A. and Mahin, S. A., "Interactive Computer Analysis Method for Predicting the Inelastic Cyclic Behavior of Structural Sections," *Report No. UCB/EERC-83/18*, Earthquake Engineering Research Center, University of California, Berkeley, CA, 1983.

12. Park, R., Kent, D. C. and Sampson, R. A., "Reinforced Concrete Members with Cyclic Loading," *Journal* of the Structural Division, ASCE, July 1972.

13. Vecchio, F. and Collins, M. P., "The Response of RC to In-Plane Shear and Normal Stresses," *Publication No. 82-03*, University of Toronto, Department of Civil Engineering, March 1982.

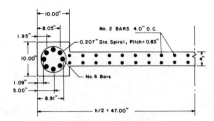

Fig. 1. Dimensions and Rebar Detailing of the Wall Section

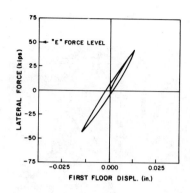

Fig. 2. Test Responses of the Wall

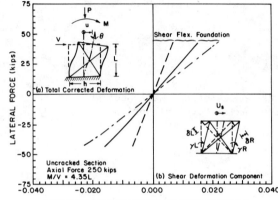

Fig. 3. Different Components of Wall Displacement

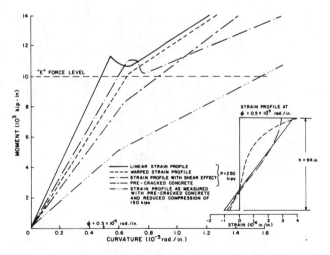

Fig. 5. Flexural Responses of the Wall Section Based on Different Assumptions on Strain Distribution and Concrete Response

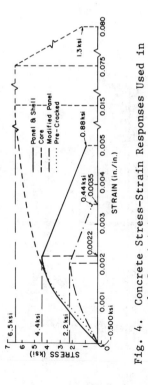

Fig. 4. Concrete Stress-Strain Responses Used in the Sensitivity Study

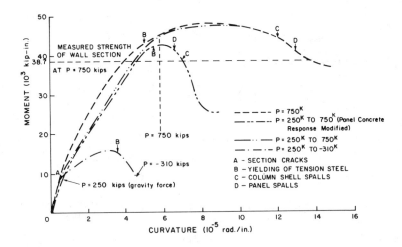

Fig. 6. Axial-Flexural Responses of the Wall Section Based on Different Axial Force Histories

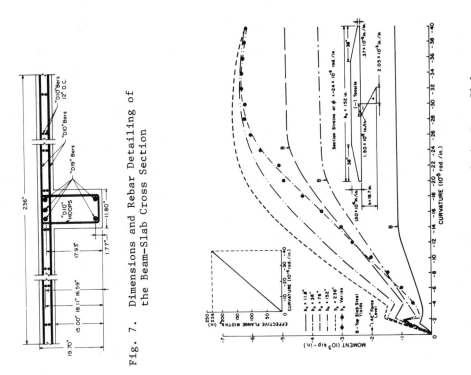

Fig. 7. Dimensions and Rebar Detailing of the Beam-Slab Cross Section

Fig. 8. Flexural Responses of the Beam-Slab Section for Different Effective Flange Widths

A HYSTERESIS MODEL FOR BIAXIAL BENDING OF REINFORCED CONCRETE COLUMNS

G. E. Ghusn, Jr.[I] and M. Saiidi[II]

ABSTRACT

A new nonlinear biaxial bending element for reinforced concrete columns was developed. The new element, called the multiple spring element (MSE), was implemented in a nonlinear static analysis program. The analytical results using the MSE compared favorably with experimental results.

The new MSE model was also implemented in program NEABS-86 which is a modified version of program NEABS (Nonlinear Earthquake Analysis of Bridge Systems). A two-span bridge structure was modeled using both the elasto-plastic yield surface element and the MSE. Comparisons of the dynamic responses showed that the MSE provided higher amplitude acceleration response and lower amplitude displacement response than the elasto-plastic element. The results from MSE were more stable and required less computer execution time.

INTRODUCTION

A three-dimensional analysis using a hysteresis model for biaxial bending of columns is required to evaluate the inelastic dynamic response of reinforced concrete structures subjected to strong bidirectional ground motions. The model must define the interaction of axial load and the two bending moment components for elastic and inelastic deformations. Modeling biaxial behavior is crucial for the accurate and realistic prediction of inelastic dynamic response of reinforced concrete structures.

A wide range of mathematical manipulations using the yield surface have been used in the past [1, 2, 3]. These techniques are cumbersome and inexact. Other methods such as finite element techniques are accurate but often require enormous computational effort for even simple structures. A multiple spring element approach that does not require the construction of a yield surface appears to be a preferred alternative. The purpose of this paper is to present such a model and the related results.

[I] Associate Staff Member, The BDM Corporation, Seattle, Washington

[II] Associate Professor, Civil Engineering Department, University of Nevada, Reno

NONLINEAR MODEL ELEMENT DESCRIPTION

The multiple spring method, first developed by Lai [4], assumes that the hysteretic behavior of a reinforced concrete column can be approximated by a zero-length nonlinear multiple spring element (MSE) at the column end. The element is formulated neglecting torsional and shear deformations. The MSE, as developed by Lai [4], consists of nine springs: four representing reinforcing steel and five representing concrete. The springs are allowed to deform only in the axial direction. The nine springs are located in five positions: one steel and one concrete spring at each corner and a concrete spring in the center (Fig. 1). The hysteretic behavior for each spring is accounted for separately. The forces and deformations in each spring are related by equilibrium and by planar strain compatibility relationships (plane sections remain plane). Thus, the MSE is capable of accounting for axial load variations during biaxial bending, which is not easily accomplished with yield surface techniques.

The nine-spring model described above has been shown to produce results which are in close agreement with experimental data [4]. Even though the model is considerably simpler than the yield surface model, it requires a relatively large memory because there are, in effect, nine subelements within every element. A new model was developed which combines the concrete and steel springs within each corner into one spring thus reducing the number of subelements to five. The new corner springs have different stiffnesses in each direction: the steel stiffness in tension and the sum of the steel and concrete stiffnesses in compression. The properties for each type of spring are calculated as described below and are then combined to determine the composite spring properties. The middle concrete spring is treated separately as in the original nine-spring configuration.

MULTIPLE SPRING ELEMENT FORMULATION

The initial properties of each spring and their positions are generally determined using the method described in Ref. 4. In developing the methodologies, it is assumed that the column has at least one axis of symmetry in terms of cross section geometry and steel distribution.

The Steel Spring

Each steel spring represents the behavior of one-quarter of the steel in the cross section. The properties of this spring incorporate the slippage of the reinforcing bars and bar elongation. The bond strength is assumed to be uniform along the development and can be approximated by

$$u = 14*(f_c')^{0.5} \tag{1}$$

in which

    u is the bond strength in psi and
    f_c' is the specified compressive strength of concrete in psi.

This equation produces bond strengths which are larger than those implied in ACI formulas [5]. The larger bond strength values lead to more realistic response histories than do the conservative values used by the ACI. The bond strength can be used to determine the development length of the steel bar at yield stress.

$$l_d = A_b f_y / (\pi * d_b * u) \tag{2}$$

in which

l_d is the development of length in inches,
A_b is the area of a steel bar,
f_y is the specified yield stress of the steel, and
d_b is the diameter of the bar.

Assuming a uniform bond stress distribution, a triangular strain distribution within the bar will result. The concentrated displacement at the joint due to bond slip is

$$d = f_y l_d / (2 * E_s) \tag{3}$$

in which

E_s is Young's modulus for steel.

Using these simple relationships, the initial elastic stiffness, k_{se}, of the spring is calculated from

$$k_{se} = 2 A_s E_s / l_d \tag{4}$$

in which

A_s is 1/4 of the total area of the longitudinal steel in the column.

The yield displacement, d_y, for the spring is given by

$$d_y = A_s f_y / k_{se} \tag{5}$$

This yield displacement is used in the hysteresis rules and in other relationships described in subsequent sections.

The Concrete Spring

The concrete spring simulates the behavior of concrete in a reinforced concrete member. The properties of this spring are determined based on the compressive strength of concrete and the balanced axial load and moment for the section.

The "yield" force in the concrete spring is

$$P_{cy} = 0.85 f'_c A_c \tag{6}$$

in which

A_c is the area of concrete represented by the spring, and

f'_c is the specified compressive strength of concrete.

The area represented by each of the corner concrete springs is the same because of symmetry. The areas are determined from the balanced condition defined by conventional flexural theory [2, 5]. The area for the center spring is the remaining area of concrete not represented by the corner springs.

$$A_{corner} = P_b/(2*0.85f'_c) \qquad (7)$$

in which

A_{corner} is the area represented by the concrete spring and
P_b is the balanced axial load for the section.

An average P_b is used for rectangular sections with different balanced loads in each orthogonal direction. The center spring area becomes

$$A_{center} = A_{gross} - 4*A_{corner} - A_{st} \qquad (8)$$

in which

A_{center} is the area represented by the center spring,
A_{gross} is the gross area of the cross section, and
A_{st} is the total area of steel in the cross section.

The initial stiffness for the corner springs is determined assuming that steel and concrete springs have the same yield displacement. The stiffness for the corner spring is

$$k_{ce} = 0.85 A_{corner} f'_c / d_y \qquad (9)$$

and for the center spring is

$$k_{cce} = 0.85 * A_{center} f'_c / d_y \qquad (10)$$

These stiffnesses are valid only for compression. The springs have no contribution in tension because the section is assumed to be cracked. This is not an unreasonable assumption because the initial cracking strength of concrete does not make a significant contribution to response. A zero postyielding stiffness is assumed for the concrete springs.

The locations of the springs are determined from the balanced condition. The moment at the yielding of opposing spring sets is assumed to be equal to the balanced moment in the corresponding direction. The distance between the springs is

$$d_{si} = 2M_{bi}/(A_{st}f_y + 2*0.85 A_{corner} f'_c) \qquad (11)$$

in which

d_{si} is the distance between the springs centered about the centroidal axis,

M_{bi} is the balanced moment computed from flexural theory, and
i is either the x or y coordinate direction.

The above relationships describe the characteristics of the components of the element. The balanced moments and axial loads are the only values that need to be calculated for the actual section in order to formulate the element properties.

ELEMENT STIFFNESS MATRIX

The element stiffness matrix translates the axial deformations of the component springs into joint rotations and axial displacement. Shear and torsion deformations are neglected. This is a reasonable assumption for columns designed based on current seismic codes, which require a high degree of confinement at joints. The derivation of the stiffness matrix is based on equilibrium of forces and planar strain compatibility. Using planar strain compatibility, the joint deformations can be written

$$\underbrace{\begin{Bmatrix} D_p \\ T_x \\ T_y \end{Bmatrix}}_{\{D\}} = \underbrace{\begin{bmatrix} 1/2 & 1/2 & 0 \\ 1/d_{sx} & 0 & -1/d_{sx} \\ 0 & -1/d_{sy} & -1/d_{sy} \end{bmatrix}}_{[T_1]} \underbrace{\begin{Bmatrix} d_1 \\ d_3 \\ d_4 \end{Bmatrix}}_{\{d\}} \quad (12)$$

in which

d_{sx} and d_{sy} are the distances between the corner springs,
d_i is the displacement in the ith spring location,
D_p is the net axial displacement defined as the displacement at the center of the section, and
T_x and T_y are the rotations about the x and y axes, respectively.

Using the equilibrium, the relationship between spring forces and the moments and the axial load is obtained

$$\underbrace{\begin{Bmatrix} P \\ M_x \\ M_y \end{Bmatrix}}_{\{P\}} = 1/2 \underbrace{\begin{bmatrix} 2(K_5/2+K_1+K_2) & 2(K_5/2+K_3+K_2) & 2(K_4-K_2) \\ (K_1+K_2)d_{sx} & (K_2-K_3)d_{sx} & (-K_4-K_2)d_{sx} \\ (K_1-K_2)d_{sy} & (-K_2-K_3)d_{sy} & (K_2+K_4)d_{sy} \end{bmatrix}}_{[T_2]} \underbrace{\begin{Bmatrix} d_1 \\ d_3 \\ d_4 \end{Bmatrix}}_{\{d\}} \quad (13)$$

in which

k_i is the stiffness at the ith spring location.

Substituting and rearranging Eqs. 12 and 13 results in

$$\{P\} = [T_1][T_2]^{-1}\{D\} \quad (14)$$

which can be rewritten as

$$\{P\} = [K]\{D\} \tag{15}$$

where $[K]$ is the element stiffness matrix.

HYSTERESIS MODELS FOR THE SPRINGS

Two hysteresis models are used for the multiple spring element: a simple one for the center concrete spring and a more complex model for the composite steel-concrete springs. The hysteretic behavior for each spring is taken into account separately resulting in a relatively complex hysteretic behavior for the overall element.

The GHYST hysteresis model (Fig. 2) for the center concrete spring is a simple approximation of the response of concrete. The model consists of only three rules. This model was originally developed by Lai [4] for use in his nine-spring element.

A new hysteresis model, AQHYST, was developed to model the response of the new composite spring. AQHYST is based on the same four rules as the QHYST model developed by Saiidi and Sozen [6, 7], except that the initial stiffnesses for compression and tension are different (Fig. 3). The AQHYST model does not incorporate the rules from GHYST, rather, the concrete spring is assumed to "follow" the hysteretic path of the steel spring. The unloading stiffnesses in the compression and tension ranges are calculated using separate decay factors for each region.

The absolute value of the yield displacement is the same for both compression and tension. The yield force in tension is the yield force for a steel spring, and the yield force in compression is the sum of the yield forces for a steel spring and a concrete spring. The postyielding stiffness for both the compression and tension regions is a fraction of the elastic stiffness of a steel spring. The details of the AQHYST model are described in Ref. 8.

Comparisons with Experimental Data

Otani and Cheung [9] performed static biaxial bending tests for two cantilever reinforced concrete columns (Fig. 4). The experimental results from Otani's specimens were utilized for parametric studies of the effects of changing the postyielding stiffness values and the unloading stiffness decay factors. The best fit results were obtained using a postyielding stiffness of two percent of the initial steel spring stiffness and a value of 0.2 for decay factors b1 and b2 (Fig. 3). The results for one of the specimens (SP-7) were found to be representative and are shown in this paper. The measured and calculated results are shown in Figs. 5 and 6. The overall shape of the curves shows a good fit in both directions. The areas within the curves generated by the analytical model are approximately the same as those from the experimental data indicating that MSE predicted similar energy dissipation. The MSE underestimates the peak load capacity for the column by about ten percent, which is acceptable for reinforced concrete elements.

SEISMIC BRIDGE ANALYSIS

The new multiple spring element was implemented as a new nonlinear element in program NEABS (Nonlinear Earthquake Analysis of Bridge Systems). The new version of the program is called NEABS-86. The original program used only an elasto-plastic yield surface model [10]. The program uses Newmark's technique to solve the equations of motion for a bridge system [11].

Comparisons of the dynamic response of a bridge modeled using both the elasto-plastic and multiple spring column elements reveal dramatic differences in the predicted responses. The Meloland Overpass in California (Fig. 7) was used as the basis for the mathematical models analyzed using NEABS-86.

Two models of the bridge were prepared: one using an elasto-plastic column element and another using a multiple spring element at the base of the column. The objective of this study was to examine the possible differences in dynamic response created by using different column models. Ordinarily, the nonlinear response is affected by the nonlinearity of the foundation, columns, hinges, expansion joints, etc. To isolate the effect of the modeling of the columns, however, it was necessary that only the columns develop nonlinear deformations. As a result, the bridge was not modeled exactly, but, rather, provided the basis for realistic structural geometries and cross section properties. Foundations were assumed to be infinitely rigid because the flexibility of the foundation can interfere with the yielding of columns. Abutments were modeled as roller supports. Therefore, the results shown in this paper represent the dynamic response of a ficticious bridge structure with realistic structural properties for the comparison of the two biaxial bending elements. The results do not represent the actual bridge response.

The Meloland Overpass is a 208-ft. long (63.4 m), two-span symmetrical reinforced concrete box girder bridge located near the Imperial Fault in southern California. The deck center is supported on a single round concrete column pier 20.5 ft. (6.3 m) high and five feet (1.5 m) in diameter and is reinforced with eighteen, #18 bars distributed around its perimeter. An elevation of the bridge is presented in Fig. 7a. The bridge was subjected to the magnitude 6.4 Imperial Valley earthquake in October of 1979 but did not experience any structural damage [12].

The Meloland Overpass was idealized as shown in Fig. 7b. The abutments were modeled as rollers in order to allow for large deformation and yielding of the column at reasonable acceleration levels. Structural damping was assumed to be five percent for all cases.

The measured free field horizontal acceleration components from the 1979 Imperial Valley earthquake were applied to the model of the Meloland Overpass. These two acceleration histories, one logitudinal and one transverse, were recorded by an instrument 200 ft. (61 m) from the centerline of the bridge. The results from the analyses are presented in Figs. 8 and 9. The solid lines represent the deck response from the MSE model, and the dashed lines represent the elasto-plastic response. Note that the early low amplitude accelerations and displacements at the pier top are identical indicating the same elastic response for both element models. Upon yielding at approximately four seconds in the history, the responses separate dramatically. The elasto-plastic column has zero stiffness upon yielding, whereas the MSE reduces stiffness gradually and allows for some effective stiffness. The elasto-plastic acceleration response is less than the MSE response, and the displacement response for the elasto-plastic element is much larger than the MSE response. The responses also show phase differences due to the different stiffnesses. A study of the response of two other bridges showed similar trends [8]. In one case, it was noted that the elasto-plastic yield surface model required a considerably shorter time interval for numerical integration to yield stable results.

CONCLUSIONS

A new relatively simple nonlinear biaxial bending element for the cyclic analysis of reinforced concrete columns was developed. The model produced excellent correlations with experimental data for a statically tested specimen. The accuracy of the MSE's dynamic response is inferred from these few data. There is little doubt that this new element produces more realistic results than the elasto-plastic element for biaxial bending of reinforced concrete columns. It has been demonstrated that reinforced concrete is not elasto-plastic. However, researchers in this field have used the elasto-plastic model, not because it is accurate but because it is simple and convenient. The MSE model provides a simpler alternative to the yield surface model and can be used in modeling the biaxial hysteretic behavior in reinforced concrete buildings and bridges. The accuracy of the MSE for dynamic analyses of reinforced concrete columns cannot be fully evaluated until more test data become available.

Nonlinear analyses using the new multiple spring element execute faster, do not become unstable, and provide for a more realistic dynamic response for reinforced concrete columns subjected to biaxial bending. The analyses of models using the MSE can offer insight into the performance of a reinforced concrete structure subjected to biaxial bending due to bidirectional earthquake loadings.

ACKNOWLEDGEMENTS

The study leading to the thesis presented here was funded by grant No. CEE-8317477 from the National Science Foundation. The analyses presented herein were performed using the CYBER 830 computer at the University of Nevada, Reno. Special thanks are due to Professor George Will of the University of Toronto and Mr. Shing Lai, a former graduate student at the University of Toronto, for providing information about their models.

REFERENCES

[1] Bressler, B., "Design Criteria for Reinforced Concrete Columns Under Axial Load and Biaxial Bending," American Concrete Institute Journal, No. 57, pp. 481-490, 1960.

[2] Park, R. and T. Paulay, <u>Reinforced Concrete Structures</u>, John Wiley and Sons, Inc., 1975.

[3] Takizawa, H. and H. Aoyama, "Biaxial Effects in Modelling Earthquake Response of R/C Structures," <u>Earthquake Engineering and Structural Mechanics</u>, Vol. 4, pp. 523-552, 1976.

[4] Lai, S.-S., "Inelastic Analysis of Reinforced Concrete Space Frame Under Biaxial Earthquake Motions," a dissertation submitted in partial fulfillment of the requirements for the degree of Doctor of Philosophy in Civil Engineering, University of Toronto, 1984.

[5] ACI Committee 318, "Building Code Requirements for Reinforced Concrete," Detroit, American Concrete Institute, 1983.

[6] Saiidi, M. and M.A. Sozen, "Simple and Complex Models for Nonlinear Seismic Response of Reinforced Concrete Structures," Civil Engineering Studies, SRS No. 465, University of Illinois at Urbana-Champaign, August 1979.

[7] Saiidi, M., "Hysteresis Models for Reinforced Concrete," <u>Journal of the Structural Division</u>, ASCE, Vol. 108, No. ST5, pp. 1077-1087, May 1982.

[8] Ghusn, G.E., "A Hysteresis Model for Reinforced Concrete Columns Subjected to Biaxial Bending," a thesis submitted in partial fulfillment of the requirements for the Master of Science degree in Civil Engineering, University of Nevada Reno, May 1986.

[9] Otani, S. and V.W.-T. Cheung, "Behavior of Reinforced Concrete Columns Subjected to Biaxial Lateral Load Reversals," Publication 81-02, Dept. of Civil Engineering, University of Toronto, February 1981.

[10] Penzien, J., R. Imbsen, and W.D. Liu, "<u>NEABS</u>: Nonlinear Earthquake Analysis of Bridge Systems (Users Manual)", NISSEE/Computer Applications, Earthquake Engineering Research Center, University of California at Berkeley, May 1981.

[11] Newmark, N.M., "A Method of Computation for Structural Dynamics," <u>Journal of the Mechanics Division</u>, ASCE, pp. 67-94, July 1959.

[12] Hart, J.D., "Nonlinear Modeling of Short Highway Bridges Subjected to Earthquake Loading," a thesis submitted in partial fulfillment of the requirements for the Master of Science degree in Civil Engineering, University of Nevada Reno, May 1984.

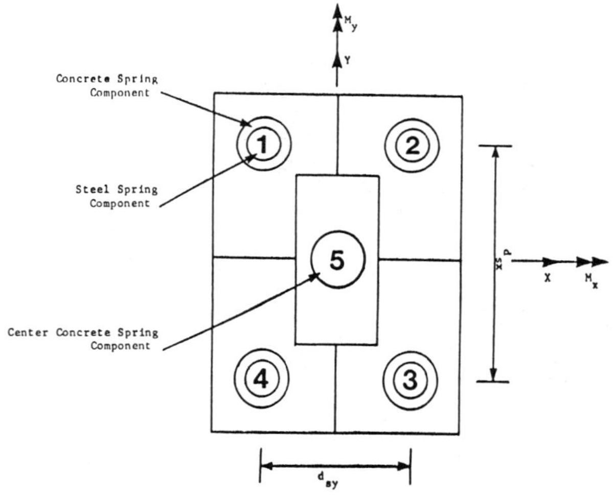

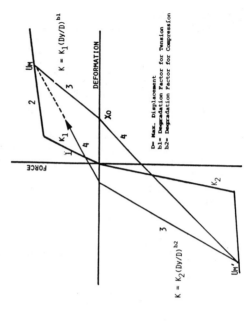

Fig. 2 Hysteresis Model for Concrete Springs

Fig. 3 AQ-HYST Model for Composite Springs

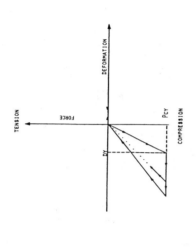

Fig. 1 Multiple Spring Model

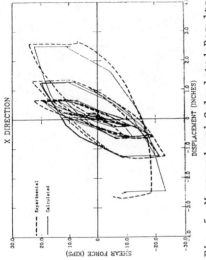

Fig. 5 Measured and Calculated Results in the X Direction (1 inch = 25.4 mm)

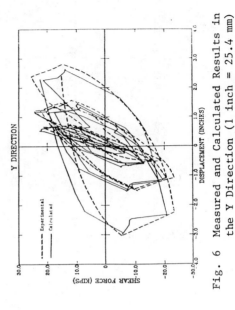

Fig. 6 Measured and Calculated Results in the Y Direction (1 inch = 25.4 mm)

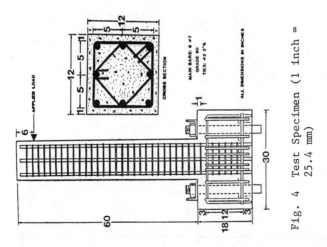

Fig. 4 Test Specimen (1 inch = 25.4 mm)

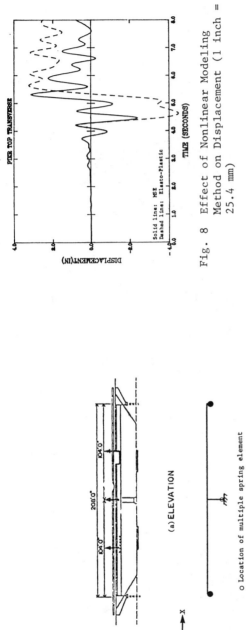

Fig. 8 Effect of Nonlinear Modeling Method on Displacement (1 inch = 25.4 mm)

Fig. 9 Effect of Nonlinear Modeling Method on Acceleration Response

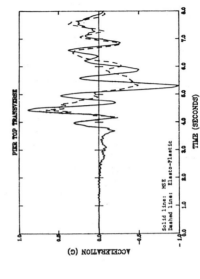

Fig. 7 Meloland Overpass and the Analytical Model (1 foot = 0.305 m)

CONSENSUS-OPINION EARTHQUAKE DAMAGE PROBABILITY MATRICES AND LOSS-OF-FUNCTION ESTIMATES FOR FACILITIES IN CALIFORNIA

C. Rojahn[I], R. L. Sharpe[II], R. E. Scholl[III],
A. S. Kiremidjian[IV], R. V. Nutt[V], and R. R. Wilson[VI]

ABSTRACT

This paper describes the results of questionnaire processes used to obtain consensus opinion damage-factor estimates (expected physical damage) and loss-of-function estimates for all types of existing industrial, commercial, residential, utility and transportation facilities in California. Damage-factor estimates are provided for Modified Mercalli Intensities (MMI) VI through XII in the form of Damage Probability Matrices. Loss-of-Function estimates, which specify the time required to restore a facility to 30%, 60%, and 100% of the pre-damage usability, are provided for each of the following seven damage states: 0%, 0-1%, 1-10%, 10-30%, 30-60%, 60-100%, and 100% damage.

INTRODUCTION

Recent studies have demonstrated that damage and consequent economic losses from a moderate to great magnitude earthquake centered near a major metropolitan area in California would be severe[1,2,3]. In view of these significant postulated losses, the Federal Emergency Management Agency (FEMA) has undertaken a comprehensive program to estimate the economic impacts of major California earthquakes on the state, region, and nation. The damage and loss estimates are calculated through the use of a computer simulation model known as the FEMA Earthquake Damage and Loss Estimation System (FEDLOSS)[4], which utilizes earthquake damage/loss estimates and inventory methodology developed by the Applied Technology Council (ATC) for existing facilities in California[5]. The economic impact estimates are based on the FEDLOSS results using another computer simulation methodology known as the FEMA Earthquake Impacts Modeling System (FEIMS), which utilizes a joint supply-side/demand-side economic impacts model[6] that involves assessment of damage to existing facilities in California as well as the economic interactions among functions housed in these facilities.

I.   Executive Director, Applied Technology Council, Redwood City, California.

II. Senior Consultant, Applied Technology Council, Redwood City, California.

III. Consultant, Redwood City, California.

IV. Consultant, Palo Alto, California.

V.   Consultant, Orangevale, California.

VI. Chief, Economic Analysis Branch, Federal Emergency Management Agency, Washington, D.C.

Because the earthquake damage and loss data required for FEDLOSS were not available in the literature, ATC and FEMA agreed that the best way to develop the required data was to draw on the experience and judgment of seasoned earthquake engineers. Accordingly, ATC established a 13-member advisory Project Engineering Panel (PEP) composed of senior-level specialists in earthquake engineering and commissioned 58 other earthquake engineering specialists to provide the required consensus damage/loss estimates.

This project involved four primary tasks:

1. Identification of the earthquake shaking characterization most appropriate for estimating earthquake damages and losses (for this project)

2. Development of facility classification scheme(s) that would account for all existing facilities within California

3. Development of earthquake damage and loss estimates in terms of the earthquake shaking characterization selected and the facility classes identified

4. Development of inventory data and methodology that are consistent with the facility classification scheme adopted as well as the inventory data currently available to FEMA

The methodologies utilized to develop the required data are described in detail in the ATC-13 Report[5].

Facility Classifications

Because of the comprehensive nature of the overall FEMA economic impacts investigation, it was essential that all types of industrial, commercial, residential, utility, transportation and other existing facilities in California be considered. These facilities were classified in two ways: (1) by Earthquake Engineering Facility Classification, which characterizes structures in terms of their size, structural system, and type (e.g., low-rise unreinforced masonry buildings), and (2) by Social Function Classification, which characterizes facilities in terms of their economic function (e.g., commercial retail trade).

The Earthquake Engineering Facility Classification is required because earthquake-induced physical damage is dependent upon structural properties. This classification contains 78 classes of structures, 40 of which are buildings and 38 of which are other structure types--bridges (3 classes), pipelines (2), dams (2), tunnels (3), storage tanks (6), roadways and pavements (3), high industrial chimneys (3), cranes (1), conveyor systems (1), on-shore towers (3), off-shore towers (1), canals (1), earth retaining structures (1), waterfront structures (1), and equipment (6--residential, office, electrical, mechanical, high technology and laboratory, and vehicles). These 78 structure classes were selected on the basis of expected dominance in the existing inventory of California structures and on the basis of expected uniqueness in seismic performance; the structure classes were not established on the basis of inventory sampling.

The Social Function Classification is required because that is the form in which structures in the existing FEMA database are listed and because this form is required as input in the economic impacts model utilized by FEMA. In addition, loss of function (or usability) is related to social function class. This classification contains 35 classes of facilitities--residential (3 classes), commercial (7), industrial (8), agriculture (1), mining (1), religion and nonprofit (1), government (2), education (1), transportation services (4), utilities (5), communication (1), and flood control (1). These 35 facility classes were selected so as to account for all facility types listed in the four digit Standard Industrial Classifications of the U. S. Department of Commerce.

EARTHQUAKE DAMAGE AND LOSS ESTIMATES

In addition to incorporating an exhaustive listing of facility types, the economic impacts model utilized by FEMA considers the following types of losses:

- The expected physical damage caused by ground shaking

- The expected losses from collateral earthquake hazards such as ground failure, inundation, and fire

- The expected percentage of loss of function or usability, including the time required to restore the facility to its pre-damage usability

- The expected percentage of population killed and injured

An overview of the methodologies used to develop the estimates for each of these loss types and examples of the resulting data are described below.

Physical Damage Caused by Ground Shaking

Estimates of percent physical damage caused by ground shaking, expressed in terms of damage factor versus Modified Mercalli Intensity scale[7], were developed through a multiple questionnaire process involving the PEP and other selected earthquake engineering specialists. The Modified Mercalli Intensity (MMI) scale, which has numerous shortcomings[5], was selected as the most appropriate earthquake shaking characterization for this project because the great preponderance of existing motion-damage data for earthquakes in the United States exist in this form.

The objective of the questionnaire process was to develop damage probability matrices (DPM's) similar in form to that suggested by Whitman, Reed, and Hong[8]. By using such DPM's, it is possible to estimate the expected dollar loss caused by ground shaking for each facility by multiplying the damage factors for the structure and its contents by the estimated replacement values for each, respectively. Shown below are the damage states and corresponding damage factor ranges defined for this project:

Damage State	Damage Factor Range (%)	Central Damage Factor (%)
1 - None	0	0
2 - Slight	0 - 1	0.5
3 - Light	1 - 10	5
4 - Moderate	10 - 30	15
5 - Heavy	30 - 60	45
6 - Major	60 - 100	80
7 - Destroyed	100	100

In Round One of this three-round questionnaire process each expert was asked to provide low, best, and high estimates of the damage factor to selected earthquake engineering facility types for Modified Mercalli Intensity levels VI through XII. For all types of facilities except pipelines, damage factor was defined as the ratio of earthquake dollar damage divided by the facility replacement value. For pipelines, each expert was asked to specify the number of breaks per kilometer. In addition to providing low, best, and high damage factor estimates, each expert was also asked to evaluate his level of experience with the facility class being evaluated and to provide a self-evaluated degree of certainty in the low, best, and high estimates. In order to remain unbiased in the Round One questionnaire, the experts were asked not to communicate with each other regarding this aspect of the project prior to making their evaluations.

The objective of the Round Two and Round Three questionnaires was to approach consensus on the damage factor estimates, and each expert was again asked to answer questions on the degree of damage at MMI levels VI through XII. The procedure for the Round Two and Round Three questionnaires, however, was slightly different from that followed in Round One in that each expert was provided with graphs showing his answers to the previous questionnaire together with the answers of all other experts (shown anonymously) responsible for the same facility class. Each expert was then asked to re-evaluate his estimates in light of the responses of others, with all other rules essentially the same as the rules for Round One.

Following the Round Three questionnaire and prior to conversion to damage probability matrices, the data were tested using Beta, normal, and lognormal probability distributions. From all the facility classes tested with these three distributions, it was felt that the Beta fitted the data uniformly better than either the lognormal or the normal. This distribution (Beta) was then used to develop DPM's for the 78 Earthquake Engineering Facility Classes considered in this project. Example DPM's are shown in Table 1.

The DPM's developed under this project apply to facilities having standard construction, which includes all facilities except those designated as special or nonstandard. Special construction includes (1) California elementary and secondary public school buildings, (2) post-1972 California hospitals, (3) railway bridges, and (4) any facility determined to have special earthquake damage control features. Nonstandard construction includes those structures that are more susceptible to earthquake damage than standard construction. The quantitative manner in which special and nonstandard construction is treated in this project is to shift the probability of a given damage state, P_{DSI}, up or down, depending on the grade or quality of design and construction.

Losses Due to Collateral Hazards

In addition to damage caused by strong ground shaking, collateral hazards such as ground failure, fault rupture, inundation, and fire can also cause serious damage to facilities. Initially, a literature review was conducted to ascertain existing quantitative information on the losses caused by these collateral hazards. On the basis of this information, plus judgment on the part of the project participants, methods for estimating damage caused by the following collateral hazards were developed:

- Poor ground/liquefaction, as it impacts surface and buried facilities

- Landslide, in terms of slope failure probability
- Fault rupture, both within the fault and drag zone
- Inundation, in terms of depth of high velocity water

The estimated damage from each of these four collateral causes is defined in terms of mean damage factor, which is the same form used to describe damage due to ground shaking. The total mean damage factor for a facility, then, is conservatively the sum of the mean damage factors for ground shaking, poor ground/liquefaction, landslide, fault rupture, and inundation.

Loss of Function or Usability

The procedure for estimating loss of function and restoration time for this project is based on the premise that loss of function and subsequent restoration time are directly related to: (1) direct damage to the individual facility and (2) direct damage to lifelines on which the facility depends. Lifeline systems considered include water supply, sanitary sewer (waste water), power/energy (electric power, natural gas, and petroleum fuels), transportation (highway, railway, air, and sea/water), and communication (telephone, radio, and television).

The methodology for evaluating the impact of lifeline failures on loss of function of particular facilities presumes that the extent to which a lifeline system is affected overall is largely dependent upon the extent of damage to the main components, distribution components, and service components. Importance factors that reflect the extent to which each of the 35 Social Function Classes will be affected by the failure of main and distribution components of the 11 spatially distributed lifeline systems considered under this project were developed and are provided in the ATC-13 Report[5].

Recognizing the paucity of statistical data, expert opinion on loss of function for the 35 Social Function Classes was solicited in a manner similar to that used in securing expert opinion on motion-damage relationships. For each social function, the PEP and 29 additional specialists were asked to estimate the time required to restore facilities to 30, 60 and 100% of their pre-earthquake usability. Restoration time was to be given for each of the seven levels of damage defined earlier. For purposes of this questionnaire, the experts were asked to consider only on-site effects, such as damage to the structure, damage to the equipment necessary for the operation of the facility, and loss of on-site utilities. In addition to providing estimates of restoration time, each expert was also asked to evaluate his level of experience with the facility class being considered. In Round One of this two-questionnaire process each expert was asked to provide estimates on the basis of his own experience, whereas in Round Two each expert was asked to re-evaluate his estimates in light of the responses of others, which were plotted anonymously on graphs included with the questionnaire.

Weighted-mean restoration times for each facility class were computed for each damage-factor level and each restoration level considered. Example restoration times are provided in Table 2. Specific application of these data is facilitated by preparing function restoration curves, which are simply plots of the time required to restore function to levels of 30, 60 and 100%. An example is shown in Fig. 1.

Using the expert-opinion restoration time estimates and the importance factors developed under this project, it is possible to determine the functionality of specific facilities by following the following steps:

1. Determine the shaking hazard in the area

2. Determine the collateral hazards for the facility

3. Determine the facility damage state, which would be the sum of percent damage due to ground shaking and percent damage due to collateral hazards

4. Prepare a function restoration curve for the facility using the expert-opinion data

5. Determine the damage state for lifeline distribution components and lifeline main components affecting the facility

6. Prepare function restoration curves for lifeline distribution and main components

7. Calculate functionality, at any time T, as the product of the facility functionality (percent functional for time T as determined from facility function restoration curve) times the functionality of the main and distribution lifelines (equations used to calculate lifeline functionality are provided in the ATC-13 Report[5])

Death and Injury Estimates

Deaths and injuries resulting from severe earthquakes in California will be principally due to the failures of man-made facilities, such as dams and buildings. In order to estimate deaths and injuries for this project, the literature was first reviewed to determine the rates of deaths and injuries as a function of damage to various facilities. On the basis of this information[2,3,9], injury and death rates were developed that are based on the total damage to a structure, including damage resulting from ground shaking and collateral causes. Estimates are provided in Table 3 for two categories of construction: (1) light steel construction and wood-frame construction and (2) all other types of construction.

CONCLUDING REMARKS

It is important that the reader and user of the data described in this paper be aware that the loss estimates for shaking, for collateral hazards, and for collateral losses and loss of function are based on judgment and were established using an iterative questionnaire process. It is also important to note that (1) the estimates provided are for facilities in California, where structures are designed to resist earthquakes, and (2) great amounts of experimental data (i.e., from actual earthquakes) are needed to verify or alter these estimates. When using the expert-opinion data developed in this study, care must be taken to recognize the limitations of the method employed in developing them. The estimates are based upon the clearly subjective judgment of individuals who have drawn on their experience history and very limited data. Although the questionniare process was performed under a highly controlled environment, biases, such as those due to conservatism, pessimism, and optimism, and subjective pressures are difficult to eliminate. As a result, the motion-damage relationships developed under this project may not reflect "real life" characteristics of structures and may not confirm field data collected after a major earthquake. At high intensities (MMI IX, X, XI, XII), for example, there are always some structures that are undamaged, or slightly damaged, and vice-versa at low intensities. Because the expert-opinion motion-damage relationships represent only average conditions, they may not reflect such occurrences. The data described in this paper, however, do reflect the best judgment of a group of highly prominent earthquake engineers and, with the exception of weak statistical data on damage for about a half-dozen types of structures, represent the only available information for the wide variety of structure types currently existing in California.

REFERENCES

(1) Steinbrugge, K. V., Algermissen, S. T., Lagorio, H. J., Cluff, L. S., and Degenkolb, H. J., "Metropolitan San Francisco and Los Angeles Earthquake Loss Studies: 1980 Assessment," U. S. Geological Survey Open-File Report 81-113.

(2) NOAA, "A Study of Earthquake Losses in the San Francisco Bay Region," prepared for the Office of Emergency Preparedness by the U. S. Department of Commerce, National Oceanic & Atmospheric Administration, 220 pp., 1972.

(3) NOAA, "A Study of Earthquake Losses in the Los Angeles, California Area," prepared for the Federal Disaster Assistance Administration by the Department of Housing and Urban Development/U.S. Department of Commerce, National Oceanic & Atmospheric Administration, 331 pp., 1973.

(4) Moore, D., Okamoto, T., Russo, J., Wilson, R., and Rojahn, C., "The FEMA Earthquake Damage and Loss Estimation System (FEDLOSS)," Proceedings of the 1985 Multiconference of the Society for Computer Simulation, San Diego, California, 1985.

(5) ATC, "Earthquake Damage Evaluation Data for California," Applied Technology Council Report ATC-13, Redwood City, California, 484 pp., 1985.

(6) Lofting, E. M., "Kern County Interindustry Modeling Study," Report to the Kern County Water Agency, Engineering-Economics Associates, Berkeley, California, 137 pp., 1982.

(7) Wood, H. O., and Newmann, F., "Modified Mercalli Intensity Scale of 1931," Seismological Society of America Bulletin, Vol. 21, No. 4, pp. 277-283, 1931.

(8) Whitman, R. V., Reed, J. W., and Hong, S. T., "Earthquake Damage Probability Matrices," Proceedings of the Fifth World Conference on Earthquake Engineering, International Association for Earthquake Engineering, Rome, Italy, 1973.

(9) Anagnostopoulos, S. A., and Whitman, R. V., On Human Loss Prediction in Buildings During Earthquakes," Proceedings of the Sixth World Conference on Earthquake Engineering, New Delhi, India, 1977.

Table 1
Damage Probability Matrices Based on Expert Opinion for
Example Earthquake Engineering Facility Classes

Central Damage Factor	Modified Mercalli Intensity						
	VI	VII	VIII	IX	X	XI	XII

Wood Frame Buildings (Low Rise: 1-3 stories)

0.00	3.7	***	***	***	***	***	***
0.50	68.5	26.8	1.6	***	***	***	***
5.00	27.8	73.2	94.9	62.4	11.5	1.8	***
15.00	***	***	3.5	37.6	76.0	75.1	24.8
45.00	***	***	***	***	12.5	23.1	73.5
80.00	***	***	***	***	***	***	1.7
100.00	***	***	***	***	***	***	***

Unreinforced Masonry Buildings with Load Bearing Frame (Medium Rise: 4-7 stories)

0.00	0.5	***	***	***	***	***	***
0.50	15.3	2.9	***	***	***	***	***
5.00	81.2	66.6	13.5	1.9	0.3	***	***
15.00	3.0	30.1	69.3	40.6	14.1	2.0	0.2
45.00	***	0.4	17.2	54.4	63.4	28.4	8.5
80.00	***	***	***	3.1	22.2	67.5	78.8
100.00	***	***	***	***	***	2.1	12.5

Moment-Resisting Non-Ductile Distributed Concrete Frame (High Rise: 8+ stories)

0.00	0.1	***	***	***	***	***	***
0.50	27.0	2.2	***	***	***	***	***
5.00	72.9	89.3	32.2	3.0	***	***	***
15.00	***	8.5	66.9	68.1	19.9	3.9	0.1
45.00	***	***	0.9	28.9	74.2	57.8	12.4
80.00	***	***	***	***	5.9	38.3	84.3
100.00	***	***	***	***	***	***	3.2

Moment-Resisting Steel Distributed Frame (High Rise)

0.00	26.8	0.5	***	***	***	***	***
0.50	60.0	22.2	2.7	***	***	***	***
5.00	13.2	77.1	92.3	58.8	14.7	5.9	0.8
15.00	***	0.2	5.0	41.2	83.0	67.1	42.3
45.00	***	***	***	***	2.3	26.9	55.7
80.00	***	***	***	***	***	0.1	1.2
100.00	***	***	***	***	***	***	***

***Very small probability

Table 1 (Continued)
Damage Probability Matrices Based on Expert Opinion for Example Earthquake Engineering Facility Classes

Central Damage Factor	Modified Mercalli Intensity						
	VI	VII	VIII	IX	X	XI	XII
Conventional Continuous/Monolithic Bridges (less than 500-foot spans)							
0.00	93.6	8.1	0.9	***	***	***	***
0.50	6.4	77.8	17.6	***	***	***	***
5.00	***	14.1	78.6	56.5	***	***	***
15.00	***	***	2.9	43.5	1.8	1.2	0.7
45.00	***	***	***	***	98.2	36.8	5.7
80.00	***	***	***	***	***	61.9	39.1
100.00	***	***	***	***	***	0.1	54.5
Earthfill and Rockfill Dams							
0.00	50.9	***	***	***	***	***	***
0.50	49.1	86.6	20.0	1.1	***	***	***
5.00	***	13.4	80.0	88.9	62.5	7.8	***
15.00	***	***	***	10.0	37.5	71.1	21.4
45.00	***	***	***	***	***	21.1	74.1
80.00	***	***	***	***	***	***	4.5
100.00	***	***	***	***	***	***	***
On Ground Liquid Storage Tanks							
0.00	94.0	2.5	0.4	***	***	***	***
0.50	6.0	92.9	30.6	2.1	***	***	***
5.00	***	4.6	69.0	94.6	25.7	2.5	0.2
15.00	***	***	***	3.3	69.3	58.1	27.4
45.00	***	***	***	***	5.0	39.1	69.4
80.00	***	***	***	***	***	0.3	3.0
100.00	***	***	***	***	***	***	***
Highways							
0.00	93.3	18.8	2.8	1.0	***	***	***
0.50	6.7	61.5	27.0	13.8	1.3	0.1	***
5.00	***	19.7	68.8	75.4	59.0	20.5	4.6
15.00	***	***	1.4	9.8	39.1	65.2	50.2
45.00	***	***	***	***	0.6	14.2	43.4
80.00	***	***	***	***	***	***	1.8
100.00	***	***	***	***	***	***	***

***Very small probability

Table 2
Loss of Function Restoration Time Based on Expert Opinion for Example Social Function Classifications

Central Damage Factor	Weighted Mean Time (in Days) to Restore To		
	30% Usability	60% Usability	100% Usability
Permanent Residential Dwelling			
0.50	0.2	0.2	0.8
5.00	0.3	1.5	3.3
15.00	1.9	5.4	10.5
45.00	15.2	30.5	71.9
80.00	57.2	93.8	146.6
100.00	105.5	152.1	211.9
Professional, Technical and Business Services			
0.50	1.2	2.4	5.8
5.00	3.4	10.2	20.0
15.00	9.8	44.6	71.0
45.00	37.0	111.6	202.7
80.00	114.7	213.7	343.1
100.00	214.8	355.9	439.3
Food and Drugs Processing			
0.50	1.0	2.2	4.4
5.00	3.0	6.4	16.1
15.00	17.5	37.3	72.7
45.00	122.8	180.9	235.6
80.00	150.5	257.6	380.7
100.00	362.5	503.8	534.1
High Technology			
0.50	0.0	0.0	1.1
5.00	4.7	5.5	16.5
15.00	36.8	55.9	111.8
45.00	136.4	198.2	258.2
80.00	198.2	281.1	429.1
100.00	365.0	548.0	612.0
Mining			
0.50	0.5	0.7	6.1
5.00	4.9	9.4	18.2
15.00	23.8	43.0	83.0
45.00	92.7	156.0	265.3
80.00	352.3	460.6	648.6
100.00	734.5	797.9	949.0

Table 2 (Continued)
Loss of Function Restoration Time Based on Expert Opinion
for Example Social Function Classifications

Central Damage Factor	Weighted Mean Time (in Days) to Restore To		
	30% Usability	60% Usability	100% Usability
Education			
0.50	2.8	4.2	5.7
5.00	6.4	11.4	15.5
15.00	21.5	43.8	72.1
45.00	80.8	125.2	183.0
80.00	177.7	267.2	362.1
100.00	312.3	386.5	562.6
Highway Systems, Major Bridges			
0.50	0.3	0.8	1.6
5.00	1.1	3.2	7.4
15.00	55.3	80.6	141.6
45.00	168.1	256.3	391.8
80.00	598.7	760.3	844.5
100.00	758.8	878.4	946.8
Airport Terminals			
0.50	0.0	0.0	0.0
5.00	0.4	2.0	18.3
15.00	20.3	73.0	112.0
45.00	83.0	178.0	264.3
80.00	160.0	303.3	530.3
100.00	240.0	455.3	638.7
Electrical Transmission Lines			
0.50	0.2	0.4	1.0
5.00	1.4	1.7	2.3
15.00	8.6	12.3	16.9
45.00	31.8	37.3	48.9
80.00	61.9	74.4	81.9
100.00	87.5	107.1	126.7
Dams			
0.50	0.0	0.0	2.6
5.00	0.0	1.3	6.4
15.00	10.7	30.0	98.0
45.00	64.7	207.2	367.6
80.00	563.4	563.4	683.9
100.00	632.3	632.3	700.8

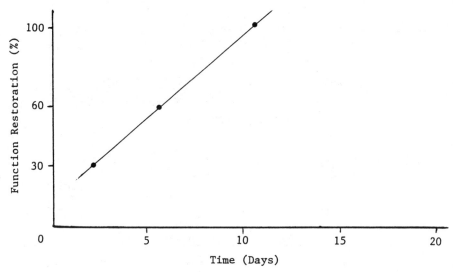

Fig. 1 Function Restoration Curve for Permanent Residential Dwellings at Damage State 4 (Central Damage Factor = 15.00).

Table 3
Injury and Death Rates*

Damage State	CDF** (%)	Fraction Injured		Fraction Dead
		Minor	Serious	
1	0	0	0	0
2	.5	3/100,000	1/250,000	1/1,000,000
3	5	3/10,000	1/25,000	1/100,000
4	20	3/1,000	1/2,500	1/10,000
5	45	3/100	1/250	1/1,000
6	80	3/10	1/25	1/100
7	100	2/5	2/5	1/5

*Estimates are based on consensus of the PEP and are for all types of construction except light steel construction and wood-frame construction. For light steel construction and wood-frame construction, multiply all numerators by 0.1.
**Central Damage Factor

ANALYSIS OF MULTISTORY STRUCTURE-FOUNDATION SYSTEMS
INCLUDING P-DELTA EFFECTS

K.S. Sivakumaran[1]

ABSTRACT

Seismic provisions in the National Building Code of Canada [1] do not include either the soil-structure interaction or the P-delta effects. ATC's tentative seismic provisions [5] include guidelines on how to incorporate the above effects in the design of a multistory building. This paper demonstrates the method of application and compares the calculated responses of two (five- and twenty-story) moment resisting steel frames, designed in accordance with the National Building Code of Canada, using (1) Applied Technology Council Equivalent Lateral Force Procedure [5] and (2) Elastic Dynamic Analysis. A method of elastic dynamic analysis of multistory structures on soft soil including P-delta effects is described in this paper.

INTRODUCTION

Construction of earthquake resistant buildings in Canada is regulated by the National Building Code of Canada (NBCC 1985) [1], issued by the National Research Council of Canada. According to NBCC 1985 [1], multistory buildings located in highly seismic zones must be designed (a) to limit lateral deflection (b) to have sufficient structural capacity and structural integrity to resist safely all loads. The interstory drift must be within the established limitation to minimize nonstructural damage. NBCC 1985 [1] recommended interstory drift limitation is $h_i/200$ under specified earthquake loads, where h_i is the height of the ith story. To prevent collision of buildings during an earthquake, the maximum lateral deflections should be limited such that the sum of the individual deflections of the adjacent structures does not exceed the separation between them. The applied loads are the stationary gravity forces from dead loads and live loads due to intended use and occupancy and transient lateral forces caused by earthquake. The loading due to earthquake motion is determined by first calculating the minimum lateral base shear force, $V = v \cdot S \cdot K \cdot I \cdot F \cdot W$ and distributing it as equivalent static forces to be concentrated at each storey. The base shear forces can be distributed either by a dynamic analysis or by considering a portion to be concentrated at the top of the structure and the remainder to have a distribution approximately triangular in shape, with apex at the base. In the calculation of base shear, v is the specified zonal horizontal ground velocity expressed as a ratio to 1 m/s, S is the seismic response factor which is dependent on the seismic zone and the period T of the structure, K is the numerical coefficient that reflects the material and type of construction, damping and ductility of the structure, I is the importance factor of the structure, F is the foundation factor and W is the specified total gravity forces.

---

[1] Assistant Professor, Department of Civil Engineering and Engineering Mechanics, McMaster University, Hamilton, Ontario, Canada L8S 4L7.

Soft soil conditions at a site have been shown to alter the structural response due to (a) amplification of seismic waves (b) flexibility and damping capacity of foundation. The former is called the site effects and have been incorporated in the seismic provisions of the NBCC 1985 [1] through the foundation factor. The latter is soil-structure interaction. Due to soil-structure interaction, the fundamental period and the effective damping of a structure are increased [2]. The significance of soil-structure interaction in structural response depends on the properties of the structure relative to the foundation. If the foundation is stiff and supported on a deep, soft soil layer considerable energy will be transferred from the structure to the soil and the response of the structure may differ drastically. Hence, in general, the soil-structure interaction can be important for buildings on soft soil and must be accounted for in an earthquake response analysis. However, soil-structure interaction effects are not considered explicitly in the NBCC 1985 [1] and presumed that neglecting the effects results in conservative design.

In general the P-delta effects produced by vertical loads acting on a structure in its displaced configuration should be taken into account in the design of structures and their structural members. The P-delta effect has been a matter of concern in the determination of the ultimate static carrying capacity and stability of frames [3] and were found to have considerable influence upon lateral strength of structures. However, results on seismic P-delta effects are rather limited [4]. The current building code [1] does not give specific recommendations to evaluate P-delta effects during earthquake. In view of this, an analytical study was undertaken to evaluate the responses, including the effects of soil-structure interaction and P-delta, of multistory buildings designed in accordance with NBCC 1985 [1] seismic provisions. This paper demonstrates the method of application and compares the calculated responses with that of the design values of two moment resisting frames using (1) Applied Technology Council (ATC) Equivalent Lateral Force Procedure [5], (2) Elastic Dynamic Analysis. The results using inelastic analysis methods are not available at this time.

CHARACTERISTICS OF STRUCTURE-FOUNDATION

In order to examine the influence of soil-structure interaction and P-delta the following structural systems, applied forces and site conditions were postulated. Two steel ($E = 30 \times 10^6$ psi (200 GPa)) buildings of five- and twenty-stories, as shown in Figure 1, are fully moment resisting unbraced frames, square in plan and have bay widths of 40 ft (12 m) and 80 ft (24 m) respectively. The floor system is considered rigid relative to the columns and carries a uniformly distributed average gravity load of 100 psf (4.8 kPa) at each floor. The buildings were designed for the seismic conditions, $v = 0.4$ (1940 ElCentro), $S = 0.22/\sqrt{T}$, $K = 0.7$, $I = 1.0$, $F = 1.0$ and to satisfy NBCC 1985 [1] strength and stiffness requirements. The moment of inertia of the columns, which were varied every fifth floor from the base, are also shown in Figure 1. The structures are doubly symmetrical in the horizontal plane; hence, the torsional effect has been eliminated.

Now, the responses of these structures are to be determined for two postulated earthquakes: (1) Major earthquake, 1940 ElCentro S 00° E component (2) Moderate earthquake, 1952 Taft S 69° E component. These earthquake records are considered to include site effects; hence, they are the resulting "free field" surface motions. The following data are also necessary in order to incorporate soil-structure interaction and P-delta effects in the analysis methods namely, ATC Equivalent Lateral Force Procedure and Elastic Dynamic Analysis. The fundamental damping ratio of the superstructures is taken to be 5%. The base of the building is a thick reinforced concrete mat, weighing 240 kip (1035 kN) and 960 kip (4265 kN) respectively, and assumed to rest at or near the ground surface of soft soil (sand and gravel, mass density $\rho = 120$ lbf/ft^3 (1925 kg/m^3), Poisson's ratio $v = 0.33$, average shear wave velocity at small strains $V_s = 450$ ft/sec (135 m/s). Shear wave velocity of the soil medium can generally indicate the flexibility of the foundation (rigid rock $V_s > 2500$ ft/sec (750 m/s)).

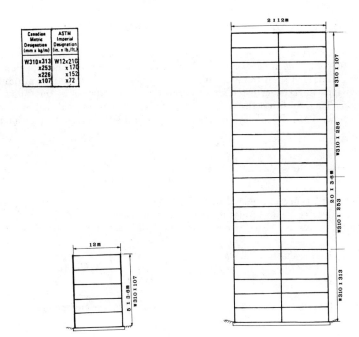

Figure 1. Schematic Elevation of Multistory Buildings

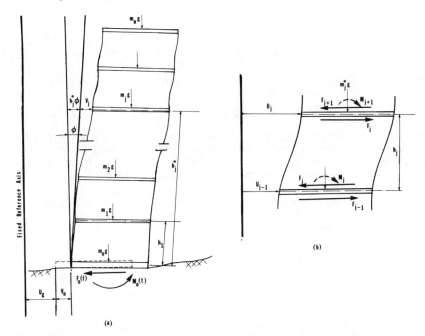

Figure 2. Structure-Foundation system including P-delta effects.

ATC EQUIVALENT LATERAL FORCE PROCEDURE

Recently, by the concerted effort of a team of experts in earthquake engineering, Applied Technology Council published a document entitled Tentative Provisions for the Development of Seismic regulations for Buildings [5]. Although, this document is not by any means a building code, when the various provisions have been tested by research and field experience they may be considered for code adoption. Therefore first the structures described here were analysed in accordance with ATC recommendations. The ATC [5] design base shear of a building on rigid foundation is given by

$$V = \frac{1.2 A_v FW}{RT^{3/2}} \leq 2.5 \frac{A_a W}{R} \tag{1}$$

In "Eq. 1", A_v is a coefficient representing Effective Peak Velocity-related acceleration, A_a is the coefficient representing Effective Peak Acceleration and R is the response modification factor. For the present site conditions, the values of coefficients using ATC [5] recommendations are

$\left. \begin{array}{l} A_v = 0.32 \\ A_a = 0.25 \end{array} \right\}$ (1940 ElCentro) $\qquad \left. \begin{array}{l} A_v = 0.17 \\ A_a = 0.15 \end{array} \right\}$ (1952 Taft)

$T = 0.75$ sec (five story) $\qquad T = 2.13$ sec (twenty story)

$F = 1.0$ $\qquad R = 4.5$

However, in the design of buildings on soft soil, incorporation of soil-structure interaction in accordance with ATC [5] provisions results in a reduced design base shear $\widetilde{V}$. The magnitude of reduction is dependent on soil conditions and structural properties. Here, computations in accordance with the ATC [5] recommendations indicated a maximum reduction of 12.5% of the ATC [5] design base shear of the twenty-story structure on soft soil ($V_s = 450$ ft/s) subjected to 1940 ElCentro earthquake. The deflections including the soil-structure interaction were determined using ATC [5] equation 6.11,

$$\text{modified deflection} = \frac{\widetilde{V}}{V} \left[\frac{M_o h_j^*}{K_\phi} + V_j \right] \tag{2}$$

where M_o is the overturning moment at the base, h_j^* is the height above the base to the floor level 'j', V_j deflection of the jth floor determined by an elastic analysis considering the building fixed at base and K_ϕ is the rocking stiffness of the foundations given by $K_\phi = 8 Gr^3/3(1-v)$. Here, G is the shear modulus of elasticity of the soil medium and r is the equivalent radius of the foundation mat.

Although ATC [5] has suggested procedures to take into account of P-delta effects, here they are taken into account in the following manner (see Figure 2(b)). The additional overturning moment M_j at each floor, due to P-delta effect is given by

$$M_j = m_j^* g (U_j - U_{j-1}) \tag{3}$$

where g is the gravitational acceleration, U_j is the total horizontal displacement of the jth floor and

$$m_j^* = \sum_{i=j}^{N} m_i$$

where m_i is the mass of the ith floor. This overturning moment can be replaced by statically equivalent lateral forces F_j for each story applied at the upper and lower end along opposite directions. These fictitious horizontal forces are equal to the overturning moment divided by the story height. Thus

$$F_j = \frac{m_j^* g}{h_j} V_j - \frac{m_j^* g}{h_j} V_{j-1} + m_j^* g \phi \qquad (4)$$

Here, the rotation of the base $\phi = M_o/K_\phi$. Using direct stiffness method and combining ATC [5] story shears S_j with these horizontal destabilizing forces, the following relations are obtained

$$\left[[K] - [K_G]\right] \{V_j\} = \{S_j\} + \{m_g\} g \cdot \frac{M_o}{K_\phi} \qquad (5)$$

In these equations and in the equations to follow $\{x_j\}$ is a Nx1 vector of quantity x_j, where x_j denotes that it is associated with the jth floor or j^{th} mode of vibration, [K] is the elastic stiffness matrix of order NxN and $[K_G]$ is the lateral geometric stiffness matrix. [K] and $[K_G]$ are tri-diagonal matrices whose non zero terms are

$$K(j,j) = 12\,(EI)_j/h_j^3 + 12\,(EI)_{j+1}/h_{j+1}^3; \quad K(j,j-1) = K(j-1,j) = -12\,(EI)_{j+1}/h_{j+1}^3$$

$$K_G(j,j) = m_j^* g/h_j + m_{j+1}^* g/h_{j+1}; \quad K_G(j,j-1) = K_G(j-1,j) = -m_{j+1}^* g/h_{j+1} \qquad (6)$$

where, $j \leq N$ and $(EI)_j$ is the sum of the flexural rigidities of the jth story columns. The static deflections of the system were obtained by simple inversion. Note that "eq. 5" and "eq. 6" contain the gravitational acceleration g, which represents the P-delta effect. The ATC [5] equivalent static responses of the two buildings for various conditions are shown in Figures 3-6.

ELASTIC DYNAMIC ANALYSIS

In considering dynamic soil-structure interaction the base of the building can be considered as a massless rigid disc on an elastic half-space. Appropriate dynamic-force-displacement relationships of a rigid disc on a half-space can be evaluated by methods of continuum mechanics [6]. Consequently, the interaction should be modelled by spring-dashpot devices whose stiffness and damping terms are functions of the frequency of excitation. In principle, such a frequency-dependent foundation model could be used directly in a frequency-domain analysis of soil-structure interaction problem. However, such a procedure requires a large computational effort, especially for structures with many degrees of freedom such as multistory buildings. To avoid this difficulty, Veletsos and Verbic [7] obtained approximate dynamic-force-displacement relationships which are independent of frequency and yet applicable for earthquake motions. Neglecting the small coupling between horizontal and rocking motions, such integro-differential force-displacement relationships for horizontal translation and rocking motion at time t are given by:

$$F_0(t) = K_v \left[V_0(t) + b_0 \frac{r}{V_s} \dot{V}_0(t) \right] \qquad (7)$$

$$M_0(t) = K_\phi \left[(1-b_1)\phi(t) + b_1 b_2 \left(\frac{r}{V_s}\right)\dot{\phi}(t) + b_3 \left(\frac{r}{V_s}\right)^2 \ddot{\phi}(t) \right.$$

$$\left. + \frac{b_1}{b_2} \frac{V_s}{r} \left\{ \int_o^t \phi(\tau) e^{-(V_s/b_2 r)(t-\tau)} d\tau \right\} \right] \qquad (8)$$

where, V_0 is the horizontal displacement of the base and $K_v = 8Gr/(2-v)$ is the horizontal stiffness of the half-space. The values of b_0, b_1, b_2 and b_3 are Poisson's ratio dependent. (For $v = 0.33$, $b_0 = 0.65$, $b_1 = 0.5$, $b_2 = 0.8$, $b_3 = 0.0$). Influences of the P-delta are incorporated again using the equivalent lateral forces F_j given by "eq. 4". The building-foundation system is now subjected to horizontal free-field ground displacement U_g (acceleration $\ddot{U}_g$). Therefore, the total horizontal displacement U_j of the jth floor with respect to a fixed vertical reference axis is

$$U_j = U_g + V_0 + h_j^* \phi + V_j \qquad (9)$$

Now, the equation of motion for any floor 'j' can be formulated by expressing the equilibrium of the effective forces associated with each degree of freedom. Thus the equation of motion in terms of the degrees of freedom V_j for each floor mass of the superstructure above the base

$$[M]\{\ddot{V}_j\} + [C]\{\dot{V}_j\} + [[K] - [K_G]]\{V_j\}$$
$$= -\{m_j\}(\ddot{U}_g + \ddot{V}_0) - \{m_j h_j^*\}\ddot{\phi} + \{m_j\} g \phi \qquad (10)$$

In addition, the equations of motion of the whole building in translation and rotation

$$m_0(\ddot{U}_g + \ddot{V}_0) + \{m_j\}^T \{\ddot{U}_j\} + F_0(t) = 0 \qquad (11)$$

$$I^* \ddot{\phi} + \{m_j h_j^*\}^T \{\ddot{U}_j\} - \{m_j\}^T \{h_j^*\} g\phi - \{m_j\}^T \{V_j\} g + M_0(t) = 0 \qquad (12)$$

where

$$I^* = \sum_{j=0}^{N} I_j \qquad (13)$$

where I_j is the mass moment of inertia of the jth floor including that of the footing, m_0 is the mass of the footing, $[M]$ is the diagonal mass matrix of order NxN, $[C]$ is the damping matrix of order NxN. When the damping matrix $[C]$ satisfies certain conditions the simplest of which is when it is linear combination of $[M]$ and $[K]$ the system has normal modes of vibration. Assuming this condition is satisfied and also assuming that the superstructure possess classical normal modes, the N coupled equations of motion of the superstructure "eq. 10" can be changed to N uncoupled equations using normal coordinates transformation. Thus the uncoupled equations in terms of normal coordinates q_j are given by

$$\ddot{q}_j + 2\xi_j \omega_j \dot{q}_j + \omega_j^2 q_j = -a_j(\ddot{U}_g + \ddot{V}_0) - \beta_j \ddot{\phi} + a_j g \phi \; ; \qquad j = 1 \text{ to } N \tag{14}$$

where

$$\{a_j\} = [\gamma]^T \{\frac{m_j}{M_j^*}\}, \quad \{\beta_j\} = [\gamma]^T \{\frac{m_j h_j^*}{M_j^*}\} \tag{15}$$

and where ω_j, ξ_j and M_j^* are the jth mode undamped frequency, damping ratio and normal coordinate generalized mass respectively and $[\gamma]$ is the mode shape matrix. Now, the jth mode-displacement and mode-acceleration can be obtained in terms of U_g, V_0, ϕ and their time derivatives by solving the set of "eq. 14". Hence,

$$q_j(t) = -\int_0^t L_j(\tau) \lambda_j(t - \tau) \, d\tau \tag{16}$$

$$\ddot{q}_j(t) = \int_0^t L_j(\tau) \mu_j(t - \tau) \, d\tau - L_j(t) \; ; \qquad j = 1 \text{ to } N$$

where

$$L_j(t) = a_j(\ddot{U}_g(t) + \ddot{V}_0(t)) + \beta_j \ddot{\phi}(t) - a_j g \phi(t)$$

$$\lambda_j(t) = \frac{1}{\omega_{Dj}} e^{-\xi_j \omega_j t} \sin \omega_{Dj} t \tag{17}$$

$$\mu_j(t) = \frac{\omega_j^2}{\omega_{Dj}} e^{-\xi_j \omega_j t} \cos(\omega_{Dj} t - \Psi_j)$$

and where

$$\omega_{Dj} = \omega_j \sqrt{1 - \xi_j^2} \tag{18}$$

$$\Psi_j = \tan^{-1}(1 - 2\xi_j^2)/2\xi_j \sqrt{1 - \xi_j^2}$$

Using "eq. 9" and its time derivatives, now the remaining governing "eq. 11" and "eq. 12" can be written respectively in the form

$$m_o(\ddot{U}_g(t) + \ddot{V}_0(t)) + \{A_j\}^T \left\{ \int_0^t (\ddot{U}_g(\tau) + \ddot{V}_0(\tau))\mu_j(t-\tau)\,d\tau \right\} + \{B_j\}^T \left\{ \int_0^t \ddot{\phi}(\tau)\mu_j(t-\tau)\,d\tau \right\}$$

$$+ m_1^* g\,\phi(t) - \{A_j\}^T g \left\{ \int_0^t \phi(\tau)\mu_j(t-\tau)\,d\tau \right\} + F_0(t) = 0 \qquad (19)$$

$$I^*\ddot{\phi}(t) + \{B_j\}^T \left\{ \int_0^t (\ddot{U}_g(\tau) + \ddot{V}_0(\tau))\mu_j(t-\tau)\,d\tau \right\} + \{A_j\}^T g \left\{ \int_0^t (\ddot{U}_g(\tau) + \ddot{V}_0(\tau))\lambda_j(t-\tau)\,d\tau \right\}$$

$$+ \{C_j\}^T \left\{ \int_0^t \ddot{\phi}(\tau)\mu_j(t-\tau)\,d\tau \right\} + \{B_j\}^T g \left\{ \int_0^t \ddot{\phi}(\tau)\lambda_j(t-\tau)\,d\tau \right\}$$

$$- \{B_j\}^T g \left\{ \int_0^t \phi(\tau)\mu_j(t-\tau)\,d\tau \right\} - \{A_j\}^T g^2 \left\{ \int_0^t \phi(\tau)\lambda_j(t-\tau)\,d\tau \right\} + M_0(t) = 0 \qquad (20)$$

where,

$$\{A_j\}^T = \left\{ \{m_j\}^T [\gamma]_j \cdot \alpha_j \right\}$$

$$\{B_j\}^T = \left\{ \{m_j\}^T [\gamma]_j \cdot \beta_j \right\} = \left\{ \{m_j h_j^*\}^T [\gamma]_j \cdot \alpha_j \right\} \qquad (21)$$

$$\{C_j\}^T = \left\{ \{m_j h_j^*\}^T [\gamma]_j \cdot \beta_j \right\}$$

Noting that the base shear $F_0(t)$ and moment $M_0(t)$ can be expressed in terms of V_0 and ϕ ("eq. 7" and "eq. 8"), thus the governing equations of the total system with $N+2$ degrees of freedom, namely $\{V_j\}$, V_0, ϕ, are reduced to two couple integro-differential equations above in terms of the base displacements V_0 and ϕ. However, close-form solutions satisfying these equations cannot yet be obtained. Here, numerical procedures [8] are used. Trapezoidal rule and average acceleration methods have been used to obtain the integrand and the derivatives respectively. Solving the equations for V_0 and ϕ using time steps of 0.02 seconds, time history for base displacements, floor deflections, and floor shear can be obtained. For brevity, details of these procedures are not presented herein. Figures 3-6 also show the responses of the two buildings obtained using the elastic dynamic analysis described herein.

RESULTS AND DISCUSSION

Figures 3 and 4 show the maximum floor displacements of the five-story building under consideration, subjected to 1940 ElCentro and 1952 Taft earthquakes, respectively. Figure 5 and 6 show the corresponding responses of the twenty-story building. The plots with symbols are the results from ATC [5] equivalent static procedure. The figures also show the NBCC 1985 [1] allowable design deflections. In these figures, the displacements at the zero story level are the maximum horizontal translation of the footing. In ATC [5] analysis procedure the base is considered to be fixed; hence, the horizontal displacement of base is zero. Using the elastic dynamic analysis, maximum base displacement of 0.242 inch and maximum base rotation of 0.0019 rad. were observed, both on the twenty-story building on soft soil subjected to the major earthquake. Here, the results are shown for the soil conditions (1) V_s = 450 ft/sec (135 m/s), (2) V_s = 2500 ft/sec (750 m/s). The in-situ shear wave velocity depends on the strain level, confining stress etc. Here, 65% and 75% of the shear wave velocity at small strains were considered to be the appropriate in-situ V_s during (a) major, (b) moderate earthquakes respectively.

From these limited analyses and the corresponding results, the following observations were made:

Base shear

ATC [5] base shear, without soil-structure interaction, is 18% and 25% higher than NBCC 1985 [1] base shear for major and moderate earthquakes respectively. The same base shear obtained using the elastic dynamic analysis were 6 times larger than NBCC 1985 [1] base shear of the five-story building and 3 times larger than that of the twenty-story building. This discrepancy is due to the fact that most building codes allow design of structures to behave inelastically during severe ground excitations. Soil-structure interaction reduces the base shear. Using ATC [5] seismic provisions reductions varying from 4% to a maximum of 12.5% of the base shear of the buildings on stiff soil (V_s = 2500 ft/s) were obtained. In the elastic dynamic analysis a reduction of the same of 9% to 25% were noted, maximum being on the twenty-story building. Due to P-delta effects the base shears were increased; however, the maximum increase was about 5%.

Deflections

NBCC 1985 [1] design deflections were obtained using a deflection amplification factor value of 3. This is the value suggested by NBCC 1985 to obtain realistic values of anticipated deflections. In accordance with ATC [5], the deflection amplification factor for ordinary moment resisting steel frames is 4. This value was used in calculating ATC [5] based deflections. Due to soil-structure interaction, the deflections of the buildings on soft soil can be either increased or decreased [2]. In accordance with ATC [5] the deflections due to soil-structure interaction were decreased in the five-story building and were increased in the twenty-story building. This characteristic was also evident in the results using elastic dynamic analysis. Inexplicably, ATC [5] predicted deflections of the five-story building were lower than that of the elastic dynamics analysis and the deflections of the twenty-story buildings were higher than that of the same.

ACKNOWLEDGEMENT

This study was supported by the Natural Sciences and Engineering Research Council of Canada Grant No. U0384.

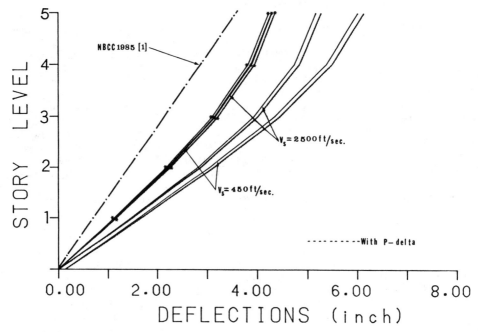

Figure 3. Maximum Story Deflections of a Five-story Building (1940 El Centro)

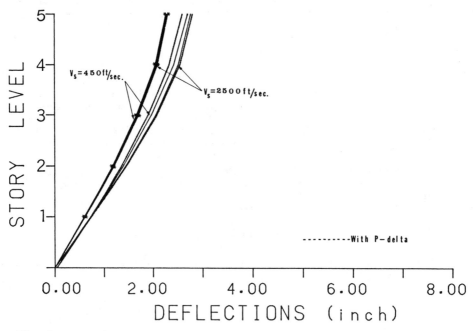

Figure 4. Maximum Story Deflections of a Five-story Building (1952 Taft)

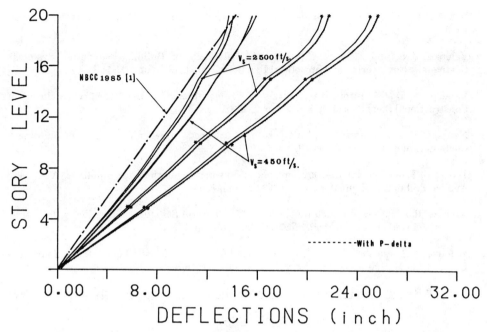

Figure 5. Maximum Story Deflections of a Twenty-story Building (1940 El Centro)

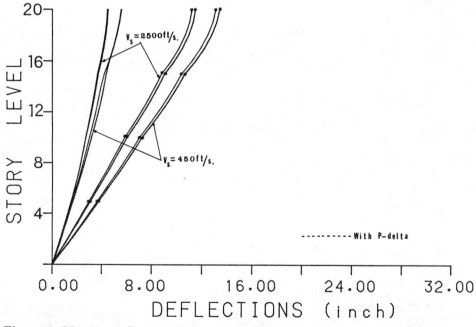

Figure 6. Maximum Story Deflections of a Twenty-story Building (1952 Taft)

REFERENCES

[1] National Building Code of Canada, National Research Council of Canada, Ottawa, Canada, 1985.

[2] Veletsos, A.S. and Meek, J.W., 'Dynamic Behaviour of Building-Foundation Systems', Earthquake Engineering and Structural Dynamics, vol. 3, 1974, pp. 121-138.

[3] Vandepitte, D., 'Non-iterative Analysis of Frames Including the P-Δ-effect', Journal of Constructional Steel Research, Vol. 2, No. 2, 1982, pp. 3-10.

[4] Neuss, C.F. and Maison, B.F., 'Analysis for P-Δ Effects in Seismic Response of Buildings', Computers and Structures, Vol. 19, No. 3, 1984, pp. 369-380.

[5] Tentative Provisions for the Development of Seismic Regulations for Buildings, ATC 3-06, Applied Technology Council, Palo Alto, California, 1984.

[6] Savuzzo, R.J., Bailey, J.L. and Raftpoulos, D.D., 'Lateral Structure Interation with Seismic Waves', Journal of Applied Mechanics, Vol. 38, 1971, pp. 125-134.

[7] Veletsos, A.S. and Verbic, B., 'Basic Response Function for Elastic Foundations', Journal of Engineering Mechanics Division, ASCE, Vol. 100, 1974, pp. 189-202.

[8] Craig, R.R., '<u>Structural Dynamics, An Introduction to Computer Methods</u>', John Wiley and Sons, New York, 1981.

Identification of Time-Dependent Dynamic Characteristics of Structures Responding to Strong Ground Motion

Vahid Sotoudeh [I]

Haresh C. Shah [II]

ABSTRACT

Major changes in the dynamic properties of structures are expected to occur and have been observed from the earthquake records obtained from several instrumented buildings.

An appropriate signal processing technique for analyzing both the earthquake records and the building response data should be capable of identifying the time variations in the amplitudes and in the frequency content.

The methodology discussed in this paper utilizes the impulse invariant transformation to obtain the discrete time equivalent of a single-degree-of-freedom oscillating system which is used as a second order component in parallel form realizations of earthquake and structural system. The overall system's input-output difference equation is then defined and its time-varying coefficients are identified.

INTRODUCTION

In general, earthquake or strong wind accelerograms obtained from instrumented structures can be considered as realizations of a random process containing information about both the dynamic excitation environment and the large-amplitude behavior of the structure.

Past earthquake building accelerograms reveal the time-varying features of the amplitudes and of the frequency content of these records. This time variation has been attributed to the expected nonstationarities in the amplitude and in the frequency content of the earthquake excitation as well as to the changes in the structure itself induced by the strong motion.

The discrete-time, earthquake-structure, time-variant model developed to study the earthquake response of a building, was initially suggested by Sato [1]. This model which is a single-input, single-output system consists of two linear, time-variant digital filters placed in series. The first filter has a white noise input and its output (which is the input to the second filter) represents the nonstationary earthquake excitation. The second filter is a multi-degree-of-freedom (MDOF) model, representing a structure with time-varying dynamic properties.

Sato, based on Lee's [2] work, used an autoregressive moving average (ARMA) model of order $(2n, 2n)$ for the response of a linear, time-invariant, n-degree-of-freedom structural system with white noise excitation. He simulated the response of a time-invariant SDOF oscillator with a band-pass filtered, amplitude-modulated white noise input. He used stationary time-series methods to identify the parameters of the combined system, for different segments of the simulated response. He also mentioned

---

(I) Ph.D. Candidate, The John A. Blume Earthquake Engineering Center Stanford University, Stanford, California
(II) Professor and Chairman, Civil Engineering Department, Stanford University, Stanford, California

the sensitivity of the estimated parameters to the size of these segments.

The two basic types of non-stationarity which are considered here are the non-stationarities in amplitude and in frequency. These are commonly observed in major earthquake records.

A non-stationary time history is treated here as a realization of a discrete-time stochastic process being the output of a *linear, time-variant*, and *response-independent* system. The dynamic characteristics of such a system are continuously changing with time and the input or the " signal generating process" is assumed to be a zero-mean white noise process. The time-varying property of this system enables it to generate a non-stationary time history.

DYNAMIC RESPONSE OF A LINEAR SINGLE-DEGREE-OF-FREEDOM (SDOF) SYSTEM

The Continuous-Time Model

The differential equation for a linear, viscously damped SDOF system is

$$m\ddot{y}(t) + c\dot{y}(t) + ky(t) = x(t) \tag{1}$$

or

$$\ddot{y}(t) + 2\xi p\dot{y}(t) + p^2 y(t) = \frac{x(t)}{m} \tag{2}$$

- $y(t)$ steady state response (displacement)
- $x(t)$ input forcing function
- m mass of the oscillator
- p undamped circular frequency = $\sqrt{k/m}$
- ξ damping ratio = $c/2mp$

The solution to this second-order, linear differential equation is

$$y(t) = h(t) * x(t) \tag{3}$$

where

$$h(t) = \frac{1}{m \cdot p_D} e^{-p\xi t} \sin(p_D t) \quad \text{for } 0 \leq \xi \leq 1 \tag{4}$$

$h(t)$ is the unit impulse response of the system, * between $h(t)$ and $x(t)$ is the convolution sign, and $p_D = p\sqrt{1-\xi^2}$ is the damped circular frequency.

The block diagram corresponding to this system in the continuous time-domain is

$$x(t) \longrightarrow \boxed{h(t)} \longrightarrow y(t)$$

Using Laplace transformation, Equation 3 can be written in the s-domain as follows.

$$Y(s) = H(s)X(s) \tag{5}$$

where $Y(s)$, $H(s)$ and $X(s)$ are the Laplace transforms of $y(t)$, $h(t)$ and $x(t)$, respectively. Specifically, for a SDOF oscillator when $x(t)$ is the input forcing function and $y(t)$ represents the displacement response, the expression for $H(s)$ is

$$H(s) = \frac{1/m}{(s + p\xi)^2 + p_D^2} \tag{6}$$

The Discrete-Time Model

To build a discrete-time equivalent of a continuous-time system several methods can be used.

As long as the frequency response of the system is bandlimited (as of a lightly damped oscillator) and large errors due to aliasing are eliminated by sampling at a sufficiently high rate, the impulse invariant transformation [3] provides a relatively accurate discretized realization of a system. This type of transformation can be described as replacing the unit impulse response of the continuous-time system by its sampled version in the discrete-time domain. The resulting system's block diagram is :

$$X(z) \longrightarrow \boxed{H(z)} \longrightarrow Y(z)$$

where, $X(z)$ and $Y(z)$ are the z-transforms of the sampled $x(t)$ and $y(t)$, and

$$H(z) = Z\left\{Th(nT)\right\} =$$

$$\frac{T}{m \cdot p_D} \frac{z^{-1} e^{-p\xi T} \sin(p_D T)}{1 - 2z^{-1} e^{-p\xi T} \cos(p_D T) + e^{-2p\xi T} z^{-2}} \qquad (7)$$

$$Y(z) = H(z) \cdot X(z) \qquad (8)$$

$H(z)$, the transfer function in the z-domain, which is obtained by taking the z-transform of the discrete-time impulse response $Th(nT)$,[4], has a complex conjugate pair of poles z_1, z_2.

The locations of z_1 and z_2 in the complex z-plane are related to the system properties as follows.

$$|z_1| = |z_2| = e^{-p\xi T} \qquad (9)$$

$$Arg(z_1) = -Arg(z_2) = p_D T \qquad (10)$$

T is equal to $\frac{1}{f_s}$, where f_s is the sampling frequency.

$H(z)$ represents the transfer function of a *stable* system if both of its poles lie inside the unit circle in the z-plane. Thus, the stability of $H(z)$ means that $e^{-p\xi T}$ is less than one. Given that p and T are both positive and real, as long as ξ stays positive and real, the stability condition is always satisfied.

To obtain the recursive relationship between the input and output of the system , for $z \neq 0$ Equation 7 can be written as

$$H(z) = \frac{Y(z)}{X(z)} = \frac{\theta z^{-1}}{1 - \varphi_1 z^{-1} - \varphi_2 z^{-2}} \qquad (11)$$

where

$$\theta = \frac{T}{m \cdot p_D} e^{-p\xi T} \sin(p_D T) \qquad (12)$$

$$\varphi_1 = 2 e^{-p\xi T} \cos(p_D T) \qquad (13)$$

$$\varphi_2 = -e^{-2p\xi T} \qquad (14)$$

Then the input-output relation in the z-domain becomes

$$Y(z) - \varphi_1 z^{-1} Y(z) - \varphi_2 z^{-2} Y(z) = \theta z^{-1} X(z) \qquad (15)$$

Taking the inverse z-transform of both sides of Equation 15 the input-output difference equation becomes:

$$y(n) - \varphi_1 y(n-1) - \varphi_2 y(n-2) = \theta x(n-1) \tag{16}$$

Equation 16 is a difference equation substitute for the differential equation 2. From this difference equation, the steady state displacement at any time-step n can be calculated from the response values at the previous two steps and the input at the immediately preceding step.

COMBINED EARTHQUAKE-STRUCTURE MODEL

For situations when the response of the structure consists of more than a single mode, or when parallel realization for the effective earthquake model are needed, systems with orders greater than four apply.

The nonstationarities of the structural response are modeled by time-dependent coefficients of the corresponding difference equation.

The normal modes approach is adopted here. It is assumed that the observed response is the weighted sum of a finite number of modal responses, and each mode is modeled by a time-variant second-order system. A block diagram of such earthquake-structure system is shown below.

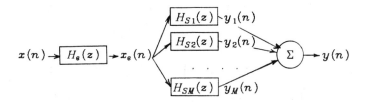

The following difference equations apply to the structural system.

$$y_1(n) - \varphi_1^{S1}(n) y_1(n-1) - \varphi_2^{S1}(n) y_1(n-2) = \theta^{S1}(n) x_e(n-1)$$

$$\cdot \quad \cdot \quad \cdot \quad \cdot \quad \cdot \quad \cdot \tag{17}$$

$$y_M(n) - \varphi_1^{SM}(n) y_M(n-1) - \varphi_2^{SM}(n) y_M(n-2) = \theta^{SM}(n) x_e(n-1)$$

where $\theta^i(n) = \sqrt{-\varphi_2^i(n)}$ and $y(n) = \sum_{i=1}^{M} B_i(n) y_i(n)$

The overall transfer function in the z-domain is

$$H(z) = \frac{Y(z)}{X(z)} = H_e(z) \sum_{i=1}^{M} B_i(n) H_{Si}(z) \tag{18}$$

The effective earthquake is represented by $H_e(z)$ which is assumed to be the transfer function of a parallel-form realization of a finite-order, linear, time-variant system, described by the following equation

$$H_e(z) = \sum_{i=1}^{I} B_i'(n) H_{ei}(z) \tag{19}$$

where I is the number of second-order blocks in the parallel realization of the earthquake model.

For the combined earthquake-structure system considered so far, the input-output difference equation can be written as

$$y(n) - \varphi_1(n)y(n-1) - \ldots \varphi_{2(M+I)}(n)y(n-2M-2I) =$$

$$a_2(n)x(n-2) + a_3(n)x(n-3) + \ldots + a_{2M+2I-2}(n)x(n-2M-2I+2)$$

(20)

The difference equation 20 is identical to the input-output relation of a nonstationary ARMA process of order $(2M+2I, 2M+2I-2)$ for which the first two moving average terms, namely $a_0(n)$ and $a_1(n)$, are equal to zero. In a standard ARMA model, $a_0(n)$ is equal to one.

In the process of system identification, the right-hand-side of Equation 20 is referred to as $e(n)$, the residuals. Since $x(n)$ is assumed to be a white noise process, $e(n)$ becomes a zero-mean process with a time-varying variance. As indicated in Equation 20, the process $e(n)$ corresponds to the combined earthquake-structure system which is a combination of cascade and parallel realizations of $M+I$ second order blocks. For this combination, there are $2M+2I-3$ terms describing $e(n)$, (Equation 20), and the autocorrelation of $e(n)$ vanishes for lags greater than $K = 2M+2I-4$.

In other words, the behavior of the residuals autocorrelation function of a best fit model is related to the overall system configuration. The two special cases are:

1. A purely parallel realization of all the $M+I$ components which corresponds to $2M+2I-1$ terms in the description of $e(n)$. For this case, the autocorrelation of $e(n)$ vanishes for lags greater than $K=2M+2I-2$.
2. A purely cascade realization of all the $M+I$ components which corresponds to only one term in the description of $e(n)$. For this case the autocorrelation of $e(n)$ vanishes for lags greater than zero.

From the preceding discussion it can be concluded that, regardless of the overall system configuration (parallel, cascade or mixed), the autocorrelation of the residuals becomes zero for lags greater than $K=2M+2I-2$. This property will be used later in the process of system identification.

The existing correlation in $e(n)$ does not affect the unbiasedness of the estimated regression coefficients. However, the variances of such estimates will be affected.

SYSTEM IDENTIFICATION

To identify the time-varying parameters of the system, first a trial order of the combined earthquake-structure system is selected as

$$P = 2M + 2I \tag{21}$$

The $\varphi_i(n)$ are expanded in terms of a set of orthogonal functions and the coefficients of these expansions are estimated through linear least squares, minimizing the sum of squares of the residuals, $e(n)$.

$$e(n) = a_2(n)x(n-2) + \ldots + a_{P-2}(n)x(n-P+2) \tag{22}$$

As discussed earlier, for an exact fit model, $e(n)$ is a zero-mean process with a time-dependent variance and its autocorrelation vanishes for lags greater than $L = P - 2$

$$E[e(n)] = 0. \tag{23}$$

$$E[e^2(n)] = a_2^2(n) + \ldots + a_{P-2}^2(n) \tag{24}$$

$$R_e(n,k) = E[e(n)e(n-k)] = 0. \quad \text{for } k > P - 2 \tag{25}$$

In the next step, the variance of $e(n)$ is stabilized by dividing it by its envelope function $g(n)$. As was suggested by Nigam [5],[6], an estimate for this envelope function can be obtained from the following relationship

$$g(n) = \sqrt{e^2(n) + \tilde{e}^2(n)} \tag{26}$$

where $\tilde{e}(n)$ is the Hilbert transform of $e(n)$.

Usually, $g(n)$, as obtained from Equation 26, is not smooth. Digital filtering is one of the recommended techniques for smoothing this estimated envelope function [7]. As an additional processing step, $g(n)$ is smoothed by passing it through a zero-phase, second-order, Butterworth lowpass filter, with a cutoff period at one-fifth of the $e(n)$ duration.

The variance-stabilized residuals, $e'(n)$, can be calculated using this processed $g(n)$ from the following relationship.

$$e'(n) = e(n) / g(n) \tag{27}$$

At this stage, it can be assumed that $e'(n)$ is a stationary, normally distributed process with the following properties.

$$E[e'(n)] = 0. \tag{28}$$

$$E[e'^2(n)] \approx \text{constant} \tag{29}$$

$$R_{e'}(k) = E[e'(n)e'(n-k)] = 0. \quad \text{for } k > P - 2 \tag{30}$$

The autocorrelation function of $e'(n)$ can be estimated as

$$\hat{R}_{e'}(k) = \frac{1}{N'} \sum_{n=k}^{N'} e'(n)e'(n-k) \tag{31}$$

where N' is the number of estimated residuals and is equal to $N - P$, (N is the number of data points in the original signal).

The following approximate statistics, known as the large-lag approximations, can be found for the estimated autocorrelation function of a process with the same statistical properties as of $e'(n)$ [8]. The term "large-lag" is used because the following statistics apply to lags which are greater than $P-2$.

$$E[\hat{R}_{e'}(k)] = 0. \tag{32}$$

$$E[\hat{R}_{e'}^2(k)] \approx \frac{1}{N'}\left\{1 + 2\sum_{i=1}^{P-2} R_{e'}^2(i)\right\} \tag{33}$$

$$E[\hat{R}_{e'}(k) \cdot \hat{R}_{e'}(k-s)] \approx \frac{1}{N'}\sum_{i=-\infty}^{\infty} R_{e'}(i) R_{e'}(i-s) \tag{34}$$

From Equation 34 together with Equation 30 it can be concluded that

$$E[\hat{R}_{e'}(k) \cdot \hat{R}_{e'}(k-s)] = 0. \quad \text{for} \quad s > P-2 \tag{35}$$

Equation 35 implies that the large-lag estimated autocorrelations which their lag differences are greater than $P-2$, are uncorrelated. This lack of correlation together with the property of being normally distributed means that these $P-2$ lag apart, large-lag autocorrelations are also statistically independent. Therefore, a set of independent, normally distributed random variables, with zero mean, can be obtained by selecting a set of estimated autocorrelations of the variance-stabilized residuals, $\hat{R}_{e'}(k)$, at lags greater than $P-2$ and more than $P-2$ lags apart. The variances of these random variables can be normalized to one, by dividing their values by the square root of their estimated variances. These can be calculated from Equation 33 in which the actual autocorrelations are replaced by their corresponding estimates.

The sum of squares of ν such normalized, independent and normally distributed estimates of large-lag autocorrelations has a Chi square distribution with ν degrees of freedom. The area α under the upper tail of this distribution (Figure 1) can be used as a measure of the goodness of the fitted model. In fact, α is the probability of all the large-lag autocorrelations being zero. The larger this probability, the better fit the model is.

The following stepwise system identification process is developed based on the procedure described so far.

1. Select the order of the model, P, and the degrees of the polynomial expansions of the coefficients.
2. Use the linear least squares regression to estimate the unknowns vector, Φ, and the residuals, $e(n)$.
3. Find the envelope of the residuals, $g(n)$.
4. Divide the residuals by the envelope function to find $e'(n)$.
5. Estimate $\hat{R}_{e'}(k)$, the autocorrelation function of $e'(n)$.
6. Select a set of ν uncorrelated, large-lag estimated autocorrelations, and find their sum of squares.
7. Divide this sum by the variance given by Equation 33, and use the result as a Chi square random variable with ν degrees of freedom, to find the probability α.
8. Try another model, until α is minimized.

It should be mentioned that when selecting a set of ν uncorrelated estimates of the autocorrelations (step 6), a local search over all the large-lag autocorrelations needs to be made to identify the set corresponding to the smallest estimated value of α. This is done to eliminate the possibility of the selection of a non-optimum model.

The order of the model discussed so far has been restricted to be an even number. However, in some cases, an odd-order model may provide a better fit to the data. Therefore, both the odd- and even-order models should be considered in the system identification process. By putting a first-order block in series with an even-order subsystem, (all together representing the nonstationary earthquake excitation), odd-order models can be incorporated in the modeling process. In this setup, the oscillatory properties of the earthquake model is still controlled by the second-order subsystem. As a result, the more general forms of Equations 30 and 35 become

$$R_e(k) = 0. \quad \text{for} \quad \begin{cases} k > P-2 & P \text{ even} \\ k > P-3 & P \text{ odd} \end{cases} \quad (36)$$

$$E[\hat{R}_e(k) \cdot \hat{R}_e(k-s)] = 0. \quad \text{for} \quad \begin{cases} s > P-2 & P \text{ even} \\ s > P-3 & P \text{ odd} \end{cases} \quad (37)$$

A CASE STUDY

The accelerogram obtained at the roof level of Imperial County Services Building, during the October 15, 1979 earthquake is studied here. The accelerogram was recorded at the center of the roof in the East-West direction. The structure was a six-story reinforced concrete frame and shear-wall building, in El Centro, California, and sustained significant structural damage during the earthquake [9].

The preliminary data processing of this record was done by CDMG, and the corrected accelerogram is used for the analysis here. Calculations of the Fourier amplitude spectrum of this record have indicated that its frequency content is primarily in the range of 0-5 Hertz. Therefore, this data is lowpass filtered, using a 5 Hertz, zero-phase, second-order Butterworth filter, and the first 40 seconds of the record, at a sampling rate of 25 samples per second, is taken for the analysis

The acceleration time history, the residuals, the variance stabilized residuals, the corresponding envelope function are shown in Figures 2.a through 2.d.

For this data, it has been observed that better models can be fitted, if Chebychev polynomials are used instead of Fourier series for expanding the model coefficients. Also, these expansions are truncated at 10% of the total record length, away from each end, due to poor convergence of this kind of polynomial expansion at the limits of expansion interval. The best-fit model, obtained for this data, is a sixth-order model with a polynomial expansion degree of fourteen. The statistical properties of the variance-stabilized residuals are given in Table .

Pre-earthquake information obtained from ambient vibration tests together with the properties of the estimated autocorrelation function of the variance-stabilized residuals can be used to identify the overall system configuration.

From the ambient vibration testing of this building which was performed in the spring of 1979 [10], the frequencies of the first two bending modes in the East-West direction were found to be 1.54 and 5.1 Hz. Time-variation of these two modal frequencies of this building, during the 1979 Imperial Valley earthquake, have also been identified by Pauschkee et al [11]. These time-varying identified modes are plotted , as solid lines, in Figure 3.a. Therefore, the second-order blocks, corresponding to the lowest two modal frequencies, as shown in Figure 3.a, can be taken as the structural modes of vibration.

autocorrelations occur only at the first two lags which implies that the numerator of the overall system transfer function is a polynomial of order 4 . Therefore, this transfer function can be written as follows.

$$H(z) = \frac{A_1 z^{-1}}{1-\varphi_1^1 z^{-1}-\varphi_2^1 z^{-2}} \left[\frac{A_2 z^{-1}}{1-\varphi_1^2 z^{-1}-\varphi_2^2 z^{-2}} + \frac{A_3 z^{-1}}{1-\varphi_1^3 z^{-1}-\varphi_2^3 z^{-2}} \right]$$

$$= H_1(z) \left[H_2(z) + H_3(z) \right] \qquad (38)$$

where all the φ_{ij} and A_i are time-dependent.

From Equation 38, it can be seen that the overall system has the following configuration.

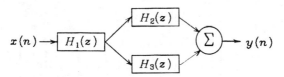

where $H_2(z)$ and $H_3(z)$ correspond to the first two modes of the building, and $H_1(z)$ can be taken as the transfer function of a second-order block representing the earthquake excitations.

CONCLUSION

In this paper, it has been shown that finite-order, discrete-time time-varying models are suitable analytical tools for system identification in the analysis of strong motion structural response.

ACKNOWLEDGEMENTS

The material presented in this paper is part of the author's Ph.D. work being done at Stanford University, which was supported by the J.A. Blume Earthquake Engineering Center, and the U.S. Geological Survey at Menlo Park.

REFERENCES

[1] Sato,T., Detection of Dynamic Properties of Structural Systems by AR-MA Process through Microtremor Observation, *Proceedings of the 7 th World Conference on Earthquake Engineering* Vol. 6, Istanbul, 1980.

[2] Lee, R.C.K., Optimal Estimation and Control, MIT Press, 1964

[3] Rabiner, L.R., Gold, B., *Theory and Application of Digital Signal Processing*, Prentice-Hall, 1975.

[4] Lee, V.W., A new fast algorithm for the calculation of response of a single-degree-of-freedom system to arbitrary load in time, *Soil Dynamics and Earthquake Engineering*, 1984, Vol.3, No.4, pp. 191-199.

[5] Nigam, N.C., Phase Properties of Earthquake Ground Acceleration Records, *Proceedings of the 8th World Conference in Earthquake Engineering*, San Francisco, 1984, Vol. II

[6] Nigam, N.C., *Introduction to Random Vibrations*, MIT Press, 1983.

[7] Bendat, J.S., Piersol, A.G., *Random Data: Analysis And Measurement Procedures*, Wiley-Interscience, New York, 1971.

[8] Box,G.E., Jenkins, G.H., *Time Series Analysis : Forcasting and Control*, San Francisco : Holden Day, 1976.

[9] Rojahn, C., Mork, P.N., An Analysis of Strong-Motion Data from A Severely Damaged Structure - The Imperial County Services Building, El Centro, California, *The Imperial Valley, California Earthquake of October 15, 1979*, USGS professional paper, 1254, 1982

[10] Pardoen, G.C., Imperial County Services Building ambient vibration test results, Christianchurch, Newzeland, University of Canterbury, report 79-14, 21p, 1979.

[11] Pauschke, J.M., Oliveira, C.S., Shah, H.C., Zsutty, T.C., A Priliminary Investigation of the Dynamic Response of the Imperial County Services Building during the October 15, 1979 Imperial Valley Earthquake, The John A. Blume Earthquake Engineering Center, Department of Civil Engineering, Stanford University, Report No. 49, 1981.

TABLE 1

	N	P	K	$\sigma_{\hat{R}(k)}$	χ^2	ν	Z	α
Imp. Cnty Bldg. TR4	1000	6	200	0.039	40.7	39	0.25	40%

N	number of sampled data points
P	order of the best-fit model
K	number of estimated autocorrelations
$\sigma_{\hat{R}(k)}$	standard deviation of large-lag autocorrelations
χ^2	chi square deviate
ν	degrees of freedom of χ^2 distribution
Z	equivalent, standard normal deviate of χ^2 for $\nu > 30$
α	area under the upper tail of the χ^2 distribution

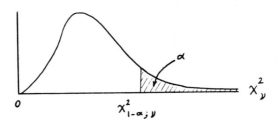

FIGURE 1- CHI SQUARED DISTRIBUTION

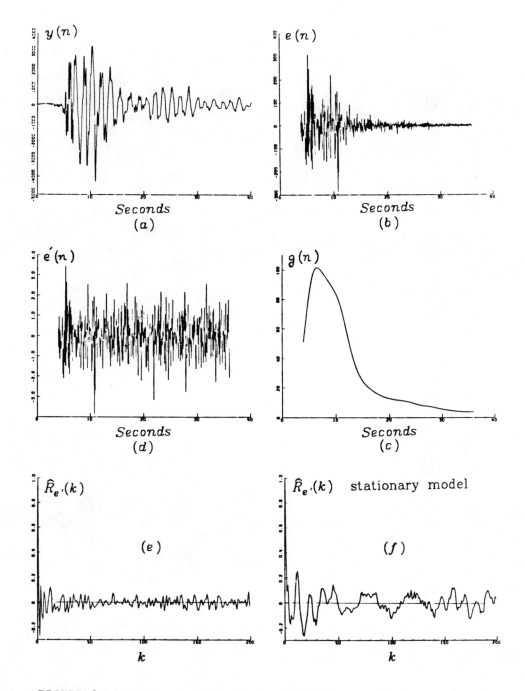

FIGURE 2- SYSTEM IDENTIFICATION OF TRACE 4, IMPERIAL COUNTY BUILDING, OCTOBER 1979 EARTHQUAKE

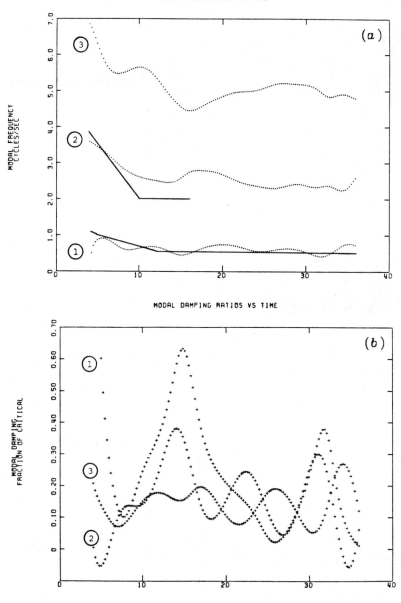

FIGURE 3- (a) MODAL FREQUENCIES
(b) MODAL DAMPINGS

COUPLED LATERAL-TORSIONAL RESPONSE OF MDOF SYSTEMS
TO NONSTATIONARY RANDOM EXCITATION

Wei-Joe Sun[I] and Ahsan Kareem[II]

ABSTRACT

A procedure for calculating coupled lateral-torsional response of a multi-degree-of-freedom system to nonstationary non-white vector-valued random excitation is developed by employing a modal time-domain approach. The nonstationary excitation is represented by an amplitude modulated stationary random process that can be expressed as a product of a stationary random process with a deterministic envelope intensity function. The response of the coupled system is obtained at any time interval using the modal impulse-response function. The envelope intensity function is approximated by a staircase unit step function. This formulation facilitates computations by a reduction in the multiplicity of the integration involved in the evaluation of the response covariance matrices. The peak structural response at any level on the structure may be obtained following the evolutionary distribution of the extreme values. As an example structure, a 10-story unsymmetrical building subjected to ground motion in any arbitrary direction is considered.

INTRODUCTION

Many random environmental load effects have nonstationary characteristics, i.e., their frequency contents and/or amplitude are time-variant. Typical examples of such loadings are atmospheric turbulence, and earthquake induced excitation. The response of structures to these load effects can be accomplished by using Priestly's concept of evolutionary spectral analysis (10). A special class of evolutionary processes that can be represented by separable nonstationary model has facilitated convenient modeling of nonstationary load effects. These random processes are represented by arbitrary broad band stationary Gaussian processes modulated by envelope intensity functions.

The response of a linear single-degree-of-freedom system to a nonstationary excitation has been addressed in a number of studies, some of these are available in references, 1, 2, 7, 11, 12, 13 and 14. The studies concerning the response of a multi-degree-of-freedom systems have been rather limited and some of them can be found in references 3, 4, 5, 6, 15 and 16. In reference 16 closed form expressions for increasing-decreasing exponential

---

I Research Associate

II Associate Professor and Director Structural Aerodynamics and Ocean System Modeling Laboratory, University of Houston, TX.

modulating functions are developed for evaluating nonstationary random response of a multi-degree-of-freedom system.

In this paper a simple time-domain approach is developed that utilizes a modal impulse-response function for any given envelope intensity function. The modulating function is approximated by a staircase unit step function. This formulation facilitates computations through a reduction in the multiplicity of the integration involved in the evaluation of response covariance matrices. An example of a coupled lateral-torsional response of a ten story building to an earthquake excitation is used to illustrate the methodology. The computationally efficient procedure presented herein may be used for analyzing a wide class of problems related to structures excited by nonstationary environmental or man-made load effects. For example, the procedure presented here is useful for structures excited by turbulence, i.e., tall buildings or long span structures under the gust loading during a hurricane. The methodology is being further extended to include nonlinearities in the system.

MATHEMATICAL FORMULATION

The equation of motion for a typical N-story rigid floor building subjected to random excitation can be expressed as

$$\underline{M}\,\ddot{d}(t) + \underline{C}\,\dot{d}(t) + \underline{K}\,d(t) = \underline{P}(t) \tag{1}$$

where $\underline{M}$ and $\underline{K}$ are mass and stiffness matrices respectively and are given by

$$\underline{M} = \begin{vmatrix} \underline{m} & & \underline{0} \\ & \underline{m}\,\underline{\gamma}^2 & \\ \underline{0} & & \underline{m} \end{vmatrix}$$

$$\underline{K} = \begin{vmatrix} \underline{K}_{xx} & \underline{K}_{x\theta} & \underline{0} \\ \underline{K}_{x\theta}^T & \underline{K}_{\theta\theta} & \underline{K}_{y\theta}^T \\ \underline{0} & \underline{K}_{y\theta} & \underline{K}_{yy} \end{vmatrix}$$

The displacement vector d has three components, x, y and θ for each lumped mass

$$\underline{d} = [\,\underline{x},\,\underline{\theta},\,\underline{y}\,]^T$$

The load p(t) corresponding to the two lateral and a rotational direction is expressed as

$$\underline{p}(t) = [\underline{p}_x(t), \underline{p}_\theta(t), \underline{p}_y(t)]^T \qquad (2)$$

If p(t) may be an earthquake loading acting at an angle α with respect to x-axis; $\underline{p}(t)$ has the form

$$\underline{p}(t) = -\underline{M}\,\ddot{d}_g\,[\,\underline{I}\cos\alpha \quad 0 \quad \underline{I}\sin\alpha\,]^T \qquad (3)$$

in which $\ddot{d}_g$ is ground acceleration, $\underline{I}$ is a unit vector. The frequencies and the associated mode shapes, may be obtained following the eigenvalue analysis. Following modal displacement technique, the displacement vector $\underline{d}(t)$ is expressed as a product of normalized mode shape $\underline{\Phi}$ and the normal coordinate q(t)

$$\underline{d}(t) = \underline{\Phi}\,\underline{q}(t) = \sum_{j=1}^{M} q_j(t)\,\underline{\phi}_j \qquad (4)$$

Introducing eq. 4 and its derivatives in eq. 1 and applying the orthogonality conditions, leads to a typical uncoupled modal equation

$$\ddot{q}_j(t) + 2\xi_j\omega_j\dot{q}_j(t) + \omega_j^2 q(t) = \underline{\phi}_j^T\,\underline{p}(t) \qquad (5)$$

In eq. 5, ω_j and ξ_j and $\underline{\phi}_j^T$ are the jth mode frequency, damping ratio and normalized mode shape respectively. The jth mode generalized force $f_j(t)$ is defined as

$$\begin{aligned}f_j(t) &= \underline{\phi}_j^T\,\underline{p}(t) \\ &= \underline{\phi}_{xj}^T\,\underline{p}_x(t) + \underline{\phi}_{\theta j}^T\,\underline{p}_\theta(t) + \underline{\phi}_{yj}^T\,\underline{p}_y(t) \end{aligned} \qquad (6)$$

In this study an earthquake excitation oriented at an angle α with respect to x-axis is considered. The nonstationary stationary ground acceleration $\ddot{d}_g(t)$, is expressed as a product of an envelope intensity function C(t) and a stationary Gaussian process W(t). In view of eq. 3, eq. 6 reduces to the following

$$f_j(t) = \phi_j^T(-\underline{M})\ddot{d}_g(t) [\underline{I} \cos\alpha \quad \underline{0} \quad \underline{I} \sin\alpha]^T$$

$$= C(t) \Gamma_j W(t) \tag{7}$$

in which the participation factor Γ_j is given by

$$\Gamma_j = \cos\alpha \sum_{n=1}^{N} m_n \phi_{xnj} + \sin\alpha \sum_{n=1}^{N} m_n \phi_{ynj}$$

The correlation between modal response components in terms of the input ground excitation is given by (15)

$$R_{q_j q_k}(t) = \int_{-\infty}^{\infty} \Gamma_j \Gamma_k \int_0^t C(\tau_1) h_j(t-\tau_1) e^{i\Omega\tau_1} d\tau_1$$

$$\times \int_0^t C(\tau_2) h_k^*(t-\tau_2) e^{-i\Omega\tau_2} d\tau_2 \, S_w(\Omega) d\Omega \tag{8}$$

Similarly, the correlation between modal response $q_i(t)$ and its time derivatives is obtained (15)

The nonstationary (or transient) character of earthquakes can be represented by the stationary motion multiplied by a deterministic intensity or envelope function as shown in eq. 7 [7,11]. It is possible to define a suitable envelope function for a given earthquake record. Any continuous arbitrary function can be approximated by a series of unit step functions as shown in fig. 1.

The envelope function, $C(\tau)$, can be expressed as

$$C(\tau) \simeq \sum_{n=1}^{N} e_n [U_{\varepsilon_n}(\tau) - U_{\varepsilon_{n+1}}(\tau)] \tag{9}$$

in which e_n is constant for $\varepsilon_n < \tau < \varepsilon_{n+1}$ and $U_{\varepsilon_n}(\tau)$ is a unit step function.

Introducing the concept of unit step function in eq. 8 and further algebraic manipulations leads to (15)

$$R_{q_j q_k}(t) = \frac{\Gamma_j \Gamma_k}{\omega_{d_j} \omega_{d_k}} \int_{-\infty}^{\infty} \sum_{n=1}^{N} e_n [R_n(S) - i\, I_n(S)]\Big|_{\varepsilon'_n}^{\varepsilon'_{n+1}}$$

$$\times \sum_{n=1}^{N} e_n [R_n(S) + i\, I_n(S)]\Big|_{\varepsilon'_n}^{\varepsilon'_{n+1}} S_w(\Omega) d\Omega \tag{10}$$

Following the above procedure, expressions for correlation between the derivatives of modal response components is obtained (15). The covariance function of the responses are given by

$$\begin{vmatrix} \Sigma_{dd}^{(t)} & \Sigma_{d\dot{d}}^{(t)} \\ \Sigma_{\dot{d}d}^{(t)} & \Sigma_{\dot{d}\dot{d}}^{(t)} \end{vmatrix} = \begin{vmatrix} \Phi \Sigma_{qq}(t) \Phi^T & \Phi \Sigma_{q\dot{q}}(t) \Phi^T \\ \Phi \Sigma_{\dot{q}q}(t) \Phi^T & \Phi \Sigma_{\dot{q}\dot{q}}(t) \Phi^T \end{vmatrix} \quad (11)$$

The absolute structural response can be formulated in terms of relative response and ground acceleration. Details of this formulation are available in reference 15.

EXAMPLE

To illustrate the proposed methodology the example used in reference 4 was selected for comparison. A typical envelope intensity function in conjunction with Kanai-Tajimi PSD function is used. In references 4 and 5 a complex formulation employing the time history state space transition matrix approach was used. All the results based on the procedure presented here are in good agreement with reference 4.

As a second example, a more complicated 10-story unsymmetrical building is considered (Fig. 2). The geometric properties of the building are listed in Table I. The damping ratio for all the modes was assumed to be 5%. Two sets of ground parameters, one for $\xi_g = 0.4$, $\omega_g = 2\pi$ and another for $\xi_g = 0.6$, $\omega_g = 5\pi$ studied. 10 sec., the intensity function shown in Fig. 1, was divided in 11 unit step segments to approximate the exact envelope shape. Three orientations of the excitation, $\alpha = 0°$, $30°$ and $45°$ were examined. In this study only the first 15 modes were included.

The building response is sensitive to the site conditions represented by the ground filter parameters in the Kanai-Tajimi spectra. The sensitivity depends on the closeness of the ground filter frequency to the system frequency. Figures 3, 4 and 5 illustrate the nonstationary RMS acceleration responses for lateral and torsional degrees of freedom, respectively for $\alpha = 0°$, $\xi_g = 0.4$ and $\omega_g = 2\pi$. The response estimates for the ground excitation at different angles are available in reference 15.

The influence of various parameters on the response estimates confirmed most of the trends that have been highlighted in previous studies by other researchers using both deterministic and probabilistic approaches (9 and 10). Some of the key conclusions are stated here. The coupling induced torsional response is amplified as the eccentricity perpendicular to the ground motion is increased. For a two-way torsional coupling the torsional response depends on the eccentricity perpendicular to the ground motion, whereas, eccentricity in the direction of motion tends to diminish it. There is additional transverse lateral response. The maximum response in various directions due to the evolutionary excitation did not occur at the same time, instead a short time lag was observed. The nonstationary response did not overshoot the corresponding stationary response except for torsional response in a few cases. The level of exceedance was limited to a couple of percent, that may be attributed to a numerical error.

CONCLUSIONS

A procedure for calculating coupled lateral-torsional response of a multi-degree-of-freedom system to nonstationary non-white vector-valued random excitation is developed by employing a modal time-domain approach. This formulation facilitates computations by a reduction in the multiplicity of the integration involved in the evaluation of the response covariance matrices. The methodology developed herein was used to analyze a coupled lateral-torsional response of a single-story and a multi-story building that led to the following conclusions.

The nonstationary lateral-torsional response estimates of a single-story structure based on the proposed procedure are in good agreement with results obtained from a time history state space transition matrix approach. For both single and multiple-story buildings the response is sensitive to the site conditions represented by the ground filter parameters in Kanai-Tajimi spectra. The multi-story building responds differently for one-way or two-way eccentricities between the mass and the stiffness centers and are sensitive to the effect of ground motion directionality.

AKNOWLEDGEMENT

The financial support for this research was provided by NSF Grant no. ECE-8352223. This support is gratefully acknowledged.

REFERENCES

[1] Caughey, T.K. and Stumpf, H.J., "Transient Response of a Dynamical System Under Random Excitation," Transactions of the American Society of Mechanical Engineers, Journal of Applied Mechanics, Vol. 28, 1961.

[2] Corotis, R.B. and Marshall, T.A., "Oscillator Response to Modulated Random Excitation," Journal of the Engineering Mechanics, ASCE, Vol. 103, EM4, 1977.

[3] DebChaudhury, A. and Gaspirini, D.A., "Response of MDOF Systems to Vector Random Excitation," Journal of Engineering Mechanics Division, ASCE, Vol. 108, No. EM2, 1982.

[4] Gaspirini, D.A., "Response of MDOF Systems to Nonstationary Random Excitation," J. of Engineering Mechanics Division, ASCE, Vol. 105, No. EM1, 1979.

[5] Gasparini, D.A., and Debchaudhury, A., "Dynamics Reponse to Nonstationary Nonwhite Excitation," Journal of the Engineering Mechanics Divisions, ASCE, Vo. 106, No. EM6, Dec. 1980.

[6] Hammond, S.K., "On the Response of Single and Multi-Degree-of-Freedom Systems to Non-Stationary Random Excitations, "J. of Sound and Vibration, Vol. 7, 1968.

[7] Hasselman, T., "Linear Response to Nonstationary Random Excitation," J. of Engineering Mechanics, ASCE, EM3, June 1972.

[8] Kan, C.L. and Chopra, A.K., "Elastic Earthquake Analysis of Torsionally Coupled Multiply Buildings," Earthquake Engineering and Structural Dynamics, Vol. 5, 1977.

[9] Kung, Shyh-Yuan and Pecknold, D.A., "Effect of Ground Motion Charactristics on the Seismic Response of Torsionally Coupled Elastic Systems," Department of Civil Engineering, University of Illinois, Tech Rep., UILU-ENG-82-2009.

[10] Priestly, M.B., "Power Spectral Analysis of Nonstationary Random Process," J. of Sound and Vibration, V. 6, 1967.

[11] Ross, B., Vanmarcke, E.H. and Cornell, C.A., "First Passage of Nonstationary Random Vibration," Journal of the Engineering Mechanics, ASCE, EM3, June, 1972.

[12] Roberts, J.B., "The Covariance Response of Linear Systems to Non-Stationary Random Excitation," Journal of Sound and Vibration, Vol. 14, 1971.

[13] Shinozuka, A.M. and Sato, Y., "Simulation for Non-Stationary Random Process," Journal of the Engineering Mechanics Division, ASCE, Vol. 93, No. EM1, 1967.

[14] Solomos, G.P. and Spanos, P-T.D., "Oscillator Response to Nonstationary Excitation," J. of Applied Mechanics, Vol. 51, 1984.

[15] Sun, W.J. and Kareem, A., "Response of Torsionally Coupled Multi-story Buildings to Random Excitation," Technical Report, Civil Engineering Dept., UHCE 86-2, University of Houston, 1986.

[16] To C.W.S.," Non-Stationary Random Responses of a Multi-Degree-of-Freedom System by the Theory of Evolutionary Spectra," J. of Sound and Vibration, Vol. 83, No. 2, 1982.

TABLE I

Properties of 10-Story Building

Story No.	m_i	k_x	k_y	k_θ	e_x	e_y	γ
1	0.3500E+01	0.4900E+04	0.5825E+04	0.2105E+06	0.3677E+02	0.5515E+02	0.1838E+03
2	0.3500E+01	0.4768E+04	0.5698E+04	0.2056E+06	0.3677E+02	0.5515E+02	0.1838E+03
3	0.3500E+01	0.4543E+04	0.5448E+04	0.1966E+06	0.3677E+02	0.5515E+02	0.1838E+03
4	0.3500E+01	0.4418E+04	0.5295E+04	0.1910E+06	0.3677E+02	0.5515E+02	0.1838E+03
5	0.2000E+01	0.4165E+04	0.4993E+04	0.1800E+06	0.2828E+02	0.3677E+02	0.1414E+03
6	0.2000E+01	0.2525E+04	0.3243E+04	0.6785E+05	0.2828E+02	0.3677E+02	0.1414E+03
7	0.2000E+01	0.2188E+04	0.2810E+04	0.5910E+05	0.2828E+02	0.3677E+02	0.1414E+03
8	0.2000E+01	0.1768E+04	0.2270E+04	0.4815E+05	0.2828E+02	0.3677E+02	0.1414E+03
9	0.2000E+01	0.1263E+04	0.1450E+04	0.3500E+05	0.2828E+02	0.3677E+02	0.1414E+03
10	0.2000E+01	0.6730E+03	0.8680E+03	0.2005E+05	0.2828E+02	0.3677E+02	0.1414E+03

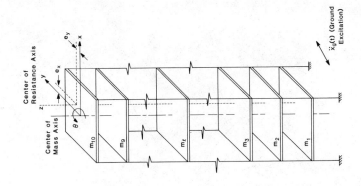

Fig 2. Example Building

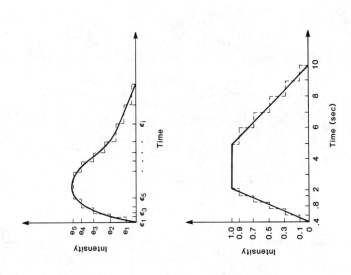

Fig 1. Envelope Intensity Function

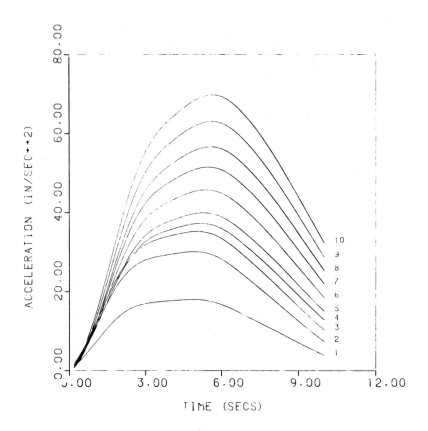

Fig 3. X-Acceleration

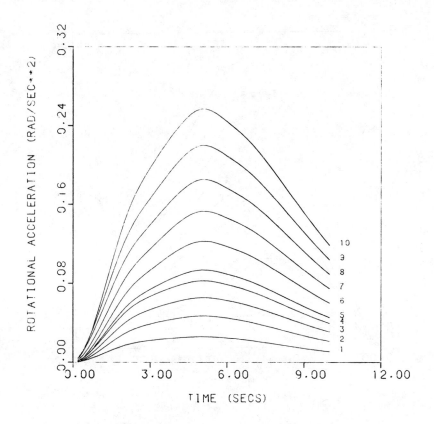

Fig 4. θ-Acceleration

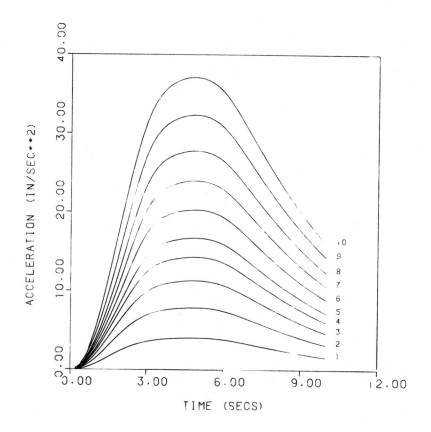

Fig 5. Y-Acceleration

PREDICTION OF THE LARGEST PEAKS IN SINGLE-DEGREE-OF-FREEDOM
OSCILLATOR RESPONSE TO STRONG GROUND MOTION

J.M. Pauschke[I] and J. Chatterjee[II]

ABSTRACT

Exponential and Rayleigh probability distributions are used to model the distributions of the largest peaks in response time histories of a linear, elastic single-degree-of-freedom oscillator subjected to earthquake excitation. Average acceleration spectra computed for the actual first, second, fifth, tenth, and twentieth largest peaks and average acceleration spectra computed for the first and tenth largest peaks predicted from these probability distributions are presented. Characterization of the largest peaks in response time histories enables prediction of expected amplitude levels of cyclic response in addition to the traditionally retained maximum amplitude of a response time history.

INTRODUCTION

For use in seismic design, response spectra are traditionally plotted for the maximum amplitude of the response time history of a linear, elastic single-degree-of-freedom (SDOF) oscillator subjected to strong earthquake ground motion. Such characterization, however, does not retain information on the amplitudes and the duration of the lesser but near maximum peaks of the response time history which contribute to the strong motion portion of the response. Therefore, the objective of this study is to characterize explicitly the amplitudes and implicitly the duration of the largest peaks during the strong motion portion of the response time history of a SDOF oscillator subjected to earthquake excitation.

A peak is defined as the maximum absolute amplitude between two consecutive zero crossings in the acceleration, relative velocity, or relative displacement time history of a linear, elastic SDOF oscillator. As shown in Fig. 1, the kth largest peak, $X(k)$, of a response time history is the kth peak in the time history when the peaks are ranked in descending order from the largest, $X(1)$, to smallest peak. The peak $X(k)$ then summarizes the amplitude above which there will be k/2 cycles of response exceeding this amplitude. The adequacy of the exponential and Rayleigh probability distributions to predict the kth largest peaks $X(k)$ in response time histories is investigated below. In a slightly different approach, Refs. [1] and [2]

I. Assistant Professor, Department of Civil Engineering, University of Pennsylvania, Philadelphia, PA.

II. Doctoral Candidate, Department of Civil Engineering, University of Pennsylvania, Philadelphia, PA.

have computed the ratio of the amplitude corresponding to N cycles of linear SDOF oscillator response to the maximum response amplitude for relative displacement and acceleration response time histories, respectively.

METHODOLOGY

Prediction of the peaks X(k) of the response time histories is investigated for a SDOF oscillator with 5% critical damping subjected to 112 horizontal ground motion records from 14 earthquakes as listed in Table 1. Of the 112 records considered, 34 are from earthquakes prior to 1971 and 34 are from the 9 February 1971 San Fernando, CA earthquake [3]. The remaining 44 records are from the 15 October 1979 Imperial Valley, CA earthquake [4]. To investigate the effects of local site conditions on the selection of the best predictive distribution of the largest peaks, the soil conditions of the recording stations are divided into two soil categories: "soil" and "rock". Site category "rock" corresponds to the rock category per Ref. [5]. Site category "soil" includes both the "stiff soil conditions" and the "deep cohesionless soil conditions" per Ref. [5]. Of the 112 records considered, 86 and 26 records correspond to soil and rock sites, respectively. While other researchers have considered more refined "soil classifications" for computation of average response spectra, the two site categories used in the present study adequately illustrate the trends in the distributions of the largest peaks in response time histories as a function of different soil conditions. The linear, elastic response of the SDOF oscillator subjected to earthquake excitation is computed using the numerical procedure outlined in Ref. [6].

The largest peaks in acceleration, relative velocity, and relative displacement response time histories of the SDOF oscillator are modelled by exponential or exponential-like probability distributions. The exponential probability density function $f_Y(y)$ of a random variable Y is given as:

$$f_Y(y) = \lambda e^{-\lambda y} \tag{1}$$

where λ is the parameter of the distribution. Ref. [7] has shown through a simple change of variable transformation that if the set $\{X\} = \{X_1, X_2,...X_L\}$ of L ordered observations is not exponentially distributed, but the set $\{Y\} = \{X_1^r, X_2^r, ..., X_L^r\}$ is exponentially distributed, then for r equal to 1/2 and 2, the set $\{X\}$ follows the Weibull and Rayleigh distributions, respectively. From the parameter λ of the exponential, Rayleigh, or Weibull distribution and N, the total number of peaks in the time history, the mean value of the expected kth largest peak, $\bar{X}(k)$, can be predicted from extremal statistics per Ref. [8] as:

$$\bar{X}(K) = \frac{1}{\lambda} [\sum_{z=k}^{N} \frac{1}{z}]^m \tag{2}$$

where m = 1, 1/2, or 2 for an exponential, Rayleigh, or Weibull distribution, respectively.

In this study, the largest peaks in the response time history of a SDOF oscillator are modelled by the following probability distributions:

1) Exponential
2) Rayleigh
3) Exponential (EHT)
4) Rayleigh (EHT)

The first two probability distributions are the well-known exponential and Rayleigh distributions per Eq. 1. These two distributions are derived using the complete set of peaks in each response time history. Hence, in Eq. 2, N corresponds to the total number of peaks in each response time history. However, a number of peaks in a response time history will consist of small amplitude peaks which do not contribute to the strong motion portion of the record. Consequently, the inclusion of these smaller amplitude peaks in the derivation of the exponential or Rayleigh distributions will greatly reduce the accuracy in prediction of the largest near maximum peaks which comprise the strong motion portion of interest.

To improve the prediction of the near maximum peaks in ground motion accelerograms without the influence of the smaller peaks, Refs. [7] and [9] developed an exponential half-tail (EHT) distribution to characterize an upper median of largest peaks in a ground motion acceleration time history. The EHT model describes, by an upper half-tail exponential, Rayleigh, or Weibull distribution, the distribution of the largest peaks above a median which is determined for each time history by a trial and error procedure. The use of the median as a cutoff point for the ranked peaks was chosen because it is a well-established statistical quantity. If the rank of the median is k_m, then the number of peaks to be used in Eq. 2 to predict $\bar{X}(k)$ is given as $N = 2k_m$ for an EHT distribution.

The EHT model is used in this study to model the distributions of the largest peaks in response time histories. The analytical criterion used by Ref. [7] to determine whether the largest peaks in a given response time history are most accurately predicted by an exponential (EHT), Rayleigh (EHT), or Weibull (EHT) distribution is based on minimizing the weighted square difference between the observed and predicted peaks in accordance with the rank of the peaks. Table 2 shows, as a function of the natural period of a SDOF oscillator with 5% damping subjected to the ground motion in Table 1, the fraction of acceleration, relative velocity, and relative displacement response time histories whose largest peaks are selected by the above criterion to be best modelled by a particular EHT distribution. Except at very low periods (less than about 0.1 second), the largest peaks in response time histories favor either the exponential (EHT) or Rayleigh (EHT) distributions. Beyond a period of about 0.05 seconds, typically less than 20% of the response records follow a Weibull distribution with the exception of relative velocity response for periods greater than 10 seconds. These same trends are observed for the response records considered separately for soil and rock sites. Hence, the third and fourth probability distributions considered in this study are the exponential (EHT) and Rayleigh (EHT) distributions. It is assumed that the largest peaks in all response records can be modelled by these two distributions regardless of the EHT distribution selected by the above criterion.

RESULTS

As a part of model verification, graphical comparisons are presented in Fig. 2 for combined soil and rock sites to illustrate the adequacy of the four considered probability distributions to predict the maximum observed peak, X(1), in acceleration response time histories for a SDOF oscillator with 5% critical damping and 1 second period. For aid in comparison, also shown in each graph is the line $\bar{X}(1) = X(1)$. The upper two graphs, corresponding to the exponential and Rayleigh distributions, show that typically the first peak is underpredicted by these distributions derived from the entire set of peaks in each response time history. However, accuracy in the prediction of X(1) is improved by modelling only the largest peaks in the response time history by the exponential (EHT) and Rayleigh (EHT) distributions as shown in the lower two graphs.

Average acceleration spectra for the actual acceleration peaks X(1), X(2), X(5), X(10), and X(20) of the response of a SDOF oscillator with 5% damping and for periods between 0.03 second and 10 seconds are shown separately for soil and rock sites in Fig. 3. Each peak X(k) has been normalized by the peak ground acceleration (PGA) of the input record. The spectra for acceleration peak X(1) is the traditionally plotted maximum acceleration response spectra. Fig. 3 indicates that the normalized average acceleration spectra for the near maximum peaks X(2), X(5), X(10), and X(20) are similar in shape to the traditional average acceleration spectra for X(1).

Figs. 4 and 5 compare, for X(1) and X(10), respectively, the average actual acceleration spectra shown in Fig. 3 to the average acceleration spectra computed for these peaks predicted from the exponential, Rayleigh, exponential (EHT), and Rayleigh (EHT) distributions. Fig. 4 shows that for both soil and rock sites, X(1) is best predicted by the Rayleigh (EHT) distribution, which tends to slightly overpredict X(1) in the period range between 0.1 second and 1.0 second; otherwise, for all other period ranges, the agreement between the actual and the Rayleigh (EHT) spectra is excellent. The average acceleration spectra for X(1) computed from the exponential (EHT) distribution tend to overpredict the average actual acceleration spectra computed for X(1). For periods less than 0.10 second, the average acceleration spectra computed on rock for the actual peak X(1) and for X(1) predicted from the exponential (EHT) and Rayleigh (EHT) distributions are almost identical. For periods less than about 10 seconds and 5 seconds on soil and rock sites, respectively, both the traditional exponential and Rayleigh distributions underpredict the average actual acceleration spectra by a factor of about 2 to 3. These same trends have also been noted for the second largest peak X(2).

However, the average actual acceleration spectra for X(5), X(10), and X(20) are bounded by the Rayleigh (EHT) distribution as an upper limit and the exponential (EHT) distribution as a lower limit. The average actual vs. average predicted acceleration spectra for X(10), shown in Fig. 5 separately for soil and rock sites, illustrate this trend. The traditional exponential and Rayleigh distributions are shown to underpredict X(10) by a factor of about 2 to 3 for periods less than 0.5 second. Hence, as shown in Figs. 2, 4, and 5, the exponential (EHT) and Rayleigh (EHT) distributions, derived from consideration of only the largest peaks in a response time history, are better predictors of the maximum and near maximum acceleration peaks than the traditional exponential and Rayleigh distributions.

CONCLUSIONS

The distributions of the largest peaks in the response time histories of a SDOF oscillator with 5% critical damping are shown to favor the exponential (EHT) and Rayleigh (EHT) distributions over the Weibull (EHT) distribution and the traditional exponential and Rayleigh distributions. Empirically derived average actual response spectra have been computed for selected peaks $X(k)$ in this study. However, the advantage of characterizing the largest peaks in terms of the exponential (EHT) and Rayleigh (EHT) distributions is that any desired peak $\bar{X}(k)$ can be predicted from the two parameters, λ and N, of these distributions. From average acceleration spectra computed for $(1/\lambda)/PGA$ and for a standardized number of peaks, N^*, derived for each oscillator period, then any desired peak response amplitude $\bar{X}(k)$ could be predicted for a SDOF oscillator subjected to a given earthquake excitation.

ACKNOWLEDGMENTS

This research was supported by National Science Foundation Grant CEE-83-07187.

REFERENCES

[1] Perez, V. and A.G. Brady, "Reversing Cyclic Demands on Structural Ductility During Earthquakes", Critical Aspects of Earthquake Ground Motion and Building Damage Potential, ATC-10-1, pp. 95-104, 1984.

[2] Prince, J., "Influence of the Number of Response Maxima on the Observed Seismic Behavior of Structures", Critical Aspects of Earthquake Ground Motion and Building Damage Potential, ATC-10-1, pp. 57-66, 1984.

[3] California Institute of Technology, Earthquake Engineering Research Laboratory, "Strong-Motion Earthquake Accelerograms - Digitized and Plotted Data", Volume II, 1973.

[4] Brady, A.G., V. Perez, and P.N. Mork, "Digitization and Processing of Main-Shock Ground-Motion Data from the U.S. Geological Survey Accelerograph Network" in The Imperial Valley, California Earthquake of October 15, 1979, U.S. Geological Survey Professional Paper 1254, pp. 385-406, 1982.

[5] Seed, H.B., R. Murarka, J. Lysmer, and I.M. Idriss, "Relationships of Maximum Acceleration, Maximum Velocity, Distance from Source, and Local Soil Conditions for Moderately Strong Earthquakes", Bull. Seis. Soc. Am. Vol. 66, No. 4, pp. 1323-1342, August 1976.

[6] Nigam, N.C. and P.C. Jennings, "Digital Calculation of Response Spectra from Strong-Motion Earthquake Records", California Institute of Technology, Earthquake Engineering Research Laboratory, June 1968.

[7] Deherrera, M. and T. Zsutty, "A Time Domain Analysis of Seismic Ground Motions Based on Geophysical Parameters", Stanford University, John A. Blume Earthquake Engineering Report No. 54, 1982.

[8] Gumbel, E., Statistics of Extremes, Columbia University Press, 1958.

[9] Zsutty, T. and M. Deherrera, "A Statistical Analysis of Accelerogram Peaks Based Upon the Exponential Distribution Model", 2nd U.S. National Conference on Earthquake Engineering, Stanford University, Stanford, CA, pp. 733-742, 1979.

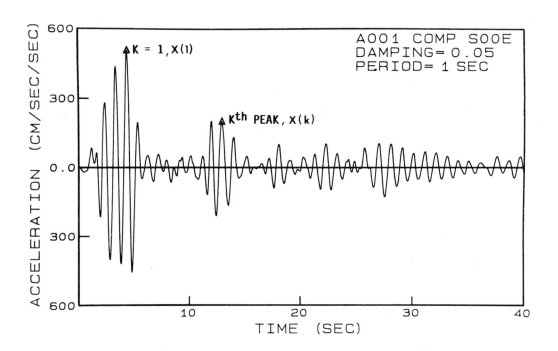

Figure 1   Definition of kth largest peak in SDOF oscillator response time history.

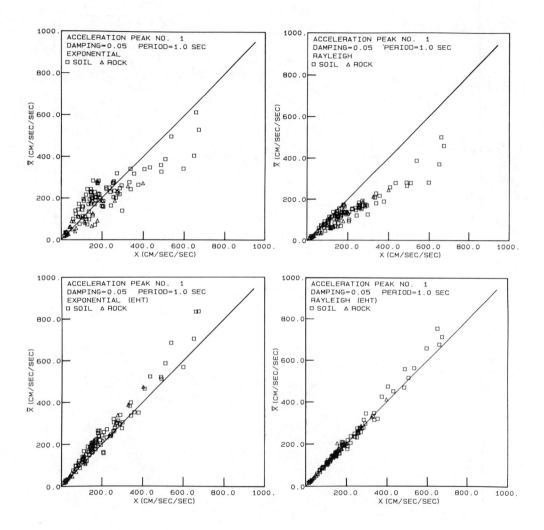

Figure 2   Observed X(1) vs. predicted X̄(1) peak acceleration from exponential, Rayleigh, exponential (EHT) and Rayleigh (EHT) distributions for SDOF oscillator response with 5% damping and 1 second period.

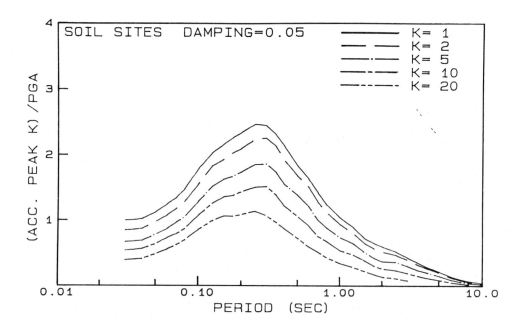

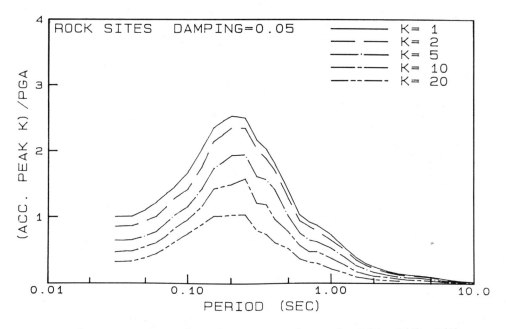

Figure 3 Average acceleration spectra for peaks X(1), X(2), X(5), X(10), and X(20) of response of SDOF oscillator with 5% damping for soil and rock sites.

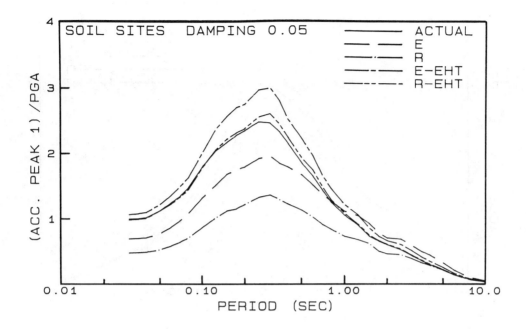

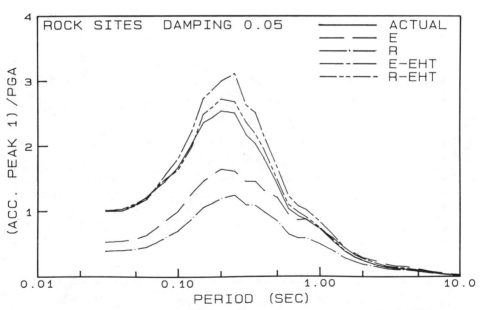

Figure 4 Average actual acceleration spectrum vs. average predicted acceleration spectra from Exponential (E), Rayleigh (R), Exponential EHT (E-EHT), and Rayleigh EHT (R-EHT) distributions for peak X(1) of the response of SDOF oscillator with 5% damping for soil and rock sites.

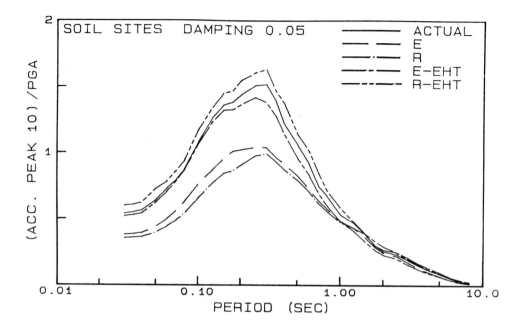

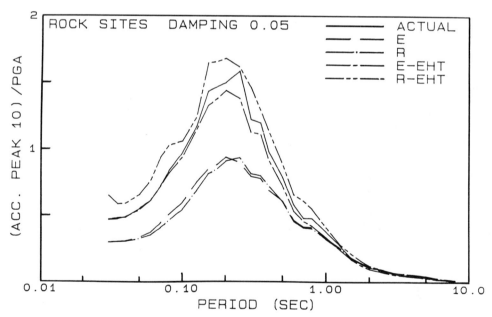

Figure 5   Average actual acceleration spectrum vs. average predicted acceleration spectra from Exponential (E), Rayleigh (R), Exponential EHT (E-EHT), and Rayleigh EHT (R-EHT) distributions for peak X(10) of the response of SDOF oscillator with 5% damping for soil and rock sites.

TABLE 1

Summary of Records Analyzed[1]

Earthquake	Date	Richter Magnitude	Recording Station	I.D. No.	Site[2] Condition
Imperial Valley, CA	05-18-40	7.1	El Centro	A001	S
Kern County, CA	07-21-52	7.7	Pasadena, CIT Athenaeum	A003	S
Kern County, CA	07-21-52	7.7	Taft, Lincoln School Tunnel	A004	R
Eureka, CA	12-21-54	6.6	Eureka Federal Bldg.	A008	S
Eureka, CA	12-21-54	6.6	Ferndale City Hall	A009	S
San Francisco, CA	03-22-57	5.3	Alexander Building	A014	S
San Francisco, CA	03-22-57	5.3	Golden Gate Park	A015	R
San Francisco, CA	03-22-57	5.3	State Building	A016	S
Lower California	12-30-34	6.5	El Centro	B024	S
Helena, Montana	10-31-35	6.0	Carroll College	B025	R
Western Washington	04-13-49	7.0	Olympia, Highway Test Lab	B029	S
Puget Sound, WA	04-29-65	6.5	Olympia, Highway Test Lab	B032	S
Parkfield, CA	06-27-66	5.3	Cholame-Shandon, Array 5	B034	S
Parkfield, CA	06-27-66	5.3	Temblor, CA, No. 2	B037	R
Borrego Mt., CA	04-08-68	6.5	San Onofre, SCE Power Plant	B040	R
San Fernando, CA	02-09-71	6.4	Pacoima Dam	C041	R
San Fernando, CA	02-09-71	6.4	Los Angeles, 8244 Orion	C048	S
San Fernando, CA	02-09-71	6.4	Castaic	D056	S
San Fernando, CA	02-09-71	6.4	LA, Hollywood Storage	D058	S
San Fernando, CA	02-09-71	6.4	LA, 3470 Wilshire	E075	S
San Fernando, CA	02-09-71	6.4	LA, 3407 West Sixth	E083	S
San Fernando, CA	02-09-71	6.4	Pasadena, Seismological Lab	G106	R
San Fernando, CA	02-09-71	6.4	Pasadena, CIT Athenaeum	G107	S
San Fernando, CA	02-09-71	6.4	Pasadena, CIT Millikan Library	G108	S
San Fernando, CA	02-09-71	6.4	Pasadena, JPL	G110	S
San Fernando, CA	02-09-71	6.4	LA, 15250 Ventura	H115	S
San Fernando, CA	02-09-71	6.4	Lake Hughes, No. 4	J142	R
San Fernando, CA	02-09-71	6.4	Lake Hughes, No. 12	J144	R
San Fernando, CA	02-09-71	6.4	LA, 3838 Lankershim	L166	R
San Fernando, CA	02-09-71	6.4	LA, Griffith Park Observatory	O198	S
San Fernando, CA	02-09-71	6.4	LA, 14274 Ventura	Q233	S
San Fernando, CA	02-09-71	6.4	LA, 3550 Wilshire	S266	S
Northern CA	12-10-67	5.8	Ferndale City Hall	U312	S
Lytle Creek, CA	09-12-70	5.4	Wrightwood, 6074 Park Drive	W334	R
Imperial Valley, CA	10-15-79	6.6	El Centro, Array 1	IV01	S
Imperial Valley, CA	10-15-79	6.6	El Centro, Array 2	IV02	S
Imperial Valley, CA	10-15-79	6.6	El Centro, Array 3	IV03	S
Imperial Valley, CA	10-15-79	6.6	El Centro, Array 4	IV04	S
Imperial Valley, CA	10-15-79	6.6	El Centro, Array 5	IV05	S
Imperial Valley, CA	10-15-79	6.6	El Centro, Array 6	IV06	S
Imperial Valley, CA	10-15-79	6.6	El Centro, Array 7	IV07	S
Imperial Valley, CA	10-15-79	6.6	El Centro, Array 8	IV08	S
Imperial Valley, CA	10-15-79	6.6	El Centro, Diff. Array	IV09	S
Imperial Valley, CA	10-15-79	6.6	El Centro, Array 10	IV10	S
Imperial Valley, CA	10-15-79	6.6	El Centro, Array 11	IV11	S
Imperial Valley, CA	10-15-79	6.6	El Centro, Array 12	IV12	S
Imperial Valley, CA	10-15-79	6.6	El Centro, Array 13	IV13	S
Imperial Valley, CA	10-15-79	6.6	Parachute Test Facility	IV14	S
Imperial Valley, CA	10-15-79	6.6	Calipatria, Fire Station	IV15	S
Imperial Valley, CA	10-15-79	6.6	Superstition Mt.	IV16	R
Imperial Valley, CA	10-15-79	6.6	Plaster City Storehouse	IV17	S
Imperial Valley, CA	10-15-79	6.6	Coachella Canal #4	IV18	S
Imperial Valley, CA	10-15-79	6.6	Bond's Corner	IV19	S
Imperial Valley, CA	10-15-79	6.6	Brawley Airport	IV20	S
Imperial Valley, CA	10-15-79	6.6	Holtville Post Office	IV21	S
Imperial Valley, CA	10-15-79	6.6	Calexico Fire Station	IV22	S

[1] Two horizontal components at each recording site analyzed

[2] S - Soil Site
R - Rock Site

TABLE 2

Fraction of SDOF Oscillator Response (5% Critical Damping) Records Whose Largest Peaks Follow Exponential, Rayleigh, and Weibull EHT Distributions
(112 Soil and Rock Sites)

Period (Sec)	ACCELERATION			RELATIVE VELOCITY			RELATIVE DISPLACEMENT		
	Exponential	Rayleigh	Weibull	Exponential	Rayleigh	Weibull	Exponential	Rayleigh	Weibull
0.03	0.37	0.34	0.29	0.44	0.30	0.26	0.30	0.28	0.42
0.04	0.32	0.37	0.31	0.51	0.25	0.24	0.28	0.28	0.44
0.05	0.39	0.33	0.28	0.57	0.28	0.15	0.38	0.31	0.31
0.06	0.44	0.35	0.21	0.52	0.33	0.15	0.40	0.27	0.33
0.07	0.42	0.36	0.22	0.45	0.40	0.15	0.35	0.31	0.34
0.08	0.50	0.32	0.18	0.51	0.40	0.09	0.46	0.30	0.24
0.10	0.41	0.44	0.15	0.49	0.42	0.09	0.45	0.39	0.16
0.125	0.39	0.47	0.14	0.39	0.55	0.06	0.41	0.46	0.13
0.15	0.34	0.52	0.14	0.33	0.60	0.07	0.35	0.50	0.15
0.175	0.42	0.46	0.12	0.45	0.45	0.10	0.52	0.26	0.22
0.20	0.32	0.51	0.17	0.32	0.57	0.11	0.33	0.50	0.17
0.30	0.32	0.55	0.13	0.32	0.56	0.12	0.30	0.56	0.14
0.40	0.37	0.47	0.16	0.37	0.52	0.11	0.35	0.48	0.17
0.50	0.28	0.57	0.15	0.23	0.59	0.18	0.27	0.58	0.15
0.60	0.37	0.49	0.14	0.31	0.50	0.19	0.36	0.48	0.16
0.70	0.31	0.50	0.19	0.30	0.53	0.17	0.29	0.53	0.18
0.80	0.23	0.62	0.15	0.29	0.59	0.12	0.24	0.61	0.15
1.00	0.32	0.53	0.15	0.35	0.53	0.12	0.31	0.68	0.16
1.25	0.22	0.68	0.10	0.36	0.54	0.10	0.22	0.62	0.10
1.50	0.25	0.62	0.13	0.30	0.52	0.18	0.25	0.62	0.13
1.75	0.29	0.59	0.12	0.38	0.47	0.15	0.29	0.58	0.13
2.00	0.25	0.60	0.15	0.35	0.46	0.19	0.24	0.60	0.16
2.50	0.26	0.65	0.09	0.24	0.60	0.16	0.26	0.66	0.08
3.00	0.21	0.74	0.05	0.23	0.61	0.16	0.18	0.77	0.05
4.00	0.15	0.84	0.01	0.21	0.66	0.13	0.16	0.83	0.01
5.00	0.08	0.90	0.02	0.23	0.66	0.11	0.07	0.91	0.02
6.00	0.15	0.81	0.04	0.28	0.55	0.17	0.11	0.86	0.03
7.00	0.08	0.88	0.04	0.24	0.59	0.17	0.11	0.86	0.03
8.00	0.13	0.83	0.04	0.33	0.47	0.20	0.13	0.84	0.03
10.00	0.07	0.86	0.07	0.27	0.44	0.29	0.08	0.85	0.07
12.50	0.16	0.77	0.07	0.30	0.36	0.34	0.12	0.83	0.05
15.00	0.25	0.69	0.06	0.38	0.33	0.29	0.21	0.75	0.04
17.50	0.26	0.66	0.08	0.38	0.29	0.33	0.24	0.73	0.03
20.00	0.30	0.55	0.15	0.36	0.28	0.36	0.20	0.78	0.02
25.00	0.22	0.60	0.18	0.37	0.25	0.38	0.12	0.84	0.04
30.00	0.27	0.53	0.20	0.34	0.25	0.41	0.12	0.79	0.09

SEISMIC BEHAVIOUR OF FRICTION DAMPED BRACED FRAMES

P. Baktash[I] and C. Marsh[II]

ABSTRACT

The objective of this investigation is to experimentally ascertain the hysteretic behaviour of friction damped braced frames subjected to dynamic loadings. A large scale steel model frame with friction devices has been tested on a shake table at Concordia University, Montreal. The experiments show the advantage of using friction devices in dissipating the energy induced to the structure during seismic loadings.

INTRODUCTION

Energy dissipation in most structural systems is achieved through ductility i.e. the inelastic yielding of the structural components. The importance of developing a mechanism for dissipating energy during earthquakes without producing permanent damage has been stressed in recent years.

The friction damped braced frame was originally proposed by Pall [1], and nonlinear inelastic dynamic analysis has shown the superiority of this system when compared to other structural systems such as moment resisting frames, concentric braced frames, and eccentric braced frames [1,2].

Shake table tests [3] have already shown the beneficial effects of mechanical energy dissipation devices in improving the seismic response of shear wall structures. To study the behavior of the friction damped braced frames compared to that predicted by analysis, experiments were conducted using a large model on a shaking table.

DESIGN OF TEST MODEL

For the experimental study it was necessary to design a frame which would fit the testing facilities at Concordia University, Montreal. The existing shaking table has dimensions of 4m x 6m (13ft x 20ft), and a top clearance of 6.5m (22ft). Another consideration was to have as many friction joints in the frame as possible in order to simulate earthquake effects on a taller structure. The model was designed to behave as a moment resisting frame (MRF), a concentric braced frame (CBF), or a friction damped braced frame (FDBF), depending on the values of the slip force in the diagonals.

I. Research Assistant, Centre for Building Studies, Concordia University, Montreal, Quebec, Canada.

II. Professor of Engineering, Centre for Building Studies, Concordia University, Montreal, Quebec, Canada.

Two four story, one bay, steel frames were connected together by a set of bracings to eliminate any torsion in the system, thus ensuring that the two frame deformed equally. The total height of the structure was 4m (158 in), each story height being 1m (39.5 in). The bay width and the distance between the frames were 1m (39.5 in). The columns and the beams consisted of the special light profile, SLP4. The members were welded together with a typical moment resistant connection. A steel mass of 910 kg (2000 lb) was added to the top floor in order to ensure that only the fundamental mode of vibration occured. The structure is shown on the shaking table in Fig. 1.

FIG. 1   OVERALL VIEW OF THE MODEL

The braces were connected to the structure by bolting through steel gusset plates with slotted holes. Brake linning pads were inserted at both sides of the plates. The details are shown in Fig. 2.

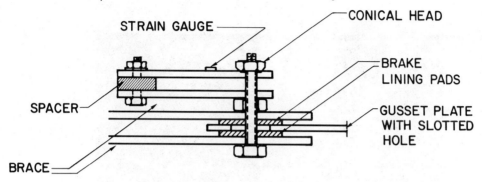

FIG. 2 DETAIL OF THE SPRING LOADING DEVICE

SLIP DEVICE

To create the pressure to control the slip force in the friction joints, the following methods were tried:

a) The turn of the nut method. In this case the start point was subjected to human error.

b) Coil springs were placed under the bolt heads and the shortening calibrated to the slip force. This system was not sufficiently accurate, as the forces were not repeatable.

c) Torque meters were used to measure the torque applied at each bolt, and calibrated with slip loads. This again proved not to be repeatable.

As none of the above methods was sufficiently accurate or repeatable a new system was developed, shown in Figs. 2 and 3.

Force is monitored by a strain gauge on the spring plates. Conical nuts were used to eliminate the hole clearance. The relationship between the slip force of the device and the strain gauge readings were calibrated by means of a load cell. The coefficient of friction was obtained from this relationship. It varied between 0.6 for small slip forces and 0.53 for higher slip forces.

By using this 'spring loading device' the slip force was sufficiently predictable.

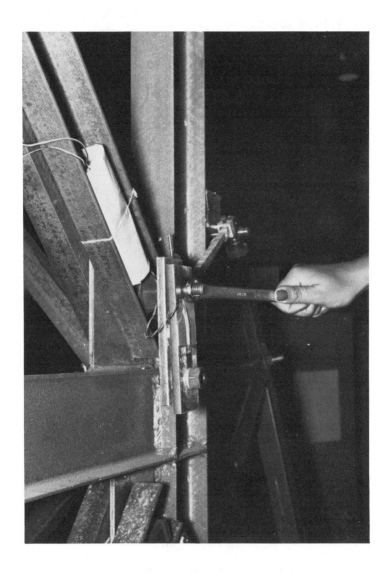

FIG. 3   FRICTION DAMPED BRACED FRAME
         INCORPORATING THE SPRING LOADING DEVICE

QUASI-STATIC TEST

Energy dissipation in any structure is measured by the area enclosed by the hysteretic loop. The larger the area the higher is the energy dissipation in the structure. In friction damped braced frames these curves are steady, repeatable, and unpinched.

Figure 4 shows the measured hysteretic loops for different slip forces. Each graph includes the hysteretic loop for the moment resisting frame (zero slip force), and for a particular value of the slip force. The other extreme case is when the slip force is very high, representing a concentric braced frame.

The hysteretic loops shown are for the sixth to tenth cycles. The loops were repeatable for many more cycles. The dashed lines indicate the theoretical results obtained by using the ADINA program [4], for the non-linear static analysis.

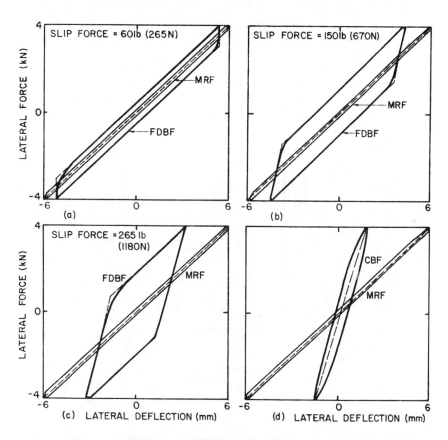

FIG. 4 LATERAL FORCE - TOP STORY DEFLECTION HYSTERETIC BEHAVIOR FOR DIFFERENT SLIP FORCE
(1 in = 25.4 mm, 1 lb = 4.448 N)

DYNAMIC TEST

For the dynamic test experiments were conducted using different slip loads.

Sinusoidal excitation was used to operate the shaking table. The excitation was based on a constant acceleration sweeping through a frequency domain of one to ten cycles per second. With this type of excitation the occurrance of the resonant frequency or pseudo resonance was passed through for all values of the slip force. The natural period of vibration decreases as the slip force increases, moving from one extreme for the moment resisting frame with natural frequency of 3.7 Hz. to the braced frame with nautral frequency of 7.4 Hz.

Electric resistance strain gauges were used to measure the strains at 32 locations in the structure.

Figure 5 shows the top story deflections for various slip forces.

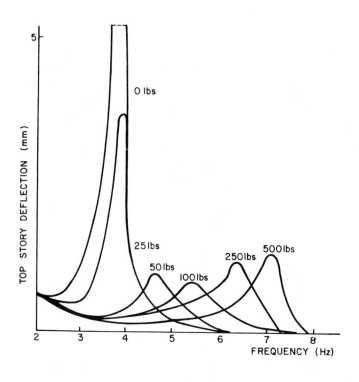

FIG. 5 TOP STORY DEFLECTION AT DIFFERENT SLIP FORCES

The optimum solution occurs [2] when the shear force causing the braces to slip is equal to the shear force causing the rigid frame to yield, giving a slip force equal to:

$$P_s = \frac{2M_p}{h \cos\alpha} \tag{1}$$

in which P_s is the slip force, M_p is the plastic moment capacity of the beam, h is the floor height, and α is the brace angle. For the present study the M_p was chosen as the bending moment in the beam for an arbitrary stress. Tests were performed for varying slip forces. It was found that the optimum is close to that suggested by the above equation. It is observed that as the acceleration of the base increases the optimum would shift to a higher value due to the larger limiting moment in the beam.

CONCLUSION

This study shows the efficiency of friction devices when used in the bracing of steel framed buildings. The energy is dissipated mechanically by friction, reducing the building accelerations, and shear forces. This kind of structure has stable, repeatable hysteretic loops. The use of spring loading device assures that the required slip force is consistently obtained in the device.

REFERENCES

[1] Pall, A.S., Marsh, C., "Seismic Response of Friction Damped Braced Frames", American Society of civil Engineers, Journal of Structural Division, ASCE, No. ST6, Proc. Paper 17175, pp. 1313-1323, June 1982.

[2] Baktash, P., and Marsh, C., "Comparative Seismic Response of Damped Braced Frames", Third ASCE Engineering Mechanics Speciality Conference on Dynamic Response of Structures, April 1986.

[3] Baktash, P., Marsh, C. and Pall, A.S., "Seismic Tests on a Model Shear Wall with Friction Joints", Canadian Journal of Civil Engineering, Vol. 10, pp. 52-59, March 1983.

[4] ADINA "A Finite Element Program for Automatic Incremental Nonlinear Analysis", Report AE.81-1, ADINA Engineering, September 1981.

MODIFICATION OF EARTHQUAKE RESPONSE SPECTRA WITH RESPECT TO DAMPING RATIO

K. Kawashima[I] and K. Aizawa[II]

ABSTRACT

Proposed is a modification coeffcient which is to be multiplied to earthquake response spectra with 5% damping ratio of critical to obtain response spectra of arbitrary damping ratio h. A modification coefficient ξ_{SA} was defined as $S_A(T,h)/S_A(T,0.05)$ in which $S_A(T,h)$ represents absolute acceleration response spectral amplitude for natural period T and damping ratio h. Studied was a variation of S_A in accordance with natural period T and frequency characteristics of ground motions represented in terms of acceleration response spectral ratio $\beta(T,0.05)$ ($\beta = S_A(T,0.05)/a_{max}$, in which a_{max} represents the peak acceleration). Horizontal strong motion records of 206 components obtained at free field sites in Japan were used for the analysis. It was concluded that coefficient ξ_{SA} can be approximated as $1.5 / (40h+1)+0.5$ for damping ratio less than 0.5 of critical.

INTRODUCTION

Earthquake response spectra can be calculated for arbitrary damping ratios if a time history of the ground motion is available. However when a design spectrum is specified for a certain damping ratio, it becomes necessary to modify it to the values corresponding to the desired damping ratios. Such a modification is also required to evaluate the respose spectra for a given earthquake magnitude, epicentral distance and damping ratio with use of attenuation equations of earthquake response spectra, because the attenuation equations generally give spectral amplitudes for only selected damping ratios.

This paper proposes a modification coefficient which is to be multiplied to earthquake response spectra with 5% damping ratio of critical to obtain the spectra with an arbitrary damping ratio h.

STRONG MOTION DATA ANALYZED

A total of 103 sets of two orthogonal horizontal components of strong motion acceleration records (206 components) were used in this study. They were recorded between January 1966 and June 1978 at 43 free field sites in

---

I. Head, Earthquake Engineering Division, Earthquake Disaster Prevention Department, Public Works Research Institute, Tsukuba Science City, Japan.

II. Assistant Research Engineer, Ground Vibration Division, ditto.

Japan, any records on structures including the first floor and basement being excluded. Only earthquakes with magnitude greater than or equal to 5.0 and with focal depth less than 60km were considered. Fig.1 shows the classification of the records in terms of earthquake magnitude M and epicentral distance Δ. Fig.2 shows the distribution of peak ground accelerations.

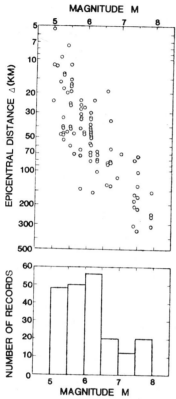

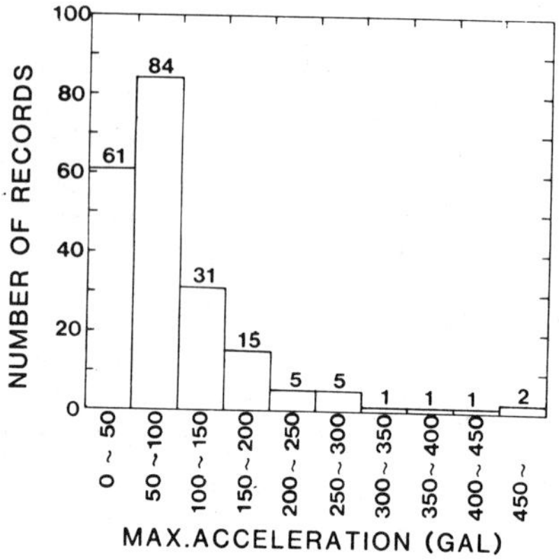

Fig.2 Distribution of Peak Ground Acceleration

Fig.1 Classification of Strong Motion Records in terms of Earthquake Magnitude and Epicentral Distance

DEFINITION OF MODIFICATION COEFFICIENT ξ_{SA}

To study the effect of damping ratio on earthquake response spectrum, a modification coefficient ξ_{SA} is defined as

$$\xi_{SA}(T,h) = S_A(T,h)/S_A(T,0.05) \qquad (1)$$

in which $S_A(T,h)$ represents an absolute acceleration response spectrum for a natural period T and a damping ratio h of critical. The damping ratio of 0.05 is assumed as a reference because general structures have damping ratio of about 0.05.

As an example of $\xi_{SA}(T,h)$, Figs.3 and 4 represent those estimated for two ground motions, i.e., the Itajima record and the Kaihoku record, which were obtained from the Bungo-suido earthquake of 1968(M=6.6, Δ =10km) and the Miyagi-ken-oki earthquake of 1978(M=7.4, Δ =100km), respectively. Earthquake response spectral ratio β (T,0.05) as defined by $S_A(T,0.05)/ a_{max}$, in which a_{max} represents the peak ground acceleration, is also presented in Figs.3 and 4 to show the frequency characteristics of two ground motions. It is seen in Figs.3 and 4 that the modification coefficient $\xi_{SA}(T,h)$ is not constant over entire natural period, i.e., it takes high and/or low values depending on damping ratio at certain natural periods. Such characteristics

(a) Acceleration Response Spectra

(b) Effect of Damping Ratio $\xi_{SA}(T,h)$

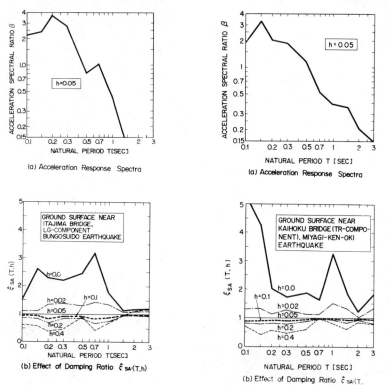

Fig.3 Modification Coefficient ξ_{SA} for the Itajima Record.

(a) Acceleration Response Spectra

(b) Effect of Damping Ratio $\xi_{SA}(T,$

Fig.4 Modification Coefficient ξ_{SA} for the Kaihoku Record

are especially significant for damping ratio of 0.0 and 0.4. The natural period where the modification coefficient ξ_{SA} takes high and/or low values appears to be consistent with the natural period where the acceleration response spectral ratio β takes peak values, i.e., in case of the Itajima record, $\xi_{SA}(T, 0.0)$ takes high values at natural periods of 0.15 and 0.7 second, where $\beta(T,0.05)$ takes a peak spectral ratio. This implies that the modification coefficient $\xi_{SA}(T,h)$ takes either large or small values in accordance with damping ratio at the predominant period of ground motion. This may be explained from the resonance.

Figs.5 and 6 represent decrease of the modification coefficient $\xi_{SA}(T,h)$ in accordance with increase of damping ratio h for the Itajima record and the Kaihoku record, respectively. Although ξ_{SA} mostly decreases with increasing h, decreasing rate of ξ_{SA} with respect to h depends on natural period T. It should be noted here that there is an exceptional case that ξ_{SA} increases with increasing damping ratio such as the case of the Itajima record for natural period of 2 seconds.

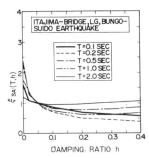

Fig.5 ξ_{SA} vs. h relation for the Itajima Record

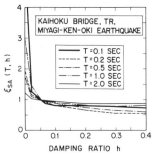

Fig.6 ξ_{SA} vs. h relation for the Kaihoku Record

ESTIMATION OF MODIFICATION COEFFICIENT ξ_{SA}

Because the modification coefficient $\xi_{SA}(T,h)$ is closely related with the acceleration response spectral ratio $\beta(T,0.05)$ as was discussed in the preceding section, a relation between $\xi_{SA}(T,h)$ and $\beta(T,0.05)$ was studied for 206 components of strong motion records. Fig.7 shows $\xi_{SA}(T,h)$ vs. $\beta(T,0.05)$ relation thus obtained. It should be noted here that $\xi_{SA}(T,h)$ vs. $\beta(T,0.05)$ relations at 10 discrete natural periods (T = 0.1, 0.15, 0.2,

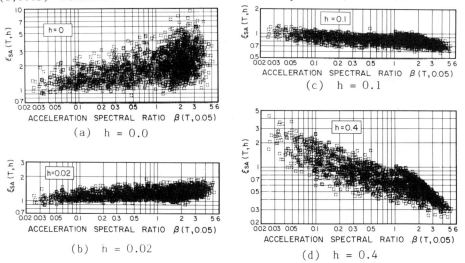

Fig.7 $\xi_{SA}(T, h)$ vs. $\beta(T,0.05)$ Relation for 206 Components of Ground Motion

0.3, 0.5, 0.7, 1, 1.5, 2 and 3 seconds) are presented for each record in Fig.7. Although scatters are significant when damping ratio becomes either smaller or larger than 0.05, it is apparent that $\xi_{SA}(T,h)$ is closely related with $\beta(T,0.05)$.

Therfore it was decided to approximate $\xi_{SA}(T,h)$ vs. $\beta(T,0.05)$ relation as

$$\log \xi_{SA}(T,h) = \log a(T,h) + b(T,h) \log \beta(T,0.05) \qquad (2)$$

in which coefficients $a(T,h)$ and $b(T,h)$ are constants to be determined for each natural period T and damping ratio h. After determining two coefficients by means of least square fitting of the date in Fig.7, $\xi_{SA}(T,h)$ vs. $\beta(T,0.05)$ relation was obtained as shown in Fig.8. Because there are upper and lower limits in the variation of $\beta(T_0,0.05)$ associated with a certain natural period T_0, $\xi_{SA}(T,h)$ vs. $\beta(T,0.05)$ relation predicted by Eq.(2) is shown in Fig.8 only for a range of β between mean value ± 1 standard deviation of $\beta(T_0,0.05)$.

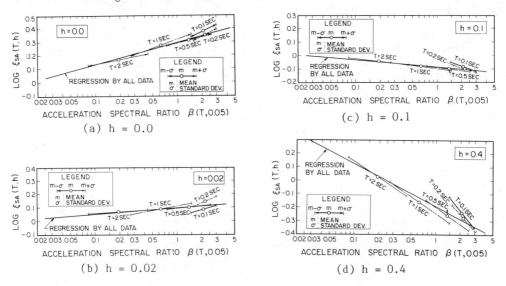

Fig.8 $\xi_{SA}(T, h)$ vs. $\beta(T, 0.05)$ Relation Prediced by Eq.(2)

It is seen in Fig.8 that the variation of $\xi_{SA}(T,h)$ vs. $\beta(T,0.05)$ relation with respect to natural period T is less sensitive, which gives credit to assume that $\xi_{SA}(T,h)$ vs. $\beta(T,0.05)$ relation is independent of natural period T. It is therefore reasonable to assume Eq.(2) as

$$\xi_{SA}(T,h) = \log a(h) + b(h) \log \beta(T,0.05) \qquad (3)$$

in which coefficients $a(h)$ and $b(h)$ are constants to be determined for each damping ratio. They were obtained as shown in Table 1 by least square fitting of the data presented in Fig.7. In Fig.8, $\xi_{SA}(T,h)$ vs. $\beta(T,0.05)$ relation predicted by Eq.(3) is also presented for comparison. It is apparent in Fig.8

Table 1   Coefficients a(h) and b(h)

Damping Ratio h of Critical	Coefficients of Eq.(2) a(h)	b(h)	Coefficients of Eq.(3) a(h)	b(h)
0	1.98	0.167	2.00	0.167
0.01	1.44	0.071	1.57	0.103
0.02	1.25	0.044	1.33	0.067
0.03	1.14	0.026	1.18	0.043
0.05	1.00	0.000	1.00	0.008
0.07	0.911	0.022	0.895	-0.019
0.10	0.824	-0.052	0.800	-0.052
0.20	0.694	-0.150	0.667	-0.157
0.30	0.655	-0.239	0.615	-0.230
0.40	0.647	-0.316	0.588	-0.312

that regression for $\xi_{SA}(T,h)$ vs $\beta(T,0.05)$ relation by Eq.(3) represents the overall relation predicted by Eq.(2).

Fig.9 shows the variation of $\xi_{SA}(T,h)$ vs. $\beta(T,0.05)$ relation by Eq.(3) with respect to damping ratio. When the damping ratio is greater than 0.05, $\xi_{SA}(T,h)$ increases with increasing $\beta(T,0.5)$, and the reverse is true for damping ratio less than 0.05. One interesting result shown in Fig.9 is that changing rate of $\xi_{SA}(T,h)$ with respect to damping ratio becomes significant in accordance with increase of $\beta(T,0.05)$. Reminding that $\beta(T,0.05)$ for a given ground motion tends to have large ratios at its predominant period, this implies that the effect of damping ratio on the modification coefficient is sensitive at the predominant period of ground motions, which is the trend discussed in Figs.3 and 4.

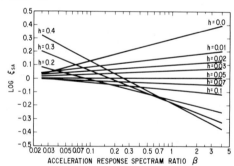

Fig.9 $\xi_{SA}(T, h)$ vs. $\beta(T, 0.05)$ Relation Prediced by Eq.(3)

The coefficients a(h) and b(h) presented in Table-1 may be approximated as

$$a(h) = \frac{1.5}{40h+1} + 0.5 \tag{4}$$

$$b(h) = \frac{1}{300h+1} - 0.8h \tag{5}$$

Fig.10 compares the coefficients a(h) and b(h) predicted by Eqs.(4) and (5) with the values determined by regression analysis. It is seen that Eqs.(4) and (5) give a reasonable approximation.

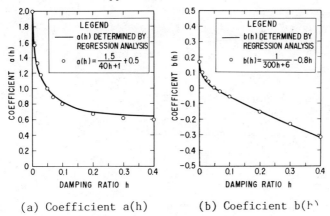

(a) Coefficient a(h) (b) Coefficient b(h)

Fig.10 Accuracy of Eqs.(4) and (5) for a(h) and b(h)

Consequently, substituting Eqs.(4) and (5) into Eq.(3), the modification coefficient $\xi_{SA}(T,h)$ can be obtained as

$$\xi_{SA}(T,h) = \left(\frac{1.5}{40h+1} + 0.5\right) \times \beta(T,0.05)^{\left(\frac{1}{300h} - 0.8h\right)} \qquad (6)$$

It should be however noted here that for infinitely large damping ratio ($h \to \infty$), the coefficients a(h) and b(h) by Eqs.(4) and (5) converge to 0.5 and $-\infty$, respectively, which, in turn, brings the modification coefficient $\xi_{SA}(T,h)$ to zero. However, because $S_A(T,h)$ approaches a_{max} with regardless of natural period for infinitely large damping ratio, ξ_{SA} defined by Eq.(1) should become.

$$\lim_{h \to \infty} \xi_{SA}(T,h) = 1/\beta(T,0.05) \qquad (7)$$

Eq.(7) is satisfied only when the coefficients a(h) and b(h) of Eqs.(4) and (5) converge to 1.0 and -1.0, respectively, for infinitely large damping ratio.

To study the upper limit of Eq.(6), the coefficients a(h) and b(h) were evaluated for large damping ratios. Fig.11 comparies those results with the one calculated by Eqs.(4) and (5). It is seen that the approximations of coefficients a(h) and b(h) by Eqs.(4) and (5) are sufficient for damping ratio less than 0.5. Therefore, the modification coefficient $\xi_{SA}(T,h)$ proposed by Eq.(6) is valid only for damping ratio less than 0.5. However because it is less frequent to assume a dapming ratio larger than 0.5 in dynamic response

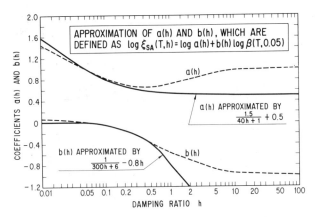

Fig.11 Applicability of Eqs.(4) and (5) for Large Damping Ratio

analysis for usual structures, Eq.(6) is expected to provide a realistic basis for assessing the earthquake response spectra with arbitrary damping ratio less than 0.5.

NUMERICAL EXAMPLES

For showing applicability of Eq.(6), absolute acceleration response spectra $S_A(T,h)$ were calculated directly from the records, and they were compared with the results determined by multiplying the modification coefficient $\xi_{SA}(T,h)$ given by Eq.(6) with $S_A(T,0.05)$. It was assumed here that acceleration response spectral ratio $\beta(T,0.05)$ is known as well as $S_A(T,0.05)$.

Figs.12 and 13 shows $S_A(T,h)$ with damping ratio of 0.0, 0.02, 0.1 and 0.3 for the Itajima record and the Kaihoku record, respectively. Also shown in Figs.12 and 13 are the simplified solution of Eq.(6) by disregarding second term of Eq.(6), which may be practically useful when $\beta(T,0.05)$ is not known. It should be noted here that to disregard the second term of Eq.(6) is equivalent to assume $\beta(T,0.05)$ equals 1.0 with regardless of natural period. It is seen in Figs.12 and 13 that the predicted spectral amplitude by Eq.(6) agrees quite well with the exact solution and that the simplified solution by disregarding the second term of Eq.(6) also gives a reasonable agreement although its accuracy is slightly poor as compared with Eq.(6). Attention should be paid to the case of zero damping ratio, in which response spectra directly calculated from the records show rapid change of spectral amplitude with respect to natural period. Therefore although overall characteristics of the response spectrum are predicted by Eq.(6), it is likely to develop difference at the natural periods where the response spectra have sharp peaks.

CONCLUDING REMARKS

The preceding pages present the modification coefficient $\xi_{SA}(T,h)$ as defined by Eq.(1), which is to be multiplied to earthquake response spectra with 5% damping ratio of critical to obtain the spectra with an arbitrary

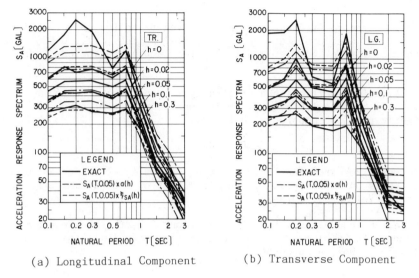

(a) Longitudinal Component   (b) Transverse Component

Fig.12   Example for Estimating $S_A(T,h)$ with Use of Eq.(6) for the Itajima Record

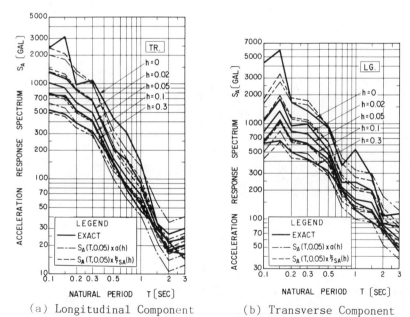

(a) Longitudinal Component   (b) Transverse Component

Fig.13   Example for Estimating $S_A(T,h)$ with Use of Eq.(6) for the Kaihoku Record

damping ratio h. From the results presented herein, the following conclusions may be deduced.

1) The modification coefficient ξ_{SA} may be given by Eq.(6) in terms of damping ratio h and acceleration response spectral ratio β (T, 0.05) with 5% damping ratio. Eq.(6) is applicable for damping ratio less than 0.5 of critical

2) The modification coefficient disregarding second term of Eq.(6) by assuming β (T, 0.05) equals 1.0 also gives reasonable result although the accurary is slightly poor as compared with Eq.(6).

REFERENCE

1) Public Works Research Institute: Strong Motion Acceleration Records from Public Works in Japan, No.1-7, Technical Note of the Public Works Research Institute, Vol.32-38, 1978-1981.

2) Kawashima, K. and Aizawa, K. : Modification of Earthquake Response Spectra with respect to Damping, Proc. of Japan Society of Civil Engineers, No.344, pp.351 - 355, 1984(in Japanese).

OBSERVATIONS ON SPECTRA AND DESIGN

W. J. Hall[I] and S. L. McCabe[II]

ABSTRACT

The earthquake engineering literature is replete with articles on the subject of response spectra and their applications in analysis and design. This paper contains some observations on spectra as viewed in their present, as well as possible future, context and also addresses related developments that are believed to be needed for improvements in seismic design procedures. Recent research pertaining to energy and damage considerations is described generally. The purpose of this conceptual presentation is to outline developments that it is believed will eventually need to find their way into the seismic design and evaluation process.

INTRODUCTION

The purpose of this paper is to present some observations about response and design spectra, and on certain selected topics pertaining to seismic design that it is believed will need to be developed in further detail if improved design procedures are to be developed. The earthquake engineering literature is replete with articles on spectra and their applications in analysis, in part because of the simplicity of the approach offered by this design tool; nonetheless, spectra as currently employed are quite limited in terms of the information contained therein. It is the authors' belief that in the future, advances in seismic design procedures will require a somewhat different assemblage of input information than is currently the case. It is the purpose of this paper to offer some observations in this regard based upon both our research and practice in this field.

RESPONSE AND DESIGN SPECTRA, AND GROUND MOTIONS

The response spectrum is defined as a graphical relationship of the maximum response of a damped single-degree-of-freedom (SDOF) elastic system to dynamic motion or forces. In the case of earthquake engineering the dynamic motion normally is applied at the base of the simple oscillator, as would be

---

I. Professor and Head, Department of Civil Engineering, University of Illinois at Urbana-Champaign, Urbana, Illinois.

II. Assistant Professor, Department of Civil Engineering, University of Illinois at Urbana-Champaign, Urbana, Illinois.

the case for a simple structure supported on the ground. The simple damped mass-spring system normally is depicted as shown in Fig. 1. When excitation is applied to the base, usually acceleration versus time in the case of an earthquake, one can calculate any number of quantities that are descriptive of the response of the simple system; normally the peak value of a response quantity, for example relative displacement between the mass and base, u, pseudo relative velocity between the mass and base, ωu, or pseudo acceleration of the mass, $\omega^2 u$, are plotted as a function of frequency. The plots may be made in any number of ways, but one useful way is that of the tripartite plot. An example of two response spectra computed and plotted in this manner is shown in Fig. 2, one for a long-duration of strong shaking, namely the Imperial Valley El Centro Earthquake of May 18, 1940 (S00E), and another earthquake with a short burst of energy, the Parkfield Earthquake of 27 June 1966, Cholane-Shandom No. 2 Station (N65E) both for 5 percent damping.

It is not the purpose of this paper to recast the theory of computing and plotting response spectra; the uninitiated reader is referred to Refs. 1-5, as well as other texts, papers and reports, for a description of the details involved. However, one must read and study the literature with care because of the great variation in nomenclature employed, and the subtle differences in intended applications that are sometimes discussed. In addition to peak ground motions, as for example peak ground acceleration, velocity and displacement, another characterization of the effects of an earthquake is provided by the spectrum; it should be obvious that the spectrum is a form of transfer function since it relates excitation (loading) and simple structural response (resistance).

On the basis of studying response spectra associated with many earthquakes, and of calculations of response from recorded strong ground motion acceleration-time histories, it was possible to estimate the shape of typical spectra or to arrive at general rules for constructing approximate representative spectra for different values of damping [1-5]. This major development, which seems quite simple at this point in time, was of great importance as it permitted one to sketch a spectrum for general analysis purposes with knowledge of only a few key parameters. Typical examples of such "smoothed" response spectrum shapes are shown in Fig. 3. If these spectra were used for analysis purposes in connection with a seismic design project they commonly would be called "design spectra". Thus, there is a very definite difference between response spectra and design spectra, the former being associated with computed response for a particular ground motion for a particular earthquake, whereas a design spectrum is a characterization of the same general shape intended to capture the essential features of the response spectra for a region and class of earthquakes, but scaled to a level believed to be applicable for design purposes.

In some articles and texts the term "effective acceleration" or "effective motion" has been employed to denote the level of acceleration that is to be used as the anchor point (or "zero period" [ZPA] acceleration) for the design spectrum, the shape of the spectrum being determined generally as just described. The effective motion values are chosen by considering many factors as for example the effect on response of sustained long-duration shaking as contrasted to a short burst of energy with a high spike of high frequency acceleration; the latter normally is not of great consequence in affecting response. Normally distant source motion has been so filtered by the ground medium that it would be considered the effective motion. Also, in some cases, the effective motion term may mean specified values of velocity and displacement controls if these are believed applicable because of site conditions or for other reasons. In any event the design spectrum (or spectra) must be established to some decided level and one can see that the reasoning behind "design spectra" and "effective motions" are intertwined and the differences in arriving at the final result are subtle at best and essentially indistinguishable. Much more could be said about these topics but in the interests of brevity the reader is referred to Ref. 1-2.

For what purposes are design spectra employed? For a simple SDOF system, or a normal or principal mode of a model of a building, for example, one can enter the spectrum with the appropriate frequency or period and read off the maximum response value and in turn employ that value in the analysis computation in an appropriate fashion [6-8]. Or, for two simple systems, entering with the appropriate frequencies one can estimate displacements of each mass and draw conclusions about relative motions. Many other uses could be cited. Clearly the design spectrum is an approximation to a set of scaled response spectra or equivalent, wherein the responses may fall above or below the design spectrum, and for these reasons one wonders at times about the extreme accuracy that is attributed to such plots by some researchers and designers!

The foregoing discussion sets the stage for the primary point to be made in this section of the paper. Up to this time, in codes and design documents it has been assumed that one design spectrum would suffice for use in design. As more earthquakes are studied in detail around the world, and as more records of ground motions are obtained, it appears that the practice of employing only one design response characterization (one spectrum), even for elastic computations without any concern about nonlinear effects, may have some obvious shortcomings; such observations have not received much attention in the literature in any organized fashion. Several examples of situations where multiple design conditions are needed follow, along with some observations on associated decisions that must be made as well.

Some time ago in Ref. 7 the senior author noted the need in some cases for another spectrum (or alternative motion definition) for a distant earthquake source. In the case cited the normal design spectrum (broad banded) was intended to handle near-field to moderate distance earthquakes from sources with relatively long-duration and strong shaking, but that spectrum clearly failed to fulfill the requirement for a distant source long-period sustained type motion that could possibly affect high-rise buildings and fluid sloshing in large diameter tanks, or other long period items. Such a spectrum is shown in Fig. 4 as No. 2; the normal design spectrum is denoted as No. 1 there. Even the use of multiple spectra can lead to deficiencies insofar as design parameter definition is concerned as will be noted a little later herein.

It has been observed over the years, especially in the eastern United States, that local earthquakes (for example those of magnitude 4 and 5, which are relatively common) are narrow banded, as illustrated by spectrum No. 3 in Fig. 4. A highly debated question is whether or not such spectra should be used for design solely in lieu of the broad banded spectrum (No. 1 in Fig. 4) that reflects sustained shaking from strong ground motion. How can one be sure that the next earthquake will be so limited in source characteristics as to insure that such a narrow banded spectrum is appropriate? Perhaps the use of several design spectra, coupled with other appropriate criteria will resolve this problem. We do know that these small earthquakes can lead to sustained nearly harmonic motion (surface waves) at great distances from the source; even though the amplitudes are measured in microns, the fact that the motions are repeated for ten or twenty cycles, for example, can lead to significant excitation for long period items, as for example high-rise buildings.

Another interesting item has been detected in connection with some of the local earthquakes, namely that they may be accompanied by some significant excitation at selected frequencies, and in some cases these frequencies are high. In such cases one often can find another peak at a higher frequency as noted by No. 4 in Fig. 4. Usually such a peak is associated with a series of moderate high frequency acceleration-type ground motions. From an acceleration or force point of view, at this high frequency the values are of little structural significance; but, they may have an effect on the building contents or equipment. Generally most equipment qualification shake table testing, by virtue of the properties of the table, leads to testing at modest to reasonably high values of acceleration in the high frequency range of excitation; however, this is not always the case and this noted situation needs to be addressed in the design criteria.

In the last few years there has been increasing interest in utilizing the Power Spectral Density (PSD) as a "monitor" of the energy at the frequencies of interest in the ground motion time-history and thereby the spectrum as well. One suspects that a specified PSD value over a given frequency range may aid in maintaining adequate frequency and amplitude control but may not be totally sufficient. By this latter statement it is meant that the variation in input motions (frequency variation, duration, effects at distance, etc.) may serve to make the PSD an incomplete additional criterion. This topic needs further study by researchers and practitioners.

It should be clear by this time that depending upon the circumstances one can have a range of conditions that may have to be considered, and it is clear that one response spectrum will not serve to characterize all situations. Even more important is the fact that the spectrum by itself may not represent many of the desired the criteria parameters. For example, in the last example given above, for equipment one may well have to specify the ground motion time history (or more likely a set of such values) that is to be used for testing or analyzing the equipment item; in other words, the spectrum by itself is not a sufficient criterion.

NONLINEAR RESPONSE, MODIFIED SPECTRA, AND DAMAGE

In those cases where limited nonlinear behavior is to be incorporated in the response through analysis it is possible to rigorously compute the response of a simple oscillator as a function of frequency and thereby arrive at modified response spectra, i.e., spectra that correspond to specific nonlinear force deformation relationships. Studies conducted thus far suggest that the use of such spectra may lead to reasonable results, albeit approximate, when the ductility (defined as the ratio of the maximum deformation to the yield deformation) is limited to ductility values of no more than 5 or 6. Details for constructing such spectra are presented in Ref. 1 and 4, and are not repeated here; a typical modified spectrum is shown in Fig. 5. This concept was developed initially for monotonic elasto-plastic resistance type functions subjected to blast-type loadings and was later extended to the earthquake engineering field. A recent definitive study in this area [9] offered suggestions for refining the bounds, but noted that the overall ductility was still bounded fairly well by the maximum deformation achieved, even in the case of cyclic motion. Thus for elasto-plastic resistance modelling the bounds were essentially the same as developed earlier [1,4,7] in terms of reductions, as a function of the ductility factor, reflecting the limit acceleration and yield displacement.

Even so it was realized that the behavior was not well represented by the approximations just noted, and a subsequent study [10] undertook to examine the energy considerations in the hysteretic behavior of a SDOF system. Briefly the energy input to a structure subjected to ground motion is dissipated in part by damping and, in part, by yielding or inelastic deformation in the structural components. Well designed and well constructed buildings should be able to absorb and dissipate the imparted energy with minimal damage, but more on that point later herein. In this investigation the energy time-history response of SDOF systems subjected to earthquakes was studied. Very briefly the energy imparted to a simple structure for a unit mass [10] is given by Eq. (1) and the energy absorbed in the simple structure is given by Eq. (2).

$$E_I = \int_0^t \ddot{y}(t)\dot{u}(t)dt \qquad (1)$$

$$E_I = \int_0^t \ddot{u}(t)\dot{u}(t)dt + 2\beta\omega \int_0^t \dot{u}(t)^2 dt + \int_0^t R(u[t])\dot{u}(t)dt \qquad (2)$$

where E is the energy, $\ddot{u}$, $\dot{u}$, and u are the relative acceleration, velocity and displacement respectively, $\ddot{y}$ is the base acceleration time-history, β is the critical damping ratio, and ω is the natural circular frequency.

In Eq. (2) the first term represents the kinetic energy, the second term represents the energy dissipated by viscous damping, and the third term represents the sum of the hysteretic energy plus the strain energy.

Two examples of the energy, time and yielding sequence are shown in Figs. 6 and 7 for a simple structure subjected to the 1940 El Centro Earthquake and the 1966 Parkfield earthquake identified earlier; in the former case there was sustained long-duration shaking and energy input, and in the latter case a short burst of energy was input into the structure. It will be observed that the energy input curve for a structure subjected to long-duration motion has a large number of peaks and troughs as compared to that for short-duration motion. Also one should note the number of yield excursions, 15 in the case of the long-duration El Centro earthquake and 4 in the case of the shorter Parkfield earthquake. Many other observations arose out of this study pertaining to energy considerations as a function of structural frequency, including such items as effective times for energy absorption, effective motions, and energy spectra; the study offered some insight into steps that might be taken to evaluate damage in manners not previously addressed by researchers. The study suggests other matters that should be studied as part of investigations involving nonlinear response of frame-type structures, and several such studies are now underway in the Department of Civil Engineering at the University of Illinois.

However, in terms of this paper and the thesis developed herein, among the more important studies underway and nearing completion is one by the second author [11] dealing with development of structural damage evaluation criteria for seismic loading situations wherein low-cycle fatigue concepts are being adapted to the cyclic deformation process that occurs in structural members during an earthquake. An intensive examination of the low-cycle fatigue damage theories suggests that several of the plastic strain rules have potential application in the seismic response field if modified to represent energy absorption instead of cycles to fracture; it is apparent that the mechanisms involved in fatigue fracture and cyclic earthquake motion are quite similar in concept. One such expression, developed by Prof. J. D. Morrow of the University of Illinois Theoretical and Applied Mechanics Department, has the following form:

$$\frac{\Delta\epsilon\rho}{2} = \epsilon_f' \, (2N)^{-0.6} \tag{3}$$

where $\frac{\Delta\epsilon\rho}{2}$ is the plastic strain amplitude, ϵ_f' is the fatigue ductility coefficient and $2N_f$ is the number of reversals to failure. And, in turn, this general form of representing cyclic fatigue behavior can be restructured into a form to reflect the structural behavior associated with monotonic plastic ductility and effective hysteretic ductility. Also it has been necessary to develop a procedure to reflect the fact that the straining cycles in the member can be tensile and compressive, and not necessarily equal cycles in each domain. And, finally, it has been possible to demonstrate the applicability in simple structures for different types of seismic excitation and to check the developed relationships against available experimental evidence.

This important development should point the way to permit us to gain some insight into how one might incorporate realistic damage concepts into design spectra or alternatively other design criteria reflecting the ability of the structure to withstand cyclic loading. Also it may provide a basis in some part for helping evaluate the damage sustained by a structure in an earthquake and in turn reduce the large number of structures that are deemed unusable following an earthquake and thereby are torn down.

In conclusion one final observation needs to be made. Rational modeling of structures for purposes of analysis, no matter how approximate, requires one to be able to predict the behavior of the elements (including also connections, components, cladding, etc.) making up the structure, as well as the global behavior of the structure; our reservoir of experimental information in this area is indeed small and more studies are needed, not only to develop the entire strength relationship, and not only to arrive at better design in the sense of use of materials and detailing, but also in providing a better basis for estimating serviceability limits and remaining margins of strength. Experimental laboratory studies, and studies of undamaged and lightly damaged structures in earthquakes, the ultimate laboratory in this case, should do much to increase our knowledge in this area. In short much remains to be done.

ACKNOWLEDGMENTS

This research was supported in large part by the National Science Foundation under Grants PFR 80-02582, CEE-8203973 and DFR-8419191 at the University of Illinois at Urbana-Champaign, for which the authors are grateful. Any opinions, findings and conclusions or recommendations expressed in this paper are those of the authors and do not necessarily reflect the views of the National Science Foundation.

REFERENCES

1. Newmark, N. M. and W. J. Hall, <u>Earthquake Spectra and Design</u>, Monograph, Earthquake Engineering Research Institute, Berkeley, CA, 103p., 1982.

2. Housner, G. W. and P. C. Jennings, <u>Earthquake Design Criteria</u>, Monograph, Earthquake Engineering Research Institute, Berkeley, CA, 140p., 1982.

3. Chopra, A. K., <u>Dynamics of Structures - A Primer</u>, Monograph, Earthquake Engineering Research Institute, Berkeley, CA, 126p., 1981.

4. Newmark, N. M. and W. J. Hall, "Vibrations of Structures Induced by Ground Motion," in <u>Shock and Vibration Handbook</u>, ed. by C. M. Harris and C. E. Crede, McGraw-Hill, Inc., Second Edition, pp. 29-1 to 29-19, 1976.

5. Nau, J. M., and W. J. Hall, "Scaling Methods for Earthquake Response Spectra", Journal of Structural Engineering, ASCE, 110:7, pp. 1533-1548, July 1984. (See also Structural Engineering Studies Structural Research Report No. 499, "An Evaluation of Scaling Methods for Earthquake Response Spectra", by the same authors, May 1982)

6. Cruz, E. F., and A. K. Chopra, "Simplified Procedures for Earthquake Analysis of Buildings,", ASCE, Journal of Structural Engineering, 112:3, pp. 461-480, Mar. 1986.

7. Hall., W. J., "Observations on Some Current Issues Pertaining to Nuclear Power Plant Seismic Design," Nuclear Engineering and Design, 69:3, pp. 365-378, 1982.

8. Applied Technology Council, "Tentative Provisions for the Development of Seismic Regulations for Buildings," Report ATC 3-06, Palo Alto, CA, 505p., 1978.

9. Riddell, R. and N. M. Newmark, "Statistical Analysis of the Response of Nonlinear Systems Subjected to Earthquakes", Civil Engineering Studies Structural Research Series Report No. 468, 291p, August 1979.

10. Zahrah, T.F., and W. J. Hall, "Earthquake Energy Absorption in SDOF Structures," Journal of Structural Engineering, ASCE, 110:8, pp. 1757-1772, Aug., 1984. (See also Civil Engineering Studies Structural Res. Series Report No. 501, "Seismic Energy Absorption in Simple Structures", by the same authors, July 1982.)

11. McCabe, S. L., Doctoral Dissertation in the Department of Civil Engineering, Graduate College, Univ. of Ill. at Urbana-Champaign, Urbana, IL 1986.

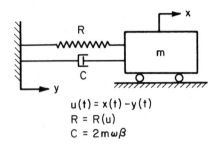

FIG. 1 SIMPLE DAMPED MASS-SPRING SYSTEM

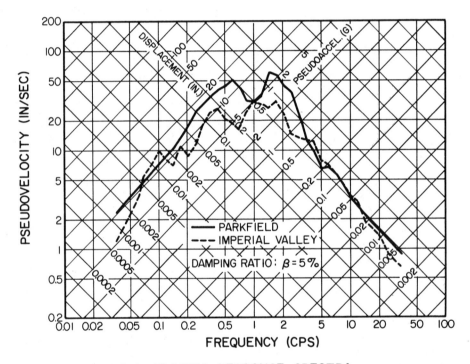

FIG. 2 ELASTIC RESPONSE SPECTRA

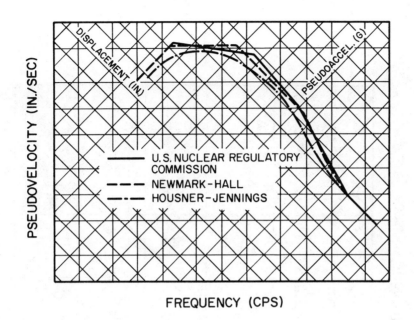

FIG. 3 TYPICAL SHAPES OF SMOOTHED DESIGN SPECTRA

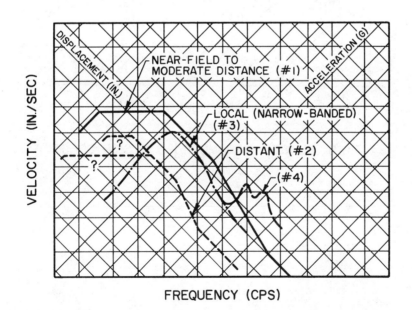

FIG. 4 DESIGN SPECTRA FOR DIFFERENT EARTHQUAKE SOURCES

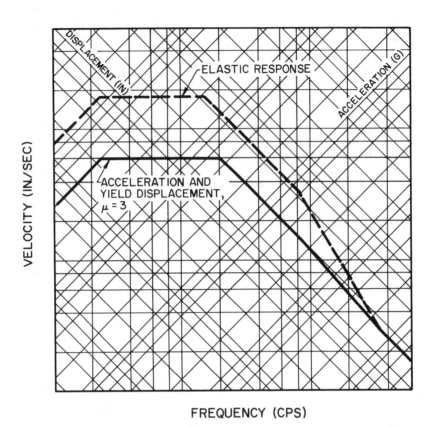

FIG. 5 ELASTIC AND MODIFIED DESIGN SPECTRA

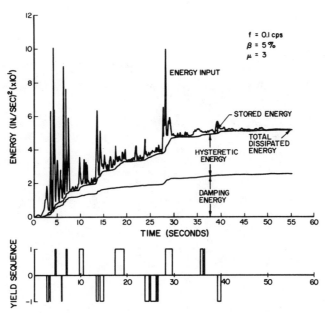

FIG. 6 ENERGY VERSUS TIME AND YIELD SEQUENCE FOR A SIMPLE STRUCTURE SUBJECTED TO EL CENTRO GROUND MOTION

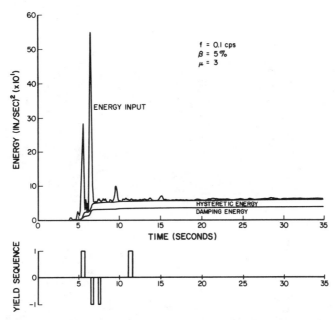

FIG. 7 ENERGY VERSUS TIME AND YIELD SEQUENCE FOR A SIMPLE STRUCTURE SUBJECTED TO PARKFIELD GROUND MOTION

SUPPLEMENTAL MECHANICAL DAMPING FOR IMPROVED
SEISMIC RESPONSE OF BUILDINGS

Robert D. Hanson[I], David M. Bergman[II] and Samir A. Ashour[III]

ABSTRACT

An earthquake resistant design philosopy which relies upon increased system damping rather than increased strength and ductility is investigated herein. The effects of additional mechanical damping on classic response spectra parameters and damping matrix proportionality as well as cyclic dynamic characteristics of viscoelastic mechanical damping devices are discussed.

INTRODUCTION

For many years it has been recognized that damping in our structures has been beneficial by limiting the maximum responses of the structures when subjected to earthquake ground motions. The fraction of critical damping has been used as an important parameter in the display of the earthquake ground motion effects as reflected by the earthquake response spectra. It has also been assumed that the effective fraction of critical damping increases with increased displacements, and that it increases more rapidly as material damage and/or material yielding occurs. Even in cases where inelastic energy absorption is included in the analyses it has been shown that increased viscous damping significantly reduces the inelastic energy demand.

While aerospace systems and mechanical systems have utilized added damping to control unwanted vibrations for many years, similar efforts have not been made for building systems. Several methods to provide supplemental mechanical damping to structural systems are available. The issues which need to be discussed are their static and dynamic characteristics, their reliability and their cost effectiveness for implementation in structural designs.

The purpose of this paper is to provide an understanding of the consequences of increasing system damping. To achieve this purpose the effect of damping on the classic response spectra parameters (displacement response, pseudo velocity and pseudo absolute acceleration) will be studied. The

---

I. Professor, Department of Civil Engineering, University of Michigan, Ann Arbor, Michigan, 48109-2125.

II. Ph.D. candidate, Department of Civil Engineering, University of Michigan, Ann Arbor, Michigan, 48109-2125.

III. Ph.D. candidate, Department of Civil Engineering, University of Michigan, Ann Arbor, Michigan, 48109-2125.

effect of supplemental mechanical damping on the proportionality (with respect to mass and/or stiffness) of the system damping matrix is also investigated and finally, the dynamic behavior of a viscoelastic mechanical damping device is discussed.

EFFECT OF DAMPING ON DISPLACEMENT RESPONSE SPECTRA

The displacement response spectra, SD, is a key parameter in estimating the maximum displacement responses in each mode of a building for past earthquakes or for a specified design spectra. This spectral displacement decreases as damping increases. Typically spectral calculations have used fractions of critical damping up to 20 percent. With supplemental damping the fraction of critical damping is no longer restricted to these low values. From the work of Ashour [1] a relationship for the change in SD with changes in damping is proposed and correlated with results obtained from existing earthquake accelerogram records.

Method of Analysis

The notion of a response spectrum is very simple. The maximum response of a linear single degree of freedom system to any given component of earthquake motion depends only on the natural frequency or period of the system and its fraction of critical damping. The natural periods, Tn, used in this study are 0.5, 1.0, 1.5, 2.0, 2.5 and 3.0 seconds which covers a representative range of natural periods. To determine the response spectra, nine earthquakes were used for excitation input. Three of the seismic records were real (El Centro 1940 N/S, Taft July 1952 S21W and Alameda Park May 1962 N10W) while the remaining six were from artificial earthquakes (B1, B2, C1, C2, D1) generated at the California Institute of Technology [2] and Gateway. The response spectra were generated by numerical solution of the equation of motion using Duhamel's integral.

Three different damping ranges were considered in evaluating Duhamel's integral. They are the underdamped, critically damped and overdamped conditions. The damping values used were 0, 2, 5, 10, 20, 30, 50, 75, 100, 125 and 150 percent of critical. The values of SD for a given Tn and damping factor were normalized with respect to the maximum value corresponding to zero damping. The normalized maximum displacements for the nine earthquakes were then averaged to obtain a mean value for each period and fraction of critical damping. In addition the standard deviation of the combination for each period was determined. Fig. 1(a) shows the resulting relation for the mean value of SD as a function of period and damping while Fig. 1(b) shows the relation of the mean plus one standard deviation. The general trend of the two curves is similar with little dispersion.

Approximate Relationship

An approximate relationship which represents the characteristics of the normalized SD as a function of period and damping is illustrated in Fig. 1 as solid continuous lines. The proposed relationship is based on a statistical approach studied previously by Caughey [3]. The average square of the displacement of an oscillator with small damping and zero initial conditions subjected to white noise with mean zero is described by Housner and Jennings [4]:

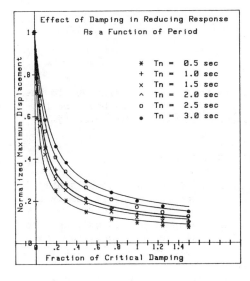

(a) Mean Displacement Response

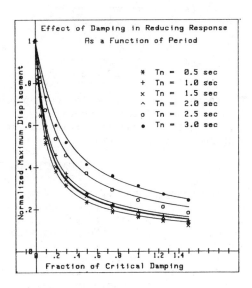

(b) Mean Plus Standard Deviation Response

Figure 1. Normalized Maximum Displacements for Nine Earthquake Accelerograms

$$X^2 = \frac{G(w)\pi(1 - e^{-2\xi w_n t})}{4\xi w_n^3} \qquad (1)$$

where $G(w)$ is the power spectral density, ξ is the fraction of critical damping and w_n is the undamped natural frequency. The ratio of the displacements of two oscillators having equal undamped natural frequencies but different amounts of damping is then given by the following equations:

$$(X_2/X_1)^2 = \frac{\xi_1 (1 - e^{-2\xi_2 w_n t})}{\xi_2 (1 - e^{-2\xi_1 w_n t})} \qquad (2)$$

and for $\xi_1 = 0$:

$$(X_2/X_1)^2 = \frac{(1 - e^{-2\xi_2 w_n t})}{2\xi_2 w_n t} \qquad (3)$$

However, because only an approximate empirical relationship is needed the previous two equations are modified to:

$$(X_2/X_1)^2 = \frac{\xi_1 (1 - e^{-2\xi_2 w_n B})}{\xi_2 (1 - e^{-2\xi_1 w_n B})} \quad (4)$$

$$(X_2/X_1)^2 = \frac{(1 - e^{-2\xi_2 w_n B})}{2\xi_2 w_n B} \quad (5)$$

The factor B is selected for each period so as to fit these relations to the data points calculated numerically. The fitting technique was based on the method of least squares. This relation is shown in Fig. 1 as the solid continuous lines.

EFFECT OF DAMPING ON PSEUDO SPECTRAL VALUES

The consequences of adding damping to structural systems was presented in terms of its effect on the displacement response spectra, SD. Typically the traditional tripartite response spectra are given as the displacement response spectra, SD, the pseudo velocity response spectra, PSV=wSD, and the pseudo acceleration, PSA=w^2SD. These relationships are valid only for low damping [5]. A comparison of the actual maximum absolute acceleration and PSA for fractions of critical damping up to 150 percent has been made by Ashour, et al [6]. Their conclusions are given here without the supporting data.

The general trend of the data showed that as damping increases the discrepancy between actual absolute Acceleration Spectra, SA and PSA increases for the Taft, the Alameda Park, and the El Centro earthquakes. Based on these limited comparisons it appears that direct integration evaluation of absolute Acceleration Spectra for large damping should be done rather than by assuming an w^2SD pseudo acceleration spectral technique which is valid with small damping.

NON PROPORTIONAL SUPPLEMENTAL DAMPING

Introducing damping to multi-degree of freedom systems requires that severe constraints be placed on the distribution of the damping if a proportional damping matrix [C] is desired. In this context a proportional damping matrix is defined as one which is uncoupled when the mass matrix [M] and the stiffness matrix [K] are uncoupled by the undamped eigenvectors (mode shapes) [7,8]. A damping matrix which is not proportional develops dynamic coupling of the modal responses and requires special considerations for a complete mathematical solution. Although supplemental damping devices can provide significant amounts of damping it is not practical to install supplemental dampers throughout the building on the basis of mass and/or stiffness proportionality. It would be convenient if the nonproportional [C] matrix could be assumed to have modal properties identical to a corresponding proportional [C] matrix. Due to the simplicity of using proportional damping for response computations [9] it is important to verify the accuracy of this assumption by comparing it to the actual dynamic response for the actual damping distribution. A comparison of the corresponding responses for two different buildings and two different earthquakes will be used to demonstrate that when subjected to certain restrictions equivalent proportional modal damping techniques can be used with confidence of achieving an acceptable degree of accuracy.

To simplify computational efforts, the example buildings are considered to be elastic rigid floor structures with building weights concentrated at the floor levels. The girders are infinitely rigid and column stiffnesses are unaffected by column axial load. Moments induced by the lateral displacement of gravity loads (P-Delta effects) are neglected. The damping is assumed to be interstory viscous damping.

Analytical results were obtained and comparisons made for direct integration of the complete set of multi-degree of freedom equations and mode superposition by integration and summation at each time step of each modal equation assuming that the undamped mode shapes and effective proportional modal damping are correct.

Nineteen Story Building

A nineteen story building, Table 1, is subjected to the 1940 El Centro N/S earthquake accelerogram. It is assumed that the building has fundamental damping proportional to mass and stiffness giving 5% in modes 1 and 2. The supplemental damping is added as equal interstory viscous dashpots in each story from the seventh level to the nineteenth level with no supplemental damping in levels one through six. To study the effect of non-proportionality of damping the building is analyzed for all the cases described by Hanson [10]. These cases are characterized by the amount of the fraction of critical damping in the fundamental undamped mode. The floor responses relative to the foundation and the story displacement responses are determined by direct integration of the equations of motion in the original coordinates and by integration of six modal equations independently with instantaneous combinations of the responses.

These results for 10 percent and 30 percent fraction of critical damping in the first mode are given in Fig. 2. It can be seen that as damping increases, the direct integration solution gives larger values than the assumed proportional damping technique in the lower floors where supplemental damping was not provided. A factor of two reduction in response in the upper levels occurs when the damping in the first mode is increased from 10 to 30 percent of critical.

TABLE 1.   NINETEEN STORY BUILDING FLOOR WEIGHTS AND
                    INTERSTORY STIFFNESSES

Level	19	18	17	16	15	14	13	12	11	10	9
Weight	12282	7571	7972	7954	8108	8108	8174	8174	8197	8197	8208
Stiffness	2204	2615	3590	3981	4312	4384	4783	5936	6542	7652	8218

Level	8	7	6	5	4	3	2	1
Weight	8259	13637	16081	29981	14954	21625	23737	28571
Stiffness	9721	11065	207806	52016	42456	50000	496191	1984249

Note: Floor weights are in kips and interstory stiffnesses are in kips/inch.
      1.0 kip = 4.45 kN and 1.0 kip/inch = 1.75 kN/cm

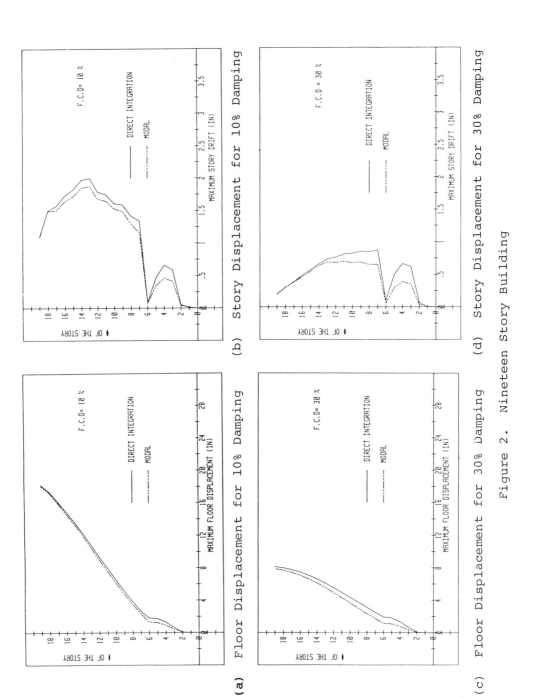

(a) Floor Displacement for 10% Damping (b) Story Displacement for 10% Damping

(c) Floor Displacement for 30% Damping (d) Story Displacement for 30% Damping

Figure 2. Nineteen Story Building

Ten Story Building

The 1952 Taft S21W earthquake accelerogram increased in amplitude by a factor of three was used as input ground motion for a ten story ductile moment frame with minimum cross braces. A single braced bay provides the lateral resistance for three bays of building weight equal to 132 kips per floor. The approximate interstory stiffnesses from the top to bottom are: 298.4, 345.7, 345.7, 417.8, 417.8, 490.9, 495.5, 544.0, 577.5 and 530.0 kips per inch. Responses for 10, 20 and 30 percent of critical first mode damping are shown in Figs. 3(a), (b) and (c). The differences between the direct integration solution and the modal integration solution is a function of the amount of damping in the system. One interesting point to be noted is that the interstory drift in the first story increased as damping increased. This may be due to the associated increase in the stiffness of the structure in all except the first story when damping was added to the upper levels. The supplemental dampers provided both velocity and displacement dependent contributions to the structural resistance.

If the same distribution of damping is used but damping and stiffness are added in the first floor, the story displacements will be as given in Fig. 3(d). It is obvious that a uniform addition of the supplemental damping provides better overall behavior as well as better accuracy of the modal integration solution.

VISCOELASTIC MECHANICAL DAMPER CHARACTERISTICS

By incorporating mechanical damping devices in a structural frame it is possible to increase the energy dissipation capabilities of the structure while decreasing its accelerations and inertia loads [11]. The end result is improved dynamic structural response.

At the University of Michigan mechanical damping devices using viscoelastic materials as energy absorbers have been tested in a hydraulic testing machine through a broad range of frequencies and amplitudes. The viscoelastic energy dissipation is accomplished by converting input energy to heat loss in the damper [12]. The results of a preliminary test program reported and analyzed by Bergman and Hanson [13] provide the damping characteristics of these devices for reducing the response of buildings to dynamic excitation.

The dynamic behavior of viscoelastic materials is strongly dependent on temperature and frequency. At high frequencies (or low temperatures) the storage modulus is large while the loss modulus is small. In this "glassy" region the behavior is like that of an ideal elastic material. At low frequencies (or high temperatures) the storage and loss moduli are both small. This "rubbery" region is not conducive to energy dissipation. The loss modulus is largest at intermediate temperatures and frequencies. Because damping is related to the loss modulus, the capacity to dissipate energy is greatest in this "transition" region.

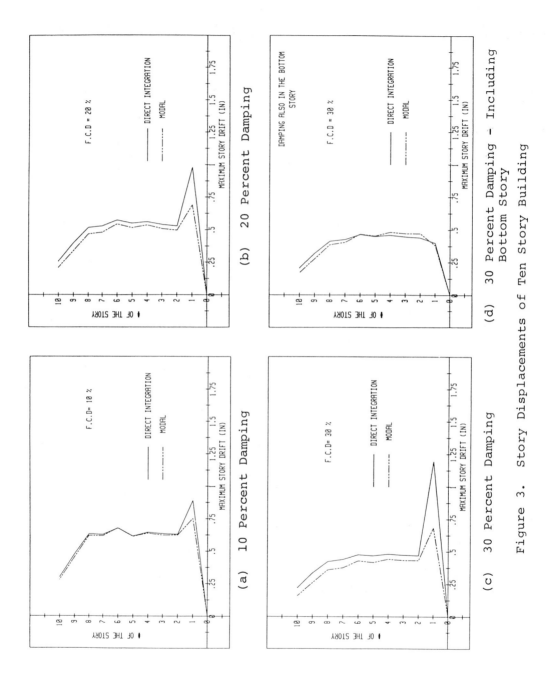

Figure 3. Story Displacements of Ten Story Building

Experimental Program and Results

The damper tests reported herein used pure shear shear dampers. This damper consists of two T-sections, one plate and two layers of a viscoelastic polymer which are sandwiched between the outer T-sections and the middle plate. The assemblage is bolted together through four slotted holes in the interior plate in order to allow relative displacement of the device's components. The damper is installed in a frame's diagonal brace so that lateral drift of the frame causes the middle damper plate to move in a direction opposite to the two outer T-sections, thus subjecting the viscoelastic material to pure shear. Two different dampers were tested in this program. Each contained two pads of viscoelastic material that were 4.0 inches wide by 10.0 inches long. The viscoelastic pad thicknesses in the first damper (D1) were 0.063 inches while those in the second damper (D2) were 0.125 inches.

Each damper was dynamically tested in an Instron hydraulic loading machine. Displacement control mode was used to cycle the devices 20 times at each of the following viscoelastic shear strain levels of +/- 40%, +/- 80% and +/- 160% sequentially for one test run. Six separate tests were conducted for each damper at frequencies of 0.10, 0.25, 0.50, 1.0, 3.0 and 5.0 cycles per second. Forty data points (load and relative displacement of the damper components) were taken for each cycle. This data was used to generate hysteresis loops for each damper to assess the effects of frequency and strain level on the viscoelastic material behavior.

As the frequency of loading increased the change in hysteresis behavior from cycle 1 to cycle 20 became greater. Although this is attributed to temperature rise in the damper material, temperature data was not recorded in these pilot tests. In order to provide some numerical interpretation of these results the cyclic energy dissipation during the last (20th) cycle at each strain level and frequency for the two devices is listed in Table 2. The twentieth cycle data was used to provide a conservative estimate of the energy dissipation capacity of these two specimens.

At each frequency and strain level the amount of energy dissipated during a cycle can be used to determine the equivalent viscous damping coefficient, c, [12,14]:

$$\text{Energy Dissipated/Cycle} = \pi * c * w * A^2 \tag{6}$$

where c has the units of force-seconds/unit displacement, w is the frequency in radians per second and A is displacement amplitude. The right side of this equation represents the energy dissipated per cycle due to viscous damping. The experimental data were used to determine values for the equivalent viscous damping coefficient (c) at each strain level and frequency as shown in Table 2.

TABLE 2. ENERGY DISSIPATED and EQUIVALENT VISCOUS DAMPING
              COEFFICIENT for CYCLE 20

Direct Shear Damper 1 (D1) Thickness=0.063 inches; Area=80.0 in}

Shear Strain		40%		80%		160%	
Frequency (Hertz)	Period (sec.)	ED	C	ED	C	ED	C
0.10	10	453	367	1481	300	4073	206
0.25	4	342	111	1337	108	4355	88
0.50	2	624	101	1584	64	4128	42
1.00	1	370	30	1234	25	3394	17
3.00	0.33	480	13	1680	11	2840	5
5.00	0.20	864	14	1152	5	2057	2

Direct Shear Damper 2 (D2) Thickness=0.125 inches; Area=80.0 in}

0.10	10	836	169	2379	120	6300	63
0.25	4	964	78	2636	53	6365	26
0.50	2	743	30	2171	22	4400	9
1.00	1	650	13	1900	10	3450	3
3.00	0.33	800	5	1450	2	2750	1
5.00	0.20	686	3	1086	1	971	0.2

Energy Dissipated, ED, (inch-pounds) (1.0 in-lb = 11.3 cm-N)
Equivalent Viscous Damping Coefficient, C, (kip-sec./inch)
            (1.0 kip-sec/in = 1.75 kN-sec/cm)

Analysis of Test Results

   The mechanical dampers tested displayed stable hysteretic behavior at all shear strain levels up to a frequency of 1.00 cycle per second. At higher frequencies the temperature of the viscoelastic material increases significantly which leads to a reduction in the values of both loss modulus and equivalent viscous damping. This indicates that the devices would be reliable over the range of frequencies expected from earthquake or wind excitation.

   Damping forces are proportional to velocity up to the 80% strain levels for frequencies of 1.00 cycles per second and less. This indicates that viscous damping could be used with excellent accuracy to model the damping characteristics of a structural frame incorporating these devices.

CONCLUSIONS

1. A relationship between displacement response spectra, SD, and fraction of critical damping has been established by using existing response spectra records and nine accelerograms for damping up to 150% of critical. These results are useful in estimating the maximum modal displacements for structures with increased levels of damping.

2. Damping has a major influence on the amplitude of response which may develop in a building subjected to seismic excitation. At the present time it is recommended that traditional pseudo spectral techniques not be used to establish spectral veolcities and spectral accelerations for systems with damping greater than 20 percent of critical.

3. The concept of representing supplemental damping in structures by independent modal damping ratios has been shown to be acceptable with some limitations. On the basis of this limited investigation, it appears that discontinuity in supplemental damping distribution should be avoided as much as possible for a reliable response determination using modal analysis. For smooth variations in damping, no significant difference in the response is obtained by direct integration or by modal combinations.

4. Direct shear dampers are capable of dissipating large amounts of energy in vibrating structural systems. This supplemental damping may be accurately modelled as viscous damping and its hysteretic behavior assured for typical buildings subjected to wind or earthquake excitation provided the strain levels in the material are limited to 100 percent strain.

ACKNOWLEDGEMENTS

This paper is based upon research supported by the National Science Foundation under Grant No. ECE-8512726. Mr. Ashour is supported by a fellowship from the Government of Saudi Arabia. The direct shear damping specimens were provided by Minnesota Mining & Manufacturing Company (3M). The opinions, findings and conclusions expressed herein are those of the authors and do not necessarily reflect the views of the National Science Foundation or 3M.

REFERENCES

[1] Ashour, S., Ph.D. Thesis, University of Michigan, in preparation.

[2] Jennings, P.C., Housner, G.W. and Tsai, N.C., "Simulated Earthquake Motions", California Institute of Technology, April, 1968.

[3] Caughey, T.K. and Stumpf, H.J., "Transient Response of Dynamic Systems Under Random Excitation", *Journal of Applied Mechanics*, ASME, December, 1961.

[4] Housner, G.W. and Jennings, P.C., "Generation of Artificial Earthquakes", *Journal of the Engineering Mechanics Division*, ASCE, February, 1964.

[5] Hudson, D.E., "Some Problems in Application of Spectrum Techniques to Strong-Motion Earthquake Analysis", *Bulletin Seismological Society of America*, Vol. 52, No. 2, 1962.

[6] Ashour, S., Hanson, R.D. and Scholl, R.E., "Effect of Supplemental Damping on Earthquake Response", *Proceedings, ATC Seminar on Base Isolation and Passive Energy Dissipation*, Applied Technology Council, Redwood City, CA, March, 1986.

[7] Caughey, T.K., "Classical Normal Modes in Damped Linear Dynamic Systems", *Journal of Applied Mechanics*, Vol. 6, No. 4, June 1960.

[8] Hansteen, O.E. and Bell, K., "On the Accuracy of Mode Superposition Analysis in Structural Dynamics", *Earthquake Engineering and Structural Dynamics*, Vol. 7, pp. 405-411, 1979.

[9] Clough, R.W. and Mojtahedi, S., "Earthquake Response Analysis Considering Non-Proportional Damping", *Earthquake Engineering and Structural Dynamics*, Vol. 4, 489-496, 1976.

[10] Hanson, R.D., "Basic Concepts and Potential Applications of Supplemental Mechanical Damping For Improved Earthquake Resistance", *Proceedings, ATC Seminar on Base Isolation and Passive Energy Dissipation*, Applied Technology Council, Redwood City, CA, March, 1986.

[11] Scholl, R.E., "Brace Dampers: An Alternative Structural System for Improving the Earthquake Performance of Buildings", *Eighth World Conference on Earthquake Engineering*, San Fransico, Vol. V, 1984.

[12] Mahmoodi, P., "Structural Dampers", *Journal of the Structural Division*, ASCE, Vol. 95, No. ST8, August 1969.

[13] Bergman, D.M. and Hanson, R.D., "Characteristics of Viscoelastic Mechanical Damping Devices", *Proceedings, ATC Seminar on Base Isolation and Passive Energy Dissipation*, Applied Technology Council, Redwood City, CA, March, 1986.

[14] Driscoll, W.A., "Linear Viscoelasticity", Vibration Control Systems, 3M, undated.

DETERMINING DESIGN RESPONSE SPECTRA FOR A SITE OF LOW SEISMICITY ON THE BASIS OF HISTORICAL RECORDS

Donald D. Hunt,[1] John T. Christian,[2] Thomas Y. H. Chang,[1] and Pedro A. Cadena[3]

ABSTRACT

Site dependent response spectra are developed for a region of very low seismicity by using two approaches: one based on accelerograms recorded on similar sites and the other on accelerograms recorded on rock and amplified through the soil profile. Differences in stiffness contrast between the recording stations and the site are accounted for by a parametric study. The computed spectra are combined statistically to give best estimates of the median spectrum and the variance of the estimate. A seismic hazard analysis gives values of the annual hazard associated with the spectra.

INTRODUCTION

Beaver Valley Power Station (BVPS) is a nuclear electric power generating plant consisting of two 800 megawatt pressurized water reactors (PWR) located in the town of Shippingport in western Pennsylvania. Design and construction of Unit 1 began in the late 1960's, and it has been in commercial operation since 1976. Unit 2 is now nearing completion and is scheduled to come on line in August, 1987. The two units are built on a flat-lying glacio-alluvial terrace of medium dense to dense sands, gravels, and silty sands and gravels on the south bank of the Ohio River. The terrace is about 115 ft. thick, with an average shear wave velocity of 1200 ft./sec., and overlies competent carbonaceous shale with a shear wave velocity of 5000 ft./sec.

The historical seismicity of the site was first investigated in 1968 for the design of Unit 1. Whitman [1] performed an analysis of the amplification characteristics of the soil profile at the site and recommended that the Safe Shutdown Earthquake (SSE - then called the Design Basis Earthquake) have a peak particle acceleration of 0.125 g at the ground surface. The final shapes of the design response spectra for both units were established through negotiation with the U. S. Atomic Energy Commission during the licensing

---

[1]Senior Geotechnical Engineer, Stone & Webster Engineering Corporation, Boston, Massachusetts.

[2]Senior Consulting Engineer, Stone & Webster Engineering Corporation, Boston, Massachusetts.

[3]Supervising Engineer, Duquesne Light Company, Pittsburgh, Pennsylvania.

process. The development of the BVPS design response spectra is summarized by Stone & Webster Engineering Corporation [2].

In 1984, during the review of the design of Unit 2 in connection with the application for an operating license, the US Nuclear Regulatory Commission asked that the seismic design basis be supported using current procedures that would account for local site conditions. Accordingly, the seismicity of the BVPS site was again studied in detail, and response spectra were developed for comparison with the BVPS-2 design response spectra. Two different approaches were employed: the site matched analysis and the soil response analysis. The site matched analysis involved establishing a suite of strong motion records observed at or near the surface of geologic profiles similar to BVPS during events having magnitudes similar to that of the SSE and evaluating the statistics of the resulting response spectra. The soil response analysis involved assembling a suite of strong motion records observed on rock during events of similar magnitude, amplifying them through a BVPS soil profile model, and evaluating the statistics of the resulting ground surface response spectra. The results of these two analyses were then combined statistically to provide a best estimate of the site dependent response spectrum. Seismic hazard analyses were performed to provide a measure of the likelihood of experiencing the design response spectrum. This paper summarizes those analyses; more details, particularly regarding the accelerograms used and the statistical and probabilistic methodology is found in reference [3]. Although they were motivated by specific questions arising during licensing of a nuclear power plant, the methodology and conclusions are useful for developing seismic design parameters for other projects.

SEISMICITY

The BVPS site lies near the center of the Appalachian Plateau tectonic province and is characterized by a low level of earthquake activity (see Fig. 1). During the past 180 years there have been few earthquakes within 200 miles of the site and only three within 50 miles. There are two areas of localized activity within the general region: one at Anna, Ohio, and the other at Attica, New York. The largest earthquakes at either source (an intensity VII-VIII (MM) event at Anna and an intensity VIII (MM) event at Attica) were barely perceptible at the site. The site has not experienced ground motion exceeding intensity III-IV (MM) from earthquakes within 200 miles. There have, however, been more severe ground motions from larger but more distant earthquakes. Examination of isoseismal maps of the large earthquakes felt in the eastern United States shows that the New Madrid events of 1811 and 1812 probably caused the maximum historical ground motion at the site, corresponding to an intensity of V (MM).

Most of the events shown in Fig. 1 cannot be associated with a particular geologic structure or causative fault. It is, therefore, postulated that similar events could occur anywhere within the same tectonic province. From this approach the maximum earthquake potential for the site is estimated to be an event of epicentral intensity VI (MM) occurring near the site.

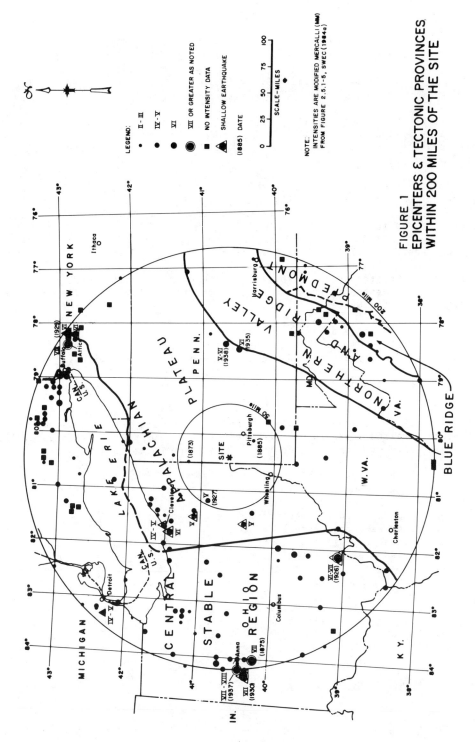

SCALING STRONG MOTION RECORDS

A correlation between the body-wave magnitude of eastern United States earthquakes and epicentral intensity is given by Nuttli and Herrmann [4] as:

$$m_b = 0.5\ I_o + 1.75 \quad (m_b \pm 0.5\ \text{units}) \tag{1}$$

where:
  m_b = body wave magnitude
  I_o = epicentral intensity (MM).
This converts the SSE intensity VI (MM) to a body wave magnitude of 4.75 ± 0.5.

Most of the available strong motion records are for earthquakes occurring in the western United States, which are most often described in units of local magnitude M_L. Chung and Bernreuter [5] developed an empirical correlation between M_L and m_b:

$$M_L\ (\text{west}) = 0.57 + 0.92\ m_b\ (\text{east}). \tag{2}$$

Therefore, an eastern m_b of 4.75 is equivalent to a western local magnitude of 4.95 ± 0.5. Events occurring in Italy were assumed to be similar to those in the western United States.

While it would be desirable to select a suite of records from events of magnitudes falling within the desired range, the present catalogues of strong motion records do not provide enough records to constitute an adequate population for statistical analysis. To increase the number and variety of usable records, a procedure was developed to scale the accelerograms to the desired magnitude.

Nuttli [6] considers the following relationship, presented by Nuttli [7] for South Carolina and by Nuttli and Herrmann [8] for the Mississippi Valley, to be generally valid for the central and eastern United States:

$$\log a_h = A + B\ m_b - 0.83\ (R^2 + h^2)^{1/2} - C\ R \tag{3}$$

where:
  a_h = peak horizontal acceleration in cm/sec²
  m_b = body wave magnitude
  R = epicentral distance in Km
  h = focal depth in Km
  A, B, C = empirical constants
  B = 0.5 for $m_b \geq 4.5$
      0.25 for $m_b < 4.5$.

Assuming all parameters and constants in Eq. 3 are held constant except for the magnitude and the acceleration leads to the following relations. For eastern earthquakes,

$$\Delta \log a_h = 0.5\ \Delta m_b \quad \text{for } m_b \geq 4.5 \tag{4a}$$

$$\Delta \log a_h = 0.25\ \Delta m_b \quad \text{for } m_b < 4.5 \tag{4b}$$

For equivalent western earthquakes,

$$\Delta \log a_h = 0.54 \Delta M_L \quad \text{for } M_L \geq 4.7 \quad (5a)$$

$$\Delta \log a_h = 0.27 \Delta M_L \quad \text{for } M_L < 4.7 \quad (5b)$$

Eqs. 4 and 5 were used to scale selected acceleration time histories either up or down to the target SSE magnitudes of 4.75 m_b for an eastern earthquake or to 4.95 M_L for a western or Italian earthquake.

SITE MATCHED RESPONSE SPECTRA

Available catalogues of strong motion records were searched to establish a suite of strong motion records caused by nearby earthquakes recorded at or near the surface of a soil profile similar to that of the site. Nine suitable recordings were found, all made in California, with M_L between 4.0 and 6.5 and R between 6 and 29 Km. The response spectrum for 5% of critical damping was computed for each of the eighteen horizontal components of time history of acceleration (two for each of nine events). Examination of the results showed that the ordinates of the spectra follow closely a log normal distribution, as would be expected. The computed site matched response spectrum was, therefore, represented by plots of the mean and of the mean plus one standard deviation of the logarithm of the spectral velocity for each frequency at which the individual response spectra were computed. These were identified as the "50th percentile" and "84th percentile" spectra, respectively.

Although the accelerograms selected were recorded at sites with soil profiles matching BVPS as closely as possible, the profiles at the recording stations did not have the same contrast between the shear wave velocities of the soil and the underlying rock. This difference can affect the motions recorded at the surface. The effect can be characterized by the velocity contrast ratio (VCR), defined as the ratio between the shear wave velocity of the rock or rock-like layer and the soil just above the rock. With a decrease in the VCR, i. e., as the shear wave velocity of the rock approaches that of the soil, more of the energy of the waves reflected at the free surface of the soil is reabsorbed by the rock, and spectral ordinates decrease. Conversely, as the VCR increases, more of the energy remains in the soil layer, and spectral ordinates increase.

The VCR at BVPS is 4.2; the average VCR for the sites of the recording stations is 2.0. A correction function was established by means of a shear wave propagation analysis of the BVPS soil profile using the SHAKE program [9]. Several rock outcrop accelerograms were amplified through the a model of the soil profile, and the stiffness of the underlying rock was varied to give values of VCR of 4.2, 2.0, and 1.0.. Ground surface response spectra for 5% structural damping were computed. The mean of the logarithms of the spectral velocities was found at each frequency for each of the three values of VCR. Fig. 2 shows the percentage change in the mean spectral ordinates caused by changing the VCR from 1.0 to 4.2 and from 2.0 to 4.2.

The 50th and 84th percentile spectra previously computed from the scaled, site-matched accelerograms were adjusted at each frequency by the ratios shown in Fig. 2 for VCR changing from 2.0 to 4.2. Fig. 3 shows the resulting curves for 5% structural damping compared with the corresponding SSE design spectrum.

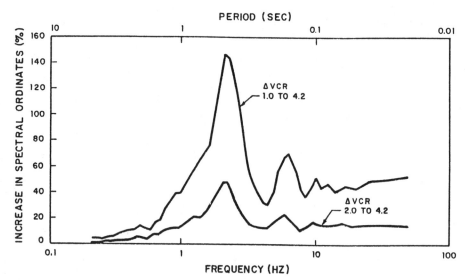

FIGURE 2. PERCENT INCREASE IN RESPONSE SPECTRA ORDINATES DUE TO CHANGE IN VELOCITY CONTRAST RATIO

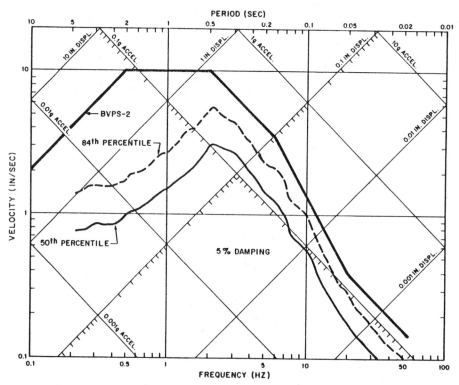

FIGURE 3. SITE MATCHED RESPONSE SPECTRA ADJUSTED FOR SHEAR WAVE VELOCITY CONTRAST

SOIL RESPONSE ANALYSIS

An alternative to the procedure using site matched records is to amplify accelerograms recorded on rock through a numerical model of the soil profile, using SHAKE [9] or some similar program. As was done for the site matched records, the catalogues of available strong motion records made on rock were searched to establish a suite of accelerograms for earthquakes with appropriate magnitudes. The conditions under which the records were obtained were thoroughly researched to ensure that they were, in fact, recorded on rock at the surface and were suitable for this analysis. Fourteen appropriate recordings were found, most made in California but including accelerograms from Helena, MT, and from the Friuli, Italy, and Miramichi, N. B., events. M_L ranged between 4.6 and 6.4, and R between 2 and 29 Km.

Each of the twenty-eight time histories of horizontal acceleration (two for each of fourteen events) was used without scaling and amplified through the model of the BVPS profile. The output was a time history of acceleration at the surface, which was then scaled, in the same way as the site matched records, to the desired magnitude. The scaling was done to the surface records instead of to the input records because the empirical relations, on which the scaling is based were derived from surface data. Ground surface scaling also made the time histories computed in the soil response analysis conceptually the same as those for the site matched analysis. The response spectrum for 5% structural damping was computed for each ground surface motion. The mean and the mean plus one standard deviation of the logarithm of the spectral velocity was computed at each frequency. Fig. 4 shows the resulting curves, labelled "50th percentile" and "84th percentile."

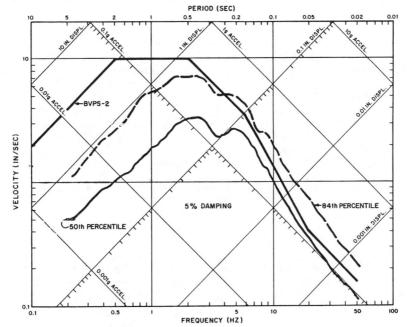

FIGURE 4. HORIZONTAL RESPONSE SPECTRA FROM SOIL
RESPONSE ANALYSIS: SCALED

COMBINED ESTIMATE OF RESPONSE SPECTRUM

The site matched records and the soil response analysis provide two separate statistical estimates of the site dependent response spectrum. Advantages and limitations can be ascribed to each approach, and it cannot be stated with certainty which of the two gives the best estimate. They are combined statistically to provide the best estimate of the site dependent response spectrum.

If v is an unbiased estimate of the underlying population mean, it is possible to compute v as a linear combination of the means estimated by the two approaches:

$$v = w_{sm}v_{sm} + w_{sr}v_{sr} \tag{6}$$

where:
  w is a weighting factor
  subscript sm indicates site matched
  subscript sr indicates soil response analysis

The weighting factors are found by first assuming that the two approaches are dealing with the same underlying population and then minimizing the variance of the estimate of the combined mean. The results are

$$w_{sm} = (1/D)(n_{sm}/s_{sm}^2) \tag{7}$$

$$w_{sr} = (1/D)(n_{sr}/s_{sr}^2) \tag{8}$$

$$D = n_{sm}/s_{sm}^2 + n_{sr}/s_{sr}^2 \tag{9}$$

where:
  n = number of points (18 for sm and 28 for sr)
  s = square root of the variance.

The weighting factors in Eqs. 7 through 9 are different for each frequency.

The variance of a sample of size n from a normal population can be described with a chi-squared distribution with (n-1) degrees of freedom. The sum of chi-squared distributions is itself chi-squared distributed, and it follows that

$$s^2 = ((n_{sm}-1)s_{sm}^2 + (n_{sr}-1)s_{sr}^2) / (n_{sm} + n_{sr} - 2). \tag{10}$$

For the numbers of records used here, this becomes at all frequencies

$$s^2 = 0.3864\, s_{sm}^2 + 0.6136\, s_{sr}^2. \tag{11}$$

Eqs. 6 through 11 were applied at all frequencies to combine the estimates from the site matched and soil response analyses. The results were expressed by the mean and the mean plus one standard deviation of the logarithm of the spectral velocity. Fig. 5 shows these curves, labelled "50th percentile" and "84th percentile," along with the SSE design spectrum for 5% damping.

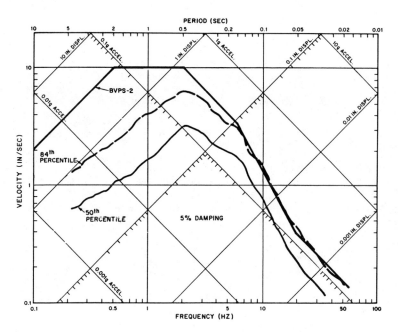

FIGURE 5. SITE DEPENDENT RESPONSE SPECTRA

PROBABILISTIC ANALYSES

Underlying the preceding analyses is the question of what hazard is associated with these spectra. Several factors must interact for the response spectra to be equaled or exceeded at the site, and there are uncertainties associated with all of them. Therefore, two sets of probabilistic analyses were performed. The first consisted of a set of conventional seismic hazard analyses using the model developed by Cornell [10] and extended by McGuire [11]. The second was a more detailed analysis of the conditions immediately around the site.

The conventional seismic hazard analysis used three models of the seismogenic regions around the site, which are described by Acharya et al. [12] in more detail than can be presented here. The first two used seismotectonic provinces and concentrated local sources developed on the basis of geologic information and historical data. Because many events occurred near the boundaries of the provinces, the alternate choices led to two distributions of seismic activity and hence to two models of seismogenic regions. The third model used source zones and background seismicity established on the basis of historical activity alone. The attenuation relation of Eq. 3 was used with a standard deviation of the error of 0.6.

The results of the seismic hazard analyses were principally that the annual hazard for a_h greater than or equal to 0.125 g (the SSE value) is between 1.3 and 2.5 X 10^{-4} and that the Appalachian Plateau Province, which surrounds the site, contributes between 92 and 96 per cent of the hazard.

This suggested that a more detailed examination of the contribution of the local region would be fruitful.

The second set of analyses considered how the design response spectra might be exceeded. The probability can be expressed as

$$P(A_2 \text{ \& } A_3) = P(A_2) P(A_3) = P(A_1) P(A_2|A_1) P(A_3) \tag{12}$$

where A_2 and A_3 are stochastically independent and the occurrences are defined as:

A_1 an earthquake occurs within the Appalachian Plateau tectonic province and with m_b equal to or greater than 4.75.

A_2 an earthquake occurs within 29 Km of the site and with m_b equal to or greater than 4.75.

A_3 an earthquake occurs within 29 Km of a site similar to that of BVPS and with m_b equal to or greater than 4.75 produces a response spectrum that exceeds a given response spectrum.

The historical seismicity of the region defined as the Appalachian Plateau tectonic province is sparse, but conventional analyses can be performed to establish recurrence relations. These yield estimates of the annual probability of equaling or exceeding an earthquake of m_b = 4.75, that is $P(A_1)$, between 0.0110 and 0.0045. Alternatively, data for the province within 200 miles of the site yield $P(A_1)$ of 0.006. These results are not sensitive to the precise location of the boundaries of the province or to the inclusion or exclusion of particular events.

The probability $P(A_2)$ is computed by multiplying $P(A_1)$ by the ratio between the areas of a circle of radius 29 Km and of the province. This ratio is 0.0193. The probability $P(A_3)$ is 0.5 for the 50th percentile spectra and 0.16 for the 84th percentile spectra.

Combining these values gives the annual probability of exceeding the mean response spectrum between 4.4×10^{-5} and 1.1×10^{-4}. For the 84th percentile spectrum the probability of exceedance is between 1.4×10^{-5} and 3.4×10^{-5}. These results do not include the contribution of smaller magnitudes. Approximate extrapolations, using Eq. 3, indicate that the smaller earthquakes may increase the annual probabilities to about 2.3×10^{-4} for the 50th percentile spectrum and 5.3×10^{-5} for the 84th percentile spectrum. The probability associated with the 50th percentile spectrum is below that usually considered acceptable for events such as the SSE, and that associated with the 84th percentile is much lower.

CONCLUSIONS

The studies of the site dependent response spectra for a site in a region of very low seismicity have provided several useful conclusions and examples for future application. The suite of earthquake records must be examined carefully to remove events recorded at stations whose conditions do not correspond to those assumed in the analyses. Records for site matched analyses and for soil response analyses by shear wave propagation can be scaled to the desired magnitude, reducing the scatter in the results and expanding the available catalogue. The effects of velocity contrast differences are accounted for by a parametric evaluation of the effects of the velocity contrast ratio. A statistical procedure is then used to combine different estimates of measures of the response spectra, leading to best estimates of the statistics of the site dependent response spectra. The example described above combines two estimates, but the techniques could be used or any number of independent methods of estimating the spectra. Finally, a seismic hazard analysis gives values of annual hazard associated with different spectra.

ACKNOWLEDGEMENTS

Many people contributed to the work described in this paper. Particularly important were the contributions of the Seismic Advisory Panel, whose members were I. M. Idriss, Chairman; J. A. Jan; C. W. Lin; D. E. Shaw; and the second author. Significant contributions were also made by the following personnel of Stone & Webster Engineering Corporation: H. K. Acharya, R. W. Borjeson, J. K. Downing, J. W. McCoy, and P. J. Trudeau. The statistical methodology relies on major contributions made by K. F. Reinschmidt of Stone & Webster Engineering Corporation.

REFERENCES

[1] Whitman, Robert V., "Effect of Local Soil Conditions upon Seismic Threat to Beaver Valley Power Station." Report prepared for Stone & Webster Engineering Corporation, 1968.

[2] Stone & Webster Engineering Corporation, "Seismic Design Response Spectra, Beaver Valley Power Station - Unit 2. Docket No. 50-334." Prepared for Duquesne Light Company, Pittsburgh, Pa., June, 1984.

[3] Stone & Webster Engineering Corporation, "Site Dependent Response Spectra, Beaver Valley Power Station - Unit 2. Docket No. 50-334." Prepared for Duquesne Light Company, Pittsburgh, Pa., February, 1985.

[4] Nuttli, O. W., and Herrmann, R. B., "State-of-the-Art for Assessing Earthquake Hazards in the United States: Credible Earthquakes for the Central United States," Miscellaneous Paper S-73-1, Report No. 12, U. S. Army Waterways Experiment Station, Vicksburg, MI., 1978.

[5] Chung, D. H., and Bernreuter, D. L., "Regional Relationships among Earthquake Magnitude Scales," Lawrence Livermore Laboratory Report 52745, prepared for U. S. Nuclear Regulatory Commission, NUREG/CR-1457, 1980.

[6] Nuttli, O. W., Personal Communication with H. Acharya of Stone & Webster Engineering Corporation, October 11, 1984.

[7] Nuttli, O. W., "Instrumental Data," included in: Nuttli, O. W., Rodriguez, R., and Herrmann, R. B., "Strong Ground Motion Studies for South Carolina Earthquakes," prepared for U. S. Nuclear regulatory Commission, NUREG/CR-3755, April, 1984.

[8] Nuttli, O. W., and Herrmann, R. B., "Ground Motion of Mississippi Valley Earthquakes," Journal of Technical Topics In Civil Engineering, ASCE, Vol. 110, No. 1, May, 1984, pp. 54-69.

[9] Schnabel, P. B., Lysmer, J., and Seed, H. B., "SHAKE: A Computer Program for Earthquake Response Analysis of Horizontally Layered Sites," Report EERC-72-12, University of California, Berkeley, 1972.

[10] Cornell, C. A., "Engineering Seismic Risk Analysis," Bulletin, Seismological Society of America, Vol. 59, No. 5, October, 1968, pp. 1583-1606.

[11] McGuire, R., "FORTRAN Computer Program for Seismic Risk Analysis," U. S. Geological Survey Open File Report 76-67, 1976.

[12] Acharya, H. K., Lucks, A. S., and Christian, J. T., "Seismic Hazard in the Northeastern United States," Soil Dynamics and Earthquake Engineering, Vol. 3, No. 1, 1984, pp. 8-18.

RESPONSE SPECTRA FOR BUILDING DESIGN

N. C. Donovan[I] and A. M. Becker[II]

ABSTRACT

Probabilistic studies are now routinely used to provide design response spectra for major structures. Following a detailed review of hazard studies in northern California simple methods have been developed to allow construction of response spectra compatible with the fully probabilistic response. The preferred spectral construction requires a probabilistically derived or mapped effective peak acceleration value. The alternate method uses average attenuation data provided from an appended figure. The major requirement found in the development of compatible spectra was the need to consistently include the parameter uncertainty one time only, either during the hazard integration or the spectral construction.

INTRODUCTION

A seismic hazard study using probabilistic analyses is generally recognized as the preferred method for developing design response spectra for major projects in seismic regions. However, simpler alternative spectra construction procedures using either estimates of the response spectra produced by the maximum event on the nearest significant fault or by anchoring a standard spectral shape to a probabilistically estimated peak acceleration are more frequently used.

Significant differences in the design spectra developed can occur when these methods are applied. This paper examines these differences and shows how they have arisen and suggests a modification to the simpler approach which provides spectra consistent with the results of more detailed hazard analyses. Design response spectra developed in accordance with the recommended methods should be adequate for preliminary studies for most projects and could also be used for final design, particularly for projects where more detailed and expensive studies are not feasible.

SPECTRAL ATTENUATION

Earthquake records obtained since McGuire [8] first derived attenuation equations for spectral values have allowed refinement of the relationships by several investigators [3, 6, 7]. With these improved relationships hazard analyses can be performed to obtain response spectra with uniform probability

---

I. Partner, Dames & Moore, 500 Sansome Street, San Francisco, California 94111.

II. Project Engineer, Dames & Moore, 500 Sansome Street, San Francisco, California 94111.

values for all spectral ordinates. Probabilistic analyses should include the uncertainty of the input parameters in the results. These uniform risk spectra differ from the simpler procedures mentioned above in that they have lower values than the anchored response spectra unless probabilities much smaller than those normally used for buildings are considered. The same trend is also apparent when they are compared with spectra anchored to the acceleration caused by the occurrence of the largest event on the closest fault. This difference, which has been observed in many studies, is demonstrated in Fig. 1 where the uniform risk spectra for several different return periods obtained using the Joyner and Boore [6] relationship based on both horizontal motion components are compared with the median level spectrum for the maximum size event on the closest approach to the controlling fault.

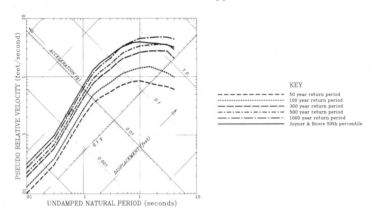

Fig. 1 Example of Probabilistic Response Spectra

The term return period is used in this paper because it appears to be a preferred term among practicing engineers. As used here it represents the reciprocal of the annnual probability and, as always for probability based items, does not preclude zero or multiple events occurring during the specified time period.

The results of the hazard study with uncertainty included only approach the median upper bound spectrum with an annual probability of 0.001 or a return period of 1000 years, a probability level too small for industrial and commercial structures. This example illustrates that the proportional spectral contributions at long periods in a hazard study increase more rapidly than the spectral values at short periods but start from much smaller values for short return periods. The frequent occurrence of smaller sized events has a controlling influence on short period response but cannot influence response at long periods. The relationship between the peak acceleration values representing different probability levels has been shown to satisfy approximate relationships by Algermissen and Perkins [1] and Joyner and Fumal [7].

SAN FRANCISCO BAY EXAMPLE

Some of the principles briefly described above can be best demonstrated by considering an example. The San Francisco Bay is chosen for this because

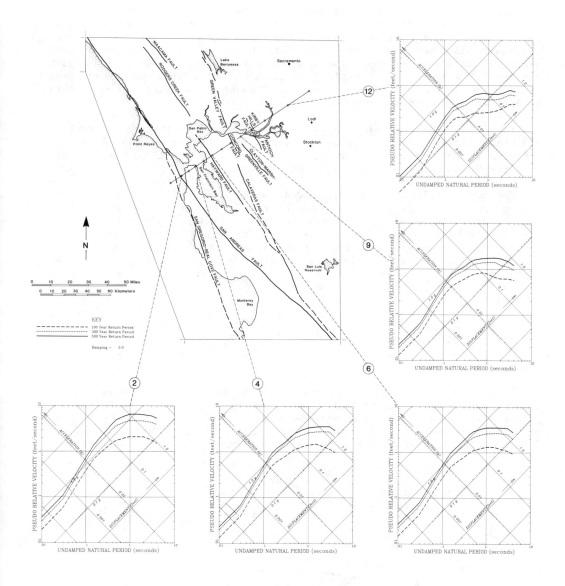

FIGURE 2. SAN FRANCISCO BAY FAULTS. RISK MODEL SECTION ACROSS ACTIVE REGION WITH SAMPLES OF HAZARD ANALYSIS RESPONSE SPECTRA.

of its familiarity to much of the earthquake engineering profession. For purposes of consistency we have used the Joyner and Boore [6] spectra attenuation equations for the mean peak component (randomly oriented) on a soil profile together with the mean peak acceleration and velocity equations given in Joyner and Fumal [7]. The error term sigma is used as given for each equation. The choice of the Joyner and Boore [6] equations was not made with any special endorsement in mind. The points made in this paper have been found to be independent of the choice of attenuation equation series so this readily available set was chosen.

In addition to the Joyner and Boore mean peak acceleration attenuation equation, in some of the following comparisons we have also used average attenuation curves derived from recent references [5]. These average peak acceleration curves are presented as an appendix.

The San Francisco Bay study area is shown together with the principal active faults in Fig. 2. Also shown on the Figure is a section line that extends from the offshore area on the Pacific tectonic plate towards the northeast and the central San Joaquin and Sacramento Valley near the city of Sacramento. The extension to the Sacramento area is of some current interest because of tentative suggestions for revising the seismic zoning map for the state of California. A detailed seismic source model prepared for a recent project was used but is not reproduced as part of this paper. It should be pointed out that the model incorporates seismic moment accrual resulting from measured creep rates such as were considered in Southern California studies by Joyner and Fumal [7]. Thirteen points were selected across the section. A complete set of probabilistic spectra was developed for each point. Figure 1 represents the results for appropriate soil profiles in downtown San Francisco.

While there is much information that can be obtained from the results of an examination of 13 individual sets of response spectra it is helpful to present some of the main observations more directly. This can be done by presenting the variation of several of the main spectral parameters along the 150 kilometer long seismic cross-section. Figure 3 shows the variation of

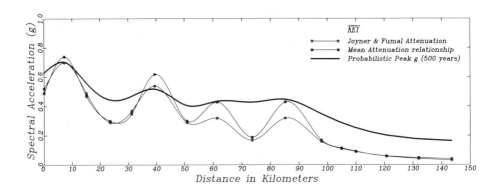

Fig. 3   Estimated Peak Acceleration Variation Along Cross-Section

peak acceleration across the section obtained from the probabilistic hazard analysis and computed deterministically by two sets of attenuation equations. (Unless noted otherwise all probabilistic results are for a 500 year return period and for a 5 percent damping ratio.) The smoothing effect of probabilistic values over the directly computed maximum values is readily evident on Fig. 3. This is the result of including the uncertainties in the probabilistic study. Including uncertainty raises the values away from fault crossings. At the fault crossings, the probabilistic accelerations are lower than the deterministic values due to the relatively lower probability of the maximum event occurring.

Spectral accelerations are compared with the peak accelerations in Fig. 4. The values shown have been normalized to 1.0g at the San Andreas Fault crossing to better demonstrate the relative attenuation with distance from the active seismic regions towards inland California. The spectral values chosen were computed at periods of 0.4 and 1.5 seconds and are believed to encompass the fundamental period of vibration of most building structures. It is recognized that major structures, which have longer periods of response, should be designed on the basis of site specific studies and therefore longer periods are not considered in this paper. A major question is posed by the results shown on Fig. 4. The ATC3-06 [2] studies were predicated on the assumption that longer period motion attenuated more slowly with distance than high frequency or short period motion. Spectral attenuation relationships were not available at the time the ATC3-06 hazard maps were prepared. The results in Fig. 4 appear to contradict this interpretation and suggest additional conservatism in ATC3-06. The inability of small to moderate sized earthquakes to produce significant structural response has not been fully recognized until now. Although California zoning is not a specific topic of this paper, the results shown in Fig. 4 suggest possible changes could be discussed.

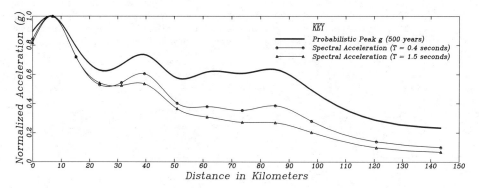

Fig. 4 Comparison Between Peak Acceleration and Spectral Acceleration

DESIGN RESPONSE SPECTRA

The detailed studies which are necessary to develop probabilistic response spectra can only be justified for the design of major structures. These studies are too expensive for most buildings so there is a distinct need for a simpler procedure to develop design spectra for these structures without performing a complete probabilistic hazard study.

A comparison of the Joyner and Boore [6] response spectra with spectra obtained using a simple extension of the Newmark and Hall [9] spectral construction method was proposed by Donovan [4]. The change recommended by Donovan was to use a v/a ratio that recognized its dependency on magnitude and distance. Using the Joyner and Boore strong motion data set with both components included, Donovan [4] derived an equation for the velocity to acceleration ratio. This relationship is as follows:

$$\log (v/a) = 0.92 + 0.065 M + 0.00127 R + 0.23 S \qquad (1)$$

where $R^2 = d^2 + 7.5^2$
$S = 1$ for soil
$ = 0$ for rock

and M is moment magnitude, and d is the shortest distance to the surface projection of the aftershock area. The v/a ratio is given in inches/second/g when distances in kilometers are used.

Comparisons of probabilistic hazard analyses are often made on the basis of peak accelerations. This is done for convenience and follows from the custom of anchoring a fixed spectral shape to an effective peak acceleration. For this study direct comparisons of the spectral values obtained by different procedures have been made. Figure 5 shows the variation of spectral acceleration across the San Francisco cross-section at a period of 0.4 seconds while Fig. 6 shows similar data for a structural period of 1.5 seconds. The heavy line on each figure represents the spectral values obtained from the complete probabilistic hazard study. The two lighter curves present the variation computed by the Newmark-Hall [9] spectral construction procedure and using the v/a ratio obtained from Eq. 1. One curve is based on spectra anchored to the 500 year return period peak acceleration from the hazard study and the other uses the deterministic values obtained from the average acceleration curves. Examination of Fig. 6 shows that both construction procedures give modest approximations to the probabilistic study values except at distances close to major fault crossings.

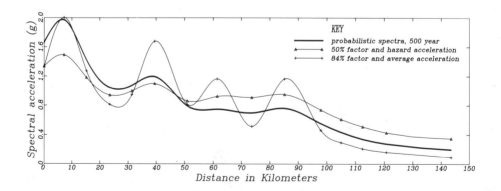

Fig. 5 Variation of the Spectral Acceleration at a 0.3 Second Period Using Alternative Spectral Construction

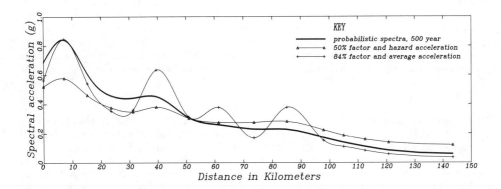

Fig. 6 Variation of the Spectral Acceleration at a 1.5 Second Period Using Alternative Spectral Constructions

Perhaps the most important conclusion to be made from the results summarized on Figs. 5 and 6 is the importance of the use of the parameter uncertainties. These must be included in any study and this incorporation must be done consistently. The uncertainty is included in each of the sets of curves. In the two sets of probabilistic results it is included in the integrations from which the values were obtained. For the spectral values anchored to the average attenuation curve, the uncertainty was incorporated by using the 84th percentile spectral amplification parameters.

SIMPLIFIED SPECTRAL CONSTRUCTION PROCEDURES

The complete response spectra for two sites, at approximately 50 and 100 kilometers along the profile, are shown on Fig. 7. From these curves it can

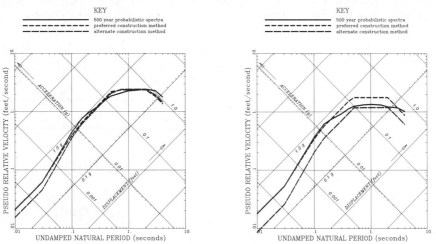

Fig. 7 Hazard Analysis and Constructed Spectra - 50 (Left) and 100 (Right) Kilometers Along Cross-Section

be seen that as a simple method of spectral construction the use of the probabilistic peak acceleration value with the mean amplification factors is the preferred approach. However, this procedure still requires the results of an acceleration hazard study for implementation. What can be suggested are two sets of spectral construction procedures ranked in order of preference. These are described in the step by step construction procedures below.

Preferred Method

This method requires the results of a probabilistic hazard analysis for peak acceleration. These can be obtained from probabilistic based hazard maps. Because there is an increasing interest in the use of this type of map for code use, they are becoming more available. The results shown above for the San Francisco Bay profile compare well with values obtained from the map given as Figure C1-3 in the ATC3-06 [2] report except in the zone close to fault crossings. The developers of the map specifically excluded use of the map close to fault crossings.

Step 1: Select the peak acceleration value for the appropriate probability (return period) for the project site.

Step 2: Using Eq. 1 compute the v/a ratio for the event which would be expected to control the deterministic approach to the design. This is usually the maximum event on the closest fault to the project site. Note that the v/a ratio is dependent on the site soil profile type. Compute the appropriate velocity using the acceleration obtained in step 1.

Step 3: Compute the maximum relative ground displacement from the dimensionless ad/v^2 term. This term may be given the value of 5 for rock sites and 4 for soil sites.

Step 4: Use the equations given by Newmark and Hall [9], included here as Table 1, to compute the median values for spectral amplification factors to be used with the values of acceleration, velocity, and displacement.

Table 1 Equations for Spectral Amplification Factors for Horizontal Motion (from Newmark and Hall [9])

Quantity	Cumulative Probability %	Equation
acceleration	50 (median)	$3.21 - 0.68\ln(b)$
velocity		$2.31 - 0.41\ln(b)$
displacement		$1.82 - 0.27\ln(b)$
acceleration	84.1 (one sigma)	$4.38 - 1.04\ln(b)$
velocity		$3.38 - 0.67\ln(b)$
displacement		$2.73 - 0.45\ln(b)$

where b is the damping ration expressed as a percentage of critical damping

Step 5: Construct the desired response spectra on a tripartite format using the values obtained in step 4 together with the peak acceleration. A description of the construction procedure is given in Newmark and Hall [9].

Alternate Method

This procedure may be used to approximate the design response spectra when the results of a probabilistic analysis or hazard maps are not available. This method is shown for convenience and should only be used for non-critical structures or when the other methods cannot be implemented. The large perturbations of the peak acceleration values along the profile shown on Fig. 3 are obtained from direct use of the attenuation equations. These can result in overestimation of spectral requirements close to a fault and possible underestimation of requirements elsewhere.

Step 1: Using an appropriate attenuation equation select the peak acceleration given by the maximum event on the fault closest to the site. The average attenuation curves given in the appendix could be used.

Step 2: Using Eq. 1 compute the v/a ratio for the event which would be expected to control the deterministic approach to the design. This will be the same event as used in step 1. Note that the v/a ratio is dependent on the site soil profile. Compute the appropriate velocity using the acceleration obtained in step 1.

Step 3: Compute the maximum relative ground displacement from the dimensionless ad/v^2 term. This term may be given the value of 5 for rock sites and 4 for soil sites.

Step 4: Use the equations given by Newmark and Hall [9] in Table 1 to compute the 84th percentile values for the spectral amplification factors to be used with the values of acceleration, velocity, and displacement.

Step 5: Construct the desired response spectra on a tripartite format using the values obtained in step 4 together with the peak acceleration.

DISCUSSION AND CONCLUSIONS

Situations frequently arise where preliminary design information may be required for a project. Detailed probabilistic hazard analyses cannot be used in such circumstances so alternate methods are required. By examining different methods of obtaining response spectra in some detail in a well studied area, it has been shown that preliminary spectra may be obtained by simple methods that should be consistent with the results of more detailed studies. The major requirement is the inclusion of the uncertainty in the spectral development in a consistent manner. This consistent inclusion requires that uncertainty be entered into the procedure just one time. In a probabilistic hazard study the uncertainties should be integrally included as part of the analysis. Thus, a peak acceleration estimate obtained from a hazard study and used to scale a response spectrum has the uncertainty included. An acceleration value obtained from the use of an attenuation relationship does not, so the uncertainty must be included in the spectral construction. The aim of the consultant should be to provide a best estimate of the design values together with a confidence bound so that the designer can include both of these in the final project evaluation.

In addition to the need to stress the importance of including the uncertainty in a consistant manner several conclusions can be made.

1. Simple approaches may be used to develop design spectra which are consistent with spectra obtained from detailed probabilistic hazard studies.

2. The simple procedures must be used with caution at distances of less than 10 kilometers from an active fault.

3. Probabilistic hazard studies suggest that the effects of velocity response from distant earthquakes may have been overestimated in some studies and map development. This difference has probably arisen by neglecting the small probability of occurrence of the large distant event.

REFERENCES

[1] Algermissen, S. T. and D. M. Perkins, "A Probabilistic Estimate of Maximum Acceleration in Rock in the Contiguous United States," United States Geologic Survey, Open File Report 76-416, 1976.

[2] Applied Technology Council, "Recommended Tentative Guidelines for Development of Comprehensive Seismic Design Provisions for Buildings," Palo Alto, California, ATC3-06.

[3] Crouse, C. B., personal communication, 1984.

[4] Donovan, N. C., "A Practitioner's View of Site Effects on Strong Ground Motion," U.S.G.S. Workshop on "Site Specific Effects of Soil and Rock on Ground Motion and the Implications for Earthquake-Resistant Design," Santa Fe, New Mexico, July 26-28, 1983.

[5] Donovan, N. C., "Ground Motion Issues for Base Isolation," ATC Seminar on Base Isolation, San Francisco, California, March 1986.

[6] Joyner, W. B. and D. M. Boore, "Prediction of Earthquake Response Spectra," Proceedings of the 51st Annual Convention of Structural Engineers Association of California, October 1982.

[7] Joyner, W. B. and T. E. Fumal, "Predictive Mapping of Ground Motion," Evaluating Earthquake Hazards in the Los Angeles Region, U.S.G.S. Professional Paper 360, October 1985.

[8] McGuire, Robin K., "Seismic Structural Response Risk Analysis, Incorporating Peak Response Regressions on Earthquake Magnitude and Distance," School of Engineering, Massachusetts Institute of Technology, Cambridge, Mass. 02139, R74-51 Structures Publication 99, 1974.

[9] Newmark, N. M. and W. J. Hall, "Earthquake Spectra of Design," Earthquake Engineering Research Institute Monograph Series, M. G. Agbabian, editor, 1982.

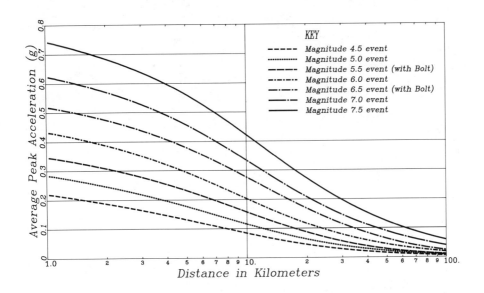

Appendix A. Acceleration Attenuation Curves for Magnitude 4.5 to 7.5 Based on Equations by Bolt, Campbell, Crouse, Donovan, Joyner and Boore, and Woodward-Clyde

HIGHER MODES CONTRIBUTION TO TOTAL SEISMIC RESPONSE

A. H. Hadjian[I] and S. T. Lin[II]

ABSTRACT

The contribution of the higher modes to the total response of three idealized beams (shear, tapered and bending) is evaluated. Both period and ductility dependency are considered. The response calculations are based on the Newmark type inelastic spectra and the modes are combined by the SRSS method. The response parameters are the base shear and base overturning moment and the shear and moment profiles throughout the beam height. It is concluded that both period and ductility tend to modify the effects of the higher modes in relation to a first mode approximation that considers all of the beam mass. Although shear and tapered beams have very similar behavior, bending beams tend to have a more complex response profile.

INTRODUCTION

The response of a multi-mode system, such as a high rise building, to earthquake excitation is governed primarily by the contribution of the first several modes of the system to the total response. Given the special frequency content characteristics of earthquake ground motions, the contribution of the higher modes to the total response is, among many other parameters, a function of the first mode frequency. Thus, for a very flexible system, the higher modes contribution could be significant and for a relatively stiff system, the contribution could be only minor. This frequency dependency of the problem will be investigated in this paper.

SYSTEM CHARACTERISTICS

In this study structures are idealized as uniform shear, uniform bending or tapered cantilever beams. The tapered beam has a fundamental mode shape given by a straight line. Bielak [1] obtained the necessary and sufficient condition for a non-uniform shear beam to exhibit a linear first mode: a shear beam of uniform cross section and mass density will have the above property if and only if the stiffness is distributed parabolically. Fig. 1 gives the eigen-parameters of the first six modes for all three beams. α is the frequency ratio of the higher modes to the fundamental mode; M_{mr} is the modal mass defined as

---

I. Principal Engineer, Bechtel Power Corp., Los Angeles, CA

II. Member of the Technical Staff, Rockwell International, Downey, CA

$$M_{mr} = \frac{\left[\{\phi_r\}^T \{m_i\}\right]^2}{\{\phi_r\}^T [M] \{\phi_r\}} \; ; \qquad (1)$$

Γ_r is the modal participation factor defined as

$$\Gamma_r = \frac{\{\phi_r\}^T \{m_i\}}{\{\phi_r\}^T [M] \{\phi_r\}} \; ; \qquad (2)$$

and H_r is the "height" of an equivalent simple oscillator such that the product of H_r and the modal mass, M_{mr}, equals the beam modal base overturning moment [2]. The sixth modal mass and participation factor have been modified such that the total beam mass is accounted for by the first six modes. In other words there is no "missing" mass. The following discrete frequencies have been assigned to the fundamental mode for each beam: f_1 = 0.2, 0.25, 0.333, 0.5, 1.0, 2.0, 2.5, 3.33, and 5.0 Hz (or alternatively T_1 = 5.0, 4.0, 3.0, 2.0, 1.0, 0.5, 0.4, 0.3 and 0.2 sec.). The frequencies of the higher modes then follow from the respective frequency ratios shown in Fig. 1. It is assumed that all modes of the structures are equally damped at 5% of critical at about yielding.

The inelastic response of structures will be approximated through the use of ductility factors. Obviously such a treatment assumes that the structure is adequately designed to respond inelastically throughout its height. Even though this is a desirable design objective, it is not easily attainable. Structures fail prematurely because inelastic response tends to concentrate in very localized areas such that the post-yield energy absorbtion capacity throughout the structure is not properly mobilized. The treatment herein implicitly assumes that this is not the case and that inelastic response occurs throughout the structure. Global ductility factors of 1, 2, 4 and 8 are used to cover a wide range of inelastic response. The K factors of the UBC code can be interpreted as allowable ductility factors [3]. Thus, for example, in the frequency range of about 0.25 Hz to 2.5 Hz, K = 0.67 (ductile moment-resisting space frame designed to resist total lateral force) can be interpreted as being equal to a uniform ductility factor of 4.

INPUT GROUND MOTION

With regards to input ground motion two obvious alternatives could be pursued: the ground motion can be characterized by an ensemble of recorded or artificially generated motions and the response parameters treated statistically; or, the variability of the ground motion can be accounted for by using smoothed response spectra. For economic reasons, the latter approach is adopted. Additionally smoothed response spectra, while retaining the general frequency characteristics of earthquake ground motions, would not significantly impact the end result due to fortuitous frequency matches between structure and ground motion. Modal responses are combined by the Square Root of the Sum of Squares (SRSS) method to estimate total system response. The SRSS method breaks down when used for the modal combination of relatively rigid structures [4]. The present results may thus possibly represent slight underestimates for rigid structures.

The Newmark-Riddle [5] inelastic mean response spectra for the above ductility factors is used as input ground motion. For the 5% damped mean response curves shown in Fig. 2, the elastic amplification values for the three segments of the response curve are:

Displacement - 1.47 x 36 = 52.9 in
Velocity - 1.55 x 48 = 74.4 in/sec.
Acceleration - 2.28 x 1 = 2.28 g

The deamplification factors, for elasto-plastic systems, as a function of the ductility factor, are given by the following formulas:

Displacement $R = 0.951\mu^{-1.07}$ (3a)
Velocity $R = (2.418\mu-1.418)^{-0.619}$ (3b)
Acceleration $R = (2.851\mu-1.851)^{-0.422}$ (3c)

BASE SHEAR AND OVERTURNING MOMENT

The total response for each beam and for each ductility is calculated from its modal responses. In order to evaluate the significance of the higher modes on the total combined responses the latter is divided by the modified response of the first mode. The use of only the first mode response implicitly assumes that the total mass of the structure is accounted for under the umbrella of the frequency and mode shape of the first mode. Thus the results of the first mode must be modified before they are used to normalize the total combined response. This modification is achieved by dividing the first mode results by the modal mass ratio of the first mode. The results from the above computations is shown graphically in Fig. 3. Obviously a ratio of greater than 1.0 measures the contribution of the higher modes to the total response if the total mass was accounted for in the fundamental mode. For example a ratio of 1.2 signifies that the higher modes contribute only 20% of the total response if the total mass was accounted for in a fundamental mode approximation.

Several observations could be made from Fig. 3. One unexpected result is that the behavior of bending beams is significantly different from those of the shear and tapered beams. The higher modes contribution for the flexible bending structures is about three times those of the other two beams. Whereas the maximum contribution (at T = 5 see and μ = 8) for the shear and tapered beams is only about 30%, for the bending beam it is more than 300%, an order of magnitude different. The weighted mode shapes of Fig. 1 cannot explain this difference. A second rather unexpected result is that, for the stiffer structures, the response ratios for all three beams are less than unity. That is, the total response due to the modal contributions would be less than the response of the first mode if the total mass is included in the first mode approximation. This behavior may partly be the result of the SRSS combination of modes used herein. However, this effect should be small since the highest fundamental frequency used in this study is only 5.0 Hz. Usually this problem becomes significant beyond about 10.0 Hz. A related observation is that the contributions of the higher modes begins to exceed unity beyond about 3 seconds period for the shear and tapered beams.

Setting aside the issue of the comparison with the first mode, it is observed that the higher modes contribution is both period and ductility dependent. Ignoring the very high ductilities and the very long periods, the following equations provide a good approximation to the curves in Fig. 3, for $0.2 \leq T \leq 4.0$ and $\mu \leq 4.0$.

Shear Beam

Base Shear Response Ratio
$$R_{VS} = 0.81 + (0.028 + 0.004\mu)T \tag{4}$$

Base Overturning Moment Response Ratio
$$R_{MS} = 0.81 \tag{5}$$

Tapered Beam

Base Shear Response Ratio
$$R_{VT} = 0.74 + (0.046 + 0.006\mu)T \geq 0.75 \tag{6}$$

Base Overturning Moment Response Ratio
$$R_{MT} = 0.75 \tag{7}$$

Bending Beam

Base Shear Response Ratio
$$R_{VB} = 0.5 + (0.22 + 0.04\mu)\,T \geq 0.61 \tag{8}$$

Base Overturning Response Ratio
$$R_{MB} = 0.6 + (0.024 + 0.005\mu)\,T \geq 0.61 \tag{9}$$

The above response ratios are based on the assumption that the total mass of the beams is accounted for by the first mode response. To obtain a corresponding set of equations where the first mode response is used as is, the above equations must be divided by the respective first modal masses of the beams (Fig. 1).

SHEAR AND MOMENT ALONG BEAM HEIGHT

Figs. 4-7 show sample results of the shear and moment profiles along beam height. Two extreme periods are represented, 0.2 sec. and 4.0 sec. The solid lines represent the response due to the first mode normalized by the modal mass to account for the total mass of the beam. The dashed lines represent the total response based of the SRSS modal combination.

Fig. 4 shows the shear and moment profiles for the 0.2 sec. tapered beam. The distribution is representative of the other two beams except of course for the values of the two parameters. The first mode approximation is always greater than the combined modal response and the response ratios at the base are given in Fig. 3. The differences tend to be smaller with height (zero at $x/L = 1.0$). For relatively rigid structures, it can be concluded that the behavior of the beams is insensitive to beam type and ductility; except of course for the absolute value of the parameters, with the bending beam being the lowest and the shear beam the highest. The differences though are relatively small.

Figs. 5-7 show the shear and moment profiles for the three type of beams of 4.0 sec. period. Unlike the relatively rigid beams, these flexible structures have a somewhat more complex response.

With respect to shear, the behavior of the shear and tapered beams is essentially the same, except maybe at the top 5% of the beams when the shear beam has slightly higher shear for all ductilities. The effect of ductilities is more significant in that as the ductility increases, the underestimate of the combined shear relative to the first mode tends to increase. Whereas for the elastic case, the shear from the combined response exceeds the first mode response at about 0.6 of the height, for $\mu=8$, the combined shear (a smoothed curve is assumed through the zig-zagged curve) begins to deviate right from the base for both shear and tapered beams. For the bending beam, the combined modal shear is more complex and, in general, significantly larger than the first mode appoximation.

With respect to moment, all three type of beams behave in a similar fashion: at the lower segment of the beams, the combined modal moment is less than the first mode approximation, and in the upper segment, it is more than the first mode approximation. This crossover takes place higher up for the tapered beam followed by the shear and bending beams. Ductility influences the location of the crossover for both the tapered and shear beams more significantly than for the bending beam, which for the latter occurs at about midheight.

CONCLUSIONS

Shear distribution tends from a parabolic shape for relatively rigid beams to almost a linear distribution as a function of increasing period and ductility. For example, the shear at the top of a shear beam with period of 4.0 sec. and ductility of 8 is about 60% of the shear at the base. Even though combined modal base shear and moment tends to be smaller than the first mode approximation for relatively flexible shear and tapered beams, the response ratio reverses up the structure. The crossover point is a function of the structure ductility. The bending beam behavior is more complex and, in general, a first mode approximation is inadequate. A more detailed evaluation is in order.

REFERENCES

[1] Bielak, J., "Base Moment for a Class of Linear Systems," Journal of the Engineering Mechanics Divisions, ASCE, Vol. 95, No EM5, pp. 1053-1062, Oct. 1969.
[2] Housner, G. W., "Design Spectrum," Chapter 5, Earthquake Engineering, R. L. Wiegel, Coordinating Editor, Prentice Hall, Englewood Cliffs, N.J., 1970.
[3] Hadjian, A. H., "A Calibration of the Lateral Force Requirements of the UBC," Proc. 8WCEE, San Francisco, July 21-28, 1984
[4] Hadjian, A. H., "Seismic Response of Structures by the Response Spectrum Method," Nuclear Engineering and Design, Vol. 66, No. 2, pp. 179-201, Aug. 1981
[5] Riddle, R. and Newmark, N.M., "Statistical Analysis of the Responses of Nonlinear Systems Subjected to Earthquakes," UILU 79-2016, Dept. of Civil Engineering, University of Illinois, Aug. 1979

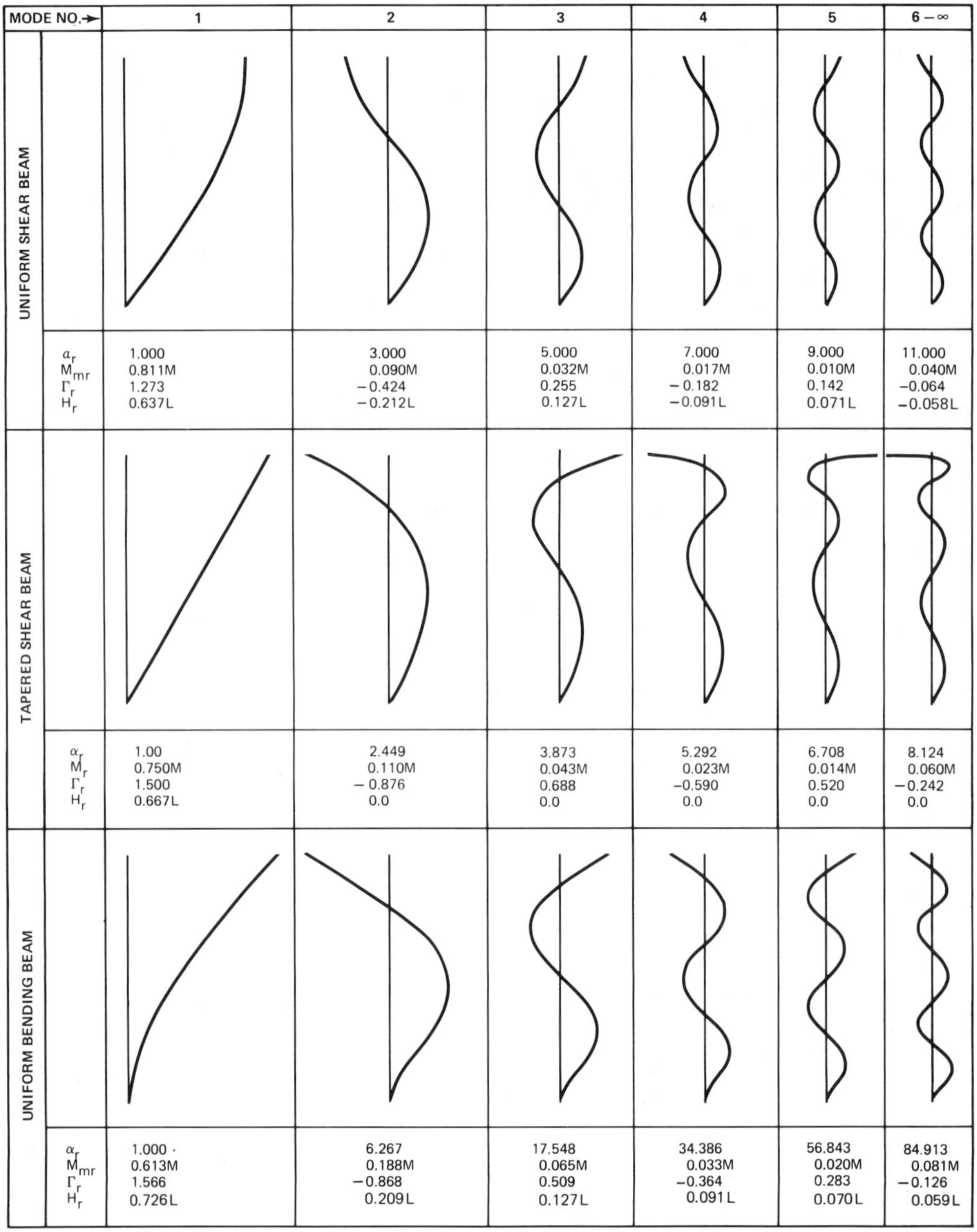

MODE NO. →	1	2	3	4	5	6 — ∞
UNIFORM SHEAR BEAM α_r M_{mr} Γ_r H_r	1.000 0.811M 1.273 0.637L	3.000 0.090M −0.424 −0.212L	5.000 0.032M 0.255 0.127L	7.000 0.017M −0.182 −0.091L	9.000 0.010M 0.142 0.071L	11.000 0.040M −0.064 −0.058L
TAPERED SHEAR BEAM α_r M_r Γ_r H_r	1.00 0.750M 1.500 0.667L	2.449 0.110M −0.876 0.0	3.873 0.043M 0.688 0.0	5.292 0.023M −0.590 0.0	6.708 0.014M 0.520 0.0	8.124 0.060M −0.242 0.0
UNIFORM BENDING BEAM α_r M_{mr} Γ_r H_r	1.000 0.613M 1.566 0.726L	6.267 0.188M −0.868 0.209L	17.548 0.065M 0.509 0.127L	34.386 0.033M −0.364 0.091L	56.843 0.020M 0.283 0.070L	84.913 0.081M −0.126 0.059L

Figure 1 EIGENPARAMETERS OF BEAMS STUDIED. (MODE SHAPES ARE WEIGHTED BY THE MODAL PARTICIPATION FACTOR)

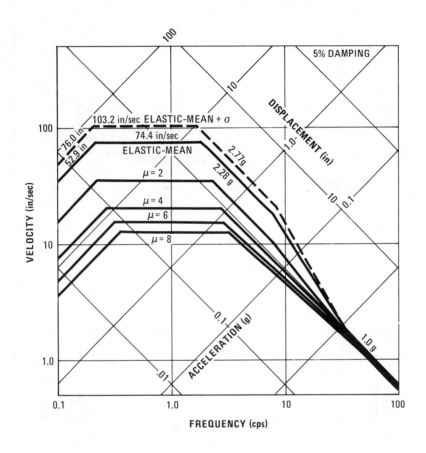

Figure 2 MEAN DESIGN SPECTRA SCALED TO 1g GROUND ACCELERATION (RIDDLE AND NEWMARK, 1979)

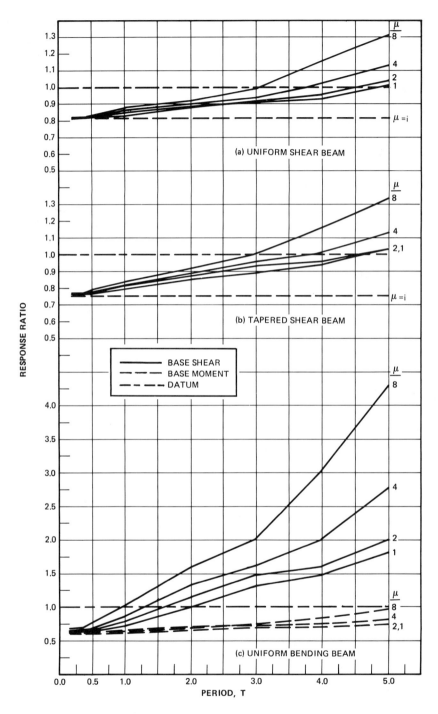

Figure 3 RATIO OF TOTAL RESPONSE TO FIRST MODE RESPONSE.

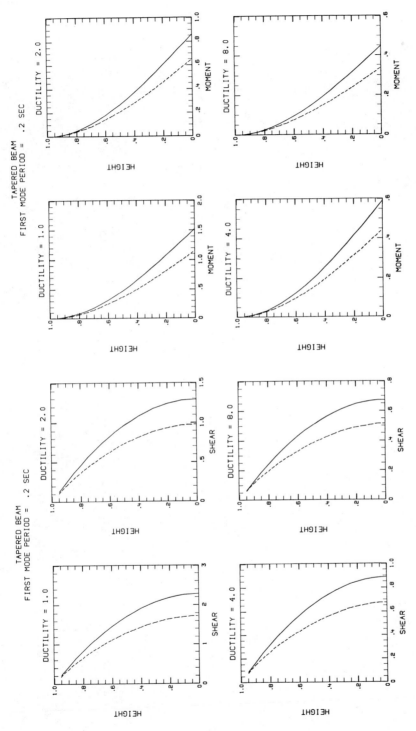

Fig. 4 SHEAR AND MOMENT PROFILES - TAPERED BEAM T_1 = 0.2 SEC.

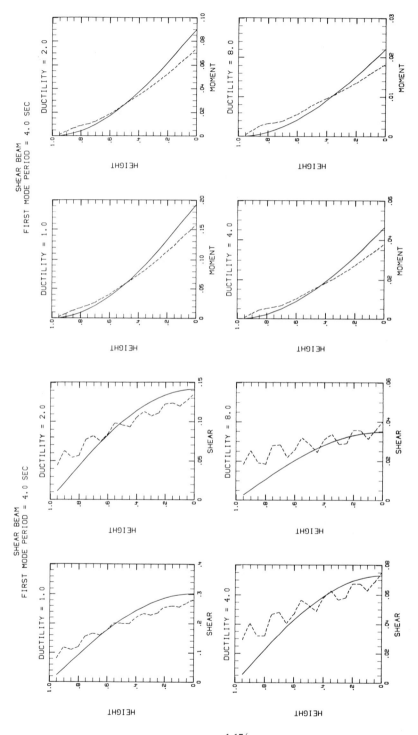

Fig. 5 SHEAR AND MOMENT PROFILES - SHEAR BEAM T_1 = 4.0 SEC.

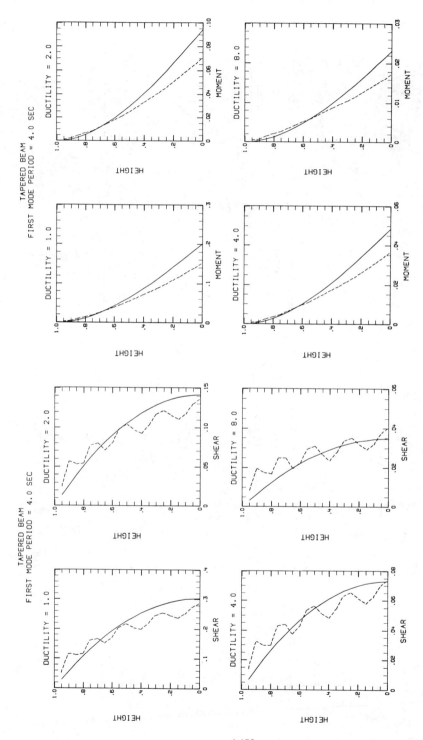

Fig. 6 SHEAR AND MOMENT PROFILES - TAPERED BEAM T_1 = 4.0 SEC.

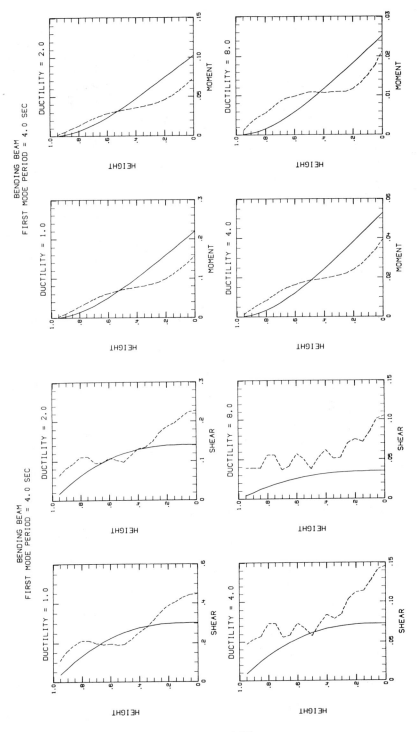

Fig. 7 SHEAR AND MOMENT PROFILES - BENDING BEAM T_1 = 4.0 SEC.

6. TESTS ON STRUCTURES AND COMPONENTS, EXPERIMENTAL METHODS

BEHAVIOR AND DESIGN OF REINFORCED CONCRETE
JOINTS USING SPECIAL MATERIALS

R. J. Craig[I]

ABSTRACT

The behavior and design of reinforced concrete joints using special materials such as polymer concrete and high strength concrete will be described. There will be a presentation of test results of fourteen beam column joints: regular concrete, polymer concrete, and high strength concrete. The benefits of adding fibers will also be shown for each different joint cementing material. All design considerations and construction techniques will be addressed. From this beam column joint test series, it was shown that, with the use of high strength and polymer concrete and/or fibers in the joint region, a more superior joint was developed than the conventional reinforced concrete joint.

INTRODUCTION

From exploratory research of reinforced ocncrete joints, it has been shown that the use of fibers and high strength concrete in the joint region is potentially superior than normal concrete.{1} From the research of fibers, high strength concrete, and polymer concrete in the joint region, we see that it has increased ductility, increased energy absorption in the area, eliminated some diagonal cracking, helped in the confinement of the joint, and changed the mode of failure as compared to the normal reinforced concrete systems used today. To show the behavior of the use of fibers, high strength concrete, and polymer concrete in the joint region, fourteen beam column joint studies will be described. The varying strength of the cementing agents in the joint region are: 1) normal strength concrete (f'_c = 4,000 psi); 2) high strength concrete (f'_c = 10,000 psi); and 3) polymer concrete (f'_c = 12,000 psi). Half of these joints contained 1½ percent by volume of hooked end fibers in the joint region.

The beam column specimen dimensions and steel arrangement will be shown. The loading cycle which was used to simulate the earthquake is that which was used by V. V. Bertero and E. P. Popov.{2} The test set up will be shown. Some of the test results of the load vs. beam tip deflection will be shown. Pictures of some of the beam column connections plus the polymer concrete beam columns will be shown. From the testing results, it was shown that the major findings when using the high strength concrete in the joint area were: 1) more ductility if properly confined; 2) decreases the diagonal cracking; 3) increases the stiffness of the member; 4) increases the energy dissipation

---

[I] Associate Professor, Civil and Environmental Engineering, New Jersey Institute of Technology, Newark, New Jersey, 07102

if properly confined; 5) increased shear capacity; and 6) increased bond capacity over normal reinforced concrete joint. With the addition of fibers into the normal and high strength concrete joint area, there was an increase in the ductility, decrease in cracking and damage, and confinement of the joint region. It will be shown that the use of fibrous polymer concrete in the joint leads to the joint behaving <u>elastically</u> while the failure occurred near the joint region in an area behaving inelastically. A specific joint which should be considered in design of reinforced concrete joints will be proposed which will give similar results as to the polymer concrete joint.

The paper will present the results of the testing program. The benefits of the addition of fibers, high strength concrete, and polymer concrete will be shown. Also, design considerations and construction techniques will be addressed.

From these fourteen beam column joint test series, it was shown that, with the use of high strength concrete and polymer concrete and/or fibers in the joint region, a more superior joint was developed than the conventional reinforced concrete joint and can be used in different seismic zones.

This paper uses some of the material and test results from references {3} and {4} in its discussion and further information can be sought from these references.

Previous Work

Normal strength concrete has been used for beam column joint construction and has been tested for its behavior through many years of research. To expand on this area would result in a lengthy discussion. In this paper, we are concerned with the new materials such as high strength concrete, polymer concrete, and fibers in normal, high, and polymer concrete.

High strength concretes represent a rather recent development which is now finding rapidly increasing use, especially in the construction of tall buildings. However, the scope and extent of the applications of these high strength concretes are still limited, partly because the knowledge of the distinct features of the mechanical behavior of high strength concretes is not clearly established. High strength concrete is a relative term depending upon the location of the concrete being made. The strength which will be considered high strength here is 10,000 psi to 12,000 psi. The economic advantages of using high strength concrete for columns and shearwalls of high-rise buildings have been demonstrated already in buildings such as the Water Tower Place in Chicago. In general, high strength concrete can be and has been advantageously used for: 1) columns and shearwalls of high-rise buildings; 2) elevated structures; 3) precast and/or prestressed products; and 4) construction where durability is critical.

There is a lot of research being done now on the mechanical characteristics (under monotonically increasing and cyclic loading) of high strength concrete - confined and unconfined. Also, there is research in improving the knowledge of structural behavior, flexure, shear, etc., of elements with consideration of static and earthquake loadings. There is examination of ACI Code methods for the prediction of behavior for design use.{5}

The answer to whether high strength concrete behaves brittlely appears to be both yes and no. At the material level, the ductility of plain concrete seems to decrease with the increasing compressive strength. At the section level, it is not necessarily the same as the material ductility. It has been shown that the ductility ratio of a reinforced concrete section subjected to flexure increases with increasing compressive strength, provided that the amount of steel is kept constant. However, if the amount of steel is maintained as a constant fraction of the balanced amount of steel, then the ductility ratio is indepednet of the compressive strength of concrete. High strength columns need to be confined for loads above the balanced conditions.{5} At the structural level, the effects of load reversals involving high shear seems to be that, as the strength of the concrete increases, the joint becomes more ductile.{4} In most cases, the high strength concrete should be confined in order to produce a ductile type of member.

The use of polymer concrete has shown some advantages over normal strength concrete. The presence of fibers will increase the strength of the cementing material. Work in polymer concrete has basically been in the material behavior and little has been done in the structural behavior. The use of reinforced polymer concrete is now in the developmental stages. Research into the behavior of reinforced polymer concrete has indicated the following benefits over regular concrete: 1) increases the tensile, flexural and compressive strengths of the material; 2) increases shear capacity of member; 3) increases moment capacity; and 4) increases bond of reinforcing. The use of polymer concrete as structural members has to be done in a cautious manner. There are disadvantages of this material which must be considered in the design procedure. These are that polymer concrete is susceptible to large creep deflections under sustained loading and, also, it is temperature sensitive and loses its increased strength properties at elevated temperatures.

From exploratory research of reinforced concrete joints, it has been shown that use of fibers and high strength concrete in the joint region is potentially superior than normal concrete.{1} From the research of fibers, high strength concrete, and polymer concrete in the joint region, it has increased ductility, increased energy absorption in the area, eliminated some diagonal cracking, helped in confinement of the joint, and changed the mode of failure as compared to the normal reinforced concrete systems used today.

Other areas of interest which will not be discussed here - but important for this area of interest - are: 1) materials; 2) bond; 3) cracking; 4) confinement; 5) member ductility-beam behavior; 6) column behavior; 7) joint behavior; and 8) subassemblage ductility. Because this information is so vast, no further attempt will be made to explain the basic behavior of high strength concrete. For a review of previous work on subassemblage behavior, the ATC 11 "Cyclic Loading of Reinforced Concrete Frame Joints" {6} is a good paper.

Experimental Program

The description of the following: 1) laboratory similation; 2) materials; 3) mixes - regular concrete, high strength concrete, and polymer concrete; 4) fibers; and 5) fabrication of the specimens can be found in references {3} and {4}.

Testing Program

In studying the behavior of joint regions, two different failure mechanisms were observed: 1) critical regions whose inelastic behavior is controlled by bending (flexural critical region); and 2) critical regions whose inelastic behavior is controlled by high shear existing in the region rather than bending. Fourteen half-scale partially modified seismic joint specimens from ACI 318 requirements: 1) six plain portland cement concrete (five normal and one high strength concrete); 2) six portland cement concrete with 1.5 percent by volume steel fibers (five normal and one high strength concrete); and 3) two polymer cement concrete (one normal and one with 1.5 percent by volume steel fibers). The actual dimensions and beam column joint details can be found in Figure 1.

Descriptions of: 1) the loading cycle which was used to simulate the earthquake; 2) the testing procedure with set up; and 3) the instrumentation and readings are explained in references {3} and {4}.

Discussion of Results

The following is a summary of the results of the fourteen beam column joints. A full description of the results is shown in references {3} and {4}.

The specimens SP1, SP2, SP3, and SP4 had an a/d ratio of 4.33, 4.33, 3.614, and 3.614, respectively. The behavior of the joints was similar to SP9 and SP10 having a flexural failure mode (fracturing of the longitudinal steel in the beam). SP1, SP3, and SP4 had diagonal cracking in the beam which became large at load stages near failure. The smaller the a/d ratio, the more damage was found. In SP2, the joint with fibers, a large crack at the face of the column developed. The presence of fibers increased the moment capacity, decreased cracking, reduced crack width, and caused less damage, and maintained more structural integrity. There was less pinching of the hysteresis loops with the presence of the fibers, also.

The specimens SP5, SP6, SP7, and SP8 failed in shear and bond in the column at the joint region. In all four cases the failure in the column was caused by large diagonal cracks with concrete severely spalling off when fibers were not present. There was little or no ductility exhibited by these joints. The fibers did help in the failure area by showing less structural damage and also exhibiting more energy dissipation. The presence of fibers in these joint specimens were not as beneficial as in joints SP2, SP4, SP10, SP12, and SP14.

Normal concrete specimens SP9 and SP10 (fibers) (a/d = 4.337) had the positive and negative steel area equal ($A_s = A'_s$). There was diagonal shear cracking in the column at joint at early ductility factors ($\mu = 1$). This cracking progressed and widths increased through $\mu = 5$. For SP9, there was cracking which indicated bond failure in the column of the longitudinal steel ($\mu = 4, 5,$ and 6). For SP10, a diagonal crack opened up on the beam and a shear failure occurred at $\mu = 7$. The crack widths were also smaller and fewer in SP10. There was no spalling of concrete in SP10 and the structural integrity of the joint was maintained. The amount of dissipated energy, which is indicated by the area of the hysteresis loops of the moment rotation diagram, was a 125 percent increase for the reinforced fiber concrete joint.

Because of the severe cracking of the regular joint, the tip deflection was greater than in the reinforced fiber concrete joint. The hysteresis loops for load vs. beam tip deflection is shown in Figure 2. There was more pinching of the loops for the regular concrete. The stiffness was also increased by the addition of fibers.

High strength concrete specimens SP11 and SP12 (fibers) showed dramatic changes from the normal concrete joints with and without fibers. There was diagonal cracking in column in both specimens at $\mu = 2$. In SP11, diagonal cracking started at $\mu = 5$, and large cracking occurred at $\mu = 7$, with the concrete finally disintegrating in the joint beam region at maximum load at $\mu = 9$. In SP12, there was a rupturing of the beam longitudinal steel at $\mu = 5$. The high strength concrete allowed the ductility factor at failure to increase 100 percent (SP9 at $\mu = 5$, and SP11 at $\mu = 10$). See Figure 3 for details. The column cracking was less in SP11 and there was no bond failure. The addition of fibers increased the joint stiffness. The ductility factor at failure was decreased from 10 to 5 for SP11 and SP12, but the amount of energy dissipated was increased by 25 percent because of fibers. The total cyclic tip deflection for SP12 was more than for SP11 (25 percent). There was an increase in dissipated energy of 27 percent because of using higher strength concrete, but the addition of fibers to the high strength concrete was not as beneficial as adding it to the normal strength concrete.

With testing of the high strength concrete joint region, there was an increase in the amount of tensile stress to cause diagonal cracking and the compressive stress/strain curve shows that the concrete becomes more brittle. The areas where the high strength concrete is to be placed should be reinforced with hoop steel in order to confine the concrete so that the material becomes more ductile and the structural integrity is maintained. Fibers may be used along with the hoop steel in order to help even more to confine the concrete in the regions of the inelastic behavior in the beams or column. Also, in the area where high strength concrete is used, the bond and anchorage strength increases, the shear strength of the concrete increases, and the shear friction mechanism across a crack changes over normal concrete. With these effects on the area where the high strength concrete is placed, the mode of failure of the normal concrete in the area may also change. And, the addition of fibers will help to increase these benefits obtained by the use of high strength concrete even more.

With polymer concrete, specimens SP13 and SP14 (fibers) demonstrated a far superior joint than the high strength and normal concrete with and without fibers. In SP13, the reinforced polymer concrete joint, a crack across the face of the column at the beam occurred at $\mu = 1$. There were essentially no other cracks which occurred in the joint region. The beam just rotated, putting cyclic compression and tension on the beam longitudinal bars at the column face with the reinforcing steel fracturing occurring at $\mu = 8$. In SP14, the crack occurred at the column face and did not propagate all the way through. The addition of fibers into the polymer concrete caused the joint to behave <u>elastically</u> during the entire test loading sequence, forcing the inelastic region to be formed away from the joint area in the normal concrete ten inches from the face of the column.[4]

The polymer concrete joints and the high concrete joints behaved similarly for loads and deflection (see Figure 4) and rotation. The cracking and failure modes (see Figure 5) were different. When looking at the data, we discover that some of the trends are: 1) significant improvement in energy dissipation capacity when properly confined; 2) increase in the ductility of the joint region; and 3) a stiffer and stronger joint is produced.

This research project has shown that there is a good potential use of high strength concrete in critical regions which are confined by the use of hoop steel or the combination of hoop steel and fibers. In further studies of the behavior of high strength reinforced concrete joints, four different cases should be examined: 1) joint controlled by flexural failure of the beam; 2) joint controlled by shear failure of the beam; 3) joint controlled by shear failure of the column; and 4) interior column test series. The research should basically look at conventional seismic joints with normal and high strength concrete in the joint region. In addition, it should be proposed for joints that cutting and bending partial steel at distance d from the column face with normal concrete, and cutting and bending partial steel at distance d from the column face with high strength concrete, be tested. These proposed joint selections are shown in Figure 6 as has been suggested by reference {6}. The cutting and bending partial steel at distance d from the column face is to try and force the location of the plastic hinge away from the column area. Also, fibers should be introduced in the test series for a few joints in order to see its behavior and possible use as confinement and improvement in ductility and structural integrity. These are the construction recommendations for the joints. If the problems of creep and loss of strength with increases in temperature would be solved for the polymer concrete joint, this system should be used because of the results which have just been explained.

Conclusions

Within the limitations of the test program, the fourteen beam column joints showed the following ocnclusions:

1) The tensile strength to cause diagonal cracking increases considerably with the addition of high strength concrete and polymer concrete. The amount of cracking was decreased in the joint area.

2) The compressive stress/strain behavior becomes more brittle as the concrete strength is increased.

3) The bond and anchorage strength increases with the increase in concrete strength.

4) The shear strength of the concrete increases with the increase in concrete strength.

5) The shear friction mechanism across a crack changes with the increase in concrete strength.

6) The mode of failure from normal concrete in the area may change with the addition of high strength concrete and polymer concrete.

7) The moment capacity of the member was increased because of the use of high strength concrete and polymer concrete.

8) The high strength concrete used in the joint region should be confined by using hoop steel or the combination of hoop steel and fibers.

9) The use of fibers in the normal, the high strength, and the polymer concrete helped the stiffness, ductility, confinement, and structural integrity of the joint.

10) Also, because of high strength concrete which was confined in the joint region, the joint (a) increased the energy dissipation of that region; (b) helped in the structural integrity of that region; and (c) increased the stiffness of that region, as compared to a joint using normal concrete.

11) More research is needed as explained in the discussion.

Acknowledgements

The NJIT graduate students who worked on the project should be recognized: S. Mahadev, C. C. Patel, M. Viteri, C. Kertesz, I. Kafrouni, J. Souaid, and H. W. Valentine.

Also, special recognition should be given to Danny DiSerio, the technician, who helped in the construction of the test frame. Thanks are also due to the Willis & Paul Corporation who helped on the design and fabrication of the test frame, as well as to the Bekaert Steel Wire Corporation for the donation of fibers for the research project. This project was funded through SBR research money at the New Jersey Institute of Technology.

REFERENCES

{1} Henager, C. H., "Steel Fibrous, Ductile Concrete Joint for Seismic-Resistant Structures," ACI Publication SP-53, "Symposium on Reinforced Concrete Structures in Seismic Zones," 1974 ACI Annual Convention, pp. 371-386.

{2} Bertero, V. V., and Popov, E. P., "Seismic Behavior of Ductile Moment Resisting Reinforced Concrete Frames," ACI Publication SP-53, "Symposium on Reinforced Concrete Structures in Seismic Zones," 1974 ACI Annual Convention, pp. 247-292.

{3} Craig, R. J., Mahadev, S., Patel, C. C., Viteri, M., and Kertesz, C., "Behavior of Joints Using Reinforced Fibrous Concrete," ACI Publication SP-81, "Fiber Reinforced Concrete - International Symposium," 1984, pp. 125-168.

{4} Craig, R. J., Kafrouni, I., Souaid, J., Mahadev, S., and Valentine, H. W., "Behavior of Joints Using Reinforced Pllymer Concrete," ACI Publication SP-89, "Polymer Concrete - Uses, Materials, and Properties," 1985, pp. 279-312.

{5} Shah, S. P., "High Strength Concrete - A Workshop Summary," Concrete International, American Concrete International, May 1981, pp. 94-98.

{6} ATC 11, "Cyclic Loading of Reinforced Concrete Frame Joints," Fall, 1983, Preliminary Draft Form.

TABLE 1 - SPECIMENS TESTED 3,4

Specimen Number	Concrete Strength, PSI Lower Column	Upper Column	Beam	Joint	Column Loading (Kips)	a/d*	A_{s_2} in.	A'_{s_2} in.	Failure Mode
SP1	4090	4300	4250	4250	80	4.33	0.88	0.40	Flexure
SP2+	4090	4300	4250	5030	80	4.33	0.88	0.40	Flexure
SP3	3540	4140	3640	3640	80	3.61	0.88	0.40	Flexure
SP4+	3540	4140	3640	3850	80	3.61	0.88	0.40	Flexure
SP5	4880	4300	5080	5080	80	2.89	0.88	0.40	Joint Shear
SP6+	4880	4300	5080	5460	80	2.89	0.88	0.40	Joint Shear
SP7	4750	4900	5050	5050	80	4.81	1.58	0.88	Joint Shear
SP8+	4750	4900	5050	5260	80	4.81	1.58	0.88	Joint Shear
SP9	4510	5320	5210	5210	80	4.33	0.88	0.88	Flexure
SP10+	4510	5320	5210	5560	80	4.33	0.88	0.88	Beam Shear
SP11	5210	4910	5210	8380	80	4.33	0.88	0.88	Flexure
SP12+	5210	4910	5210	10500	80	4.33	0.88	0.88	Flexure
SP13	5530	4820	5530	10700	80	4.33	0.88	0.88	Flexure
SP14+	5530	4820	5530	11400	80	4.33	0.88	0.88	Beam Shear

* a = distance from load to face of column

d = structural depth

\+ = joint had fibers - 1.5 percent by volume of concrete

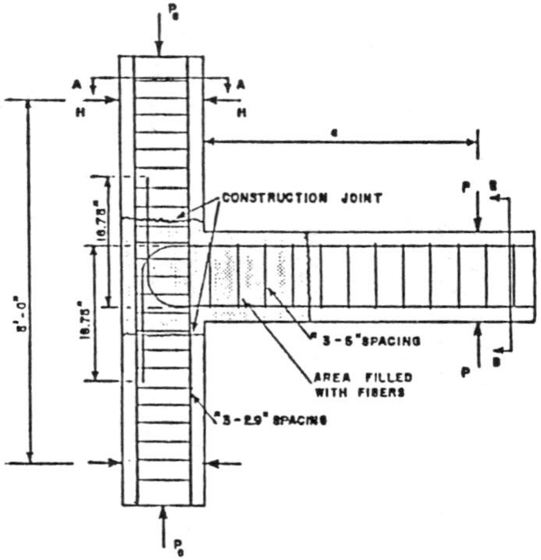

REINFORCED BEAM-COLUMN JOINT TEST SPECIMEN

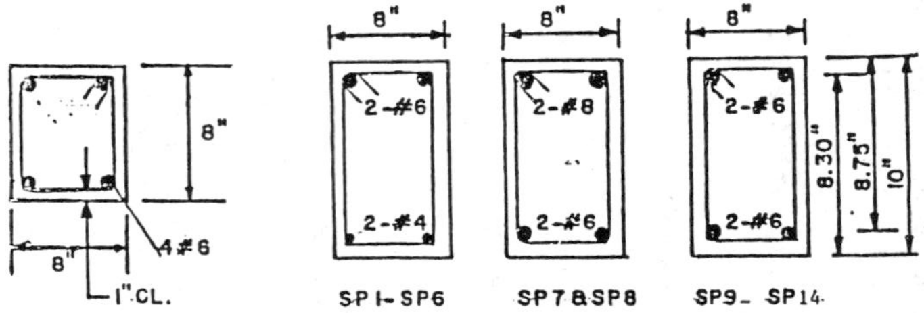

SECTION A-A COLUMN SECTION B-B BEAM

FIGURE 1 BEAM COLUMN JOINT SPECIMEN AND SECTIONS FOR TESTS 3,4

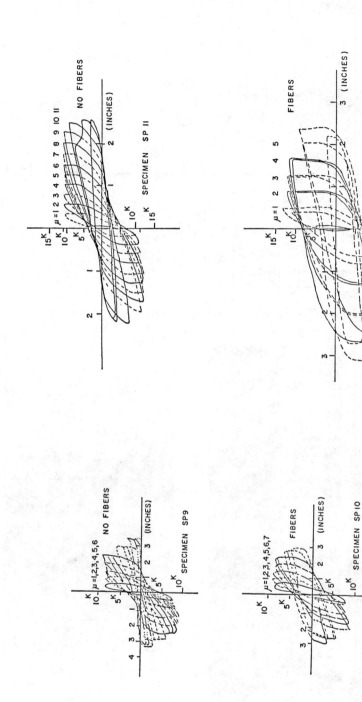

FIGURE 3 BEAM TIP LOAD VERSUS BEAM TIP DEFLECTION, SPECIMENS SP 11 AND SP 12

FIGURE 2 BEAM TIP LOAD VERSUS BEAM TIP DEFLECTION, SPECIMENS SP 9 AND SP 10

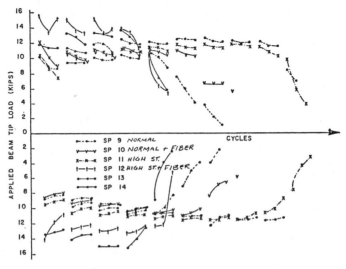

FIGURE 4 BEAM COLUMN APPLIED BEAM TIP·LOAD VERSUS CYCLE

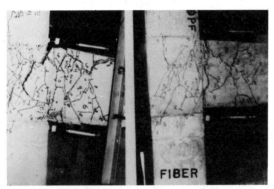

a) SPECIMENS SP 9 & SP 10 (NORMAL)

b) SPECIMENS SP 11 & SP 12 (HIGH STRENGTH)

c) SPECIMENS SP 13 & SP 14 (POLYMER)

FIGURE 5 BEAM COLUMN SPECIMEN FAILURES

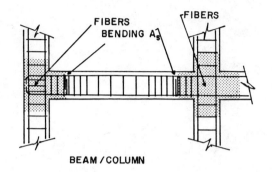

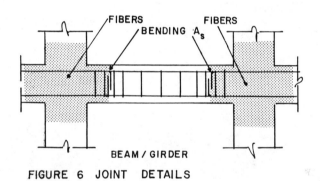

FIGURE 6 JOINT DETAILS

T-BEAM EFFECT IN STRUCTURES SUBJECTED TO LATERAL LOADING

Catherine Wolfgram French [I] and Ali Boroojerdi [II]

ABSTRACT

Three one-half scale reinforced concrete beam-column-slab subassemblages were tested under reversed cyclic lateral loading. The objective of the tests was to determine the influence of transverse beam elements on the effective width of slab participating as a tension flange to the longitudinal beam. The effect of initial damage due to loading in the transverse direction was also investigated.

INTRODUCTION

In reinforced concrete structures, the width of slab to be considered effective as a girder flange has been studied and is discussed in the ACI 318-83 Code [1] for the case of gravity loading. The effective slab width to be used for the case of structures subjected to lateral loading such as earthquake, however, has not been explicitly addressed in design codes. As a result designers often ignore the contribution of floor slabs to lateral load resistance, or they might assume an effective slab width as defined in the codes for the case of gravity loading.

Ignoring the slab contribution to the stiffness and strength of the girders will likely result in a significant underestimation of lateral load resistance. This phenomenon was observed in the test of a full-scale reinforced concrete structure conducted at the Building Research Institute in Tsukuba, Japan [2,3,4]. In this case, the calculated base shear varied by 80-percent depending on whether or not the floor slab reinforcement was considered to act as tension reinforcement to the beam [5].

Before testing, the contribution of the floor slabs was ignored in estimating the base shear capacity of the structure. Based upon this assumption the structure was expected to have a base shear capacity of 0.21 W (where "W" is the total weight of the structure). The structure actually developed a base shear capacity of 0.36 W. A significant portion of this increase in capacity may be attributed to slab reinforcement acting as tension reinforcement for the beams. Making the assumption that the full width of slab acted as tension reinforcement for the beams results in a

---

I. Assistant Professor, Department of Civil and Mineral Engineering, University of Minnesota, Minneapolis, Minnesota.

II. Research Assistant, Department of Civil and Mineral Engineering, University of Minnesota, Minneapolis, Minnesota.

calculated base shear capacity of 0.39 W. Due to the complexity of the structure, the actual amount of slab reinforcement to be considered as tension reinforcement for the beams is uncertain.

The question of how much the slab contributes to the lateral strength of structures is a very important one for design. As seen in the discussion above, it is possible to significantly underestimate the strength of a structure by ignoring the slab contribution. This has advantages as well as disadvantages. For example, it is conservative in design to underestimate the total shear and moment capacity of a structure. However, a piece of sensitive equipment placed on an upper floor in the structure might be damaged by the unexpectedly high acceleration forces. In addition, the failure mechanism of the structure might be different than that anticipated by ignoring the contribution of the slab.

One parameter which affects the amount of slab participation is the transverse beam element. A series of tests was conducted to investigate the slab contribution to lateral load resistance as the torsional stiffness of the transverse beam element was varied. The tests were conducted on three one-half scale reinforced concrete subassemblages designated EW1, EW2 and EW3. The transverse beam element was varied in the models from no beam in model EW1 to a beam which was identical to the longitudinal beam in model EW3.

DESCRIPTION OF THE MODELS

Three one-half scale reinforced concrete models were constructed to represent an interior beam-column-slab subassemblage. The ends of the beams and columns were modelled to simulate inflection points in a laterally loaded frame. The plan and elevation views of the models are shown in Fig. 1.

The nominal dimensions for the longitudinal beams (oriented in the direction of lateral loading) and columns were 150 mm x 250 mm (6 in. x 10 in.) and 250 mm x 250 mm (10 in. x 10 in.), respectively. The nominal slab thickness was 60 mm (2.5 in.). The only parameter that was varied in the models was the transverse beam element (oriented transverse to the direction of lateral loading). Model EW1 did not have a transverse beam element. Models EW2 and EW3 had 150 mm x 150 mm (6 in. x 6 in.) and 150 mm x 250 mm (6 in. x 10 in.) beams, respectively. The beam and column cross sections are shown in Fig. 2.

Grade 60 deformed bars with nominal yield strengths of 415 MPa (60 ksi) were exclusively used in the models. The column was reinforced with two No. 5 and one No. 4 bars on each face. The beam reinforcement comprised three No. 3 top bars and two No. 3 bottom bars. The slab was reinforced with two layers of No. 2 bars nominally spaced at 190 mm (7.5 in.) on center as shown in Fig. 3.

Concrete compressive strengths were measured at the time of test by testing 150 mm x 300 mm (6 in. x 12 in.) cylinders which were cast with the models. The average compressive strengths were 35, 35, and 48 MPa (5000, 5080, and 7000 psi) for models EW1, EW2 and EW3, respectively.

DESCRIPTION OF THE TEST PROGRAM

A schematic view of the test setup is shown in Fig. 1. Pinned-ended rigid links attached to the ends of the longitudinal beams and the pin at the base of the column simulated inflection points assumed to be located at midspan of the beams and midheight of the columns in a laterally loaded frame. The lateral load was applied with a hydraulic ram attached to the top of the column.

The beam and slab reinforcement was instrumented with strain gages along the column faces and along beam lines in both directions. Linear Voltage Differential Transducers (LVDT's) were used to monitor relative rotations between the column and longitudinal beam, vertical and horizontal displacements of the slab, and relative twists of the transverse beam measured with respect to the column. The lateral load was measured with a load cell in the hydraulic ram. Load cells were incorporated in the rigid links to measure the distribution of the load in the subassemblage.

The typical load history used in the tests is shown in Fig. 4. The load history was based on a predetermined value of the yield deflection which was calculated assuming nominal strength values of 415 MPa (60 ksi) and 35 MPa (5 ksi) for the steel and concrete, respectively. Strain hardening of the reinforcement was also taken into consideration. The yield deflection was defined as the deflection that would cause yielding in the bottom reinforcement of the beam for the case of slab in compression. The imposed deflection levels represent inter-story drift values (ratio of lateral story deflection to story height) ranging from 0.5-percent to 8.0-percent. Deflections were amplified progressively to the stroke limit of the actuator which was 125 mm (5 in.) in each direction.

To obtain an understanding of structural behavior associated with lateral loading, it is important to take into consideration the fact that earthquake loads do not occur in a single direction. Damage resulting from lateral loading in the transverse direction may have a significant effect on the behavior in the longitudinal direction. It is particularly important to take this effect into account when investigating the influence of a transverse element on the structural behavior in the longitudinal direction. To investigate this effect, model EW3 which had identical longitudinal and tranverse beams was loaded initially in the transverse direction to a displacement of twice the nominal yield displacement (Test EW3-I). The model was then rotated 90 degrees and tested in the longitudinal direction to failure (Test EW3-II).

SUMMARY AND DISCUSSION OF TEST RESULTS

A brief summary and discussion of the test results is presented in the following sections. The data presented include load-deflection and moment-rotation curves, relative twists of the transverse beam measured with respect to the column, and strain distributions measured in the longitudinal and transverse directions. The response histories of one of the tests, EW3-II, are shown in Figs. 5, 6, 7, 8, 9, and 10. For clarity, only the values of the first peaks for each displacement level are plotted. In general, the response histories of the three models, EW1, EW2 and EW3, were similar. The primary differences in the behavior of the models are described below.

Load-Deflection Response

Figure 5 shows a plot of the load versus deflection response of model EW3-II. In the figure, the lateral load applied to the top of the column is plotted with respect to the corresponding horizontal displacement measured at the top of the column. As seen in the figure, the model exhibited a very ductile behavior. The structure was displaced to the limiting stroke of the hydraulic actuator which represented an inter-story drift of approximately 8.0-percent. The maximum capacity of the structure was reached at an interstory drift of 4.0-percent (six times the nominal yield deflection).

As mentioned previously, model EW3 was first subjected to four cycles of the load history in the transverse direction. As a result, the model appeared to be weaker during the initial load cycles when subsequently tested in the longitudinal direction. During the first load cycle in the longitudinal direction, the stiffness of the structure was 75-percent of that measured in the first test. As the imposed deflection amplitudes increased in test EW3-II, the initial damage in the transverse direction had a negligible effect on the lateral load resistance of the structure.

A comparison of the loads measured for the same top column deflection indicates that the loads carried by models EW1 and EW2 were on the order of 80-percent and 95-percent, respectively, of that carried by EW3.

Moment-Rotation Response

The plot of moment versus rotation is shown in Fig. 6 for model EW3-II. The moment in the plot corresponds to the moment in the east longitudinal beam at the column face. The rotation is the relative rotation of the beam measured with respect to the column face. Positive moments in the figure correspond to the slab-in-tension case; negative moments correspond to the the slab-in-compression case. As seen in the figure, the slab-in-tension case has more flexural capacity than the slab-in-compression case due to the participation of the slab steel and the additional top bar in the beam.

The difference in moments corresponding to the slab-in-compression case is less than 10-percent for all three models. These moments are primarily controlled by yielding of the tension reinforcement in the bottom of the beam, and therefore, should be similar for all three models. The moments corresponding to the slab-in-tension case give an indication of the amount of slab participation. A comparison of the slab-in-tension moments for the same measured top column displacement shows that EW3 had a larger slab participation. At a particular load step, the values of the moments of EW1 and EW2 were on the order of 75-percent and 95-percent, respectively, of that measured for EW3.

A comparison of the moments measured in EW3-I to those measured in EW3-II confirms the observations made from the load-deflection histories described in the previous section.

Table 1 lists the ultimate flexural capacities of the slab-beam element based on nominal material properties and dimensions for the case of slab in tension. The assumed slab widths acting as the effective tension flanges to the beam are shown in Fig. 7. The three cases shown are: Beam Width, ACI Width and Full Width. The assumed effective width of slab acting as a tension flange to the beam was zero in the Beam-Width case. In the ACI-Width case, the effective flange width defined for the case of gravity loading was used [1]. The Full-Width case represents the upper bound of the ultimate flexural capacity in which case the entire width of slab was assumed to act as a tension flange to the beam. The slab contribution for the Beam-Width and ACI-Width cases was calculated based on the assumption that the remaining portion of the slab acted independently of the beam.

A comparison of the maximum measured moments to the calculated values indicates that all three models, EW1, EW2 and EW3, had an effective slab participation between the ACI-Width case and the Full-Width case. For EW3, the difference between the measured and calculated maximum moment assuming the full width of slab to be effective was within 10-percent.

Twist of the Transverse Beam

The relative twist of the transverse beam with respect to the column was measured at four locations along the transverse beam (75, 450, 825 and 1200mm from the column face). Relative twist measurements give an indication of the effectiveness of the slab in resisting lateral load. A zero relative twist measured across the transverse beam would indicate that the transverse beam and column rotated as a rigid body; in such a case, the entire width of the slab would act as an effective tension flange to the beam. The lowest values of relative twist were measured in the tests on model EW3 (Fig. 8). As observed in the load-deflection and moment-rotation responses, the values of relative twist indicate that, compared to the other models, the slab of EW3 had the greatest contribution to the lateral load resistance of the structure.

Strain Distribution in Longitudinal Reinforcement

Strains in the slab reinforcement were measured across the slab along the beam lines and column faces. A plot of the strain distribution measured along the east column face is shown in Fig. 9 for the slab-in-tension case of model EW3-II. The solid lines, in the plot, indicate strains measured in the top slab bars; dashed lines indicate strains measured in the bottom slab bars.

Plots of strain distribution for EW1 and EW2 indicate that the strain in the reinforcement decreased with distance from the face of the column. In EW3, however, the longitudinal reinforcement yielded across the entire width of slab as shown in Fig. 9.

Strain Distribution in Transverse Reinforcement

The strains measured in the reinforcement oriented transverse to the loading direction are shown in Fig. 10 for the slab-in-tension case of EW3-II. The gages were located along the north column-face line. As mentioned above, solid lines represent strain measurements for top slab bars; dashed lines represent strains measured in bottom slab bars. The measured strains increase with distance from the column face. This phenomenon was also observed in previous tests and was discussed by Suzuki, et. al. [6,7].

CONCLUSIONS

The following preliminary conclusions may be made from these tests:

1. The effective slab participation was greater for the models with increased transverse torsional stiffness. This phenomenon was observed in load-deflection, moment-rotation, and strain distribution measurements.

2. A reduction in the initial stiffness was observed in the second test of model EW3. This reduction may be attributed to the damage caused by loading the structure in the transverse direction first. This effect became negligible after the structure was displaced beyond the maximum deformation level imposed during the first test.

3. The flexural resistance measured for model EW3 at ultimate was within 10-percent of the nominal calculated value assuming the full width of slab to be effective as a tension flange to the beam.

ACKNOWLEDGMENTS

This research investigation was sponsored by the National Science Foundation under Grant No. CEE-8404732 and was carried out at the Civil and Mineral Engineering Structures Laboratory, University of Minnesota. The valuable assistance of M. Moeller in the experimental phase of the project is greatly appreciated.

REFERENCES

[1] ACI Committee 318, "Building Code Requirements for Reinforced Concrete (ACI 318-83)," Detroit, 1983.

[2] Yoshimura, M. and Y. Kurose, "U.S.-Japan Cooperative Earthquake Research Program, Phase I: Reinforced Concrete Structure", <u>Earthquake Effects on Reinforced Concrete Structures--U.S.-Japan Research</u>, ACI SP-84, 1984.

[3] Otani, S., T. Kabeyasawa, H. Shiohara and H. Aoyama, "Analysis of the Full-Scale Seven-Story Reinforced Concrete Test Structure", <u>Earthquake Effects on Reinforced Concrete Structures--U.S.-Japan Research</u>, ACI SP-84, 1984.

[4] Okamoto, S., S. Nakata, Y. Kitagawa, M. Yoshimura and T. Kaminosono, "A Progress Report on the Full-Scale Seismic Experiment of a Seven Story Reinforced Concrete Building - Part of the US-Japan Cooperative Program," Building Research Institute Paper No. 94, ISSN 0453-4972, Tsukuba, Japan, March 1982.

[5] Wolfgram, C. E., "Experimental Modelling and Analysis of Three One-Tenth Scale Reinforced Concrete Frame-Wall Structures," Doctoral Dissertation, Graduate College, University of Illinois, Urbana, Illinois, January 1984.

[6] Suzuki, N., S. Otani and H. Aoyama, "The Effective Width of Slabs in Reinforced Concrete Structures," <u>Transactions of the Japan Institute</u>, Vol. 5, 1983.

[7] Suzuki, N., J. K. Halim, S. Otani and H. Aoyama, "Behavior of Reinforced Concrete Beam-Column Subassemblages with and without Slab," Department of Architecture, Faculty of Engineering, University of Tokyo, March 1984.

[8] Joglekar, M. R., P. A. Murray, J. O. Jirsa and R. E. Klingner, "Full Scale Tests of Beam-Column Joints," <u>Earthquake Effects on Reinforced Concrete Structures--U.S.-Japan Research</u>, ACI SP-84, 1984.

[9] Morrison, D. and M. Sozen, "Response of Reinforced Concrete Plate-Column Connections to Dynamic and Static Horizontal Loads," <u>Structural Research Series</u>, No. 490, April 1981.

Table 1 Nominal Flexural Capacities (Slab in Tension)

	M_{ult}, kN-m		
	Beam Contribution	Slab Contribution	Total
Beam Width	21	12	33
ACI Width	41	9	50
Full Width	82	-	82

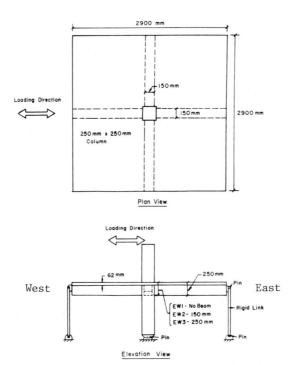

Fig. 1 Nominal Dimensions of Subassemblages

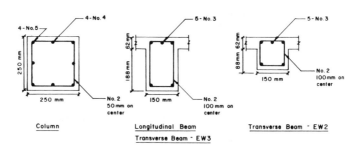

Fig. 2 Member Cross Sections

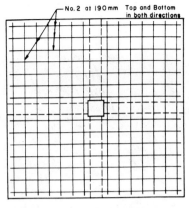

Fig. 3 Slab Reinforcement

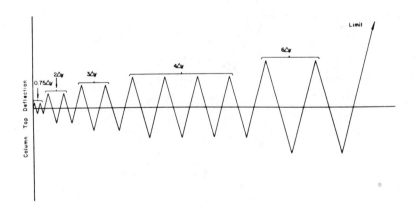

Fig. 4 Load History

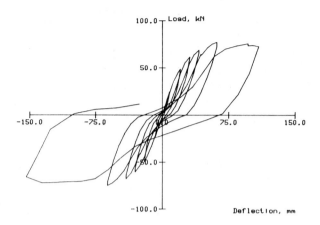

Fig. 5 Load-Deflection Curve - EW3

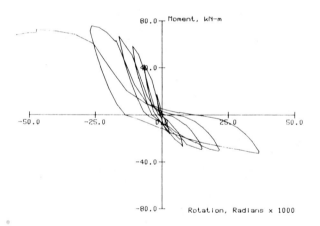

Fig. 6 Moment-Rotation Curve - EW3

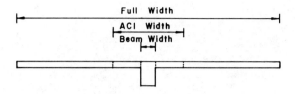

Fig. 7   Effective Slab Width for Analysis

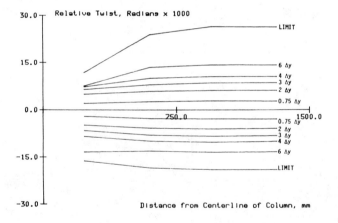

Fig. 8   Relative Twist of Transverse Beam
         with respect to Column – EW3

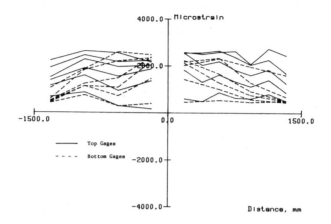

Fig. 9 Strain Distribution of Longitudinal Slab Reinforcement - EW3

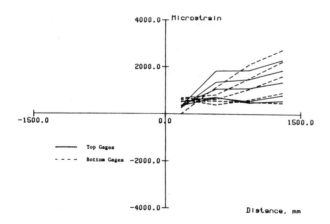

Fig. 10 Strain Distribution of Transverse Slab Reinforcement - EW3

BEHAVIOR OF A STRENGTHENED REINFORCED CONCRETE FRAME

T.D.Bush[I], C. E. Roach[II], E. A. Jones[III] and J. O. Jirsa[IV]

ABSTRACT

A two-third scale model of two bays and two stories of an exterior moment resisting frame of a reinforced concrete building was constructed and retrofitted with two different seismic strengthening schemes. The ductility and strength of the existing frame were deficient because of low column shear capacity. The strengthened frames were tested under reversed cyclic lateral loading. The first strengthening scheme involved the addition of reinforced concrete piers around each column to increase the shear capacity. The second scheme was a structural steel bracing system mounted on the exterior of the frame. The bracing system was designed to carry all seismic lateral forces.

INTRODUCTION

Background. There are a number of existing reinforced concrete buildings in high risk earthquake areas which are in need of seismic strengthening. These structures, although designed and constructed in compliance with earlier building codes, rely on lateral-load resisting systems which are found deficient when evaluated by stricter and more complex modern building codes. The purpose of seismic strengthening generally is to increase the strength and/or the ductility of the structure under lateral loads. Some of the more common methods of seismic strengthening for reinforced concrete buildings are infilling walls, adding wing walls to existing columns, encasing columns or attaching structural steel braces. In seismic strengthening, the details of the connections of the strengthening elements to the existing structure are of particular importance and greatly affect the overall response of the strengthened structure.

Objective. The objective of this research was to investigate the performance of two different schemes for strengthening a reinforced concrete framing system considered typical of many built in California in the 1950s

---

I. Graduate Research Assistant, University of Texas at Austin.

II. Graduate Research Assistant, University of Texas at Austin.

III. Magadini-Alagia Associates, Phoenix Arizona.

IV. Ferguson Professor of Civil Engineering, University of Texas at Austin.

and 1960s. The reinforced concrete structural system investigated was a seven story, flat plate system with shear walls carrying lateral forces in one direction and an exterior, moment resisting frame in the other direction. The exterior frame had deep, stiff spandrel beams framing into flexible columns having low shear strength. An evaluation of the failure mechanism of the structure pointed to a column shear failure in the exterior frame at loads much lower than the 1982 UBC earthquake design loads for the structure. Retrofitting the exterior frame was needed to improve both strength and ductility. The two schemes adopted for this investigation were: (1) the addition of reinforced concrete piers around each exterior column to increase the shear capacity of the columns, and (2) a structural steel bracing system mounted on the exterior of the frame to act as the primary lateral-load carrying system.

The research was carried out in three phases. In the first phase, a 2/3 scale model of a portion of the existing exterior frame was constructed and tested to levels below those expected to produce shear failure in the columns. In the second phase, reinforced concrete piers were added to improve the shear capacity of the columns and change the failure mode to one involving flexure in the beams. In the last phase, the piers were removed and the frame was retrofitted with a structural steel bracing system designed to carry the entire lateral shear force.

Organization. This research program had two features which distinguished it from previous research on seismic strengthening in the United States. First, the large scale model decreased or eliminated the effects of scale in construction and detailing of the connections. Second, the investigation was a cooperative effort between a design engineering group from H. J. Degenkolb Associates and a research team from The University of Texas at Austin. Further information on the research is contained in a companion paper written by the design team at H. J. Degenkolb Associates.

ORIGINAL FRAME

The test structure reproduced, in 2/3 scale, two bays between the third and fifth stories of the structure. The frame was designed in full scale using material strengths and design techniques considered typical of those used in the 1950s and 1960s. The model was built in place on the floor-wall reaction system in the laboratory. The construction sequence was typical of that expected in the field. The original frame model had two columns and three beams with a 4 ft (1.22 m) width of slab included at each level. Fig. 1 shows the original frame model located on the floor-wall system and Fig. 2 shows the dimensions and reinforcement details of the columns and beams.

Vertical struts were pin connected to the ends of the beams to model the vertical deflection restraint and approximate moment conditions which would exist at the midspan of the beams of a frame under lateral loads. The column bases were supported by neoprene pads. Out-of-plane bracing was connected at the third and first level of the frame but did not restrain lateral movement.

TESTING

Testing of the experimental model involved subjecting the frame to reversed cyclic lateral deformations. In an actual building, interstory shear due to earthquake motions is distributed to lower levels through the columns. For this experiment, introducing high magnitude lateral loads into the columns of the experimental frame would have required considerable change in the column design in the area of loading. To avoid this problem, the loads and the reactions were introduced to the slab near the two columns. The four rams were located on either side of the two columns with one ram active at each column location during loading in one direction and the other set of rams active when the loading was reversed. In general, testing involved three cycles of reversed loading to incrementally increasing levels of interstory drift. The forces and reactions applied to the frame, lateral and vertical deflections of the spandrel beams, stresses at critical sections, and slip between the original structure and the strengthening elements were monitored.

PIER STRENGTHENING SCHEME

Reinforced concrete piers were cast around the exterior three sides of the columns along the entire height of the structure. The piers were designed to resist the lateral loads necessary for the spandrel beams to develop full flexural capacity. Besides increasing the lateral strength of the frame, the retrofitting changed the failure mode from a non-ductile column shear failure to a ductile flexural beam failure mechanism. The pier design increased the column width from 12 in. to 60 in. (305 mm to 1524 mm) and the thickness from 12 in. to 13-1/3 in. (305 mm to 339 mm). Reinforcement details are illustrated in Fig. 3. The design was based on the new concrete acting monolithically with the existing concrete. Transfer of force between the original concrete and the new elements was assumed to occur through adhesive bond, lug action within the window opening, and shear-friction across the spandrel beam face. Adhesive bond was increased by sandblasting the surface of the existing frame. The connection design also included dowels epoxy-grouted into the beam faces and the sides of the columns as shown in Figs. 3 and 4.

Construction. After sandblasting the frame, holes for the dowels were drilled, brush cleaned, and vacuumed as recommended in Ref. 1. The dowels were grouted using a non-sag gel epoxy. The construction of the reinforcement cages was complicated because the transverse reinforcement was closely spaced, bars had to be threaded between previously epoxy-grouted dowels, and the cage had to be tied directly against existing concrete. The piers were cast in three lifts, with each lift extending from bottom of spandrel to bottom of spandrel on the floor above. The excellent bond between the piers and the original frame was confirmed after testing was completed by the difficulty of removing the concrete piers for the next testing phase. The only new/old concrete interface which did not have complete contact was the interface at the beam soffit where the concrete hydraulic head during casting was not sufficient to adequately force the fresh concrete against the beam soffit. The completed pier strengthened frame can be seen in Fig. 5.

Results. The pier strengthened frame was subjected to three cycles each of reversed cyclic deformation to increasing levels of drift as seen in Fig. 6. The load on the frame versus the drift, in inches, between the third and first levels of the frame was plotted. Fig. 6 shows the envelopes for each set of three cycles to the four successive drift levels. Also in Fig. 6 is a load-drift envelope for the unstrengthened or original frame test. A comparison of the curves indicates that the initial stiffness of the pier strengthened frame was approximately three times the initial stiffness of the original frame. The test was stopped at 0.5% drift to prevent excess damage to the frame. The envelope of the load-drift plot was approaching a horizontal tangent indicating that the frame was near its ultimate capacity. It is likely that slightly higher lateral loads could have been applied. The calculated shear capacity of the original column was approximately 31 k (138 kN). The maximum lateral load on the frame was 310 k (1380 kN), five times the capacity of the two original columns.

The crack patterns observed and the data from instrumented reinforcement indicated that flexural hinging had occurred at all critical beam cross sections. The longitudinal bars in the beams were yielding and, in some cases, strain hardening. The flexural cracks in the beams reached 1/8 to 1/4 in. (3.2 to 6.4 mm) widths and some spalling and concrete splitting was noted. In comparison, the pier exhibited little distress. There were some hairline flexural cracks on the exterior face of the pier and some shear cracking on the interior face with widths under 0.02 in. (0.51 mm). The maximum longitudinal reinforcement stresses in the pier were 30 ksi (207 MPa), one-half of yield. There were some pier stirrup and dowel stresses approaching 30 ksi (207 MPa) where the steel crossed shear cracks.

The frame was loaded to develop beam hinging throughout the structure but loading was stopped before high levels of distress were created which might have affected the use of the frame for retrofitting with the steel bracing scheme. The original columns were protected to the extent that only minor cracking occurred. The piers were removed and the frame was restored to its original dimensions for the steel bracing test.

Behavior of the Pier. Analysis of the instrumented areas of the pier indicated that the pier was acting monolithically with the frame. The data indicated no slip at the new/old concrete interface at low drift levels and only small slip at high drift levels. The outside corner beam dowels acted as shear-friction reinforcement, showing stresses of approximately 25 ksi (172 MPa) at drift levels where concrete slip was observed. The column dowels which crossed shear cracks in the window segment of the pier showed stresses of approximately 30 ksi (207 MPa). All other column and beam dowels showed little or no stress. The crack patterns and reinforcement stresses showed bending along the pier in the exterior 8 in. thickness where the longitudinal reinforcement was continuous. In the interior 5-1/3 in. (135 mm) thickness of the window segment there was no continuous longitudinal reinforcement and there were no dowels into the top and bottom of the spandrel beams. Crack patterns indicated that shear was transferred through a compression strut originating at the top of the window segment close to the original column and terminating at the bottom of the window segment at the edge of the pier as shown in Fig. 7. It was likely that the poor contact at the top interface of the window segment contributed significantly to the width and orientation of the compression strut. Further discussion of the pier strengthened frame test is contained in Ref. 2.

STEEL BRACING SCHEME

The steel diagonal bracing system formed a vertical truss in each bay as shown in Fig. 8. Vertical steel sections (channels) were attached to the column with epoxy grouted dowels, horizontal members were dowelled to the spandrels at slab level to aid in "collecting" the shear forces, and diagonal X-braces framed into the joints. The system was designed to transfer the lateral loads from the frame to the bracing, thus bypassing the weak columns. Bracing capacity was governed by buckling and tension yielding of the diagonals.

<u>Construction</u>. Connection details for the bracing scheme are shown in Fig. 9. Epoxy-grouted 5/8 in. (16 mm) diameter mild steel bolts were used to attach the horizontal collector sections (WT3x6) and vertical channel sections (MC6x15.1) to the concrete frame. Dowel spacings were based on dowel shear capacity and were occasionally altered to avoid interference with existing reinforcement in the frame. The general construction procedure involved first installation of the column channels, then the collector tees, and finally the diagonal brace members. Holes were drilled in the channel and tee sections and the members were clamped to the frame to be used as templates for drilling dowel holes into the concrete frame. All dowels were inserted with the steel member in place on the frame. Dowel grouting procedures and hole cleaning techniques followed the recommendations of Ref. 1. With the channels and tees in place all dowel bolts were tightened to a uniform torque of 75 ft-lb (102 N-m) and the tees were welded to the channels. An additional short tee section was welded to the in-place tee in the brace joint regions to form a wide flange section the same depth as the braces and channels. The braces consisted of W6x9 sections with the flanges trimmed to a width of 3-1/2 in. (89 mm); a standard rolled shape with the appropriate section properties for the scale model was not available. Brace members were positioned in the steel system with erection bolts, and the ends butt-welded to the joint regions. Clearances between the steel members and the frame did not allow use of backing plates which are normally required for butt welds made from one side of the member.

<u>Testing</u>. The model was subjected to five sets of load cycles as shown in the left portion of Fig. 10. The test was stopped twice during the 0.4% drift cycles to replace defective welds on brace members. Prior to the final set of cycles, F1 through F3, all brace welds were inspected and several were strengthened.

The first observance of brace buckling occurred in the upper braces during cycle F1 at a second to third interstory drift of about 0.5%. By the end of the load cycles, all upper level braces had buckled, with no buckling in the lower level braces. The final failure mechanism involved fracture of brace welds due to a combination of high local stresses and suspect weld quality. Local flange buckling of the braces occurred at the brace to tee connections of the middle bay. When the load was reversed and the member attempted to straighten, high localized stresses caused fracture of the flange welds, leading to failure of the remainder of the weld due to overstress. The sudden loss in brace capacity and shift of load into the columns led to concrete shear failure in the columns. This failure occurred after considerable deformation with upper level interstory drifts in excess of 1%. The column shear failure was accompanied by a limited amount of dowel pullout.

Results. Load-drift envelopes of each set of cycles and the final three cycles are shown in Fig. 10. The maximum load of 357 k (1588 kN) applied to the specimen represents a strength increase of nearly six times the calculated capacity of the original frame. The braced frame was approximately one and one-half times stiffer than the original frame, however the spandrels were heavily cracked at the beginning of the test due to damage from the pier strengthening test. The distribution of load carried by braces and columns when loading in the north direction can be seen in Fig. 11. The data points represent the horizontal component of the brace loads at the peak of each cycle through F1 (brace loads could not be determined reliably after buckling). The portion of shear carried by the braces for loading in either direction increased slightly with more severe load cycles but remained between 60% and 70% throughout the test. At an applied load of 300 k (1334 kN) it can be estimated that the column members were carrying approximately 50% more load than the calculated column shear failure load of 62 k (276 kN) for the original frame, due to the strengthening effect provided by the steel channels. The system showed good redistribution characteristics in that as buckling occurred, load was transferred to alternate braces with reserve tensile capacity.

The steel to concrete and steel to steel connections had a large influence on the behavior of the system. Dowel performance was good as no shear failures occurred and signs of pullout were limited to a small region where the channel in a third level brace to column joint pulled away from the frame at failure. This region was congested with gusset plates which made dowel placement difficult and also coincided with an area where some dowels were omitted due to interference with frame reinforcement. Lack of clearance for backing plates combined with the problem of aligning members composed of thin plates made butt welding at the connections difficult. Improvements in weld quality and connection design could have led to better performance with respect to overall system ductility. Further discussion of the steel bracing strengthened frame test is contained in Ref. 3.

CONCLUSIONS

The pier strengthening and steel bracing schemes successfully increased the strength, ductility, and stiffness of the original frame. Load-drift envelopes contained in Fig. 12 summarize the overall behavior of the strengthened frames:

The concrete pier strengthening system increased the strength 5 times and the initial stiffness 3 times as compared to the original frame.

The steel bracing system increased the strength 6 times and the initial stiffness 1-1/2 times (despite the precracked spandrels) as compared to the original frame.

Results of the pier strengthened frame test indicated that:

1. The failure mechanism was shifted to a ductile hinging of the spandrel beams with the piers experiencing minor distress.
2. Spandrel hinging occurred throughout the frame resulting in uniform interstory drifts.
3. The piers acted monolithically with the original frame as evidenced by low dowel stresses and low slip at new/old concrete interfaces.
4. Global behavior was dominated by frame action with the piers bent in double curvature rather than acting as rigid shear elements.
5. Lateral shear forces in the inside portion of the piers were transmitted by a diagonal compression strut.

Results of the steel bracing strengthened frame test indicated that:

1. Capacity was governed by buckling of the upper level braces, resulting in larger interstory drifts between the second and third levels than between the first and second levels.
2. The epoxy-grouted dowels performed well as evidenced by limited pullout and no dowel shear failure.
3. Weld quality and connection detail improvements could have led to better overall ductility characteristics.
4. The steel channels attached to the concrete columns increased their capacity by at least 50% as compared to the calculated capacities of the original columns.

ACKNOWLEDGEMENTS

This study was conducted at the Phil M. Ferguson Structural Engineering Laboratory of The University of Texas at Austin in a cooperative program with H. J. Degenkolb Associates, San Francisco. The assistance of other students in construction and testing, especially Marc Badoux for his laboratory and analytical assistance, is gratefully acknowledged. The work was supported under Grant No. CEE-8201205 from the National Science Foundation.

REFERENCES

1. Luke, Chon, and Jirsa, "Strength and Behavior of Reinforcing Bar Dowels Epoxy-grouted in Concrete," PMFSEL Report, 85-1, Dec. 1985.

2. Roach, C. E., "Seismic Strengthening of a Reinforced Concrete Frame Using Reinforced Concrete Piers," Thesis submitted for M.S. degree, The University of Texas at Austin, May 1986.

3. Jones, E. A., "Seismic Strengthening of a Reinforced Concrete Frame Using Structural Steel Bracing," Thesis submitted for M.S. degree, The University of Texas at Austin, Dec. 1985.

4. Bass, R. A., "An Evaluation of the Interface Shear Capacity of Techniques Used in Repair and Strengthening Reinforced Concrete Structures," Thesis submitted for M.S. degree, The University of Texas at Austin, Aug. 1984.

5. Weiner, D. F., "Behavior of Steel to Concrete Connections Used to Strengthen Existing Structures," Thesis submitted for M.S. degree, The University at Austin, Aug. 1985.

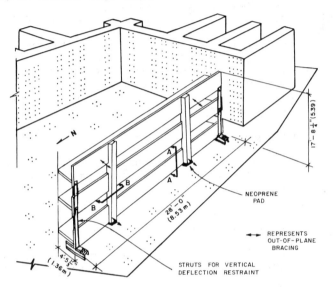

Fig. 1. Frame location in laboratory.

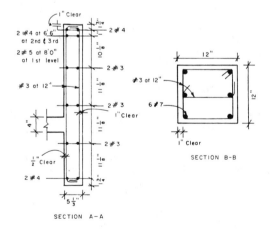

Fig. 2. Details of reinforcement of original frame.

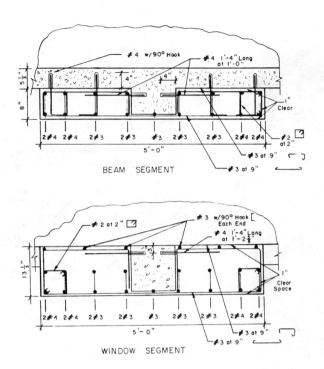

Fig. 3. Reinforcement details for pier strengthening.

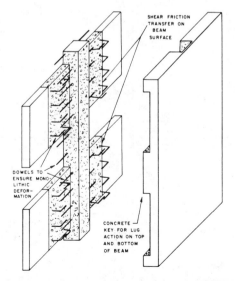

Fig. 4. Load transfer and connection mechanisms for pier strengthening.

Fig. 5. Pier strengthened frame.

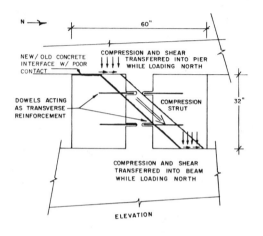

Fig. 7. Strut mechanism in interior 5-1/3 in. of window segment.

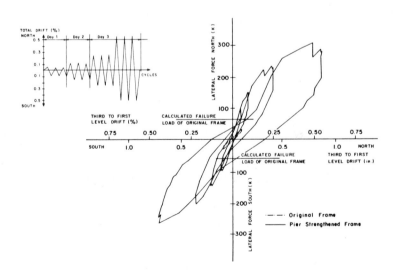

Fig. 6. Load-drift relationship - pier strengthened frame.

Fig. 8. Steel braced frame.

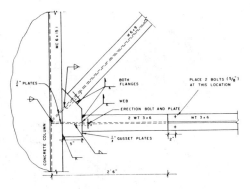
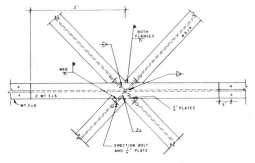

Fig. 9. Connection Details.

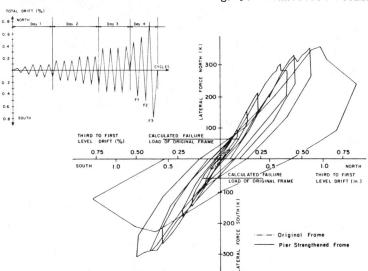

Fig. 10. Load-drift relationship - steel braced frame.

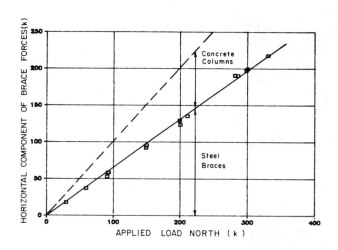

Fig. 11. Load distribution between braces and columns.

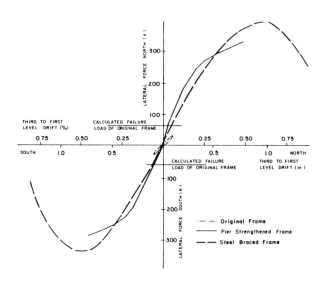

Fig. 12. Load-drift envelopes of strengthening schemes.

DESIGN AND BEHAVIOR OF A STRENGTHENED
REINFORCED CONCRETE FRAME

Loring A. Wyllie, Jr.[1], Chris D. Poland[1],
James O. Malley[2], and Maryann T. Wagner[2]

ABSTRACT

A large number of existing concrete frame buildings have details which proved to be hazardous in past earthquakes. A large scale experimental program was performed to investigate techniques for strengthening these frames. This paper outlines the design procedures and assumptions used in the experiments. A discussion of the test results and implications for Design Engineers is presented.

INTRODUCTION

Non-ductile concrete frame buildings have proven to be vulnerable to severe damage or even collapse when subjected to severe earthquake ground motions. As a result, present building codes include provisions which are intended to result in ductile seismic performance of these structures. But, between the late 1940s and the implementation of the ductile provisions in the early 1970s, a large number of non-ductile concrete frame structures were constructed in seismic areas. Many of these structures are still in use today.

One particularly vulnerable form of non-ductile concrete frame structure occurs where the frames consist of deep, stiff spandrel beams and short, weak columns. This system can be especially hazardous because it may result in a situation where the column shear capacity is the weak link in the lateral force resisting system. Concrete column shear failures are often brittle and may damage the vertical load carrying system. Deep spandrel, weak column systems were often utilized on exterior elevations where the windows occupied the area between the deep spandrels, such as that shown in Figure 1.

Because of the seismic hazard posed by non-ductile concrete frame buildings, a number of research investigations have been performed to study appropriate seismic strengthening techniques. These investigations included small scale tests of various strengthening schemes, such as infilling concrete or masonry in the frame openings, adding concrete wing walls to

---

1. Structural Engineer, H.J. Degenkolb Associates, Engineers, San Francisco, California.
2. Civil Engineer, H.J. Degenkolb Associates, Engineers, San Francisco, California.

strengthen the columns, jacketing the columns, or adding steel bracing. While these tests demonstrated the feasibility of the different strengthening schemes, the small scale test limited the information which could be useful to design engineers. Large scale tests which employ connections and details that can be used on actual structures were therefore necessary.

This paper presents a discussion of a recently completed investigation which included three tests on a 2/3 scale model of a deep spandrel, weak column exterior concrete frame. The first experiment tested the non-ductile concrete frame at low levels to estimate its elastic stiffness properties. The second test studied the effects of strengthening the columns with concrete piers. The final part of the investigation included a test of the specimen strengthened by the addition of a steel bracing system.

This project is being performed as a joint venture of H.J. Degenkolb Associates, Engineers and the University of Texas at Austin. This paper will discuss the basic features of selecting and designing the prototype structure and the two strengthening schemes, and evaluating the test results. A companion paper [1] presented by the University of Texas research team includes more detail concerning the construction and testing of the specimens. A set of detailed final reports are also being prepared [2,3,4].

SELECTION AND DESIGN OF THE TEST SPECIMEN

The seven-story prototype building investigated in this project is depicted in Figure 2. The experimental investigation centered on the exterior deep spandrel, weak column frames (See Figure 1). These frames form the primary lateral force resisting system in the longitudinal direction. Lateral forces in the transverse direction are resisted by concrete shear walls. The majority of the gravity loads are supported by interior flat plate and column frame action. Because of their flexibility, these interior frames do not contribute significantly to the lateral force resisting system.

Typical construction practice for the exterior frames aligned one face of the spandrels with the columns. This practice creates an eccentricity since the columns are typically wider than the spandrels. Matching the exterior column faces results in a flush surface which facilitates the attachment of architectural veneers. Matching the interior face results in vertical column lines on the building exterior. Similar concepts will be required to retrofit a building with either of these conditions, though the details will vary. Figure 1 shows that the condition which aligns the interior faces was chosen for this project.

Although the prototype building was designed to represent typical practice in the 1950s, some liberties were taken to accommodate the testing program. Specifically, an 18 inch (457 mm) square column size was selected primarily to correlate with previous 2/3 scale tests on 12 inch (305 mm) square columns. A rectangular or 24 inch (610 mm) square column with proportionally larger bay spacing may have been more typical for this type of system, but this variance was felt to have no direct bearing on the test results.

Analysis and design of the prototype building was based on accepted practice in the 1950s. The structure was designed for a base shear of 9% g in accordance with the 1955 edition of the Uniform Building Code. The portal method was used to analyze the frame. Member design and detailing was in accordance with the 1951 American Concrete Institute Building Code Requirements. Strength calculations indicated that the shear capacity of the columns limits the frame resistance. Once the design of the prototype building was completed, the 2/3 scale model could be detailed. These details are shown in Figure 1 of Reference 1.

Economic and technical constraints required that a subassemblage of the exterior frame be tested. The subassemblage configuration was chosen to isolate elements critical to the frame response and to properly account for the boundary conditions. The resulting subassemblage was a two-story, two-bay structure which included a 48 inch (1219 mm) width of slab at each floor level. Details of the boundary conditions and the instrumentation of the specimens can be found in the companion paper [1] and the project reports [2,3,4].

Before strengthening measures were taken, the bare specimen was tested. This was done to check the instrumentation and to estimate the stiffness of the bare frame. Two cycles to drifts of 0.05% were performed which caused some minor cracking in the spandrels near the joints. The stiffness of the bare frame was approximately 400 k/in. (70050 N/mm). No further testing was performed on the bare frame so that it would be essentially undamaged for the subsequent tests.

DESIGN AND TESTING OF THE CONCRETE STRENGTHENING SCHEME

The major objectives in the seismic retrofitting of a deep spandrel, weak column frame are to increase both the strength and ductility such that collapse is not probable. These objectives can be achieved by increasing the column strength such that a weak spandrel, strong column condition exists. Such strengthening measures with reinforced concrete can be achieved by adding piers or wingwalls, or encasing the columns. In this investigation, the columns were strengthened by the addition of full height reinforced concrete piers.

The strengthening system was designed to increase the lateral force resistance of the prototype building to the level required by the 1982 Uniform Building (UBC). This resulted in a design base shear of 18.6% g, approximately twice that required in the original design. This criteria was felt to be conservative since strengthening to current code force levels may provide capacity beyond that required for life safety concerns.

The layout and configuration of the prototype frame made possible the design of a strengthening pier constructable from outside the building. Such a system offers considerable advantages over a similar scheme constructed from the interior since the building can generally function with only relatively minor inconveniences to its occupants and contents during construction.

The new piers were designed and detailed in accordance with the provisions for structural walls in Appendix A of ACI 318-83. The length of the piers was selected to limit their nominal shear stress to about $4\sqrt{f'_c}$ and to sufficiently stiffen the building to satisfy code drift requirements without excessively reducing the window openings or causing the spandrel strength to be governed by shear. A length of 7 feet 6 inches was chosen for the prototype pier. The thickness of the new piers was governed by the dimensions of the prototype, the desire for one continuous curtain of steel over the height of the columns, and the need to provide lug action between the new and existing concrete to assist in the transfer of forces. A prototype pier thickness of 12 inches (305 mm) was used at the spandrels and 20 inches (508 mm) was used between spandrels. The pier details for the 2/3 scale model are shown in Figure 3 of Reference 1. The 28-day design compressive strength of that new concrete was 3000 psi. All reinforcement was Grade 60 except for the Grade 40 bars used for the boundary element hoops.

The piers were detailed to ensure monolithic behavior of the structural system. To assist in the transfer of force between the existing spandrel and new pier, epoxied dowels were designed to mechanically connect the two. The dowels were considered as shear friction reinforcement with embedment depth selected in accordance with the results of tests conducted at the University of Texas [5]. At low loads, the transfer of force was anticipated to occur through the bond between the new and existing concrete. Once the strength of this adhesive bond was exceeded, the load transfer was assumed to be achieved by three different mechanisms: lug action between the new concrete key acting on the top and bottom of the existing spandrel, shear friction developed over the sandblast interface of the new and existing concrete, and, after some degradation of the concrete, dowel action. The dowel details are also shown in Figure 3 of Reference 1.

The strengthening measures were constructed using procedures typical for the retrofit of actual structures. The specimen was loaded in sets of three cycles at each of the following displacement levels: 0.05%, 0.125%, 0.25% and 0.5% drift. No cycling beyond 0.5% drift was performed so that the specimen would not be too severely damaged for the final test. This discussion will include only the general response features. More detailed discussions, including the hysteretic response, is given in the companion paper [1] and the project reports [2,3,4].

The initial frame stiffness was 1250 k/in. (218900 N/mm), a factor of slightly over three times that of the bare frame. Spandrel cracking began in the first set of cycles, and increased in number, length, and width throughout the test. The first sign of stiffness degradation occurred in the cycles to 0.25% drift. This drift level also caused the initiation of diagonal pier cracking. During the final loading to 0.5% drifts, a peak load of 308 k (1370 kN), was reached. At this level, the cracking in the spandrels caused some spalling and splitting along the longitudinal reinforcement. The crack patterns in Figure 3 and the strain gage measurements demonstrated that the response of the strengthened frame was limited by the flexural yielding of the spandrels.

The strength of the bare frame model, limited by the shear strength of its columns, was estimated at 66 kips (294 kN) in accordance with ACI design equations using actual material strengths without capacity reduction factors. The strength of the retrofitted frame was limited by the flexural capacity of the spandrels. Although the peak load reached during testing was 308 kips (1370 kN), the positive slope of the load-drift envelope indicates that the capacity of the frame was not yet reached. Extrapolation of the load-drift envelope suggests a capacity of approximately 310 to 320 kips (1379 to 1423 kN). The estimated capacity of the test frame was based on the formation of plastic hinges in the spandrels at each face of each pier and an actual yield strength of 62 ksi (428 N/mm^2) in the spandrel reinforcement, was 305 kips (1357 kN). Good correlation between analytical predictions and experimental results was expected and achieved because flexure governed the specimen capacity.

At high loads, the transfer of force between the piers and frame was assumed to include direct bearing of the concrete lugs on the spandrels and shear friction between the piers and spandrels. Examination of the crack pattern in the specimen provides some insight into the actual shear transfer mechanism in the piers. At relatively low loads, horizontal cracks formed along the joints between the new piers and the spandrel soffits. As the shear increased and exceeded the frictional resistance along these cracks, significant slip was measured and inclined shear cracks formed in the piers. It is likely that at this stage the shear force was transferred to the spandrels through inclined compression struts in the piers.

Because of this effective lug action, the dowels which were provided for shear friction reinforcement between the spandrels and piers did not play a significant role in the force transfer. The dowels in the window opening, however, acted as transverse shear reinforcement and restrained the diagonal shear cracks from widening. It should be noted that the role of the dowels in the transfer of force is dependent on the exact geometry and detailing of the piers. If lugs are not provided, for example, detailing of the shear friction reinforcement would be critical to the specimen response. Because no back-up shear transfer mechanism would be available, conservatism in the design of such dowels would be recommended.

DESIGN AND TESTING OF THE STEEL STRENGTHENING SCHEME

Another method for the seismic strengthening of non-ductile concrete frame structures is to add steel bracing. Steel bracing schemes offer the possible advantages of minimization of disruption to building occupants and reduction of construction time. One possible disadvantage of such schemes is that the exterior architectural features will be altered. This method of bracing was studied in the final portion of this investigation.

The base shear and distribution of lateral forces used in the design of the steel bracing were identical to those used in the concrete strengthening scheme. The steel bracing was designed to take the entire lateral load and to limit deflections in order to protect the non-ductile columns from damaging displacements.

The configuration of the diagonal bracing elements is shown in Figure 4. This configuration of concentric X-bracing over two stories was chosen to avoid the problems of Chevron-braced frames, while restricting the unbraced length of the diagonals. In an effort to ensure acceptable hysteretic behavior of the diagonal braces, the KL/r values were limited to 100. To make use of this ductile hysteretic behavior, the ultimate capacity of the connections and the other elements were designed to ensure that they were capable of buckling the compression diagonals.

Providing connection details which develop the members but could be easily fabricated using typical construction practices became a major consideration in the design of the strengthening scheme. The choice of structural sections was also affected by the requirement that the collector and column sections would be dowelled to the existing exterior concrete frame. The use of efficient closed sections such as square tubes were eliminated because of the difficulties which would be encountered in providing acceptable connections and proper attachment to the existing concrete. The detailing scheme chosen for the steel strengthening is shown in Figure 5. As this figure shows, wide flange braces, channel columns, and tee collectors were chosen for the frame elements. The channel and tee sections were chosen to facilitate the dowel installation procedure, while the wide flange braces were selected to facilitate the connections between the elements. To eliminate connection eccentricities, all members were the same depth and the collector was changed from a tee to a wide flange section at the joints. Full penetration welds were used to connect the elements since less expensive bolted connections would not provide the required capacity and/or ductility. Gusset plates 1/4 inch (6.4 mm) thicker than the brace flanges were used to provide fit-up tolerance for the full penetration field welds.

The success of strengthening concrete structures with braced steel frames is critically dependent on the proper interaction of the new and existing elements. This is accompanied by the dowels which transfer both diaphragm shears and column overturning forces between the two systems. The value used to ensure that the bolts would have enough capacity to develop the bracing member strengths were one-half of the average ultimate capacity of previous dowel tests performed for this project [6].

After removing the pier elements added for the concrete strengthening scheme, the steel bracing elements were attached to the test frame using procedures similar to that which could be used in retrofitting an actual structure. It is important to note that the steel channels and collectors were in place before the dowels were epoxied. The loading history was similar to that of the concrete strengthening scheme with three cycles at each displacement level, until the final cycles. The companion paper [1] and the project report [2,3,4] provide more detail than the general discussion of response which will be presented here.

The initial frame stiffness was approximately 1.5 times that of the bare concrete frame. It should be noted that the contribution of the concrete frame to this stiffness had been reduced by the cracking caused by previous tests. During the initial cycles, the column shear cracks formed during previous tests were reopened. Virtually elastic response occurred until the 0.4% drift level, where the concrete cracking began to increase. At this level, a faulty brace weld fractured at a load near 300 k (1334 kN). After repairs were made, testing continued until another flange weld failed. After a third failure occurred, all welds were visually inspected and almost all were strengthened.

During the first displacement beyond 0.5% drift, the compression braces in the top story began to buckle at about 300 k (1334 kN). A peak load of 352 k (1566 kN) was reached. During the load reversal, the other top story braced buckled before a weld failed at the previously buckled brace. This weld probably failed due to a combination of lack of weld penetration and local stress concentrations caused by the previous buckling. In the next cycle, at a peak load of 360 k (1600 kN), two weld failures in the top story braces led to an explosive shear failure of the top story columns. The final loading caused two other weld failures and column shear failures to occur. Note that almost all distress in the frame was concentrated in the top story of the test specimen.

The first item to consider in evaluating the response of the specimen is the interaction of the steel and concrete systems. Before the test, it was assumed that the percentage of load carried by the steel bracing would increase as the test proceeded due to the cracking and subsequent stiffness deterioration of the concrete elements. But, the percentage of shear taken by the steel stayed almost constant at approximately 65% throughout the test. This result is probably due to the fact that the concrete frame was already cracked during the previous tests. Once brace buckling commenced, the interaction could not be accurately estimated since strain gage readings became unreliable.

The dowels between the steel and concrete elements performed quite well. This was demonstrated by the linear force variation in the collector elements and the small measured slip between elements (all less than 1 mm). This performance was certainly aided by provision of proper embedment, hole preparation, and extra epoxy to fill any gaps between the steel and concrete and the space between the bolt shaft and the steel column.

Another important design consideration is the ability to predict the ultimate capacity of the strengthened system. An accurate estimate of the specimen capacity was important both in designing the structural elements, and in ensuring that the loading apparatus was strong enough to yield the test frame. The shear capacity of the steel bracing elements was estimated to be the sum of the horizontal components of the fully yielded tension braces and the buckled compression diagonals. A K factor of 0.5 was used because of the fixity produced by the fully welded connections. The yield stress, F_y was assumed to be 45 ksi (310 N/mm^2) for the A36 material. Using the AISC Specification to estimate the column strength resulted in an estimated shear capacity of 279 kips (1241 kN) for the steel bracing elements. No shear capacity was estimated for the steel channel sections which were bolted to the concrete columns.

This capacity was added to the 66 k (298 kN) concrete column capacity to conservatively estimate the total capacity to be 345 kips (1535 kN). While it was recognized that these capacities are not strictly additive, this was felt to be appropriate in this case since the relative stiffness of the two systems were of the same order of magnitude. The maximum load resisted by the test specimen was 360 kips (1600 kN), approximately 10% above the estimated maximum, a factor of five larger than the bare concrete frame, and over three times the code level base shear.

It should be noted that the shear in the top story columns was almost 100 kips (448 kN) at the maximum load. This value is almost double the capacity calculated using code shear equations. The reasons for this discrepancy are two-fold. First, the actual concrete shear stress capacity is well above the $2\sqrt{f_c'}$ level assumed in the code equation. Second, and more important, were the steel channels which were attached to the concrete columns. These channels increased the effective shear area of the columns significantly since the closely spaced dowels allowed for composite action of these elements. In addition, these elements helped to confine the concrete and delay any spalling of the columns.

The final item of importance is the proper detailing of connections. The choice of member sections and connection details depends on the configuration of the existing concrete frame. Providing connections capable of developing the strength of the bracing members is critical to the proper inelastic response of the steel strengthening scheme. In this test, the strength and/or ductility of the full penetration welds which connected the brace to the column and collector elements failed before larger interstory drifts could be reached. If full penetration field welds are employed, the designer should ensure that the welds can be fabricated without great difficulty. The weld locations should be easily accessible. Gusset plates used to connect frame elements should be thicker than the member plates to insure that the full thickness of member plates can be welded. Special inspection and testing of field full penetration welds should be employed because of the possible access problems involved with attaching the frame to an existing concrete structure.

CONCLUSION

Both the concrete pier and steel braced frame strengthening schemes increased the stiffness, strength and ductility of the non-ductile concrete test frame. Both strengthening schemes, which were designed for the force level of the 1985 UBC, rely heavily on the proper interaction of the new and existing elements. In concrete strengthening schemes, the goal is to provide "monolithic" response between the new and existing concrete through mechanisms such as bond, shear friction, lug action, and dowel action. In steel strengthening schemes, a sufficient number of properly installed dowels are required to transfer the lateral forces into the new steel bracing elements.

Great care should be taken in designing a strengthening scheme which will provide the necessary strength and ductility in the most economical fashion. The existing configuration and condition of the concrete frame therefore becomes crucial to the selection of an appropriate bracing scheme. This project demonstrated the effectiveness of both steel and concrete bracing schemes for the seismic strengthening of non-ductile concrete frame structures.

ACKNOWLEDGMENTS

This study was performed as a joint effort of H.J. Degenkolb Associates and the University of Texas at Austin. The project was funded by the National Science Foundation, Grant No. CEE-8201205.

REFERENCES

1. Bush, T.D., Roach C.E., Jones, E.A., and Jirsa, J.O., "Behavior of a Strengthened Reinforced Concrete Frame", Proceedings of the Third U.S. National Conference on Earthquake Engineering, Charleston, South Carolina, August 1986.

2. Roach, C.E., "Seismic Strengthening of a Reinforced Concrete Frame Using Reinforced Concrete Piers," Thesis submitted for M.S. degree, The University of Texas at Austin, May 1986.

3. Jones, E.A., "Seismic Strengthening of a Reinforced Concrete Frame Using Structural Steel Bracing," Thesis submitted for M.S. degree, The University of Texas at Austin, December 1985.

4. H.J. Degenkolb Associates, "Seismic Strengthening of Reinforced Concrete Frames," 1986.

5. Luke, Chon, and Jirsa, "Strength and Behavior of Reinforcing Bar Dowels Epoxy-grouted in Concrete," PMFSEL Report, 85-1, December 1985.

6. Weiner, D.F., "Behavior of Steel to Concrete Connections Used to Strengthen Existing Structures," Thesis submitted for M.S. degree, The University of Texas at Austin, August 1985.

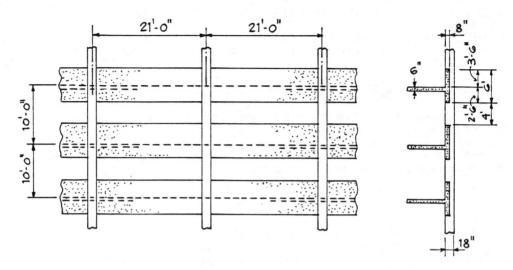

Fig. 1 Elevation and Section of Typical Deep Spandrel, Short Column Frame

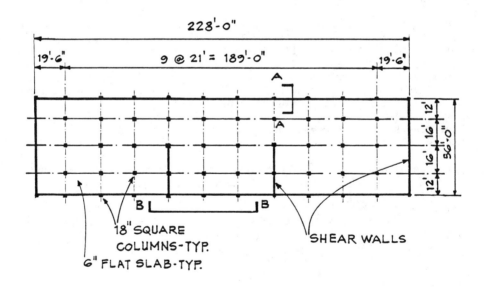

Fig. 2 Typical Floor Plan of Prototype Structure

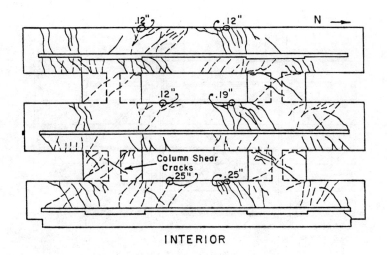

Fig. 3　Crack Pattern on Interior Face at Conclusion of Pier Strengthening Test

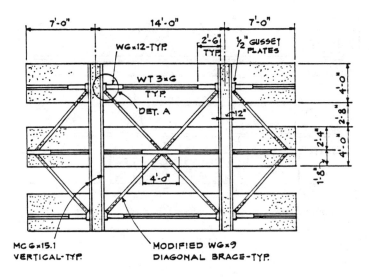

Fig. 4　Configuration of Steel Braced Specimen

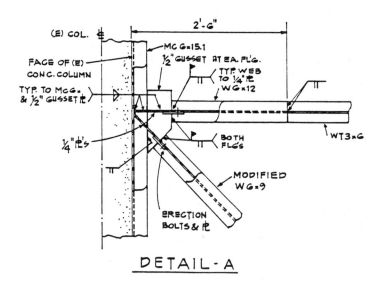

Fig. 5　Typical Connection Detail for Test Specimen

INFLUENCE OF TRANSVERSE REINFORCEMENT ON
SEISMIC PERFORMANCE OF COLUMNS

L.S. Johal,[I] D.W. Musser,[II] and W.G. Corley[III]

ABSTRACT

This paper describes a test program to develop design criteria for transverse reinforcement in reinforced concrete columns for seismic resistant buildings. Variables include level of axial load, amount and type of transverse reinforcement, and details of transverse reinforcement. Results indicate that flexural capacity of a column increases with axial load but deformability reduces substantially. Reduction in the amount of transverse reinforcement results in lower deformability. Details of transverse reinforcement including hook bends and hook extensions can be further simplified.

INTRODUCTION

Columns in building frames are normally designed to prevent hinging, crushing, or otherwise losing their capacity to support the building. However, columns in buildings subjected to an earthquake may sometimes be subjected to forces that cause unintended hinging. The possibility of yielding occurring at the column ends makes it important to ensure that columns are capable of behaving in a ductile manner.

Performance of concrete structures in recent earthquakes has clearly demonstrated the need for adequate information on behavior and design of columns [1-3]. Inadequate transverse column reinforcement has resulted in severe damage to structures. Current design provisions [4,5] for transverse reinforcement are based on providing confinement to increase concrete strain capacity. However, confinement reinforcement may not always be the governing criteria for columns. Transverse reinforcement is also needed to prevent premature buckling of vertical reinforcement and to provide shear resistance at the potential plastic hinge regions. As part of the experimental program being carried out at the Construction Technology Laboratories (CTL) of the

---

I. Senior Structural Engineer, Structural Experimental Section, Construction Technology Laboratories, a division of the Portland Cement Association, Skokie, Illinois.

II. Manager, Structural Engineering Department, Portland Cement Association, Skokie, Illinois.

III. Executive Director, Engineering and Resource Development, Construction Technology Laboratories, a division of the Portland Cement Association, Skokie, Illinois.

Portland Cement Association (PCA), test columns with differing amounts and details of transverse reinforcement and under different levels of axial load are subjected to moment reversals at increasing inelastic deformations. Observed behavior from the tests of seven columns is described and a summary of the test results is presented.

EXPERIMENTAL PROGRAM

Test column dimensions and reinforcement details of a representative test specimen are illustrated in Figs. 1 and 2. Design compressive strength of concrete was 6,000 psi (41.4 MPa). Specified yield stress of vertical and transverse reinforcement was 60 ksi (414 MPa). Vertical reinforcement consisted of eight No. 8 bars providing a reinforcement ratio of 0.0195.

Test Specimens

Seven full-scale column specimens have been tested. The test portion of each specimen represents the column extending upward from the beam-column connection to approximately the point of inflection. Cross-sectional dimensions are 18x18 in. (457x457 mm) and height is 10.5 ft (3.20 m) between horizontal load points. The beam portion was simulated by a short stub that also provided a loading point for the lateral load. Test specimens were designed and detailed in a manner to force hinging into the upper column.

Reinforcement Details and Materials

Transverse reinforcement details designated A, B, C, D, and E are shown in Fig. 3, and listed in Table 1. Transverse reinforcement for test specimen

Table 1 Details of Test Variables

Specimen Designation	Vertical Load		Transverse Reinforcement			
	P_v/P_o*	kips	Detail**	Bar Size	A_{sh} (in.^{2})	Percent
NC-1	0.30	570	A	No. 4	0.68	2.19
NC-2	0.20	380	B	No. 4	0.68	2.19
NC-3	0.40	780	B	No. 4	0.68	2.19
NC-4	0.30	580	B	No. 3	0.38	1.26
NC-5	0.30	575	C	No. 4	0.68	2.19
NC-6	0.30	520	D	No. 4	0.40	1.29
NC-7	0.30	540	E	No. 4	0.40	1.29

*$P_o = 0.85 f'_c (A_g - A_{st}) + A_{st} f_y$
P_v = Vertical column load

Metric Equivalents:
1 kip = 4.45 kN
1 in. = 25.4 mm

**Details of transverse reinforcement shown in Fig. 3.

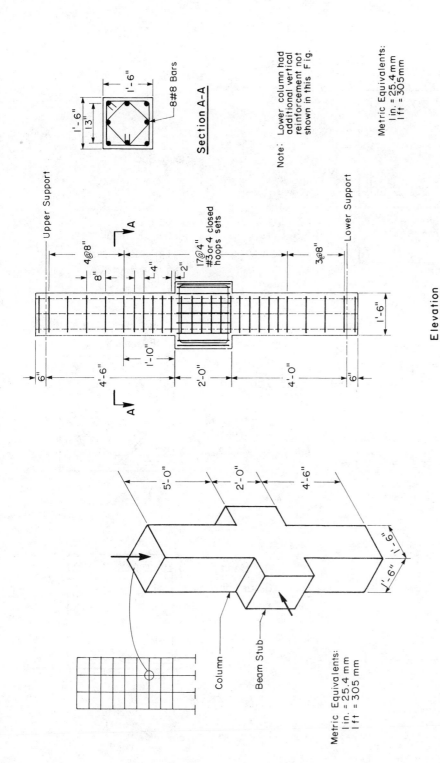

Fig. 2 Reinforcement Details

Fig. 1 Test Specimen

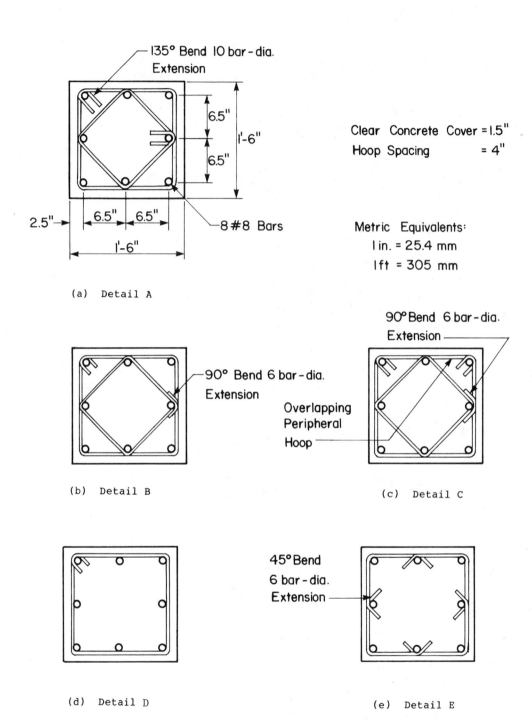

Fig. 3 Details of Transverse Reinforcement

NC-1, as shown in Fig. 3(a), was designed in accordance with the provisions of Section A.4.4 of the ACI Building Code [4]. This required 135 degree hook bends with 10 bar-diameter extensions for both inner and peripheral confining hoops. For all other specimens, hook extensions were reduced to six bar-diameter lengths. In addition, in Specimens NC-2, NC-3, NC-4, and NC-5, hook bends for the inner hoops were reduced to 90 degrees as shown in Fig. 3(b). Specimen NC-5 also used overlapping peripheral hoops as shown in Fig. 3(c). Specimen NC-6 used single peripheral hoops with 135 degree hook bends and six bar-diameter extensions as shown in Fig. 3(d). Transverse reinforcement for Specimen NC-7 also consisted of single peripheral hoops. Each of these hoops was formed with four identical ties as shown in Fig. 3(e).

The length of column confined by transverse reinforcement was kept constant at 22 in. (0.56 m). As shown in Fig. 2, hoops were spaced 4 in. (100 mm) on centers in the confined region. Transverse reinforcement in the unconfined region of column was designed to carry maximum shear stress. Clear cover was maintained at 1.5 in. (38 mm) in upper and lower columns.

Test Setup

Test setup and loading arrangement are shown schematically in Figs. 4 and 5. A one-million-lb (4448 kN) capacity testing machine was used to apply the vertical compressive force. Lateral load was applied with hydraulic rams pushing against reaction frames.

Instrumentation

Several types of instruments were required to obtain load-displacement, moment-curvature, vertical bar strain profiles, confining hoop strains, plastic hinge lengths, and maximum concrete compressive strains. External measurements were used to determine column axial and lateral loads, horizontal column displacements, column rotations, vertical reinforcement strains, and concrete strains. Internal measurements included strain on the vertical and transverse column reinforcement in the potential plastic hinge region.

Horizontal displacements were measured at three locations along the height of the test specimen. The measurement taken at the top of the stub was considered representative of upper column displacement. Groups of linear potentiometers were used to measure column rotations over gage lengths above the beam stub. By considering a pair of potentiometers measuring longitudinal displacement over the same height on each side of the column, and knowing the distance between them, a strain distribution across the column section can be obtained. Determination of the neutral axis depth from the strain gradient allows curvatures to be determined. Electric resistance strain gages were used to obtain reinforcing steel strain profiles.

Test Procedure

Each test was started by applying vertical load to the column. During a test, this load was kept constant at a predetermined level. Horizontal force was applied in increments alternately first in one direction and then

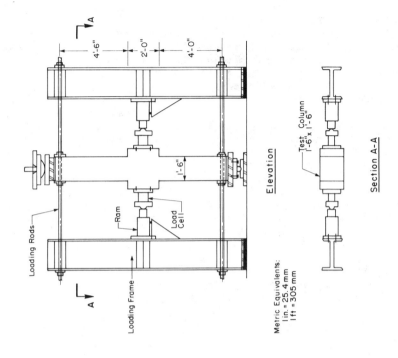

Fig. 5 Loading Arrangement

Fig. 4 Test Setup

in the opposite direction. The specimen was loaded to initial yielding in about three increments of horizontal force. Subsequent to initial yielding, loading was controlled by deflection increments.

Basic loading cycles were generally applied as follows: two cycles before yield, one cycle at yield, one cycle at ductility between 1 and 2, two cycles each at ductility 2, and subsequent ductilities. Testing was stopped at a stage when the specimen could not sustain the vertical load under increasing lateral displacement.

SUMMARY OF TEST RESULTS

Behavior of Specimens

Photographs of hinging regions of four specimens after testing are shown in Fig. 6. Observed length of hinging regions varied from approximately 10 to 14 in. (254 to 365 mm) for the seven specimens. Hysteresis loops of lateral load versus horizontal displacement obtained for three tests are shown in Figs. 7 through 9. A downward sloping line indicating maximum theoretical horizontal load, P_h, obtained from the calculated moment capacity is also shown in these figures. The downward slope indicates additional moment due to the P-Δ effect. Maximum horizontal displacement ranged from five to eight times yield displacement. A representative plot of moment versus curvature for Specimen NC-5 is shown in Fig. 10. This plot includes total moment at the upper column-stub interface versus curvature obtained over the first 4 in. (100 mm) from the beam stub. The calculated moment capacity, M_u, indicated in this figure, was determined using provisions of the ACI Building Code [4]. Capacity reduction factor, ϕ, was taken as 1.0. For a displacement ductility of seven in Specimen NC-5, maximum curvature ductility exceeded 20 as shown in Fig. 10. Maximum measured horizontal displacement, calculated flexural strength, and measured flexural strength values are listed in Table 2.

Table 2 Test Results

Specimen Designation	Measured Displacement Ductility	Flexural Strength, kip-in.		$\dfrac{M_2}{M_1}$
		Calculated, M_1	Measured, M_2	
NC-1	6	5184	6102	1.18
NC-2	8	4814	5773	1.20
NC-3	5	5198	6514	1.25
NC-4	5	5268	6152	1.17
NC-5	7	5226	6365	1.22
NC-6	5	4800	3651	0.76
NC-7	5	4950	5873	1.19

Metric Equivalent: 1 kip-in. = 0.113 kN·m

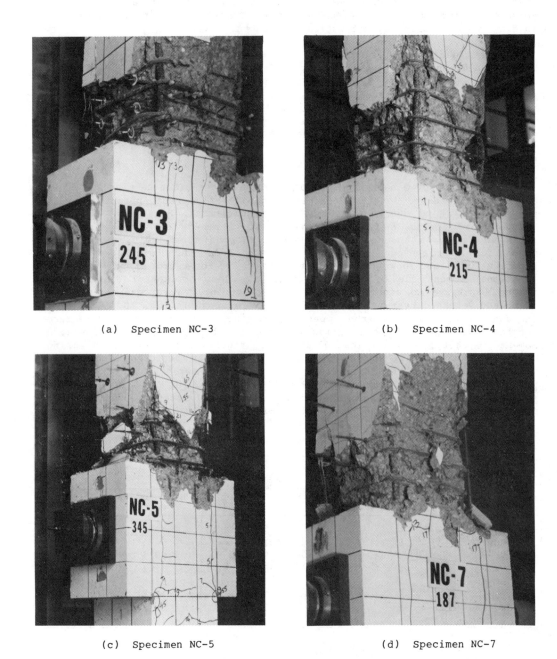

Fig. 6 Column Specimens after Test

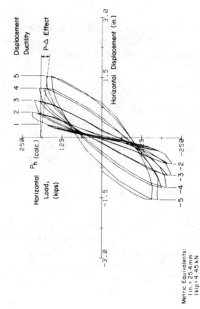

Fig. 7 Horizontal Load versus Displacement for Specimen NC-2

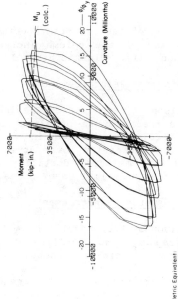

Fig. 8 Horizontal Load versus Displacement for Specimen NC-4

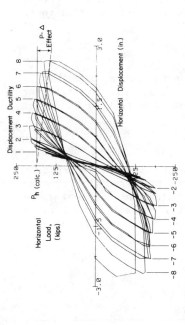

Fig. 9 Horizontal Load versus Displacement for Specimen NC-5

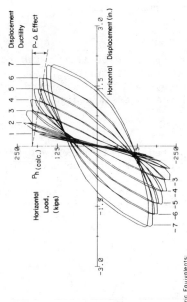

Fig. 10 Moment versus Curvature for Specimen NC-5

Effect of Variables

Axial Load

A comparison of results from Specimens NC-2 and NC-3 indicates that the flexural capacity of the column increased with axial load but ductility reduced substantially. Plastic hinge length increased with axial load.

Amount of Transverse Reinforcement

A comparison of results from Specimen NC-1 with those of NC-4, NC-6, and NC-7 indicates that the use of almost 50% less transverse reinforcement in these three specimens resulted in slightly lower ductility. Maintained strength was also generally lower at all load stages. It should be noted, however, that the measured ductility in Specimens NC-4, NC-6, and NC-7 exceeded that generally implied by codes.

Details of Transverse Reinforcement

A comparison of test results from Specimens NC-1 and NC-5 indicates that the flexural capacity and ductility of Specimen NC-5 was not reduced by the use of overlapping peripheral hoops. Also, flexural capacity and ductility were not reduced by the use of special hoops shown in Fig. 3(e) as indicated by a comparison of results from Specimens NC-4 and NC-7. However, the use of single peripheral hoops in Specimen NC-6 resulted in lower flexural strength.

Hook Bends of Inner Hoops

All specimens except NC-1 used 90 degree hook bends for inner hoops. Behavior of these specimens indicated that a standard 90 degree hook on inner hoops did not reduce ductility.

Hoop Extensions

Test results indicate that a ten bar-diameter extension as required by Section A.1 of the ACI Building Code [4] is not needed. Six bar-diameter extensions used in all specimens produced displacement ductilities exceeding those generally assumed in design.

ACKNOWLEDGMENTS

Work reported in this paper was based on research jointly supported by the National Science Foundation under Grant No. CEE 83-14226 and the Portland Cement Association. Any opinions, findings, and conclusions or recommendations are those of the authors and do not necessarily reflect the views of the National Science Foundation.

REFERENCES

[1] Zeris, C.A. and Altmann, R., "Implications of Damages to the Imperial County Services Building for Earthquake-Resistant Design," <u>Proceedings of the Eighth World Conference on Earthquake Engineering</u>, Vol. 4, San Francisco, California, July 21-28, 1984, pp. 823-830.

[2] Kreger, M.E. and Sozen, M.A., "A Study of the Causes of Column Failures in the Imperial County Services Building During the 15 October 1979 Imperial Valley Earthquake," a report to the National Science Foundation, University of Illinois, Urbana, Illinois, August 1983.

[3] "Reconnaissance Report, Imperial County California Earthquake, October, 1979, "Earthquake Engineering Research Institute, Berkeley, California, February 1980.

[4] ACI Committee 318, Building Code Requirements for Reinforced Concrete (ACI 318-83), American Concrete Institute, Detroit, Michigan, 1983.

[5] Uniform Building Code, International Conference of Building Officials, Whittier, California, 1985.

SEISMIC BEHAVIOR OF PRECAST WALLS

Peter Mueller[I]

ABSTRACT

This paper briefly reviews the results from tests on the behavior of precast walls typical of U. S. construction practice subjected to cylic seismic and constant axial load. The walls exhibit a relatively low flexural ductility due to concentration of inelastic deformations into connections and a completely altered shear transfer meachanism. The behavioral differences to monolithic walls clearly establish the need for code provisions that specifically address precast walls.

INTRODUCTION

Structural walls are one of the most efficient lateral force resisting systems. Increased usage of prefabrication cannot be avoided if the U. S. construction industry is to become more competitive. However, prefabricated structural walls are still rather an exception, particularly in seismic regions. A primary reason for this situation is insufficient knowledge on their behavior and a corresponding lack of specific code provisions. The NSF sponsored research project "Hysteretic Behavior of Precast Walls" addresses this problem through an experimental and analytical investigation centering on the two main concerns: the effect of horizontal connections on the ductility and energy dissipation capacity on the one hand, and their effect on the shear transfer capacity on the other [1].

TEST SPECIMEN

The 20 ft. (6 m) high wall specimens (Figs. 1 to 3) represent a scaled-down model of the lower half of a ten-story precast wall with details typical of large panel construction. 8 ft. (2.4 m) long, 2 ft. - 8 in. (0.8 m) high, and 4 in. (0.1 m) thick story-sized wall panels and 4 in. (0.1 m) deep hollow-core floor plank stubs are stacked on top of each other and connected with the platform-type horizontal connection shown in Fig. 5. The lower three stories represent the test region proper, that is the potential plastic hinge region. A transition panel at the fourth story is post-tensioned to a monolithic reusable top block to achieve the proper moment-shear ratio.

The connection details (Fig. 5) are selected from PCI publication Ref. 2. The panel edges are completely smooth (cast against steel) and the floor planks sit on smooth 1/16 in. plastic bearing strips. There are no shear keys or shear connectors. Each connection contains two times six

---

I. Associate Professor, Department of Civil Engineering, Lehigh University, Bethlehem, Pennsylvania.

individually cast hollow-core floor plank stubs. The vertical continuity steel is connected by grouted splice sleeves or for the prestressed walls, by post-tensioning bar couplers. These connectors are the only details in Ref. 2 which are free of unfavorable eccentricities that would adversely affect seismic performance. The wall panels are reinforced with vertical continuity steel along vertical edges, edge reinforcement serving as shear reinforcement along horizontal edges, and 0.1% of wire mesh in a single layer for crack control (Fig. 4). In addition, wire mesh is wrapped around all edges to control panel splitting along horizontal edges and to form a semblance of a boundary element along vertical edges. Panels and planks were cast in a local precasting plant and assembled in the lab using grouting and drypacking procedures followed in practice.

The parameter variation includes two reinforcement ratios, $200/f_y$ (corresponding to ACI minimum reinforcement) and $100/f_y$; two axial force nominal stress levels, 533 psi (3.7 MPa) and 350 psi (2.4 MPa); and prestress/no prestress. The low-high combination (PW1) and the high-low combination (PW2) achieve in different ways the same lateral strength permitting comparisons regarding ductility and shear strength. The latter (PW2) is at the critical limit for shear slip. The high-high combination (PW3) probes a diagonal crushing failure mode of the horizontal connection. This paper reports on these three non-prestressed specimens. Three companion specimens are prestressed. All walls have a moment-shear ratio of $M/Vl_w = 2$, corresponding to the average value obtained from inelastic dynamic analysis of prototype ten-story precast walls [3]. It is obvious that these walls would have difficulties passing the detailing requirements for zones of major seismicity. However, the objective of these tests is to determine the seismic performance of precast walls detailed for ease of fabrication and erection according to good current practice in zones of lesser seismicity.

TEST SETUP AND PROCEDURE

Figure 1 shows the test setup. The heavy steel reaction frame is designed to resist a lateral load of 150 kips (667 kN) at 19.5 ft. (6 m) and a gravity load of 250 kips (1110 kN). Lateral (seismic) and gravity load are applied to the wall through the monolithic top block. Gravity load simulators keep the gravity load vertical while the wall sways. The wall is fixed to the test bed through a longitudinally post-tensioned baseblock. The base block contains holes and fixtures such that it can be reused; only a small insert for reinforcement anchorage must be cast again for each new specimen. The bracing elements introduce the lateral bracing forces not directly into the wall panels, but more realistically over steel framing into the floor system. Care has been taken that this steel framing does not restrain the floor panels in the longitudinal direction of the horizontal connection.

The 235 kips (1045 kN) double-acting actuator for the seismic load as well as data acquisition are controlled from a lab computer. The instrumentation of the wall permits measurement of lateral displacements at each floor level, average curvature, elongation, and shearing of each wall panel and each connection, strains in the reinforcement at many locations, lateral and gravity load, and bracing forces. The test procedure consists of subjecting the walls to displacement cycles of increasing amplitude, three at each multiple of the yield displacement, until failure. Between each of these sets of three large amplitude cycles, three small amplitude cycles below the yield load are inserted.

TEST RESULTS

The following brief review of the test results centers on the effects that the horizontal connections have on energy dissipation capacity and ductility, on the one hand, and on shear strength and shear transfer mechanisms on the other hand.

Energy Dissipation and Ductility

Figure 6 shows the seismic (lateral) resistance versus top deflection hysteresis loop for specimen PW1 with a reinforcement ratio equal to half the ACI minimum ratio for flexure and a nominal axial load stress of 533 psi (3.7 MPa). The displacement amplitude of the three cycles shown is three times the yield displacement. It is obvious that the loops are not particularly fat, i.e. that the energy dissipation capacity is rather low. However, it is felt that this is not a particular characteristic of precast walls, but must primarily be attributed to the low reinforcement ratio in comparison with the axial load. Lateral resistance may be considered stemming from two sources: the resistance provided by the steel in flexure without axial load and that provided by the stabilizing effect of the axial force. The energy dissipated at a given displacement amplitude, on the other hand, primarily stems from the first source only. Therefore, if a major part of the resistance is provided by the second source, the hysteresis loops will be narrow. In other words, a similarly reinforced and loaded monolithic wall would not show much fatter loops. Accordingly, the hysteresis loops of PW2, which achieves the same resistance with doubled reinforcement ratio and lower axial load, are fatter.

The hysteresis loops shown in Fig. 6 are for the largest amplitude that could be sustained over three cycles. Specimen PW1 failed in the second half of the first cycle to four times the yield displacement through crushing and splitting of the compression zone. Figure 7 shows the failure zone after the specimen had been pushed still to six times the yield displacement in the non-failed direction. Specimen PW2 on the other hand failed in the second half of the third cycle to four times the yield displacement through bond failure in the splice sleeves in the first panel. The most likely cause for the bond failure is strain-hardening of the steel crossing the base connection beyond the steel stress of 90 ksi (620 MPa) that can be accommodated by the splice sleeves used. Figure 8 shows the first panel after the specimen had still been cycled to five times the yield displacement. The measured usable top displacement ductility ratios for PW1 and PW2 are thus 3 and 4, respectively. It must be noted though that these ratios relate to the top displacement of the test specimens which corresponds roughly to the fifth-story displacement of the prototype wall. Converting to top displacement ductility ratios of the ten-story prototype wall, values of roughly 2 to 2.5 result.

These low ductility ratios can be attributed to two causes: concentration of inelastic deformations into the connections and lack of effective boundary elements in the bottom panel. The compressive strength of a platform-type connection is only 50% to 70% of that of the panel. The connections are also weaker than the panel in tension, because the wire mesh is not connected. Although the wire mesh ratio is only 0.001, this is still not negligible relative to the low ratios of vertical continuity steel. As a result, both the base connection and the platform connections exhibit moment-curvature diagrams which are sufficiently lower than those for the panels to concentrate much of the inelastic deformations into the connections. Measured average curvature distributions clearly show that. The connections themselves are unexpectedly ductile. At four times the yield displacement, the first platform connection of PW2 had reached an average curvature ductility ratio of 8 without failure in spite of severe slip and ensuing spalling of panel corners. The problem is that these connection deformations occur over a small volume and that the inelastic deformation capacity of the panels which account for most of the wall volume, is not sufficiently activated. The panel cracks never open significantly. The severe concentration of crack opening into the connections, particularly the base connection, puts severe localized strain demands on the compressive corner of the bottom panel. The tendency of this region to split might be enhanced by wedging action of the splice sleeves. For this reason and to achieve a confined boundary element, a not yet tested specimen contains spiral reinforcement dropped over the splice sleeves.

Shear Transfer and Shear Strength

Figures 2 and 3 show crack patterns which are certainly quite different from those of monolithic walls. The development of the crack pattern of a wall panel with increasing moment can be nicely followed by comparing different panels (Fig. 3 top to bottom). First the connections open up. Then one or two flexural cracks form above the bottom corners of the panel around the splice sleeves. The next cracks that form are the steeply inclined cracks in the panel center. With increasing moment and opening of the connection above the panel, more steeply inclined cracks form towards the compression edge of the panel following the movement of the neutral axis of the connection. Already existing steeply inclined cracks progress towards the bottom compression corner. Also, new flexural and flexure-shear cracks form. The steeply inclined cracks open more (PW1) or as much (PW2) as the flexural cracks. Although both specimens have the same strength, PW2 is more thoroughly cracked than PW1 due to the higher reinforcement ratio and lower axial load.

Regarding shear transfer and strength, two issues must be clearly separated: shear transfer mechanism and strength of the connection and the effect that the shear transfer mechanism of the connection has on shear transfer mechanism and strength of the panel. As this unusual crack pattern clearly shows, the presence of the connections completely alters the shear transfer mechanism of the wall panel in comparison to a monolithic wall. While a monolithic wall transfers shear over the whole length of the wall, particularly also over the tension zone, through a field of compression diagonals, this is not possible in the present precast walls. When the connections open up, as shown in Fig. 9, the total shear force must be transferred over the compression zone of the connection, because little or no aggregate interlock is possible in the open tension zone due to the smooth panel edges and bearing pads. As shown in Fig. 9, in the truss model for a precast wall the compression diagonals must "bypass" the open connections. The inclination of the diagonals must roughly coincide with the panel diagonal rather than with the 45 degrees assumed in the ACI Code. For this reason and because the empirical constant V_c has not been determined for such walls, shear design according to the ACI Code was considered inadequate. The concentrated shear force at the top compression corner of the panel tends to tear off this corner resulting in the potential failure mechanism shown in Fig. 10 (top), which explains the observed crack pattern. Obviously, the most effective way to reinforce against this potential failure mechanism is to concentrate transverse reinforcement for shear at the top edge of the panel, a conclusion which also follows from the truss model of Fig. 9. When connections open up as shown in Fig. 9, individual panels may be considered as cantilevers fixed in the compression zone of the wall and loaded by the differential tensile force in the flexural reinforcement, as shown in Fig. 10 (bottom). The cantilevers would form cracks as indicated, explaining again the observed crack pattern, and we would reinforce them for flexure with concentrated reinforcement along the top edge. In effect, the tested precast walls represent an ideal realization of Kani's shear teeth model!

Anticipating this behavior, the wall panels were reinforced for shear, in addition to the minimal wire mesh, with concentrated transverse reinforcement at the panel edge using the truss model shown in Fig. 9. The compression members of the truss model represent the resultants of concrete compression fields, while the tension members represent the reinforcement. The inclination of and the force in the inclined compression chord were estimated by determining the location and magnitude of the concrete compressive resultants in each connection by a strain compatibility analysis. The shear resisted by the horizontal component of the inclined compression chord was used as a rational estimate for V_c. The remainder of the required shear strength was supplied by the edge transverse reinforcement along with the wire mesh. Although this resulted in twice the shear reinforcement required for a similar monolithic wall, it was clearly needed as in both PW1 and PW2 the edge reinforcement yielded in the two lower panels.

Specimen PW2 was designed to probe shear slip behavior of the connections. With a shear-axial force ratio of 0.6, it indeed developed more than 1/2 in. (12 mm) slip above the first panel, while PW1 with a ratio of 0.4 did not slip under the same shear force. The slip movement of PW2 gradually pushed off all the cover of the top corners of the first panel (Fig. 8) leaving only the section within the wire mesh cage and pointing out the importance of the mesh wrapped around all edges. In spite of this slip and damage, the connection developed more than the design flexural and shear strength and did not fail. The slip movement started also to tear apart the individual floor planks. It is doubtful whether under such slip conditions tie reinforcement for strucural integrity placed according to usual practice between floor planks, would remain sufficiently bonded. From the results presented in this paper, particularly the behavioral differences between precast and monolithic walls, it can be concluded that there is a clear need for code provisions that address precast walls specifically.

ACKNOWLEDGEMENTS

The research project "Hysteretic Behavior of Precast Panel Walls" is being conducted at Lehigh University, Fritz Engineering Laboratory, and is sponsored by the National Science Foundation under Grant CEE-8206674. The views expressed in this paper are those of the writer and do not necessarily reflect the position of the National Science Foundation.

REFERENCES

1. Mueller, P.
   "Behavioral Characteristics of Precast Walls," Proceedings, ATC-8 Seminar on Design of Prefabricated Concrete Buildings for Earthquake Loads, Los Angeles, April, 1981.

2. Martin, L. D. and Korkosz, W. J.
   "Connections for Precast Prestressed Concrete Buildings; including Earthquake Resistance," Technical Report No. 2, Prestressed Concrete Institute, Chicago, 1982.

3. Becker, J. M., Llorente, C., and Mueller, P.
   "Seismic Response of Precast Concrete Walls," Earthquake Engineering & Structural Dynamics, Vol. 8, No. 6, November-December, 1980.

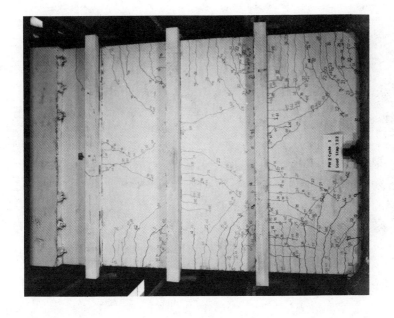

Fig. 2 Specimen PW2

Fig. 1 Test Setup

Fig. 4 Reinforcement Detail of Panel Corner Showing Splice Sleeve (PW1)

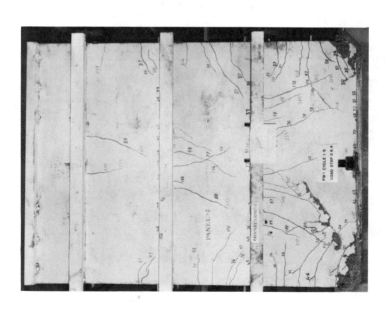

Fig. 3 Specimen PW1

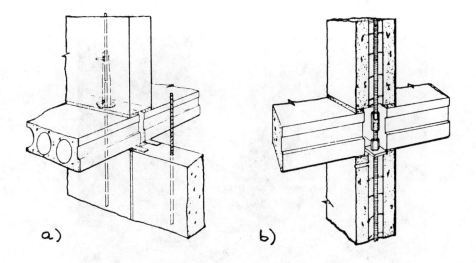

Fig. 5 Platform-type Horizontal Connection
with (a) Grade 60 Reinforcement   (Ref. 2)
      (b) Post-tensioning Bars

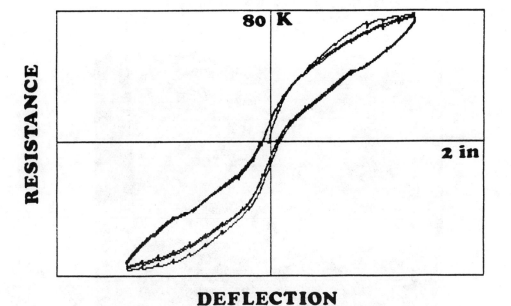

Fig. 6 Lateral Resistance vs.
        Top Deflection Hysteresis Loop

Fig. 7 Failure Zone Specimen PW1

Fig. 8 Specimen PW2 After Failure

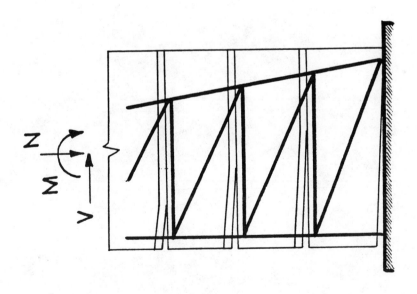

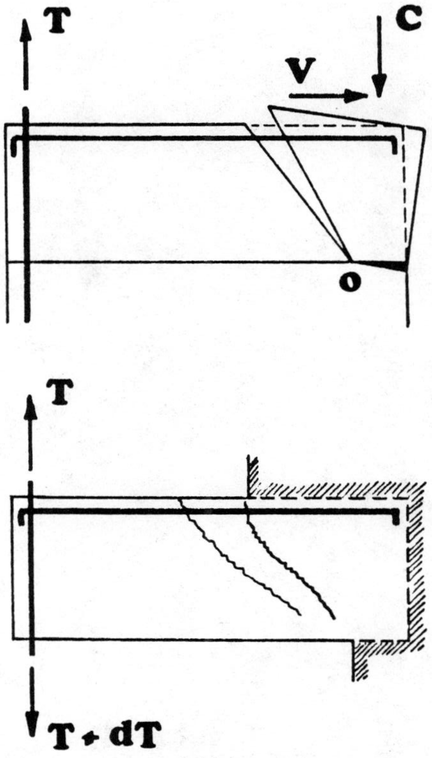

Fig. 10 Panel Shear Failure Mechanisms

Fig. 9 Shear Transfer in Precast Walls

RESULTS OF FULL-SCALE TESTS OF STEEL-DECK-REINFORCED CONCRETE FLOOR DIAPHRAGMS

M. L. Porter[I] and W. S. Easterling[II]

ABSTRACT

Results of 31 full-scale tests of steel-deck-reinforced concrete floor diaphragms are presented. These systems consist of cold-formed steel deck fastened to steel framing, with either arc spot welds, headed shear fasteners or some combination of the two, and a layer of concrete cover. In-plane cyclic loading was used for all specimens, and six of the specimens were subjected to a combination of gravity and in-plane loads. A wide range of test parameters have been used, including deck type, deck thickness, fastener type, fastener frequency, aspect ratio and concrete thickness.

INTRODUCTION

The use of steel-deck-reinforced concrete floor systems in steel frame buildings has become prevalent in recent years. Original interest, and therefore much of the subsequent research, focused on the gravity load carrying capacity of these floor systems. A significant amount of research on steel-deck-reinforced slabs has taken place at Iowa State University (ISU). The research at ISU has, in part, lead to the recent publication of specifications and commentary pertaining to the design and construction of composite slabs [1,2]. These documents do not consider diaphragm action that is a result of lateral loading of the structure, perhaps because diaphragm action of steel-deck-reinforced concrete floor systems has not been investigated to the extent that rational analytical and design procedures are available.

In order to pursue experimental and analytical information regarding the diaphragm behavior of composite slabs, a research projected is currently underway at ISU. The current study is an extension of a previous research program conducted at ISU [3]. To date a total of 31 diaphragm specimens have been tested in both research programs. The principal aim of this paper is to describe the test arrangement and present experimental results regarding the strength and stiffness of the steel-deck-reinforced concrete diaphragms.

I. Professor, Department of Civil Engineering, Iowa State University, Ames, Iowa.

II. Research Assistant, Department of Civil Engineering, Iowa State University, Ames, Iowa.

TEST ARRANGEMENT

The testing was performed in the Structural Engineering Laboratory at ISU utilizing a previously designed test frame [4]. Figure 2 shows the test frame configuration, which is a cantilever diaphragm frame with a fixed edge. The fixed edge consists of three large reaction blocks that are anchored to the structural tie-down floor. The remaining sides consist of W24 x 76 steel members fastened together with flexible tee connections.

The loading is applied via two hydraulic actuators attached to the north frame member (see Fig. 2). A quasi-static cyclic loading was used for all tests. Figure 3 shows a typical displacement history, which was selected, not with intent of representing a particular natural event, but rather, with the intent of studying the basic behavior of composite diaphragms under cyclic loading. In addition to the cyclic loading, six of the diaphragm specimens were subjected to vertical load prior to the cyclic load being applied. The selected vertical load was maintained during the cyclic loading. A schematic of the vertical load mechanism is shown in Fig. 4 [5].

Data collection devices included electrical resistance strain gages, direct current linear variable displacement transducers (DCDT), load cells, mechanical dial indicators and still photography. Continuous load versus displacement plots were made during the test procedure. An example of those plots is shown in Fig. 1.

TEST PARAMETERS AND RESULTS

The 31 test specimens were constructed using a variety of parameters. In particular, parameters such as deck type, deck thickness, fastener type, fastener frequency, aspect ratio and concrete thickness were varied during the test program. Table 1 lists the major components for each of the specimens. An additional note should be that specimens 1-21 were tested on a 15 ft. x 15 ft. (4.57 m x 4.57 m) frame and specimens 22-31 were tested on a 15 ft. x 12 ft. (4.57 m x 3.66 m) frame. The frame members perpendicular to the applied in-plane load direction were the ones shortened.

Test results, consisting of initial stiffness, ultimate capacity and failure mode, are listed in Table 2. The initial stiffness values were calculated by determining the slope of the load-displacement curve between the points corresponding to the origin and the first nominal displacement of 0.025 (0.635 mm) inches. A DCDT attached at the northeast corner in line with the load was used to determine the displacement value for the initial stiffness calculation. Ultimate loads were taken as the maximum applied in-plane load during the load history. Load cells located at both actuator locations were summed to obtain the in-plane load values. Each of the tests has been described with a type of failure that is associated with the ultimate load obtained by the slab. A complete listing and description of failure modes associated with composite diaphragms has been made previously [3]. Three failure modes, diagonal tension, shear transfer mechanism and stud zone failure, have proven to be the most prevalent in the experimental program, thus only these three are discussed herein.

Diagonal tension failure is characterized by the occurrence of cracks that form diagonally across the surface of the concrete. These cracks are typically at angles of approximately 45 degrees and extend over a large portion of the diaphragm. The ultimate load occurs just prior to the development of the diagonal tension crack. An upper limit with respect to capacity, and therefore the upper limiting failure mode is associated with a diagonal tension failure of the diaphragm.

The load transfer path of composite diaphragms is such that the load must pass from the loading frame through the fasteners (arc spot welds) and steel deck into the concrete. This transfer of load through the connectors and steel deck is primarily through shear transfer and has been shown to take place in a finite edge zone [3,6]. When degradation of the composite system occurs to a degree such that load cannot be transferred adequately to the concrete, failure of the shear transfer mechanism is said to have occurred.

When headed shear studs are used as fasteners, failure of the concrete around the stud is classified as a stud zone failure. If this occurs prior to the attainment of diagonal tension capacity of the diaphragm, the ultimate load is said to be associated with a stud zone failure. This is similar to the shear transfer mechanism failure, which is associated primarily with the use of arc spot welds.

DISCUSSION OF TEST RESULTS

This section of the paper consists of a discussion of the tests results that were presented in the previous section. The main focus of the discussion centers around the various test parameters. In particular, the way in which the parameters influence the modes of failure is addressed. The discussion that follows is based on the observations made during the experimental program and the recorded experimental data.

To begin the discussion, the influence of steel deck type and steel deck thickness will be addressed. At the outset of the original composite diaphragm research project at ISU, the investigators recognized that, due to the large variety of steel deck types and thicknesses on the market, investigating the influence of deck type and thickness should be a high priority. In view of this, 11 different steel deck cross sections or thicknesses, from five different manufacturers have been used in the experimental program. Test results show that varying the steel deck thickness has a distinct effect on the behavior of the composite diaphragms, while specific deck type influences are not yet isolated. When using only arc spot welds, 20 ga. (0.036 in., 0.914 mm) decks failed in the shear transfer mechanism mode. This failure mode is generally associated with relatively lower strength values. The 16 ga. (0.06 in., 1.524 mm) decks, when fastened with arc-spot welds, generally failed in the diagonal tension mode. If headed studs are used as fasteners, the deck thickness, as well as deck type, influence appears to diminish significantly.

The influence of fastener type and frequency can be seen from the data in Tables 1 and 2. When only arc spot welds are used with a frequency of approximately one per foot, the failure mode is of the shear transfer mechanism type. If the welds are placed with a frequency of four per foot, the failure mode for the 20 ga. decks is still by the shear transfer mechanism mode, but as discussed in the previous paragraph, the 16 ga. decks failed in the diagonal tension mode. Specimens that were fastened with headed studs or a combination of studs and welds failed by either a stud zone failure or a diagonal tension failure. When the number of studs was small, the force being transferred from the frame member to the diaphragm was concentrated at a few points, thus the stud zone failure occurred. If enough studs were used such that local stress concentrations did not reach magnitudes high enough to cause cracking around the studs, then a diagonal tension failure was observed.

The influence of a varying aspect ratio was investigated to a limited extent. Two frame sizes were used during the project. As previously mentioned, the first 21 tests were made on a 15 ft x 15 ft (4.57 m x 4.57 m) frame and the remainder of the tests were made on a 15 ft x 12 ft (4.57 m x 3.66 m) frame, with the frame members perpendicular to the in-plane load direction being the ones that were shortened. Based on this somewhat limited data there appears to be no effect on the failure modes when the aspect ratio is changed. As is expected, with the decrease in length the stiffness values generally increase.

Neilson [5] investigated the influence of combined gravity and in-plane loading and considered test specimens prior to and including specimen 18. His study showed that gravity load has no significant effect on initial stiffness or strength in general, although in one comparison of specimens with and without gravity load (slabs 15, 16), the applications of vertical load apparently caused the failure mode to be changed from one of shear transfer mechanism to one of diagonal tension. Correspondingly an increase in strength occurred.

SUMMARY

A test arrangement and methodology has been described for an experimental investigation that has been performed at ISU to determine behavioral characteristics of steel-deck-reinforced concrete floor diaphragms. Test results were presented for 31 diaphragms, and consisted of initial stiffnesses, ultimate capacities and failure mode classification. The influence of various parameters on the test results was discussed.

ACKNOWLEDGEMENTS

The authors are grateful to the National Science Foundation for the sponsorship of this research under contract no. CEE-8209104. Mssrs. S. M. Dodd, M. K. Neilsen, M. D. Prins and D. L. Wood have contributed significantly to the current research effort and their assistance is gratefully acknowledged.

REFERENCES

[1] "Specifications for the Design and Construction of Composite Slabs", ASCE Standard, Published by American Society of Civil Engineers, New York, New York, October, 1985.

[2] Porter, M. L., "Commentary on Specifications for the Design and Construction of Composite Slabs", ASCE Standard, Published by American Society of Civil Engineers, New York, New York, October, 1985.

[3] Porter, M. L. and Greimann, L. F., "Seismic Resistance of Composite Floor Diaphragms", Final Report, ERI-80133. Engineering Research Institute, Iowa State University, Ames, Iowa, May, 1980.

[4] Arnold, V. E., Greimann, L. F. and Porter, M. L. "Pilot Tests of Composite Floor Diaphragms". Progress Report, ERI-79011. Engineering Research Institute, Iowa State University, Ames, Iowa, September, 1978.

[5] Neilsen, M. K., "Effects of Gravity Load on Composite Floor Diaphragm Behavior." Unpublished Master's Thesis, Iowa State University, Ames, Iowa, 1984.

[6] Prins, M. D., "Elemental Tests for the Seismic Resistance of Composite Floor Diaphragms". Unpublished Master's Thesis, Iowa State University, Ames, Iowa, 1985.

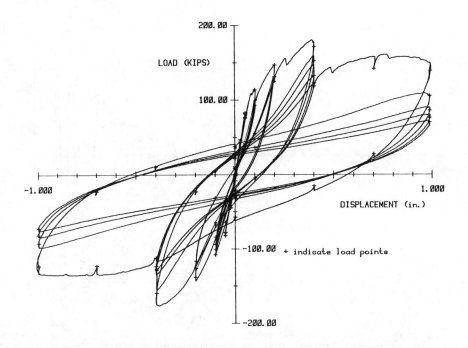

Fig. 1. Load-displacement diagram.

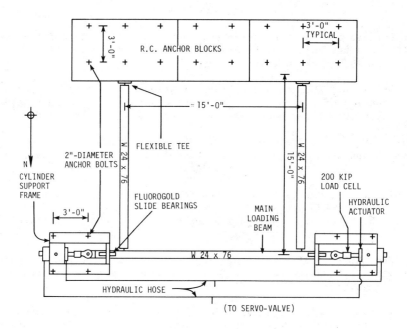

Fig. 2. Diaphragm test frame schematic.

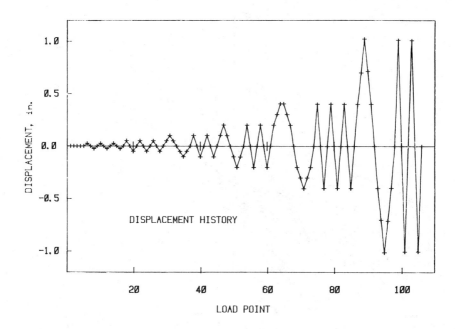

Fig. 3. Typical in-plane displacement history.

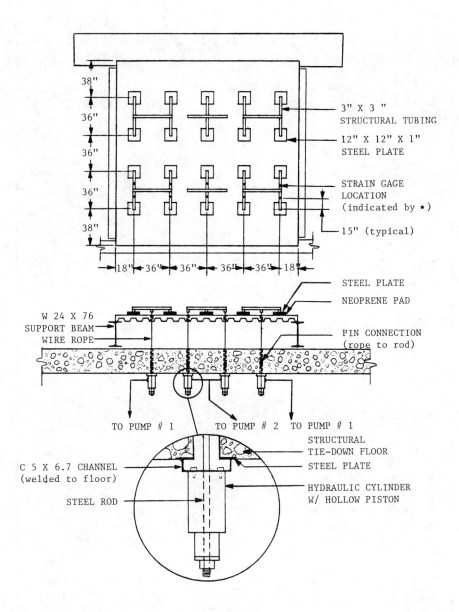

Fig. 4. Vertical load mechanism schematic.

Table 1. Summary of slab parameters

Slab Number	Concrete Parameters			Steel Deck Parameters				
	Nominal Thickness (in.)	Actual Thickness (in.)	f_c' (psi)	Deck Type	Thickness (in.)	Yield Strength (ksi)	Ultimate Strength (ksi)	Connections Per Side
1	5 1/2	5.38	5634	1	0.034	41.7	53.4	30 studs
2	5 1/2	5.50	5250	1	0.034	41.7	53.4	30 studs
3	5 1/2	5.65	4068	1	0.034	41.7	53.4	60 welds
4	5 1/2	5.28	3849	1	0.034	41.7	53.4	60 welds
5	3 1/2	3.53	2966	2	0.062	48.2	60.7	30 welds
6	7 1/2	7.44	4549	2	0.062	48.2	60.7	60 welds
7	5 1/2	5.40	5435	3	0.058	49.7	61.1	60 welds
8	5 1/2	5.47	3345	1	0.035	41.7	53.4	4 studs (N-S side) 6 studs (E-W side)
9	5 1/2	5.48	5412	4 (Pan) 4 (Pan)	0.058 0.057	51.8 52.4	63.2 64.9	60 welds
10	5 1/2	5.53	3311	5	0.062	40.4	53.2	60 welds
11	5 1/2	5.72	3533	6	0.047	89.7	93.7	60 welds
12	5 1/2	5.59	3412	5	0.062	40.4	53.2	60 welds
13	5 1/2	5.53	6187	4 (Pan) 4 (Pan)	0.058 0.057	51.8 52.4	63.2 64.9	60 welds

Table 1 (cont'd) Summary of slab parameters

	Concrete Parameters			Steel Deck Parameters				
Slab Number	Nominal Thickness (in.)	Actual Thickness (in.)	f'_c (psi)	Deck Type	Thickness (in.)	Yield Strength (ksi)	Ultimate Strength (ksi)	Connections Per Side
14	8	8.2	3699	5	0.062	40.4	53.4	60 welds
15	4	4.21	2844	7	0.047	89.7	93.6	60 welds
16	4	4.18	2952	7	0.047	89.7	93.6	60 welds
17	7 1/2	7.44	4261	2	0.062	46.0	54.4	60 welds
18	5 1/2	5.55	3052	5	0.062	40.4	53.4	60 welds
19	5 1/2	5.75	2681	8	0.062	49.4	55.5	60 welds
20	5 1/2	5.55	3973	9	0.037	48.6	56.2	40 welds
21	5 1/2	5.67	3638	5	0.062	40.4	53.4	15 welds
22	5 1/2	5.68	3301	5	0.062	40.4	53.4	60 welds (N-S side)
								49 welds (E-W side)
23	5 1/2	5.75	3496	9	0.037	48.6	56.2	40 welds (N-S side)
								34 welds (E-W side)
24	5 1/2	5.63	4047	8	0.062	49.4	55.5	48 welds
25	5 1/2	5.69	4672	5	0.062	40.4	53.4	14 studs (N-S side)
								9 studs (E-W side)

Table 1 (cont'd) Summary of slab parameters

	Concrete Parameters			Steel Deck Parameters				
Slab Number	Nominal Thickness (in.)	Actual Thickness (in.)	f'_c (psi)	Deck Type	Thickness (in.)	Yield Strength (ksi)	Ultimate Strength (ksi)	Connections Per Side
26	4 1/2	4.72	3462	10	0.036	92.8	93.6	6 welds (W) 10 studs (E) 8 studs, 14 welds (N-S side)
27	5 1/2	5.66	2883	9	0.037	48.6	56.2	9 welds (E-W side) 8 studs, 16 welds (N-S side)
28	5 1/2	5.60	3611	9	0.037	48.6	56.2	15 welds (all sides) 6 studs (E-W side) 8 studs (N-S side)
29	5 1/2	5.55	2887	11	0.035	86.9	89.8	16 studs (N-S side) 11 studs (E-W side)
30	5 1/2	5.68	3565	11	0.035	86.9	89.8	12 studs, 4 welds (N-S side) 7 studs (E-W side)
31	5 1/2	5.75	3336	11	0.035	86.9	89.8	23 welds (N-S side) 13 welds (E-W side)

Table 2. Summary of experimental results

Slab Number	Vertical Load (psf)	Initial Stiffness (KIPs/in.)	V_u(KIPs)	Failure Mode
1	0	1800	168	diagonal tension
2	0	2000	186	diagonal tension
3	0	1600	98	shear transfer mechanism
4	0	1300	88	shear transfer mechanism
5	0	1700	116	diagonal tension
6	0	2600	147	shear transfer mechanism
7	0	1500	137	shear transfer mechanism
8	0	1100	54	stud zone failure
9	0	1900	220	diagonal tension
10	0	1700	161	diagonal tension
11	0	1600	95	shear transfer mechanism
12	61	1800	180	diagonal tension
13	200	1900	250	diagonal tension
14	135	1900	208	shear transfer mechanism
15	0	1300	103	shear transfer mechanism

Table 2 (cont'd) Summary of experimental results

Slab Number	Vertical Load (psf)	Initial Stiffness (KIPs/in.)	V_u(KIPs)	Failure Mode
16	35	1300	124	diagonal tension
17	100	2200	146	shear transfer mechanism
18	135	1700	161	diagonal tension
19	0	1300	147	diagonal tension
20	0	1300	95	shear transfer mechanism
21	0	1200	122	shear transfer mechanism
22	0	2100	169	diagonal tension
23	0	1700	106	shear transfer mechanism
24	0	2100	168	diagonal tension
25	0	1900	180	diagonal tension
26	0	1700	87	diagonal tension
27	0	2000	91	stud zone failure
28	0	2000	119	stud zone failure
29	0	2300	137	diagonal tension
30	0	1900	115	stud zone failure
31	0	1500	65	shear transfer mechanism

DYNAMIC RESPONSE OF R/C FRAMES
WITH IRREGULAR PROFILES

Sharon L. Wood[I]

ABSTRACT

Two small-scale multi-story reinforced concrete structures with irregular profiles were tested using the University of Illinois earthquake simulator. The objective of the tests was to investigate the influence of setbacks on the earthquake response of planar structures. Measured response is presented and evaluated with respect to the expected dynamic behavior of regular structures.

INTRODUCTION

Traditionally, building codes in the United States have separated buildings into two classes for earthquake resistant design [1,2,3]. Buildings with nearly uniform distributions of story strength, stiffness, and mass over the height of the structure are classified as "regular". Simplified design procedures, such as the Equivalent Lateral Force method, may be used to proportion members in these structures. "Irregular" structures are believed to be more susceptible to damage during earthquakes, and more stringent design procedures, such as linear dynamic analysis, are required to anticipate the demands placed on structural elements.

This investigation [4] was concerned with the dynamic response of a particular type of irregular structure: planar reinforced concrete frames with setbacks. Setback structures are characterized by abrupt reductions in floor area in the upper stories of the building. Story mass, strength, and stiffness also decrease with height above the base, but not necessarily at the same rate.

Two building profiles, one symmetrical and one asymmetrical relative to the center of the base, were selected for investigation. This paper provides a brief description of the experimental study and a summary of the measured response.

I. Assistant Professor of Civil Engineering, University of Illinois at Urbana-Champaign, Illinois.

DESCRIPTION OF THE TEST STRUCTURES

Two small-scale reinforced concrete model structures were subjected to strong ground motions. Each structure comprised two identical, planar, nine-story frames (Figs. 1 and 2). The test structures were not models of prototype structures, however, dimensions corresponded to approximately one-fifteenth scale.

The primary experimental parameter was the building profile. The Tower Structure (Fig. 1) comprised a seven-story tower and a two-story base. The Stepped Structure (Fig. 2) included an asymmetrical arrangement of a three-story tower, a three story middle section, and a three-story base. A tall first story was selected for both structures.

A stiff base girder was cast monolithically with each frame. Frames were secured to the simulator platform with bolts post-tensioned through the base girders, creating a nearly fixed-base condition. Story weights of approximately 360, 720, and 1100 lb (160, 320, and 500 kg) were supported at one, two, and three-bay levels, respectively. The weights acted as rigid diaphragms and coupled the frames at each level. A system of channels transferred the inertial and vertical forces from the story weights to the frame joints without eccentricity. The dead load at each level was distributed equally to the columns at that level.

The test structures were constructed using small-aggregate concrete with steel wire used as reinforcement. Longitudinal reinforcement was continuous throughout the structure, welding or splicing of reinforcement was not required. Anchorage of longitudinal reinforcement was provided in short cantilever members which extended beyond exterior joints. First-story column reinforcement was anchored within the base girder. The distribution of longitudinal reinforcement is summarized in Table 1. Sufficient transverse reinforcement was provided to avoid shear failure of any member. Mean material properties are summarized in Table 2.

EXPERIMENTAL PROGRAM

Each test structure was subjected to a series of simulated earthquake motions of varying intensity. Low-amplitude free-vibration tests were conducted before and after each earthquake simulation to measure changes in the natural frequency of the structures. After the completion of dynamic testing, the structures were subjected to a series of lateral-load tests to establish story strengths under static conditions.

Earthquake motion for all simulations was modeled after the north-south component of the 1940 El Centro ground motion. The time-scale of the original earthquake record was compressed by a factor of 2.5. The design-level earthquake for both structures corresponded to a peak base acceleration of 0.4g. A representative base acceleration record for a design simulation is shown in Fig. 3a. Acceleration response spectra calculated with a damping ratio of 0.10 are shown in Fig. 3b for the design simulations of both structures.

Absolute accelerations and relative displacements in the direction of the base motion were measured continuously at each level of the test structures during the earthquake simulations. Vertical and transverse accelerations were also recorded at the top level. Load was applied incrementally during the static tests. Horizontal displacements were recorded for each load increment. Crack patterns and crack widths were recorded after each earthquake simulation and during the static tests.

OBSERVED RESPONSE

The dynamic response of the setback structures may be interpreted through representative response histories and distributions of response over the height of the structure. Data presented in this paper is limited to the design earthquake simulations.

Top-level displacement, top-level acceleration, and base shear response histories for the Tower and Stepped Structures are shown in Figs. 4 and 5. The displacement response for both structures is characterized by a smooth waveform, indicating a dominance of the apparent fundamental mode. Displacement distributions during cycles of maximum top-level response are shown in Fig. 6. The overall drift ratio was approximately 1% for both structures. The observed maximum deflected shape of the structures corresponded closely to the first mode shape for the gross-section model.

Maximum interstory drifts occurred in the fifth story of the Tower Structure and in the first story of the Stepped Structure. Displacement concentrations were not observed in stories with significant changes in strength and stiffness.

In contrast with the displacement response histories, the acceleration waveforms displayed a jagged appearance, indicating the participation of higher modes. To a lesser extent higher mode participation could also be observed in the base-shear waveforms, however, the first mode appeared to govern the response. Shear distributions during two cycles of maximum response for the Tower Structure are shown in Fig. 7. The magnitude of the base shear at 1.62 and 2.91 sec were essentially equal, however, the participation of the higher modes was perceptively different. Maximum shears in the middle and upper stories generally occurred during cycles in which the first mode dominated the shear response (Fig. 7b).

Because the displacement response of the structures was governed by the first mode, the top-level displacement histories were used to estimate changes in the fundamental mode of the structures. The time interval between relative maxima was interpreted to be the effective period of the structure. The variation of the apparent frequency during the design simulation for the Tower Structure is shown in Fig. 8. A general decrease in apparent frequency with time may be observed (Fig. 8a), corresponding to a softening of the structure with the number of cycles of response.

The apparent frequency is compared with the magnitude of the displacement response during the cycle in Fig. 8b. A trend of decreasing frequency with increasing damage, or increasing displacement, may be observed. However, a decrease in apparent frequency in cycles after the cycle of maximum displacement was also observed. The structures appeared to soften without an increase in displacement.

CONCLUDING REMARKS

The dynamic behavior of the two test structures with irregular profiles did not differ from the expected behavior of frames with regular vertical configurations. The participation of the higher modes did not significantly influence the response. Displacement and shear response was governed by the first mode, and amplification of shears or displacements in stories with significant changes in stiffness was not observed. The observed behavior of the two planar multi-story setback frames provided no evidence that the design procedure for such frames should be different from that of regular frames.

ACKNOWLEDGEMENTS

This experimental study was part of the investigation of Earthquake Response of Reinforced Concrete Frames conducted in Newmark Civil Engineering Laboratory at the University of Illinois, Urbana under grant CEE-8114977 from the National Science Foundation. The research was supervised by M. A. Sozen. Appreciation is due to P. Doak and P. Stork for their assistance in the experimental phase of the investigation.

REFERENCES

1. Applied Technology Council, <u>Tentative Provisions for the Development of Seismic Regulations for Buildings</u>, ATC 3-06, June 1978.

2. International Conference of Building Officials, <u>Uniform Building Code</u>, Los Angeles, California, 1985.

3. Structural Engineers Association of California, <u>Recommended Lateral Force Requirements and Commentary</u>, 1980.

4. Wood, S.L., "Experiments to Study the Earthquake Response of Reinforced Concrete Frames with Setbacks," Thesis submitted to the Graduate College of the University of Illinois, Urbana, Illinois, January 1986.

Table 1. Reinforcement Ratios (%)

Level/Story	Tower Structure			Stepped Structure		
	Interior Columns	Exterior Columns	Beams	Interior Columns	Exterior Columns	Beams
9	2.26		1.73	2.26	2.26	1.73
8	2.26		1.73	2.26	2.26	1.73
7	2.26		1.73	2.26	2.26	1.73
6	2.26		1.73	2.26	2.26	1.73
5	2.26		1.73	2.26	2.26	1.73
4	2.26		1.73	2.26	2.26	1.73
3	3.39		2.59	2.26	2.26	1.73
2	3.39	1.13	2.59	3.39	3.39	2.59
1	3.39	2.26	2.59	4.52	3.39	2.59

Note: Reinforcement ratios are based on gross sections for columns and effective depth for beams.

Table 2. Mean Material Properties

CONCRETE		
Compressive Strength	6 ksi	40 MPa
Secant Modulus	3900 ksi	27000 MPa
STEEL		
Column - No. 13 Gage		
Yield Strength	56.3 ksi	388 MPa
Beam - No. 7 Gage		
Yield Strength	55.2 ksi	381 MPa

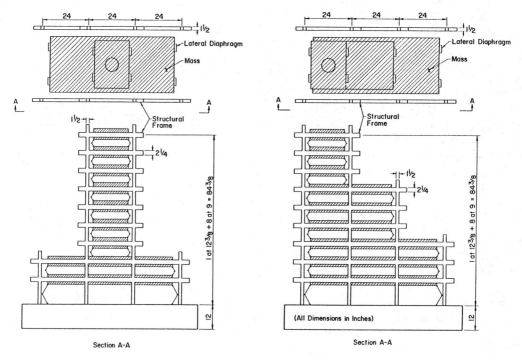

Fig. 1 Tower Structure

Fig. 2 Stepped Structure

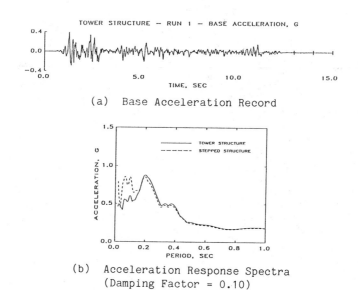

(a) Base Acceleration Record

(b) Acceleration Response Spectra
(Damping Factor = 0.10)

Fig. 3 Characteristics of Simulated Ground Motion

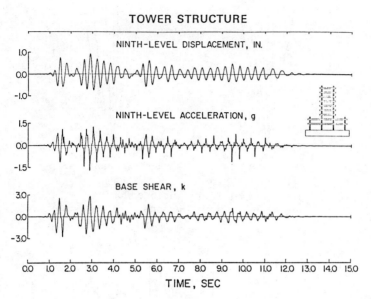

Fig. 4 Representative Response Histories - Tower Structure

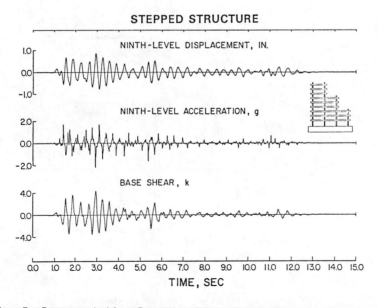

Fig. 5 Representative Response Histories - Stepped Structure

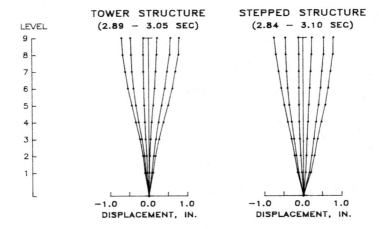

Fig. 6 Displacement Distributions During Cycles of Maximum Response

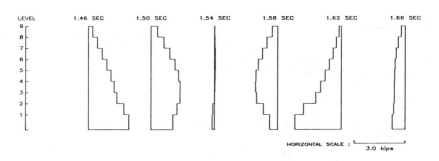

(a) Higher Mode Contribution

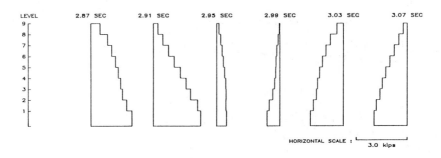

(b) First Mode Domination

Fig. 7 Shear Distributions During Cycles of Maximum Response - Tower Structure

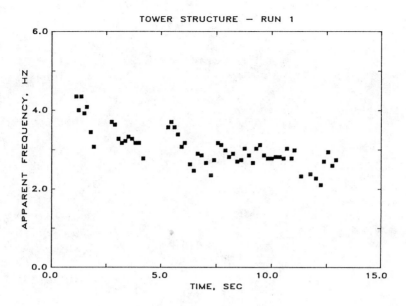

(a) Variation with Number of Cycles

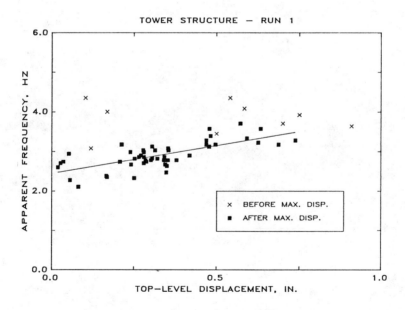

(b) Variation with Top-Level Displacement

Fig. 8 Variation of Apparent Frequency During Design Earthquake Simulation

STRONG-MOTION INSTRUMENTATION OF STRUCTURES IN CHARLESTON, SOUTH CAROLINA AND ELSEWHERE

M. Celebi[I] *and R. Maley*[I]

ABSTRACT

Instrumentation of structures is part of earthquake hazard mitigation programs of many institutions, including the United States Geological Survey (USGS). The USGS Strong-Motion Instrumentation of Structures Program is designed to complement other programs and to implement its own, within budget and other constraints. This paper reviews the overall national effort, cites examples of structures implemented and describes progress made to date. A recent example of instrumentation of an eight-story building in Charleston, South Carolina is documented.

INTRODUCTION

The main objective of any instrumentation program for structural systems is to improve the understanding of the behavior, and potential for damage caused under seismic loading. The acquisition of response data during earthquakes is essential to confirm and/or develop further methodologies used for analysis and design of earthquake resistant structural systems. This objective can best be realized by selectively instrumenting structural systems by acquiring strong ground motion data, and by recording the responses of structural systems (buildings, components, lifeline structures, *etc.*) to that strong ground motion. As a long term result, one may expect design and construction practices to be modified to minimize future earthquake damage [1].

Various codes in effect in the United States, whether nationwide or local, recommend different quantities and schemes of instrumentation. For example the Uniform Building Code (UBC) [2] recommends for Seismic Zones 3 and 4, a minimum of three accelerographs be placed in every building over six stories in height with an aggregate floor area of 60,000 square feet or more, and every building over 10 stories in height regardless of floor area. The City of Los Angeles adopted the above recommendation in 1966 but in 1983 revised this requirement to only one accelerograph. Experience from past earthquakes shows that the instrumentation guidelines given by the UBC code, although adequate for the limited analyses projected at the time, do not now provide sufficient data to perform the model verifications and structural analysis demanded by the profession.

On the other hand, valuable lessons have been derived from the study of the data obtained from a well-instrumented structure, the Imperial County Services Building, during the moderate size Imperial Valley earthquake ($M_s=6.5$) of October 15, 1979 [3].

[I] U.S. Geological Survey, Menlo Park, CA 94025

To reiterate, it is expected that a well-instrumented structure for which a complete set of recordings has been obtained would provide useful information to:

o   check the appropriateness of the dynamic model (both lumped mass and finite element) in the elastic range,
o   determine the importance of non-linear behavior on the overall and local response of the structure,
o   follow both the spreading non-linear behavior throughout the structure as the response increases and the effect of the non-linear behavior on frequency and damping,
o   correlate the damage with inelastic behavior,
o   determine ground motion parameters that correlate well with building response damage, and
o   make recommendations eventually to improve seismic codes.

The USGS effort in instrumentation of structures is based on an adopted policy that aims to complement programs in seismically active regions or to establish new programs in regions where there is little or no instrumentation. For example, in California, the USGS complements the Strong Motion Instrumentation Program (SMIP) of the California Division of Mines and Geology (CDMG). Therefore, the USGS program in California has concentrated on non-typical structures since the intent of the SMIP is to instrument typical structures.

THE OVERALL PROGRAM

The USGS structural instrumentation program has proceeded since 1983 on a basis that is heavily dependent on recommendations of advisory committees established in seismically active regions. A general location map of present and future regions for structural instrumentation is provided in Figure 1. The regional committees are comprised of academicians, practicing engineers, and state and local government officials.

The objectives of the committees can be specified as follows:

o   develop a list of structures in the designated region within the objectives of the USGS program
o   develop priorities for the list of structures,
o   coordinate the effort on instrumentation of structures with other programs and organizations,
o   communicate to the public and private sectors the importance of programs for instrumentation of buildings,
o   extend the program to other regions as required,
o   enhance the maintenance of instruments in a coordinated way, and
o   provide guidance and develop methodologies related to instrumentation of structures.

The committees currently formed in different regions are shown in Figure 2. Two committees, San Francisco and San Bernardino, have completed their deliberations and issued a final report; four others; Charleston, Boston, Anchorage, and New Madrid, have reports near completion.

The primary factors in selecting structures for instrumentation within the general USGS program results in structures that are representative of systems and materials that are likely to be repeated and those that have sufficient long-term engineering interest. Further detailed structural and site parameters included:

A) Structure parameters: The material of construction, structural system, geometry, discontinuities, importance, age, and interest.

B) Site parameters: Probability of occurrence of a large earthquake, proximity to faults, soil conditions, and expected damage and loss.

These various parameters are weighted by the committee members in order of perceived importance to develop a ranked list of structures as candidates for extensive strong-motion instrumentation. The high-priority structures are considered first.

Implementation of the USGS strong-motion instrumentation program is conducted within the Branch of Engineering Seismology and Geology (ES&G), a unit in the Office of Earthquakes, Volcanoes, and Engineering (OEVE).

When one of the Instrumentation Advisory Committees (IAC) formulates its recommendations, the USGS engineering staff obtains instrumentation permits for the high-priority selected structures and gathers information pertinent to the project, including the design calculations, structural plans, and model information. The staff ultimately directs structural evaluation, and if necessary, a ambient response study. This integrated data then is used as a basis for determining transducer locations that will adequately define the response of the structure during a strong earthquake. After the sensor locations have been selected, the installation team, an owner representative, and an electrical contractor jointly plan the actual cable and transducer placement. Following installation, a documentation report will indicate transducer location and orientation, characteristics of total system response, and other important details of the instrumentation.

RECENTLY INSTRUMENTED UNIQUE STRUCTURES

In Figure 3, the cross section of the Transamerica Building in downtown San Francisco and the associated instrumentation at various locations of the structure are shown. The instrumentation is designed to monitor the translational and torsional motions at different levels, and torsional and rocking motions of the overall structure.

Currently the JCG Wilshire Finance Building, a unique structure in Los Angeles, is being instrumented jointly by the owner and the USGS. The general instrumentation scheme is shown in Figure 4. The objective of this scheme is to monitor the translational and torsional motions at different levels (particularly at the levels of abrupt change of stiffness) and the rocking motion at the base of the structure.

As demonstrated by these two examples, the USGS intends to instrument unique and non-typical structures in California, thus complementing the SMIP program (of the State of California) by which typical structures are instrumented.

INSTRUMENTATION PROGRAM IN THE SOUTHEASTERN UNITED STATES

The USGS, in collaboration with the Technology Transfer and Development Council, established an instrumentation advisory committee for the southeastern United States, with a focus on the Charleston, South Carolina area. The advisory committee developed a preliminary list of potential structures which were deemed important such that, if instrumented, the engineering community would benefit from studying the data acquired during strong earthquakes. The structures were then rated according to the committee's criteria. The top priority candidate is the Charleston Place, a major convention and commerical center being constructed in the historic downtown area of Charleston, South Carolina.

The Charleston Place structural system consists of a reinforced concrete frame and post-tensioned concrete floor slabs. The superstructure is supported on a pre-stressed concrete pile system extending down through approximately 60 feet of soft inorganic clay and fine sand into a stiff brownish-green calcareous clay known locally as "marl". Figures 5 and 6 show the plan view and elevation of the building complex. The Charleston Place has an eight-story tower section containing three tower elements and a low-rise hotel section consisting of four stories, featuring east and west open interior courtyards [4].

The eight-story tower will be instrumented by installing 12 accelerometers at selected locations within the structure. The strong-motion signals from these sensors will be transmitted by cable to a central location where they will be recorded on a single strip of photographic film. Locations of accelerometers, shown in Figures 6 and 7, are planned as follows:

1) three vertical and two horizontal accelerometers on the first floor to record horizontal base motion as well as potential foundation rocking or independent vertical inputs,

2) three horizontal sensors at the sixth floor to record both translational and torsional motions, and

3) four horizontal sensors at the ninth level (roof) with a configuration comparable to that on the sixth floor. The additional horizontal sensor is provided to measure diaphragm in-plane motions.

In conjunction with this structural system, a time synchronized free-field accelerograph will be located north of the building.

ACKNOWLEDGMENTS

The authors gratefully acknowledge the contributions of many participating members of the advisory committees described in the paper.

REFERENCES

[1] Celebi, *et al.*, 1984, Report on recommended list of structures for seismic instrumentation in the San Francisco Bay Region: *U.S. Geological Survey Open-File Report 84-488*.

[2] _____, *Uniform Building Code, International Conference of Building Officials,* Whittier, CA, 1970, 1976, 1982 editions.

[3] Rojahn, C., and Mork, P.N., 1982, An analysis of strong motion data from a severely damaged structure--The Imperial County Services building, El Centro, California *in* The Imperial Valley, California, earthquake of October 15, 1979: *U.S. Geological Survey Professional Paper 1254*.

[4] Lindbergh, C. (Chairman), 1985, The U.S. Geological Survey Instrumentation Advisory Committee, Southeastern United States: *Preliminary Report*, 21 p.

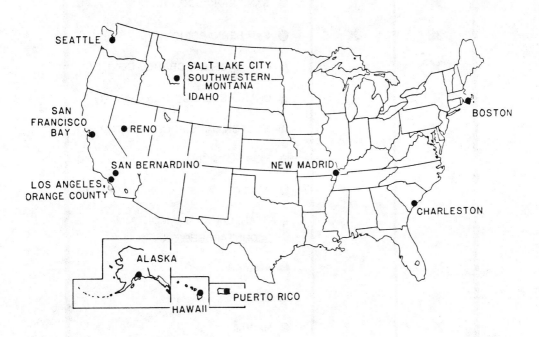

Fig. 1. Target Regions for Instrumentation of Structures Program.

ADVISORY COMMITTEES FOR STRUCTURAL INSTRUMENTATION

COMMITTEE FORMED	REPORT COMPLETED	REGIONS CONSIDERED
X	X	● SAN FRANCISCO AREA
X	X	● SAN BERNARDINO
X		● LOS ANGELES, ORANGE COUNTY
X		● CHARLESTON, SC (SOUTHEAST)
X		● BOSTON, MASS. (NORTHEAST)
X		● NEW MADRID
		● SEATTLE, WASH. (NORTHWEST)
		● UTAH, IDAHO, SW MONTANA (MOUNTAIN REGION)
X		● ALASKA
		● RENO
X		● HAWAII
		● PUERTO RICO

Fig. 2 Current Status of Advisory Committees.

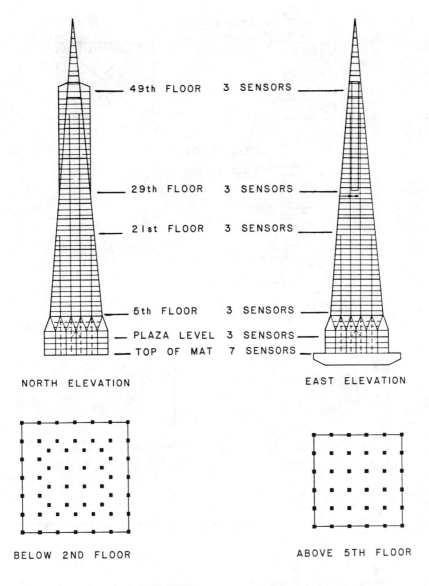

Fig. 3. Instrumentation Scheme of the Transamerica Building.

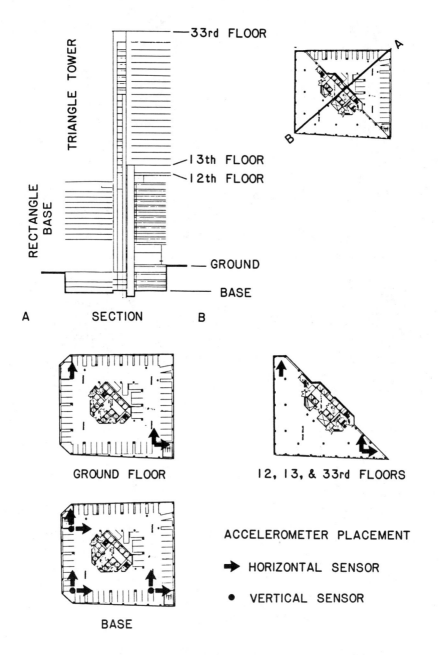

Fig. 4. Instrumentation Scheme of the Wilshire Finance Building.

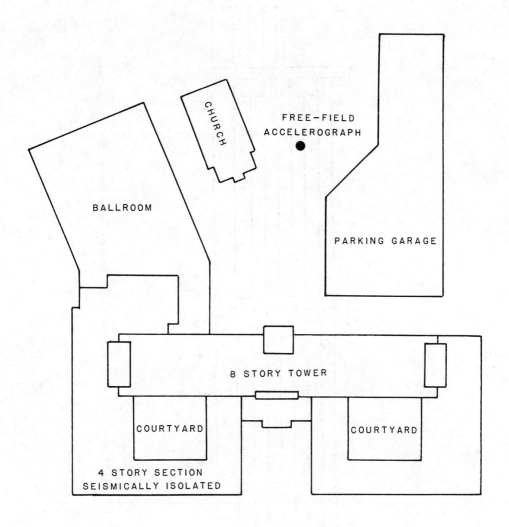

Fig. 5. General Plan View of the Charleston Place.

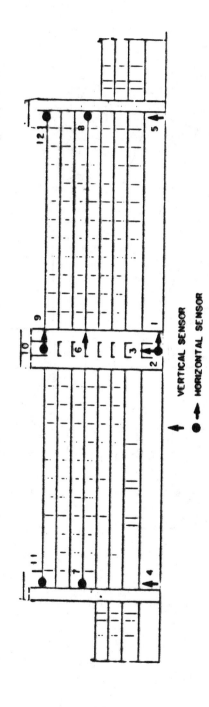

Fig. 6. Section of the Charleston Place Showing Location of Accelerometers.

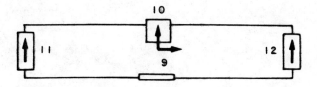
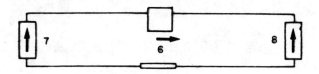
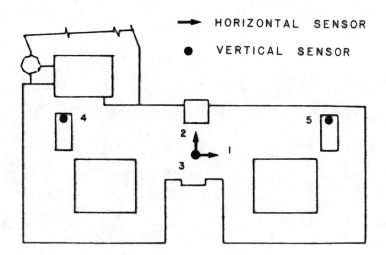

Fig. 7. Plan View of Accelerometer Deployment Scheme at Three Levels of Charleston Place.

LOW-CYCLE FATIGUE OF SEMI-RIGID
STEEL BEAM-TO-COLUMN CONNECTIONS

J. B. Radziminski[I] and A. Azizinamini[II]

ABSTRACT

This paper presents the results of a study of the performance of semi-rigid steel beam-to-column connections under cyclic loadings. Fatigue crack propagation behavior and hysteretic energy absorption characteristics are reported for connections subjected to constant and variable amplitude displacement histories. Empirical low-cycle fatigue life relationships established by the constant amplitude tests have been applied to a linear damage accumulation model; the results of several variable amplitude block cyclic tests are compared with damage summations predicted by the model.

INTRODUCTION

The importance of beam-column connections to the satisfactory performance of building structures under earthquake ground motions is well established. Moment resisting steel connections are generally well suited for seismically induced loads because of their high ductility and the energy absorbing capabilities they provide. The purpose of this investigation has been to evaluate the static and cyclic behavior of bolted, semi-rigid beam-to-column connections typical of those used for wind moment design requirements in the Eastern U.S. From the static tests, the geometric parameters which most significantly affect connection performance have been quantified, and compared with analytical models formulated to predict the initial stiffness and complete non-linear moment-rotation behavior of the connections. The results of this phase of the investigation are reported in detail in other reports [1-3].

With connection geometries determined from the static test parametric study, a series of cyclic tests was then conducted, with the test members subjected to constant amplitude and variable amplitude displacement excursions. The hysteretic energy absorption capabilities of the connections have been determined, and the mechanisms of failure (primarily low-cycle fatigue in the connection elements) identified. The results of the constant amplitude tests have provided the reference data for the application of low-cycle fatigue and of fracture mechanics concepts to damage accumulation models. Using these models, the lives of the specimens subjected to the variable amplitude loading sequences were predicted and compared with the test data.

I. Professor, Department of Civil Engineering, University of South Carolina, Columbia, SC.

II. Structural Engineer, Construction Technology Laboratories, Skokie, IL.

The results of crack propagation predictions using fracture mechanics principles are reported elsewhere [4]; cumulative damage estimates based on semi-empirical low-cycle analysis procedures are reported herein.

DESCRIPTION OF TEST PROGRAM

The test members consisted of a pair of beam sections attached to a centrally positioned stub column. The connection elements were comprised of top and seat angles bolted to the flanges of the beams and supporting column, together with double web angles bolted to the beam web and to the column flanges. The top and seat angles were the same for a given test, and the web angles were centered on the beam. ASTM A36 steel was used for the members and connection elements; all fasteners were 3/4-inch or 7/8-inch diameter ASTM A325 heavy hex, high strength bolts. A325 hardened washers were used under the turned elements.

Two beam sizes, W14X38 and W8X21, were used in the test program. For the 14-inch sections, the overall beam length was 20 feet, and for the 8-inch sections, 12 feet, so that the span-to-depth ratio was slightly less than 20 in each case. The stub column for the W14X38 beams was a W12X96 section; a W12X58 column section was used with the W8X21 beams. Heavy column sections were selected to minimize column panel zone deformation, thereby confining the hysteretic behavior to the connection elements. Complete details of the W14X38 and W8X21 connections are shown in Figs. 1 and 2, respectively.

A pair of duplicate specimens was tested simultaneously using the framing arrangements shown in Fig. 3. In all of the tests, the controlled input variables were the rate and magnitude of actuator displacement. For the cyclic tests, loading was imposed sinusoidally about a zero mean displacement using the general constant amplitude or variable amplitude block sequences illustrated in Fig. 4. Complete details of the test procedures, including specimen preparation, and types of instrumentation used, are reported in Ref. [1].

The static test program [1-3] demonstrated that the geometric parameters that most significantly affect the initial stiffness and monotonic moment-rotation performance of the semi-rigid connections investigated are: the depth of the beam sections to which the connections are framed, the thickness of the top and seat flange angles, and the gage (including the effect of bolt diameter) in the leg of the flange angle attached to the column. Consequently, these were the variables that were altered during the subsequent cyclic test program.

TEST RESULTS

Low-to-High Amplitude Cyclic Tests

Seven specimens were tested under low-to-high amplitude block displacement histories (Fig. 4a); the purpose of this testing sequence was to determine the characteristics of the hysteresis loops and to quantify the energy absorption capabilities of the connections at progressively increasing displacement amplitudes. Details of the specimen geometries are presented in Table 1; symbols used in the table are identified in Fig. 5.

An initial displacement amplitude (one-half the total range of displacement, about a zero mean) of 0.2 inches was selected, which was intended to produce a hysteresis loop with minimal non-linear response. Testing proceeded by applying a few (1-5) individual cycles at a frequency of 0.1 or 0.25 Hz, followed by an additional 10 continuous cycles at 0.25 Hz. After each block loading sequence, the displacement amplitude was increased by 0.2 inches, and the procedure repeated.

Stable hysteresis loops were established, for the 14-inch beam specimens, within a few cycles after an increase in amplitude was imposed relative to the preceding displacement under the block-type loading. For the 8-inch beam, a continual, though small, softening (loss of moment) was noted for each progressive cycle at a constant displacement amplitude; however, succeeding hysteresis loops were otherwise similar in appearance.

For each of the specimens, the moment-rotation (M-ϕ) behavior was characterized by hysteresis loops of continually decreasing slope for relatively small displacements in the non-linear range. In contrast, the loops exhibited a moderate "pinching" effect at larger amplitudes, the degree of pinching being more pronounced in the W14X38 beam connections than in the W8X21 members. An example of the hysteresis loop traces at each displacement amplitude for a typical test specimen is shown in Fig. 6.

Each of the tests culminated in the formation and subsequent propagation of fatigue cracks at the toe of fillet in one or more of the beam flange angles. Generally, two distinct cracks would initiate in the horizontal leg of a flange angle in the regions between the bolts in the leg of the angle attached to the column, and the corresponding bolts in the leg attached to the beam flange. These cracks would then propagate both through the thickness and along the width of the flange angle. The tests were terminated when cracking had progressed across all or most of the surface of a least one flange angle; no tests were extended to the point of rupture of a connection element. No slip was observed during the tests, nor was there any local buckling of the connection elements.

In general, for the low-to-high amplitude block tests, it was found that, with the exception of the first cycle following an increase in displacement amplitude, the hysteretic energy absorbed per cycle remained reasonably constant at each amplitude. Further, the ductile behavior of the connections was demonstrated by the increase in hysteresis loop area with each succeeding increase in displacement amplitude, even with pinching evident at the larger amplitudes. The average hysteresis loop area at each displacement amplitude is presented in Tables 2 and 3 for the 14-inch and 8-inch beam specimens, respectively.

High-to-Low Amplitude Cyclic Tests

Two specimens, 14B1 and 8B1, were tested under high-to-low amplitude displacement histories (Fig. 4c); details of the specimens are presented in Table 1. The high-to-low amplitude cyclic tests were intended to follow, essentially, the reverse time-displacement histories of those used in the low-to-high amplitude tests for specimens of duplicate geometry. At each amplitude, 2 individual cycles were applied at 0.25 Hz, followed by an additional 10 continuous cycles, also at the 0.25 Hz frequency. The displacement amplitude was then decreased in 0.2-inch increments, repeating the sequence of 2 initial cycles followed by 10 additional cycles. For specimen 8B1, which had not failed at the time cycling was completed at the minimum amplitude of 0.2 inches, an attempt was made to repeat the entire loading sequence; failure (pronounced crack extension) occurred during the next cycle at the 1.2-inch amplitude, and testing was concluded.

The test results for specimens 14B1 and 8B1, including loading histories and average hysteresis loop areas, are reported in Table 4. Representative hysteresis loops at each displacement amplitude for 8B1 are illustrated in Fig. 7. Although the appearances of the hysteresis loops are similar to those for the specimens tested under the low-to-high amplitude block loadings, the fatigue crack propagation behavior was notably different. At low-to-high amplitudes, once cracking had begun, the crack growth rate would increase with each succeeding increase in displacement amplitude. In the high-to-low amplitude tests, however, cracking initiated during the first block of cycles at the largest displacement amplitude, but in some instances exhibited retardation (i.e., temporary cessation of growth) as the displacement amplitudes were progressively decreased. These differences in crack growth rate between the low-to-high and the high-to-low amplitude tests serve to illustrate the dependence of cyclic performance, under random or earthquake-type loadings, on sequencing history as well as amplitude of load or displacement.

Constant Amplitude Cyclic Tests

As a consequence of the fatigue failures exhibited by the connections tested under the variable amplitude cyclic loadings, a series of constant amplitude cyclic tests (Fig. 4b) was conducted to establish bench-mark low cycle fatigue life relationships for application to damage accumulation models. Sixteen specimens were tested in this phase of the investigation, the details of which are presented in Table 1.

With the exception of several specimens that were initially subjected to a number of half cycles, the constant amplitude tests were conducted using full reversal of controlled displacement at a cyclic frequency of 0.25 Hz. The displacement amplitudes chosen resulted in fatigue lives ranging from nine to approximately 3500 cycles to "failure" (defined as the number of cycles at which the longest fatigue crack had extended over approximately three-fourths of the width of the flange angle). A summary of the test results, including average hysteresis loop areas and corresponding fatigue lives, is presented in Table 5.

As with the variable amplitude tests, the specimens tested in the constant displacement amplitude series exhibited fatigue cracking that initiated at the toe of the fillet in one or more of the beam flange angles. The fatigue cracks were generally first detected as a series of fine, hairline cracks aligned with slight irregularities in the surface of the flange angle at the toe of the fillet. With continued cycling, these individual cracks would eventually coalesce into a single crack, which then propagated more rapidly across the surface of the angle. However, the hysteresis loops remained quite stable throughout each test, with only modest loss of maximum moment evident from the time fatigue cracking was noticed to the termination of a test, when the cracks had progressed to some depth through

the thickness of the flange angle. Individual hysteresis loops at selected
cycles are shown in Fig. 8 for a representative specimen in the constant
amplitude test series.

ANALYSIS OF TEST RESULTS

Analysis of Constant Amplitude Test Data

Regression analysis of each of the four test sets in the constant
amplitude series (two thicknesses of flange angle framed to both W14X38 and
W8X21 beam sections) demonstrated that linear log-log relationships could
be expressed between the cyclic range of rotation of the connection, 2ϕ, and
the resultant total fatigue life, N_f. In order to develop a single expression
capable of predicting constant amplitude fatigue lives for specimens of varying
geometry, a non-dimensionalized nominal flange angle chord rotation index, R,
was calculated for each of the cyclic test specimens:

$$R = 2\left[\frac{(d+t)\tan\phi}{g - \frac{d_w}{2} - t}\right] \quad (1)$$

where the geometric parameters d, t, g, and d_w are as shown in Fig. 5. This
index is representative, proportionally, of the deformation of the tension
flange angle at a particular displacement amplitude (see Fig. 9). The chord
rotation index is reported in Tables 2, 3, 4, and 5 for each of the test
specimens at the various displacement amplitudes.

Figure 10 shows a plot of R versus the number of cycles to failure, on
a log-log scale, for all of the constant amplitude cyclic test data. From
a linear least-squares fit of the data, the following relationship was
obtained:

$$N_f = 1.868(R)^{-3.2531} \quad (2)$$

This equation was then used, as described subsequently, to predict
cumulative damage in the specimens tested under variable amplitude loadings.

At large displacement amplitudes, plastic strain in the connection
elements is the predominant means of energy dissipation in the semi-rigid
connections. In this study, the amount of dissipated energy is approximated
as the area of the hysteresis loop under the cyclic moment-rotation curve.
This average energy per cycle (i.e., area of an individual hysteresis loop)
is reported in Tables 2, 3, 4, and 5 for all of the test specimens. Since
the hysteretic behavior of the connections remained nearly constant throughout
the constant amplitude tests, an "average" energy per cycle was taken as the
area of the hysteresis loop at approximately mid-life for those specimens.

The cyclic hysteretic energy was compared to the number of cycles to
failure for each of the constant amplitude test specimens. As shown in
Fig. 11, a linear log-log expression was found to offer a reasonably good
relationship between energy per cycle and total fatigue life, for each of
the beam sizes individually:

$$N_f = 844.9(E)^{-1.20} \qquad \text{(W14X38 beam section)} \qquad (3)$$

$$N_f = 298.65(E)^{-1.2639} \qquad \text{(W8X21 beam section)} \qquad (4)$$

The parameter, E, may be used as an alternative to the chord rotation index, R, to predict damage accumulation for variable amplitude test specimens, as discussed below.

Prediction of Cumulative Damage in Variable Amplitude Tests

To predict cumulative damage in the specimens tested under the low-to-high and high-to-low amplitude cyclic loadings, the chord rotation index, R, was first calculated for each test displacement. Equation (2) was then used to estimate, individually, the total number of cycles to failure at each displacement amplitude. Using Miner's linear cumulative damage model [5], the summation n_i/N_{f_i} was then calculated, where, n_i represents the number of applied cycles at displacement amplitude, i, and N_{f_i} the predicted constant amplitude fatigue life at that amplitude. As reported in Table 6, cumulative damage summations ranging from 0.428 to 1.17 were obtained for the nine variable amplitude test specimens. (A summation of 1.0 represents perfect correlation between failure predicted by the damage accumulation model and actual specimen test behavior).

Cumulative damage in the variable amplitude cyclic tests was also predicted using energy dissipation. The average energy per cycle, E, measured at each displacement amplitude was used with Equation (3) or (4), as appropriate, to estimate the total fatigue life at that displacement amplitude. Again, Miner's Rule was applied to predict accumulated damage in the test specimens. Summations ranging from 0.586 to 1.28 were obtained, as seen in Table 6. Specimen 14C4 ($\Sigma = 0.586$) appears to be an anomaly in this test series, as its total accumulated cyclic energy at failure (i.e., summation of all hysteresis loop areas) was only about one-half that achieved by the other 14-inch beam specimens. Further discussion of this comparative behavior is presented in detail in Ref. [6].

Comparison of the results in Table 6 indicates that somewhat better predictions of cumulative damage are obtained when cyclic energy, E, rather than the chord rotation index, R, is used to estimate constant amplitude fatigue life. The differences between calculated damage summations and the test results may be explained in part by the inability of a simple linear summation index based on total fatigue life to satisfactorily model the complex behavior of the connections. No distinction is made between the crack initiation and propagation phases of fatigue in this model, nor does it account for the effect of immediate prior history (e.g., crack retardation or acceleration), on the rate of damage accumulation.

CONCLUSIONS

The variable amplitude, block-type cyclic tests have demonstrated the stable behavior of the semi-rigid connections studied, and have enabled quantification of their cyclic energy absorption capacities under large inelastic deformations. The tests have shown, also, that unless major connection slip or general frame instability intervenes, the effectiveness of the bolted connections will be limited by low-cycle fatigue under multiple excursions of displacement typical of those encountered in seismic loading.

Reasonably good correlations were observed, between the number of cycles to failure exhibited by the specimens tested at constant displacement amplitudes, and fatigue lives predicted by either a nominal flange angle chord rotation index, R, or by the average energy per cycle, E. Estimates of variable amplitude cyclic behavior based on Miner's linear damage accumulation model, and using fatigue lives predicted by the parameter, R, were found to be less than satisfactory for the large inelastic connection deformations considered in the investigation. Damage summations based on cyclic energy dissipation, E, appear to hold greater promise for predicting cumulative damage under conditions of low-cycle fatigue.

ACKNOWLEDGEMENT

The support provided for this investigation by the Earthquake Hazard Mitigation Program of the National Science Foundation, under Grant No. CEE-8115014, is gratefully acknowledged.

REFERENCES

[1] Altman, W. G., Jr., Azizinamini, A., Bradburn, J. H., and Radziminski, J. B., "Moment-Rotation Characteristics of Semi-Rigid Steel Beam-Column Connections," Structural Research Studies, Department of Civil Engineering, University of South Carolina, Columbia, S.C., June 1982.

[2] Radziminski, J. B., and Bradburn, J. H., "Experimental Investigation of Semi-Rigid Steel Beam-Column Connections," Proceedings, Eighth World Conference on Earthquake Engineering, San Francisco, CA, July 21-28, 1984.

[3] Azizinamini, A., Bradburn, J. H., and Radziminski, J. B., "Static and Cyclic Behavior of Semi-Rigid Steel Beam-Column Connections," Structural Research Studies, Department of Civil Engineering, University of South Carolina, Columbia, S.C., March 1985.

[4] Azizinamini, A., and Radziminski, J. B., "Cumulative Damage in Semi-Rigid Steel Beam-to-Column Connections," Proceedings, Southeastern Conference on Theoretical and Applied Mechanics, SECTAM XIII, Columbia, S.C., April 17-18, 1986.

[5] Miner, M. A., "Cumulative Damage in Fatigue," Transactions, ASME, Vol. 12, No. 3, September 1945.

[6] Azizinamini, A., "Cyclic Characteristics of Bolted Semi-Rigid Steel Beam to Column Connections," Ph.D. Dissertation, University of South Carolina, Columbia, S.C., May 1985.

TABLE 1. SCHEDULE OF CYCLIC TEST SPECIMENS

Specimen Number	Type of Test*	Beam Section	Bolt Diameter (inches)	Top and Bottom Flange Angles**				Web Angles**	
				Angle	Length, "L" (inches)	Gage in Leg on Column Flange, "g" (inches)	Bolt Spacing in Leg on Column Flange, "p" (inches)	Angle	Length, "L_c" (inches)
14C1	LH	W14x38	3/4	L6x4x3/8	8	2 1/2	5 1/2	2L4x3 1/2x1/4	8 1/2
14C2	LH	W14x38	3/4	L6x4 1/2	8	2 1/2	5 1/2	2L4x3 1/2x1/4	8 1/2
14C3	LH	W14x38	7/8	L6x4x1/2	8	2 1/2	5 1/2	2L4x3 1/2x1/4	8 1/2
14C4	LH	W14x38	7/8	L6x4x3/8	8	2 1/2	5 1/2	2L4x3 1/2x1/4	8 1/2
8C1	LH	W8x21	3/4	L6x3 1/2x5/16	6	2	3 1/2	2L4x3 1/2x1/4	5 1/2
8C2	LH	W8x21	3/4	L6x3 1/2x3/8	6	2	3 1/2	2L4x3 1/2x1/4	5 1/2
8C3	LH	W8x21	7/8	L6x3 1/2x3/8	6	2	3 1/2	2L4x3 1/2x1/4	5 1/2
14F1	CF	W14x38	7/8	L6x4x3/8	8	2 1/2	5 1/2	2L4x3 1/2x1/4	8 1/2
14F2	CF	W14x38	7/8	L6x4x3/8	8	2 1/2	5 1/2	2L4x3 1/2x1/4	8 1/2
14F3	CF	W14x38	7/8	L6x4x3/8	8	2 1/2	5 1/2	2L4x3 1/2x1/4	8 1/2
14F4	CF	W14x38	7/8	L6x4x3/8	8	2 1/2	5 1/2	2L4x3 1/2x1/4	8 1/2
14F5	CF	W14x38	7/8	L6x4x1/2	8	2 1/2	5 1/2	2L4x3 1/2x1/4	8 1/2
14F6	CF	W14x38	7/8	L6x4x1/2	8	2 1/2	5 1/2	2L4x3 1/2x1/4	8 1/2
14F7	CF	W14x38	7/8	L6x4x1/2	8	2 1/2	5 1/2	2L4x3 1/2x1/4	8 1/2
14F8	CF	W14x38	7/8	l6x4x1/2	8	2 1/2	5 1/2	2L4x3 1/2x1/4	8 1/2
14F9	CF	W14x38	7/8	L6x4x3/8	8	2 1/2	5 1/2	2L4x3 1/2x1/4	8 1/2
8F1	CF	W8x21	7/8	L6x3 1/2x3/8	6	2	3 1/2	2L4x3 1/2x1/4	5 1/2
8F2	CF	W8x21	7/8	L6x3 1/2x3/8	6	2	3 1/2	2L4x3 1/2x1/4	5 1/2
8F3	CF	W8x21	7/8	L6x3 1/2x3/8	6	2	3 1/2	2L4x3 1/2x1/4	5 1/2
8F4	CF	W8x21	7/8	L6x3 1/2x3/8	6	2	3 1/2	2L4x3 1/2x1/4	5 1/2
8F6	CF	W8x21	7/8	L6x3 1/2x5/16	6	2	3 1/2	2L4x3 1/2x1/4	5 1/2
8F7	CF	W8x21	7/8	L6x3 1/2x5/16	6	2	3 1/2	2L4x3 1/2x1/4	5 1/2
8F8	CF	W8x21	7/8	L6x3 1/2x5/16	6	2	3 1/2	2L4x3 1/2x1/4	5 1/2
8B1	HL	W8x21	7/8	L6x3 1/2x3/8	6	2	3 1/2	2L4x3 1/2x1/4	5 1/2
14B1	HL	W14x38	7/8	L6x4x1/2	8	2 1/2	5 1/2	2L4x3 1/2x1/4	8 1/2

*LH: Low to high amplitude block loading, Fig. 4a
CF: Constant amplitude loading, Fig. 4b
HL: High to low amplitude block loading, Fig. 4c
**See Nomenclature, Fig. 5

TABLE 2. SUMMARY OF LOW-TO-HIGH AMPLITUDE CYCLIC TESTS W14x38 BEAM SPECIMENS

Actuator Displacement Amplitude (inches)	Specimen 14C1			Specimen 14C2			Specimen 14C3			Specimen 14C4		
	Number of Applied Cycles	Nominal Chord Rotation Index, R	Average Hysteresis Loop Area, E (k.-in.)	Number of Applied Cycles	Nominal Chord Rotation Index, R	Average Hysteresis Loop Area, E (k.-in.)	Number of Applied Cycles	Nominal Chord Rotation Index, R	Average Hysteresis Loop Area, E (k.-in.)	Number of Applied Cycles	Nominal Chord Rotation Index, R	Average Hysteresis Loop Area, E (k.-in.)
0.2	18	0.037	0	13	0.040	0	12	0.046	0.032	12	0.041	0.020
0.4	15	0.073	0.091	13	0.081	0.203*	14	0.091	0.092	12	0.081	0.103
0.6	15	0.110	0.324	13	0.121	0.345	12	0.137	0.570	12	0.121	0.405
0.8	15	0.146	0.891	15	0.162	0.950	12	0.182	1.637	12	0.162	1.026
1.0	15	0.183	1.848	13	0.202	2.27	12	0.228	3.48	12	0.202	2.09
1.2	15	0.219	3.26	13	0.243	4.08	12	0.273	6.30	12	0.243	3.64
1.4	15	0.256	5.05	17**	0.283	6.22	13	0.319	8.95	12	0.283	5.65
1.6	16	0.292	6.58	14	0.324	8.58	12	0.364	12.02	12	0.324	7.99
1.8	15	0.329	8.54	13	0.364	10.90	14	0.410	14.97	9	0.364	10.44
2.0	11	0.365	10.47	4	0.405	13.79	4	0.455	16.93			

*Data questionable
**For Specimen 14C2, two additional cycles were applied at 0.5 in. displacement amplitude; avg. hysteresis loop area: 0.343 k.-in.

TABLE 3. SUMMARY OF LOW-TO-HIGH AMPLITUDE CYCLIC TESTS
W8x21 BEAM SPECIMENS

Actuator Displacement Amplitude (inches)	Specimen 8C1			Specimen 8C2			Specimen 8C3		
	Number of Applied Cycles	Nominal Chord Rotation Index, R	Average Hysteresis Loop Area, E (kip-inches)	Number of Applied Cycles	Nominal Chord Rotation Index, R	Average Hysteresis Loop Area, E (kip-inches)	Number of Applied Cycles	Nominal Chord Rotation Index, R	Average Hysteresis Loop Area, E (kip-inches)
0.2	11	0.055	0.038	11	0.059	0.294	12	0.070	0.020
0.4	12	0.109	0.187	12	0.118	0.275	12	0.140	0.245
0.6	13	0.164	0.809	12	0.177	1.280	12	0.210	1.050
0.8	14	0.219	2.36	13	0.236	2.93	12	0.280	2.67
1.0	14	0.273	4.27	12	0.294	5.00	12	0.350	5.04
1.2	12	0.328	6.09	12	0.353	7.10	10	0.420	7.84
1.4	2	0.382	7.78	4	0.412	9.56			

TABLE 4. SUMMARY OF HIGH-TO-LOW AMPLITUDE CYCLIC TESTS
W14x38 AND W8x21 BEAM SPECIMENS

Actuator Displacement Amplitude (inches)	Specimen 14B1			Specimen 8B1		
	Number of Applied Cycles	Nominal Chord Rotation Index, R	Average Hysteresis Loop Area, E (kip-inches)	Number of Applied Cycles	Nominal Chord Rotation Index, R	Average Hysteresis Loop Area, E (kip-inches)
1.8	12	0.410	16.08			
1.6	12	0.364	10.71			
1.4	12	0.319	7.64			
1.2	12	0.273	5.16	12	0.420	8.23
1.0	12	0.228	3.24	12	0.350	4.96
0.8				12	0.280	2.69
0.6				12	0.210	1.196
0.4				12	0.140	0.347
0.2				12	0.070	0.029
1.2				1	0.420	6.62

TABLE 5. SUMMARY OF CONSTANT AMPLITUDE CYCLIC TESTS

Specimen Number	Thickness of Flange Angle (inches)	Actuator Displacement Amplitude (inches)	Nominal Range of Rotation, $2\phi^*$ (radians x 1000)	Nominal Flange Angle Chord Rotation Index, R	Average Hysteresis Loop Area, E (kip-inches)	Fatigue Life (cycles)
14F3	3/8	2.8	49.1	0.569	25.5	9
14F2	3/8	2.0	35.1	0.407	13.2	58
14F1	3/8	1.6	28.1	0.325	8.1	72
14F9	3/8	1.0	17.5	0.203	2.4	230
14F4	3/8	0.5	8.8	0.102	0.22	3450
14F7	1/2	2.5	43.9	0.570	26.1	26
14F5	1/2	1.5	26.3	0.341	8.7	59
14F6	1/2	1.0	17.5	0.228	3.0	316
14F8	1/2	0.7	12.3	0.160	0.90	1031
8F8	5/16	1.5	45.5	0.481	9.8	16
8F7	5/16	1.0	30.3	0.320	3.8	62
8F6	5/16	0.7	21.2	0.224	1.4 est.[†]	173
8F1	3/8	1.5	45.5	0.525	13.1	10
8F4	3/8	1.0	30.3	0.350	4.9	56
8F3	3/8	0.7	21.2	0.245	1.8	147
8F2	3/8	0.5	15.2	0.175	0.59	560

*Twice actuator displacement amplitude divided by distance from support to column face
[†]Malfunction in data recording system until cycle 170

TABLE 6. PREDICTION OF CUMULATIVE DAMAGE IN VARIABLE AMPLITUDE TEST SPECIMENS

Specimen Number	Thickness of Flange Angle (inches)	Testing Sequence	Predicted Cumulative Damage Index $\Sigma n_i/N_{f_i}$	
			Based on Chord Rotation Index, R	Based on Average Cyclic Hysteretic Energy, E
14C1	3/8	Low-to-High Amplitude	0.852	0.907
14C2	1/2	Low-to-High Amplitude	0.891	0.930
14C3	1/2	Low-to-High Amplitude	1.17	1.28
14C4	3/8	Low-to-High Amplitude	0.585	0.586
14B1	1/2	High-to-Low Amplitude	0.894	0.968
8C1	5/16	Low-to-High Amplitude	0.428	0.961
8C2	3/8	Low-to-High Amplitude	0.584	1.26
8C3	3/8	Low-to-High Amplitude	0.683	0.956
8B1	3/8	High-to-Low Amplitude	0.778	1.13

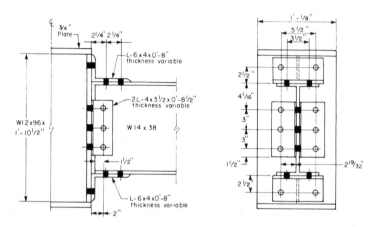

FIG. 1 DETAILS OF CONNECTION FOR W14x38 BEAM

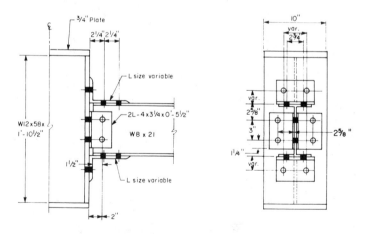

FIG. 2 DETAILS OF CONNECTION FOR W8x21 BEAM

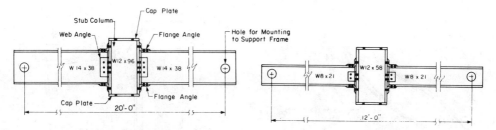

FIG. 3 GENERAL CONFIGURATIONS OF TEST SPECIMENS

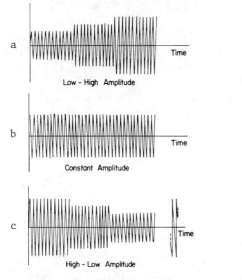

FIG. 4 TYPICAL TIME-DISPLACEMENT HISTORIES

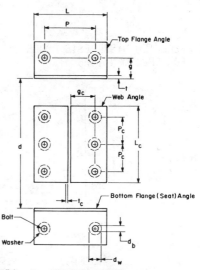

FIG. 5 NOMENCLATURE FOR CONNECTION ELEMENTS

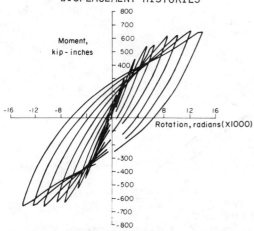

FIG. 6 STABLE HYSTERESIS LOOPS FOR 14C4

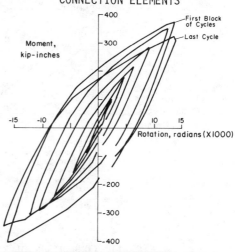

FIG. 7 STABLE HYSTERESIS LOOPS FOR 8B1

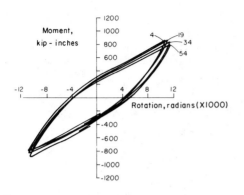

FIG. 8 SELECTED HYSTERESIS LOOPS FOR SPECIMEN 14F5

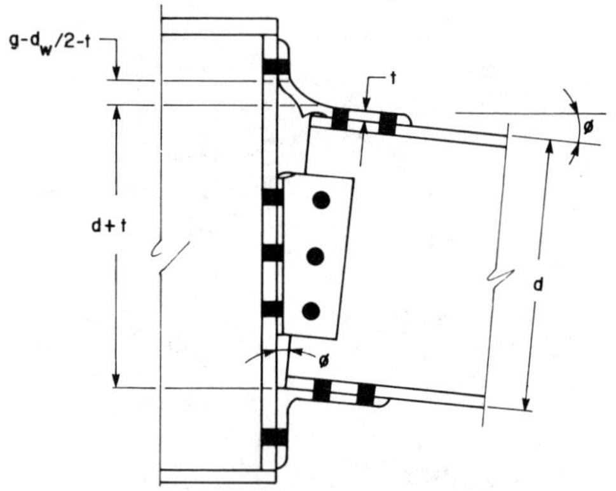

FIG. 9 DEFLECTED SHAPE OF CONNECTION AT MAXIMUM DISPLACEMENT

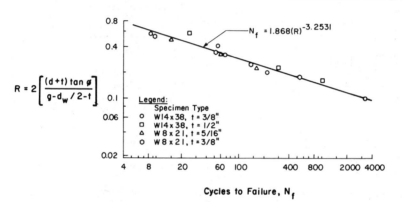

FIG. 10 COMPARISON OF TOTAL FATIGUE LIFE WITH NOMINAL FLANGE ANGLE CHORD ROTATION

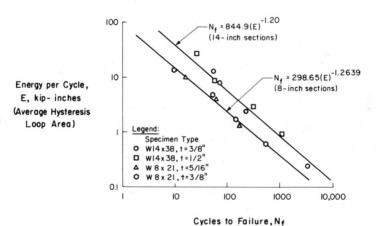

FIG. 11 COMPARISON OF TOTAL FATIGUE LIFE WITH AVERAGE AREA OF INDIVIDUAL HYSTERESIS LOOP

EARTHQUAKE RESISTANCE AND BEHAVIOR OF
WOOD-FRAMED BUILDING PARTITIONS

Satwant S. Rihal[I]

ABSTRACT

This paper presents the results of an experimental investigation into the dynamic behavior of building partitions during earthquakes. Cyclic in-plane racking tests of full-height (eight feet wide and approximately ten feet high) wood-stud framed building partitions, with representative facing panels (e.g., gypsum wallboards and plywood), fasteners and other details, were carried out to investigate their dynamic behavior, earthquake resistance and thresholds of damage. Current practices for design, detailing and installation of building partitions as well as applicable provision of the UBC have been evaluated. One conclusion is that the earthquake resistance of wood-stud framed partitions in certain classes of buildings can be relied upon, provided sound detailing practices (e.g., use of holdowns or equivalent devices and fastener layouts) are implemented. Test results show that the frequency of input motions definitely affects partition dynamic behavior and damage; and that there is a need to further investigate the provided factors of safety under severe dynamic actions, as compared to those implied in seismic design provisions of building codes which are based primarily on static test results.

INTRODUCTION

It is now widely recognized that building partitions in a broad class of building types, not only constitute a significant percentage of the overall building cost but also influence the overall behavior and performance of buildings during earthquakes. In new and existing residential buildings as well as in the large stock of existing unreinforced masonry buildings, there are significant number of wood-framed partitions of different layouts, the effects of which on the overall earthquake resistance and behavior of such buildings need to be investigated. A study of damage data from the Coalinga, California earthquake of May 2, 1983 [3],[14],[15] has shown that in many buildings, interior building partitions seemed to be intact while the rest of the low-rise masonry building structure suffered extensive damage. Furthermore it appears there is evidence that a well-designed and constructed wood-framed low-rise building with gypboard as facing panels can survive a destructive earthquake e.g., Coalinga, 1983 with only minor damage [3],[4]. It is also necessary to evaluate the effectiveness of the design provisions of the regulatory standards e.g., Uniform Building Code, through correlation with recent dynamic test results, as many of the current seismic

---

I. Professor, Architectural Engineering Department, California Polytechnic State University, San Luis Obispo, California.

design provisions applicable to building partitions are based on static test results carried out in the late 1950's.

SCOPE AND OBJECTIVES OF THE TESTING PROGRAM

The objectives of the cyclic racking tests program are to investigate the following:

o The behavior of full-height wood-framed building partitions with and without holdowns.

o The behavior and contribution of gypsum wallboard facing panels to the earthquake resistance of full-height wood-framed building partitions.

o The behavior and earthquake resistance of full-height wood-framed building partitions with plywood on one side and gypsum wallboard on the opposite side.

REVIEW OF CURRENT DESIGN AND CONSTRUCTION PRACTICES

Building partition assemblies currently in use are dictated by building type based on occupancy (e.g., commercial, residential, institutional) and governing regulations for fire resistance and separation as well as applicable acoustic criteria and economics.

Wood-Stud Framed Partitions with Gypsum-Wall Board, Plywood or Stucco as Facing Materials

These construction assemblies are typically found in residential construction and possibly in some buildings of other occupancies. Such partition components typically consist of 2 x 4 wood studs spaced every 16 inches or 24 inches, positioned between top and bottom wood-plates fastened to the main structure and faced with different types of facing panels as follows:

Interior Partitions: Gypboard facing panels applied vertically or horizontally on both sides.

Exterior Walls/Partitions: . Gypsum facing panels applied vertically or horizontally on inside face and

. Plywood and/or stucco finish on the exterior face.

In the seismic design and detailing of wood-framed building systems, the designer assumes that the entire design lateral force caused by earthquake or wind will be resisted by a few strategically placed walls designated as shear walls. These wood-stud framed walls may have the following combination of finishing materials.

. Interior walls - gypsum wallboard panels on both sides or gypsum wallboard on one side and plywood panels on the opposite side.

. Exterior walls - gypsum wallboard panels on the inside face with plywood panels with or without stucco finish on the exterior face.

Furthermore, these walls are provided with holdown devices (if necessary) to resist the overturning actions of earthquake motions [1],[4], [5],[9],[16],[18].

In these types of buildings, there are a large number of other non-structural walls or partitions that the designer assumes, do not participate in the overall system of earthquake resistance.

Because of the nature of existing construction practices for wood-framed buildings, it is likely that under certain conditions, many of the non-structural walls or partitions will participate in the overall earthquake resistance of such buildings. Therefore it was decided to explore the possible contribution of wood-framed building partition walls, to the overall earthquake resistance of buildings.

DYNAMIC TESTING PROGRAM

Description of Test Specimens

A detailed description of the full-height partition test specimens is presented elsewhere [15].

All partitions are wood-stud framed and are 8 feet wide and approximately 10 feet high. Installation of building partitions is done according to accepted current practices.

The partition dynamic test specimens may be categorized as follows:

Specimen No. PD3-V Run No. 1

Wood-stud framed partition with 1/2 inch gypboard panels applied vertically on both sides without any holdowns, nail spacing as per UBC Table 47-I.

Specimen No. PD3-V Run No. 2

Wood-stud framed partition with 1/2 inch gypboard panels applied vertically on both sides with holdowns, nail spacing as per UBC Table 47-I.

Specimen No. PD3-H

Wood-stud framed partitions with 1/2 inch gypsum panels applied horizontally on both sides with holdowns, nail spacing as per UBC Table 47-I.

Specimen No. PD-4 Run No. 1/Run No. 2

Wood-stud framed partitions with 1/2 inch gypboard panels applied on the inside face and 3/8 inch plywood panels on the outside face with holdowns, nail spacing as per UBC Table 47-I and Table 25-K.

Dynamic Testing Method

The overall test set-up including location and orientation of measurement transducers is shown in Figure 1. Complete details of the test set-up, equipment and instrumentation are presented elsewhere [14],[15].
The dynamic testing scheme basically consists of a steel-framed grid simulating a horizontal floor diaphragm that is free to roll on a wheel/bearing assembly. The full-height partitions are attached to the steel grid at the top and also fastened at the bottom to a precast concrete based bolted to the laboratory floor. The partition test specimens are subjected to cyclic racking motions at the center line of the loading grid using an MTS electro-hydraulic closed-loop system. The input excitation is sinusoidal and full-height partition specimens are subjected to cyclic displacements at controlled magnitudes and frequencies.

Test Equipment and Instrumentation

Data acquisition of dynamic test control and specimen responses are provided by transducers e.g., load-cell, LVDT's, strain gages, accelerometer. A detailed description of the six measurement transducers for all partition test specimens is presented in an earlier report [15].
Each transducer output was conditioned by a pre-amplifier module in the Honeywell Visicorder Model 1858, which provided an almost immediate hard copy of each sensor's output.
From the Visicorder's buffered output drives, the signals on all seven channels are sent through a parallel-to-series multiplexer (MUX) to give three channels of test data and one timing signal for recording. The analog dynamic test data is recorded on an HP 3960A four channel, three-speed, instrumentation tape recorder using FM recording of signal from DC up to 5000 Hz at 15 ips. This provides a permanent record of all dynamic test data on one-quarter inch magnetic tape, for further processing.
The retrieval of any transducer's response signal is provided by playing back the magnetic tape through a series-to-parallel demultiplexer (DEMUX). Additional hard-copy recordings of a sensor's response were obtained by means of a strip chart recorder (e.g., B & K Graphic Recorder Model 2309).
A Block Diagram of Dynamic Testing Equipment and Instrumentation is shown in Figure 2.
Furthermore, a x-y recorder provided instantaneous hard copy plots of load-cell vs. LVDT mounted at the centerline of the loading grid for each test run.

Dynamic Testing Procedure

After consideration of testing procedures used by other investigators [6] and in accordance with the testing procedures developed earlier in this research project [14],[15] it was decided to subject the partition specimens to Block Cyclic Tests.
During each test run, frequency is fixed and specimens are subjected to several complete cycles of loading for each increasing level of peak command horizontal displacement starting with 1/8, 1/4, 3/8, 1/2, 3/4, 1, 1-1/4, 1-1/2 ---- inches.
During each test a log sheet was kept showing all pertinent details of the dynamic test, measurement transducers, calibration of measuring instruments and recording devices. Complete details of dynamic test control

parameters for all specimens tested to date are presented in a detailed previous report [15].

DYNAMIC TEST RESULTS

Observed Behavior and Partition Performance

The behavior of each partition specimen was observed and recorded during each test sequence of the Block Cyclic Tests. In addition photographic record was kept of the specimen response and performance during each test sequence. A detailed summary of observed partition damage level and corresponding motion parameters; as well as a photographic record of specimen performance on a specimen-by-specimen basis is presented in a previous report [15].

Peak Dynamic Responses

In the absence of any sophisticated signal processing equipment the available analog test data was manually analyzed using the time-histories of test data on Honeywell Visicorder rolls of paper. The peak responses of all transducer channels for each test run for each partition test specimen were documented. For every specimen tested, graphs are plotted between peak-load and peak-displacement of each block of cyclic motions.

In addition, plots of peak-command-displacement vs. measured peak-grid-displacement at top of the partition are presented for all test specimens.

Unique efforts to record behavior of typical holdown devices [18] under cyclic motions were successfully made. Peak values of measured holdown forces during each block cyclic test were successfully obtained for specimens PD-4 through the use of internally-gaged-threaded-studs made by Strainsert Corporation. Graphs between peak load-cell output and peak holdown forces were plotted.

From the cyclic load-displacement curves obtained for each block cyclic test for each specimen, an estimation of the modulus of rigidity was made as suggested by Freeman [6] as follows:

$$\text{Rigidity} = (\text{Load}/(\text{Length} \times \text{Thickness}))/(\text{Displacement}/\text{Height})$$

For all partition test specimens, graphs between the estimated rigidity and peak-command-displacement, for each block cyclic test are presented in the previous report [15].

Dynamic responses of the four partition test specimens were manually analyzed and a partial summary of test results is presented in Table 1. For each partition test specimen, the maximum peak lateral load reached and the corresponding peak measured horizontal displacement at top of the partition during the block cyclic tests are summarized. The associated frequency of the imposed blocks of cycles of loading is also shown in Table 1. The peak lateral shears defined as the peak lateral load divided by the width (8 feet) of the partitions, as well as the UBC allowable lateral design shears [21] are also presented in Table 1.

Summary of Test Results

A summary of peak measured holdown forces, corresponding peak loads reached and calculated holdown forces for partitions specimen PD-4 for each block cyclic test are presented in a previous report [15]. The calculated holdown forces were based on maximum peak load, dimensions of the partitions and principles of structural equilibrium.

For all partition test specimens the measured peak acceleration at center line of the loading grid, i.e., at top of the partitions varied between 0.006 g and 0.32 g. Measured peak horizontal displacement at top of partitions varied between ± 0.05 inch (Specimen No. PD3-V Run No. 2, Test No. A-1, Peak Command Displacement $\pm 1/8$ inch) and ± 1.55 inches (Specimen No. PD3-H, Test No. A-11, Peak Command Displacement ± 2.8 inches) for all the block cyclic tests with frequencies of 0.5 Hz, 0.7 Hz and 1.0 Hz.

Relationship between motion parameters, i.e., frequency and amplitude of displacements, forces, etc., and level of partition damage is systematically presented in a previous report [15]. Except for partition Specimen No. PD3-V Run No. 1, without any holdowns, initial partition damage appears to start at a peak horizontal displacement at top of the partition of ± 0.20-0.25 inch @ frequency of 0.5 Hz. Severe partition damage takes place at peak horizontal displacement at top of partition of ± 0.85-1.35 inches.

Figure 3 shows the result of an effort to plot the relationship between partition damage level and peak measured horizontal displacement @ top of partition, for all test specimens.

DISCUSSION OF TEST RESULTS AND CONCLUSIONS

Results of cyclic racking tests to-date have shown that the strength, behavior and performance of building partitions is influenced by motion parameters e.g., magnitude and frequency of block cyclic displacement levels, and number of cycles in each test sequence. The relationship between partition damage level and peak measured horizontal displacement @ top of partition, for all test specimens is presented graphically in Fig. 3. The damage levels shown may be divided into the following classifications: Low/Minor Damage - Level I-III; Moderate Damage - Level III-V; Severe Damage - Level V-VIII. Detailed descriptions of damage levels are documented in a previous report [15]. In all test specimens the damage is initiated @ ends of the partition and edges of facing panels e.g., crumbling of gypboard around nail-heads. Eventually through working of the nails under cyclic motions, the facing panels start slipping relative to the stud framing and become loose, at which point popping of nail heads in clearly visible. Severe damage consists of failure of taped joints in gypboard and further crumbling of gypboard around nail-heads. In the partition with 3/8 inch plywood on one face (Specimen PD-4), damage on the plywood face essentially consisted of slippage @ edges of plywood panels, damage of plywood @ nail-heads and eventual crushing of plywood @ bottom left corner of panel. Under large cyclic displacements, severe partition damage also consisted of tearing of edge studs and the ends of the bottom sill plate. It could be reasoned that a good part of this damage is repairable damage. Studies need to be carried out to investigate effectiveness of techniques of repairing earthquake damage to these building components. Comparison of test results for the specimens with and without holdowns, shows that without holdowns failure takes place very early in the initial few cycles of motion @ lower ends of the partitions. It can be clearly concluded that without holdowns or equivalent restraints, partition components cannot have dynamic stability under the severe motions that can be expected during earthquakes. Therefore earthquake resistance of wood-framed building partitions can be relied upon, provided dynamic stability is provided for, through the use of holdowns or equivalent devices. A study of test results shows that the peak load-level reached for every test is definitely related to the frequency of the input block cyclic displacements. Peak load-levels for test runs @ 0.5 Hz are significantly higher than those for test runs @ 0.7 Hz and 1.0 Hz. A study of the results shows that in general panel rigidity keeps on decreasing with increasing peak amplitudes and frequencies of blocks of cyclic displacements. Except for the specimen without any holdowns, the maximum peak lateral shears reached for all other specimens were greater than those allowed by the Uniform Building Code, and it is concluded that further studies should be carried out to investigate and compare the factors of safety for walls/partitions obtained through dynamic racking tests with those obtained through static racking tests. It should be noted that allowable lateral shears found in most regulatory codes are based on static racking tests only. It can be stated that for the first time the dynamic behavior of holdowns typically used in wood-framed buildings has been documented as one useful product of this pilot testing program.

ACKNOWLEDGEMENTS

This research is based upon work supported by the National Science Foundation under Grant No. CEE-81-17965. This support of the National Science Foundation is gratefully acknowledged. Any opinions, findings, conclusions or recommendations expressed in this paper are those of the author and do not necessarily reflect the views of the National Science Foundation. The author acknowledges the contribution and assistance of Dr. Gary Granneman of the Engineering Technology Department during the testing phase of this research project.

REFERENCES

1. Adams, N. R., Plywood Shear Walls, American Plywood Association, Tacoma, Washington, February 1976.

2. Anderson, R. W., Investigation of the Seismic Resistance of Interior Building Partitions, Phase I Report for National Science Foundation Contract No. PFR-8009921, Agbabian Associates, El Segundo, California, February 1981.

3. Coalinga, California Earthquake of May 2, 1983, Reconnaissance Report, Earthquake Engineering Research Institute, Berkeley, California 1984.

4. Drywall Proves Its Worth in Coalinga Quake, Interior Building Systems, Vol. 17, No. 2, Spring 1983.

5. Forest Products Laboratory, Contribution of Gypsum Wallboard to Racking Resistance of Light-Framed Walls, Prepared for HUD, Washington, D. C., September 1982.

6. Freeman, Sigmund A., Progress Reports of Racking Tests of Wall Panels, URS/J. A. Blume and Associates, Engineers, San Francisco, California (First, Second, Third and Fourth), 1966-1974.

7. McCutcheon, W. J., Racking Deformations in Wood Shear Walls, Journal of Structural Engineering, ASCE, Vol. III, No. 2, Feb. 1985, pp. 257-269.

8. Meehan, J. f., "Public School Buildings," *The San Fernando, California Earthquake of February 9, 1981*, National Oceanic and Atmospheric Administration, Washington, D. C., pp. 667-884, Vol. I, Part B, 1973.

9. Mendes, Stan, Personal Communication (1984, 1985).

10. The 1984 Morgan Hill, California Earthquake, California Division of Mines and Geology, Special Publication 68, Sacramento, California, 1984.

11. PARABOND Capsule Anchors, Molly Fastener Group, Temple, Pennsylvania, 1980.

12. Rawl Masonry Anchors, The Rawl Plug Company, Inc., New York, 1983.

13. Rihal, Satwant S., Racking Tests of Non-Structural Building Partitions, Final Technical Report, Submitted to the National Science Foundation (Grant No. PFR-78-23085); Architectural Engineering Department, Report ARCE R80-1, California Polytechnic State University, San Luis Obispo, California, December 1980.

14. Rihal, Satwant S. and Granneman, Gary, Experimental Investigation of the Dynamic Behavior of Building Partitions and Suspended Ceilings During Earthquakes, Interim Progress Report Submitted to the National Science Foundation (NSF Grant No. CEE 81-17965) Report No. ARCE R84-1. Architectural Engineering Department, California Polytechnic State University, San Luis Obispo, California, June 1984.

15. Rihal, Satwant S., and Granneman, Gary, Dynamic Testing of Wood-Framed Building Partitions, Final Technical Report, Submitted to the National Science Foundation (NSF Grant No. CEE 81-17965) Report No. ARCE R85-1, Architectural Engineering Department, California Polytechnic State University, San Luis Obispo, California, August 1985.

16. Shapiro, Okino, Hom and Associates, The Home Builder's Guide for Earthquake Design, Prepared for the Applied Technology Council, SEAOC, June 1980.

17. State of California Title 21 (Schools) and Title 24 (Hospitals), Sacramento, California.

18. Strong-Tie Connectors Catalog, Simpson Company, San Leandro, California, 1924.

19. Structural Engineers Association of California, Lateral Force Requirements and Commentary, 1980.

20. Tentative Provisions for the Development of Seismic Regulations for Buildings, ATC 3-06, Applied Technology Council, Structural Engineers Association of California, 1978.

21. Uniform Building Code, International Conference of Building Officials, Whittier, California, 1982.

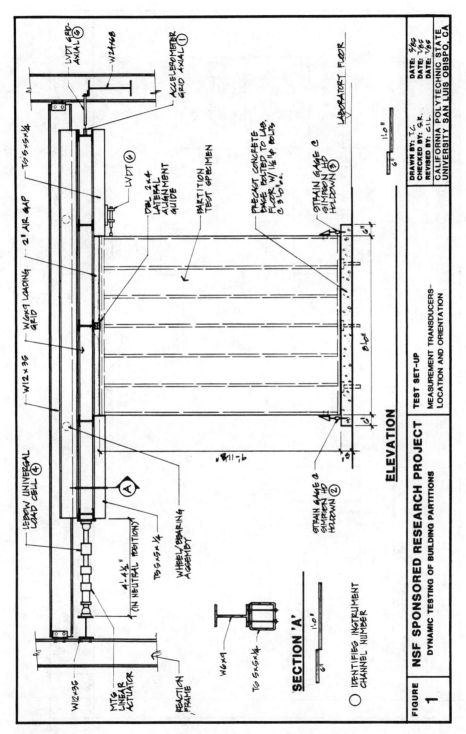

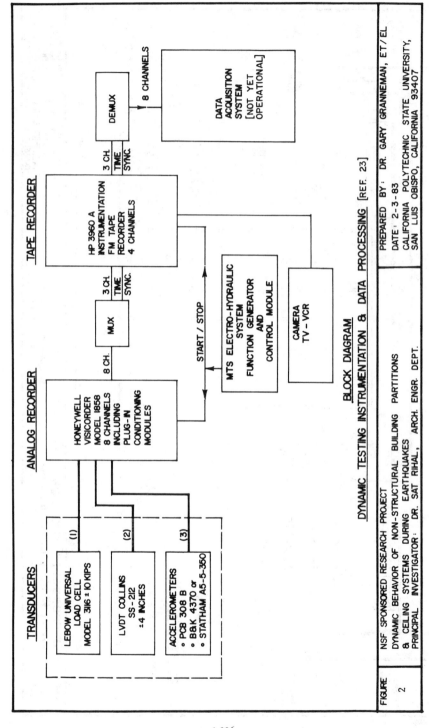

NSF SPONSORED RESEARCH PROJECT: DYNAMIC TESTING OF BUILDING PARTITIONS

TABLE 1: SUMMARY OF DYNAMIC TEST RESULTS

SPECIMEN NO.	DESCRIPTION All Specimens 8'-0" Wide X 10'-0" High	FREQUENCY (Hz)	PEAK LATERAL LOAD LBS.	PEAK LATERAL DISPLACEMENT INCHES	PEAK LATERAL SHEAR LBS./LIN. FT.	UBC ALLOWABLE DESIGN SHEAR LBS./LIN. FT.
PD3-V Run No. 1	2 X 4 Wood Studs at 16" O.C. 1/2" Gypsum Wall-Board Applied Vertically Each Side No Holdowns	0.5	±620	±0.55	77.5	200 (Table 47-I)
PD3-V Run No. 2	2 X 4 Wood Studs at 16" O.C. 1/2" Gypsum Wall-Board Applied Vertically Each Side Simpson HD-2 Holdowns at Base at Each End	0.5	±2200	±1.20	275.0	200 (Table 47-I)
PD3-H	2 X 4 Wood Studs at 16" O.C. 1/2" Gypsum Wall-Board Applied Horizontally Each Side Simpson HD-5 Holdowns at Base at Each End	0.5	±2200	±1.55	275.0	200 (Table 47-I)
PD-4 Run No. 1 Run No. 2	2 X 4 Wood Studs at 16" O.C. 1/2" Gypsum Wall-Board Applied Horizontally On Side and 3/8" Opposite Side Simpson HD-5 Holdowns at Base at Each End	0.5	±2380	±1.35	297.5	264 (Table 25-K)

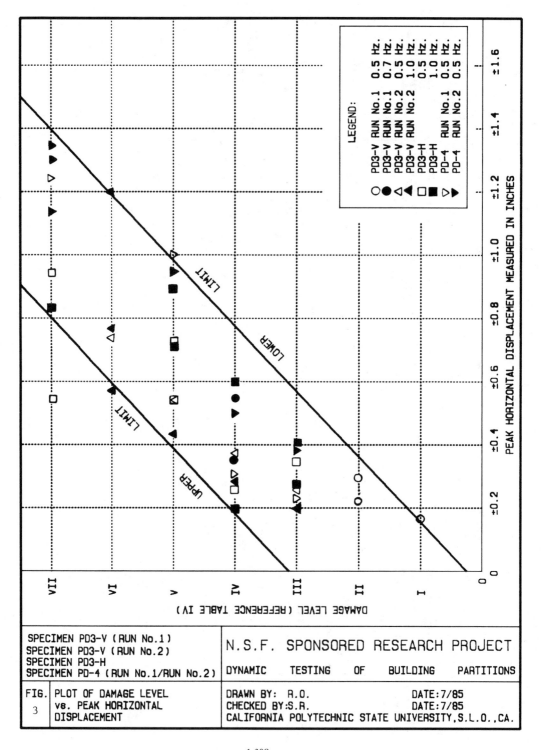

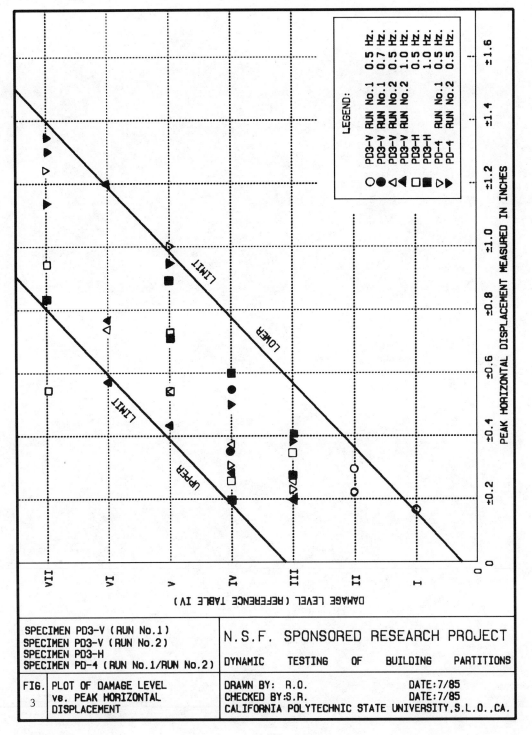

PSEUDO-DYNAMIC TESTING AND MODEL IDENTIFICATION

J. L. Beck[I] and P. Jayakumar[II]

ABSTRACT

The pseudo-dynamic test method is a recently developed procedure for subjecting full-scale structures in a laboratory to simulated earthquake response by means of on-line computer control of hydraulic actuators. Such pseudo-dynamic data from a full-scale six-story steel-frame structure tested in a special laboratory in Japan is being analyzed using a simple hysteretic model and a nonlinear system identification technique, with the ultimate objective of improving the ability to predict the dynamic response of steel structures during earthquakes. Both the time-histories of the responses and the hereditary nature of the force-deformation relationship of the structure are predicted very well by the hysteretic model employed in this study.

INTRODUCTION

A full-scale, six-story, two-bay, steel-frame structure with eccentric bracing (Fig. 1) was tested by the pseudo-dynamic method at the Building Research Institute, Tsukuba, Japan in 1984. This represented Phase II of the steel program under the U.S.-Japan Cooperative Earthquake Research Program Utilizing Large Scale Testing Facilities. All three frames of the structure in the direction of loading were moment resisting and, in addition, the north-bay of the interior frame was braced with eccentric K-bracing (Fig. 1). The floor system consisted of a formed metal decking with cast-in-place light-weight concrete, acting compositely with the girders and floor beams, as shown in Fig. 2. No nonstructural component was attached to the frame system.

The principal pseudo-dynamic tests consisted of one performed at low amplitudes to give a nominally elastic test, one at moderate amplitudes to excite the structure into the inelastic range, and three large amplitude tests using a sinusoidal ground acceleration pulse to explore the ultimate capacity of the structure. The uni-directional loading in the elastic and inelastic tests was produced by an early digitized version of the Taft S21W component from the 1952 Kern County, California earthquake, scaled to peak accelerations of 6.5%g and 50%g respectively.

The nominally elastic test data was analyzed by the authors using linear structural models and system identification techniques, primarily to examine the validity of the pseudo-dynamic test method [2,3,4]. Using single-input single-output and multiple-input multiple-output system identification methods, it was shown that the structural properties could be estimated reli-

---

I. Assistant Professor,   II. Graduate Student,
Earthquake Engineering Research Laboratory, California Institute of Technology, Pasadena, California 91125.

ably by isolating the feedback errors which are inherent in the experiment. The feedback errors act as excitations to the structure in addition to the prescribed earthquake ground motion and their effect on the apparent damping of the structure was estimated. The optimal estimates for the structural properties were in excellent agreement with those from a 0.3 scale model of the full-scale structure tested on a shaking-table facility at the University of California, Berkeley.

The dynamic response of the test structure during the inelastic test is currently being studied using a simple hysteretic model and a nonlinear system identification technique. This paper is a report of what the initial stages of the investigation have revealed.

PSEUDO-DYNAMIC TESTING

The pseudo-dynamic test method is a recently developed quasi-static procedure [5] for subjecting full-scale structures in a laboratory to simulated earthquake response by means of on-line computer control of hydraulic actuators. The inertial effects of the structure are modeled in an on-line computer, but in contrast to the usual quasi-static test procedures the relation between the interstory forces and deformations is not prescribed <u>a priori</u>. Instead, feedback from displacement and load transducers is used to force the appropriate earthquake behavior on the structure in an interactive manner as the experiment proceeds. Hence, full-scale structures can be tested at strong-motion amplitude levels while making no assumptions about the stiffness and damping characteristics of the structure. Furthermore, it is relatively inexpensive to test full-scale structures by the pseudo-dynamic method compared with the construction and operation of a big shaking table facility.

Description of the Test Procedure

The equation of motion of the multi-degree-of-freedom system shown in Fig. 3 is

$$M\ddot{x} + C\dot{x} + R(x) = F(t) = -M\ddot{z}(t)\mathbf{1} \tag{1}$$

where R is the restoring force and $F(t)$ is the excitation due to earthquake accelerations $\ddot{z}(t)$. The $C\dot{x}$ term represents the viscous damping added artificially by the on-line data processing computer during the tests. There is no rate-dependent damping from the structure itself, since the tests are carried out quasi-statically. The numerical integration of Eq. 1 using the central-difference method gives:

$$[M + C\Delta t/2]x_{i+1} = [F_i - R_i] + 2Mx_i + [C\Delta t/2 - M]x_{i-1} \tag{2}$$

where the subscript i denotes time, $t_i = i\Delta t$, $i = 0,1,\ldots,N$.

The mass matrix is prescribed from the known mass distribution of the test structure, so that the on-line computer can simulate its inertial effects, and the damping matrix is set equal to that derived from the preliminary free and forced vibration tests of the structure at low amplitudes. From the knowledge of restoring forces and displacements at the previous time steps, the displacement at any time step is calculated using Eq. 2 in the data processing computer which is part of the testing facility (Fig. 4). This displacement is first converted to a voltage change in the servo-controller which then regulates the flow of high pressure hydraulic fluid to the actuators. The actuators in turn force the structure to deflect to the required position. When the desired displacement is achieved, the load cells mounted on the actuators measure the restoring forces, and the displacement transducers on the structure measure the final displacements achieved. This information is fed back to the on-line data processing computer to calculate the displacement to be imposed at the next time step, using Eq. 2.

HYSTERETIC MODEL

During strong ground motions, structures dissipate energy through hysteresis and the modeling of this phenomenon has been a challenge to researchers. Most of the hysteretic models used are either simple and unrealistic, such as elasto-plastic and bilinear hysteretic models, or very complex, as in the case of Takeda's model for reinforced concrete [7]. In this paper, an attempt is made to predict the dynamic response of the multi-degree-of-freedom system in Fig. 1 by modeling its interstory force-deformation characteristics using a simple nonlinear hysteretic model. This model was first proposed by Bouc [8] and later generalized by Wen [9]. The model is further modified in this study to incorporate physical entities, such as the elastic stiffness and ultimate strength, as model parameters.

The interstory restoring force r is modeled by the following differential equation:

$$\dot{r} = K_0 \dot{x} \left[1 - \text{sgn}(\dot{x}) \, (r/r_u) \, |r/r_u|^{n-1} \right] \quad (3)$$

where $\quad x$ = interstory drift
sgn (signum function) = +1 (if $\dot{x} > 0$)
= −1 (if $\dot{x} < 0$)
= 0 (if $\dot{x} = 0$)

and the model parameters are

K_0 = small-amplitude "elastic" stiffness
r_u = ultimate strength (assumed here to be the same for $x > 0$ and $x < 0$)
n = an index (>0).

The effect of the model parameters on the resulting hysteresis loops can be seen in Fig. 5. The model represents an elastic system when r_u is very large and an elasto-plastic system when $n \to \infty$. Hence, it has the ability to model a wide range of structural response characteristics.

SYSTEM IDENTIFICATION TECHNIQUE

An output-error approach for system identification is used in conjunction with the modified Gauss-Newton method to determine optimal estimates of the parameters for the hysteretic model from experimental data.

Output-Error Method

If the outputs of the real system and the model, for the same input, are y and $\underset{\sim}{m}$ respectively, then the output-error $\underline{e}$ is given by:

$$\underline{e}(\underline{\theta}) = \underset{\sim}{y} - \underset{\sim}{m}(\underline{\theta}) \tag{4}$$

where $\underline{\theta}$ is a vector of model parameters. If x_0, v_0, a_0 and r_0 are the displacement, velocity, acceleration and restoring force for the real system, and x, $\dot{x}$, $\ddot{x}$ and r are the same response quantities for the model, then

$$\underset{\sim}{y} = [x_0, v_0, a_0, r_0]^T \quad , \quad \underset{\sim}{m} = [x, \dot{x}, \ddot{x}, r]^T \tag{5}$$

and

$$\underline{e} = [x_0-x, v_0-\dot{x}, a_0-\ddot{x}, r_0-r]^T \tag{6}$$

The optimal values for the model parameters are calculated by minimizing a measure-of-fit $J(\underline{\theta})$ which is defined in terms of the output-error by:

$$J(\underline{\theta}) = K_1 V_1 \int_{t_s}^{t_e} (x_0-x)^2 dt + K_2 V_2 \int_{t_s}^{t_e} (v_0-\dot{x})^2 dt + K_3 V_3 \int_{t_s}^{t_e} (a_0-\ddot{x})^2 dt$$

$$+ K_4 V_4 \int_{t_s}^{t_e} (r_0-r)^2 dt \tag{7}$$

The K_i are either 0 or 1, so that any combination of response quantities may be chosen in order to estimate the model parameters. Each V_i is chosen to normalize the corresponding integral so that comparison can be made between the optimal values of J for different response quantities and for different time segments of the response records considered. For example,

$$V_1^{-1} = \int_{t_s}^{t_e} x_0^2 \, dt \tag{8}$$

The method is an extension of that developed by Beck and Jennings [10] for structural identification using linear models.

Optimization Algorithm

The modified Gauss-Newton method [11] is chosen, because of its simplicity and computational efficiency over some other gradient methods, to determine the optimal estimates for the k model parameters which minimize the measure-of-fit $J(\underline{\theta})$.

J is approximated locally by a quadratic function and this approximate function is minimized exactly. If $\underline{\theta}_i$ is the i-th estimate for the model parameters, J can be approximated by the truncated Taylor series near $\underline{\theta}_i$ as:

$$J(\underline{\theta}) \simeq J(\underline{\theta}_i) + [\nabla J(\underline{\theta}_i)]^T (\underline{\theta}-\underline{\theta}_i) + \frac{1}{2}(\underline{\theta}-\underline{\theta}_i)^T \nabla\nabla J(\underline{\theta}_i) (\underline{\theta}-\underline{\theta}_i) \qquad (9)$$

Then $J(\underline{\theta})$ is minimized at

$$\underline{\theta}_{i+1} = \underline{\theta}_i - [\nabla\nabla J(\underline{\theta}_i)]^{-1} \nabla J(\underline{\theta}_i) \qquad (10)$$

If $J(\underline{\theta})$ is twice continuously differentiable, then the algorithm is well defined near the optimal estimates $\hat{\underline{\theta}}$ since the Hessian matrix $\nabla\nabla J(\underline{\theta})$ is positive definite. But for estimates remote from $\hat{\underline{\theta}}$, this is not so, and the algorithm must be modified in order to guarantee convergence.

Since J is not quadratic in $\underline{\theta}$ (Eqs. 3 and 7), a step size $\rho > 0$ is introduced to ensure that the value of J decreases at every iteration step. The selection of step size is based on a one-dimensional line search developed by Bard [12]. Also, the Hessian matrix may not be positive definite, or its inverse may not even exist, to yield a direction of descent in the multi-dimensional $J - \underline{\theta}$ space. This is resolved by using an approximate Hessian $\widetilde{\nabla\nabla} J(\underline{\theta})$ which is positive semi-definite. These modifications to Eq. 10 give:

$$\underline{\theta}_{i+1} = \underline{\theta}_i - \rho_i [\widetilde{\nabla\nabla} J(\underline{\theta}_i)]^{-1} \nabla J(\underline{\theta}_i) \qquad (11)$$

where ∇J, $\widetilde{\nabla\nabla} J$ can be calculated from Eq. 7 as follows:

$$(\nabla J)_j = -2K_1 V_1 \int_{t_s}^{t_e} (x_0-x) \frac{\partial x}{\partial \theta_j} dt - \ldots \qquad j = 1,2,\ldots,k \qquad (12)$$

$$[\widetilde{\nabla\nabla} J]_{j\ell} = 2K_1 V_1 \int_{t_s}^{t_e} \frac{\partial x}{\partial \theta_j} \frac{\partial x}{\partial \theta_\ell} dt - \ldots \qquad j,\ell = 1,2,\ldots,k \qquad (13)$$

in which only the terms corresponding to the displacement output-error are shown for briefness. The integral terms containing the second derivatives of response quantities have been dropped from the Hessian matrix so that the resulting matrix is positive semi-definite. To ensure positive definiteness, a multiple λ of the identity matrix is added to the approximate Hessian whenever it is singular or nearly singular, where λ is a very small positive number. Figure 6 illustrates the process of parameter estimation in a flow diagram.

IDENTIFICATION RESULTS

The interstory restoring force between two adjacent floors of the test structure is represented by the hysteretic model presented earlier in this paper, and then the optimal model parameters are estimated using the system identification procedure explained in the previous section. In this case, $\underline{\theta}=\{K_0, r_u, n$ for each story$\}$. The restoring forces measured from the test structure during the inelastic test are matched with the model predictions to estimate the parameters. The parameters are estimated successively for each story, starting with the top story, although it is eventually planned to simultaneously estimate the parameters for all stories. The results of the investigation are presented in Table 1 and the resulting time-histories of responses at the roof level, mid-height (floor 4) and at floor 2 are compared in Fig. 7. The output-errors in accelerations and forces are given instead of the response histories, because of the presence of high frequencies which masks the quality of the response comparison. The model predictions are excellent except at floor 2, which is evident from Fig. 7 and the optimal values of J in Table 1.

The model response was observed to be very sensitive to changes in the stiffnesses K_0, suggesting that these parameters should be estimated reliably from the test data. The values for the ultimate strength r_u, however, appear to be unreliable since they should decrease with height. It is clear that for the top stories, which behave almost elastically, an empirical determination of the ultimate strength is an ill-conditioned process, since large changes in r_u make small changes in the response. It turns out, however, that even for the inelastic deformations of the lower stories, the parameter r_u is not well-determined. The effect of a large relative increase in r_u can be almost balanced by a small relative decrease in the parameter n. It may be better to theoretically determine the value of r_u and then use the test data to determine the parameters K_0 and n for each story.

Hysteresis curves obtained from the test are compared with the curves predicted by the optimal hysteretic models, in Fig. 8. Except at the top story, the hysteresis predictions are very good, which is further reinforced by the comparison of total hysteresis energy dissipation in the test and the model (Table 2). The excellent agreement of the hysteresis curves suggest that the stiffness estimates are close to the actual stiffnesses of the test structure. For the top story, the test indicated several "nonphysical" negative hysteresis loops which resulted in a very small total energy dissipation, and also the test peak-displacements are larger than that of the model. This suggests that the test results for the top story, which behaved almost elastically, may have been more affected by the feedback of experimental errors. These errors were shown to add energy in the higher modes of the structure during the elastic test [4].

CONCLUSIONS

The interstory restoring force-deformation relationship for a multi-degree-of-freedom "shear-building" system is represented by a simple hysteretic model whose parameters are estimated using a nonlinear system identification technique applied to inelastic pseudo-dynamic test data. The optimal hysteretic model predicts the time-histories of the responses and also the hysteresis curves very well. This suggests that the mathematical form of the model used may be useful for response predictions during design, particularly since the parameters K_0 and r_u may be estimated theoretically. In fact, an analytical estimation of the interstory stiffness and strength will be made to compare with the empirical values obtained from the model using system identification. In addition, further tests of the prediction capabilities of the hysteretic model are planned.

ACKNOWLEDGMENTS

This work was partly supported by the National Science Foundation under Grant No. CEE-8119962.

REFERENCES

[1] Goel, S.C. and M.K. Boutros, "Analytical Study of the Response of an Eccentrically Braced Steel Structure," *Proc. ASCE Structural Engineering Congress '85*, Chicago, September 1985.

[2] Beck, J.L. and P. Jayakumar, "Analysis of Elastic Pseudo-Dynamic Test Data from a Full-Scale Steel Structure Using System Identification," *Proc. 6th JTCC Meeting*, U.S.-Japan Cooperative Research Program Utilizing Large-Scale Testing Facilities, Maui, Hawaii, June 1985.

[3] Beck, J.L. and P. Jayakumar, "Application of System Identification to Pseudo-Dynamic Test Data from a Full-Scale Six-Story Steel Structure," *Proc. Int. Conf. Vibration Problems in Engrg.*, Xian, China, June 1986.

[4] Beck, J.L. and P. Jayakumar, "System Identification Applied to Pseudo-Dynamic Test Data: A Treatment of Experimental Errors," *Proc. 3rd ASCE Engineering Mechanics Specialty Conference on Dynamic Response of Structures*, UCLA, April 1986.

[5] Takanashi, K. et al., "Non-Linear Earthquake Response Analysis of Structures by a Computer-Actuator On-Line System," *Bull. of Earthquake Resistant Structure Research Center*, No. 8, Institute of Industrial Science, University of Tokyo, Japan, 1975.

[6] Okamoto, S. et al., "Techniques for Large Scale Testing at BRI Large Scale Structure Test Laboratory," Research Paper No. 101, Building Research Institute, Ministry of Construction, May 1983.

[7] Takeda, T., Sozen, M.A. and N.N. Nielsen, "Reinforced Concrete Response to Simulated Earthquakes," *J. Struct. Div.*, ASCE, Vol. 96(12), 2557-2573, December 1970.

[8] Bouc, R., "Forced Vibration of Mechanical Systems with Hysteresis," Abstract, *Proc. 4th Conf. Nonlinear Oscillation*, Prague, Czechoslovakia, 1967.

[9] Wen, Y.K., "Method for Random Vibration of Hysteretic Systems," *J. Engrg. Mechanics Div*, ASCE, Vol. 102(2), 249-263, April 1976.

[10] Beck, J.L. and P.C. Jennings, "Structural Identification Using Linear Models and Earthquake Records," *Earthq. Engrg. and Struct. Dyn.*, Vol. 8(2), 145-160, 1980.

[11] Luenberger, D.G., *Introduction to Linear and Nonlinear Programming*, Addison-Wesley Publishing Co., 1973.

[12] Bard, Y., "Comparison of Gradient Methods for the Solution of Nonlinear Parameter Estimation Problems," *SIAM J. Numer. Anal.*, Vol. 7(1), 157-186, March 1970.

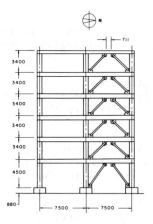

Fig. 1 Elevation of the Interior Frame of Test Structure (All dimensions in mm)

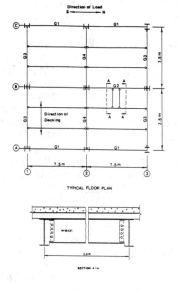

Fig. 2 Plan of Test Structure [1]

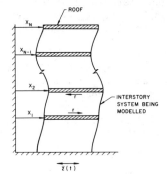

Fig. 3 Multi-Degree-of-Freedom System Excited by Earthquake Ground Accelerations

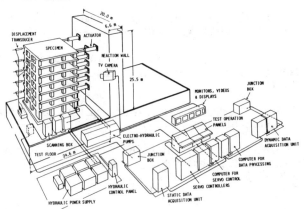

Fig. 4 Pseudo-Dynamic Testing Facility at the Building Research Institute, Tsukuba, Japan [6]

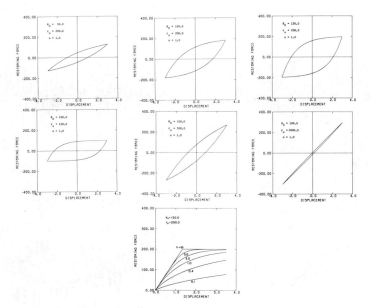

Fig. 5 Effect of Model Parameters on Hysteresis Loops

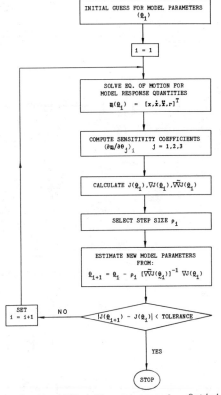

Fig. 6 Simplified Flow Diagram for Optimization

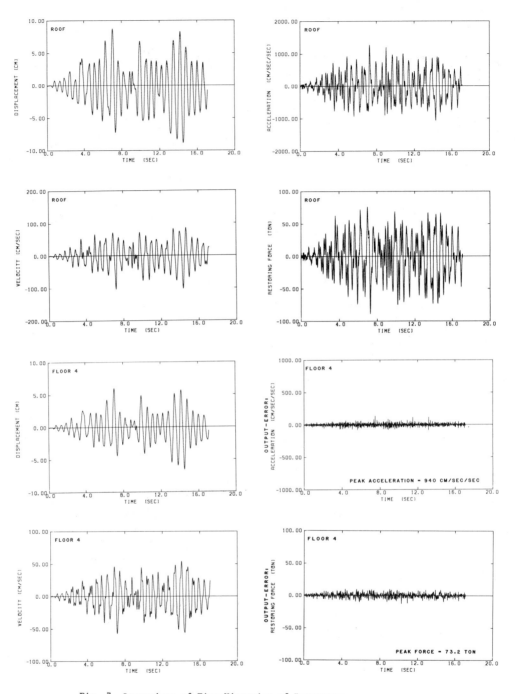

Fig. 7 Comparison of Time-Histories of Responses
(―― Pseudo Dynamic Test; ---- Hysteretic Model)

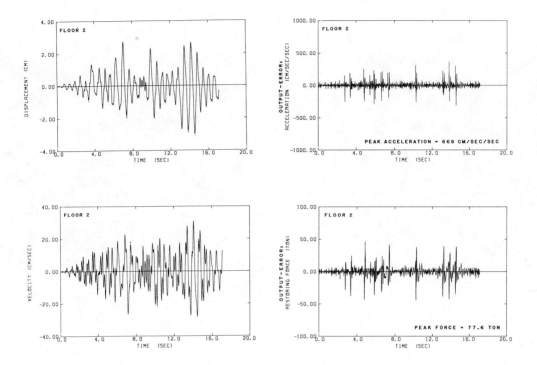

Fig. 7 (Continued)
Comparison of Time-Histories of Responses
(—— Pseudo Dynamic Test; ---- Hysteretic Model)

Table 1 Optimal Model Parameters Estimated from Restoring Force Records Using Nonlinear Hysteretic Model

Story	Model Parameters			J (%)
	K_0 (ton/cm)	r_u (ton)	n	
6	119.1	425	1.04	1.70
5	155.7	228	2.67	1.14
4	186.8	340	2.39	2.07
3	200.7	305	2.13	1.65
2	210.0	273	2.24	6.29
1	161.8	347	1.51	21.2

Table 2 Total Hysteresis Energy Dissipation During the Test and for the Model

Story	Total Hysteresis Energy Dissipation (ton-cm)	
	Pseudo-Dynamic Test	Hysteretic Model
6	9.3	165.8
5	113.1	206.1
4	175.4	256.6
3	727.9	861.0
2	1862.8	2165.9
1	4386.0	4348.1

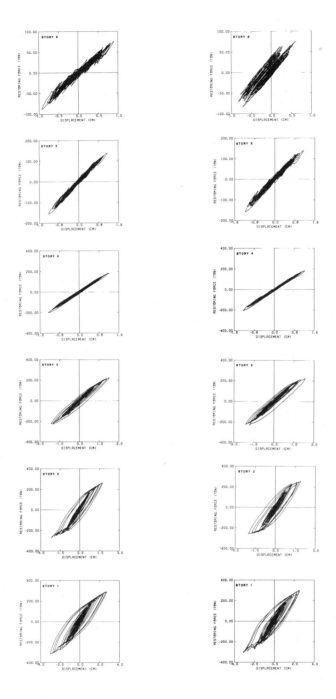

Fig. 8 Hysteresis Behavior: Observed in the Test (Left) and Predicted by the Model (Right)

SEISMIC STRENGTHENING OF STRUCTURAL MASONRY WALLS
WITH EXTERNAL COATINGS

S. P. Prawel[I], A. M. Reinhorn[II], and S.K. Kunnath[III]

ABSTRACT

The risk represented by the building stock comprising old masonry buildings that were originally designed with little or no provision for lateral loading, in earthquake prone regions, has been recognized as one of the major problems facing the earthquake engineer today. While some methods of upgrading, such as the addition of new structural frames or shear walls, have been proven to be effective in actual earthquakes, they have been either too costly or restricted in use to only certain types of structures.

Recent research has shown that a relatively thin layer of external coating applied to the wall surface has excellent properties as a retrofit/repair method. This paper highlights procedures involving thin reinforced coatings and, in particular a new process using a thin layer of heavily reinforced ferrocement. Results of testing carried out at the State University of New York at Buffalo indicate that the ferrocement developed substantial ductility and strength for the seismic strengthening of masonry walls. Examples of the use of ferrocement for various structures using both the ATC and UBC codes are also included.

INTRODUCTION

It is well recognized that masonry structures are a major source of hazard during a seismic event, particularly when unreinforced. These structures are usually constructed from brick, concrete block and, in older examples, stone. In each case, the units are tied together by a cement mortar mixture and in some cases, steel or other reinforcement. When there are cavities in the masonry, they are often grouted.

While there are several types of masonry structural elements within a building, the most important element that is subject to earthquake damage is the load bearing wall. These elements are solid planes which are designed primarily to carry the vertical loads within the structure. In a seismic event, however, they must also carry any in-plane or out-of-plane horizontal loads resulting from the earthquake.

The high risk of earthquake damage to such older masonry walls,

---

[I] Associate Professor, State University of New York at Buffalo, NY 14260.
[II] Assistant Professor State University of New York at Buffalo, NY 14260.
[III] Graduate Student, State University of New York at Buffalo, NY 14260.

particularly when unreinforced and the potential for a great loss of life has made the masonry structure the subject of a wide range of ongoing research. One area that is very active is that of defining methods whereby these structures can be strengthened or upgraded in lateral load capacity to satisfy modern seismic design codes. These retrofit measures are, in many cases, equally effective in the repair of damaged masonry.

SEISMIC BEHAVIOR OF STRUCTURAL MASONRY WALLS

The behavior of masonry buildings under strong dynamic loading such as an earthquake is brittle with little or no ductility and both structural and nonstructural parts suffer local cracking, crushing or complete failure. Moreover, overstressing can lead to various types of damage ranging from invisible cracking to crushing and eventual disintegration.

Nonbearing masonry walls are used to separate space and not intended to support vertical or lateral loads. Such walls are built with little or no attention to their structural integrity and are most often damaged by cracking or crushing at the corners or at the top connector line. These local damage effects, however, do not effect the overall stability of the building unless deformations are large enough to cause direct action between parts.

Load bearing walls, on the other hand, are solid planes which are designed to carry the vertical thrusts of the structure but in a seismic event, must also carry simultaneously with any parallel walls, both in-plane and out-of-plane lateral earthquake effects. The vertical loads act as a prestressing force within the wall and reduces the tendency to slip laterally along the bed joints, and to separate along the joints when the load is normal to the wall. This stiffening effect most often leads to crushing of the edges of the supports due to bending or to axial tension or compression. Stand-alone load bearing walls and masonry infill walls in a lateral load system develop similar damage when overstressed. This usually consists of crushing of corners, a horizontal or vertical open split due to excessive horizontal load or diagonal cracking due to the high shear forces in the wall. Most of the characteristic horizontal or vertical cracking is associated with an arching effect within the wall which results when the vertically loaded wall is subjected to out-of-plane loads [1].

REVIEW OF RETROFITTING TECHNIQUES FOR MASONRY WALLS

Several procedures have been found to be effective in the retrofit of load bearing walls. For damage in the form of relatively small cracks, injections of epoxy and epoxy-ceramic foams have been found to be effective [2,3]. Epoxies, however, are quite sensitive to temperature variations and can become brittle at low temperatures and flow at high temperatures [4]. The epoxy is also quite expensive and when used to repair large cracks, particularly in concrete masonry units, large amounts of epoxy can pass into the voids in the masonry where it is lost as a repair agent. This wasted material can also develop high bond within the void and lead to unwanted stiffness concentration which can be detrimental to the surrounding masonry. A less expensive method uses a cement-mortar grout which can be injected into large cavities. The repair of several masonry block walls using this technique are reported [2,5,6,7] and testing indicates a near complete restoration of initial strength.

For upgrading, two procedures have received considerable attention. The first of these involves transfering most of the load to a new composite wall consisting partly of the old masonry wall and partly of a new reinforced concrete wall. The new wall is constructed by removing one or more wythes of masonry and replacing the volume with a heavy coat of pneumatically applied concrete [8,9,10,11,12,13]. While this type of upgrading is effective, it requires a great deal of preparation work and often adds considerably to the weight of the structure which, in turn, may lead to foundation adjustments.

The second upgrading procedure involves a thin coat of reinforced cement or other material bonded to one or both sides of the masonry unit. These overlay procedures have been found to be very effective with the resulting composite wall developing at least the original in-plane shear strength of a damaged unit. When used with undamaged masonry the shear strength is usually doubled. The out-of-plane resistance is also substantially improved as is the composite ductiliy. Surface coating procedures are also attractive from a construction standpoint requiring very little surface preparation, very little forming and skilled labor [14]. These procedures are detailed in the next section.

SEISMIC RETROFIT USING EXTERNAL COATINGS

A major advantage of surface treatment methods is that it minimizes the construction effort since it does not usually require the removal of existing masonry and in most cases, requires no special high skill work force to complete the job. The following review on the research status of surface treatment methods emphasises the experimental efforts to determine the composite behavior of structural elements.

Numerous studies [7,15,16] on the in-plane strength of shear walls using mesh-reinforced mortar coatings indicate that the coated specimens were able to develop atleast twice the strength of the unreinforced wall, regardless of mesh size. Jabarov et al [15], however, noted that the load corresponding to first cracking was the same for both strengthened and unstrengthened specimens. Also, a reduction in the stiffness characteristic was observed when measuring the natural frequency of vibration for both cases. The decrease shows that while the strength was significantly increased by the coating, the upgraded specimen will show larger deformations than specimens not so treated.

Gulkan et al.[17] reported the use of glass fiber reinforced surface bonding cements (in layers not exceeding 1/8 inch) as a repair agent for damaged walls. When the walls were retested under simulated eartquake loads after repair, they showed acceleration capacity 75% greater than the unstrengthened wall. Earlier, Cagley [14] had studied the properties of such a surface bond cement and based on several static shear tests on 8 inch thick wall panels determined the ultimate resistance of the coated panels to be of the order of 30 to 50 psi.

Meli et al. [18] performed shaking table studies on 1:2.5 scale model adobe walls using welded wire meshes nailed to both faces of the wall and covered by 1 inch of cement mortar. The ultimate resistance of the strengthened walls improved two-fold. Meli also notes the weathering protection provided by the coating as well as the debris containment at failure.

Tso et al.[19] used expanded metal sheets bonded to the side of a block wall by a one inch thick layer of mortar. Standard 1-1/2 inch, No. 16 expanded metal sheets were used. To hold the sheets in place, 1/4 inch diameter bolts with 2x2 inch plates welded to the ends were driven through the joints in the masonry wall. The assembly was tested under cyclic monotonic load. A specimen coated on both sides showed lateral resistance nearly three times greater than an unreinforced wall and the effect of the cyclic load was a continuously increasing energy dissipation. Another specimen coated on one side only was found to be two times stronger than the unreinforced case. The ductility of the second specimen was also far less than that of the first. The doubly coated specimen 'contained' the cracked masonry and was effective in preventing deterioration with increasing deflection cycles. All specimens tested however behaved similarly in stiffness degradation and energy dissipation. Tso also considered out-of-plane bending and recorded the load-deformation curve for the center of the masonry wall. Following a long elastic range, the panel yielded to develop a lateral deformation of almost one inch without loosing its integrity. The cracking pattern that finally resulted was characteristic of yield lines for plate bending. The same type of reinforcement was employed by Kahn [13] along with a 1-1/2 inch layer of sprayed concrete.

Similar yield line type patterns were noted by Scrivner quoted in ref. [20] when testing panels bent out-of-plane under cyclic loading. He observed a narrow hysteric behavior which was probably due to the location of the reinforcement.

Brokken and Bertero [21,22] upgraded a 1:3 scaled brick masonry infill with external reinforcement consisting of welded wire fabric two inches on center and coated by a 5/8 inch cement mortar layer on each side of the wall. The assemblege was subjected to increasing displacment, cyclic loading and the frame load-displacement was recorded. The strengthened masonry infill developed about 1.4 times the frame response and an improved hysteric behavior, dissipating more energy per cycle. Apart from its structural contribution, the coating significantly reduced the production of debris.

A comparison of the behavior of various surface coating procedures based on diagonal tension tests was recently presented by Hutchinson et al. [23]. The procedures used to strengthen brick panels included prestressed walls, sprayed concrete, glass reinforced concrete coated walls, fiber reinforced concrete coated walls and ferrocement coated walls. It was concluded that based on these tests that the sprayed walls, the GRC and FRC walls performed the best. It was recommended that further tests had to conducted on ferrocement since its full capacity could not be tested due to a very premature bond separation between the coating and the masonry. The need to correlate the strength and bonding when using very thin coatings was born out in the tests by Reinhorn, Prawel and Jia [24] (and are discussed in greater detail in the next section).

Much information on the behavior of externally reinforced masonry walls can be drawn from the behavior of regularly reinforced masonry. Static loading tests were performed on masonry walls by Priestly and Bridgitan [25], while cyclic loading tests by Meli [26], and Gulkan et al [17] showed the dependance of the characteristic properties on the reinforcement. Failure mechanisms, in shear or flexure, are described in some detail by Gulkan et al. where it is shown that the amount of reinforcement can be used to control the failure mechanism. It was also shown by Mayes and Clough [27] that reinforced masonry panels possessed more ductility in a flexural failure than if the failure was by shear. It was further observed that the shear strength is not greatly dependent on the amount of reinforcement, but the spacing of the reinforcement has an effect on the ductility [28].

Out-of-plane behavior is governed by the development of arching after the masonry cracks along patterns defined by yield line theory [1, 29,30,31]. Reinforced masonry [31] is, therefore, stronger than unreinforced masonry in out-of-plane behavior. Surface treated masonry must show even better out-of-plane response since the tensile reinforcement is in a more advantageous position [19].

When externally coated masonry walls are subjected to shear forces, it is apparent that the shear resistance depends to a large degree upon the capacity of the coating to withstand diagonal tension [16]. The efficiency of such an assemblage can be determined by diagonal compression split tests of square specimens as suggested by Blume [16,27,28].

Particular attention must be paid to the interaction between the coating materials and the masonry wall. The wide variety of connectors used and their arbitrary distribution indicates a lack of precise information on the actual performance of the connectors. Although studies on the efficiency of such anchors were made by Manos et al [29] and by Kelly et al [quoted in Ref.16], the information provided is not sufficient for the design of the connection between the surface coating and the masonry.

In the following chapter, ferrocement, a highly-reinforced mortar material, is presented as a possible alternative for surface coating of structural walls in large structures.

FERROCEMENT AS A COATING MATERIAL

Among the available coating procedures a thin ferrocement overlay has been suggested for use with unreinforced masonry walls that need enhanced in-plane and out-of-plane strength and ductility. Ferrocement, as it is usually used, is an orthotropic composite material having a high strength cement mortar matrix and reinforced with layers of fine steel wires in the form of a mesh. The mortar strength results in a high composite compressive strength of 5000 to 8000 psi as measured in flexure which is dependent on the volume ratio of reinforcement (.5% to 5%), the mesh type and its orientation. The most significant properties of the composite are its ductility and its high tensile strength which ranges from 500 to 2000 psi for most commercially available meshes.

A study made by the authors [24] showed the effeciency of ferrocement coated masonry walls subjected to in-plane shear forces. The diagonal split or "Blume" test was chosen for the experiment. The test set-up is shown in Fig.1. The specimens tested included ferrocement plates, bare masonry and coated masonry units. The results of the ferrocement plate tests showed that after the split of the mortar matrix, all diagonal tension is carried by the reinforcement. In most cases, the specimens ultimately failed in compression by crushing the concrete at the loaded corners. Because of this, the full range of diagonal tension capacity was not realized. The cracking load in most cases was, however, 50% to 70% greater than that expected from composite materials theory.

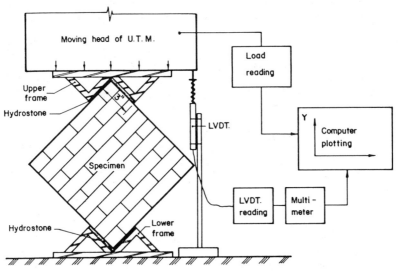

Fig.1. Test set-up.

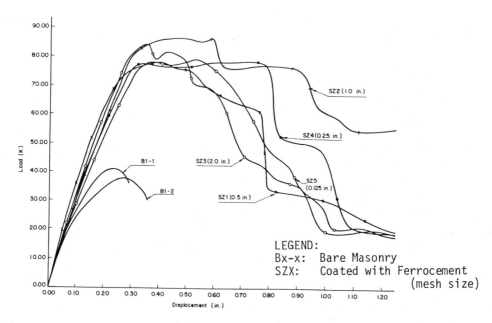

Fig.2. Load deflection curves (measured) [24].

It is clear from the load deformation curves in Fig.2 that each of the coated specimens developed a maximum strength which was approximately double that of the bare masonry specimens and that the strength reached was almost independent of the reinforcement spacing. In the ultimate stages, the ferrocement reached only one-third of its ultimate capacity and failed by local crushing of loaded corners. This, therefore, did not allow for the development of a full diagonal tension failure as in specimen SZ5 (see Fig.2).

Based upon these observations, and the stress computations, it is apparent that two different failure mechanism are possible. The first is a diagonal tension split of the entire composite section which involves yielding of the tensile steel and leads to large inelastic deformations. The second mechanism involves local crushing of the ferrocement coating at the loaded corners. This mechanism is the result of a total transfer of load from the composite section to the coating after the coating has separated from the masonry. Such a failure is characterized as a 'bond' failure. The bond failure is brittle and sudden drops in load capacity are noted when the individual tie wires joining the ferrocement to the masonry fail. If the plates remain bonded to the masonry, as in specimen SZ5, the load transfer is over the entire interface surface which allows for the development of a diagonal tension failure mechanism. It is apparent therefore, that the anchors play a key role in the effective use of the ferrocement coating.

Although the actual failure mechanism did not allow for the full utilization of the potential of the ferrocement coating, the composite structure did develop a limited ductility which was totally absent in the bare masonry walls. For medium size meshes, maximum ductility values range from 1.7 to 2.4 while at 50% of the ultimate the values are from 3.3 to 5.2.

It was also noted that the stiffness degradation of the two uncoated specimens was much more rapid than those that were coated. At a deflection of 0.3 inches the stiffness of the coated specimens has double the value of the bare masonry and at .6 inches, which was double the maximum deflection of the bare masonry, the stiffness of the coated specimens is seen to be reduced to the same value as that of the uncoated walls.

ILLUSTRATIVE EXAMPLE OF RETROFIT USING FERROCEMENT COATING

A typical perimeter wall masonry structure is shown in plan in Fig.3. The structure is constructed of unreinforced masonry walls. The thickness of the walls is variable. The weights and natural periods for the various elements with the same floor plan is shown in Table 1.

Using code provisions for various seismic zones, the base shear forces and the stresses resulting from these forces were computed using both the UBC and ATC-3 provisions. It can be seen from Table 2 that low medium height buildings in UBC zones 1 and 2 can support the seismic load with no assistance. However, structures in zone 3 and 4 and tall structures in zone 2 do not satisfy the UBC requirements for allowable stress. The ATC-3 requirements are more severe, and tall structures in the lowest seismic areas have stresses exceeding the allowable values. Using characteristic material properties, it is determined that a ferrocement overlay on both sides, 1/2 inch thick will supply an extra shear force of 262 psi x 1 in = 262 lb. For a 12 inch wall, this represents an

```
STORY HEIGHT:          10'-0"
SOLID BRICK WALLS:
     THICKNESS:        1'-0" (FOR LESS THAN 6 STORIES)
                       1'-6" (FOR MORE THAN 6 STORIES)
WOODEN FLOORS (WT):    20 PSF
PARTITIONS (WP):       20 PSF
WALLS WEIGHT (WW):     140 PSF

LOADS EQUALLY DISTRIBUTED TO EXTERNAL WALLS.
PERIOD OF VIBRATION: T = .05 H √ D
```

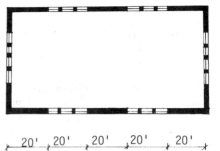

Fig.3. Typical floor plan.

Table 1. Properties of sample building.

	W(kips)	T(sec) Long side	T(sec) Short side
(1)	(2)	(3)	(4)
1 Story	350	0.05	0.07
2 Story	800	0.10	0.14
3 Story	1250	0.15	0.21
4 Story	1700	0.20	0.28
6 Story	2600	0.30	0.42
9 Story	4250	0.45	0.64

equivalent additional strength of 22 psi. If a 1-inch layer on each side was used, the additional strength from the ferrocement would be 44 psi. With a material of this strength, it is not necessary to treat the entire wall surface, and only a percentage need to coated to achieve the required overall strength level. For example, a structure with four stories in UBC zone 3 requires a strength of 33 psi at the basis of the masonry wall. Since the masonry stress cannot exceed 20 psi, the remaining 13 psi is obtained by coating a part of both sides of the wall by a 1/2 inch layer of ferrocement. If it is assumed that the load distribution is not effected by the additional stiffness, 59% of the total wall length (13 psi: required divided by 22 psi: provided by the coating) must be coated to obtain the required strength. The same treatment is required for a one-story building in ATC-3 0.4 g areas. Smaller percentages would be required at higher floor levels.

CONCLUSIONS

The class of retrofit procedures referred to as coating methods have shown all of the characteristics required to be an effective and economical method to provide masonry with needed upgrading in strength and ductility. In

Table 2 - Design Forces and Shear Stresses at Base of Masonry Walls*

	UBC 79									ATC-3			
	ZONE 4		ZONE 3		ZONE 2		ZONE 1			$A_{max}=.40$		$A_{min}=.05$	
No. of Stories	S. Force kips	Stress psi	S. Force kips	Stress psi	S. Force kips	Stress psi	S. Force kips	Stress psi		S. Force kips	Stress psi	S. Force kips	Stress psi
(1)	(2)	(3)	(4)	(5)	(6)	(7)	(8)	(9)		(10)	(11)	(12)	(13)
1	66.5	10	49.0	7	24.5	4	14.0	2		234.5	33	28.0	4
2	152.0	22	112.0	16	56.0	8	32.0	5		536.0	75	64.0	9
3	237.0	33	175.0	25	87.5	13	50.0	7		837.5	117	100.0	14
4	323.0	45	238.0	33	119.0	17	68.0	10		1139.0	159	136.0	19
6	494.0	69	364.0	51	182.0	26	104.0	15		1742.0	242	208.0	29
9	807.5	75	595.0	55	297.5	28	107.0	16		2847.0	264	734.0	32

* Hard lines divide between structures overstressed assuming well inspected reinforced brick masonry.
  VC,all= 20 psi

** Base Shear. (UBC)
   $V = ZKCSIW$ $(CS=0.14) = Z(1.33)(0.14)(1.01)W;$ $V = 0.1862 ZIW$

*** Base Shear (ATC-3)
   $V = C_W$; $C_S = 1.2 A_S / RT^{2/3}$
   $R=1.2$ unreinf. masonry. (Table 2-A,PI-10) S=1.5 Type C Soil (Table 2-C,PI-13); 0.05 < A < 0.40 (Table RA-1, PRA-1)

particular, preliminary testing using a ferrocement overlay has shown it to be versatile and cost-effective because of minimum added weight, improved strength and ductility, ease of placement and ability to assume any shape without formwork. The latter is extremely important when architectural preservation is required.

REFERENCES

[1] Lefter, J. and Colville, J., "Reinforcing Existing Buildings to Resist Earthquake Forces", Proc. of the U.S. National Conference on Earthquake Engineering, Ann Arbor, Michigan, June 1975, pp. 226.

[2] Warner, J., "Methods of Repairing and Retrofitting (Strengthening) Existing Buildings", Proc. of Workshop on Earthquake-Resistant Reinforced Concrete Building Construction (ERCBC), University of California, Berkeley, July 1977.

[3] Plecnik, J.M., Amrhein, J.E., Jay, W.H. and Warner, J., "Epoxy Repair of Structures", Proc. of the International Symposium on Earthquake Structural Engineering, Missouri, 1976, pp.1023-1036.

[4] Plecnik, J.M., Amrhein, J.E., Warner, J., Jay, W.H. and Chelapati, C.V., "Repair of Earthquake Damage Concrete Masonry Systems Subjected to Static and Dynamic Loads and Elevated Temperatures," Proc. 6th WCEE, New Delhi, 1977.

[5] Benedetti, D. and Castellani, A., "Comparative Tests on Strengthened Stone-Masonry Buildings", Proc. 2nd U.S. National Conference on Earthquake Engineering, pp. 793, Stanford University, Aug. 1979.

[6] Benedetti, D. and Castellani, A., "Experimental Determination of the Seismic Resistance of Repaired Masonry Structures", Proc. 7th WCEE, Vol. 6, pp. 159, Istanbul, 1980.

[7] Sheppard, P. and Terceli, "The Effect of Repair and Strengthening Method for Masonry Walls", Proc. 7th WCEE, Vol. 6 pp. 255, Istanbul, 1980.

[8] Kahn, L.F., "Repair and Strengthening of Masonry," Proc. 2nd Seminar on Repair and Retrofit of Structures, pp. 247, Ann Arbor, May 1981.

[9] Barnes, S.B., "Strengthening and Repair of Existing Structures", ASCE Fall Convention and Exhibit, Preprint 2936, San Francisco, California, Oct. 1977.

[10] Lee, L.A., Lee, H.H. and Nicoletti, J.P., "Seismic Rehabilitation of the California State Capitol West Wing", Preprint, ASCE Convention San Francisco, California, Oct. 1977.

[11] Wyllie, L.A., "Strengthening Existing Concrete and Masonry Buildings for Seismic Resistance," Proc. 2nd Seminar on Repair and Retrofit of Structures, pp. 322, Ann Arbor, May 1981.

[12] Khan, L.F., "Shotcrete Retrofit for Unreinforced Brick Masonry," Proceedings 8th WCEE, San Francisco, 1984, Vol. 1.

[13] Kahn, L.F., "Shotcrete Strengthening of Brick Masonry Walls," ACI Concrete International, Vol. 6, No. 7, July 1984.

[14] Cagley, J.R., "Seismic Hardening of Unreinforced Masonry Walls through a Surface Treatment", U.S. Department of Commerce, National Technical Information Service, Pb-278,930, May 1978.

[15] Jabarov, M., Kozharinov, S.V. and Lunyov, A.A., "Strengthening of Damaged Masonry of Reinforced Mortar Layers", Proc. 7th WCEE, Vol. 4, pp. 73, Istanbul, 1980.

[16] Schneider, R.R. and Dickey, W.L., Reinforces Masonry Design, Prentice-Hall, Englewood Cliffs, New Jersey, 1980.

[17] Gulkan P., Mayes, R.L. and Clough, R.W., "Shaking Table Study of Single-Story Masonry Houses, Volume 1: Test Structure 1 and 2," Earthquake Engineering Research Center Report No. UCB/EERC-79/23, University of California, Berkeley, California, Sept. 1979.

[18] Meli, R., Hernandez, D., Padilla, M., "Strengthening of Adobe Houses for Seismic Actions", Proc. of 7th WCEE, Vol. 4, pp. 265. Istanbul, 1980.

[19] Tso, W.K., Pollner, E. and Heidebrecht, A.C., "Cyclic Loading on Externally Reinforced Masonry Walls", Proc. 5th WCEE, pp. 1177, Rome Italy, 1974.

[20] Warner J., "Restoration of Earthquake Damaged Concrete and Masonry", Proc. 5th WCEE, pp. 882, Rome, 1973.

[21] Bertero, V.V. and Brokken, S.T., "Effects of Infills in Seismic Resistant Buildings", Proc. 2nd Seminar on Repair and Retrofit of Structures, Ann Arbor, pp. 377, May 1981.

[22] Brokken, S.T. and Bertero, V.V., "Studies on Effects of Infills in Seismic Resistant R/C Construction", Earthquake Engineering Research Center Report No. UCB/EERC-81/12, Univ. of California, Berkeley, Oct. 1981.

[23] Hutchinson, D.L., Yong, P.M.F. and McKenzie, G.H.F., " Laboratory Testing of A Variety of STrengthening Solutions for Brick Masonry Wall Panels," Proceedings 8th WCEE, San Francisco, 1984, Vol. 1.

[24] Reinhorn, A.M., Prawel, S.P., and Jia, Z.H., "Experimental Study on External Ferrocement Coating for Masonry Walls", Journal of Ferroecment, Vol.15, No.3, 1985.

[25] Pristley, M.J. and Bridgitan, D.O., "Seismic Resistance of Brick Masonry Walls", Bulletin of the New Zealand National Society of Earthquake Engineering, Vol. 7, No. 4, December 1974.

[26] Meli, R., "Behavior of Masonry Walls Under Lateral Loads", Proceedings of 5th World Conference on Earthquake Engineering, pp. 853, Rome 1972.

[27] Mayes, R.L. and Clough, R.W., "State-of-the-Art in Seismic Shear Strengthen of Masonry -- An Evaluation and Review", Earthquake Engineering Center Report No. UCB/EERC-75/21, Univ. of Califnornia, Berkeley, Oct. 1975.

[28] Hidalgo, P., Mayes, R.L., McNiven, H.D., and Clough, R.W., "Cyclic Loading Tests on Masonry Single Piers", Report No. UCB/EERC-79/12, Earthquake Engineering Research Center, University of California, Berkeley, May 1979.

[29] Manos, G.C., Clough, R.W., and Mayes R.L., "Shaking Table Study of Single-Story Masonry Houses". Report No. UCB/EERC-83/11, Earthquake Engineering Research Center, University of California, Berkeley, July 1983.

[30] Omote, Y., Mayes, R.L., Chen, S.W., and Clough, R.W., "A Literature Survey - Transverse Strength of Masonry Walls", Report No. UCB/EERC-77/07, Earthquake Engineering Research Center, University of California, Berkeley, March 1977.

[31] Amrhein, J.E., "Reinforced Masonry Engineering Handbook -- Clay and Concrete Masonry", Masonry Institute of America, Fourth Edition, 1983.

SHEAR STRENGTH OF REINFORCED MASONRY WALLS
UNDER EARTHQUAKE EXCITATION

P. Hidalgo[I] and C. Luders[I]

ABSTRACT

The shear strength of reinforced masonry walls is analyzed using the results of two experimental research programs that have placed their emphasis on the shear mode of failure. The effect of several parameters on this strength is studied; this permits to propose an analytical prediction of the shear strength of reinforced masonry piers. A short discussion on the inelastic behavior associated with the shear mode of failure is also presented.

INTRODUCTION

The shear strength of walls plays a significant role in the earthquake-resistant design of reinforced masonry buildings. Some controversy has arisen among researchers over the past decade whether it is possible to increase this shear strength by using substantial amounts of properly detailed shear reinforcement, in orden to force a ductile flexural mode of failure in the shear walls under unusually strong earthquake excitations. New Zealand researchers sustain this can be achieved as shown by tests performed on masonry walls with high steel percentages, and have translated their findings into a code based on ultimate strength design concepts[1]. However, the use of such amounts of reinforcing steel may not be economically feasible even in countries like the United States; moreover, the idea of reducing the shear force to bending moment ratio in the walls using flexible links between them instead of rigid coupling beams, may impose architectural limitations that are not easy to comply. This situation may force engineers to design masonry buildigs recognizing that shear walls will eventually fail in shear under severe ground motions; in such a case it is important to be aware of the amount of energy dissipation that may take place in those structural elements under the shear mode of failure, and increase the level of the earthquake forces if the code has not properly considered the structural type and the material of the structure in the definition of the base shear force due to the earthquake action.

In any of the situations mentioned above, it is important to have a good assessment of the ultimate shear strength of reinforced masonry walls, including the effect of the various parameters that influence such strength. The development of experimental research programs that provide consistent sets of data are essential to derive an analytical expresion of this shear strength. This paper presents an evaluation of the results obtained in two of

------
I. Professor, Department of Structural Engineering, Catholic University, Santiago, Chile.

these experimental programs, that developed at the University of California, Berkeley [2,3,4,5,6], and the program in progress at the Catholic University in Santiago, Chile[7,8]. Both programs have tested single piers and have a common philosophy; the chilean experimental program was derived from the Berkeley program, with changes only in the test setup, masonry units and workmanship. The main parameters in both programs were the quality of masonry, measured by its compressive prism strength, the aspect ratio of the piers, the amount of horizontal reinforcement and the axial compression force present during the test; several types of masonry units and both partial and full grouting have been used in both research programs.

This paper has placed more emphasis on the shear strength than the inelastic behavior shown by the walls tested. While quantitative results on the shear strength are included in this study, only a qualitative analysis of the inelastic behavior associated with the shear mode of failure is presented. From the 124 tests carried out on single piers, (95 American and 29 Chilean), only those that developed a shear mode of failure have been considered in this study, (78 American and 25 Chilean). Special care has been taken in discarding those tests where the ultimate strength was controlled by the flexural capacity of the piers or a sliding mode of failure prompted by flexural cracks. Further description of the different modes of failure that take place in this kind of tests may be found elsewhere[2,5,8].

TEST PROGRAM CHARACTERISTICS

The basic difference between both test programs was the test setup used to induce the in-plane shear force. The Berkeley program used a test setup that prevented rotation at both ends of the specimen; this was achieved by using two rigid vertical links or columns located at the sides of the specimen, as shown in Fig. 1. However, as the lateral deformations increased, the rigid columns induced increasing amounts of axial compression on the specimen, changing in some cases a flexural mode of failure into a shear mode. This effect was corrected in the last series of tests, replacing the columns by vertical actuators that were programed to keep the inflection point of the bending moment diagram at midheight of the pier, while maintaining a constant and controlled axial force on the specimen throughout the test. The Chilean program used the test setup shown in Fig. 2; this was devised as a cantilever system, with the horizontal shear force not directly applied to the specimen but at a higher level through a bending moment and a shear force action; vertical compression forces may also be incorporated in the test setup but have not been used until now. This system was originally thought as a simple cantilever system with a reinforced concrete beam cast on top of the specimen to distribute the shear force[9]; the improved system shown in Fig. 2 avoids the use of the top beam and its influence on the crack pattern of the specimen, thus reproducing in a better way the behavior of one half of a wall with both ends prevented against rotation.

The other test characteristics were quite the same in both programs. The shear force was applied through a sequence of cycles with monotonically increasing deformation amplitudes, until failure was attained; two or three cycles at the same amplitute were used at each stage of the sequence. The load was applied at a rate of 0.02 cycles per second in the Berkeley tests, and at a somewhat lower rate in the Chilean tests. The basic experimental information used in this study were the hysteresis curves, (load vs deformation curves throughout the test), and the crack patterns developed in the walls. There is another difference between both programs that does not have to do with the test procedure itself but with the design of the specimen to obtain a shear mode of failure. While the Berkeley tests used the amount of axial compression to induce the shear mode of failure, the Chilean tests have been performed until now under no compression and preventing the flexural failure by increasing the bending moment capacity through the use of larger amounts of vertical reinforcement in the outer cells of the specimen.

In addition to the parameter represented by the axial compression stress present at failure, that varied in the Berkeley tests between zero and 580 psi (3.96 MPa), the main parameters included in these research programs were the amount of horizontal reinforcement and the aspect ratio of the walls; the horizontal reinforcement ratio varied from zero to 1.02%, with both reinforcing bars and joint reinforcement (truss or ladder type wire mesh) being used to provide this type of reinforcement. The aspect ratio may be represented by M/Vd, where $\underline{M}$ and $\underline{V}$ are the bending moment and shear force at the base of the pier, respectively, and $\underline{d}$ is the length of the wall; this ratio varied from 0.25 to 1.05. Different types of masonry units were used to construct the walls; clay brick blocks, concrete blocks and double wythe grouted walls were used in the Berkeley tests, while the Chilean program used clay brick and concrete block units. The clay brick units typical of Chilean construction are quite different from those used in the United States[9] and the behavior of the walls constructed with this type of masonry units drifts away from the general behavior observed in the rest of the piers, as it will be shown later in the paper. The quality of masonry is represented here by f_m', the compressive strength of a prism having a height-to-thickness ratio of the order of 5 and constructed with the same materials and under the same conditions used in the walls. Both fully and partially grouted piers were used in these programs; nine of the Berkeley tests and 13 of the Chilean tests reported in this paper were partially grouted.

TEST RESULTS

Shear strength

The processing of the experimental data was done from the raw data reported for the Berkeley tests[3,4,5,6] and the Chilean tests[7,8]. In the figures that follow v_u, the ultimate shear stress was obtained by dividing the ultimate shear force by the gross cross-sectional area of the wall; in the case of the partially grouted walls, the net contact area between layers was used instead of the gross cross-sectional area. The same definition applies to σ_u, the axial stress present on the specimen when the ultimate shear force was attained. The horizontal reinforcement ratio ρ_h was obtained by dividing the area of the reinforcing bar or joint reinforcement by the product of the vertical distance between bars and the thickness of the pier.

The influence of the prism compression strength on the shear strength of masonry is shown in Fig. 3 for the case of walls with no horizontal reinforcement and M=Vd. Though it has been recognized the prism strength is not an optimum indicator of the shear strength of masonry, no other simple test that provides a better correlation with the shear strength has been found yet. The set of triangular points with v_u not larger than 50 psi (0.34 MPa) represents the Chilean tests of clay brick walls. It becomes clear from Fig. 3

the increase in shear strength produced by the axial compression force, effect that is consistently present in the following figures.

The influence of the aspect ratio of the walls is shown in Figs. 4,5 and 6 for different amounts of horizontal reinforcement. The increase in shear strength for squat piers with M/Vd ratios smaller than one is clearly shown in these graphs. The line represented by the equation

$$v_u = \frac{1}{3}(4 - \frac{M}{Vd})\sqrt{f'_m} \tag{1}$$

corresponds to one of the expressions of the allowable shear stress specified by the Uniform Building Code[10] when masonry takes all the shear. Comparison of the experimental points with this line in Fig. 4 gives an indication of the safety factor implied by the allowable shear stresses of Eq. 1. The effect of the compression stresses is also apparent from these figures.

In order to isolate the effect of horizontal reinforcement on the shear strength, the values of v_u have been divided by v_m, the allowable shear stress specified by the UBC when masonry takes all the shear, and plotted against the horizontal reinforcement ratio in Fig. 7. The variation of v_m as a function of f'_m and M/Vd is shown in Fig. 8; comparison of Fig. 8 with the graphs of Figs. 3,4,5 and 6 indicates that v_m adequately reflects the influence of the parameters f'_m and M/Vd. The graph of Fig. 7 shows that the shear strength increases as the horizontal reinforcement ratio increases; the favorable effect of horizontal reinforcement is not as significant as in the case of reinforced concrete due to anchorage problems of the reinforcing bars that prevented them from attaining yielding. The effectiveness of the horizontal reinforcement is enhanced by the use of special anchoring devices, such as those used in the last series of the Berkeley tests[6]. It is also worth to mention the effectiveness of the joint reinforcement (truss or ladder type wire mesh) when used with structural purposes as horizontal reinforcement; the Chilean tests were able to produce rupture of both wires before anchorage problems became evident; however, due to the lack of ductility of the cold drawn wire, it is not possible to obtain a 100% contribution from the joint reinforcement distributed through the height of the wall[8].

The graph of Fig. 7 also permits to draw an expression to predict the ultimate shear stress that can be attained in reinforced masonry walls. If the effect of the axial force is neglected, as in the UBC[10] and the New Zealand Code[1], the straight line shown in Fig. 7 represents a conservative estimation of such strength. The analytical expression of this line is

$$v_u = 2.5\, v_m (1 + \frac{\rho_h}{0.006}) \tag{2}$$

This expression should be used for values of ρ_h under 0.006, since there are only two experimetal points beyond that value. Very few points have been left below this line; the points corresponding to the Chilean tests using clay brick units, ($\rho_h = 0$ and v_u/v_m under 2.0), are among them.

Finally, Fig. 9 shows the influence of axial compression force on the shear strength of masonry walls. In spite of the beneficial effect of axial compression that is evident from this figure, codes have not considered it; the fact that this axial force may be overcome by the overturning moment effect, and its detrimental effect on ductility, may be among the reasons that have precluded the inclusion of the effect of this parameter in the code expressions.

Inelastic behavior

The recognition that the shear mode of failure cannot always be avoided has led the researchers to look for ways of improving the behavior of reinforced masonry walls after the ultimate shear stress has been attained. The results obtained so far are not conclusive; only qualitative results are available from the experimental research programs reviewed in this paper. It has been shown that the use of special anchoring devices for the horizontal reinforcing bars, such as the use of end plates or 180° anchoring hooks, and the use joint reinforcement, improve this inelastic behavior[6]. However, one has to be very careful when looking into this problem since it is common that another resistance mechanism develops after the shear strength has been attained[8]; this second resistance mechanism has usually less strength than the first one, and might lead to a ductile mode of failure. More research is needed to study what happens in the walls after the shear strength has been attained. Meanwhile, the use of factors to reduce the earthquake design elastic forces for reinforced masonry buildings should be limited to values of the order of 1.5[11].

CONCLUSIONS

1. The wide scatter shown in the correlation charts of this paper might be due to the following reasons:

   a) the possibility that not all the essential parameters that govern the physical phenomenon have been properly considered in the analytical model to predict the shear strength of reinforced masonry;
   b) the influence of workmanship in masonry construction and the importance of quality control.

2. The UBC 85 allowable stresses for shear design when masonry takes all the shear reflect reasonably well the tendency of the influence of the parameters f'_m and M/Vd on the shear strength.

3. In spite of the problems to analytically predict the shear strength of reinforced masonry, a simple but conservative expression is proposed.

4. More research and study are needed to comprehend the behavior of reinforced masonry walls after the ultimate shear stress has been attained. This would eventually permit to design and construct these walls so that improved energy dissipation characteristics are obtained.

ACKNOWLEDGEMENTS

The University of California, Berkeley, research program reported in this paper was jointly funded by the National Science Foundation, the Masonry Institute of America, the Western States Clay Products Association and the Concrete Masonry Association of California and Nevada. The research program developed at the Catholic University, Santiago, Chile, was jointly funded by the Research Division of the University, the Chilean Cement and Concrete Institute, and the PRINCESA Industry. The sponsorship of these institutions is gratefully acknowledged. Many individuals have also contributed to the development of these research programs; particularly, the authors want to acknowledge the steering work of Professors R. Clough and H. McNiven, and Dr. R. Mayes, and the contribution of Mr. B. Sveinsson, to the development of the Berkeley research program.

REFERENCES

[1] Standards Association of New Zealand, "Code of Practice for the Design of Masonry Structures", NZS 4230P:1985, Wellington, New Zealand, 1985.

[2] Hidalgo P. and McNiven H., "Seismic Behavior of Masonry Buildings", Seventh World Conference on Earthquake Engineering, Vol. 7, pages 111-118, Istanbul, Turkey, September 1980.

[3] Hidalgo P., Mayes R., McNiven H. and Clough R., "Cyclic Loading Tests of Masonry Single Piers, Volume 1 - Height to Width Ratio of 2", EERC Report N°78/27, University of California, Berkeley, November 1978.

[4] Chen J., Hidalgo P., Mayes R., Clough R. and McNiven H., "Cyclic Loading Tests of Masonry Single Piers, Volume 2 - Height to Width Ratio of 1", EERC Report N°78/28, University of California, Berkeley, December 1978.

[5] Hidalgo P., Mayes R., McNiven H. and Clough R., "Cyclic Loading Tests of Masonry Single Piers, Volume 3 - Height to Width Ratio of 0.5", EERC Report N°79/12, University of California, Berkeley, May 1979.

[6] McNiven H., "Cyclic Loading Tests of Reinforced Masonry Piers", EERC Report in preparation, University of California, Berkeley.

[7] Hidalgo P. and Luders C., "Resistencia al Esfuerzo de Corte de Muros de Albañilería Armada", Dept. of Structural Engineering Report N°82-1, Catholic University, Santiago, Chile, May 1982.

[8] Luders C., Hidalgo P. and Gavilán C., "Comportamiento Símico de Muros de Albañilería Armada", Dept. of Structural Engineering Report N°85-3, Catholic University, Santiago, Chile, April 1985.

[9] Hidalgo P. and Luders C., "Earthquake-Resistant Design of Reinforced Masonry Buildings", Eighth World Conference on Earthquake Engineering, Vol. VI, pages 815-822, San Francisco, California, July 1984.

[10] International Conference of Building Code Officials, "Uniform Building Code", Chapter 24, Whittier, California, 1985.

[11] Hidalgo P., Luders C. and Jordán R., "Seismic Design Provisions for Reinforced Masonry Buildings in Chile", Paper accepted for presentation at the Eighth European Conference on Earthquake Engineering, Lisbon, Portugal, September 1986.

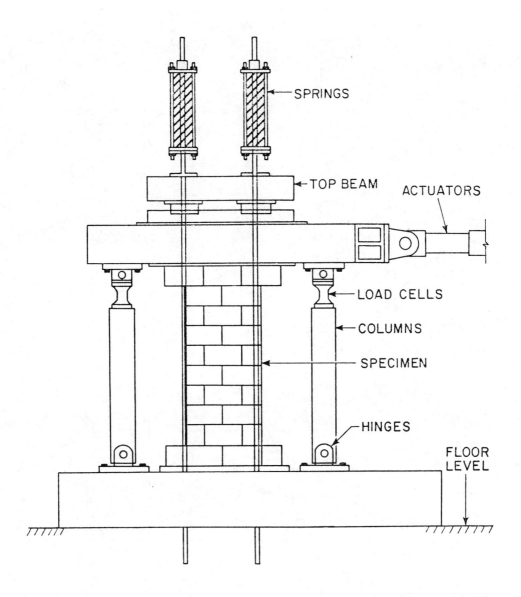

FIG. 1 UNIV. OF CALIFORNIA, BERKELEY, TEST SETUP

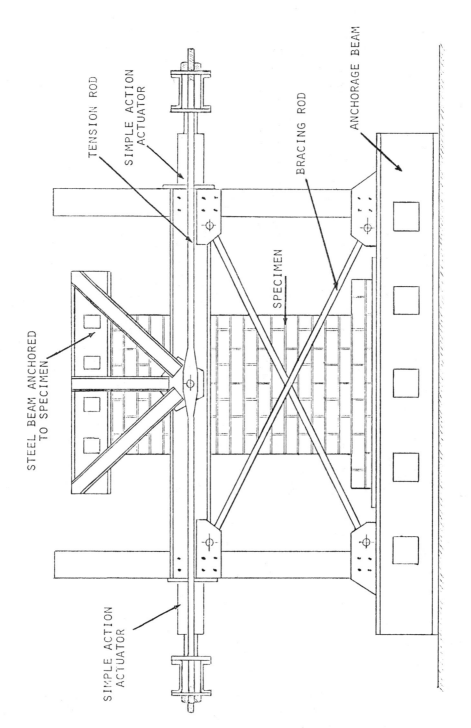

FIG. 2 CATHOLIC UNIVERSITY TEST SETUP

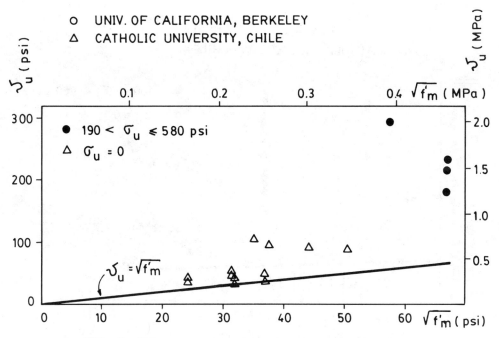

FIG. 3 INFLUENCE OF MASONRY PRISM STRENGTH ON SHEAR STRENGTH FOR $\rho_h=0$ AND $M=Vd$

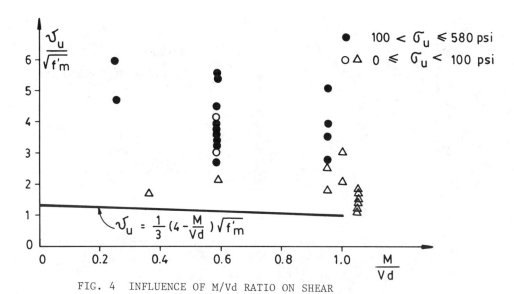

FIG. 4 INFLUENCE OF M/Vd RATIO ON SHEAR STRENGTH FOR $\rho_h=0$

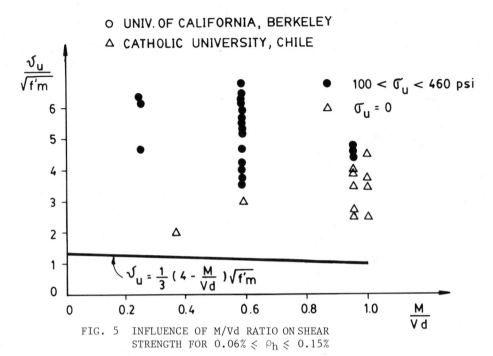

FIG. 5 INFLUENCE OF M/Vd RATIO ON SHEAR STRENGTH FOR $0.06\% \leq \rho_h \leq 0.15\%$

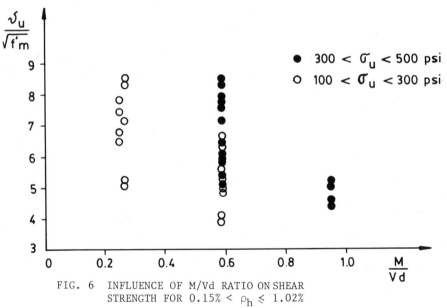

FIG. 6 INFLUENCE OF M/Vd RATIO ON SHEAR STRENGTH FOR $0.15\% < \rho_h \leq 1.02\%$

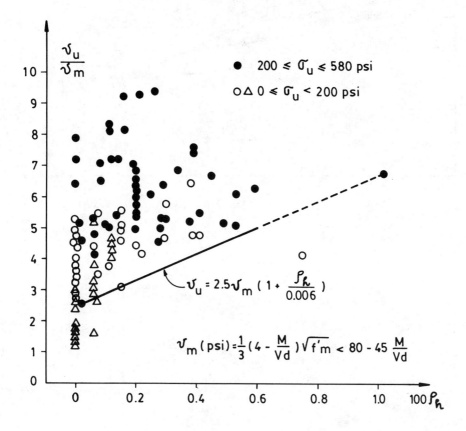

FIG. 7 INFLUENCE OF HORIZONTAL REINFORCEMENT ON SHEAR STRENGTH

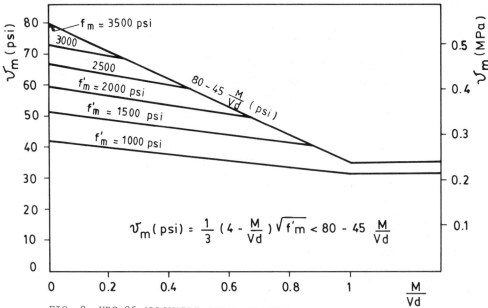

FIG. 8 UBC 85 ALLOWABLE SHEAR STRESSES
MASONRY TAKES ALL THE SHEAR

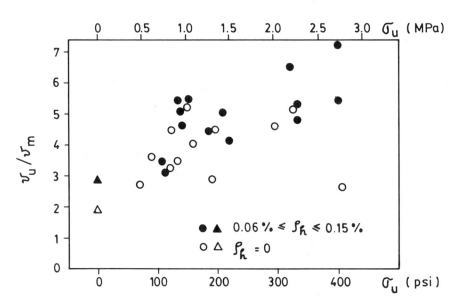

FIG. 9 INFLUENCE OF AXIAL COMPRESSION STRESS
ON SHEAR STRESS FOR M=0.58 Vd

SEISMIC BEHAVIOR OF PRECAST CONCRETE SHEAR WALLS -
CORRELATION OF EXPERIMENTAL AND ANALYTICAL RESULTS

V. Caccese[I] and H. G. Harris[II]

ABSTRACT

The seismic behavior of precast concrete shear walls is described.
The behavior is characterized by nonlinear mechanisms of rocking
and shear slip motion. Results of experimental testing of 4 five
story small scale models are discussed along with analytical studies
performed at MIT and UC Berkeley. An independent mathematical
analysis using Drain 2-D was also performed. The analytical models
portrayed the same general response that was observed in the
experimental models.

INTRODUCTION

Precast concrete wall systems are ideal for multistory residential structures where a high resistance to fire and sound is essential. These systems are composed of precast concrete walls connected to precast concrete floor planks or roof planks forming a box-like arrangement with high lateral stiffness. Unlike monolithic concrete, the precast concrete system has known regions of reduced strength and stiffness where cracking usually occurs. The regions are the connections where junction of the precast elements is made. For this reason, when compared to cast-in-place shear wall buildings, the analysis of a precast wall system is somewhat simplified owing to the fact that cracking chiefly occurs in the connection region. Thus, a typical precast concrete shear wall system can be thought of as a deep concrete beam with known regions (connections) where nonlinear behavior will occur. When subjected to high lateral forces induced by earthquakes, the nonlinear response of the connection is affected by a combination of shear slip and rocking motion.

Slip will occur when the shear stress, caused by seismic forces assumed in the plane of the stack of walls, exceeds the shear strength at the slip plane. The shear resistance is first influenced by an initial bond between the joint grout and the precast components. Once this initial bond is broken, adhesion no longer exists, and the shear strength is derived from the interface shear transfer mechanism (shear friction, and aggregate interlock) combined with dowel action contributions from the vertical tie reinforcing that is transverse to the joint. Since the loading occurs cyclically, the effect of aggregate interlock may be degraded and damage to concrete adjacent to vertical reinforcing will cause a reduction in joint stiffness. When the

---

I. Assistant Professor, Department of Mechanical Engineering, University of Maine, Orono, Maine.

II. Professor, Department of Civil Engineering, Drexel University, Philadelphia, Pennsylvania.

induced lateral force exceeds the resistance available from the interface shear transfer mechanism, the integrity of the joint is sustained by the vertical tie.

The nonlinear rocking mechanism is attributed to an overturning moment originating from any inplane lateral force acting on panels above. Since the joint is generally weak in tension, uplift of the panel may occur if the tensile stress due to bending is greater than the superimposed compressive stress due to the active gravity loads and any vertical prestressing provided. Once the tensile strength is exceeded a gap will form between the end of the wall panel and the horizontal joint.

To investigate the behavior of precast wall systems, 4 five story small scale models were tested on a shaking table housed in the Structural Dynamics Laboratory at Drexel University. Results of the experimental study are compared to analytical models developed by Becker et. al. [1,2,3] at MIT, by Schricker et. al. [4] at UC Berkeley, and by Caccese and Harris [5,6,7] at Drexel.

MIT ANALYTICAL MODEL

Becker et. al. [1,2,3] studied the seismic response of a stack of precast concrete simple shear walls using a mathematical model capable of handling both the rocking and slip phenomenon. A cross-wall Large Panel system, shown in Fig. 1, was used at the basic design unit. Large Panel precast wall systems tend to use horizontal joints of the wet or semi-wet type. The most prevalent type of horizontal joint used in North American practice is the platform joint shown in Fig. 2. The analytical connection model was based on this type of joint. For the most part, ten story buildings were studied. However, several analyses of five story buildings were performed and are compared to the experimental study at Drexel.

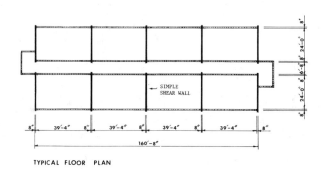

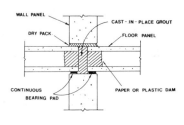

Fig. 1 - Cross-Wall Building Plan.

Fig. 2 - Platform Type Horizontal Joint.

In the analysis several assumptions were made about the behavior of the system and are given as: 1) Connections are regions of all nonlinear and inelastic behavior, 2) Floors and foundations behave rigidly, 3) Horizontal connections act as pre-cracked planes, 4) Solid wall panels remain linearly elastic, 5) Floor systems are assumed to behave rigidly.

The nonlinear inelastic rocking and slip mechanisms associated with the horizontal connections were modelled by combining a 4 node interface or contact element with elasto-plastic properties in shear and a 4 node plane stress rectangular connection element with zero tension material properties. Shear behavior was coupled with normal stress so that slip does not remain constant but varied with the overturning moment, however, it was assumed to be constant for each time step. The vertical tie tensile reinforcement was modelled using elasto-plastic truss elements at the panel edges. Vertical steel was assumed to provide vertical continuity only and had no shear transfer capability.

The MIT finite element model is shown in Fig. 3. The technique of substructuring was employed to condense out the internal degrees of freedom of the wall panel. The initial frequencies and mode shapes of the structure were determined with a linear elastic model where the five story structure had a fundamental period of 0.17 sec.

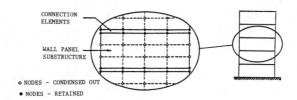

Fig. 3 - MIT Finite Element Model [3].

The loading input to the five story model was an artificial earthquake designed to match the Newmark-Blume-Kapur response spectra for 2% damping. Fig. 4 shows a comparison of the response spectra of the artificial earthquake to the response spectra of the North-South component of the El-Centro 1940 earthquake and the Taft 1952 earthquake.

Post tensioned (PT) and reinforced concrete (R/C) walls of reinforcement ratios of 0.25%, 0.5% and 1% were analyzed. The friction coefficient (μ) was taken as 0.2 or 0.4 for the PT walls and 0.4 only for the R/C walls. A summary of the results is given in Table 1.

According to Becker and Llorente [1,2,3], in the reinforced cases, global slip occurred in the upper floors only. Slip has the effect of limiting seismic forces as energy is dissipated. Rocking results in softening of the connection with a minimal amount of energy dissipation. It affects the fundamental period of the structure as the joint stiffness decreases. As the period of the structure lengthens, the spectral components of the forcing function determine whether or not the shift will increase or decrease force levels in the structure. When rocking and slip occur together, shear must be transferred to the unopened portion of the connection as shown in Fig. 5. Rocking and slip lead to high force concentrations in the panel corners. Deterioration of the panel and connection ends was predicted (Fig. 6).

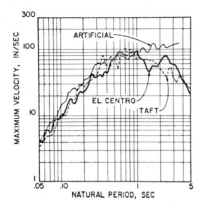

Table 1 - MIT Reinforcement Study, Five Story Wall, Artificial Earthquake, 25 per cent g [2].

Case	Shear (kN)	Moment (kN m)	Deflection (cm)	Cracked (%)	Length level	Global (mm)	Slip level
Linear elastic	1,533	12,598	0.54	–	–	–	–
PT. $\mu=0.2$	1,126	9,422	0.57	13.9	0	0.9	1
PT. $\mu=0.4$	1,467	11,751	0.53	29.0	0	–	–
R/C, 0.25% $\mu=0.4$	1,211	10,490	0.88	38.1	1	3.4	4
R/C, 1.0% $\mu=0.4$	1,185	9,590	0.81	24.3	0	3.4	4
R/C, 0.5% $\mu=0.4$	1,242	9,778	0.83	31.6	0	3.4	4

Fig. 4 - Response Spectrum 5% damping, 1g peak.

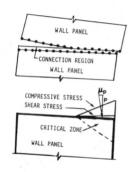

Fig. 5 - Nonlinear Connection Behavior [2].

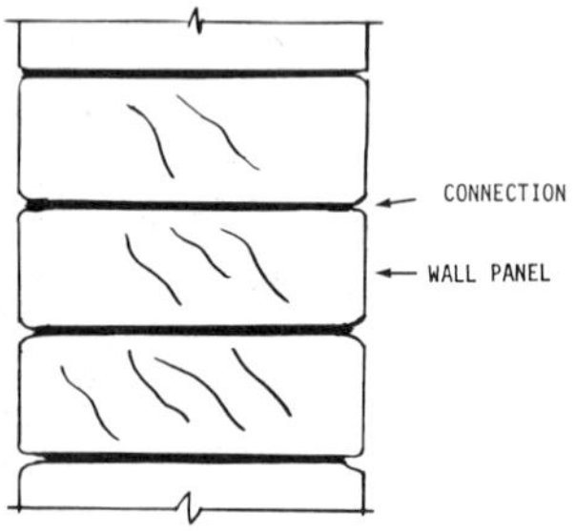

Fig. 6 - Typical Damage to Wall Panel [2].

UNIVERSITY OF CALIFORNIA AT BERKELEY
ANALYTICAL WORK

Schricker and Powell [4] idealized the wall panels in precast LP systems either by elastic beam-type elements (Fig. 7a), by a finite element model (Fig. 7b) or by a combined beam-finite element model (Fig. 7c). The horizontal joints were modelled by nonlinear spring elements with variable force deformation relationships.

Gap elements and shear elements in a variety of combinations were used. A typical gap element was nonlinear with zero tensile strength and stiffness. Shear elements were either elastic-plastic, degrading, or of the shear friction type. The elements were added to a modified version of the Drain 2-D [8] computer code.

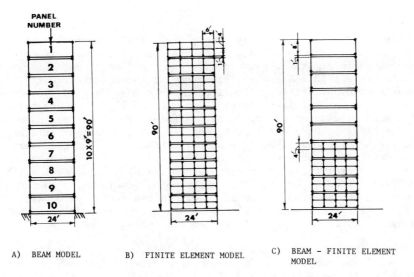

Fig. 7 - U C Berkeley Analytical Models [4].

A ten story single bay large panel wall with horizontal joints was studied. The purpose of the investigation was, to determine the extent to which forces in the structure are reduced when joint slip and gap opening are permitted, to study the distribution of joint deformations for different vertical loads, to investigate the effect of the joint force-deformation relationship on the response of the wall, and to investigate whether the response is sensitive to the assumed ground motion.

The dimensions and mass distribution of the wall were identical to those of the wall studied by Becker et al. [2] presented in the previous section. Mass was lumped at panel corners. Rayleigh damping corresponding to 5% of critical in the first mode was assumed. Post tensioning bars were modelled by elastic spring elements connecting the roof to the foundation with no intermediate connections. A time step of 0.001 seconds was used for all inelastic analyses. Four earthquake records matching the El Centro 1940 N-S component, the Pacoima Dam 1971 S16E component, an artificial earthquake AA, and an artificial earthquake generated to match the Newmark-Blume-Kapur response spectrum with 2% damping were used.

Conclusions from the analysis showed that more experimental data are needed on the behavior of Large Panel joints. When ground motions were varied, the computed response for the given mathematical models were qualitatively similar. The computed response is very sensitive to the assumed post-yield or post-slip behavior. When a joint looses strength after slip, analyses indicate a tendency for slip deformations to be concentrated in the joint that slipped first. If a particular joint is initially weaker than the other joints, slip will tend to be concentrated in this weaker joint. The response can be sensitive to computational procedure and the time step used. Time steps should be varied to insure that consistent results are obtained. Slip and gap opening at joints are effective in reducing the force induced in the panels of LP structures.

EXPERIMENTAL STUDY

An experimental study of a five story precast simple shear wall system similar to the structure modelled by Becker et al. [2,3] was performed on a small shaking table at Drexel University. The small scale modelling technique [9] was used and a 1/32 scale wire reinforced microconcrete model (Figs. 8 and 9) was constructed and tested. Similitude of the mass density was accomplished by adding lead ballast plates (Figs. 10 and 11) as the technique of artificial mass simulation was employed. A total of 4 small scale models were tested. The first 3 Models, 1-A, 1-B, and 1-C, contained a brittle,high strength, 710 GPa (103 ksi) yield strength, vertical tie,and the last Model, 2-A, contained a lower strength, 289 GPa (42 ksi) yield strength, annealed tie. Initial natural frequencies were determined by displacement pullback tests and varied from 44 Hz to 55 Hz in the model domain which corresponds to periods of 0.13 sec to 0.10 sec respectively in the prototype domain. The period of the five story small scale model was slightly lower than that reported by MIT (0.17 sec), thus indicated stiffer small scale models.

After the initial properties were determined,a random vibration test where the model was subjected to a series of simulated earthquakes was undertaken. To increase the amount of nonlinear behavior, several magnitudes of base acceleration were input. Models 1-A, 1-B, and 2-A were subjected to base motions representative of the El Centro 1940 earthquake whereas Model 1-C was subjected to the El Centro, Taft and Pacoima Dam earthquakes. During most tests, displacements measured using LVDT's, were recorded at each floor level and were typically measured at the top of the wall panels. Figs. 12, and 13 show the level 5 horizontal displacement time-history for Model 1-C subjected to the El Centro, and Taft earthquakes, respectively. In this case the earthquakes were scaled to a maximum base acceleration of 0.29g before input. Fig. 14 shows the frequency spectrum of the Model 2-A level 5 horizontal displacement for two different levels of El Centro input. The maximum value of the spectral displacements were scaled to unity for this comparison. Observed, as expected, is a drop in frequency of model response as the base motion is increased in magnitude. This behavior occurs for basically two reasons. Firstly, the model, before it undergoes the higher Magnitude 3 input, was at a state of higher damage level due to accumulated damage from any previous shocks. The model, in the test depicted as Magnitude 3 had undergone 2 previous shocks, thus, the model was initially less stiff and more prone to amplification when subjected to an earthquake such as El Centro. Secondly, the increased magnitude of base motion causes higher levels of inertia to be induced thereby increasing the nonlinear mechanisms of slip and rocking in the model. These mechanisms act to reduce the stiffness and thereby the response becomes more prone to amplification at lower frequencies.

ANALYTICAL STUDY

Description of the Finite Element Method

A mathematical finite element analysis of the simple shear wall system was carried out using a modified version of the Drain 2-D [8] computer code. The finite element mesh, shown in Fig. 15, consisted of wall panels that were idealized by twelve, linear elastic 8 D.O.F. plane stress rectangles each of 20 cm (8 in) thickness. A modulus of elasticity of 31.1 GPa (4500 ksi) and a Poisson's ratio of 0.15 were chosen to match the properties of the micro-

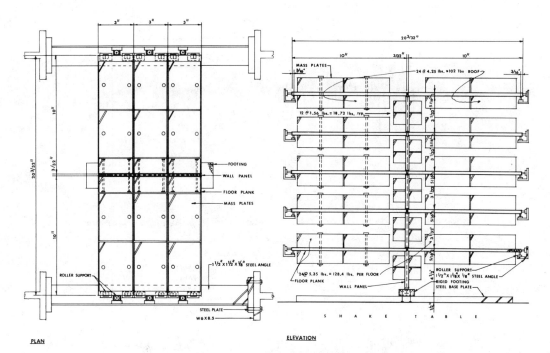

Fig. 8 - Plan View of the Small Scale Model Assembly.

Fig. 9 - Elevation View of the Small Scale Model Assembly.

Fig. 10 - First Wall Panel With Lead Ballast Attached.

Fig. 11 - Small Scale Model With Lead Ballast and Instrumentation.

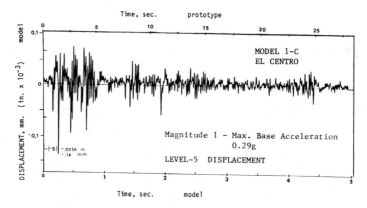

Fig. 12 - Model 1-C, Magnitude 1, El Centro, Level 5 Displacement Vs. Time.

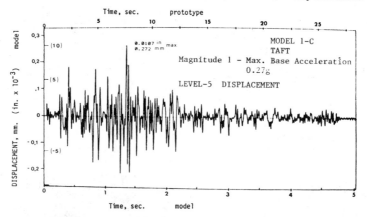

Fig. 13 - Model 1-C, Magnitude 1, Taft, Level 5 Displacement Vs. Time.

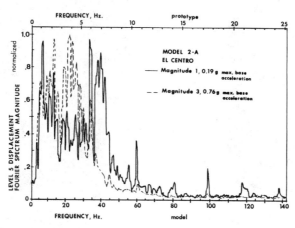

Fig. 14 - Model 2-A, Magnitude 1 and 3, El Centro, Level 5 Displacement Frequency Spectrum.

concrete wall panel. Rayleigh damping corresponding to an initial damping ratio of 5 percent in the first mode and an integration time step of 0.001 seconds were employed in the analysis.

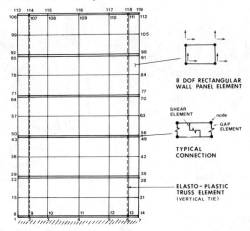

Fig. 15 - Drexel Finite Element Model.

The vertical tie was modelled by a truss element with elasto-plastic behavior. In all cases the tie was assumed to be confined and buckling was not permitted. The continuous ties were connected to node lines 2 to 114 and 6 to 118 corresponding to the location of the vertical tie in the small scale model. Elasto-plastic shear connection elements of two different material types were used across each horizontal joint. The first element type modelled the frictional resistance of the grout/concrete interface. Its strength was given as the coefficient of friction (0.3 or 0.6 were used) times the normal force. The shear stiffness, assumed to be reduced, was taken as 10, 20, or 40 percent (α) of the uncracked shear stiffness. The second shear element modelled the steel vertical tie that acts as a dowel when the joint starts to slip. Its strength was taken as the shear strength of the tie used. A nonlinear gap element modelled the extensional properties of the horizontal joint where the compressive properties of the microconcrete were used and zero strength and stiffness were assumed in tension. In all of the nonlinear studies, the slip was assumed to occur at the top of the lower wall panel (Fig. 2). To place the horizontal inertia effect in the proper location, the structural mass was lumped at the upper nodes of each connection. Joints 1 through 7 were base joints that moved uniformly and in phase during the transient earthquake.

Analysis Results

In general, the analysis was performed as a correlation to the small scale model tests, therefore, the compilation of results dealt most heavily with comparing the displacement patterns of the small scale model and the finite element analysis. Fig. 16 shows the horizontal displacement at level 5 for the Drain 2-D model with a coefficient of friction of 0.3 and an El Centro input of 0.24g maximum. Notice that when this lower coefficient of friction is used, at a time of 2.25 sec. the response becomes biased toward the negative side of the curve. This indicates a global slippage of the system that was not recovered. This type of behavior was not seen in the small scale model response curves (Fig. 12). However, due to the type of LVDT used to measure displacements, this action may have been filtered. The finite element analysis with a 0.6 coefficient of friction, shown in Fig. 17, does not show the bias due to global slip for the 0.24g El Centro case.

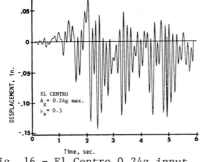
Fig. 16 - El Centro 0.24g input $\mu = 0.3$.

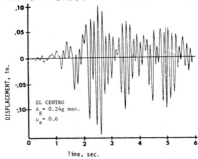
Fig. 17 - El Centro 0.24g input $\mu = 0.6$.

Fig. 18 shows the rocking displacement for the Drain 2-D analysis with a coefficient of friction equal to 0.6 and a base acceleration of 0.36g maximum. The rocking displacement shown is the relative displacement in the vertical direction across the level 1 joint at the rear of the model. This behavior is characterized by an increase in displacement as the joint opens, as indicated on the negative side of the curve, and much smaller displacement on the positive side as compression occurs and stiffness increases. This same type of behavior was shown in the small scale model.

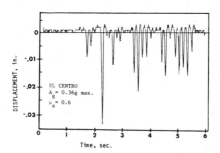

Fig. 18 - Rocking Displacement El Centro 0.36g, $\mu = 0.6$.

Figs. 19 and 20 give a comparison of the maximum response predicted by the finite element model to the maximum response measured in the small scale model tests scaled up to the prototype domain. As shown in Fig. 19, an elastic analysis will grossly underpredict the overall maximum displacement, even for base accelerations lower than 0.1 g maximum which was less than 1/3 of the actual El Centro earthquake. Also, the finite element analysis using a coefficient of friction of 0.3 or 0.6 was able to predict the measured experimental values for Models 1-B, 1-C, and 2-A up to a base acceleration of 0.3g. Higher magnitude tests showed much more scatter especially since accumulated damage in the small scale model would alter its initial conditions during these tests. The Drain 2-D model still gave a reasonable prediction at the higher magnitude for these models. The response of Model 1-A was underpredicted by the mathematical analysis. This model however was accidently damaged prior to the first earthquake input and its behavior will not be represented by the mathematical model used. The maximum level 5 displacement for the MIT analysis of a five story precast wall was more than double any Drain 2-D prediction. The MIT model, however, was initially less stiff than the Drain 2-D or the small scale model and was subjected to an artificial earthquake where a greater response would be expected.

Maximum level 1 rocking displacement is compared for the Drain 2-D analysis to the small scale model tests in Fig. 20. The rocking displacement is the relative movement in the vertical direction across a joint. Again, Model 1-A was damaged and the results may not be representative for this model. For Models 1-B and 1-C the FEM model using a $\mu = 0.6$ greatly overestimates the rocking response and the lower coefficient of friction of 0.3 gives a better prediction. In predicting the maximum horizontal displacement as well as the maximum rocking displacement the mathematical model with a coefficient of friction equal to 0.3 is indicated from these studies.

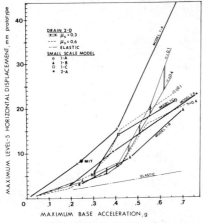

Fig. 19 - Max. Level 5 Displacement Analytical and Experimental

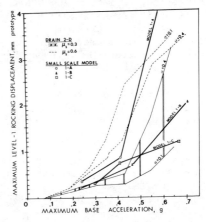

Fig. 20 - Max. Rocking Displacement Analytical and Experimental

CONCLUSIONS

More experimental work is needed if the behavior of precast shear walls is to be quantified. The comparison of small scale model tests to a computer analysis has shown similar horizontal floor-level displacement and rocking displacement time-histories. The computer study was able to predict, within reason, maximum values of displacements especially for initial tests with maximum values of base acceleration lower than 0.3 g.

The overall behavior of the system can not be predicted by an elastic analysis even for low levels of base acceleration. Some type of nonlinear analysis must be performed to predict response. For design purposes the complex analytical and experimental models presented in this report are prohibitive with respect to cost. More simplified, less costly, methods of analysis are necessary. Simplified methods can be attained through a full understanding of structural behavior which can be accomplished with further full scale, small scale, and analytical studies.

ACKNOWLEDGMENTS

The work reported herein was conducted in the Dept. of Civil Engrg., Drexel University, under Grants No. PFR-7924723 and CEE-8342561 from the National Science Foundation. The cognizant NSF Program official for these grants was Dr. John B. Scalzi.

REFERENCES

1. Becker, J.M. and Llorente, C., "Seismic Design of Precast Concrete Panel Buildings," Workshop on Earthquake Resistant Reinforced Concrete Building Construction, July 11-16, 1977, University of California, Berkeley, sponsored by NSF.

2. Becker, J.M., and Llorente, C., "The Seismic Response of Simple Precast Concrete Panel Walls," Proceedings of the U.S. National Conference on Earthquake Engineering, Stanford University, California, August, 1979.

3. Becker, J.M., Llorente, C., and Mueller, P., "Seismic Response of Precast Concrete Walls," <u>Earthquake Engineering and Structural Dynamics</u>, John Wiley & Sons, Inc., NY, Vol 8, 1980, pp. 545-564.

4. Schricker, V., and Powell, G.H., "Inelastic Seismic Analysis of Large Panel Buildings," Report No. EERC 80-38, University of California, Berkeley, CA, September, 1980.

5. Caccese, V., and Harris, H.G., "Seismic Behavior of Precast Concrete Large Panel Buildings Using a Small Shaking Table, Report 3 - Correlation of Experimental and Analytical Results," Drexel University, Structural Dynamic Laboratory, Report No. D85-01, Department of Civil Engineering, Drexel University, Philadelphia, Penna., June 1985.

6. Harris, H.G., and Caccese, V., "Seismic Behavior of Precast Concrete Large Panel Buildings Using a Small Shaking Table," Proceedings of the 8th World Conference on Earthquake Engineering, July 21-28, 1984, San Francisco, California.

7. Caccese, V., and Harris, H.G., "Nonlinear Behavior of Precast Concrete Shear Walls Under Simulated Earthquake Loading," Proceedings of the 3rd ASCE Engineering Mechanics Specialty Conference on Dynamic Response of Structures, University of California, Los Angeles, March 31, 1986.

8. Kanaan, A.E., and Powell, G.H., "General Purpose Computer Program for Dynamic Analysis of Inelastic Plane Structures," Report No. EERC 76-6, Earthquake Engineering Research Center, University of California, Berkeley, California, April, 1973.

9. Sabnis, G.M., Harris, H.G., White, R.N., and Mirza, S.M., <u>Structural Modeling and Experimental Techniques</u>, Prentice Hall, Inc., Englewood Cliffs, New Jersey, 1983.

STRESSES IN COMPOSITE MASONRY SHEAR WALLS
DUE TO EARTHQUAKE LOADS

S. C. Anand[I] and M. A. Rahman[II]

ABSTRACT

The behavior of brick-block composite masonry walls which act as shear walls and are subjected to horizontal in-plane loads due to earthquakes is a subject that has received little attention. Of particular interest in these walls are the shearing stresses created in the collar joint due to transfer of the vertical and horizontal loads from the loaded block wythe to the brick wythe. The authors have previously presented analyses to estimate these shear stresses utilizing a 2-D finite element model for a single story wall. In this paper, collar joint shear stresses due to vertical and horizontal in-plane loads for a two story composite masonry wall are presented.

INTRODUCTION

A composite wall, as shown in Fig. 1, consists of one wythe of brick and another of concrete block, with the cavity between the two wythes (called the collar joint) filled with grout. The loads, which may be vertical due to gravity, or horizontal due to wind or earthquakes, are generally transmitted only to the block wythe by the floor slab as shown in Fig. 2. These loads, nevertheless, are partially transferred to the brick wythe and create shear stresses in the collar joint. If the shear stress becomes too large at the brick-grout or block-grout interface, delamination may occur and the load could no longer be transferred to the brick wythe and the wall may experience distress. It is, therefore, important that a correct estimate of these interlaminar shearing stresses in the collar joint is made.

The senior author of this paper, together with his graduate students and colleagues, have been engaged in research during the last few years to determine the magnitude and variation of shear stress in the collar joint in composite masonry walls subjected to vertical and horizontal in-plane loads. This investigation has been carried out both analytically and experimentally, results of which have been presented in Proceedings of the various national and international conference [1-2]. Besides this research, only a few other studies have been undertaken to estimate the magnitudes of shear stresses in composite masonry walls [13-16].

I. Professor, Department of Civil Engineering, Clemson University, Clemson, SC, 29634-0911

II. Graduate Research Assistant, Department of Civil Engineering, Clemson University, Clemson, SC, 29634-0911

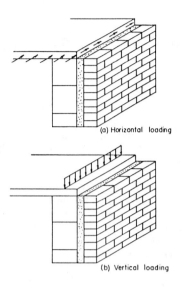

Fig. 1. Loads on a Block Wythe

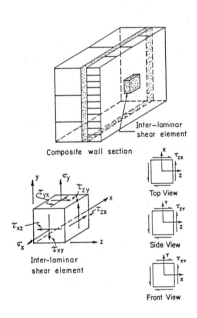

Fig. 2. Shear Stresses in the Collar Joint

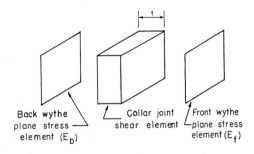

Fig. 3. Components of A Composite Element

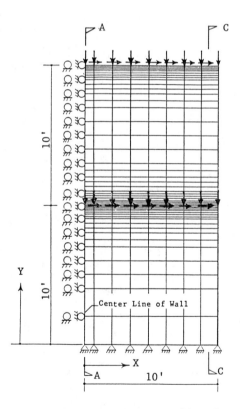

Fig. 4. Finite Element Model and Loads

The previous research by this author and his colleagues has led to the development of a two-dimensional finite element model that is capable of predicting the normal and shear stresses in the collar joint. This model has primarily been utilized to compute collar joint shear stresses due to vertically applied loads on the block wythe. However, recently the authors used this 2-D model to estimate shear stresses in the collar joint of a one-story composite masonry shear wall subjected to horizontal in-plane loads due to earthquake effects [11]. It was found in this investigation that the magnitudes of the horizontal interlaminar shear stresses were of the same order of magnitude as the vertical shear stresses due to vertical loads and could cause delamination of the wythes from the collar joint. It was also shown that the basic shape of the shear stress distribution due to horizontal and vertical loads is quite similar, although horizontal loads lead to larger shear stresses in the collar joint because of smaller in-plane horizontal stiffness of the block wythe.

In this paper, the previously developed 2-D model is utilized to estimate shear stresses in the collar joint of a two-story composite masonry shear wall subjected to gravity and earthquake loads. It is shown that both the in-plane shear stress τ_{xy} and out-of-plane shear stresses τ_{xz} and τ_{yz} in the collar joint can have substantial magnitudes, and may cause delamination of the collar joint.

2-D FINITE ELEMENT MODEL

The special two-dimensional finite element model, that is capable of predicting the out-of-plane shear stresses τ_{xz} and τ_{yz} (as shown in Fig. 2) in the collar joint of a composite masonry wall when subjected to vertical and horizontal in-plane loads, has been described in detail elsewhere [1-2, 5-6, 8-9, 11]. However, a short description of this model is given here for completeness. In the development of the 2-D model, a new "composite" element is created. In this new element, the front and back wythes are each modelled as plane stress elements. These two elements are joined together by a collar joint shear element. These three elements are shown in Fig. 3, a combination of which forms the composite element with eight nodes, four for each wythe. As each node has two degrees of in-plane displacement freedom, there are in total sixteen degrees of freedom for the composite element. The shear stresses that act in the collar joint shear element have previously been shown in Fig. 2.

The stiffness matrix of the proposed composite element is formed by combining the stiffness matrices of the two wythe elements with the collar joint stiffness matrix to yield

$$[k_{ce}] = \left[\begin{bmatrix} [k_f]_{8\times 8} & [0] \\ [0] & [k_b]_{8\times 8} \end{bmatrix} + [k_{cj}]_{16\times 16} \right]_{16\times 16} \quad (1)$$

in which $[k_{ce}]$ = composite element stiffness matrix, $[k_f]$ = front wythe element stiffness matrix, $[k_b]$ = back wythe element stiffness matrix, and $[k_{cj}]$ = collar joint stiffness matrix.

Detailed expressions for these matrices may be found in Refs. [1-3]. It should be noted that following assumptions have been made in the development of Eq. (1): 1) All materials are considered as linearly elastic, homogeneous, and isotropic; 2) Displacements are assumed to vary linearly between the nodes in an element; 3) Out-of-plane bending effects in the wall are ignored; and 4) The collar joint as well as the two wythes are unreinforced.

Any composite masonry wall subjected to vertical and/or horizontal in-plane loads can be analyzed using this model. Each wythe is divided into an appropriate finite element mesh, noting that the mesh distribution on each wythe is exactly the same. Thus, the complete wall becomes an assemblage of the above developed composite elements. The solution for the unknown nodal displacements is obtained by the standard stiffness method. The values of the nodal displacements are then used in the appropriate strain-displacement matrices to compute the element normal and shear strains in the wythes and collar joint. Expressions for these matrices may be found in Ref. [11].

ANALYSIS OF A TWO-STORY COMPOSITE SHEAR WALL

As a composite masonry wall could be used as shear wall in a two-story construction, it is of interest to estimate shear stresses in the collar joint due to vertical (gravity) and horizontal (earthquake) loads. The composite shear wall is assumed to be made of an 8 in (0.2 m) thick wythe of concrete block that is attached to a 4 in (0.1 m) thick wythe of clay brick through a 2 in (0.05 m) thick grouted collar joint. The wall is 20 ft (6.1 m) high and is subjected to vertical and horizontal loads from the floor slabs at 10 ft (3.05 m) and 20 ft (6.1 m) heights, respectively.

It is assumed that the composite wall is a 20 ft (6.1 m) long end-wall for a two-story covered area of the size 20 ft x 60 ft (6.1 m x 18.3 m). The 60 ft (18.3 m) long sides are provided with columns in the middle on which rest beams each 30 ft (9.15 m) long. The floor slabs at each level, thus, behave as 20 ft x 30 ft (6.1 m x 9.15 m) two-way slabs as far as the vertical loads are concerned, and one 20 ft (6.1 m) side of each slab rests on the concrete block wythe of the composite masonry shear wall. The vertical load at the roof level is based on a 5 in (0.13 m) concrete slab with 2 in (0.05 m) covering and no live loads whereas at the 10 ft (3.05 m) level, it is based on a 10 in (0.25 m) concrete slab, 2 in (.05 m) floor finish and a live load of 100 psf (4.79 kN/m^2) As the slab is supported on all four sides, the total load on the composite wall is calculated from the appropriate tributary area. This leads to the total vertical loads on the block wythe of 8 k (35.6 kN) and 24.8 k (110.4 kN) at the roof and 1st floor levels, respectively. Assuming a parabolic distribution for the vertical loads yields the corresponding maximum load intensities at the center of the wall of 0.63 k/ft (9.2 kN/m) and 1.95 k/ft (28.5 kN/m).

The horizontal earthquake loads that act on the composite masonry shear wall at each floor level are computed using the base shear formula given in ANSI A58.1-1982 [17]. It is assumed in this calculation that the shear wall supports the horizontal load due to a tributary area of 20 ft x 30 ft (6.1 m x 9.15 m). It is further assumed that the building is located in earthquake zone 3 and has an importance factor of one. Since the collar joint is unreinforced, even if the composite wall itself is reinforced, the value for the numerical coefficient, K, in the base shear formula is taken as four.

Performing standard calculations for the base shear formula and dividing the total load for the two floors appropriately leads to horizontal loads of 50.5 k (225 kN) and 42 k (187 kN) at the roof and 1st floor levels, respectively. These loads further yield the corresponding uniform horizontal load intensities on the block wythe of 2.52 k/ft (36.8 kN/m) and 2.10 k/ft (30.6 kN/m).

It is not quite clear if the above mentioned horizontal loads would act with a uniform intensity on the block wythe. The actual distribution of this load would depend upon the aspect ratio of the floor slab and the relative stiffnesses of the shear wall to the rest of the structure. It could be surmised, however, that the center of the wall would resist the largest horizontal load. Accordingly, a load distribution that is parabolic in shape, and leads to the same total load at each floor level, has also been assumed in these analyses. The maximum horizontal load intensities in this case are equal to 3.97 k/ft (57.9 kN/m) at the roof level and 3.30 k/ft (48.1 kN/m) at the first floor at the center of the wall and zero at the ends.

As a wall with the horizontally applied loads can be assumed to be in a state of antisymmetry about an axis through the midpoint along its length, only half the length of the wall is considered in the analysis. The wall is considered pinned at the base and the antisymmetric condition can be modelled by providing horizontal rollers at the midpoint of the wall. Similarly, for the vertically applied loads, only half the length of the wall with vertical rollers at the center line can be considered due to symmetry. These boundary conditions along with the finite element mesh used in the analysis are shown in Fig. 4. It should be noted that the loads at each floor level are applied only on the block wythe. The finite element mesh consists of 768 quadrilateral elements and 882 nodal points with a half-bandwidth of 44. A relatively fine mesh is utilized near the points of load application as it is known from the previous experience that large stress changes in the collar joint occur near these points. The values for the elastic modulus and Poisson's ratio utilized in the analysis are based upon the formulas recommended by the Brick Institute of America and the National Concrete Masonry Association, and ultimate strengths of the various material components measured in the laboratory. These calculations may be found in Appendix A of Ref. [3] and lead to modulii values of 1040 ksi (7.18×10^6 kN/m^2) and 2000 ksi (13.8×10^6 kN/m^2) for the block and brick wythes, and 1600 ksi (11.04×10^6 kN/m^2) for the collar joint, respectively. The corresponding values for the Poisson's ratio are 0.25, 0.25 and 0.20.

RESULTS AND DISCUSSION

Because the strength of the collar joint is critical in the composite action of the wall, shear stresses in the collar joint at two different locations, defined by Sections A-A and C-C in Fig. 4, are shown and discussed. In addition, the normal strains and normal stresses in the brick and block wythes, and collar joint are investigated along the length of the wall. The above mentioned stresses and strains are shown for a uniform as well as a parabolic horizontal load distribution at these sections. In addition, shear stress distribution in the collar joint due to vertical loads is also presented.

Horizontal Shear Stress τ_{xz} in the Collar Joint

The horizontal loads due to earthquake are applied to the block wythe through the slabs and are transferred to the brick wythe through the collar joint [11]. This load transfer produces horizontal shear stresses in the collar joint which are functions only of the x-displacements of the individual wythes as the proposed model does not include the out-of-plane displacements in the z-direction in its formulation.

The horizontal shear stress distribution in the collar joint due to the uniformly distributed horizontal earthquake loads of 2.1 k/ft (30.6 kN/m) at the 2nd floor level and 2.52 k/ft (36.8 kN/m) at the roof level is shown in Fig. 5. It can be noted that the shear stress magnitude near the roof level is very much higher compared to its value at the 2nd floor level [40 psi vs. 16 psi (276 N/m^2 vs. 110 N/m^2)] although the load intensity at the roof is only slightly larger than at the 2nd floor level. This phenomenon can be attributed to the fact that the load in the block wythe at the 2nd floor level can be transferred to the brick wythe through the collar joint in a region both above and below the slab. On the other hand, the load at the roof level can be transferred to the brick wythe only below the roof slab. Since the stiffness of the collar joint at the 2nd floor level to resist the horizontal in-plane loads is double than that at the roof level, smaller displacements, strains and stresses are caused at this level. The shear stresses in the collar joint drop to zero within a distance of 10-12 inches (0.25 m - 0.305 m) from the slab level, a phenomenon similar to the one observed in the analysis of a single story composite wall [11]. This indicates that the relatively large shear stiffness of the collar joint is much more predominant in the load transfer mechanism between the two wythes.

The horizontal shear stress distribution in the collar joint at the two levels due to a parabolic distribution of the horizontal earthquake loads is shown in Fig. 6. As the maximum load intensity at each level is at Sec. A-A near the center of the wall, the maximum shear stresses also occur at this section with a value of 60 psi (414 N/m^2) at the roof level and 25 psi (173 N/m^2) at the 2nd floor level. These magnitudes are approximately 1.57 times larger than those due to the uniform load. It is of interest to note that the ratio between the maximum horizontal shear stress due to the parabolic load to that due to the uniform load at each level is the same as the corresponding ratio between the loads at any point along the length of the wall. This indicates once again that the stiffness of the collar joint in transferring the load across the collar joint is of major importance instead of the total longitudinal stiffness of the composite wall. It can be seen in Fig. 6 that the shear stress magnitudes at Sec. C-C are much smaller than at Sec. A-A in this case. This is so because the horizontal load intensity varies parabolically at each floor level and has a much smaller magnitude at Sec. C-C.

It should be noted here that the vertical loads from slabs at the two levels act symmetrically with respect to the center line along the length of the wall and, thus, do not cause any horizontal shear stresses in the collar joint irrespective of the shape of the load distribution.

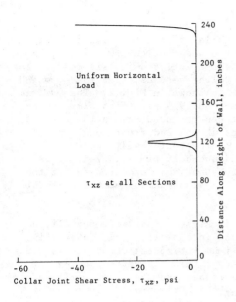

Fig. 5. τ_{xz} in the Collar Joint due to Uniform Horizontal Load

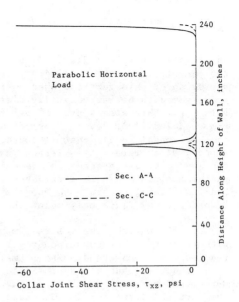

Fig. 6. τ_{xz} in the Collar Joint due to Parabolic Horizontal Load

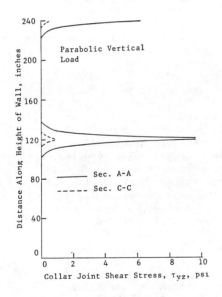

Fig. 7. τ_{yz} in the Collar Joint due to Vertical Load

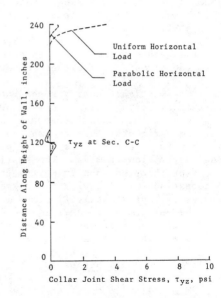

Fig. 8. τ_{yz} in the Collar Joint due to Horizontal Load

Vertical Shear Stress τ_{yz} in the Collar Joint

The vertical shear stress, τ_{yz}, in the collar joint is a function only of the relative vertical displacements between the nodes of the two wythes, as the out-of-plane horizontal displacements are not included in this model. The variation of this vertical shear stress due to the vertical slab loads, which are assumed to be parabolically distributed, is shown in Fig. 7. The maximum values for this stress are 6.5 psi (44.8 N/m^2) and 9.5 psi (65.5 N/m^2) at the roof and 2nd floor levels, respectively. It can be noted, as in the case of the horizontal loads, that although the loads at the 2nd floor level are approximately three times larger than those at the roof level, the corresponding shear stresses are only one and one-half times larger. This phenomenon can again be attributed to the fact that the load transfer at the 2nd floor level occurs both above and below the slab whereas at the roof this transfer takes place only below the slab level.

If the maximum shear stress magnitudes in Figs. 6 and 7 are normalized with respect to the corresponding maximum load intensities, some interesting results can be observed. The normalized maximum horizontal shear stresses due to earthquake loads are approximately 50% larger than the corresponding normalized vertical shear stresses. This is so because the wythe stiffness in the vertical direction is much larger than in the horizontal direction, which in turn reduces the vertical load transfer to the brick wythe. Hence, the shear stresses in the collar joint are smaller in this case.

Due to the antisymmetric behavior of the composite wall about its center line along the length when subjected to in-plane horizontal earthquake loads, the maximum vertical displacements occur near the wall end Sec. C-C and are zero at the center line Sec. A-A. Thus, the vertical shear stresses are largest at Sec. C-C and zero at Sec. A-A. As the vertical displacements in the two wythes become equal to each other at approximately 20 in (0.51 m) away from the slabs, τ_{yz} vanishes to zero at this height.

The vertical shear stress distribution in the collar joint due to the horizontal earthquake loads is shown in Fig. 8. It is obvious in this figure that the maximum shear stress τ_{yz} occurs at the roof level. Its value for a uniform horizontal load assumption is equal to 3.7 psi (25.5 N/m^2) at Sec. C-C. This stress reduces to 0.7 psi (4.8 N/m^2) at this section if a parabolic load assumption is made. The smaller value for the parabolic load is due to the fact that the load intensity at this section is much smaller in this case. A comparison of τ_{yz} with τ_{xz} suggests that the maximum horizontal shear stress is approximately 11 times larger than the maximum vertical shear stress for the uniform horizontal load. For parabolic loading, however, the horizontal shear stress is much larger at the center line of the wall (Sec. A-A) where the vertical shear stress is zero.

Collar Joint Shear Stress τ_{xy}

The collar joint shear stress τ_{xy} for the uniform and parabolic horizontal loads is shown in Fig. 9. Although at first glance, these distributions appear to be rather difficult to grasp, a little effort can lead to a better understanding of this shear stress variation if one notes that τ_{xy} in the collar joint is a function of the horizontal displacement gradient along the height and vertical displacement gradient along the length of the wall.

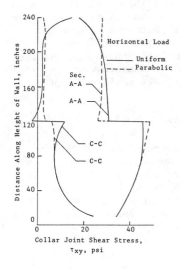

Fig. 9. τ_{xy} in the Collar Joint due to Horizontal Load

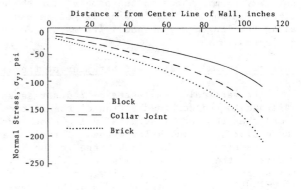

Fig. 10. σ_y at the Base of the Wall

It can be anticipated that the value of the shear stress τ_{xy} at the top of the wall due to the uniform horizontal load would be the same at all points along its length. This can be seen in Fig. 9 where τ_{xy} is equal to 17 psi (117 N/m²) at Secs. A-A and C-C at the roof level. At Sec. A-A near the center of the wall, τ_{xy} increases from 17 psi (117 N/m²) at the top to 30 psi (207 N/m²) just above the second floor level. This can be attributed to the fact that Sec. A-A of the wall could be construed as the neutral axis of a cantilever wall fixed at the base and subjected to a uniform horizontal load at the roof level. Shear stress τ_{xy} at the neutral axis in a beam is naturally maximum. Similarly, Sec. C-C could be regarded as the top (or bottom) fiber in the cantilever wall where shear stress is zero. This can also be seen in Fig. 9 where τ_{xy} at Sec. C-C just above the second floor level is approximately zero. As additional horizontal load intensity is applied at the second floor level, shear stress τ_{xy} takes a sudden jump and its magnitude becomes equal to 45 psi (310 kN/m²) and 12 psi (83 kN/m²) just below the second floor level at Secs. A-A and C-C, respectively. The total horizontal in-plane load along the length of the wall is transferred uniformly to the foundation and produces a uniform τ_{xy} in the collar joint of 32 psi (221 kN/m²) at the wall base.

If the horizontal load is assumed to have a parabolic load distribution along the length of the wall, the shear stress τ_{xy} at the roof level will be proportional to the load intensity at any point. This is the case in Fig. 9 where τ_{xy} is equal to 28 psi (193 kN/m^2) at Sec. A-A and 5 psi (35 kN/m^2) at Sec. C-C. These values remain approximately constant up to just above the second floor level where additional horizontal load with a parabolic distribution is applied. This produces a sudden increase of shear stress which becomes equal to 48 psi (331 kN/m^2) and 8 psi (55 kN/m^2) at Sec. A-A and Sec. C-C, respectively. As in the case of the uniform load distribution, the magnitude of the collar joint shear stress τ_{xy} at the bottom of the wall becomes equal at all sections along its length. This magnitude is equal to 32 psi (221 kN/m^2) which is the same as for the uniform load. Although the shear stress τ_{xy} in the collar joint is not zero due to a non-uniform vertical load along the length of the wall, its magnitude is relatively small and can be neglected.

Vertical Normal Strains and Stresses at the Wall Base

The vertical normal strains and stresses were computed at the base of the wall due to the combined action of the vertical and horizontal loads. It was found that the normal strains in the block wythe were the same as those in the collar joint and brick wythe. This is due to the fact that most of the load transfer from the block wythe to the brick wythe through the collar joint occurs in the top short distance of the wall. It is also for this reason that the shape of the load intensity does not have any effect on the strain distribution at the wall base, and the normal strains for the uniform and parabolic horizontal loads are the same.

The normal stresses at the base in various materials of the composite wall are shown in Fig. 10. The maximum stresses, on the compression side of the center line of the wall, in the brick wythe, collar joint, and block wythe are 210 psi, 170 psi and 105 psi (1.45 x 10^3, 1.17 x 10^3 and 0.72 x 10^3 kN/m^2), respectively. It is evident from this figure that the stress variation along the length of the wall is nonlinear. This shape can be attributed to the large length to height ratio of the wall, making the wall behave like a very deep cantilever beam.

CONCLUSIONS

The maximum horizontal shear stress in the collar joint for a two-story wall subjected to horizontal earthquake loads (assuming a parabolic load distribution) occurs near the center line of the wall at the roof level and has a magnitude of 60 psi (415 kN/m^2). This shear stress reduces to zero at a distance of approximately 20 inches (0.51 m) from the top of the wall. The horizontal shear stress at the second floor level, on the other hand, is much smaller.

The vertical shear stress, τ_{yz}, due to all loads is rather small. Its maximum value due to vertical loads occurs at the center line of the wall and is equal to 9.5 psi (66 kN/m^2) at the second floor level and 6.5 psi (39 kN/m^2) at the roof level. The corresponding maximum vertical shear stress due to the horizontal loads is much smaller and it occurs at Sec. C-C where its value due to the vertical loads is small. This suggests that Sec. A-A at the center of the wall length is a more critical section.

The shear stress, τ_{xy}, on planes perpendicular to where τ_{xz} acts, is quite significant with its maximum value of 48 psi (331 kN/m^2) just below the second floor level. It appears, therefore, that both locations, one at the roof level and the other just below the second floor level are critical for a failure of the collar joint. Accordingly, it is recommended that failure criteria based on the normal and shear stresses in the collar joint at these locations should be developed and utilized to predict the progressive failure of composite masonry walls subjected to gravity and earthquake loads.

ACKNOWLEDGMENTS

The research reported in this paper was supported by Grant No. ECE-8410081 from the National Science Foundation. Computations were carried out on VAX-11/780 at the Computation Laboratory of the College of Engineering at Clemson University. The financial support of NSF and the cooperation of the Clemson University Computer Center are gratefully acknowledged.

REFERENCES

[1] Anand, S.C., and Young, D.T., "A Finite Element Model for Composite Masonry," Journal of the Structural Division, ASCE, Vol. 108, No. ST12, 1982, pp. 2637-2648.

[2] Anand, S.C., Young, D.T., and Stevens, D.J., "A Model to Predict Shearing Stresses Between Wythes in Composite Masonry Walls due to Differential Movement," Proceedings, 2nd North American Masonry Conference, University of Maryland, College Park, MD, August 9-11, 1982, pp. 7.1-7.16.

[3] Anand, S.C., and Gandhi, A., "A Finite Element Model to Compute Stresses in Composite Masonry Walls Due to Temperature, Moisture, and Creep," Proceedings, 3rd Canadian Masonry Symposium '83, University of Alberta, Edmonton, Alberta, June 6-8, 1983, pp. 34.1-34.20.

[4] Anand, S.C., and Dandawate, B., "Creep Modelling for Composite Masonry Walls," Proceedings, 5th ASCE-EMD Specialty Conference, University of Wyoming, Laramie, Wyoming, Aug., 1-3, 1984, pp. 432-437.

[5] Anand, S.C., and Stevens, D.J., "Computer-Aided Failure Analysis of Composite Concrete Block-Brick Masonry," Proceedings, International Conference on Computer-Aided Analysis and Design of Concrete Structures, Split, Yugoslavia, Sept. 17-21, 1984, pp. 649-661.

[6] Anand, S.C., and Stevens, D.J., "A Simple Model for Shear Cracking and Failure in Composite Masonry," Proceedings, 6th International Congress On Fracture, New Delhi, India, Dec. 4-10, 1984, pp. 2915-2922.

[7] Anand, S.C., and Dandawate, B., "A Numerical Technique to Compute Creep Effects in Masonry Walls," Proceedings, 3rd North American Masonry Conference, University of Texas at Arlington, Arlington, TX, June 3-5, 1985, pp. 75.1-75.14.

[8] Anand, S.C., and Stevens, D.J., "Shear Stresses in Composite Masonry Walls Using a 2-D Model," <u>Proceedings, 3rd North American Masonry Conference</u>, University of Texas at Arlington, Arlington, Tx, June 3-5, 1985, pp. 41.1-41.15.

[9] Anand, S.C., "Shear Stresses in Composite Masonry Walls," <u>Session Proceedings</u>, entitled "New Analysis Techniques for Composite Masonry," <u>ASCE Structures Congress'85</u>, Chicago, IL, Sept. 16-18, 1985, published by ASCE, New York, NY, pp. 120-131.

[10] Anand, S.C., and Rahman, M. A., "Creep Modelling for Masonry Under Plane Strain," <u>Proceedings, 9th Conference on Electronic Computation</u>, ASCE, University of Alabama in Birmingham, Feb. 23-26, 1986, pp. 629-643.

[11] Anand, S.C., and Rahman, M. A., "Stresses in Composite Masonry Shear Walls," <u>4th Canadian Masonry Symposium</u>, Fredricton, B.C., Canada, June 2-4, 1986, 15 pp.

[12] Brown, R.H., and Cousins, T.E., "Shear Strength of Slushed Composite Masonry Collar Joints," <u>Proceedings, 3rd Canadian Masonry Symposium '83</u>, University of Alberta, Edmonton, Alberta, June 6-8, 1983, pp. 38.1-38.16.

[13] Grimm, C.T., and Fowler, D.W., "Differential Movement in Composite Load-Bearing Masonry Walls," <u>Journal of the Structural Division</u>, ASCE, Vol. 105, No. ST7, 1979, pp. 1277-1289.

[14] Williams, R.T., and Geschwindner, L.F., "Shear Stress Across Collar Joints in Composite Masonry Walls," <u>Proceedings, 2nd North American Masonry Conference, University of Maryland</u>, College Park, MD, Aug. 9-11, 1982, pp. 8.1-8.17.

[15] Self, M.W., "Design Guidelines for Composite Clay Brick and Concrete Block Masonry. Part I - Composite Masonry Prisms," <u>Department of Civil Engineering, University of Florida</u>, Gainesville, FL, Research Report, April 1983.

[16] Porter, M., Ahmed, M., and Wolde-Tinsae, A., "Preliminary Work on Reinforced Composite Masonry Shear Walls," <u>Proceedings, 3rd Canadian Masonry Symposium '83</u>, University of Alberta, Edmonton, Alberta, June 6-8, 1983, pp. 15.1-15.12.

[17] "Minimum Design Loads for Buildings and Other Structures," ANSI A58.1-1982, <u>American National Standards Institute Inc.</u>, New York, NY.

MEASURED HYSTERESIS IN A MASONRY BUILDING SYSTEM

D. P. Abrams[I]

ABSTRACT

Results of a laboratory study are presented which was done to investigate resistance of a two-story masonry building system to repeated and reversed lateral forces. The test structure was constructed at full scale of reinforced concrete masonry, and consisted of two symmetrical C-shaped walls with window openings. Design of the specimen was based on the newly revised Chapter 24 of the 1985 Uniform Building Code for masonry structures in Seismic Zones III or IV. The measured hysteretic character of the test structure is discussed in terms of present engineering methods for estimating strength and deflection of masonry building systems.

INTRODUCTION

Design or analysis of masonry buildings subjected to light, moderate or strong earthquake motions requires a substantial amount of interpretation of how such a structure will behave under repeated and reversed lateral forces. Present U.S. codes of practice are founded on a scant amount of field or laboratory data, and still remain largely empirical in this respect. Results of the experimental program described in this paper provides new information regarding the actual behavior of such a structural system.

Primary lateral-force resistance in a typical masonry building system is provided by wall elements bending within their plane. To estimate strengths or nonlinear deflections of such an element under bending or shear force reversals requires solution of a complex structural mechanics problem. If door or window openings are present, the difficulty of the problem and certainty of the solution lies outside the realm of current engineering practice.

In lieu of the analytical approach, strengths and deflections can be approximated for design using code provisions such as the Uniform Building Code [1]. For example, shear strength of a wall with openings is computed as the product of an allowable stress and the sum of cross-sectional areas for all piers comprising the wall. It is assumed with little justification that highly stressed elements will deform inelastically with no loss of strength, and thus, will shed stress to lightly stressed elements. Despite the reliance on inelastic behavior, code provisions suggest that lateral deflections be estimated assuming linear elastic behavior of homogeneous

---

[I] Associate Professor, Department of Civil Engineering, University of Illinois, Urbana, Illinois

elements. Furthermore, no reference is made to the effects of cyclic or reversed forces for which deflections of masonry materials and walls have been shown [2,3,4] through previous experiments to be quite relevant.

The purpose of the experimental study described in this paper was to study nonlinear hysteretic behavior of a two-story masonry building system subjected to reversals of lateral force. A sample structure was chosen which contained two parallel and symmetrical C-shaped walls with openings (Fig. 1). The test specimen was designed in accordance with provisions of the 1985 Uniform Building Code [1] for buildings in Seismic Zones III or IV. Lateral forces were applied so that the structure would drift equal amounts back and forth, and for progressively increasing motions, reveal the hysteretic character of the building system.

SPECIMEN DESIGN

The layout of window openings (Figs. 1 and 2) was established so that the central pier between openings would be square, and significantly stiffer than the external piers to the outside of each opening. With this arrangement, shear force would be attracted initially to the central pier, and its inelastic deformability to redistribute lateral shear to adjacent external piers could be examined.

It was of interest to investigate the width of the end flange that was effective in resisting tensile or compressive stress for different amplitudes of lateral force. The width of the end flanges were sized three-units long past the face of the in-plane wall. The present UBC prescribes an effective flange width of six times the unit thickness which is equal to this amount.

For structures in Seismic Zones III or IV, the UBC requires a total area of both vertical and horizontal steel equal to 0.002, and a minimum in either direction equal to 0.0007 of the wall gross area. For an 8-inch block, this translates to one No. 4 bar at 16 inches in the vertical direction. One No. 4 bar was placed above and below, and along side of each opening. Horizontal reinforcing bars were placed in bond beam units. Holes were cut in the bottom of these channel units for protrusions of vertical reinforcement. To complement the horizontal bars, No. 9 gage truss-type joint reinforcement was placed at 16 inches. This was the only horizontal steel in the piers.

Walls were coupled at each of two levels with a reinforced concrete floor slab. The slab and the slab-wall connection were made sufficiently strong so that deformation would occur primarily in the masonry walls.

FABRICATION OF THE TEST SPECIMEN

Construction of the specimen was started in early November of 1985 when a concrete "grade" beam was cast on the test floor. Vertical dowel reinforcing bars were anchored in the beam which was stressed to the test floor with 20 one-inch diameter bolts. Concrete was cast within lips of floor holes to insure that slip of the beam would not occur. Readings from dial gages during testing confirmed this.

Block for the first story was laid by a crew of eight union bricklayers in two days. After three days, the first-level concrete slab which had been cast previously on the floor, was lifted and placed on top of the walls. Vertical wall reinforcement was placed through four-inch diameter holes in the slab. The lower story was then partially grouted up to the midheight of the top course.

The masons returned two weeks later to lay block for the second story. Following erection of the second-story walls which were topped with a continuous bond beam, vertical reinforcement was placed and the walls were grouted. The top-level slab was lifted and placed on top of the walls a few days later. Bolts were anchored in the grouted bond beam through four-inch diameter holes in the top slab. This provided 38 shear keys of grout with the bond beam. Construction of the structure was completed the first week in January, 1986.

MATERIALS

Type S mortar (1.0:0.5:4.5 parts Portland Cement, lime and sand) was mixed adjacent to the specimen in a mixer which was borrowed from a local masonry contractor. Compressive strengths of five 4-inch diameter cylinders averaged 1790 psi with a c.o.v. of 14%.

Concrete block was obtained locally. Flatwise compressive strength on the net area averaged 3960 psi with a c.o.v. of 11% for five specimens. Two-unit prisms were made during construction of the test specimen. Tests of three ungrouted prisms resulted in an average compressive strength equal to 3450 psi on the net area with a c.o.v. equal to 4%.

Each batch of grout consisted of 2 bags of Portland Cement, 6.0 cubic feet of surface-dried masonry sand, and 3.2 cubic feet of pea gravel. An expansive agent was added to the grout as it was mixed in a concrete mixer located within the laboratory. Grout prisms were cast using block lined with paper towels as formwork. Average compressive stress was 2630 psi with a c.o.v. equal to 8% for three specimens.

Reinforcement consisted of No. 4 Grade 60 bars. Tension tests of 10 samples resulted in a mean yield stress equal to 70.9 ksi with a c.o.v. of 4%, and an ultimate stress equal to 112 ksi with c.o.v. of 2%.

DESCRIPTION OF THE EXPERIMENTAL PROGRAM

The test structure was subjected to lateral sways within its plane. Forces from each hydraulic actuator (Figs. 2 and 3) were controlled in accordance with the same displacement signal. Because of fluctuations in resistance of each parallel wall, lateral forces applied to each wall varied slightly. However, since each wall softened with an increase in lateral force, this effect was self-correcting. Forces and deflection maxima for each half cycle of loading are presented in Table 1.

The lateral force distribution was idealized with a concentrated force at the top level. This resulted in a moment-to-shear ratio at the base typical of a three-story building subjected to a triangular force distribution. Variation in lateral force distribution with localized damage was not modeled because the effect is not significant for low-rise structures. The sequence of lateral sways was established so that the general hysteretic character of the structure could be revealed. The test was done at a static rate of loading because of flow limitations of the hydraulic power supply system.

Relations between lateral force and top-level displacement were monitored continuously using analog x-y plotters. Unloading points were chosen based on this information. Data from 76 channels were digitized and stored on diskette at specified load points. Instrumentation consisted of load cells to detect lateral forces applied to each wall, and displacement transducers to detect lateral deflection at each of the two levels, and flexure and shear distortions of each pier on the north wall. In addition, 38 strain gages were mounted to surfaces of concrete block to detect distributions across the width of the wall. Cracks were marked as they appeared with black or red felt tip markers for each direction of loading. At points of maximum displacement, high resolution photographs were taken to track the relative movement of targets marked on the north face of the test structure. Targets were marked on an eight-inch grid to detect slippage along head and bed joints. As a backup, the entire test procedure was recorded on video tape.

OBSERVED BEHAVIOR OF TEST SPECIMEN

A summary of the loading history is given in Table 1. For the system, lateral force is the sum of each actuator force, and deflection is the average of slab movements measured along the axis of each wall. Force-deflection relations for the first story of the north wall are shown in Fig. 4. These curves were similar in shape for the south wall, and for the overall system.

The relation between story shear and diagonal extension of each pier at the first story of the north wall is shown in Fig. 5. Diagonal extension is the sum of extension in one direction and shortening in the opposite direction as measured across the entire pier width and height. The correspondence in the shape and nature of the force-deformation relation for the central pier and the overall structural system suggests that behavior of the system was governed by the central piers.

Crack patterns observed at the end of testing are presented in Fig. 6.

The First Three Cycles (Fig. 4a)

Maximum lateral force for either wall was limited to nominally 40 kips (180 kN) for the first three cycles. This represented a force level equal to approximately 70% of the maximum observed for all testing. During the first cycle, the structure responded essentially linear with the exception of a noticeable slip at the extreme point of the first westwardly cycle (positive force on plot). The slip was also observed in the record of the displacement transducer located vertically at the base of the east side of the north wall. The specimen responded with a reduced stiffness for the second and third cycles corroborating that flexural cracking had occurred during the first cycle. Because the structure was deflected to a new maximum for the second cycle, the structure was slightly softer upon loading for the third cycle.

It is interesting to note that the lateral drift of the specimen for these three cycles was still quite low: the ratio of top-level deflection to height was less than 1/1500. Another interesting point was the hairline cracks first appeared during the third cycle along the mortar joints in the central piers at the first story. Two previous cycles to nominally the same force level did not reveal these cracks.

The Fourth and Fifth Cycles (Fig. 4b)

Despite the fact that the maximum deflection for the third cycle was essentially the same as for the second, the loading stiffness for the fourth cycle was less than that for the previous loading. This may be attributed to cracking of the central piers during the third cycle. The specimen was observed to stiffen during loading in each direction of the fourth cycle, suggesting that cracks from the previous half cycle were closing.

Upon reaching near the maximum force for the fourth cycle, the specimen softened appreciably. This was matched with a rapid increase in crack widths along the bed joints and across the head joints in the central pier. It was obvious that the joint reinforcement was either slipping or yielding. It is of interest to note that the story was resisting additional force during this time, inferring that the central pier was deforming inelastically but not losing strength as force was then distributed to external piers. Shear distortions of piers verify this action.

Loading in the fifth cycle showed a marked reduction in stiffness from the previous cycle. As for hysteresis in reinforced concrete elements, the force-deflection curve for the masonry structure tended to approach the point of previous unloading.

The Sixth and Seventh Cycles (Fig. 4c)

Similar tendencies as observed for the previous two cycles were observed for the sixth and seventh cycles. The structure continued to soften significantly because of progressively increasing deflections and not forces.

Maximum lateral forces were observed during the sixth cycle. A maximum force applied to the north wall equal to 57.1 kips (250 kN) is equal to 3.6 times that prescribed by UBC on the basis of an allowable shear stress equal to 35 psi and the net area of all three piers. For the south wall, this ratio was 3.3.

Slippage of units along bed joints and across head joints continued to occur. Cracks continued to propagate even when the top-level deflection was held constant for data recording. The system acted as a continuum of jointed masses that was inertly dissipating energy when no additional energy was being input. Cracks were predominantly in mortar joints and not across masonry units. Widths of cracks were minor except for those within the central piers at large deflections.

The central piers continued to deform inelastically while the story force increased. External piers were observed to behave inelastically to a much lesser degree than for the central piers. Ductility of the central pier with joint reinforcement as the only horizontal steel was sufficient to develop strengths of external piers.

Behavior of the external piers was influenced by the direction of the lateral force. Piers were observed to deform inelastically when resisting axial compression. Although shear force resisted by each pier could not be measured, it may be inferred that pier force is related to the amount of inelastic deformation. Thus, the distribution of lateral force was not symmetrical for the symmetrical structure.

The Eighth and Ninth Cycles (Fig. 4d)

These cycles were the first that showed a substantial loss of stiffness and strength for the first story. Substantial damage had occurred during the seventh cycle which was reflected in these subsequent cycles. Despite no increase in deflection or force maxima, the specimen responded with a limp and weak character.

Maximum deflections were an order of magnitude larger than those of the initial cycle. Strength of the specimen was reduced to 50% of the maxima observed during the sixth cycle.

After reducing the lateral force to zero, the cracks closed. It was very interesting to observe the lack of apparent visual damage. The mortar joints had lost bond with the block units which resulted in substantial slippage under load, but did not alter the appearance of the wall.

CONCLUSIONS

Data from the experimental study have shown that the strength and stiffness of a reinforced masonry structure can each be very large. However, after a few large-amplitude cycles of loading, the stiffness and strength of the test structure deteriorated substantially despite a rather normal appearance. This deterioration which is not reflected in the code would have a significant effect on lateral drifts, fundamental period of vibration and energy dissipation.

Specific conclusions are noted.

1. Resistance of overall structure was governed by nonlinear behavior of first-story central pier.

2. Ductility of central pier was sufficient to redistribute story shear to adjacent piers. Strengths of external piers were developed.

3. Maximum shear strength of walls was at least three times that of UBC allowable values, however, a severe reduction in strength was observed with cycling.

4. Initial lateral stiffness of walls reduced by an order of magnitude with repeated and reversed cycles. Lateral drifts ranged from 1/1500 to 1/150 of the structure height.

ACKNOWLEDGEMENTS

Concrete block and other masonry materials were furnished by the Illinois Masonry Institute in Park Ridge, Illinois. Construction of the test structure was done by Local 17 of the International Union of Bricklayers. Research support was provided by the Campus Research Board and the Department of Civil Engineering at the University of Illinois. Appreciation is extended to Jeffrey Hamera, Walter Kania and Janet Paluga, Civil Engineering students, for their assistance with the research project.

REFERENCES

1. The Uniform Building Code, International Conference of Building Officials, Whittier, California, Chapter 24, 1985.

2. Abrams, D. P., J. L. Noland, R. H. Atkinson, and P. D. Waugh, "Response of Clay-Unit Masonry to Repeated Compressive Forces," _Proceedings of the 7th International Brick Masonry Conference_, Melbourne, Australia, 565-576, February 1985.

3. Williams, D. and J. C. Scrivener, "Response of Reinforced Masonry Shear Walls to Static and Dynamic Cyclic Loading," _Proceedings of the 5th World Conference on Earthquake Engineering_, Rome, Italy, June 1973.

4. Priestley, M.J.N., "Seismic Design of Masonry Buildings - Background to the Draft Masonry Design Code DZ4210," _Bulletin of the New Zealand National Society for Earthquake Engineering_, Vol. 13, No. 4, December 1980.

Table 1

Force and Deflection Maxima per Cycle

Cycle	North Wall		South Wall		System	
	Force (kips)	Deflect. (inches)	Force	Deflect.	Force	Deflect.
1w*	39.4	0.114	33.5	0.094	72.9	0.104
1e	41.0	0.093	15.9	0.072	56.9	0.083
2w	38.5	0.137	35.1	0.120	73.6	0.129
2e	38.8	0.127	33.8	0.122	72.6	0.125
3w	38.8	0.151	32.2	0.133	71.0	0.142
3e	39.1	0.138	37.9	0.133	77.0	0.136
4w	47.6	0.274	38.1	0.244	85.7	0.259
4e	48.0	0.224	42.0	0.207	90.0	0.215
5w	47.6	0.313	38.7	0.280	86.3	0.297
5e	48.5	0.256	43.0	0.236	91.5	0.246
6w	55.3	0.599	44.7	0.539	100.0	0.569
6e	57.1	0.509	49.4	0.492	106.5	0.501
7w	54.1	0.916	44.9	0.848	99.0	0.882
7e	51.8	0.693	47.5	0.661	99.3	0.677
8w	42.6	1.345	45.5	1.268	88.1	1.307
8e	41.4	1.333	51.0	1.286	92.4	1.310
9w	24.7	1.531	30.3	1.452	55.0	1.492

Note: * w signifies westward movement, e eastward movement
1.0 inch equals 25.4 millimeter
1.0 kip equals 4.46 kilonewtons

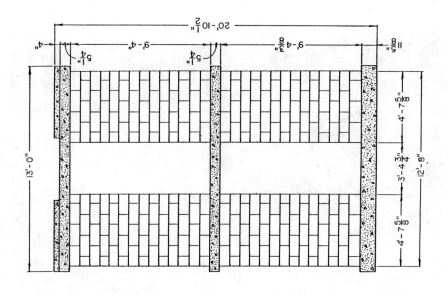

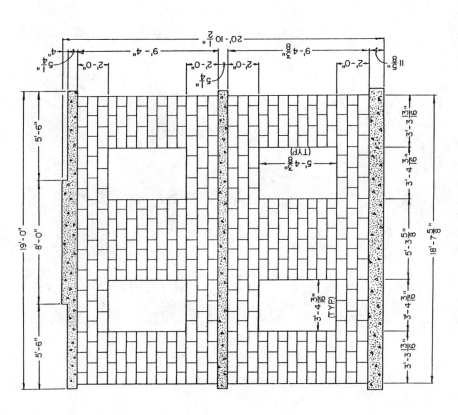

Fig. 1 Description of Test Specimen

Fig. 2 Experimental Arrangement

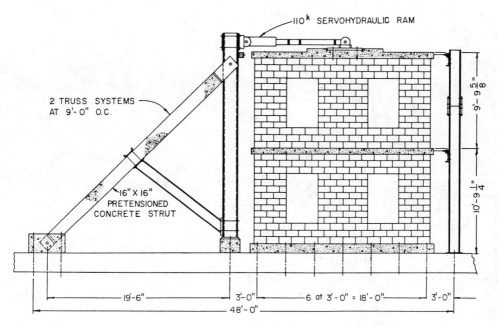

Fig. 3 Reaction Structure and Test Specimen

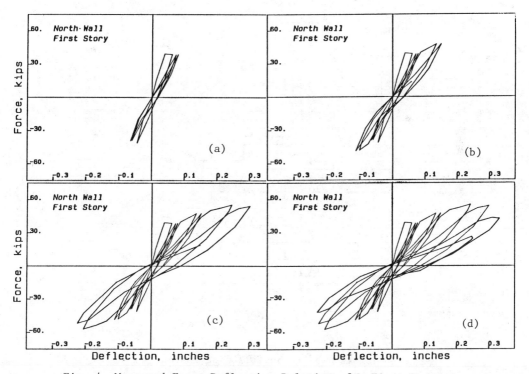

Fig. 4 Measured Force-Deflection Relations for First Story

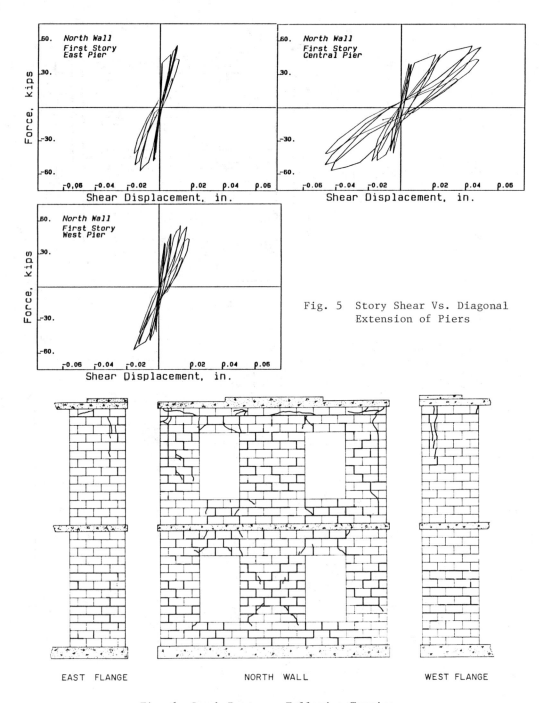

Fig. 5 Story Shear Vs. Diagonal Extension of Piers

Fig. 6 Crack Patterns Following Testing

EARTHQUAKE SIMULATED PERFORMANCE OF AN UNREINFORCED
AND A REINFORCED MASONRY SINGLE-STORY HOUSE MODEL

G. C. Manos[I] and R. W. Clough[II]

ABSTRACT

In order to determine the reinforcing requirements of prototype single story masonry dwellings built at areas of United States Uniform Building Code Seismic Zone 2 an experimental investigation was conducted with five single-story masonry models subjected to simulated earthquake input. Whereas the masonry walls of the first four models were subjected to either in-plane or out-of-plane horizontal inertia forces, the walls of the fifth model were simultaneously subjected to in-plane and out-of-plane horizontal excitations together with the vertical input motion. Some of the results obtained from the fifth model are presented together with a discussion of the fifth model earthquake performance. Comparisons are also made between the fifth model performance and the performance of the previously tested models. Finally, design recommendations are presented together with conclusions valid for prototype single-story masonry houses.

INTRODUCTION

Some of the results obtained by a research program entitled "Laboratory Studies of the Seismic Behavior of Single-Story Masonry Houses in Uniform Building Code Seismic Zone 2 of the U.S.A." are presented in this paper. The objective of this program was to evaluated the U.S. Department of Housing and Urban Development (HUD) seismic design and construction requirements for single-story masonry dwellings in Seismic Zone 2 areas by investigating the earthquake performance of single-story masonry models, constructed with full-scale components and subjected to simulated earthquake motions. In this way, by studying the performance of five models subjected to a large number of earthquake excitations it was aimed to determine, extrapolating the model performance to prototype conditions, the maximum earthquake intensity that could be resisted by an unreinforced masonry house and then to evaluate the additional resistance that could be provided to the structure by partial reinforcement. The research program was supported by the U.S. Department of Housing and Urban Development. After the completion of all the tests for the first four models tentative recommendations were made based on these results. The tests were conducted on the Earthquake Simulator of the University of California at Berkeley, which is capable of simultaneous vertical and one dimensional horizon-

---

I. Assistant Professor of Structural Engineering, Aristotle Univerity of Thessaloniki, GREECE, formerly Ass. Research Engineer, University of California, Berkeley, Earthquake Engineering Research Center, CA, U.S.A.

II. Nishkian Professor of Structural Engineering, University of California, Berkeley, California, U.S.A.

tal motion. However, because the effect of the combined in-plane and out-of-plane inertia forces on the earthquake performance of the masonry walls of these four models was not studied by these tests a fifth model was studied, basically the same as the previous models but oriented in such a way on the shaking table that its masonry walls were simultaneously subjected to two horizontal (in-plane and out-of-plane) and the vertical component of the input motion (Fig. 1 and refs. 1,2,3 and 4).

TEST PROGRAM

Features of the models

All five tested houses were full-scale models and consisted of four wall panels with standard size door and window openings (A, A1 B, B1); they were square in plan with wall length equal to 16 ft (4.88 m) and wall height 8 ft 8 in. (2.64 m). Each model had a wooden roof loaded with an additional 12,000 lb (53,392 Nt) weight in order to simulate roof load conditions of typical prototype masonry houses. For the fifth model (House 5) as well as for Houses 1,2 and 4 concrete blocks were used with nominal dimensions 4 x 6 x 16 in. (102 x 152 x 406 mm) whereas for House 3 hollow clay bricks were used with nominal dimensions 4 x 6 x 12 in. (102 x 152 x 305 mm). The compressive strength of the mortar used

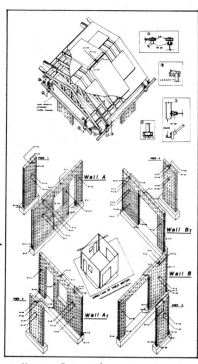

Fig. 1 House 5 testing arrangement

for House 5 was 2229 psi (15.4 Nt/mm^2) whereas the compressive and the diagonal tension strengths of masonry specimens built together with the masonry walls were 2200 psi (15.2 Nt/mm^2) and 69 psi (0.48 Nt/mm^2), respectively. Whereas Houses 1,2 and 3 had two walls unreinforced (A and A1) and two walls partially reinforced (B and B1), Houses 4 and 5 were initially unreinforced and after the completion of a sequence of tests they were partially reinforced, by placing vertical reinforcing bars inside the masonry cells at predetermined positions that were next filled

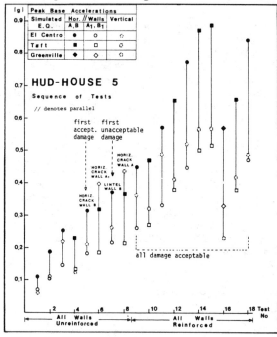

Fig. 2 House 5 test sequence

with grout, which was poured from the top of the walls. Full details of all
five models are included in references 1,2,3 and 4).

Type of excitation

All models were attached firmly to the shaking table concrete slab. For
the majority of the test the earthquake motion provided by the shaking table
was based on accelerograms recorded at El Centro and Taft (Kern County) during
the earthquakes of May 1940 and June 1952, respectively. All simulated earthquake records had no time scaling and had both the horizontal and the vertical
component. Instrumentation was provided to measure the acceleration and displacement response, both in-plane and out-of-plane, for all four masonry walls at
different levels along the wall height. In addition, vertical uplift displacements were measured at each end of every wall or at each end of every door opening in order to record the formation of cracks along the wall height or the
seperation of the masonry from the concrete foundation during the tests. All
five models were subjected to simulated motions of continously increasing
intensity; this was done by adjusting the displacement control commands of the
shaking table hydraulic actuators in a way that it resulted in base accelerations with continously increasing values. The sequence of tests both for the unreinforced and the partially reinforced House 5 is shown in figure 2. The
abscissae represent the sequential test numbers whereas the ordinates are the
peak values of the three base acceleration components. The intensity of each
test is in this way portrayed by plotting these peak acceleration values of
the three components of excitation acting simultaneously on the House 5 masonry
walls (in-plane, out-of-plane and vertical). In order to provide a more realistic means of evaluating the severity of the base motion for each shaking
table test, apart from the peak
value of horizontal base acceleration, the effective peak acceleration
(E.P.A.) was calculated for each
test from the recorded time history
of the base motion. It has been long
recognised that the effective peak
acceleration, based on the spectral
acceleration values for a relatively
wide period range (0.1 to 0.5 sec.),
it relates rather better to the
damage potential of a given earthquake than the corresponding peak
ground acceleration. Sharp spikes
in accelerograms tend to overempha-
size the peak acceleration value
while the importance of long period
motions due to distant earthquakes
may be underestimated by the peak

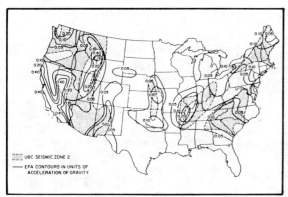

Fig. 3 Effective Peak Acceleration
Contour Map.

ground acceleration. Moreover, by characterising the base motion intensity for
each test in terms of effective peak acceleration a basis of comparison is also
provided in order to evaluate the performance of all five models that were
subjected to various earthquake records of continously increasing intensity
despite changes in the input motion that were introduced by variations of the
shaking table performance and of the table-structure interaction. Finally, the
concept of effective peak acceleration forms a basis of comparison between the
shaking table tests and the expected prototype conditions, as the prototype

ground motion expected intensity can also be expressed in terms of effective peak acceleration, as shown in figure 3 (ref. [5]).

Test results

Tables I and II list the measured peak acceleration and displacement response for the House 5 masonry walls, which was observed during tests No. [7,8] for the unreinforced House 5 and during tests [13,14] for the partially reinforced House 5, respectively. Table III lists the observed minimum and maximum acceleration amplification values during the tests of both Houses 4 and 5 (sharp spikes excluded), together with an approximate average amplification value; these values are based on the measured acceleration response at the roof level and at the top level of the mainly in-plane walls (A and B) as well as at the top level of the mainly out-of-plane walls. This is done first for the unreinforced masonry walls, with the peak horizontal base acceleration not exceeding 0.4 g (0.3 g E.P.A.), and then for the partially reinforced masonry walls, with peak horizontal base acceleration not exceeding the value of 0.9 g (0.52 g E.P.A.). In both cases the base horizontal acceleration component and the measured acceleration response were in the direction parallel to walls A and B. Figures 4 and 5 depict the observed House 5 amplification factors for the masonry wall acceleration response measured at the top of the walls level. Although the base motion for all House 5 tests, as already mentioned, resulted in two horizontal components acting simultaneously on all four masonry walls, the plotted in figures 4 and 5 values represent the ratio of the measured top of the wall peak acceleration response (in-plane for walls A and B and out-of-plane for walls A1 and B1) to the base peak horizontal acceleration component in the same direction (parallel to walls A and B).

Table I. Peak Response of Unreinforced House-5

Base Motion	Acceleration Components of Horizontal Motion (g)		Wall Name	Peak Wall Accelerations (g)		Peak Wall Displacements (in. /mm)	
				I.P.	O.P.	In-Plane	Out-of-Plane
Test No. 7 El Centro H = 0.43 g V = 0.26 g	Parallel to walls	P.B.A. | E.P.A.	A	0.690	0.550	0.034 /0.86	0.304 /7.72
			B	0.853	0.512	0.359 /9.12	0.146 /3.71
	A , B	0.368 | 0.260	A1	0.483	0.827	0.296 /7.52	0.361 /9.17
	A1, B1	0.214 | 0.150	B1	0.423	0.765	0.180 /4.57	0.239 /6.07
Test No. 8 Taft H = 0.43 g V = 0.44 g	Parallel to walls	P.B.A. | E.P.A.	A	0.857	0.732	0.080 /2.03	0.392 /9.96
			B	1.023	0.448	0.410 /10.4	0.323 /8.20
	A , B	0.368 | 0.300	A1	0.559	1.002	0.369 /9.37	0.421 /10.7
	A1, B1	0.213 | 0.170	B1	0.451	0.992	0.292 /7.42	0.284 /7.21

Notes: A,B were loadbearing walls whereas A1,B1 were nonloadbearing walls
H = Peak Horizontal Acceleration   V = Peak Vertical Acceleration
P.B.A.= Peak Base Acceleration   E.P.A.= Effective Peak Acceleration

Table II. Peak Response of Partially Reinforced House-5

Base Motion	Acceleration Components of Horizontal Motion (g)		Wall Name	Peak Wall Accelerations (g)		Peak Wall Displacements (in. /mm)	
				I.P.	O.P.	In-Plane	Out-of-Plane
Test No.13 El Centro H = 0.90 g V = 0.52 g	Parallel to walls	P.B.A. | E.P.A.	A	0.777	0.524	0.013 /0.33	0.146 /3.71
			B	0.833	0.534	0.043 /1.09	0.156 /3.96
	A , B	0.775 | 0.400	A1	0.447	1.195	0.042 /1.07	0.425 /10.8
	A1, B1	0.448 | 0.230	B1	0.477	1.096	0.045 /1.14	0.368 /9.35
Test No.14 Taft H = 1.01 g V = 0.57 g	Parallel to walls	P.B.A. | E.P.A.	A	0.876	0.664	0.021 /0.53	0.191 /4.85
			B	1.007	0.865	0.075 /1.91	0.206 /5.23
	A , B	0.872 | 0.520	A1	0.586	1.602	0.045 /1.14	0.805 /20.5
	A1, B1	0.504 | 0.300	B1	0.563	1.300	0.053 /1.35	0.557 /14.2

Notes: A,B were loadbearing walls whereas A1,B1 were nonloadbearing walls
H = Peak Horizontal Acceleration   V = Peak Vertical Acceleration
P.B.A.= Peak Base Acceleration   E.P.A.= Effective Peak Acceleration

Table III. Observed Amplification Factors

Calculated from observed acceleration response at:	House - 4						House - 5					
	Unreinforced			Partially Reinforced			Unreinforced			Partially Reinforced		
	Min. Value	Max. Value	Appr. Aver.	Min. Value	Max. Value	Appr. Aver.	Min. Value	Max. Value	Appr. Aver.	Min. Value	Max. Value	Appr. Aver.
Top of the timber Roof	1.69	2.72	2.11	1.29	2.15	1.73	1.50	4.64	2.67	1.13	4.17	2.16
Top of the walls A , B	1.00	2.72	1.75	1.00	3.33	1.73	0.90	2.38	1.48	0.95	1.18	1.10
Top of the walls A1, B1	1.59	4.33	2.72	1.41	2.16	1.78	1.00	2.26	1.55	1.16	1.85	1.50

Note: Sharp spikes were excluded from the measurements.
Appr. Aver. = An approximate average of all the measured values.

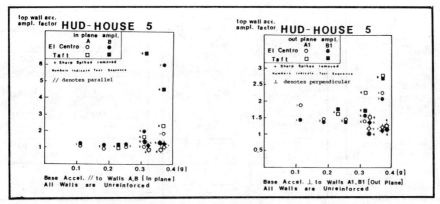

Fig. 4 Unreinforced House 5 amplification factors.

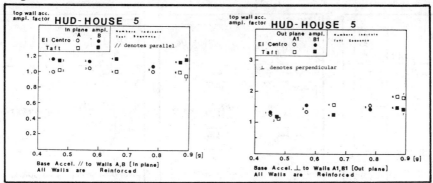

Fig 5 Partially reinforced House 5 amplification factors.

The top of the wall acceleration and displacement House 5 response is depicted in figure 6 for test No. 8 with the masonry walls being unreinforced, and in figure 9 for test No. 14 with the masonry walls being partially reinforced. These figures present the measured response in graphic form at particular instants selected to coincide either with peak values of the displacement response, which is magnified and is plotted with dashed line against the solid line that represents the undeformed masonry wall plan, or coincide with peak acceleration response, which is plotted using arrows. Time history plots of the measured at the top of the masonry walls displacement and acceleration response are presented in figures 7 and 8 for unreinforced walls A1 and B during test No. 8 and in figures 10 and 11 for partially reinforced walls A1 and B during test No. 14. These plots, which also contain vertical (uplift) displacement response for the same tests, show the most significant part of the total observed displacement or acceleration response that had a duration of over 35 seconds; this significant part in the case of the Taft simulated base motion is placed between the 5th and the 12th second from the beginning of the earthquake motion. For each wall the following parameters are presented:

- At the top row of each figure the in-plane displacement (left) and the in-plane acceleration (right) response whereas at the middle row the out-of-plane displacement (left) and the out-of-plane acceleration (right) response.

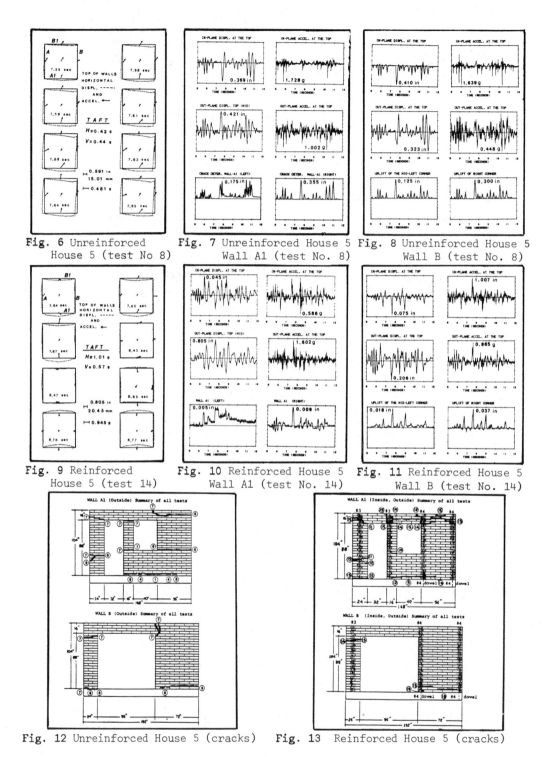

Fig. 6 Unreinforced House 5 (test No 8)
Fig. 7 Unreinforced House 5 Wall A1 (test No. 8)
Fig. 8 Unreinforced House 5 Wall B (test No. 8)
Fig. 9 Reinforced House 5 (test 14)
Fig. 10 Reinforced House 5 Wall A1 (test No. 14)
Fig. 11 Reinforced House 5 Wall B (test No. 14)
Fig. 12 Unreinforced House 5 (cracks)
Fig. 13 Reinforced House 5 (cracks)

- The vertical (uplift) displacement response is plotted at the bottom row in these figures. These vertical displacements record for wall A1 the formation of cracks at the various horizontal mortar joints all along the wall height whereas for wall B they record the formation of cracks at the first and second horizontal mortar joints near the foundation level. The time history plots for these vertical displacements correspond to the left and right end of each masonry wall (A1 and B) , respectively.

Finally, a summary of the formation and the propagation of the cracks for all House 5 tests are presented in figure 12 for unreinforced walls A1 and B, and in figure 13 for partially reinforced walls A1 and B; circled numbers indicate in these figures the sequential test number during which the particular crack formed for the first time or propagated.

DISCUSSION OF THE TEST RESULTS

From the observed performance of the unreinforced House 5 the following points can be made:

a1. During test No. 5 (El Centro base motion with peak horizontal base acceleration equal to 0.36 g) the first structural crack appeared at the bottom of the main shear resisting element of loadbearing wall B (Fig. 12). Following the formation of this first structural crack, the house response was dominated by uplift displacements of wall B at the crack location inducing large in-plane displacements for wall B and large out-of-plane displacements for wall A1 (Figs. 6,7 and 8). These large in-plane and out-of-plane displacements resulted in further cracking of the unreinforced masonry walls, mainly for the nonloadbearing walls A1 and B1 (Fig. 12). The observed peak crack opening was 0.300 in. (7.62 mm) for wall B and 0.355 in. (9.02 mm) for wall A1 during test No. 8 (Figs. 7,8 and 12).

a2. The asymmetric cracking of the main shear resisting masonry elements of House 5, which was built with an initial asymmetric configuration similar to the rest three models (Houses 2,3 and 4), influenced the response of the masonry house as a whole and resulted in the development of significant torsional and distortional displacement response (Fig. 6). However, House 5 performed satisfactorily during all tests from No. 1 to No. 6; the first unacceptable damage appeared during test No. 7 in the form of partial loss of support for the door lintel beam of wall B (Fig. 12). The peak horizontal acceleration at the base during this test was 0.43 g (E.P.A. component parallel to walls A and B was equal to 0.26 g). Despite the formation and propagation of various structural cracks in a cumulative way for the masonry walls of the unreinforced House 5, this model was far from the collapse stage at the end of test No. 8 with peak horizontal acceleration at the base equal to 0.43 g (E.P.A. component paralle to walls A and B was equal to 0.30 g) (Fig. 2 and Tables I and II).

From the observed performance of the partially reinforced House 5 the following points can be made:

b1. The presence of partial reinforcement in House 5 kept the in-plane displacement response of the loadbearing walls A and B at low levels, even during very high intensity tests with peak horizontal base acceleration equal to 1.01 g (E.P.A. component parallel to walls A and B equal to 0.52 g). As can be seen when comparing figures 8 and 11, this must be attributed to the effective way that partial reinforcement controls the damage by preventing the formation of excessive cracking. The observed peak uplift displacement for wall B was 0.037 in. (0.94 mm) during test No. 14 and 0.021 in. (0.53 mm) for wall A1 during test No. 16 (Figs. 6 to 11 and Tables I and II). Despite the large out-of-plane displacements of nonloadbearing walls A1 and B1, the partial reinforcement was again very effective in controlling the damage in these walls and succeeded in preventing extensive crack propagation, even during very high intensity tests (Figs. 7,10 and 13 and Table II).

b2. The response of the partially reinforced House 5 did not exhibit any pronounced torsional or distortional response (Figs. 6 and 9).

From the observed acceleration response amplification factors for both the unreinforced and the partially reinforced House 5 response the following points can be made:

c1. The amplification factors that were observed for the in-plane wall A and B and out-of-plane wall A1 and B1 acceleration response are larger for the unreinforced House 5 than for the partially reinforced House 5. For both the unreinforced and the partially reinforced House 5 the amplification factors at the top of the walls are in general lowere than those at the top of the roof. This may be explained by the spectral characteristics of the input motion and the frequency characteristics of the house and roof system (Figs. 4 and 5. The test sequence numbers of figure 5 start with the No. 1 test of partially reinforced House 5, which corresponds to test No. 9 of figure 2).

c2. The amplification factor at the top of the roof observed for Zone 2 intenseity tests (E.P.A. $\leq$ 0.20 g) of the unreinforced House 5 has a maximum value of 2.50, which is in agreement with the ATC-3 definition of effective peak acceleration. The amplification factor at the top of the roof for the partially reinforced House 5 and for intensities well beyond those expected in Zone 2 (E.P.A.= 0.52 g) and reaching intensities expected in Zone 4, has a maximum value of 4.17 and an approximate average value of 2.16.

c3. The amplification factors for the out-of-plane acceleration of walls A1 and B1 are larger than the corresponding values for the in-plane acceleration of walls A and B for all cases except for the unreinforced House 5 in its postcracking state (Figs. 4 and 5 and Table III).

EXTRAPOLATION OF TEST RESULTS TO PROTOTYPE CONDITIONS

The following parameters, which are believed to have a significant influence on the earthquake performance of single-story masonry houses, were considered when the validity of the extrapolation of the test results to prototype conditions was evaluated as briefly discussed below. Because all models were constructed with full scale masonry components no model scaling influence was considered in the observed house earthquake performance.

<u>Seismic Input</u>

The used shaking table input motions were realistic reproductions of recorded Western United States ground motions, which for the masonry walls of House 5 had two horizontal and one vertical component, as already explained.

The main objective of the research program was to evaluate the single-story masonry house earthquake performance for Seismic Zone 2 of the U.S. Uniform Building Code areas. As explained before the extrapolation of the shaking table test intensity to prototype conditions was made in terms of effective peak acceleration values. As can be seen from figure 3 UBC Zone 2 includes a wide range of E.P.A. values, from less than 0.05 g to 0.20 g. Because of this wide range of E.P.A. Seismic Zone 2 values, it was thought not to be reasonable to impose design requirements on all parts of Zone 2 from extrapolating the test results based on the maximum E.P.A. value of .20 g. Thus two subzones were defined within Zone 2; subzone 2A includes the areas of Zone 2 that according to the ATC-3 map of figure 3 has E.P.A. values of 0.10 g or less whereas subzone 2B includes the areas of Zone 2 with E.P.A. values from 0.10 g to 0.20 g (refs. 4 and 5).

Roof Load and Roof Flexibility

Extra load was placed on the roof in order to simulate prototype conditions for a 48 ft (14.63 m) square house with an assumed 20 lb/sqft (98 kg/m^2) prototype roof load. Thus the load per unit length for the loadbearing masonry walls of the model was 458 lb/ft (684 kg/m) whereas the corresponding load per unit length for the prototype masonry loadbearing walls was assumed to be 480 lb/ft (717 kg/m).

The roof system used for all five masonry models was less flexible than the corresponding prototype roofs. In order to investigate the influence of the roof flexibility on the masonry walls earthquake performance additional masonry piers were built having very flexible top support conditions (Fig. 1); these piers were subjected to all the tests that the masonry walls of House 5 were subjected to.

Geometry - Torsional Response

House 5 as well as Houses 2,3 and 4 had four walls with typical house wall components having standard size door and window openings forming a structure with an asymmetric plan configuration. Thus the studied masonry house earthquake performance included the influence of torsional response arising from moderately asymmetric prototype masonry house configurations.

Foundation - Preexisting State of Stress

All model masonry houses had a rigid concrete foundation thus simulating good soil prototype conditions. They did not simulate any preexisting state of stress, other than from gravity load, arising from foundation settlement, temperature changes or shrinkage. However, it must be noted that due to the continuously increasing intensity test procedure the model structures were gradually damaged thus they were in a weakened condition when they were subjected to the most intense simulated earthquake motions.

Extrapolation of test results - Unreinforced Masonry

Taking into account the previously discussed parameters it can be concluded that it is valid to extrapolate the test results to prototype single-story masonry house conditions by expressing the seismic intensity in terms of effective peak acceleration provided that the prototype foundation, roof system, roof load and masonry wall geometry together with the prototype

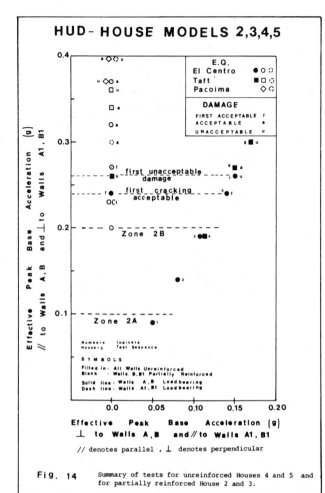

Fig. 14 Summary of tests for unreinforced Houses 4 and 5 and for partially reinforced House 2 and 3.

material strength and construction details do not differ to any considerable degree from those of the model structures.

Figure 14 depicts a summary of the earthquake performance for unreinforced Houses 4 and 5 as well as for Houses 2 and 3 with walls A,A1 unreinforced and walls B,B1 partially reinforced. The intensity of each test is expressed in terms of effective peak acceleration. The abscissae in this figure are the effective peak acceleration values for the horizontal components parallel to walls A1 and B1 whereas the ordinates are the effective peak acceleration values for the horizontal components parallel to walls A and B. In the same figure the levels of maximum expected intensity of the two subzones, namely 2A and 2B, are also shown in terms of effective peak acceleration. The letter "u" next to the symbol that represents each test indicates that the observed damage was unacceptable.

The following observations can be made from the depicted unreinforced masonry house earthquake performance:

d1. The first acceptable damage is observed for two simultaneous horizontal components of effective peak acceleration of 0.24 g and zero for unreinforced House 4 or for 0.24 g and 0.14 g for unreinforced House 5. The first unacceptable damage is observed for two simultaneous horizontal components of effective peak acceleration of 0.26 g and zero for unreinforced House 4 or 0.26 g and 0.15 g for unreinforced House 5. These effective peak acceleration values for the first damage (either acceptable or unacceptable) are higher than the expected maximum effective peak accelerations in any part of Zone 2.

d2. The simultaneous action of in-plane and out-of-plane inertia forces on all unreinforced House 5 masonry walls did not result in any noticeable increase of the damage, when this is compared with the corresponding damage of House 4 where this combined in-plane and out-of-plane action was not present.

d3. Taking into account the preconditions that were previously stated with regard to the validity of the extrapolation of the test results to prototype conditions the above observations will similarly apply to prototype single-story masonry houses.

d4. The 6 ft (1.83 m) long shear resisting unreinforced masonry panels, which were part of loadbearing walls A and B, performed satisfactorily beyond the level of earthquake intensities expected in UBC Seismic Zone 2, which for the purpose of extrapolating the test results to prototype conditions was expressed in terms of effective peak acceleration (E.P.A. ≤ 0.20 g).

DESIGN RECOMMENDATIONS

Because the effect of the combined in-plane and out-of-plane forces acting simultaneously on the masonry walls of Houses 1,2,3 and 4 was not studied the tentative recommendations that were made after the completion of all tests of these four models (ref. 3), assumed that the earthquake resistance capacity of the masonry walls would be significantly reduced by the action of the combined in-plane and out-of-plane forces in comparison with the masonry wall resistance capacity for horizontal forces in only one direction (either in-plane or out-of-plane). This reduction was estimated to be 30 to 50 percent for unreinforced masonry walls and 20 percent for partially reinforced walls. This led to the recommendation that partial reinforcement may be deemed necessary in certain areas of Seismic Zone 2. However, the observed earthquake performance of House 5 with all its masonry walls being subjected to combined in-plane and out-of-plane horizontal forces led to the conclusion, as already stated in the previous paragraph (d2), that this combined in-plane and out-of-plane action of the inertia forces did not result in any noticeable increase of the damage that was sustained by the House 5 unreinforced masonry walls, when this is compared with the corresponding damage sustained by essentially the same unreinforced masonry walls (House 4) where this combined in-plane and out-of-plane inertia force action was not present. In this way, the recommendation for a reduction by 30 to 50 percent of the horizontal earthquake resistance capacity of unreinforced masonry walls that was observed in the first four masonry house models, represents an overconservative assumption. Based on the observed earthquake performance of all five models the following recommendations were made for Seismic Zone 2 single story masonry houses (ref. 4):

<u>Definition of Structural Components (Shear Resisting Masonry Elements)</u>

e1. An unreinforced shear panel is an unreinforced masonry wall element of certain length, defined by the design recommendations, that extends from floor to ceiling without any penetrations, openings or discontinuities.

e2. A partially reinforced shear panel is a masonry wall element of a certain length, defined by the design recommendations, that has a No. 4 vertical reinforcing bar fully grouted at each end of the panel. The No. 4 reinforcing bar is dowelled at the floor level by a No. 4 dowel to provide the necessary anchorage into the foundation. All masonry and steel parts of this panel extend from floor to ceiling without any penetrations, openings, or discontinuities.

<u>Design Recommendations</u>

f1. All exterior walls must have a shear-resisting element that can resist the lateral seismic forces, with the specified length. This length is determined by a procedure that is included in the recommendations, which is not presented here due to space limitations (ref. 4).

f2. For Zone 2A the minimum length of an unreinforced shear resisting element is 6 ft (1.83 m); the minimum length for a partially reinforced shear resisting element is 5 ft (1.52 m).

f3. For Zone 2B the minimum length of an unreinforced shear-resisting element is 9 ft (2.74 m); alternatively there may be two 6 ft (1.83 m) elements. The minimum length of a partially reinforced shear-resisting element is 5 ft (1.52 m).

CONCLUSIONS

1) Despite the simultaneous action of three component simulated earthquake input of moderate to moderately high intensity and the large observed acceleration and displacement response the unreinforced House 5 performed satisfactorily.

2) Following the observed performance of all five models, it was recommended that unreinforced masonry could be used in Seismic Zone 2 areas if certain minimum length and shear capacity requirements are fullfilled by the main shear resisting masonry elements, provided that the prototype conditions do not introduce any considerable deviations in the important characteristics of the prototype masonry houses from those of the tested models.

3) The formation of structural cracks at the unreinforced House 5 shear resisting masonry elements introduced significant increase in the displacement response together with significant torsional and distortional contributions for the response of the masonry house as a whole.

4) Partial reinforcement in the shear resisting elements, when provided with dowel connections to the foundation, resulted in low level in-plane displacement response and it was very effective in controlling the damage even for tests of very high intensity, keeping at the same time the overall house response without any significant torsional or distortional contributions.

REFERENCES

1. Gulkan P., Mayes R.L., and Clough R.W. "Shaking Table Study of Single-Story Masonry Houses - Vol. 1: Test Structures 1 and 2", Earthquake Engineering Research Center Report No. UCB/EERC-79/23, 1979.

2. Gulkan P., Mayes R.L., and Clough R.W. "Shaking Table Study of Single-Story Masonry Houses - Vol. 2: Test Structures 3 and 4", Earthquake Engineering Research Center Report No. UCB/EERC-79/24, 1979.

3. Clough R.W., Mayes R.L., and Gulkan P. "Shaking Table Study of Single-Story Masonry Houses - Vol. 3: Summary, Conclusions and Recommendations", Earthquake Engineering Research Center Report No. UCB/EERC-79/25, 1979.

4. Manos G.C., Clough R.W., Mayes R.L. "Shaking Table Study of Single-Story Masonry Houses - Dynamic Performance under Three Component Seismic Input and Recommendations", Earthquake Engineering Research Center Report No. UCB/EERC-83/11, 1983.

5. " Tentative Provisions for a Development of Seismic Excitations for Buildings", Applied Technology Council Publications ATC3-06, (NSF Publication 78-8, NBS Special Publication 510) U.S. Government Printing Office, June 1978.

TESTS OF SQUAT SHEAR WALL
UNDER LATERAL LOAD REVERSALS

S. Wiradinata[I] and M. Saatcioglu[II]

ABSTRACT

This paper presents the results of an experimental investigation on seismic response of squat shear walls. The experimental program includes tests of large scale squat walls with height-to-width ratios of 1/2 and 1/4. The observed behavior of wall response is discussed with reference to crack patterns and the hysteretic force-deformation relationships. The results indicate that a different mode of failure may govern the capacity of a squat shear wall, depending on the wall height-to-width ratio, the presence of a top beam, and the percentage of horizontal and vertical reinforcement.

INTRODUCTION

Reinforced concrete walls are commonly used in earthquake resistant construction to provide lateral strength and stiffness. In the event of a strong earthquake, these walls generally behave beyond the elastic range and may develop significant inelastic deformations. It has been experimentally shown that, tall structural walls can be designed to respond in a ductile manner[1,2]. This can be achieved by promoting flexural yielding prior to shear failure. In low-rise structural walls or squat walls, it is difficult to attain the same mechanism of load resistance due to the high level of shear stresses associated with shorter shear spans.

Squat walls with height-to-width ratios of less than 2.0 behave predominantly in the shear mode. These walls are susceptible to brittle shear failures when subjected to strong ground excitations.

The objective of this paper is to present the results of large scale wall tests conducted under inelastic load reversals. The test procedure and the observed behavior of wall specimens are discussed in detail. Further details of the test program can be found in Ref.[3].

I. Structural Engineer, P. T. Dacrea, Design and Engineering Consultants, Jakarta, Indonesia.

II. Associate Professor, Department of Civil Engineering, University of Ottawa, Ottawa, Canada.

TEST SPECIMENS

Two squat walls with rectangular cross-sections were constructed for testing. Each wall specimen was reinforced with 0.8% vertical and 0.25% horizontal reinforcement. The height-to-width ratio of Wall 1 was 1/2 and Wall 2 was 1/4. Figure 1 illustrates the specimen geometry and the reinforcement detailing.

Each wall specimen was built with heavily reinforced top and bottom beams. The top beam was used to apply the horizontal load uniformly. The bottom beam was used to provide full fixity at the base. Each wall was constructed in two stages. Concrete in the bottom beam was cast first. An exposed aggregate construction joint was provided between the wall and the bottom foundation beam. The rest of the wall was cast at once, approximately a week later, in the second stage of construction.

The predicted flexural capacity of the walls was 660 kN.m (487 ft-k). This value corresponds to $0.56\sqrt{f'_c}$ MPa ($6.72\sqrt{f'_c}$ psi) and $1.08\sqrt{f'_c}$ MPa ($12.96\sqrt{f'_c}$ psi) of shear stress in Walls 1 and 2 respectively. The average concrete strength, determined on the day of test, was 25 MPa (3.6 ksi) for Wall 1 and 22 Mpa (3.2 ksi) for Wall 2. The reinforcement yield strengths were 435 MPa (63.0 ksi) and 425 MPa (61.6 ksi) for vertical and horizontal reinforcement respectively.

TEST PROCEDURE

An identical procedure was followed for both tests. The description of the test set-up, loading program and the instrumentation are discussed below.

Test Set-up

The test set-up is illustrated in Fig. 2. The specimens were post-tensioned to the Laboratory strong floor. The horizontal load was applied by means of two 1000 kN (225 kip) capacity hydraulic jacks, supported by the Laboratory strong wall. Swivel heads were attached to the jacks to ensure the allignment of load with the jack axis. The load was applied through the Loading Assembly which was post-tensioned to the specimen . The Loading Assembly consisted of two steel " I " beams and Dywidag post-tensioning bars. The stability of specimens during testing was ensured by the use of two steel " A " frames as shown in Fig. 2.

Loading Program

The specimens were subjected to displacement controlled horizontal load cycles. The displacements were applied as increments of yield displacement. The yield displacement is defined in Fig. 3.

The loading program consisted of three elastic cycles at approximately one half of the yield displacement, followed by incrementally increasing inelastic deflection cycles. Three cycles were imposed on the specimens at each deformation level. A small deformation cycle was also applied prior to each increase in the deformation level. The magnitude of deformations were increased until a significant drop was observed in the load resistance of the specimen. Figure 4 illustrates the loading program used in the tests.

Instrumentation

Vertical and horizontal deflections were measured by Linear Variable Differential Transformers (LVDT). Horizontal sliding of the specimen at wall base was also measured by means of LVDTs. Electrical resistance strain gauges were placed on selected vertical, horizontal, and tie reinforcement. All strain gauge and LVDT voltages were fed into a Hewlett-Packerd (HP) 3052A Automatic Data Acquisition System. All voltage excitations, including those from the load cells were stored as micro-strains, forces and displacements by an HP 9816 micro computer.

Additional data were taken on the concrete surface to determine shear deformations. A " Zurich " gauge was used for this purpose. A total of fifteen brass targets were glued on the west face of Wall 1 for " Zurich " gauge readings. These targets formed 200 mm square grids in the middle portion of the wall. The same grid size necessitated the use of nine targets in Wall 2.

OBSERVED BEHAVIOR AND TEST RESULTS

Wall 1

The load versus top deflection hysteretic relationship for Wall 1 is shown in Fig. 5. The specimen was initially subjected to elastic load cycles at approximately 250 kN (56 kip). Some hairline cracks along the construction joint were observed at this load stage. The first inclined shear crack occured at a shear stress of $.44\sqrt{f'_c}$ MPa ($5.3\sqrt{f'_c}$ psi) prior to reaching the flexural yield point. The flexural yield point was reached at 470 kN (105 kip) of horizontal load and 2.5 mm (0.1 in) of top displacement. Additional flexural cracks were observed at this load stage, on both sides of the wall, extending approximately up to the mid wall height. Shear cracks with inclinations ranging between 30 to 40 degrees with the horizontal were also observed. A major diagonal shear crack was noticed between the North-East lower corner and the mid-span of the top loading beam when the yield displacement was reached.

The peak load of 575 kN (129 kip) and the corresponding top deflection of 10 mm (0.4 in) were recorded at the first cycle to four times the yield displacement. This load corresponds to a shear stress of $0.56\sqrt{f'_c}$ MPa. (6.7 $\sqrt{f'_c}$ psi). The specimen showed a ductile behavior up to this load stage. Progressively increasing strength and stiffness degradation was observed in the subsequent load stages. A second set of major shear crack became appearent at the end of the cycles at six times the yield displacement. These cracks formed diagonally between the two opposite corners. The failure of the specimen was attributed to these two diagonal shear cracks, crossing each other approximately in the central region of the wall specimen. Figure 6 shows the crack pattern at the end of the cycles at eight times the yield displacement.

Wall slip along the construction joint was carefully monitored during testing. The load versus wall slip hysteretic relationship is shown in Fig. 7. It is evident from this figure that wall slip forms only a small portion of the total horizontal displacement.

The amount of vertical bar slip in the foundation beam was also measured using 15 mm stroke LVDTs. Figure 8 shows the hysteretic variation of vertical bar slip. It was observed that the vertical bar slip within the foundation beam contributed very little to the overall top displacement. This contribution was computed to be approximately 1% at two times the yield disaplacement.

Wall 2

The load versus top deflection hysteretic relationship for Wall 2 is shown in Fig. 9. During the first three elastic cycles, the specimen was subjected to a maximum load of 450 kN (101 kip). The first crack occured diagonally, starting from the top beam corner, where the load was applied, propagating towards the mid wall height. No horizontal flexural crack was observed during the elastic cycles. The first inclined shear crack occured at a shear stress of approximately $0.37\sqrt{f'_c}$ MPa ($4.4\sqrt{f'_c}$ psi).

The next doformation level, in the loading program, was the yield point. The predicted yield level of 2.5 mm (0.1 in) was reached at a horizontal load of 680 kN (153 kip). This load level was significantly lower than the predicted yield load of 1015 kN (228 kip). The first horizontal crack occured during this load stage. More inclined cracks originated from the top beam. A major crack formed along the construction joint at the wall base.

At the next deformation level, the specimen was subjected to two times the predicted yield displacement. It was noted that the recorded load of 615 kN (138 kip) at this load stage was lower than the previous peak load. Only minor additional cracks formed at this load stage. However the base crack along the construction joint became more pronounced and propagated towars the centerline of the wall section. When the force was reversed the base crack became continuous. Sliding of the wall with respect to the foundation beam became appearent in the subsequent load cycles and the crack pattern remained unchanced. The wall slip became visibly noticeable during cycles at four times the predicted yield displacement. The concrete particles dislodged by grinding along the base, "lubricating" the interface. Strength and stiffness degradation became severe at this deformation level.

The sliding of the wall relative to the foundation beam was continuously recorded througout the test using four LVDTs. Fig 10 shows the hysteretic force versus sliding shear displacement relationship. The hysteretic force-displacement relationship given in Fig. 11 is obtained by subtracting the sliding shear displacement from the total top displacement. Examination of Figs. 9 through 11 clearly indicates that Wall 2 could not develop its flexural strength due to excessive sliding. The failure of Wall 2 was attributed to the sliding shear which started at a shear stress of $0.72\sqrt{f'_c}$ MPa ($8.7\sqrt{f'_c}$ psi). Most of the shear cracks observed on the specimen were hairline cracks with no sign of diagonal tension or compression failures. Figure 12 shows the crack pattern of the specimen at two times the yield displacement. Based on the observed behavior of Wall 2, it may be concluded that the increase in the amount of horizontal reinforcement over the minimum provided, as required by the current Building Codes would not improve the wall behavior.

CONCLUSIONS

Based on the results of this investigation, the following conclusions can be made:

- The failure mode of a squat wall, subjected to reversed cyclic loading is affected by height-to-width ratio of the wall. Walls with low height-to-width ratios may fail in sliding shear prior to flexure or diagonal tension or compression failures. The wall specimen with a height-to-width ratio of 1/4, tested in this investigation, failed in sliding shear at $0.72\sqrt{f'_c}$ MPa ($8.7\sqrt{f'_c}$ psi) of shear stress.

- Shear strength of a squat wall, when governed by diagonal tension, increases with decreasing height-to-width ratio. However, this observation is valid only if the 45-degree diagonal tension failure is prevented by a top tie beam or a strong floor slab, forcing the wall to fail through the corner-to-corner failure plane.

- Vertical reinforcement is highly effective in resisting shear. The wall specimen considered in this investigation, with a height-to-width ratio of 1/2, showed a ductile behavior up to four times the yield displacement, even though it contained only a minimum amount of horizontal reinforcement.

- The contribution of vertrical bar slip within the foundation beam to total top displacement can be relatively insignificant when compared with the other components of deformation. While flexure and shear deformations in squat walls are both significant, the shear deformations become very significant after the onset of the strength degradation.

ACKNOWLEDGEMENT

This investigation was sponsored by the National Science and Engineering Research Council of Canada under Grant No. U0040. The experimental program was performed at the Structures Laboratory of the University of Toronto, Professor Peter Marti, Manager.

REFERENCES

[1] Oesterle, R. G., Fiorato, A. E., Johal, L. S., Carpenter, J. E., Russell, H. G. and Corley, W. G.,"Earthquake Resistant Structural Walls-Tests of Isolated Walls",Portland Cement Association, Skokie, Nov. 1976

[2] Vallenas, J. M., Bertero, V. V., Popov, E. P., "Hysteretic Behavior of R/C Structural Walls", EERI, University of California, Berkeley, California, August, 1979.

[3] Saatcioglu, M. and Wiradinata, S., "Response of Squat Shear Walls to Inelastic Load Cycles", Department of Civil Engineering, University of Ottawa, Ottawa, Canada, May 1986.

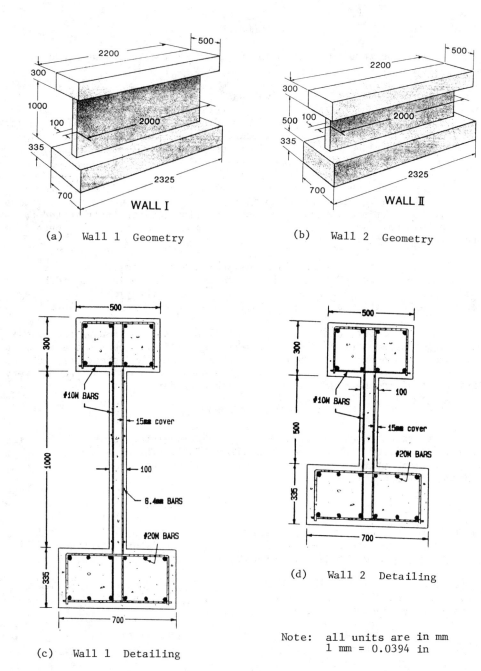

Fig. 1 Specimen Geometry and Reinforcement Detailing

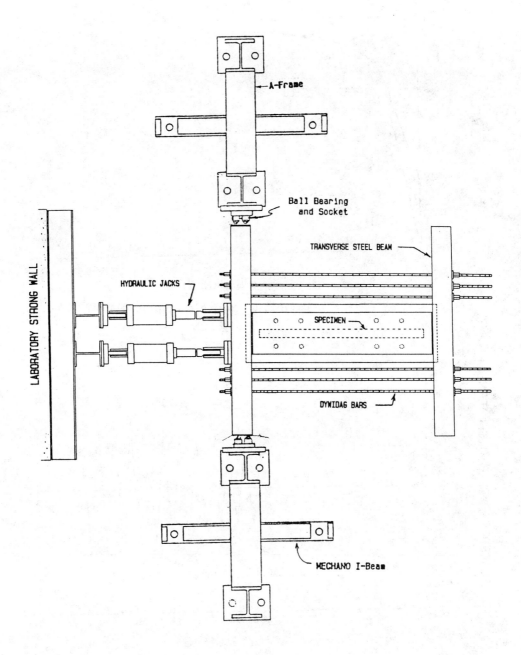

Fig. 2   Test Set-Up

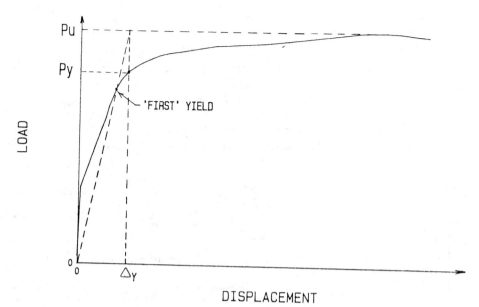

Fig. 3 Definition of Yield Displacement

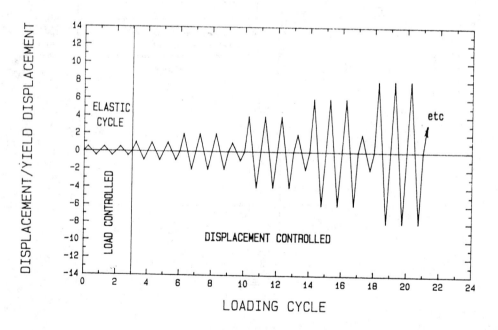

Fig. 4 Loading Program

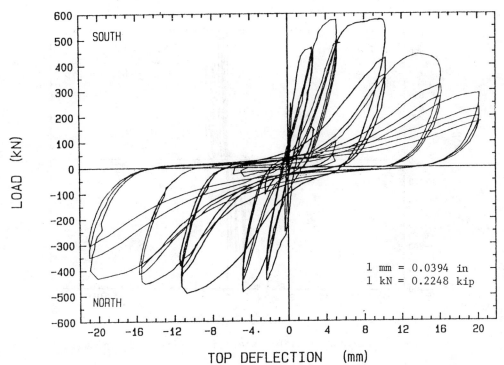

Fig. 5　Load-Top Deflection Hysteretic Relationship for Wall 1

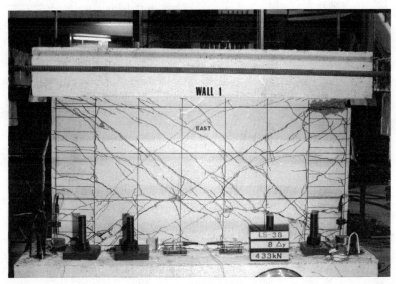

Fig. 6　Crack Pattern of Wall 1 at the End of Cycles at Eight Times the Yield Displacement

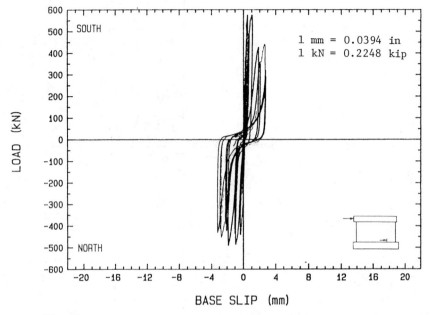

Fig. 7   Load-Wall Slip Hysteretic Relationship for Wall 1

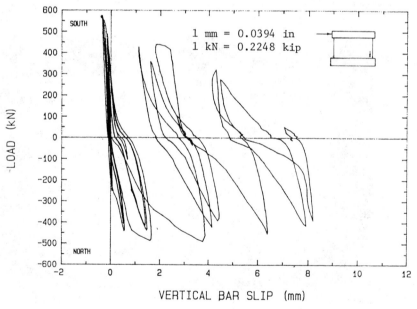

Fig. 8   Load-Vertical Bar Slip Hysteretic Relationship

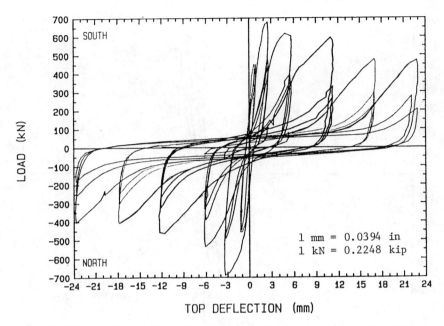

Fig. 9   Load-Deflection Hysteretic Relationship for Wall 2

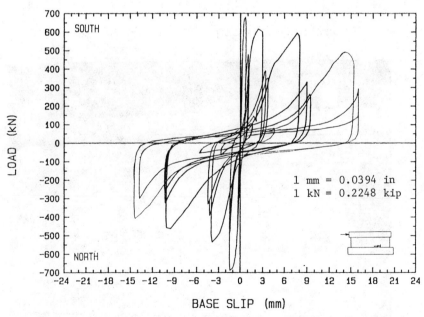

Fig. 10   Load-Wall Slip Hysteretic Relationship for Wall 2

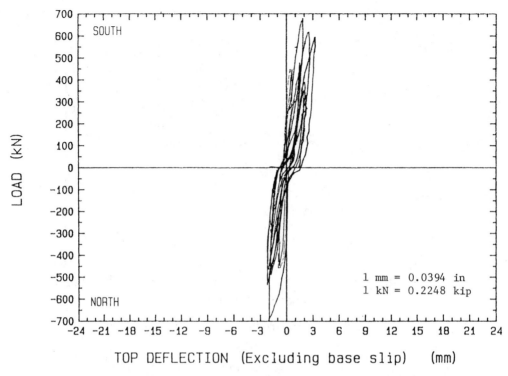

Fig. 11   Hysteretic Relationship Excluding the Wall Slip (Wall 2)

Fig. 12   Crack Pattern of Wall 2 at Two Times the Yield Displacement

MEASURED DYNAMIC RESPONSE OF HYPERBOLIC SHELL

A. N. Lin[I], H. Mozaffarian[II], and M. Helpingstine[III]

ABSTRACT

The dynamic response of a hyperbolic concrete cooling tower to ambient vibration was measured. Analysis of the data in the frequency domain led to identification of response frequencies and mode shapes. Response frequencies and mode shapes were also calculated from a free vibration analysis using axisymmetric shell elements. The experimentally obtained frequencies were more separated than the analytical results, although this may be a degenerate case. The observed and calculated mode shapes showed limited qualitative agreement. In general, however, the complex dynamic response of the tower may defy total description based solely on ambient vibration data.

INTRODUCTION

Natural draft hyperbolic cooling towers are large thin-shell structures that are subject to dynamic loads from wind and strong ground motion. Over the past two decades, the maximum height of these structures has increased by over twenty percent to the present height of 170 m (550 ft) [1]. The shell thickness is nominally 20 to 30 cm (8 to 10 in). Present trends in tower design and construction point toward greater height and reduced wall thickness.

Experimental studies of cooling tower response have been concentrated on (a) determining the pressure distribution due to wind, and (b) the buckling behavior of the shell. The impetus for these efforts stems primarily from the wind-induced failures in 1965 of three tower structures at Ferrybridge, England and in 1973 of a tower at Ardeer, Scotland. In 1978, a tornado and collapsing tower crane caused serious damage to a nearly completed structure at Port Gibson, Mississippi. The structure was later repaired. In 1985, a tower in the United KIngdom collapsed. This event has led to a program of structural inspection.

In general, experimental studies on cooling tower behavior have been conducted at three scale levels. Small scale models (1/250 to 1/500) constructed of a variety of materials and fabrication techniques have been

---

I. Assistant Professor, Department of Civil Engineering, Kansas State University, Manhattan, Kansas.

II. Graduate Research Assistant, Department of Civil Engineering, Kansas State University, Manhattan, Kansas.

III. Graduate Research Assistant, Department of Civil Engineering, Kansas State University, Manhattan, Kansas.

subjected to wind tunnel testing to determine the effect of multiple structure interaction, surrounding structures and topography on wind loading [2,3]. Buckling of elastic and brittle models under external pressure has also been examined [4,5]. Laser interferometry has been used to determine mode shapes and resonant frequencies under free vibration [6,7].

At the next scale level, a 1/40 scale microconcrete and reinforcing steel model was constructed and subjected to internal vacuum to simulate uniform external pressure. The shell was buckled to failure, repaired, and retested [8]. A second more refined model is currently under construction. A slightly smaller model, constructed of glass reinforced polyester has also been tested under similar conditions [9].

Testing of prototype cooling towers has primarily consisted of measurements of the pressure differential across the wall of the tower. Prototype towers for which such measurements have been made include Martin's Creek, PA [10,11,12,13], and Weisweiler [14] and Schmeehausen [15], Germany.

In each case, the wind pressure differential across the tower wall was measured around the circumference of the tower at a fixed height. There was no measurement of the wind pressure distribution with height. Measurements have been recorded at large time steps for static analysis and at small time steps (0.25 sec to 1.5 sec) for dynamic analysis. Data were recorded using self-triggering devices that activated when wind velocities exceeded a threshhold limit. The results of these experimental investigations form the current basis of analysis and design for wind loading [16].

The other form of full scale tower investigation has been the visual inspection for cracks and other signs of service induced distress. The TVA's Paradise Steam Power Plant's two 133 m (437 ft) towers were inspected in 1968 and 1969 [1], and a saltwater tower, completed in 1974 with a height of 63 m (208 ft) near Atlantic City, NJ, was inspected in 1983 [17].

Current analytical efforts are directed toward the development of refined finite element models for analysis. Present analysis capabilities include free vibration, time history, and response spectra analysis [18]. Additional refinements have included inclusion of foundation flexibility [19]. The use of time domain analysis and Monte Carlo simulations of wind time histories have also been presented [20].

This paper describes the results of an ambient vibration study of the 152 m (499 ft) cooling tower at the Trojan Nuclear Power Plant at Rainier, Oregon. The dynamic response measurements were made during a three week period in August, 1985. Finite Fourier Transform of the data led to identification of seven response frequencies. The lateral response mode shapes have also been plotted. The results of the experimental program are compared to the results of a finite element program that uses axisymmetric shell elements. Other aspects of the present study have been presented elsewhere [21]. A final report will be prepared at the completion of the study.

DESCRIPTION OF TOWER

The subject tower is located at Portland General Electric's Trojan Nuclear Power Plant at Rainier, Oregon. The structure was completed in May 1973. The tower is an offset hyperboloidal shell of revolution. The meridional curve equation is:

$$(r - c)^2/a^2 - z^2/b^2 = 1 \qquad (1)$$

where c is the throat offset distance of 24.82 m (81.37 ft), and the hyperboloidal parameters a and b have values of 10.56 m and 39.3 m (34.63 ft and 128.82 ft), respectively.

The structure is 182 m (499 ft) in height, including the tower legs, of which there are 44 pairs. Each leg is 12.5 m (41 ft) high and 1.1 m (3.5 ft) in diameter. The tower base is supported on rock. The tower has a base diameter of 117 m (385 ft), a throat diameter of 70.7 m (232 ft) and a cornice diameter of 76.2 m (250 ft). Nominal shell thickness of 0.25 m (10 in). A 1.5 m (4.5 ft) wide stiffening ring (cornice), also used for access, is provided at the top of the tower, with a shell thickness of 0.35 m (14.8 in) at the top. The shell thickness is gradually increased to 1.15 m (3.75 ft) at the lintel. The tower dimensions and the finite element model are shown in Figure 1. A photograph of the tower is shown in Figure 2.

Access ladders are provided at ninety degree intervals around the circumference of the shell. One ladder, referred to here as the principal access, extends to the top, while the remaining three extend only to the throat. Intermediate landings are located at approximate elevations of 46 m (150 ft), 82 m (270 ft), at 128 m (420 ft). An additional landing is provided on the principal access ladder beneath the cornice, at elevation 148 m (485 ft).

EQUIPMENT

The test instrumentation system is shown in Figure 3. The motion transducers were Model FBA-11 Uniaxial Force Balance Accelerometers, manufactured by Kinemetrics, Inc., of Pasadena, CA, having a nominal natural frequency of 50 Hertz, and sensitivity of 2.5 Volts/g. Twelve volt DC power and calibration signals to each accelerometer were provided by a Kinemetrics SMA-3 Strong Motion Accelerograph. The accelerometer output was band pass filtered and amplified by a Kinemetrics SC-1 Signal Conditioner. The system has been widely used for both ambient and forced vibration testing of full scale structures, including buildings, bridges, dams, and storage tanks.

Data were recorded on magnetic tape by a Bruel and Kjaer 7006 four-channel FM recorder. Visual monitoring of the data during the tests used a Brush 2400 four-channel strip recorder.

A two channel Hewlett-Packard 3582A Real Time Spectrum Analyser was used to calculate Fourier amplitude spectra for single channel data, and cross spectrum and coherence for two channels of data simultaneously. In the latter case, one channel was the reference accelerometer.

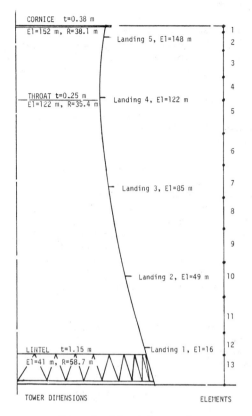

Fig. 1 Dimensions of prototype tower and finite element model.

Fig. 2 Photograph of prototype tower.

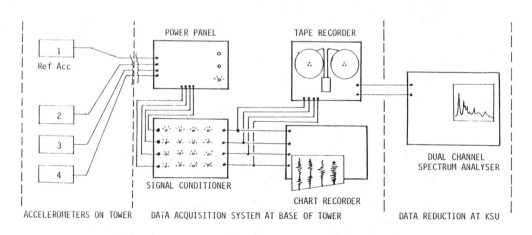

Fig. 3 Schematic diagram of test instrumentation

PROCEDURE

The test program was designed to provide a complete characterization of the structure's response subject to several conditions. First, only four channels of data acquisition and signal conditioning were available. Second, the accelerometers could only be placed at discrete locations on the tower; namely at the access ladder positions. Third, on-site analysis of the data was not possible; it was necessary to collect all of the data for reduction at KSU. Finally, decoupling and identification of the radial and lateral responses would be complicated by the axisymmetry of the structure.

The result was an extensive plan of ambient vibration measurements. Significant features of the instrumentation plan are described in the following:

(a) One accelerometer remained at the uppermost landing, oriented perpendicular to the shell surface. Relative amplitude and phase at other locations of the tower for mode shape identification were obtained by calculating the cross spectrum with respect to this so-called reference accelerometer.

(b) There were essentially three types of measurements. The first type provided information on the relative magnitude and phase of orthogonal motions at any point on the principal access ladder. Therefore, the three accelerometers were oriented in mutually orthogonal directions (normal and tangential to the shell, and vertical) and installed sequentially at each landing on the principal access ladder.

(c) The second type defined the lateral response of the tower. The three accelerometers were oriented normal to the shell surface and installed at different elevations on the principal access ladder. Accelerometers were placed at each landing and also on the ladder between landings. There was considerable redundancy in measurements, resulting in multiple values for some of the mode shape values.

(d) A third type of measurement was used to define the circumferential distribution of motion. The accelerometers were placed on three of the ladders; at the base, first, and second landing platforms, and sequentially in the three directions.

(e) The ground motions near the tower were also measured. These measurements determined the frequency content of the ambient ground motion.

(f) The sequence of accelerometer placement and testing was optimized to minimize climbing of the tower and relocation of the accelerometers and connecting cables.

As a result of these considerations, a total of 15 sets of measurements were made. The typical procedure for each set of measurements consisted of the following: (a) install accelerometers at a desired location and orientation, (b) send calibration signal to insure proper operation of system, (c) set low pass filter at 20 Hertz, (d) set amplifier

gains for each channel for maximum recordable signal strength, and
(e) record data on magnetic tape for 10, 15, or 20 minutes as required. To
maximize the dynamic range by eliminating potential high frequency noise,
steps (d) and (e) were repeated for filtering frequencies of 10 and 5
Hertz.

During data reduction, it was found that there were no identifiable
modes above 5 Hertz. Therefore, all of the spectra were computed from the
records obtained with a low pass filter frequency of 5 Hertz, to improve
the resolution on the frequency axis. Each spectrum is calculated from 16
arithmetic averages of 512 amplitude and phase points. The A/D sampling
rate was 0.08 seconds per sample.

EXPERIMENTAL RESULTS

Fourier amplitude spectra were calculated for each accelerometer
record. Of particular interest was the response of the reference
accelerometer. One of the resulting Fourier spectra is shown in Figure 4.
Some of the significant response frequencies and modal damping ratios are
shown. Modal damping ratios were calculated by the half-power method when
response peaks were isolated. The spectra remained essentially unchanged
from one test to another.

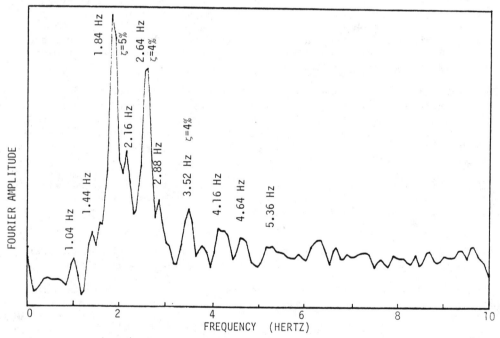

Fig. 4 Fourier Amplitude Spectrum for reference accelerometer, test 5.

Cross spectra for relative amplitude and phase with respect to the
reference accelerometer were calculated for records obtained at the other
measurement points to define the mode shapes. The mode shapes in the

w-component (normal to the shell) along the principal ladder are plotted in Figure 5 for the first seven identifiable frequencies. Each mode shape is normalized to the amplitude of the reference accelerometer. Error bars are used to indicate the range in value of multiple measurements, the number of which is shown in parenthesis ( ). In mode 6, at a frequency of 2.88 Hertz, the relative phase of one of the stations was ambiguous, and is therefore plotted with dashed lines. It does not change the number of zero crossings, however.

ANALYTICAL RESULTS

A free vibration analysis using the finite element program SHORE-III [18] was conducted for a model consisting of 12 axisymmetric shell elements and 1 axisymmetric element representing the ring columns. The model is based on a uniform shell thickness, which does not represent the true case, and will be revised in the near future.

The resonant frequencies for the first four modes of harmonics 0, 1, 2, 3, 4, and 5 are shown in Table 1. The harmonic number refers to the circumferential wave number, defined by $\cos(n\theta)$. The w-component of the mode shapes corresponding to the seven lowest frequencies are plotted in Figure 6, with nodal amplitudes normalized to the displacement of the top element.

TABLE 1. Resonant frequencies of first four modes of harmonics 0 through 5, by SHORE-III.

HARMONIC NUMBER n	MODE NUMBER m			
	1	2	3	4
0	5.50	6.90	10.52	11.04
1	2.16	3.95	7.20	8.43
2	1.19	2.21	4.24	6.61
3	0.87	1.46	2.55	4.35
4	0.77	1.30	1.88	2.94
5	0.79	1.33	1.84	2.31

DISCUSSION AND CONCLUSION

The following comments can be made about the results of the experimental and analytical portions of the study.

(a) Compared to the free vibration analysis results, the corresponding observed modal frequencies are higher. While the spectrum in Figure 4 does show some small peaks below 1.04 Hertz, no mode shape was found at the lower frequency.

(b) The experimental frequencies appear to be relatively separated, while some of the analytically obtained frequencies are close together. It would be virtually impossible, however, to experimentally distinguish between two modes with frequencies of 0.77 and 0.79 Hertz, or 1.30 and 1.33 Hertz, such as found for the first and second, and sixth and seventh modes, respectively.

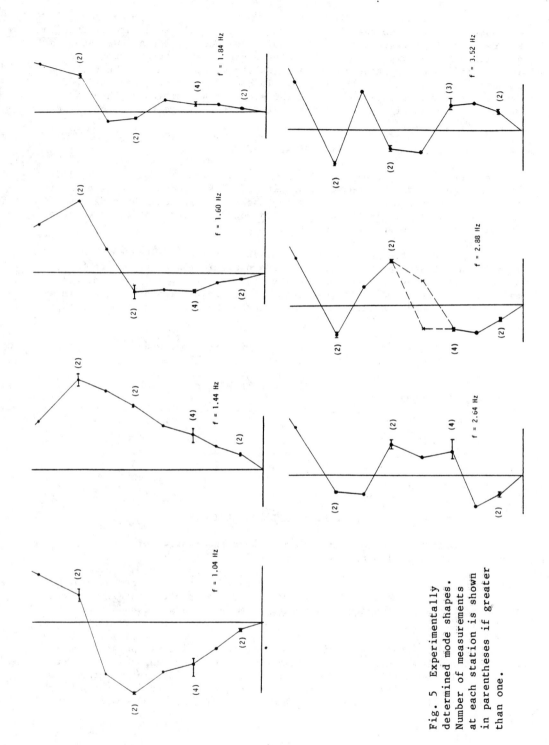

Fig. 5 Experimentally determined mode shapes. Number of measurements at each station is shown in parentheses if greater than one.

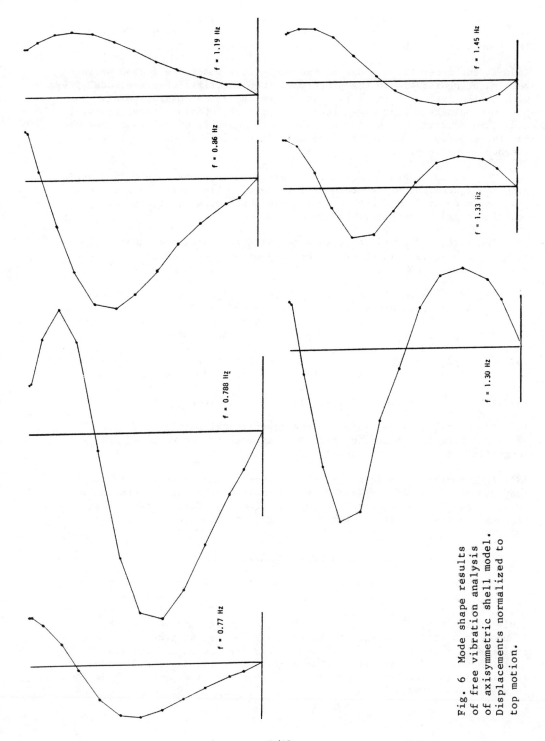

Fig. 6 Mode shape results of free vibration analysis of axisymmetric shell model. Displacements normalized to top motion.

(c) The analytically obtained mode shapes for the first, second and third modes are similar; that is they have one zero-crossing point. Similarly, the fourth and fifth modes have two zero-crossing points.

(d) If the first three modes of the analytical model are indeed present but indistinguishable in the prototype, then the first observed mode agrees with the analytical results in that it has one zero-crossing point. The harmonic numbers of the first three analytically obtained modes are 4, 5, and 3, respectively. It is not possible to identify the harmonic numbers for the experimental results since only four points around the circumference are accessible.

(e) The second observed mode at 1.44 Hertz has a comparable mode shape with the fourth calculated mode at 1.19 Hertz. Similarly, the fourth observed mode at 1.84 Hertz is similar in shape with the fourth and (particularly) fifth calculated modes.

(f) The fifth and sixth observed modes have the three zero-crossing points each and are similar in shape. There may be a higher calculated mode having similar properties.

(g) The finite element model used for this analysis has constant wall thickness. Improved agreement may result when the actual wall profile is incorporated. It was found that the frequencies and mode shapes are more dependent upon the wall thickness profile than the number of elements.

(h) Other features of the prototype tower that have not been considered in the analysis include cracking in the meridional direction leading to anisotropy and reduced stiffness [22], and the effect of foundation compliance, although the prototype is founded on rock.

(i) Additional work remaining on the experimental portion includes examination of the foundation response, as well as the vertical and tangential response of the shell.

(j) In general, it appears that the behavor of the tower is more complex than can be reasonably explained on the basis of the ambient vibration data. Plans for future work include an attempt at forced steady state harmonic excitation using vibration generators attached to the tower or to the base.

ACKNOWLEDGMENTS

We gratefully acknowledge the cooperation of Portland General Electric, which granted access to the tower and permission to conduct the tests. We also appreciate the loan of nearly 3000 feet of instrumentation cable by the Earthquake Engineering Research Laboratory at Caltech, and by Kinemetrics, Inc. Professor G. W. Housner provided helpful and generous advice during the planning stages which resulted in an efficient and thorough instrumentation plan. The assistance of B. Hull, KSU graduate student, in the installation of the accelerometers and cables on the tower is also appreciated.

Support for the project was provided by Grant No. CEE-8503993 by the National Science Foundation. Opinions and findings expressed herein are those of the authors, and do not necessarily represent those of the National Science Foundation.

REFERENCES

(1) Billington, D. F. "Two Decades of Cooling Tower Design in the United States of America" *Proceedings of the Second International Symposium on Natural Draft Cooling Towers*, Ruhr-University, Bochum, Germany, September 1984.

(2) Armitt, John, "Wind Loadings on Cooling Towers," *Journal of the Structural Division*, ASCE, Vol. 106, No. ST3, Mar. 1980.

(3) Der, T. J. and R. Fidler, "A Model Study of the Buckling Behavior of Hyperbolic Shells," *Proceedings*, Institution of Civil Engineers, Vol. 41, Sept. 1968.

(4) Veronda, D. R. and V. I. Weingarten, "Stability of Pressurized Hyperboloidal Shells," *Journal of the Engineering Mechanics Division*, ASCE, Vol. 101, No. EM5, Oct. 1975.

(5) Mungan, Ihsan, "Buckling Stress States of Hyperboloidal Shells," *Journal of the Structural Division*, ASCE, Vol. 102, No. ST10, Oct. 1976.

(6) Lu, W., Y. Cheng, J. Peng, and T. Wang, "The Finite Element Solution and Model Tests of Free Vibration of Hyperbolic Cooling Tower," *Proceedings of the Second International Symposium on Natural Draft Cooling Towers*, Ruhr-University, Bochum, Germany, Sept. 1984.

(7) Lashkari, M., V. I. Weingarten, and D. S. Margolias, "Vibration of Pressure Loaded Hyperboloidal Shells," *Journal of the Engineering Mechanics Division*, ASCE, Vol. 98, No. EM5, Oct. 1972.

(8) Swartz, S. E., A. Nikaeen, and H. Mozaffarian, "On the Validity of Reinforced Concrete Models of Cooling Tower Shells," *Proceedings of the Second International Symposium on Natural Draft Cooling Towers*, Ruhr-University, Bochum, Germany, September 1984.

(9) Ting, B. Y., "Buckling Test of a Hyperbolic Shell Subjected to Uniform Pressure," The Marley Co., Mission, Kansas (unpublished report).

(10) Sollenberger, N. J. and R. H. Scanlan, "Pressure Differences Across the Shell of a Hyperbolic Natural Draft Cooling Tower," *Proceedings of the International Conference on Full Scale Testing of Wind Effects*, London, Ontario, Canada, June 1974.

(11) Sollenberger, N. J., R. H. Scanlan, and D. P. Billington, "Wind Loading and Response of Cooling Tower," *Journal of the Structural Division*, ASCE, Vol. 106, No. ST3, March 1980.

(12) Steinmetz, R. L., D. P. Billington, and J. F. Abel, "Hyperbolic Cooling Tower Dynamic Response to Wind," *Journal of the Structural Division*, ASCE, Vol. 104, No. ST1, Jan. 1978.

(13) Scanlan, R. H. and N. J. Sollenberger, "Pressure Differences Across the Shell of a Hyperbolic Natural Draft Cooling Tower," *Proceedings of the Fourth International Conference on Wind Effects on Buildings and Structures*, Cambridge University Press, Cambridge, England.

(14) Niemann, H. J., "Wind Pressure Measurements on Cooling Towers," *Proceedings of IASS Conference on Tower Shaped Structures*, The Hague, Netherlands, April 1969.

(15) Niemann, H. J. and H. Propper, "Some Properties of Fluctuating Wind Pressures on a Full Scale Cooling Tower," *Symposium on Full Scale Measurements of Wind Effects on Tall Buildings and Other Structures*, University of Western Ontario, Canada, June 1974.

(16) Basu, P. K., and P. L. Gould, "Cooling Towers Using Measured Wind Data," *Journal of the Structural Division*, ASCE, Vol. 106, No. ST3, March 1980.

(17) Warner, M. E. and M. R. Lefevre, "Salt Water Natural Draft Cooling Towers Design Considerations," *American Power Conference*, Chicago, IL, April 1974.

(18) Basu, P. K. and P. L. Gould, *SHORE III Shell of Revolution Finite Element Analysis Program User's Manual*, Research Report No. 49, Structural Division, Washington University, St. Louis, MO, 1977.

(19) Dumitrescu, J. A., James G. Croll, and D. P. Billington, "Cooling Towers on Flexible Foundations," *Journal of Structural Engineering*, ASCE, Vol. 109, No. 10, Oct. 1983.

(20) Kapania, R. K. and T. Y. Yang, "Time Domain Random Wind Response of Cooling Tower," *Journal of Engineering Mechanics*, ASCE, Vol. 110, No. EM10, Oct. 1984.

(21) Lin, A. N., "Measurement of Prototype Cooling Tower Ambient Vibration," *Proceedings*, SEM Fall Conference on Experimental Mechanics, Grenelefe, FL, Nov. 1985.

(22) Abel, J. F. and D. P. Billington, "Effect of Shell Cracking on Dynamic Response of Concrete Cooling Towers," *Proceedings*, IASS World Congress on Space Enclosures, Montreal, Canada, July 1976.

CONFINEMENT REQUIREMENTS FOR PRESTRESSED CONCRETE COLUMNS SUBJECTED TO SEISMIC LOADING

A. J. Durrani[I] and H. E. Elias[II]

ABSTRACT

The results of an experimental investigation in which eleven one-third scale prestressed concrete columns were tested under concentric load is presented. Primary variable of this study was the amount of lateral confining reinforcement. The load-deformation behavior at high axial strains as encountered during an earthquake was found to be significantly different than that of non-prestressed columns. The concept of strain ductility as a measure of column ductility is proposed. The experimental results indicate that the ACI code limit on the spacing of eight bar diameter may be too restrictive for prestressed concrete columns which usually have a small diameter longitudinal reinforcement.

INTRODUCTION

The survival of buildings against collapse during earthquakes depends to a large degree on the performance of columns. This is particularly true for buildings with precast components because of their limited load redistribution capability. The use of precast columns in low-rise buildings in regions of low to moderate seismic risk is not very uncommon. During earthquakes, these columns may be required to undergo large axial strains. Unlike reinforced concrete columns, precast prestressed concrete columns generally have a much smaller amount of longitudinal reinforcement, and therefore, behave differently at large strain levels. Thus, for adequate strength and ductility, it is important to understand the load-deformation characteristics of precast prestressed concrete columns.

PRESTRESSED CONCRETE COLUMNS

Precast columns are prestressed mainly to resist any tensile stresses during handling, transportation and erection. ACI code [1] specifies a minimum effective prestress over the gross section of not less than 225 psi. This provision indirectly sets a minimum longitudinal steel ratio. The prestressing reinforcement is usually provided in the form of small diameter strands or bars made from high strength steel and the columns are commonly prestressed to an average stress of 500 psi. At axial strains

---

[I] Assistant Professor, Department of Civil Engineering, Rice University, Houston, Texas.

[II] Graduate Student, Department of Civil Engineering, Rice University, Houston, Texas.

which result in spalling of the cover, the prestressing reinforcement because of its small diameter is more susceptible to buckling than longitudinal reinforcement in non-prestressed columns.

Lateral reinforcement in prestressed columns performs several functions. It confines the concrete core, thus increasing ductility and toughness. It also increases the transverse shear strength, and restrains the longitudinal reinforcement against buckling. It becomes active in confining the column core only when axial strain is sufficiently high to induce lateral bulging of the core thus setting up tensile stresses in the confining reinforcement.

Closely spaced circular spiral is most effective in confining the column core because it resists lateral forces in direct tension but in the case of rectilinear ties, the lateral forces in the core are resisted by a combination of direct tension and bending of tie bars. At corners of ties where direct tension is dominant, the confinement is most effective [2,3]. In between the corners, the ties resist lateral forces in bending, thus creating regions of lower or no confinement as shown in Fig. 1.

Prestressed columns are fabricated from relatively high strength materials; steel having an ultimate strength of 250-270 ksi and concrete having a 28 days compressive strength usually in excess of 5000 psi. For square sections, a simple configuration of lateral reinforcement in the form of rectilinear ties is a common practice. The ACI code [1] is not very explicit on the use of prestressed columns in regions of high seismic risk. However, it does contain provisions for adequate strength and ductility in regions of moderate seismic risk.

The objective of this study was to investigate the strength and ductility behavior of prestressed columns under high axial strains as encountered in moderate earthquakes. Knowing the load-deformation characteristics of typical prestressed concrete columns, the performance of buildings with precast columns could be predicted more realistically.

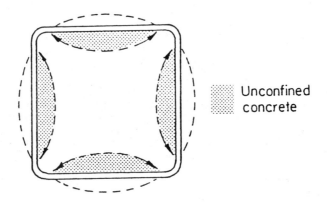

Fig. 1: Confinement by Rectilinear Ties
(From Park and Paulay 1975)

EXPERIMENTAL INVESTIGATION

The main objective of the experimental program was to investigate the effect of various amounts of lateral confining reinforcement on the load deformation behavior of short prestressed concrete columns. The testing program consisted of columns which were loaded to failure under axial load. The test columns had a cross section of 5.25 in. x 5.25 in. and were 26 inches long. With these dimensions, the test units may be regarded as one-third scale model of real size columns. Two of the columns did not have any lateral reinforcement. In the remaining nine columns, the lateral reinforcement was provided in the form of square ties which had 135 degrees hooks with ten bar diameter extensions. The volumetric ratio of the lateral reinforcement to the concrete core, ρ, was varied between 1.09 and 2.18 percent. The concrete core is defined as the area enclosed by the center line of the lateral reinforcement. The column designation, spacing and number of columns tested in each series are given in Table 1.

Table 1 - Detail of Test Columns

Series	No. of Columns	Spacing (inches)	Lateral Reinforcement Ratio
A1	2	---	---
B1	3	4	1.09%
B2	3	3	1.46%
B3	3	2	2.18%

All columns with lateral ties were designed to satisfy the recommendations of chapter 18 of the ACI code. In addition, columns in series B3 also satisfied the recommendations of Appendix A for regions of moderate seismic risk except for the spacing limitation of $8d_\ell$, where d_ℓ is the diameter of the longitudinal reinforcement. This provision would have severely limited the spacing of ties because of the relatively smaller size of the prestressing reinforcement compared to the reinforcement in non-prestressed columns. The ratio of the core area to the gross area was maintained at about 0.66 which is close to that of the prototype columns. The concrete used in preparing the columns had an average 28 days compressive strength of 5600 psi. The effective prestress over the gross section of each column was 500 psi. During the test, vertical strains were measured with two LVDT's attached to the columns over a gage length of ten inches. Columns were tested in a 400 kips closed loop Tinius-Olsen machine under strain control mode which assured accurate measurements of strains in the post-peak loading region. Figure 2 shows typical details of the test columns.

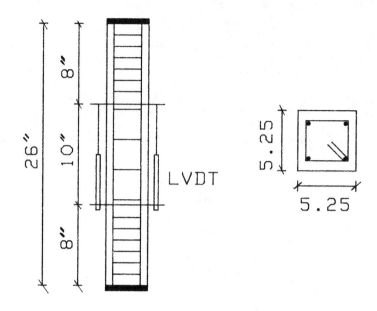

Fig. 2: Typical Details of a Test Column

TEST RESULTS

During the test, the load-strain relationship was continuously plotted and the strain in the lateral and longitudinal reinforcement was recorded at discrete intervals. At or close to the maximum load, vertical cracks started to appear in the cover. This was always the first sign of distress in the columns. These cracks spread quickly causing the concrete cover to become ineffective. This was particularly evident in the case of columns with closely spaced ties. It is believed that closely spaced ties create a plane of weakness between the core and the cover thus promoting the separation of the cover. The spalling off of the cover was completed at an average longitudinal strain of about .004 to .0045. Thereon, the mechanism of core confinement was in the form of arching between the ties and the longitudinal reinforcement. Buckling of the prestressed bars was observed at an average longitudinal strain of about .006 to .008. In most of the columns, failure occured either by the formation of an inclined plane or by the development of a shear cone. Tie fracture or anchorage failure did not occur in any of the tests. Typical load-strain plots of each series are shown in Fig. 3. The abcissa represents the average longitudinal strain over ten inch gage length. The ordinate represents the load applied to the column normalized with respect to the gross area and the concrete strength.

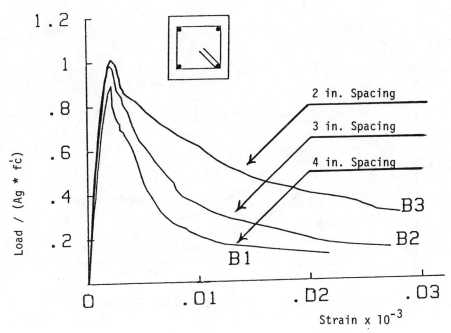

Fig. 3: Experimental Load Strain Curves

The load strain behavior depends to a large extent on the stress-strain of the concrete core, the cover, and the prestressing reinforcement. The stress-strain curve of the concrete core can be obtained by subtracting the load-strain of the cover and the prestressing reinforcement from the total load-strain curve. The remaining load is then normalized by the core area. Typical stress-strain behavior of a reinforced and a prestressed concrete column core are shown in Fig. 4. In both cases the ascending branches are almost identical. The maximum stress is generally slightly higher in the reinforced concrete than in the prestressed concrete column core. This is primarily due to the fact that the prestressed concrete columns have relatively smaller amounts of longitudinal reinforcement and the confinement is essentially provided by the lateral reinforcement. The rapid drop in stress in the post-peak region of the prestressed concrete column core is also attributed to the same reason.

The currently available models [3,4] for the prediction of the core stress-strain curve are based mainly on test results reinforced concrete columns and seem to overestimate the stress level in the post-peak descending branch. Such overestimation of column stress at high axial strains as during earthquakes could be critical to the safety of precast buildings.

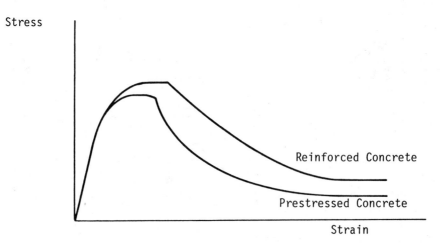

Fig. 4. Prestressed Concrete and Reinforced Concrete Core Stress Strain Curves

Stress-strain relationship of the core curve can be used to define the toughness of columns. If the area under the load-strain curve of the column up to a certain strain value is taken as a measure of column toughness the test results indicate that an increase in the amount of lateral reinforcement increased the column toughness. In the absence of any other definitive criteria to measure the ductility of columns, a strain value corresponding to a load 80 percent of the maximum load can be regarded as a reasonable estimate for the column ductility. The strain ductility can then be calculated as

$$SDF_{(80\%)} = \frac{\varepsilon_{80} - \varepsilon_{max}}{\varepsilon_{max}} \qquad (1)$$

where ε_{max} is the strain at maximum load, and ε_{80} is the strain at 80% of the maximum load as shown in Fig. 5.

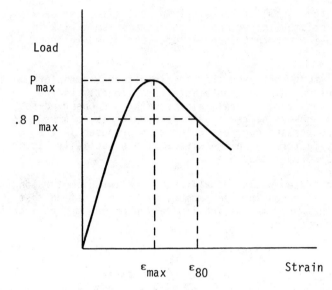

Fig. 5. Strain at Eighty Percent of the Maximum Load

Strain ductility factors for each test series calculated from the load-strain plots of columns is given in Table 2.

Table 2 - Strain-Ductility Factors

Series	$SDF_{(80\%)}$
A1	32%
B1	38%
B2	53%
B3	83%

The values shown are the average value within each series. It is noticed that in series B1 the $SDF_{(80\%)}$ did not increase appreciably over that of plain concrete columns in series A1. However, the ductility increased rapidly as the tie spacing decreased in series B2 and B3.

SUMMARY AND CONCLUSIONS

Experimentally observed load-strain behavior of several prestressed concrete columns is presented. Some of these columns satisfied the provisions for the seismic design of columns in regions of moderate seismic risk in addition to the general design of prestressed columns contained in the ACI code.

The results of this study indicated that the longitudinal reinforcement in prestressed concrete columns did not contribute significantly to the ductility of the columns. However, lateral reinforcement was observed to considerably improve the toughness of the prestressed concrete columns. A measure of ductility of prestressed concrete columns is proposed in the form of a strain ductility factor. A column with a strain ductility factor of 75% or more may be considered acceptable.

Since prestressed concrete columns usually have smaller diameter longitudinal reinforcement, the ACI code tie spacing requirement of $8d_\ell$ may be too restrictive. The test results suggest that the tie spacing limitation of $8d_\ell$ may be increased to $10d_\ell$ for prestressed concrete columns.

REFERENCES

[1] ACI Committee 318, "Building Code Requirements for Reinforced Concrete (ACI 318-83)," American Concrete Institute, Detroit, 1983, 111 pp.

[2] Park, R. and Paulay, T., Reinforced Concrete Structures, John Wiley and Sons, New York, 1975, 769 pp.

[3] Sheikh, S.A. and Uzumeri, S.M., "Analytical Model for Concrete Confinement in Tied Columns," Journal of the Structural Division, ASCE, Vol. 108, No. ST12, 2703-2722, December 1982.

[4] Park, R., Priestly, M.J. and Gill, W., "Ductility of Square Confined Conrete Columns," Journal of the Structural Division, ASCE, Vol. 108, No. ST4, 928-950, April 1982.

ANALYTICAL PREDICTION OF THE BIAXIAL RESPONSE OF A REINFORCED CONCRETE SHAKE TABLE MODEL

C. A. Zeris [†], S. A. Mahin [‡]

Analytical correlations of experimental test results is necessary in order to verify and improve seismic design procedures. Analytical studies indicate that use of simplified analytical tools, though economic, may fail to describe the physical behavior of reinforced concrete members realistically. A refined finite element is introduced that is capable of simulating experimentally observed behavior of reinforced concrete beam-columns under axial load and biaxial bending in a rational rather than phenomenological manner. In this study emphasis is placed on verification of the model against experimental test results of a single cantilever beam-column and a frame tested on the earthquake simulator. The response of this frame under uniaxial and biaxial support excitation are compared.

INTRODUCTION

The nonlinear mechanical characteristics of reinforced concrete beam-columns under simulataneous variation of axial load and components of biaxial bending depend on several factors, including the state of the structure, the load history and the member detailing. Furthermore, cracking and nonlinearity of the concrete can affect the effective axial and flexural stiffnesses well inside the maximum capacity surface. Member post-yield mechanical characteristics and hysteretic behavior are strongly influenced by Bauschinger effects and hardening of the steel, spalling of the cover and extend of crushing in the core material. Further, bond and shear deterioration and strain rate effects also contribute to the spread of damage along and across the member.

For the sake of computational efficiency and economy, available analytical models rarely attempt to take all of these factors into account in simulating both the hysteretic response and the change in strength of a member. Instead, they rely on phenomenological simplifications to represent the overall global behavior. The behavior of adequately confined beams can often be considerably simplified with satisfactory results. However, for columns the above phenomena are clearly coupled and behavior may often deviate from common idealisations, especially where poor confinement and/or high axial loads are encountered. Because of this the reliability of simplified analytical predictions may be questionable.

A new model is proposed, that can rationally simulate flexural behavior of reinforced concrete beam-columns with the minimum amount of a priori assumptions from the point of view of the user. The ability of the proposed model to simulate experimental behavior is demonstrated. For this purpose, the behavior of a single beam-column tested under controlled conditions is compared with analytically predicted behavior. Increasing the level of complexity, a simple frame tested under uniaxial excitation on a shake table is considered. Finally, in order to demonstrate the significance of the biaxial response the frame is reanalyzed under an excitation of the same intensity but different

[†] Research Engineer, Department of Civil Engineering, Univ. of California, Berkeley, California 94720.

[‡] Associate Professor, Department of Civil Engineering, Univ. of California, Berkeley, California 94720.

orientation with respect to the principal axes of the building. The reliability of the proposed model in simulating the response is discussed to identify its limitations as well as to assess the importance of biaxial effects in building response.

DESCRIPTION OF THE FORMULATION

Nonlinear modeling of beam-column elements has evolved from single component series models [1] to multi component parallel [2,3] and distributed damage multi-spring models. For behavior under constant axial load, modeling can be simplified by assigning global nonlinear behavior characteristics estimated from the section properties and some assumption regarding the flexibility distribution along the member. Special end springs can simulate the pinched hysteretic response associated with fixed end rotations characteristic of bond induced slip [4,5]. For beams detailed according to modern seismic code specifications, modeling of the flexural behavior with multilinear hardening-plastic components may be sufficient if distributed gravity loads are not significant. If the stiffness distribution along the length of the member varies considerably, or the axial load fluctuations are high, the simplified models may be rendered unreliable. Variable axial loads increase modeling complexity. In the earliest models, axial load dependence of the stiffness has been ignored and only the strength dependence was included. The flexural axial interaction has been explicitly considered by use of an appropriate bounding surface and a generalised flow rule [6]. Addition of more bounding surfaces translating and rotating in the load space results in multilinear behavior and stiffness degradation with cycling [7,8].

To model both the stiffness and strength dependence under varying axial load, the column section is refined into fibers or finite filaments. Under the basic assumption that plane sections remain plane, hysteretic behavior and resistance are obtained from uniaxial stress strain relations of the constituent materials. Typically, displacement formulation fiber models make use of cubic Hermite polynomials [9,10,11]. Modeling the entire section with fibers often requires excessive machine storage; an alternative approach of lumping the section material with few springs was adopted in [12]. A lumped nonlinearity series model, with two nonlinear end springs and an elastic component was introduced in [13], using equivalent end spring idealization. In this case, force-elongation characteristics were obtained assuming uniform bond stresses between the steel and concrete over one development length. For improved internal damage representation, flexibility based models have been introduced, interpolating the section flexibilities [14]. Furthermore, standard Hermite polynomials (exact only for a linear member) are abandoned in favor of higher order interpolations that reflect the flexibility distribution along the column. This approach was introduced in [15] for modeling circular steel members and was extended to reinforced concrete columns in [16].

Extension of the displacement formulations using fiber modeling of reinforced concrete sections has been limited due to numerical instability and lack of convergence triggered by internal unbalances within the element. In order to improve the reliability of the fiber model, and to extend its scope of application, an alternative approach has been proposed and verified against experimental test results [17]. The satisfactory performance of this formulation in uniaxial bending motivated its extension to biaxial response.

Theory of the Biaxial Element

The proposed model is a distributed flexibility line element. Internal degrees of freedom are included at an arbitrary number of slices distributed along the element length. Each section is subdivided in a two-dimensional grid of fibers defined on a system of orthogonal axes normal to the longitudinal axis of the element. A minimum of two slices at the ends of the member are required. The degrees of freedom considered are the curvatures with respect to the section axes and an axial reference strain. Global degrees of freedom are the end rotations in the two principal axes (oriented similarly as the section axes) and the axial extension of the member. Small displacements are assumed in all geometry transformations, while global P−Δ effects are formulated assuming the geometric stiffness of a simple truss element.

Flexibilities along the element length are linearly interpolated between slices. The element flexibility is obtained by a weighted integration of the flexibility function and is subsequently inverted to define the local stiffness. Following virtual work formulation, the equilibrium transformation is used as a weight for a linear bending moment and a constant axial load within the member. To avoid numerical instability, determination of internal section states for a given deformation increment is performed following a modified event to event strategy, adopted in [17].

Under the proposed scheme, equilibrium rather than compatibility is enforced at the interior slices. This is due to the demonstrated favorable behavior of this scheme for negative definite behavior in any crirical section. Compatibility is only enforced at the two end slices of the element. For nonlinear positive definite behavior the two approaches are found to be equivalent. At the state determination, conventional techniques were considered for satisfying equilibrium at the internal slices by directly minimizing the section unbalance vector norm. Such techniques failed to converge near section instability, hence a three dimensional patterned search scheme has been adopted instead.

The element is incorporated in ANSR-I [18], a general purpose nonlinear analysis program for the static and dynamic analysis of three dimensional space systems. Material models in the element include a bilinear and a nonlinear with exponential unloading (proposed in [14]) steel model and a multi-linear concrete model with cracking and strength degradation. For the solution of the dynamic problem a Newmark $\beta - \gamma$ integration scheme is used. Various iterative schemes can be used.

ANALYSIS OF A CANTILEVER BEAM-COLUMN IN BIAXIAL BENDING

Experimental Setup

Several experimental results are available in the literature for columns under biaxial bending dominated displacement control (eg. [19],[7]) under constant axial load. For the verification of the element, the beam-column cantilever test reported in [20] is selected. The cantilever is 54. in. (1.37 m) long with a shear span to depth ratio of 4.5. Shear deterioration effects during the test were small. The column section is 12 in. (30.5 cm) square, and symmetrically reinforced about the two axes with a total of eight 0.875 in. (2.2 cm) diameter bars. Transverse confinement consists of 0.25 in. (6 mm) diameter ties at 1.7 in. (4.4 cm.) spacing. The dimensions of the specimen and reinforcement details are shown in Fig. (1). The particular specimen considered here (SP8) was tested under self weight and an imposed tip displacement control. The prescribed displacement path involved 52 cycles of tip displacement reversal in either of the two plan principal directions of the member, in a square pattern. Typically the sequence of deformation consisted of the applications of two square excursions followed by smaller amplitude inner loops. Displacements were progressively increased to a maximum of five times the uniaxial bending yield displacement. This was estimated to be 0.55 in. (14 mm)[20].

For reasons of cost, all cycles leading to a smaller than previously attained absolute displacement are disregarded. Thus, prior square displacement patterns are not duplicated. The displacement pattern for which the model is analyzed is shown in Fig.2; in all, 24 displacement excursions were imposed. The cantilever model analyzed consists of a single beam-column element. The element is refined into four unequally spaced slices, each slice being modeled by two hundred concrete and eight steel fibers. Different material properties are specified for the unconfined cover and the concrete core, while a bilinear hardening steel model is used.

Results and Correlations

The variation of the base shear with imposed tip rotation for the two principal directions of the column is compared with the experimental values in Fig.3. It can be observed that the obtained pattern of load-deformation correlates well with the experimental results. All duplicated excursions are omitted. The effect of the biaxial action on the flexural strength of the column is evident. Sequential biaxial motion of the tip along two orthogonal directions (e.g. EW and then NS motion) causes the neutral axis of the section, initially aligned with the NS direction to rotate towards the EW. At extreme deformations in the normal sense the neutral axis rotation is a full ninety degrees, resulting in significant cracking of the previously compressed concrete and subsequent drop in the EW resistance. This drop is equally significant for all the displacement squares, and generally increases with increasing absolute deformation. Further, since part of the strains at the corner steel and compression concrete are due to the out-of-plane deformation, a reduction in capacity also occurs at the reloading direction also. Cyclic resistance deterioration is evident in all excursions to larger ductility in both directions, as a result of the intermediate out-of-plane cycle. This implies that the initiation of yielding bounding surface in the moment plane translates along the direction of deformation.

The effect of biaxial bending on the flexural stiffnesses is also quite significant. Considering the patterns in the EW bending, it can be seen that the unloading stiffnesses towards the negative EW displacement degrade progressively with each cycle. Although the same maximum resistance is obtained in the small ductility sense, progressive stiffness deterioration of the member is more evident.

The axial elongation variation at the base of the column with imposed EW bending is shown in Fig.4. The effect of biaxial displacements results in a cumulative increase in the centroidal extension of the member under no axial load. This extension is a result of irrecoverable plastic strains of the reinforcement with cycling. This results in an accumulation of crack opening at the base section de to the absence of axial load on the member. The extent of yielding along the member is evident in Fig.5, where the NS moment curvature characteristics for two end slices 5 in. (12.5 cm.) apart are compared. For the first three cycles of deformation (maximum displacement ductility of 2) the upper section is cracked but does not exhibit any residual curvatures. In the final cycle of reversal leading to a ductility of 5, the upper slice also deforms irrecoverably, with higher residual negative curvatures than the end slice. However, the incremental curvature reversal is twice as high for the already yielded base.

UNIAXIAL TEST CORRELATION FOR A R/C FRAME

Experimental Setup

The test structure considered in this study was a two storey reinforced concrete bare frame tested on a earthquake simulator. The structure is a 1/1.4 scale model of a two story office building designed as a moment resisting frame. The proportions and detailing of the frame (Fig. 6) were chosen so that the inelastic deformation would be primarily confined to the columns. Columns are rectangular, 8.50 by 5 3/4 in. (21.6 by 14.6 cm) reinforced with four 0.62in (16 mm) diameter bars (2.5% the gross section area). The girders are T sections with unequal amounts of top and bottom reinforcement (Fig.6). Maximum material strengths were experimentally determined to be on average 41 ksi (282 Mpa) and 4.4 ksi (30.3 MPa) for the steel and concrete, respectively. To avoid out-of-plane effects the frame was restrained in the transverse direction by lateral braces. The bare frame was tested under a longitudinal excitation input [21,22] under several ground motion intensities. In this study the test run W2 is considered; for this test the Taft earthquake was applied with a peak acceleration scaled to 0.52 g. The strength of the structure determined experimentally by static load to collapse after the tests was 67% of the total weight of the building.

To obtain correct dynamic similitude the mass of the structure is increased by concrete blocks bolted at the slabs. The total added weight is 16 k (71.2 KN) and 8 k (35.6 KN) at the bottom and top stories respectively. The uncracked periods of vibration for the fixed base model were estimated by free and forced vibration tests to be 0.1 and 0.26 seconds for the first two translational modes. After the first test run, a 0.07g scaled Taft excitation, the periods of the structure increased to 0.11 sec and 0.32 sec respectively. The modal damping at the first mode was measured to be 4.5% of critical. The above periods and damping values are identically simulated by the formulated numerical model of the frame.

The model of the uniaxially tested frame consists of ten fiber elements, namely six beams and four columns. The experimentally determined material characteristics are specified for the uniaxial stress-strain properties of the steel and concrete. Different behavior is assumed for the confined core and the unconfined cover. All members are refined using five unequally spaced slices, and a total of 800 concrete and 20 steel fibers per element. Conditions at the base are assumed to be fixed. The actual foundation of the test structure was a reinforced concrete footing bolted on the table. Some fixed end rotations would therefore be expected at this location. The same test specimen was also considered in [17], in order to verify the accuracy of the original uniaxial beam-column element.

Correlations and Results

The structure is reanalyzed with the new element. The static load is initially imposed in order to obtain the cracked pattern on the structure. The time step used for the dynamic analysis is 0.006 seconds. The biaxial elements are specified with comparable out-of-plane stiffnesses to avoid any numerical roundoff problems in the inversion of the section stiffnesses. The first eight and a half seconds of the recorded response are analyzed, starting from second 2.0 of the record. The recorded roof displacements correlate well with the analytically predicted values both in amplitude and in periodicity, failing only to estimate correctly the large peak of the response at the eighth second (Fig.7). Due to improved simulation of dead loads, the correlations are better than those reported in [17]. Reasons for the remaining lack of correlation are possibly 1) the effect of the table rocking motion, which is not included in the excitation nor in correcting the relative roof displacements and 2) the lack of modeling of the foundation conditions of the columns. It is expected that fixed end rotations of the columns at the base due to pullout of the rebars can give rise to additional drift at the roof of the specimen. In order to include this effect it is necessary to add a concentrated spring with degrading slip characteristics at the base of each member. Improvements of the formulation towards this modeling idealization are currently being pursued.

The internal force and deformation histories are briefly described. The variation of axial load at the ground story columns is given in Fig.8a. As expected, axial loads in the two members oscillate symmetrically about the dead load values. At the arrival of the first strong pulse of the Taft record axial loads in Column 1 almost become tensile. This results in a considerable reduction in the stiffness of this member, both in resisting axial load and in resisting base shear. Furthermore, it reduces the flexural capacity of the column while increasing its deformability. The above behavior can be detected in Fig.8b , where the shears resisted by the two ground columns are compared. At the 6.0 second, Column 2 resists four times the dynamic shear of Column 1. For the low amplitude cycles up to about 5.5 seconds the two columns resist equal shears.

The bending-base rotation characteristics of Column 1 are shown in Fig.9. Initial small amplitude vibrations induce the characteristic spindle shaped hysteretic curves under crack opening and closing during cycling, without any hysteretic energy absorbtion. Upon yielding of the steel the characteristics of the member are strongly dependent on the axial load level and the stress history of the metarials. Overall, since the maximum axial load on the column is below 30% the balance load (equal to 100 kips -445 KN [22]) the behavior of the member is ductile. In the positive bending sense the column is under increased compression, resulting in pinching of the hysteretic loop, while in the opposite sense the reduction of axial load causes a) earlier yielding of the member, but b) a more ductile transition of the hysteresis to yield, governed primarily by the steel characteristics. For the extent of the ground motion analyzed, the top of the column undergoes smaller nonlinear excursions, being mostly cracked under moment reversal.

In summary the analytical model of the shaking table specimen is satisfactory in estimating both the initial cracked state of the structure as well as the transition to nearly a complete mechanism response of the frame during the strong motion. The effects of changing axial load in the members both on stiffness as well as on strength and hysteretic shape are modelled in a satisfactory manner. The correlations are good, with some inability as yet to include the fixed base rotation behavior of the base-column joint. Rocking of the table is important for this test and should be considered in subsequent analyses.

BIAXIAL TEST CORRELATION OF A R/C FRAME

The discussions so far indicate that the behavior of a reinforced concrete beam-column under three dimensional excitation is complex and strongly load history dependent. Analysis of the biaxial response shows that the capacities and stiffnesses of the members in the two principal directions are coupled and depend also on the axial load resisted. In order to investigate the significance of these phenomena to the response of a real structure, the simple two bare story frame considered in the uniaxial correlation studies is reanalysed under the same support excitation. In this case, however, the ground motion is assumed to occur at an incident angle of 25 degrees relative to the longitudinal axis of the structure. This will excite a biaxial structural response. The characteristics of the structure, materials and ground motion involved are identical to the uniaxial case. The results of the analysis are compared to the previous analysis. Again, the significance of bond deterioration is not taken into account. The model of the frame consists of eight fiber elements to represent the columns. Two component elements are used to model the beam-slab subassemblage. The slab is assumed rigid in its own plane. Beam stiffnesses are taken as an average of the cracked stffnesses of the fiber element beams in the uniaxial analysis, and are assumed independent of the dead and seismic load distribution on the beam. Columns are refined into five unequally spaced slices using a total of 500 concrete fibers per column.

The global dislacements in the longitudinal direction for the uniaxial and the biaxial analysis are compared in Fig.10a. The two histories are markedly different both in periodicity and in magnitude. The biaxial frame is overall more flexible, due to the reduction of the flexural stiffness of the columns with out-of-plane response. As a result of this stifness reduction, the predicted displacement in the maximum positive displacement peak is slightly smaller than the uniaxial model response which is generally stiffer. Following the peak, the biaxial model indicates considerable stiffness degradation in both principal directions and a corresponding reduction in the displacement amplitudes in the longitudinal sense, until the second energy burst, when the displacement peaks are again comparable. In the more flexible transverse direction, the peak is actually reached in the second energy burst of the motion (Fig. 10b).

The base moment rotation demands for Column 1 in the two orthogonal rotations are shown in Fig. 11. Comparing to the response of the member under uniaxial excitation (Fig. 9), it is apparent that the member suffers a reduction in longitudinal capacity by about 30%, due to the concurrent biaxial demands. The influence of axial load fluctuation with bending sense changes the capacity of the member in the positive and negative bending. The flexural stiffness also reduces considerably (as also evident in the cracked period of the response), while the hysteretic behavior is no longer symmetric. Pinching is now less evident, due to the fact that the neutral axis rotates with crack closing, making the transition from cracked to uncracked section less sudden than in the uniaxial case. Rotational demands in the longitudinal sense are comparable with the uniaxial case, even though the elastic beam stiffness are invariant during the analysis.

CONCLUSIONS

Summarising, the following conclusions can be drawn. Biaxial effects in reinforced concrete beam-columns are significant both from the points of view of strength and of stiffness. Simplified models often disregard either of the two, and may underestimate localised demand predictions. A reliable fiber model is proposed that can incorporate rationally the observed mechanical behavior of such members with minimum assumptions. The model is able to predict both uniaxial and biaxial test results. In comparing the biaxial versus uniaxial response of a particular frame, it is observed that biaxial response tends to elongate the apparent period of the structure more than the uniaxial response, resulting (for the cases considered) in comparable drifts yet a notable srength and stiffness reduction with respect to the two principal axes of the building. For ductile moment frames, where columns should be proportioned not to be the primary members for energy absorbtion, unlike the test frame considered here, the reduction in strength should be considered.

ACKNOWLEDGEMENTS

The authors would like to acknowledge the financial support of the National Science Foundation during the course of this study. We are thankful to graduate students I. Khatib and V. Pantazopoulou for assistance with the programs and to J. Dimsdale for assistance in performing the analyses.

REFERENCES

[1] Giberson M., The Response of Nonlinear Multi-Storey Structures Subjected to Earthquake Excitation, *Earthqu. Eng. Res. Lab*, Pasadena, May 1967.

[2] Clough R. and Benuska L., Nonlinear Earthquake Behavior of Tall Buildings, *Jnl of Am. Soc. of Civil Eng.*, **93**, EM 3, June 1967, 129-146.

[3] Clough R. and Johnston S., Effect of Stiffness Degradadtion on Earthquake Ductility Requirements, *Trans. Japan Earthqu. Eng. Symposium*, Tokyo, Oct. 1966, 195-198.

[4] Otani S. and Sozen M., Behavior of Multistorey Reinforced Concrete Frames During Earthquakes, Univ. of Illinois, Str. Res. Series Rep. No 408, Urbana, July 1974.

[5] Banon H., Biggs J. and Irvine M., Seismic Damage in Reinforced Concrete Frames, *Jnl. Am. Soc. of Civil Engrs.*, **107**, ST 9, Sept. 1981, p. 1713-1729.

[6] Riahi A., Row D. and Powell G., Three Dimensional Frame Elements for the ANSR-I Program, *Earthqu. Eng. Res. Center.*, Rep. No EERC-78/06, Berkeley, 1978.

[7] Takizawa H. and Aoyama H., Biaxial Effects in Modelling Earthquake Response of R/C Structures, *Earthqu. Eng. and Str. Dyn.*, **4**, p. 523-52, 1976.

[8] Chen P.F., Generalized Plastic Hinge Concepts for 3D Beam-Column Elements, PhD Disrtn, Univ. of California, Berkeley, 1981.

[9] Kang Y., Nonlinear Geometric, Material and Time Dependent Analysis of Reinforced and Prestressed Concrete Frames, Univ. of California, Rep. UC-SESM No. 77-1, Berkeley, Jan. 1977.

[10] Mari A., Nonlinear Geometric, Material and Time Dependent Analysis of Three Dimensional Reinforced and Prestressed Concrete Frames, Univ. of California, Rep. UC-SESM No. 84-12, Berkeley, June 1984.

[11] Aktan A., Pecknold D. and Sozen M., Effects of Two-Dimensional Earthquake Motion on a Reinforced Concrete Column, *Civil Eng. Studies*, Rep. 399, Univ. of Illinois, Urbana-Champaign, 1973.

[12] Suharwardy M. and Pecknold D., Inelastic Response of Reinforced Concrete Columns Subject to Two-Dimensional Earthquake Motions, *Civil Eng. Studies*, Rep. 455, Univ. of Illinois, Urbana-Champaign, 1978.

[13] Lai S., Will G. and Otani S., Model for Inelastic Biaxial Bending of Concrete Members, *Jnl. Am. Soc. of Civil Engrs.*, **110**, ST 11, Nov. 1984, p. 2563-2584.

[14] Menegotto M. and Pinto P., Slender RC Compressed Members in Biaxial Bending, *Jnl of Am. Soc. of Civil Eng.*, **103**, ST 3, March 1977, 587-605.

[15] Mahasurevachai M., Inelastic Analysis of Piping and Tubular Structures, Ph D Disrtn, Department of Civil Engineering, Univ. of Calif., Berkeley, Nov. 1982.

[16] Kaba S. and Mahin S., Refined Modelling of Reinforced Concrete Columns for Seismic Analysis, *Earthqu. Eng. Res. Center.*, Rep. No EERC-84/03, Berkeley, 1984.

[17] Zeris C. and Mahin S., Significance of Nonlinear Modeling of R/C Columns to Seismic Response, *Am. Soc. of Civil Eng.*, Third Specialty Conf. on Dynamic Response of Structures, Unic. of California, Los Angeles, April 1986.

[18] Mondkar D. and Powell G., ANSR-I General Purpose Program for Analysis of Nonlinear Structural Response, Rep. EERC 75-37, *Earthqu. Eng. Res. Center*, Univ. of California, Berkeley, 1975.

[19] Okada T., Seki M. and Asai S., Response of Reinforced Concrete Columns to Bi-Directional Horizontal Force and Constant Axial Force, Tech. Note, *Bull. ERS*, **10**, p. 30-36, 1976.

[20] Otani S. and Cheung V., Behavior of Reinforced Concrete Columns Under Biaxial Lateral Load Reversals, (II) Test Without Axial Loads, Publ. 81-02, Univ. of Toronto, Feb. 1981.

[21] Hidalgo P. and Clough R., Earthquake Simulator Study of a Reinforced Concrete Frame, *Earthqu. Eng. Res. Center.*, Rep. No EERC-74/13, Berkeley, 1974.

[22] Clough R.W. and Gidwani J., Reinforced Concrete Frame 2:, Seismic Testing and Analytical Correlation, *Earthqu. Eng. Res. Center.*, Rep. No EERC-76/15, Berkeley, 1976.

[23] Oliva M., Shaking Table Testing of a Reinforced Concrete Frame With Biaxial Response. *Earthqu. Eng. Res. Center.*, Rep. No EERC-80/28, Berkeley, 1980.

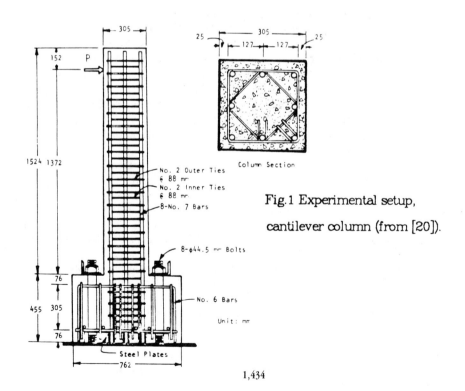

Fig.1 Experimental setup, cantilever column (from [20]).

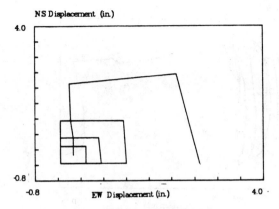

Fig.2 Tip deformation pattern imposed

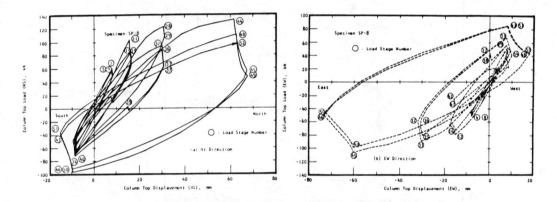

Fig.3a Experimental response of cantilever column (SP8, from [20]).

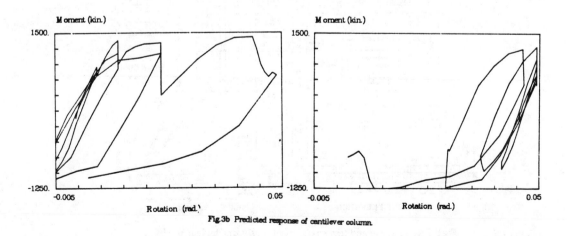

Fig.3b Predicted response of cantilever column.

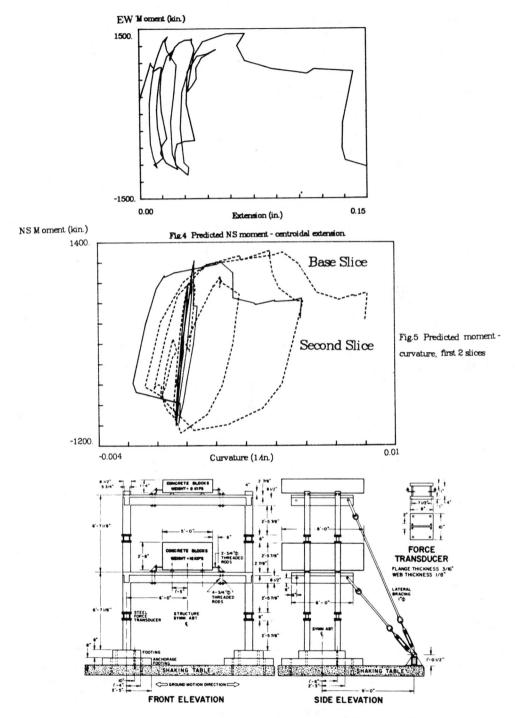

Fig.4 Predicted NS moment - centroidal extension.

Fig.5 Predicted moment - curvature, first 2 slices

Fig.6 Test structure and test arrangement on shaking table (from [22]).

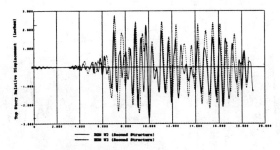

a) Recorded (from [22])

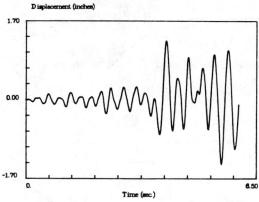

b) Predicted

Fig.7 Comparison of predicted and recorded roof response.

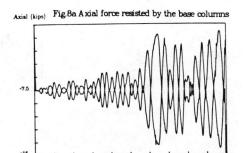

Fig.8a Axial force resisted by the base columns

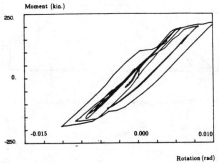

Fig.9 Predicted moment-rotation characteristics, base

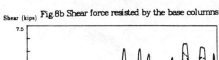

Fig.8b Shear force resisted by the base columns

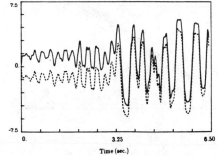

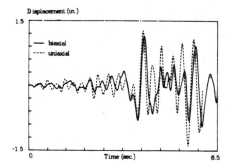
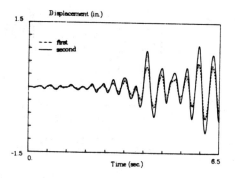

Fig.10a Comparison of uniaxial and biaxial longitudinal roof response

Fig.10b Biaxial transverse response

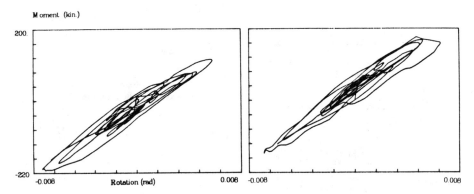

Fig.11a Predicted EW moment-rotation characteristics, base

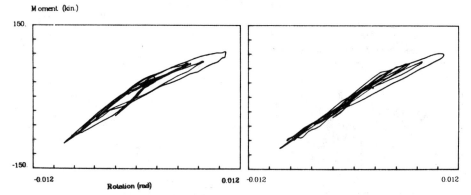

Fig.11b Predicted NS moment-rotation characteristics, base

MECHANISMS OF STIFFNESS AND STRENGTH DETERIORATION IN BIAXIALLY LOADED R.C. BEAM-COLUMN JOINTS

Roberto T. Leon (I)

ABSTRACT

Fourteen reinforced concrete beam-column joint subassemblages were tested to investigate the effects of axial loads, beam geometry, and slabs on joint response under large cyclic load reversals. The shear strength and hysteretic behavior were studied by quantifying the contributions of several mechanisms to shear and stiffness deterioration. Horizontal joint shear, bar slip and column inelastic behavior were found to contribute more to degradation of behavior with cycling than previously thought. It was found that biaxial effects and floor member geometry can have a significant impact on joint behavior, and that bond conditions and column-to-beam flexural capacity ratio control their behavior.

INTRODUCTION

The current design procedures for reinforced concrete beam-column joints in ductile moment-resisting frames are based on limiting the horizontal shear in the joint area and satisfying some minimum requirements for anchorage of the beam reinforcement and flexural overstrength of the column [1,2]. These design recommendations assume that most of the shear will be transferred by a large compressive strut forming diagonally across the joint, and that some deterioration of behavior will occur with cycling due to bar slippage and shear cracking. These provisions are intended for design of plane frames, and do not account for the contribution of the floor slabs to the strength and stiffness of the frame. The influence of floor slabs and bidirectional effects cannot be safely ignored because both can drastically increase the horizontal joint shear, decrease the column overstrength, and shift the hinging regions from the beams to the columns.

An extensive experimental study of beam-column joints subjected to bidirectional cyclic loading was recently completed at the U. of Texas. The primary objectives were to verify the ultimate shear strength of the joints and to assess the adequacy of current design provisions. Among the unique features of this test series is the large scale of the models used, the bidirectional loading applied, the low percentage of transverse joint reinforcement, the measurement of beam reinforcement slip, and the extensive instrumentation used to monitor the hysteretic deterioration.

I. Assistant Professor, Dept. of Civil and Mineral Engineering, U. of Minnesota, Minneapolis, MN 55455.

EXPERIMENTAL STUDY

To investigate the main parameters affecting the shear strength and hysteretic performance of beam-column joints a series of fourteen full-scale tests was devised, utilizing the subassemblage shown in Figure 1. The geometry and reinforcement were varied throughout the test series, and the details are given in Table 1. More complete details on the entire series have been published elsewhere [3,4], and only eight tests will be discussed here.

The beam-column joints in this test series were specifically designed so that the joint core strength would be the controlling factor. Although the specimen were designed to have proportions similar to those typical of joint elements in multi-story concrete frames, they were not intended to model an actual structure. It was anticipated that joints would show severe shear strength and bond deterioration. The joints were designed based on the shear strength equations proposed by Meinheit and Jirsa [5], and the joint dimensions were kept small in order to increase the joint shear stress and decrease the anchorage length of the beam bars.

The specimen were loaded through the beams using double-acting hydraulic rams which allowed for the application of both dead and racking loads. The first load stage consisted of deflecting the beam ends downward 0.1 in. to account for the dead load present in a typical frame (Fig. 2). The specimen were then subjected to an elastic cycle to precrack the beams and simulate the effects of the first small cycles imposed by a major earthquake. After this the specimen were subjected to three cycles at three progressively larger deflection levels. The first deflection level corresponded to first yielding in the beam bars at the joint face. The loads were applied simultaneously in both principal axes. The second and third deflection levels were multiples of this first level. The deflection levels corresponded roughly to drifts of 1.5%, 3.0%, and 4.5% in each principal direction.

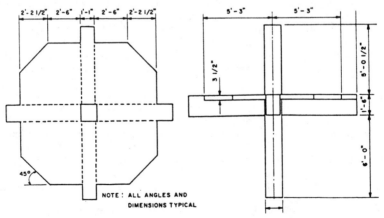

Figure 1 - Typical test specimen (BCJ9)

Table 1 - Summary of Test Series

Test	Column (1)	Beam (-) (2)	Beam (+) (2)	Mom. Ratio (3)	Variable
BCJ5	12 # 9	3 # 8	3 # 6	1.27	P(col)=300 k
BCJ7	12 # 9	3 # 8	3 # 6	1.28	Joint rein.
BCJ8	12 # 9	3 # 8	3 # 6	1.26	P(col) = 0
BCJ9	12 # 9	3 # 8	3 # 6	1.16	Slab
BCJ9A	12 # 9	3 # 8	3 # 6	1.16	Slab
BCJ10	12 # 9	3 # 8	3 # 6	1.13	Wide beams
BCJ11	8 #11	2 #10	2 # 8	1.25	Nar. beams
BCJ12	12 # 9	3 # 8	3 # 6	1.13	Wide beams

(1) Typical columns 15" by 15", # 4 ties at 4" o.c.
(2) Typical beams 13" by 18", with # 3 ties at 4" o.c.
(3) Column moment vs. beam capacities utilizing actual material properties at yield

The specimen were extensively instrumented with strain gages, load cells, LVDT's, and slip wires. The steel strains at critical sections near the joint, the deformations of the beams and column, the joint shear strain and the slip of the main longitudinal reinforcement in the beams were monitored throughout the test. All measurements were taken with respect to inserts anchored in the confined section of the members, and isolated from the cover which typically spalled early in the tests. In total 96 channels of data were recorded utilizing a VIDAR Data Acquisition System, and were processed by a DG Nova-3 minicomputer.

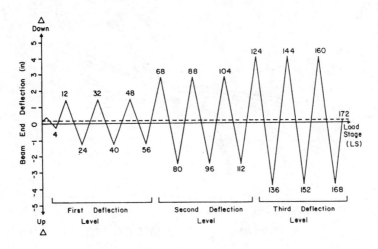

Figure 2 - Load history

EXPERIMENTAL RESULTS

The typical behavior of these specimen is shown in Fig. 3, which shows the joint shear for BCJ8 as the load history progressed. While very large shear stresses were attained, large losses occured with cycling. The ultimate shear strengths measured ranged from $27\sqrt{f'_c}$ to $33\sqrt{f'_c}$, based on an uniaxial effective area of the column. They decreased between 10% and 15% with cycling at any deflection level, but even at the end of the test the damaged joints could sustain about $15\sqrt{f'_c}$ to $18\sqrt{f'_c}$.

The hysteresis loops became very pinched with cycling due primarily to shear cracking in the joint area and slip of the beam longitudinal reinforcement. It should be noted that plastic hinges were not clearly observed in the beams; while the strain gages at the beam-column interface indicated that significant yielding took place, this yielding seems to have spread mostly into the joint, resulting in severe bond deterioration and pull-through of the reinforcing bars by the second deflection level.

Joints in ductile moment-resisting frames (DMRF) designed by the weak beam-strong column method are intended to remain essentially elastic during seismic loading, since the joint is a part of the column and any significant damage in this area can result in stability problems and collapse of the structure. It is assumed that the hinges will form in the beams adjacent to the joint region, but that the joint itself will be a rigid element. Thus all the deformations should come from elastic deformations in the beams and columns, and from inelastic beam rotations or beam plastic hinges.

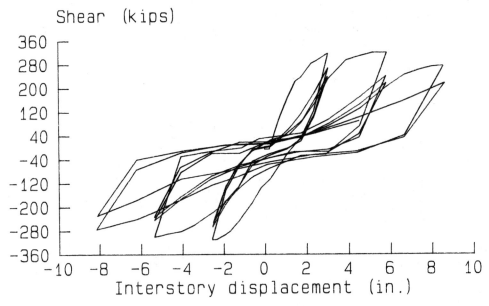

Figure 3 - Joint shear vs. interstory displacement for BCJ8

Previous research [4,6,7] indicates that in order to maintain such
stable hysteretic performance large anchorage lengths (30 to 35 bar
diameters for Grade 60 bars), substantial column overstrength ratios (column
to beam capacities in the range of 1.8 to 2.5), and low levels of shear
stress (12 $\sqrt{f'_c}$ to 18 $\sqrt{f'_c}$) must be used. These requirements can be
satisfied through the use of large columns which decrease the nominal joint
shear stress, increase the anchorage length of beam bars, and typically
result in an increase of column capacity. However, the anchorage
requirements result in excessively large columns if one considers that using
8 bars in the beams would result in a column at least 30 in. wide. Thus,
current design codes do not contain such stringent provisions, and assume
that under large cyclic load reversals some loss of bond and shear cracking
can be accepted. The study described herein represents an upper bound to
the damage and loss of energy dissipation capacity in beam-column joints
because the shear stresses imposed were on the order of 20 $\sqrt{f'_c}$ uniaxially
and 28 $\sqrt{f'_c}$ biaxially, the anchorage lengths varied from 15 to 20 bar
diameters, and the flexural overstrength was about 1.16 to 1.28.

To study the behavior of these specimen it was decided to separate the
different components contributing to the stiffness and strength
deterioration observed. The stiffness deterioration was quantified by
defining four mechanisms, or components, contributing to the beam end
deflections. The first was the elastic deformation of the beams and column,
which could be calculated from the measured forces and the cracked section
properties. The second is the beam inelastic deformation, to which both
yielding of the reinforcement and longitudinal bar slip contribute. This
quantity was determined from the rotation measurements taken at a distance
of one-half the beam depth away from the joint, the area where most of the
damage and rotation were concentrated. The bar slips were measured on both
sides of the joint by attaching very stiff wires to the reinforcing bars,
and measuring the displacements of these wires with a spring-loaded
potentiometers. Typical results for this instrumentation are shown in Fig.
4 for a top # 8 bar.

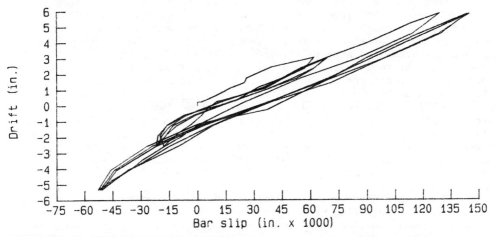

Figure 4 - Bar slip for top #8 bar (first two deflection levels).

The third is the joint shear strain, measured directly by attaching a frame to the joint region. Typical measurements are shown in Fig. 5. The last mechanism, called column inelastic rotation, comprises all other modes of deformation, including column yielding and column bar slip. The contribution of each of mechanism was computed based on the measurements taken, including corrections for geometry and the interaction of mechanisms. The different contributions were added, and the result compared to the measured beam end deflection at each load stage. Very good agreement (within 10%) was found between the calculated and measured deformations for the upper half of each load cycle; the agreement was considerably worse near the zero deformation points since it was impossible to calibrate the devices for good resolution both at the peaks and near zero.

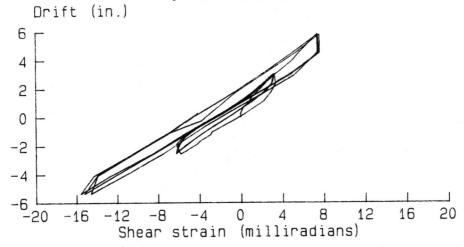

Figure 5 - Joint shear strain for first two deflection levels of BCJ8

The individual contributions, in percentage of the total deformation, of each mechanism to the interstory East-West beam deflection for BCJ8 are shown in Figs. 6 through 9. The contribution of the elastic deformation decreases as the deformation increases because the bottom beam reinforcement yields early in the load history (Fig. 6). The beam was reinforced with 3 # 8 Grade 60 bars at the top and 3 # 6 bars at the bottom. The first peak, at an interstory displacement of about 2 in., corresponds to the yield of the top bars. Because both beams were deflected equally to simulate the sidesway mechanism, the bottom bars yielded at about two-thirds of deflection required to yield the upper bars. During the second cycle the elastic contribution was lower because tensile cracks were open and the beam bars were slipping; only near the peaks, when the cracks closed and the bond transfer improved did the elastic contribution for each cycle tend to equalize.

The difference between the elastic contributions during the first two cycles was made up primarily by an increase in the joint shear strain and the inelastic beam rotations. The beam rotations were particularly large at the beginning of each reloading stage because of the bar slippage; the contribution of these rotations decreased, or remained constant, as the peaks were approached (Fig. 7). The amount of beam rotation remained essentially constant with interstory displacement, with an average value of 2.5 milliradians per inch of interstory drift. Since the anchorage lengths for the reinforcement across the column varied from 15 bar diameters for the top bars to 20 bar diameters for the bottom ones, bond deterioration and slippage are not surprising. As previously mentioned, the inelastic beam rotation accounts for both yielding and slippage. From the stress profiles obtained from the strain gages and the slip wire measurements, it was estimated that about 40% of the inelastic beam rotation at the first deflection level was due to slippage, and that this increased to about 60% at the second deflection level.

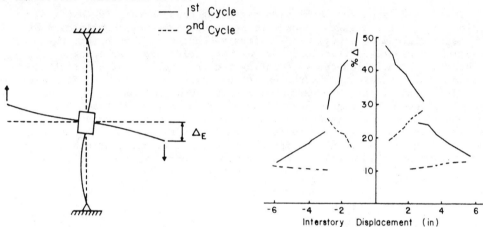

Figure 6 - Elastic contribution to total beam end deflection.

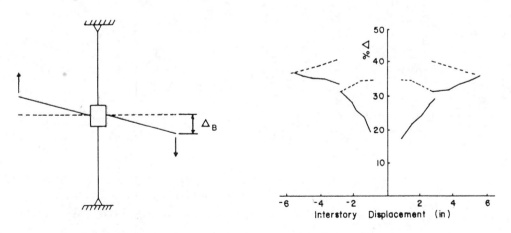

Figure 7 - Beam rotation contribution to beam end deflection.

The joint shear strain measurements lead to two important qualitative conclusions (Fig. 8). The first is that while very large shear strengths can be attained, they are associated with unacceptably large shear strains. At the first deflection level, the joint shear strain at the first peak was about 3.0 milliradians in the positive direction of loading, and about 5.5 milliradians in the negative. This pronounced skewness was observed in all tests, except BCJ7 which had a very large transverse reinforcement in the joint (10 # 4 ties vs. 2 # 4 ties for all other specimen). Even for BCJ7, however, the magnitude of the shear strains was about 80% of that measured for all other tests, indicating that packing the joint with transverse reinforcement may not be the best solution.

The second important conclusion regarding shear strain is that its contribution decreases as the peaks are approached because the yielding of the beam reinforcement represents a softer mechanism at this point. When loaded in the negative direction, however, the joint has been cracked, and its shear strength is reduced, accounting for the doubling of shear strains measured. The nominal shear stresses in the joint were about 28 $\sqrt{f'_c}$, which represents about 20 $\sqrt{f'_c}$ in each principal direction, or about the value currently allowed by design codes. The amount of shear deformation and associated damage to the joint panel zone observed in these tests is not intended to be associated with current design provisions, and thus it would seem unsafe to extend current design specifications to bidirectional loading without lowering the nominal shear stress in each principal direction.

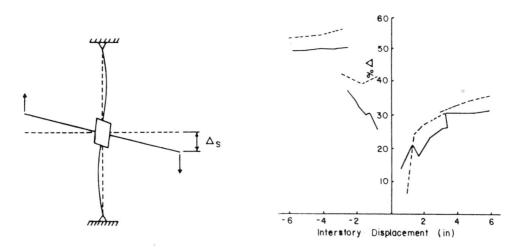

Figure 8 - Joint shear strain contribution to beam end deflection.

The fourth mechanism, column inelastic rotation, seems to be skewed towards the original direction of loading probably for two main reasons (Fig. 9). First, crushing of the column corners in compression was observed in the original direction of loading. This is due to the very large compressive strains required of this relatively small column under the action of bidirectional loading. This crushing resulted in losses of section at the joint corners, and reduced the contribution of the column bars farthest away from the neutral axis. It should be noted that poor bond performance was observed not only in the beam but also in the column longitudinal reinforcement. This reinforcement did not reach yield typically until the third deflection level, but bond deterioration seems to occur even in the elastic range. A second reason for the skewness of this contribution is that during the loading to the first peak at each deflection level the joint was softened considerably by shear cracking. When the load was reversed most of the deformation, as shown by the joint shear strain readings, took place in the joint itself rather than in the framing members.

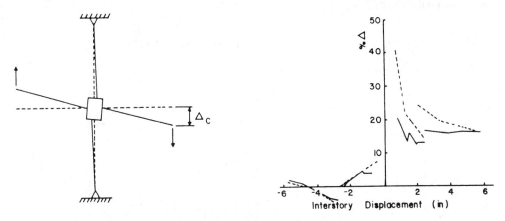

Figure 9 - Column rotation contribution to beam end deflection.

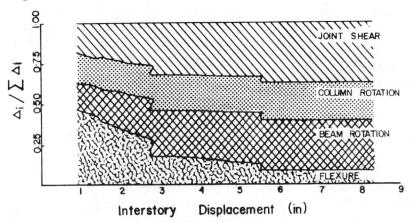

Figure 10 - Overall contributions to beam end deflection.

The average contribution of each mechanism for the East-West direction of loading is shown in Fig. 10. The total contributions were normalized assuming the measurement errors were equal for all components. The large contributions of two undesirable mechanisms, joint shear strain and column inelastic deformation are clearly shown. The values obtained for the other tests are shown in Table 2, and will lead to the following conclusions:

a) Comparison of BCJ5 and BCJ8, identical specimen except for the axial load of 300 kips (balance point) carried by BCJ5, does not indicate a substantial effect of compressive axial loads on joint behavior. The "clamping" force provided by the axial load would be expected to reduce the column elastic and inelastic deformations, and possibly decrease the joint shear strain. In fact, the specimen without axial load exhibited larger column rotations and slightly lower shear strains than the one with axial load, but both the elastic deformations and beam rotations remained the same.

Table 2 - Contributions to Interstory Drift

Test	Drift Level (%)	Elastic Def. (%)	Beam Rot. (%)	Column Rot. (%)	Joint Shear (%)
BCJ5	0.5	47	22	11	20
	1.0	39	25	14	22
	1.5	29	30	9	30
BCJ7	0.5	45	25	10	20
	1.0	39	30	9	22
	1.5	31	34	13	22
BCJ8	0.5	46	17	17	20
	1.0	36	27	15	22
	1.5	29	31	15	25
BCJ9	0.5	40	15	34	11
	1.0	31	23	31	17
	1.5	24	30	26	20
BCJ11	0.5	50	22	26	1
	1.0	41	21	32	6
	1.5	32	20	37	11
BCJ12	0.5	58	14	27	1
	1.0	47	17	32	4
	1.5	34	18	38	10

b) Comparison of BCJ5 and BCJ7, identical except that BCJ7 had 5 times as much transverse joint reinforcement, did not indicate any significant difference except that BCJ7 exhibited larger beam rotations and slightly lower joint shear strains. This indicates that additional joint confinement in the form of ties will improve the shear transfer by providing the boundary conditions necessary for a panel truss mechanism and by improving the bond conditions inside the joint.

c) Comparison of BCJ8 and BCJ9, identical except for the 4 in. thick floor slab in BCJ9, indicated that the addition of the floor slab significantly increased the column rotations, primarily at the expense of the elastic deformation and joint shear strains. It should be noted that the slab specimen had a flexural overstrength ratio (biaxial column capacity divided by the sum of the beam plus slab capacities at ultimate, utilizing actual material properties) of 1.16 versus 1.26 for that without slabs. Thus a column failure, or at least such large contribution from this component, was not anticipated. It also points out that the current 1.2 overstrength required by ACI 318-83 Appendix A may be insufficient, and that the 1.4 recommended by ASCE-ACI Committee 352 may be closer to the actual margins of safety implicit in the codes. Interestingly, although the joint shear increased by about 10% due to the presence of nominal reinforcement in the slab (#4 at 12" o.c. in both directions), the joint shear strains actually decreased. The additional area provided by the slab as well as the transfer of forces through torsion by the framing beams may help explain this phenomenon.

d) A comparison of BCJ5, BCJ11, and BCJ12 serves to illustrate the effect of framing beam size. BCJ5 had beams 13" wide by 18" deep, while BCJ12 had 18" by 18" beams (wide), and BCJ11 had 9" by 18" beams (narrow). The contributions for BCJ11 and BCJ12 are very similar, although the behavior was significantly different. BCJ11 suffered an early failure due to spalling of the joint corners, with the resulting loss of section and bond of the corner bars. Thus the hinging shifted to the column immediately above and below the joint, the strength of the beams was not fully mobilized, and the joint did not undergo large shear strains. In the case of BCJ12, the beams did reach yield but the inelastic deformations in the column increased. Because the area available to resist horizontal shear within the joint itself increased, the joint did not suffer extensive shear cracking and the yielding was shifted from the joint to the column. BCJ12 was able to carry the large shears imposed, however, because the size of the joint permitted the formation of a large compression strut, which could be maintained with cycling.

CONCLUSIONS

The results of this investigation indicate that use of small columns, which result in poor bond conditions for the beam longitudinal reinforcement and low ratios of column overstrength capacity, is not advisable for ductile moment resisting frames. It also points out that the current design recommendations available cannot be applied directly to biaxially loaded joints, and that designers should adjust their design values downwards by at least a factor of $\sqrt{2}$ when such conditions arise. Current design recommendations should be used with care, and where possible anchorage lengths in the order of 24 to 28 bar diameters as well as overstrength ratios greater than 1.4 for uniaxailly loaded frames and 2.0 for biaxially loaded ones should be used.

ACKNOWLEDGEMENTS

The work described in this paper was carried out at the Ferguson Structural Enginnering Laboratory, The University of Texas at Austin under the sponsorship of the National Science Foundation Grants ENV77-20816 and PRF-7720816. The help of the laboratory staff and Dr. James O. Jirsa, Ferguson Professor of Engineering at the U. of Texas is gratefully acknowledged. The conclusions and opinions presented here are those of the author alone.

REFERENCES

[1] American Concrete Institute, Building Code Requirements for Reinforced Concrete (ACI 318-83), American Concrete Institute, Detroit 1983.
[2] ACI-ASCE Joint Committee 352, Design of Beam-Column Joints in Monolithic Reinforced Structures, ACI Journal, Vol. 82, N. 3, May-June 1985.
[3] Leon, R.T., and Jirsa, J.O., Bidirectional Loading of R.C. Beam-Column Joints, Earthquake Spectra, Vol. 2, N. 3, May 1986.
[4] Leon, R.T., The Influence of Floor Members on the Behavior of R.C. Beam-Column Joints Subjected to Severe Cyclic Loads, Ph.D. Dissertation, The U. of Texas at Austin, Austin, Dec. 1983.
[5] Zhang, L. and Jirsa, J.O., A Study of Shear Behavior of R.C. Beam-Column Joints, PMFSEL Report 82-1, Dept. of Civil Eng. The University of Texas at Austin, Feb. 1982.
[6] Park, R., and Paulay, T., Joints in Reinforced Concrete Frames Designed for Earthquake Resistance, Research Report 84-9, Dept. of Civil Engineering, University of Canterbury, Christchurch, New Zealand, June 1984.
[7] Bertero, V.V., Seismic Behavior of Structural Concrete Linear Elements and Their Connections, CEB Bulletin 131/132, Rome 1979.

STRAIN RATE EFFECT ON THE STRESS STRAIN BEHAVIOR OF
STRUCTURAL STEEL UNDER SEISMIC LOADING CONDITIONS

K.C. Chang and G.C. Lee[I]

ABSTRACT

The strain rate and strain rate history effects on the inelastic stress-strain behavior of annealed A-36 structural steel at room temperature is examined experimentally under monotonic and cyclic loading conditions. The strain rates used are between 10^{-6}/sec and 10^{-1}/sec, which cover the range of the strain rate that might be experienced by steel building frames during earthquake ground motions. An axial-torsional electro-hydraulic servo-controlled testing system with strain control is used for all the tests. Test results show relatively more significant strain rate sensitivity of the material for monotonic loading, and less significant strain rate sensitivity for cyclic loading. Strain rate history effect is not observed to be significant in most of the tests. The results suggest that rate-independent plasticity theories should be modified to account for the strain rate effect in certain earthquake engineering applications.

INTRODUCTION

In classical plasticity theories the assumptions of the existence of a yield surface and rate-independent plastic deformation are generally made. They may not be appropriate for applications to structures subjected to dynamic or nonproportional loading conditions. Strictly speaking, theories of dynamic plasticity should be used for dynamic loading because they include the effect of strain rate. For structures subjected to severe earthquake ground motions, the strain rate at critical sections may be as high as 10^{-1} to 10^{0}/sec.[1]. Furthermore, the strain rate at the critical sections in the structure are not constant. Before dynamic plasticity theory is established for earthquake analysis, there is the question of the accuracy of using the rate-independent material models. This is the purpose of the present study.

A large number of research efforts have been devoted to the effect of high strain rate (10/sec and above)on material properties [2] under impact type monotonic loading in the past few decades. There is still very little information available about the rate sensitivity of structural steel subjected to earthquake type cyclic loading under changing strain rates. This is probably due to the lack of satisfactory testing systems that can accurately control the abrupt changes of rate of loading, and the testing process, and obtaining data with high speed.

I. State University of New York at Buffalo, Buffalo, New York

Recently, with the introduction of the servo-controlled electro-hydraulic testing machine, experimental observations of material behavior under monotonic and cyclic loading conditions with changing rates become possible. In this paper the question or rate-sensitivity of A-36 structural steel subjected to a number of selected strain paths under variable strain rates at room temperature was examined by utilizing such a testing system. This strain rate range covers most of the spectrum that may be expected in typical steel structures subjected to strong earthquake ground motions. Considerable strain rate effects are observed for A-36 structural steel. The results also show that it is possible to modify the rate-independent plasticity theories to include the strain rate effect in certain earthquake engineering applications if more accurate analytical predictions are desired without using theories of dynamic plasticity.

EQUIPMENT AND EXPERIMENTAL SET-UP

In developing the theory of "dynamic plasticity," special testing techniques such as the split Hopkinson bar [3] are often used to generate very high strain rate (10/sec and above). However, only under monotonic loading conditions can this technique be used. With the advent of a servo-controlled testing system [4] accurate control of the rate-of-loading under quasi-static or dynamic conditions is possible. The level of strain rates using this servo-controlled system is, however, substantially lower than that, using the split Hopkinson bar. The amount of strain rate the servo-controlled system can reach is generally dependent on the flow rate and pressure of the hydraulic pump, the area of the cylinder of the hydraulic actuator and the flow rate of the servo valve. The MTS testing system used in this study consists of two major components: (1) the closed-looped servo-controlled axial-torsional machine and (2) a PDP-11 computer system with the MTS Processor Interface to control the entire testing process and data acquisition. Up to 10^0/sec strain rate can be reached by using this system. The closed-looped servo-controlled system is described by a block diagram in Fig. 1.

SPECIMENS AND EXPERIMENTAL PROCEDURES

The test specimens were machined from commercially available A-36 structural steel tubings with 1-3/8 in. O.D. and 7/8 in. I.D. The dimensions of the test specimen are shown in Fig. 2. Following machining, the specimens are annealed in a furnace at 1200°F for half an hour and then furnace cooled except for the monotonically loaded specimens of Fig. 3a (1600°F for one hour and then furnace cooled).

Strain control is used throughout the test by using an MTS axial-torsional extensometer. The strain rates tested are between 10^{-6}/sec and 10^{-1}/sec. In general, all specimens are first loaded in the axial direction with variable or constant strain rates up to about 4% axial strain. Then the specimens are cycled with prescribed loading paths with variable or constant strain rates within the ranges between 1.6% and -1.6% axial strain or between 0% and 2% axial strain.

OBSERVATIONS

Strain Rate Sensitivity During Monotonic Loading

Fig. 3a shows the monotonic loading curves from three specimens with strain rates 10^{-3}/sec, 10^{-4}/sec and 10^{-5}/sec. Some strain rate sensitivity can be observed with maximum lower yield stress difference of 14%. In addition, the faster the strain rate the longer the plastic plateau.

Fig. 3b shows a monotonic loading curve with rate change between 10^{-5}/sec and 10^{-2}/sec up to axial strain 4.5%. At the end of each slow loading (10^{-5}/sec) an elastic unloading was followed with the same slow strain rate. A fast strain rate loading (10^{-2}/sec) is then used for the plastic range. It can be seen that significant strain rate effect is found (27% increase in lower yield stress). Also shown in this figure is the coincident of the unloading and reloading curves in the elastic range. The Young's modulus is not observed to be sensitive to strain rate change. After the material reached the strain hardening range the strain rate sensitivity is less profound.

Strain Rate Sensitivity During Cyclic Loading

Fig. 4a shows the cyclic loading curves corresponding to different constant strain rates 10^{-2}/sec and 10^{-4}/sec. After monotonic loading the specimens are subjected to cyclic loading in the axial direction and stabilized with strain rate 10^{-2}/sec. It is then followed by cyclic loading with different constant strain rates. Because of the complexities of these curves only the stress-strain curves corresponding to the two strain rates mentioned above are shown. Some strain rate sensitivity can also be observed (approximately 8%). This is considerably less than those displayed by the monotonic loading curves.

One phenomenon worth noting is that before the application of cyclic loading with strain rate 10^{-1}/sec, the cross-section of the specimen remained unchanged. After cyclic loading with strain rate 10^{-1}/sec, the specimen is locally buckled. However, the cyclic stress-strain curve remained stable until very severe local buckling of the specimen takes place.

Fig. 4b shows a cyclic loading cuirve with abrupt strain rate change between 10^{-2}/sec and 10^{-5}/sec during cyclic loading. Strain rate sensitivity can be observed through the sudden drop of stress level from high strain rate to low strain rate (approximately 10%).

Strain Rate History Effect During Monotonic Loading

Fig. 5a shows a monotonic loading curve (curve A) with changing strain rate bewteen 10^{-5}/sec and 10^{-2}/sec and two monotonic loading curves with constant strain rates 10^{-2}/sec and 10^{-5}/sec (curves B and C) for annealed specimens. In the range of plastic plateau very little strain rate history sensitivity is observed. In the work hardening range, however, the tangent modulus is greater for strain rate 10^{-5}/sec in curve A. This may be the contribution of strain rate history effect.

Fig. 5b shows a monotonic loading curve with changing strain rate between 10^{-5}/sec and 10^{-2}/sec for a specimen which was prestrained during previous tests and then reduced to zero strain (the extensometer was re-zeroed). Less strain rate history sensitivity is observed from this monotonic loading curve.

Strain Rate History Effect During Cyclic Loading

Fig. 6a shows two different stabilized cyclic loading curves. Curve A represents the change of strain rates (from 10^{-2}/sec to 10^{-4}/sec) at 1.2% and -1.2%. Curve B shows the change of strain rate (from 10^{-4}/sec to 10^{-2}/sec) at 0.4% and -0.4%. It can be observed that the stabilized cyclic loading curve can always be obtained despite strain rate changes during each loading cycle. Furthermore, the portion of the curves with the same strain rate coincide despite different strain rate histories. Loading curves are unique to specific strain rates despite different previous strain rate history.

Fig. 6b first shows cyclic loading curves from four continuous strain histories between 0% and 2% axial strain. The strain rate histories for those four strain histories are 10^{-2}/sec, 10^{-4}/sec, 10^{-2}/sec and 10^{-4}/sec.

Following the above loading history, 3 loading cycles between 0% and 2% strain with alternately changing strain rate for each half cycle followed (Fig. 6b). The strain rate for those 3 continuous cycles are 10^{-2}/sec, 10^{-4}/sec, 10^{-2}/sec, 10^{-4}/sec, 10^{-2}/sec and 10^{-4}/sec respectively. As was observed, cyclic loading curves are unique to the specific strain rate despite the different strain rate histories.

The Static Loading Curve

In conventional static loading tests, the load is stopped after each increment while maintaining the strains at those load points. The load is permitted to drop until stabilized. The static stress-strain curve is constructed based on those stablized load levels.

Fig. 7a shows a monotonic loading curve with variable strain rate between 10^{-6}/sec and 10^{-2}/sec. During the loading process the load is stopped at specified strains each for 10 minutes.

Fig. 7b shows a static loading curve with constant strain rate of 10^{-4}/sec. Again, 10 minutes waiting time is used after each load increment. These two figures show tht the stress relaxation depends on the previous strain rate. At the strain hardening region the stress relaxations are less than that of the plastic plateau range. Furthermore, under changing strain rate a unique stress-strain curve corresponding to the conventional static loading curve cannot be established.

SUMMARY AND CONCLUSION

Experimental study on the strain rate effect of A-36 steel subjected to monotonic loading and cyclic loading conditions with strain rate change between 10^{-1}/sec and 10^{-6}/sec is carried out. Based on the results of all the experiments described in this paper the following may be summarized.

1. The monotonic loading curves show quantitatively strain rate sensitivity of A-36 structural steel. Faster strain rate corresponds to higher yield stress and longer plastic plateau.

2. The slopes of elastic unloading and reloading curves for different strain rates are identical to that of the original elastic loading curve.

3. When there is an abrupt change in strain rate within a prescribed strain range under cyclic loading, a unique stabilized curve can be obtained, that is, there is no apparent strain rate history effect under cyclic loading.

4. The cyclic loading curves of different strain ranges for different strain rate and different strain history are identical except for the portion of the elastic range and the peak stress value which are larger for higher strain rate.

5. The strain rate effect is more visible for the case of monotonic loading than for cyclic loading, especially when the load was cycled with constant rates.

6. Although the strain rate history during monotonic loading in the plastic plateau range is insignificant, it may be important in the strain hardening range with large strains. If the strain rate changes from high to low, the value of the tangent modulus corresponding to the low strain rate is higher than the value of no strain rate change.

7. The conventional static loading curve is different for different strain rates.

The experiments show considerable rate-sensitivity of A-36 structural steel at room temperature. Therefore, using the classical time-independent plasticity theory for structural analysis under strong ground motions may be inappropriate for certain structures (e.g. high-rise slender structures). In such situations the theory of plasticity describing the material behavior of structural steel may be modified to include the strain rate effect. This is currently under investigation.

One of the more important issues concerning the modification of the classical plasticity theory for dynamic loading is the effect of strain rate history. Certain studies [5,6] have addressed the strain rate history effect in dynamic plasticity formulations. Recent studies on AISI type 304 stainless steel by Krempl [7] showed no strain rate history effect. In this current study on A-36 strucutral steel the strain rate history effect is observed to be insignificant for cyclic loading and monotonic loading in the plastic plateau range. However, there is some strain rate history effect in the strain hardening range for an annealed specimen. For prestrained specimens, no significant strain rate history effect is concluded. However, this conclusion does not automatically apply to general cyclic, multi-axial, nonproportional loading with variable strain rate changes.

ACKNOWLEDGMENT

This research study has been supported by grants from the National Science Foundation (CEE-8213437 and ECE-8516471).

REFERENCES

[1] Lee, G.C., "Pilot Study on the Dynamic Behavior of Steel Gable Frames," WRC Progress Report, pp. 24-38, Sept. 1984.

[2] Malvern, L.E., "Experimental and Theoretical Approaches to Characterization of Material Behavior at High Strain Rates of Deformation," presented at 3rd Conf. Mech.Prop.High Rates of Strain, Oxford, 1984.

[3] Clifton, R.L., "Dynamic Plasticity," J. Appl. Mech. Vol. 50, p. 941, Dec. 1983.

[4] Lindholm, U.S., L.M. Yeakley and A. Nagy, AFML-TR-149 (Wright-Patterson Air Force Base,Ohio, Air Force Material Laboratory) 1971.

[5] Marsh, K.J. and J.D. Campbell, "The Effect of Strain Rate on the Post-Yield Flow of Mild Steel," J. Mech. Phys. Solids, Vol. 11, p. 49, 1963.

[6] Frang, R.A. and J. Duffy, "The Dynamic Stress-Strain Behavior in Torsion of 1100-0 Aluminum Subjected to a Sharp Increase in Strain Rate," J. Appl. Mech., Vol. 39, p. 939, 1972.

[7] Krempl, E., "An Experimental Study of Room Temperature Rate-Sensitivity, Creep, and Relaxation of AISI Type 304 Stainless Steel," J. Mech. Phys. Solids, Vol. 27, p. 363, 1979.

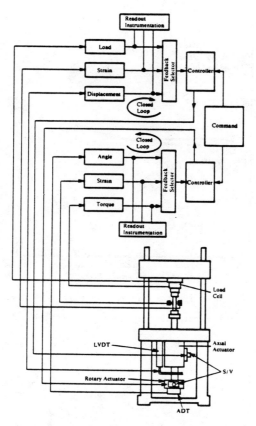

Fig. 1 Block diagram of the closed-looped servo-controlled system used in this study.

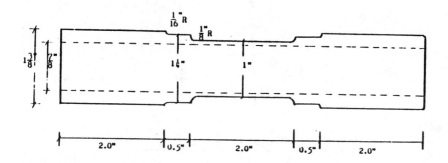

Fig. 2. Test specimen used in this study.

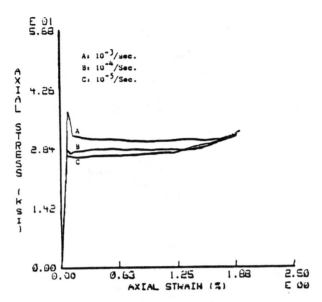

Fig. 3a. Three monotonic loading curves with different strain rates. Some strain rate sensitivity was observed with maximum lower yield stress difference being about 14%.

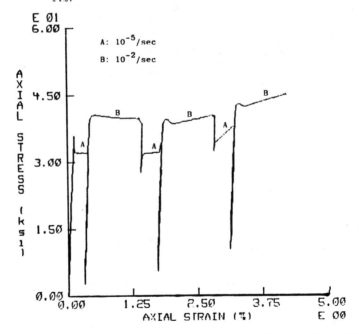

Fig. 3b. Monotonic loading curves with rate change between 10^{-2}/sec and 10^{-5}/sec with elastic loading and unloading. Up to 27% difference in lower yield stress was observed. The elastic loading and unloading curves with different strain rate are coincident.

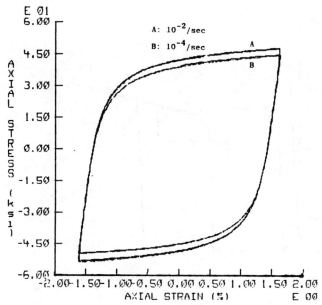

Fig. 4a. Cyclic loading curves with constant strain rates 10^{-2}/sec and 10^{-4}/sec. Some strain rate sensitivity was observed (8%) but less sensitivity than the monotonic loading curve.

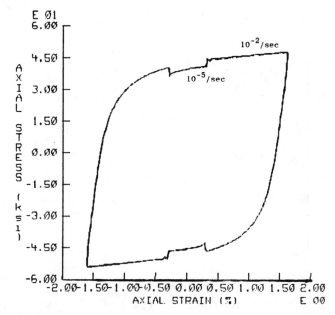

Fig. 4b. Cyclic loading curves with strain rate change within each cycle from 10^{-2}/sec to 10^{-5}/sec. Strain rate sensitivity can be observed (about 10%) but also less than the monotonic loading case.

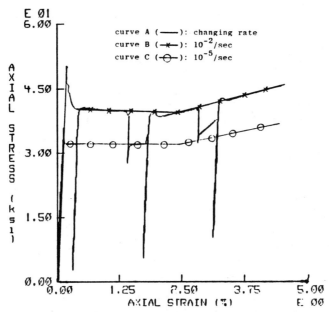

Fig. 5a. Strain rate history effect during monotonic loading of annealed specimen. Very little strain rate history sensitivity can be observed except in the strain hardening range for strain rate 10^{-5}/sec.

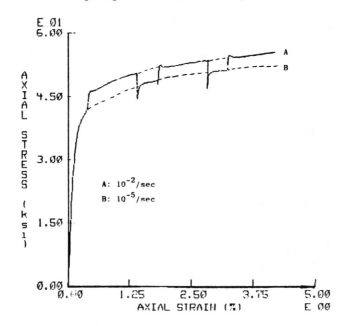

Fig. 5b. Strain rate history effect during monotonic loading of prestrained specimen. Very little strain rate history effect can be observed.

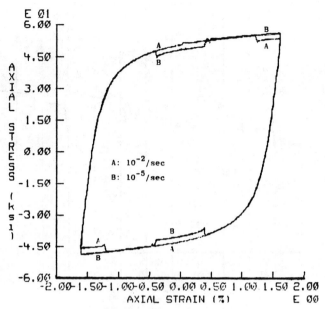

Fig. 6a. Strain rate history effect during cyclic loading with strain rate change within each cycle. Curve A and curve B coincide with each other in the portions of the same strain rate. Very little strain rate history can be observed.

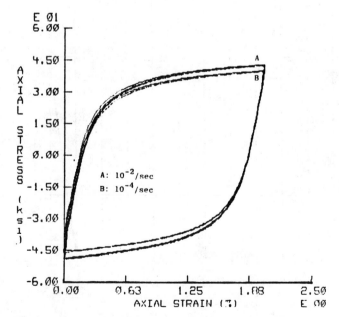

Fig. 6b. Strain rate history effect during cyclic loading with strain rate change for each cycle or each half cycle. No strain rate history effect can be observed.

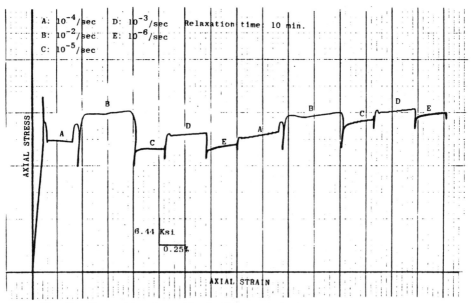

Fig. 7a. Conventional static loading curve cannot be obtained by connecting the relaxed loading points with strain rate change between 10^{-2}/sec and 10^{-6}/sec.

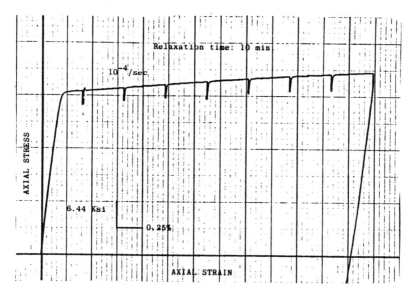

Fig. 7b. Conventional static loading curve can be obtained by connecting the relaxed loading points with constant strain rate 10^{-4}/sec.

HIGH STRENGTH CONCRETE FRAMES SUBJECTED TO
INELASTIC CYCLIC LOADING

M. R. Ehsani[I], A. E. Moussa[II], and C. R. Vallenilla[II]

ABSTRACT

Test results of four reinforced concrete beam-column subassemblies are reported. The specimens were constructed with concrete having compressive strength of approximately 9500 psi. The column portions of the specimens were subjected to a constant axial load while cyclic displacements were imposed on the free end of the beam. Based on the test results it is concluded that modification of the current codes are needed for application to high strength concrete cases. For concrete having compressive strength of 9500 psi, a maximum allowable joint shear stress of $10\sqrt{f'_c}$ (psi units) is recommended.

INTRODUCTION

The behavior of reinforced concrete ductile moment resisting frame structures has been studied since the early 1960's. Major emphasis has been placed on providing proper anchorage conditions for the beam longitudinal reinforcement so that the ultimate flexural capacity of the beam can be developed. The philosophy of "strong column-weak beam" has therefore been the major goal in all studies.

The first recommendations for the design of beam to column connections were published in 1976 and followed an approach similar to that for the design of beams subjected to shear [1]. A portion of the joint shear stress was assigned to the concrete. The area of the joint transverse reinforcement was calculated to resist the remaining joint shear stresses.

The results of the recent research studies [2-6] were considered in the revision of the recommendations in 1985 [7], which will be hereafter referred to as the Recommendations. According to the Recommendations, maximum allowable joint shear stresses are specified for different joint configurations. If the joint shear stress exceeds this limit, the area of

---

I. Assistant Professor of Civil Engineering and Engineering Mechanics, The University of Arizona, Tucson, Arizona.

II. Graduate Student, Department of Civil Engineering and Engineering Mechanics, The University of Arizona, Tucson, Arizona.

the joint must be increased rather than providing additional joint reinforcement to resist the excess shear.

The Recommendations are based on experimental results for frames constructed with ordinary strength concrete having compressive strength of 3500 psi to 5500 psi. Therefore, in applying the Recommendations, the maximum compressive strength of concrete is limited to 6000 psi. However, many concrete structures have been constructed in recent years, in which the compressive strength of the concrete used in the construction of the columns exceeds 12,000 psi. The study presented here was undertaken to investigate the behavior of frames constructed with high strength concrete in light of the above limitations.

EXPERIMENTAL STUDY

The primary objective of this study was to examine the shear capacity of beam-column connections constructed with high strength concrete. Four large-scale corner-type connections were tested. The testing program is described in the following sections.

Design of the Specimens

Four corner-type reinforced concrete beam-column connections were constructed and tested. The specimens were designed according to the Recommendations [7] and the ACI 318-83 Building Code [8]. Detailed description of the specimens and the findings of this study are presented in Ref. 9. The configuration of the specimens is shown in Fig. 1 and their physical dimensions are listed in Table 1.

The column and beam longitudinal reinforcement as well as the transverse reinforcement used within the joint region were Grade 60 steel. The column and beam ties were Grade 40. The concrete compressive strength at the time of testing for Specimens 1 and 3 was 9380 psi and that for Specimens 2 and 4 was 9760 psi.

Primary Variables

The primary variable in this study was the joint shear stress. According to the Recommendations [7], the joint shear stress for corner connections is limited to $12 \sqrt{f'_c}$ (psi units). The constant multiplier, 12 in this case, is referred to as the joint shear stress factor, γ. For the specimens tested, the joint shear stress was varied between $7.52 \sqrt{f'_c}$ and $12.84 \sqrt{f'_c}$. The flexural strength ratio, M_R, defined as the ratio of the flexural capacities of the columns to that of the beam, was almost constant and greater than 1.4 for all specimens.

All joints were reinforced with three layers of hoops. Each layer of hoop consisted of a square- and a diamond-shaped tie as shown in Fig. 1. The minor differences in concrete compressive strength were assumed to have no significant influence on the behavior of the subassemblies. The design parameters for all specimens are listed in Table 2.

Test Setup and Instrumentation

Specimens were tested in the steel reaction frame shown in Fig. 2. Using threaded rods which were cast in the specimen, steel clevices were attached to each specimen prior to the testing. These clevices were designed to represent the beam and column inflection points. The column portion of the specimens was placed horizontally in the testing frame. A small axial load, as given in Table 2, was applied to each column and kept constant throughout the test.

The free end of the beam was subjected to the displacement controlled loading history shown in Fig. 3. The maximum displacement in each cycle of loading was gradually increased from 0.5 in. during the first cycle to 4.5 in. during the last cycle of loading.

Approximately twenty four electrical resistance strain gages were attached to the reinforcing steel in the joint region. In addition, displacement transducers were attached to the end of the beam near the column. These transducers were used to measure the elongation of the hinging zone of the beams over a 20 in. gage length. During each cycle of loading, at 0.25 in. displacement increments, the application of the load was temporarily stopped until all strain gages and transducers were automatically scanned and recorded and the cracks in the specimens were marked.

DISCUSSION OF TEST RESULTS

Load vs. displacement plots are perhaps the most important results obtained from the tests. They provide useful information on the load carrying capacity, energy absorption, and the loss of stiffness of the specimens. The load vs. displacement plots for all four specimens are shown in Fig. 4. The plots for Specimens 1, 2, and 3 which failed in flexure show increasing load carrying capacity with additional displacements. A reduction in the maximum load carried and a significant loss of stiffness for Specimens 2 and 3 was observed only during the last cycle of loading. In contrast, the hysteretic response of Specimen 4 was accompanied by significant loss of stiffness near the zero displacement region, which is a common indicator of loss of bond and shear failure. The maximum load carried by this specimen was reached during the fifth cycle of loading followed by a significant loss of strength in subsequent cycles.

Because all specimens were symmetrically reinforced, the positive and negative parts of the hysteresis loops were almost symmetrical. The overall hysteretic response of the specimens were mainly influenced by a combination of (a) the curvilinear behavior of the reinforcing steel due to the "Bauschinger Effect," (b) the opening and closing of residual cracks, and (c) the slippage or loss of bond between the reinforcement and concrete.

Failure Mechanism

Based on the observations made during the tests regarding the cracking pattern of the specimens, and the analysis of the strain gage data, two distinct modes of failure were recognized. Specimens 1 and 2 failed due to the formation of flexural hinges at the fixed end of the beam. Failure of Specimen 4 was clearly due to the shear failure of the joint. The failure mechanism for Specimen 3 was a combination of the two mechanisms. The joint of this specimen was providing adequate anchorage for the beam longitudinal reinforcement through the sixth cycle of loading. Beyond that point, however, the joint was not capable of resisting the larger shear stresses caused by the strain hardening of the beam reinforcement.

The failure mechanisms can be demonstrated with the aid of Fig. 5 which is a plot of the elongation of the beam hinging region at the end of each cycle of loading. It is shown that the beams in Specimens 1 and 2 elongated considerably due to the formation of wide cracks associated with the formation of plastic hinges. On the other hand, very little elongation of the beam for Specimen 4 was measured since the majority of the damage for this specimen was concentrated in the joint rather than the beam. The beam of Specimen 3 elongated through the sixth cycle of loading, beyond which a reduction in the beam elongation was recorded due to bond failure of the joint.

It is noted that, the reduction in beam elongation of Specimens 1 and 2 after cycles 8 and 7 respectively was due to the buckling of compression reinforcement in the beams of these specimens. Lateral confinement for the No. 6 longitudinal steel of these specimens was provided by No. 3 stirrups placed at 3 in. spacing. Near the end of the test, when the cover concrete in the beam hinging region had spalled off, the compression reinforcement in these specimens started to buckle. Similar behavior has been observed by Scribner and Wight [10]. Figure 6 shows the buckled bars of Specimen 1. The joint failure of Specimen 4 is demonstrated in Fig. 7.

Energy Dissipation

A major role of the plastified regions of ductile structures during large seismic displacements is to dissipate the input energy of the earthquake. The area under the load-displacement curves for each cycle is a measure of the dissipated energy for that cycle of loading. These results for all specimens are shown in Fig. 8.

The best performance was observed in Specimens 1 and 2 which failed in a flexural mode as a result of low joint shear stresses. Specimen 4, which had the highest joint shear stress, failed in shear and dissipated the lowest amount of energy. The response of Specimen 3, which failed in a combination of flexure and shear, is between the two previous groups.

Joint Shear Stresses

As mentioned earlier, joint shear stress was the primary variable in this study. According to the Recommendations, the maximum allowable joint shear stress for this type of connection is $12 \sqrt{f'_c}$ (psi units), with the concrete compressive strength not to exceed 6000 psi. For the specimens tested, the joint shear stresses varied between $7.52 \sqrt{f'_c}$ and $12.84 \sqrt{f'_c}$ (psi units).

Excessive joint shear stress can cause premature failure of the subassembly due to loss of anchorage of the beam reinforcement in the joint. Figure 9 shows the strains measured on the joint transverse reinforcement for all four specimens. In Specimens 1 and 2, which had low joint shear stresses, the measured strains in the hoops are small and with the elastic range. The strains in Specimen 4 and later cycles of Specimen 3 are large, indicating significant cracking of the joint region. The formation of these cracks prevent the beam longitudinal reinforcement, which is anchored in the joint, from developing its full capacity. The accompanying bond and shear failure is characterized with severe loss of stiffness and energy dissipation.

Slippage of Bars

Slippage of longitudinal reinforcement is a phenomenon often observed in tests of beam-column subassemblies [4,5] and can result in significant loss of stiffness. Data from strain gages were used in determining slippage of bars. Because the specimens were subjected to larger displacements with each additional cycle of loading, it is expected that the measured strains in the longitudinal steel would also increase. If the maximum strain measured at the end of each cycle of loading does not increase in subsequent cycles, slippage of the bar has occurred.

No slippage of beam reinforcement was observed. However, as shown in Fig. 10, the column longitudinal bars in Specimens 3 and 4 started to slip during the third and fourth cycles of loading respectively. A minor slippage was recorded for Specimen 2 during the fifth cycle. The Recommendations require the beam depth to be larger than twenty column bar diameters to prevent slippage of bars [7]. The beam depths were 19.8 and 17.3 times the column bar diameters for Specimens 3 and 4, respectively.

CONCLUSIONS

The study presented is the first reported investigation of the behavior of reinforced concrete frames constructed with high strength concrete. Clearly, much more data is needed before complete design recommendations can be developed. Within the limitations of the study, the following can be concluded:

1. The maximum allowable joint shear stress factor, γ, is a function of the concrete compressive strength.

2. Modification of the maximum allowable joint shear stress factor, γ, as given by the Recommendations [7] is needed before they can be safely applied to high strength concrete frames.

3. For concrete having compressive strength of 9500 psi, the maximum allowable joint shear stress is $10\sqrt{f'_c}$ (psi units).

4. High joint shear stresses reduce the energy absorption capacity of the subassembly, even at the presence of large flexural strength ratios.

5. In specimens with low joint shear stresses, compression buckling of beam longitudinal reinforcement was observed at very large drifts, after flexural hinges were fully developed at the end of the beam and the cover concrete had spalled off. However, changes in reinforcing details to eliminate the buckling of the bars under such circumstances are not recommended.

ACKNOWLEDGEMENTS

The study reported was conducted at the structural Engineering Laboratory of the University of Arizona. Funding for the project was provided by the College of Engineering and the University of Arizona Foundation.

REFERENCES

[1] ACI-ASCE Joint Committee 352, "Recommendations for Design of Beam-Column Joints in Monolithic Reinforced Concrete Structures," ACI Journal, Vol. 73, No. 7, July 1976, pp. 375-393.

[2] Popov, E. P., "Seismic Behavior of Structural Subassemblages," Journal of the Structural Division, ASCE, V. 106, No. ST7, July 1980, pp. 1451-1474.

[3] Meinheit, D. F. and J. O. Jirsa, "Shear Strength of R/C Beam-Column Connections," Journal of the Structural Division, ASCE, V. 107, No. ST 11, Nov. 1982, pp. 2227-2244.

[4] Ehsani, M. R. and J. K. Wight, "Effect of Transverse Beams and Slab on Behavior of Reinforced Concrete Beam-to-Column Connections," ACI Journal, V. 82, No. 2, March-April 1985, pp. 188-195.

[5] Ehsani, M. R. and J. K. Wight, "Exterior Reinforced Concrete Beam-to-Column Connections Subjected to Earthquake-Type Loading," ACI Journal, V. 82, No. 4, July-August 1985, pp. 492-499.

[6] Durrani, A. J. and J. K. Wight, "Behavior of Interior Beam-to-Column connections Under Earthquake-Type Loading," ACI Journal, V. 82, No. 3, May-June 1985, pp. 343-349.

[7] ACI-ASCE Joint Committee 352, "Recommendations for Design of Beam-Column Joints in Monolithic Reinforced Concrete Structures," ACI Journal, V. 82, No. 3, May-June 1985, pp. 266-283.

[8] ACI Committee 318, "Building Code Requirements for Reinforced Concrete (ACI 318-83)," American Concrete Institute, Detroit, 1983, 111 pp.

[9] Moussa, A. E., C. R. Vallenilla, and M. R. Ehsani, "Experimental Study of Reinforced Concrete Beam-column Subassemblies Constructed with High Strength Concrete," Department of Civil Engineeirng and Engineering Mechanics, The University of Arizona, report to the completed by April 1986.

[10] Scribner, C. F. and J. K. Wight, "Strength Decay in R/C Beams Under Load Reversals," <u>Journal of the Structural Division</u>, ASCE, Vol. 106, No. ST4, April 1980, pp. 861-876.

TABLE 1. Parameters of the Specimens

Specimen Number	1	2	3	4
h_c (in.)	13.4	13.4	11.8	11.8
d_{1c} (in.)	11.4	11.4	9.8	9.8
d_{2c} (in.)	6.7	6.7	5.9	5.9
A_{s1c}	2#7+1#6	2#7+1#6	2#7+1#6	2#8+1#7
A_{s2c}	2#6	2#6	2#6	2#7
h_b (in.)	18.9	18.9	17.3	17.3
b_b (in.)	11.8	11.8	10.2	10.2
d_{1b} (in.)	16.9	16.9	15.4	15.4
d_{2b} (in.)	15.0	15.0	13.4	13.4
A_{s1b}	2#6+1#5	3#6	3#6	3#7
A_{s2b}	2#5	2#6	2#5	2#5

TABLE 2. Design Parameters

Specimen Number	1	2	3	4
Column Axial Load (Kips)	30	76	86	73
M_{col} (K-in)	1638	1865	1582	1917
M_{BEAM} (K-in)	1729	2041	1663	2290
M_R = Flexural Strength Ratio	1.89	1.83	1.90	1.67
Joint Shear (Kips)	123.0	150.0	133.3	164.7
Joint Shear Stress (Psi)	729	888	1027	1269
Concrete Compressive Strength (Psi)	9380	9760	9380	9760
γ = Joint Shear Stress/ $\sqrt{f'_c}$	7.52	8.99	10.61	12.84

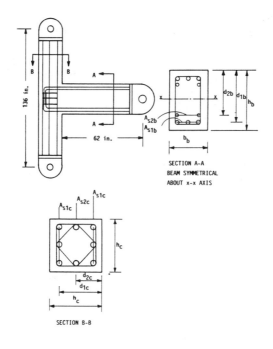

Fig. 1 Configuration of Specimens

Fig. 2 Test Frame

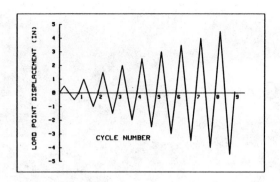

Fig. 3 Loading Sequence

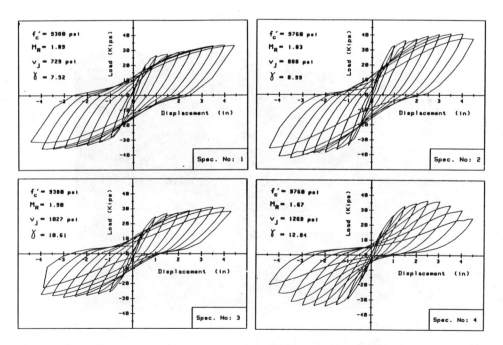

Fig. 4 Load vs. Deflection Response for Specimens

Fig. 6 Buckling of Beam Longitudinal
Reinforcement in Specimen 1

Fig. 7 Failure of Joint in Specimen 4
at the Conclusion of the Test

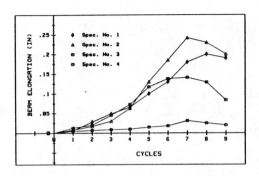

Fig. 5 Elongation of Beam Hinging Region During Different Cycles of Loading

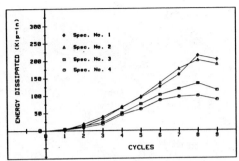

Fig. 8 Energy Dissipated During Each Cycle of Loading

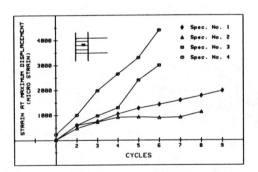

Fig. 9 Strain in Joint Reinforcement

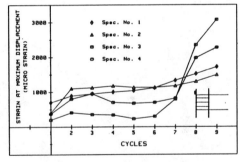

Fig. 10 Strain in Column Longitudinal Steel

BEHAVIOR OF REINFORCED CONCRETE COLUMNS SUBJECTED TO LATERAL AND AXIAL LOAD REVERSALS

Michael E. Kreger[I] and Leo Linbeck, III[II]

ABSTRACT

Recent study of reinforced concrete structures responding to strong motion earthquakes indicates columns can be exposed to large variations in axial as well as lateral loads. An experimental investigation of reinforced concrete columns subjected to various combinations of lateral and axial load reversals is described. The effect of axial load reversals on the lateral stiffness of reinforced concrete columns is observed well into the inelastic range, and implications for analytical modeling of this behavior are examined.

INTRODUCTION

Early experimental investigations of reinforced concrete columns subjected to cyclic loads considered only the application of moment reversals. Axial loads, if included, were held constant at a level considered representative of expected gravity loads [1]. More recent experimental work has examied the response of columns to changes in both axial loads and end moments [2].

The significance of this type of loading on a structure can be understood with the aid of Fig. 1. End columns of laterally loaded moment resisting frames (such as the one in this illustration) are subjected to variations in axial loads. The axial load variations are equal to shear forces developed at ends of beams which frame into the end columns. Unlike interior columns in a frame, beam reactions carried by these columns are not offset by equal and opposite reactions from a beam in the next bay. Instead, unbalanced shear forces are carried as additional axial loads by the end columns. When a frame, such as that shown in Fig. 1, is subjected to lateral load reversals, the end columns are subjected to axial load reversals. For end columns on the lowest story, which receive these axial load reversals from all floors above, this effect can be quite significant. Changes in axial load and changes in end moments will be approximately proportional as long as the frame is not forced well into the nonlinear range of response. This is not, however, the only source of axial load reversals which could be expected when a structure is subjected to seismic loads.

I. Assistant Professor, Department of Civil Engineering, University of Texas at Austin, Austin, Texas.

II. Graduate Research Assistant, Department of Civil Engineering, University of Texas, Austin, Texas.

Evaluation of the recent failure of a reinforced concrete building during a strong motion earthquake revealed a column load history which was significantly different from the loading described above [3]. Essential to the understanding of this unusual load history is an understanding of the way in which lateral-load-resisting systems transmit forces to columns. The building mentioned above had two distinctly different lateral-force-resisting systems: moment-resisting frames in one direction and a complex wall system in the other. The wall system consisted of four interior walls from the ground to the first floor and two walls from the first floor to the roof which traversed the full length of the building. The critical end of this wall system is shown in Fig. 2. This figure illustrates the stagger between one of the massive end walls and the nearest interior wall in the first story. In addition, arrows on the perspective show how shears developed by lateral forces on the structure are transmitted down the end wall, through the floor at the top of the first story to interior walls, and down through interior walls to the foundation. However, the flexibility of the horizontal diaphragm at the top of the first story did not allow efficient transfer of overturning moments between the staggered walls. As a result, end columns located beneath the massive end wall were required to develop overturning moments as illustrated in Fig. 3.

Corner columns at the end of the structure were subjected to severe changes in axial load was well as moment reversals due to response of the frame system in the direction perpendicular to the wall system. Because the two lateral-force-resisting systems responded at different frequencies, columns at the corners of the structure experienced changes in axial load and changes in end moments which varied at different frequencies. The third test described in this paper is for a column subjected to axial load reversals which vary independently of variations in end moments.

EXPERIMENTAL PROGRAM

The experimental program consisted of three identical column specimens, each subjected to a different loading program. Specimens were 10 in. (254 mm) square columns, 60 in. (1.52 m) tall, with a large endblock cast monolithically on the top and bottom of each column to provide anchorage of longitudinal reinforcement and to facilitate placement of specimens in the loading frame. Concrete used in specimens was made with Type I portland cement and 3/8 in. (9.5 mm) maximum size aggregate, and the slump of the fresh concrete was 6 in. (150 mm) to allow for ample workability when fabricating the specimens. Design strength of the concrete was 4.5 ksi (31.0 MPa), and cylinders cast with the specimens had compressive strengths between 5 and 6 ksi (34.4 and 41.3 MPa). Columns were reinforced with ten #4 bars, for a total steel area of 2 sq. in. (1290 sq. mm) and a reinforcement ratio of 2 percent. Transverse reinforcement was stirrups made of 1/4 in. (6.4 mm) diameter undeformed bar spaced at 2-1/4 in. (57 mm) on center over the entire length of the column. All longitudinal steel was Grade 60 with an average tensile yield stress of 65 ksi (448 MPa). Reinforcing was designed in accordance with ACI 318-83, Appendix A [4], for a column supporting reactions in a building with discontinuous stiffness (A.4.4.5). A schematic of the specimen is shown in Fig. 4.

All three tests were performed at the Phil M. Ferguson Structural Engineering Laboratory at the Balcones Research Center of The University of Texas at Austin. The test setup consisted of two primary hydraulic rams which were used to apply the lateral and axial loads, and six secondary rams which were used to control the relative movement of the endblocks. Secondary rams restrained the top endblock from rotating and twisting relative to the bottom endblock, assuring the specimen deformed in double curvature about the major axis of the column section. In this way, the test procedure simulated the deformation of a column with very stiff restraining members. The axial ram was rated at 320 kips (1408 kN, or approximately 60 percent of the axial capacity of the column) and the lateral ram was rated at 150 kips (660 kN), which gave it the capability of producing 4500 kip-in. (503 kN-m) of moment on the column. A schematic showing load application points and the expected deformed shape of the specimen is shown in Fig. 5.

Data for each test were recorded with an HP-3497A Data Acquisiton System and stored on a floppy disk with the aid of an HP-86 microcomputer. The data were taken from three sources: load cells on two primary and three secondary rams, displacement transducers placed on the column and endblock, and strain gages attached to the longitudinal and transverse reinforcement. Load cells were used to monitor the applied and restraining loads delivered to the specimen, displacement transducers were used to monitor the movement of the column at 10 locations on the specimen, and strain gages were used to measure the strain of the reinforcing steel at critical sections of the column. A layout of the instrumentation is shown in Fig. 6.

Each specimen was subjected to a different load program. In the first two tests, designated C-LA and C-HA, the primary rams were electronically coupled so lateral and axial loads applied to the specimen remained proportional. For these tests, the lateral ram operated under displacement control with feedback from the lateral load cell serving as the command signal for the axial ram, which operated under load control. In this way, the specimen could be loaded by forcing an increase or decrease in lateral displacement, resulting in a change in lateral load which caused a corresponding change in axial load. For these two tests, the relationship between the lateral and axial load can be written as the equation of a straight line: $P = \alpha H + \beta$. Since β was 70 kips (309 kN) for both specimens C-LA and C-HA, the only difference between the loading program for the tests was the ratio of axial load change to lateral load change, α. Test C-LA had an α of 4.1 and C-HA had an α of 8.6. Therefore, changes in axial load were more pronounced for Test C-HA than Test C-LA. The loading (or displacement) programs for these columns are shown in Figs. 7 and 8.

The third column, I-HA, used a loading program which was significantly different from the first two tests. In this test, the axial and lateral rams were uncoupled, with each ram operating independently under displacement control. A loading program was used which was based on results of an analytical investigation of a building with two distinct structural systems oriented in orthogonal directions [3]. Axial loads and end moments were applied at different rates (see Fig. 9) to represent forces which are developed in a column which is part of two orthogonal lateral force resisting systems that respond to lateral loads at two different frequencies. Maximum axial load coincided with maximum lateral load (which is directly related to end moments) at only one time during the test. The loading program for this column is shown in Fig. 9.

PRESENTATION OF TEST RESULTS

The experimental program provided ample data for analysis. Load-displacement response for each specimen will be examined, and comparisons and contrasts will be made for the group of test specimens.

Specimen C-LA

Specimen C-LA was subjected to end moment and axial load reversals which occurred in phase with each other. Examination of the lateral load-horizontal displacement plot (Fig. 10) indicates that stiffness was reduced when the specimen was subjected to a new maximum lateral displacement. In addition, the load-displacement path followed during loading to a new maximum displacement passed slightly below the previously experienced maximum load. This was true for loads applied in either the positive or negative direction.

The load-deformation response in the positive direction (first quadrant) indicates the maximum lateral load resistance was reached at a drift of approximately 1.5 percent of the column height, and later cycles taken to larger deformations were beginning to define the descending portion of the load-deformation curve. Comparisons of secant stiffnesses in the positive and negative loading direction show that the column had more stiffness when axial load was increasing than when decreasing; more of the section was effective when the axial load was large.

The area bounded by a hysteresis loop is related to the energy dissipated by the specimen while it is being deformed. A qualitative comparison of the area bounded by the hysteresis for loading in the positive and negative directions indicates the energy dissipation in the two loading directions was very similar.

Specimen C-HA

Specimen C-HA was subjected to lateral load reversals in phase with large axial load reversals. As for Specimen C-LA, lateral stiffness of the specimen deteriorated as higher levels of displacement were reached (see Fig. 11). The deterioration in stiffness was most evident in later cycles when the peak positive lateral load used in the first load cycle was repeated. Also, loading to a new maximum lateral load resulted in the load-deformation response passing near to the previously experienced maximum load level.

Ductility of the specimen was reduced in the positive loading direction compared with the negative direction. Lateral load capacity in the positive direction was reached at a displacement approximately equal to 1 percent of the column height. After the maximum lateral load capacity was realized, part of the descending portion of the load-deformation curve was defined.

Stiffness in the positive loading direction was substantially greater than for the negative direction as a result of the very different axial loads associated with the two lateral load directions. Loading in the positive direction was accompanied by increases in axial load up to 300 kips, while large negative lateral loads were accompanied by net tension in the column.

Specimen I-HA

Specimen I-HA was subjected to uncoupled lateral and axial loads. Examination of the lateral load-displacement response (Fig. 12) indicates the behavior of the column is very dependent of the load history. In fact, Fig. 12 displays two hysteresis loops which are thought by most researchers to never occur. The most impressive of the two occurrences, marked with an arrow in Fig. 12, indicates there was a decrease in lateral deflection while the lateral load was increased. This apparent anomaly can be explained by the increase in stiffness of the column due to the large change in axial force during this load sequence. The change from axial tension, when the concrete contributes virtually no stiffness to the column, to high compression, when the concrete section behaves as if it were uncracked, caused a large increase in the lateral stiffness of the column. Therefore, the lateral load required to maintain the maximum displacement was greater than the load that was actually applied, and as a result, the column deflection decreased. It is hoped that a more complete understanding of this behavior can be developed in the future by considering a three-dimensional plot which includes the changes in applied axial load.

Common Trends

When behavior of all three specimens is examined, some similarities can be noted. In general, as the compressive axial load on the specimens increased, so did their stiffness. This is very apparent when contrasting the behavior of Specimen C-HA at high and low axial loads. In the tension range, stiffness of the column approached zero, and as axial load was added, lateral stiffness increased. The reason for this increase appears to be that as more compression is added to the section, the neutral axis moves to a new location at which more of the section is effective. The increase in stiffness, then, can be directly attributed to an increased contribution of the concrete to the stiffness of the section.

Another common trend for Specimens C-LA and C-HA is overall stiffness of the section was reduced after each excursion to a new maximum deflection. In addition, the response curve for a new maximum deflection passed slightly below the previous maximum response.

Contrasting Trends

The major difference between behavior at high and low axial load ranges concerns the energy dissipation characteristics of the column. For purposes of this discussion, energy dissipation capacity of the column is assumed to be proportional to the area bounded by the lateral load-displacement response curve. In Specimen C-HA, much more energy was dissipated in the positive load direction (accompanied by increasing axial compression) than in the negative load direction. In Specimen C-LA, however, energy dissipated in the positive and negative directions were nearly equal, with slightly more being dissipated in the negative direction.

SUMMARY AND CONCLUSIONS

Three approximately half-scale reinforced concrete columns were subjected to various combinations of lateral and axial load. The following general observations were made:

1. Stiffness of columns increased with increases in axial load.

2. Each time the column was subjected to a greater drift, stiffness of the column was reduced.

3. Energy dissipation characteristics of the column were greatly dependent on the axial load history.

4. Response of a reinforced concrete column subjected to reversals in lateral and axial load is significantly different from the response of a reinforced concrete column subjected to reversals in lateral load under constant axial load. Therefore, new models are needed to predict inelastic behavior of reinforced concrete columns subjected to axial and lateral load reversals.

ACKNOWLEDGEMENTS

This investigation was supported by the National Science Foundation under its Research Initiation Program (Grant Number CEE 8404578). The conclusions and opinions expressed in this paper are solely those of the authors and do not necessarily represent the views of the sponsors.

REFERENCES

[1] Ramirez, H. and J. O. Jirsa, "Effect of Axial Load on Shear Behavior of Short Reinforced Concrete Columns under Cyclic Lateral Deformations," PMFSEL Report No. 80-1, The University of Texas at Austin, June 1980.

[2] Bedell, R. and D. P. Abrams, "Scale Relationships of Concrete Columns," Structural Research Series No. 8302, Univ. of Colorado, Boulder, January 1983.

[3] Kreger, M. E. and M. A. Sozen, "A Study of the Causes of Column Failures in the Imperial County Services Building During the 15 October 1979 Imperial Valley Earthquake," Structural Research Series No. 509, Univ. of Illinois, Urbana, August 1983.

[4] Building Code Requirements for Reinforced Concrete (ACI 318-83), American Concrete Institute, Detroit, 1983.

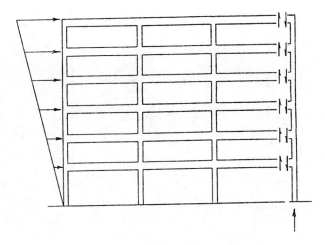

Fig. 1   Moment Resisting Frame Subjected to Lateral Forces

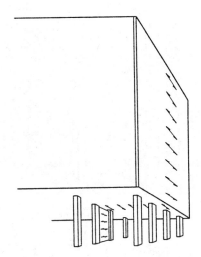

Fig. 2   Schematic of a Staggered Structural Wall System

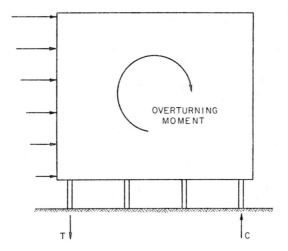

Fig. 3   Column Axial Forces Resulting from Overturning of a Staggered Structural Wall System

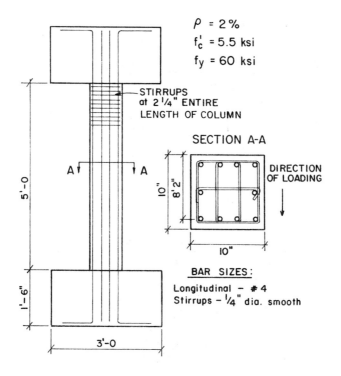

Fig. 4   Details of the Column Specimen

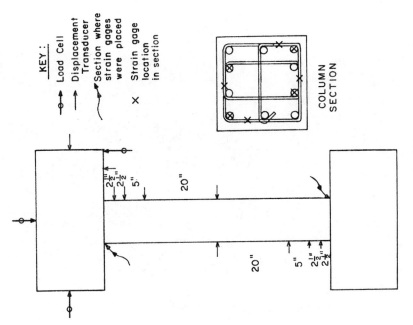

Fig. 6　Locations of Load Cells, Displacement Transducers, and Strain Gages

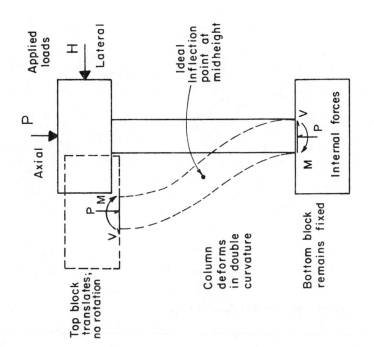

Fig. 5　Schematic of the Test Specimen under Load

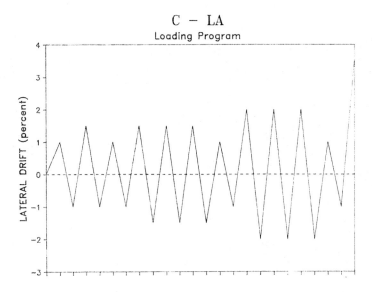

Fig. 7   Displacement History for Specimen C-LA

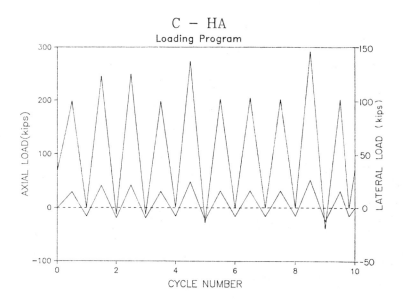

Fig. 8   Axial and Lateral Load History for Specimen C-HA

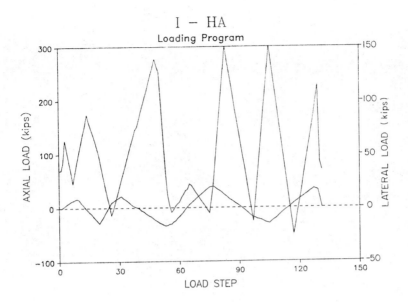

Fig. 9   Axial and Lateral Load History for Specimen I-HA

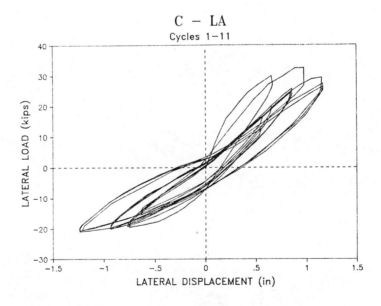

Fig. 10   Lateral Load-Displacement Response for Specimen C-LA

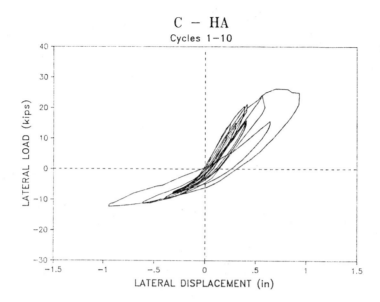

Fig. 11 Lateral Load-Displacement Response for Specimen C-HA

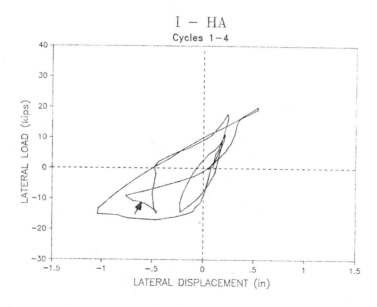

Fig. 12 Lateral Load-Displacement Response for Specimen I-HA

LOW-CYCLE FATIGUE BASED SEISMIC DESIGN OF STEEL STRUCTURES

Bahman Lashkari[I]

ABSTRACT

This paper outlines a methodology to incorporate low-cycle fatigue in seismic design of steel structures. The method is based on a simple cumulative damage model in which damage due to each plastic excursion is approximated by a Coffin-Manson type relationship. The parameters in this model are treated as random variables with distributions that are obtained from studies on the nonlinear seismic behavior of bilinear single-degree-of-freedom oscillators. The method is used to obtain an equivalent "single-excursion" ductility demand which represents the same damage state as the low-cycle fatigue model. This equivalent ductility demand will present a more reliable design requirement than the presently used ductility factor. The proposed approach will provide a basis for improving design codes and can be easily incorporated into current seismic codes.

INTRODUCTION

Current building design codes permit structures to be designed for a lateral-force level considerable lower than that anticipated during a severe earthquake. The reasoning behind this approach is that during severe seismic loading inelastic deformations will occur but the structure will not collapse or cause loss of life. The reduction in elastic strength demand depends on the type of material, structural system and the required ductility.

Existing codes are primarily concerned with the largest-excursion failure mechanism and do not account for low-cycle fatigue. This failure mechanism is incorporated into existing codes through an index known as ductility factor (i.e., ratio of maximum deformation to yield deformation) which accounts for largest expected plastic deformation. The basis for the ductility factors used in design are studies of force recudtions, ductility damand for bilinear SDOF systems [1].

In recent years, however, researchers and engineers are paying more attention to fatigue failure and recent studies, such as those by Mizuhata, et al., Ref. [2], and Lashkari, Ref. [3], have indicated that the use of cumulative fatigue as an index for measuring seismic performance of steel structures is more appropriate than the single-excursion ductility factor.

[I] M.EERI, Jack Benjamin & Associates, Inc., Mountain View, California

In this paper, a methodology to obtain a modified design ductility μ^* which includes low-cycle fatigue failure effects is outlined. First a cumulative damage model is presented, then the procedure to determine μ^* is described and application examples are presented.

CUMULATIVE DAMAGE MODEL

To assess the performance of structures under cyclic loading conditions, it is necessary to evaluate the failure modes that lead to cyclic deterioration in strength, stiffness and energy dissipation capacity. Provided that deterioration and failure in a structural component is caused by material fatigue, an extension of material damage models to structural components can be justified. For other failure modes, such as local buckling, the validity of such an extension can only be hypothesized and future experimental verification is needed.

Since structures subjected to major ground motions will undergo a relatively small number of inelastic excursions and since the inelastic excursions are usually of considerable magnitude, the concepts of low-cycle fatigue are applicable and the effects of elastic excursions on accumulated damage can usually be neglected. For elements of steel structures, experimental studies have shown that the total cumulative damage in the low-cycle fatigue regime can be evaluated approximately by using a Coffin-Manson relationship and assuming linear damage accumulation [4]. Coffin and Manson have postulated that the relationship between a constant plastic deformation range $\Delta\delta_p$, defined in Fig. 1, and the number of excursions to failure N_f can be expressed by the following equation

$$N_f = C^{-1} (\Delta\delta_p)^{-c} \tag{1}$$

in which C and c are structural resistance parameters. The parameter c can be assumed to have a value of 2.0 based on recent experimental studies on welded steel beam-to-column connections [5].

Assuming linear damage accumulation (Miner's rule), the damage caused during one excursion of range $\Delta\delta_{pi}$ can be expressed as $1/N_{fi}$, and the total damage D during N different excursions will be:

$$D = \sum_{i=1}^{N} \frac{1}{N_{fi}} = C \sum_{i=1}^{N} (\Delta\delta_{pi})^2 \tag{2}$$

where $\Delta\delta_{pi}$ is the plastic deformation range of the ith excursion. In employing Miner's rule for damage calculations the mean effect and sequence effect are ignored. Although these effects may be of some importance, they are neglected in this study to keep the damage model as simple as possible.

STATISTICAL EVALUATION OF DAMAGE

For a given severity level (e.g., peak ground acceleration and spectral shape defined by the ATC-3-06 [6]) to estimate the cumulative damage, it is necessary to know the statistical description of the number and magnitudes of all plastic deformations (i.e., expected value and standard deviation of N and $\Delta\delta_{pi}$). This information can be obtained from Reference [3] which is a pilot study to obtain the statistical description of $\Delta\delta_{pi}$ and N for steel structures using six California ground motion records. The study was performed for bilinear nondegrading single-degree-of-freedom systems with different stiffness-hardening ratios[1] and force reduction factors[1]. The study found $\Delta\delta_{pi}$ to be lognormally distributed.

Assuming $\Delta\delta_{pi}$ to be lognormally distributed with median $\bar{m}_{\Delta\delta_{pi}}$ and standard deviation in logspace of $\sigma_{\ln\Delta\delta_{pi}}$ then the $(\Delta\delta_{pi})^2$ is also lognormally distributed with median $(\bar{m}_{\Delta\delta_{pi}})^2$ and standard deviation in logspace of $2\sigma_{\ln\Delta\delta_{pi}}$. The mean value (expected value) of $(\Delta\delta_{pi})^2$ can be expressed as follows:

$$m_{(\Delta\delta_{pi})^2} = (\bar{m}_{\Delta\delta_{pi}})^2 \, e^{2\sigma^2_{\ln\Delta\delta_{pi}}} \tag{3}$$

From Equation (2) the expected value of accumulated damage can be expressed as follows (Ref. [7]):

$$E(D) = C \, E(N) \, E((\Delta\delta_{pi})^2) \tag{4}$$

Substituting for $E((\Delta\delta_{pi})^2)$ from Equation (3) and assuming N has a mean value (expected value) of m_N and standard deviation of σ_N; then:

$$E(D) = C \, m_N \, (\bar{m}_{\Delta\delta_{pi}})^2 \, e^{2\sigma^2_{\ln\Delta\delta_{pi}}} \tag{5}$$

The study by Lashkari [3] also reported a type I extreme-value distribution fit for maximum plastic deformation range $(\Delta\delta_p)_{max}$ with a mean value of $m_{(\Delta\delta_p)_{max}}$ and standard deviation of $\sigma_{(\Delta\delta_p)_{max}}$.

---

[1] The stiffness-hardening ratio α is the ratio of the post yield stiffness to the initial elastic stiffness. Force reduction factor R is the ATC-3-06 response modification factor.

PROCEDURE TO OBTAIN MODIFIED DUCTILITY μ^*

The modified ductility μ^*, which takes into consideration the effects of low-cycle fatigue, can be calculated from $\mu^* = \gamma\mu$ where μ is the conventional ductility and γ is a correction factor defined as follows:

$$\gamma = \frac{(\Delta\delta_p)_{eq}}{{}^m(\Delta\delta_p)_{max}} \quad (6)$$

where

$\quad {}^m(\Delta\delta_p)_{max}$ = mean value of maximum plastic deformation range

$\quad (\Delta\delta_p)_{eq}$ = equivalent single plastic excursion representing the low-cycle fatigue damage state

To determine $(\Delta\delta_p)_{eq}$, first, the expected cumulative damage $E(D)$ for the specified structural system and prescribed seismic event should be evaluated using Equation (5) and then the $(\Delta\delta_p)_{eq}$ can be obtained from Equation (4) with $E(N) = 1$, i.e.,

$$(\Delta\delta_p)_{eq} = \sqrt{\frac{E(D)}{C}} = \sqrt{m_N\,(\bar{m}_{\Delta\delta_{pi}})^2\,e^{2\sigma^2_{\ln\Delta\delta_{pi}}}} \quad (7)$$

CASE STUDY

Consider design of a steel structure based on the ATC-3-06 design philosophy. Assume the building has a fundamental period of 0.5 seconds with a yield level corresponding to a response modification factor R of 8 and stiffness-hardening ratio of 0.1 located in an area of high seismicity (effective peak ground acceleration of 0.4g).

For this case the following information (all normalized to yield deformation) can be obtained from Reference [3]:

$\bar{m}_{\Delta\delta_{pi}}$ = 1.95

$\sigma_{\ln\Delta\delta_{pi}}$ = 0.934

${}^m(\Delta\delta_p)_{max}$ = 14.63

m_N = 18 for strong motion duration of 7.5 seconds

m_N = 67 for strong motion duration of 20 seconds

Using Equations (6) and (7), the value of γ for the aforementioned case is calculated to be 1.35 for a strong motion duration of 7.5 seconds (short duration) and 2.6 for a strong motion duration of 20 seconds (long duration).

This means that the structure should be designed for a ductility requirement of 1.35 times or 2.6 times (depending on expected strong motion duration) the conventional ductility factor in order to account for all inelastic excursions (low-cycle fatigue).

CONCLUSIONS

A methodology to incorporate low-cycle fatigue as part of the ductility requirement in seismic design of steel structures was presented in this paper. As expected, the correction factor γ for the conventional ductility (i.e., ratio of maximum deformation to yield deformation) is proportional to strong motion duration and structural system characteristics (i.e., the expected number and distribution parameters of plastic excursions). For the example presented in this paper, γ was calculated to be between 1.35 to 2.6 for a strong motion duration of 7.5 seconds to 20 seconds. However, a comprehensive study is needed to quantify γ for different structural systems and ground motion input.

The proposed approach presents a realistic means of seismic risk evaluation and will provide a basis for improving design codes, and establishing safer and more reliable steel-frame design requirements.

ACKNOWLEDGEMENTS

The author would like to thank his collegues at Jack Benjamin & Associates (JBA) and at the Structural Engineering Department, Stanford University for their valuable suggestions. Especial thanks are due to Ashish Karamchandani, Steve Winterstein and Hassan Hadidi-Tamjed of Stanford and Kim Warren of JBA.

REFERENCES

[1] Shah, H. C., "Earthquake Engineering and Seismic Risk Analysis CE282B," John Blume Earthquake Engineering Center, Stanford University, Stanford California, 1982.

[2] Mizuhata, K., Y. Gyoten, and H. Kitamura, "Study on Low Cycle Fatigue of Structural Frames Due to Randomly Varying Load," Preprints, 6th World Conference on Earthquake Engineering, New Delhi, India, January 1977.

[3] Lashkari, B., "Cumulative Damage Parameters for Bilinear Systems Subjected to Seismic Excitations," Ph.D. Dissertation, Department of Civil Engineering, Stanford University, August 1983.

[4] Lashkari, B., J. Iihara, and A. Karamchandani, "Fatigue Assessment and Reliability in Steel Structures Subjected to Seismic Excitations," 4th International Conference on Structural Safety and Reliability, Kobe, Japan, May 1985.

[5] Zohrei, M., "Cumulative Damage in Components of Steel Structures Under Cyclic Inelastic Loading," Ph.D. Dissertation, Department of Civil Engineering, Stanford University, 1982.

[6] Applied Technology Council, <u>Tentative Provisions for the Development of Seismic Regulations for Buildings</u>, ATC Publication ATC-3-06, June 1978.

[7] Parzen, E., "<u>Stochastic Processes</u>," Holden-Day Inc., 1962.

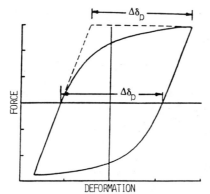

Fig. 1 Definition of Plastic Deformation Range, $\Delta\delta_p$

EARTHQUAKE DAMAGE PREDICTION FOR BUILDINGS
USING COMPONENT TEST DATA

Onder Kustu[I]

ABSTRACT

Statistics of damage threshold deformations observed in laboratory tests of building components are used to determine conditional probabilities of specific damage states for these components. Procedures are described to utilize this information in making detailed earthquake damage and loss predictions for specific buildings.

INTRODUCTION

Extensive test data exists in the literature on the behavior of various building components subjected to simulated earthquake loading. This paper describes procedures for utilizing these data in the prediction of building damage in earthquakes.

Empirical procedures are available which utilize damage statistics from past earthquakes to predict damage behavior of structures in future earthquakes. This approach is appropriate for estimating gross damage to large numbers of structures and requires adequate damage data bases for the general structure types of interest.

Damage prediction for specific buildings is not practicable using the empirical procedures if a reasonable accuracy and detail regarding the character of the damage is required. In such cases a mathematical model of the building can be used to compute its response to the estimated ground motion. The total damage or damage to specific building components is then computed based on the calculated response of the building.

Predicting damage to specific buildings on a component-by-component basis offers certain advantages. For example, the knowledge of whether the expected damage is largely structural or architectural would be very valuable in selecting seismic strengthening strategies for existing buildings or in selecting structural systems for new buildings. Building components referred to in this paper include all structural and non-structural components, such as beams, columns, shear walls, partitions, ceiling systems, cladding, electrical and mechanical systems, and other components for which data are available or can be generated through tests.

Procedures for component-by-component evaluation of damage to buildings based on calculated building response have been previously proposed by others (1, 2, 3, 4, 5, 6, 7, 8). These procedures either relate a global response parameter (e.g. base shear) or a computed local response parameter (e.g. interstory drift) to specific component damage. A major problem hindering the application of these procedures to practice has been the lack of reliable

---

I. Technical Director, Applied Technology Council, Redwood City, California

component damage functions (i.e. functions relating local response to component damage); in most procedures expert opinions or theoretically derived damage functions are used in lieu of damage functions based on factual data.

The study reported in this paper (9) developed component damage functions from published test data for a number of building components. These functions were used in example building loss evaluations (10). This paper describes the procedures to develop building component damage functions and to utilize these functions in seismic evaluation of buildings. For detailed data on specific components, the reader is referred to Reference 9.

PROCEDURE

General Procedure

In order to predict building damage by components an inventory of the components that make up the building is required. This inventory should identify all the significant components at each floor level and estimate their replacement values. Building cost statistics shown in Table 1 was developed for quickly estimating for several building types the value of building components relative to the total building cost.

The next step in the procedure is to determine the peak building response in terms of local response parameters, i.e. peak interstory drift and floor acceleration values. An appropriate structural analysis procedure needs to be used to estimate these values. For most cases, a spectrum analysis would be sufficient.

Once the peak local building responses are determined, the damage state of each component at a given story is determined using the component test data or the component damage functions derived from such test data. The total building damage is computed by summing the damage to component types at all floor levels, and adding up the damage from all floors.

This general procedure accounts for the ground motion and building response using common engineering procedures, and hence, parameters affecting structural response (e.g. peak ground acceleration, local geology and soil conditions, building stiffness, mass distribution, and damping) are explicitly considered. The component test data and the derived damage functions form the most significant parts of the procedure.

Applicable Data

The data of interest are those obtained from laboratory tests of building components similar to the "real-world" components to be evaluated. The degree of similarity between the test components and the real-world components would have a direct effect on the reliability of the damage prediction. The data should include the relationship between applied forces and displacements, such as force-displacement hysteresis curves, and should allow for the determination of degree of damage to the component at various times during the test. Pictures or drawings showing the extent of cracking, spalling, etc. at various stages of the test, or verbal descriptions of the incurred damage at specific times, would be highly desirable.

Significant Local Response Parameters

The deformation and damage behavior of each building component can be described by at least one local response parameter. Interstory drift (relative horizontal deflection between consecutive floors) closely correlates with deformation and damage behavior of most building components, such as full height partitions, structural frames, shear walls, infill walls, and exterior walls. In some cases, such as for individual beams or columns, the significant local response parameter (beam/column deflection or deflection angle) can be computed from the interstory drift. For those components which are attached only to one floor, such as mechanical equipment and suspended ceiling systems, the most appropriate local response parameter would be the floor acceleration. When screening test data for use in this procedure it should be ensured that the significant local response parameter for the component is reported for the whole test or it can be computed from other test parameters.

Significant Damage States and Damage Thresholds

All building components exhibit specific damage patterns and failure modes when subjected to cyclic deformations beyond their elastic capabilities. Each distinct damage pattern typical of a specific component can be identified as a damage state (DS) for that component. For example, for a reinforced concrete beam, four damage states may be identified: null (or undamaged), cracked, yielding, and failed. If the component damage is expressed in terms of discrete damage states the local response-component damage relationship becomes a step function in which distinct damage states are delineated by limiting values of the relevant local response parameter, or the "damage thresholds". Hence, assigning appropriate values to the damage thresholds would be sufficient to establish the component response-damage relationships for building components. For example, for a reinforced concrete cantilever beam, the tip deflections at first cracking, first yielding, and at failure would be sufficient to delineate four damage states (undamaged, cracked, yielding, and failed) and relate the deformation of the beam to these damage states.

Average Threshold Values

Damage threshold statistics, i.e. mean values and variabilities of damage thresholds, are the simplest practical forms of presenting component test data for use in damage prediction.

To obtain damage threshold statistics threshold values for each test sample are first determined. For example, for reinforced concrete beams the deflections corresponding to first cracking, first yielding, and failure are determined from the hysteresis curves (d_c, d_y, d_f). These deflections are then expressed in terms of a nondimensional parameter by dividing with an equivalent cantilever length, i.e. $D_c = d_c/L$, etc. Hence, for each test specimen in the data base, a set of threshold values are obtained. Using these values, the mean value and the variability (expressed in terms of a standard deviation) of each threshold are computed.

The average threshold values (and the associated variabilities) are very useful for quick evaluations of buildings. A simple method to map the component damage thresholds on response spectra plots is described later. Table 2 gives the average damage threshold values for a number of building components.

Probability Distributions of Damage Thresholds

Damage incurred by a building during an earthquake is a function of a large number of parameters, such as the characteristics of the ground motion, building geometry, design details, and construction quality. Because these parameters are highly variable, a probabilistic approach to damage prediction is appropriate.

This study is concerned with the variabilities of the damage thresholds. The variability of the computed local response caused by the uncertainties in the ground motion definition and analyses procedures are not addressed. Hence, it is assumed that the local response parameters (e.g. interstory drift and floor acceleration) are computed for given ground motion using deterministic analysis procedures. However, the uncertainties in the local response-damage relationships are represented by the probability distributions of the damage thresholds.

Fig. 1 shows the relative frequency histograms of damage thresholds for reinforced concrete beams. Using these histograms, or the theoretical probability density functions derived to fit the data, the probability of a component being in a specific damage state can be computed for any given local response parameter value.

Probabilities of Damage States

When the significant local response, D, for a component can be estimated the probability of the component being in a certain damage state can be computed using the probability density functions of the damage thresholds. However, for mathematical convenience, assumptions need to be made regarding the statistical dependence between the damage thresholds.

The probability density functions which can be derived from the relative frequency histograms shown in Fig. 1 are marginal probability density functions of individual damage thresholds. Assuming that the damage states DS0, DS1, DS2, etc. delineated by damage thresholds T1, T2, etc. are in increasing order of severity; i.e. DS1 represents more severe damage than DS0 and T2 is larger than T1, and that the thresholds are statistically independent, the joint probability density function is the multiplication of the marginal functions; i.e.,

$$f_{T_1 T_2}(x,y) = f_{T_1}(x) f_{T_2}(y) \qquad (1)$$

where, $f_{T_i}(x)$ is the marginal density function for threshold i. Integrating this joint density function over appropriate areas, the following probability values are obtained:

$$P(DS_0|D) = P(D<T_1) = 1 - F_{T_1}(D) \qquad (2)$$
$$P(DS_1|D) = P(T_1<D<T_2) = F_{T_1}(D)(1 - F_{T_2}(D)) \qquad (3)$$
$$P(DS_2|D) = P(T_1<D \cap T_2<D) = F_{T_1}(D) F_{T_2}(D) \qquad (4)$$

where, $F_{T_i}(D)$ are the marginal cumulative distribution functions of thresholds. For four damage states the probabilities are given as follows:

$$P(DS_0|D) = 1 - F_{T_1}(D) \qquad (5)$$
$$P(DS_1|D) = F_{T_1}(D)(1 - F_{T_2}(D)) \qquad (6)$$
$$P(DS_2|D) = F_{T_1}(D) F_{T_2}(D) (1 - F_{T_3}(D)) \qquad (7)$$
$$P(DS_3|D) = F_{T_1}(D) F_{T_2}(D) F_{T_3}(D) \qquad (8)$$

It should be noted that the above equations were derived by integrating the joint probability density function within the limits from zero to infinity for each threshold. Hence, the probability expressions may include a small error attributable to the volume of the joint density function over the areas where the sequential relationships of damage thresholds (i.e. $T_0<T_1<T_2<T_3$) is not true.

If a strong correlation between damage thresholds is indicated, the probability of damage states may be computed based on perfect correlation assumptions. Equations to compute the probabilities of damage states for an example case are given in Reference 9.

Damage Functions

A damage function is defined here as a function which describes the relationship between a local response parameter and the expected damage. If damage is quantified in terms of damage factor (called damage ratio by some investigators--ratio of repair cost to replacement value), the damage function takes the form of a relationship between the significant local response parameter and the expected value of damage factor.

The expected value of damage factor, given local response D, can be found as follows:

$$E(DF) = \sum_i P(DS_i|D) \, DF_i \qquad (9)$$

where DF_i is the damage factor corresponding to damage state i.

APPLICATION TO BUILDING EVALUATIONS

Procedures are outlined above for processing raw test data for building components to develop the basic component damageability information, i.e. average values of damage thresholds, probability distribution functions of damage states, and damage functions. The rest of this paper describes the various ways of utilizing this information in practical building evaluations.

Quick Qualitative Evaluations

In many cases, a quick qualitative evaluation of what might happen to a building when subjected to a certain ground motion level is required. In such instances, the local building response for the postulated seismic loading can be estimated and compared to the mean threshold values.

The following simple procedure can be used for quick damage evaluations:

1. Estimate the fundamental building period using approximate formulas. For this purpose, the formulas given in codes (e.g. in the Uniform Building Code (UBC)) or in the technical literature can be used. (For example, for

a 10-story reinforced concrete building the period is approximately 1 second, using the T=0.1N relationship given in the UBC.)

2. Determine the spectral acceleration, S_a, using the appropriate response spectrum for the site.

3. Estimate the spectral displacement from spectral acceleration using the relationship $S_d = (T/2\pi)^2 S_a$, or if the spectrum is available on tripartite paper, determine the spectral displacement directly.

4. Estimate the maximum roof displacement (assuming that the fundamental mode displacement approximates the maximum displacement) by modifying the spectral displacement for the fundamental mode with the modal participation factor. Sample participation factors for a straight line fundamental mode shape are shown in Table 3.

5. Compute the average interstory drift angle by dividing the roof displacement with the total building height.

6. Compare the computed drift angle, or the corresponding relevant response parameter, with the threshold values given in Table 2 to determine which components may suffer damage if subjected to the earthquake under consideration.

The threshold values given in Table 2 can also be mapped onto a tripartite spectrum plot by making certain assumptions for the relationships between building height and period, and building height and participation factor (10). Fig. 2 shows an example plot of damage thresholds for reinforced concrete shear walls superimposed onto the response spectrum of the 1940 El Centro earthquake. The curves on this plot correspond to the mean values of damage thresholds, i.e. the shear walls of buildings falling into frequency ranges where the spectrum exceeds a threshold curve would have over 50 percent likelihood of being damaged to the extent indicated by the applicable damage state.

Damage Prediction Based on Detailed Structural Analysis

More detailed, quantitative damage evaluations (e.g. predicting damage factors or repair costs) require more accurate determination of local building response. For low ground motion levels elastic spectrum analysis may be adequate. For high ground motion levels nonlinear analyses may be required. In lieu of non-linear time history analyses, an approximate iterative approach may be used where member stiffnesses may be adjusted according to the indicated damage.

Quantification of damage in terms of monetary losses would require the following additional data:

1. Total building value.

2. Estimate of the replacement value of each component as a percentage of the total building value.

3. Repair cost estimate as a percentage of replacement value (damage factor) corresponding to each damage state of all components.

The total building value relevant to this study is the "new construction cost"; i.e. how much the same building would cost if it were to be built today. This value can be estimated based on average unit costs for the type of construction and the building area or computed from the original construction cost (if it is known) with appropriate adjustment for inflation.

Specific data on how much the structural frame, partitions, etc. are worth as percentages of the total cost may not be available. In this case, the statistical data presented in Table 1 can be used to estimate the relative values of the various components.

The most difficult data to obtain are the repair cost estimates of damage states. There has been no systematic effort in the past to follow-up on buildings damaged in earthquakes and report the actual repair costs. Such data are not generally included in laboratory test reports either. Hence, the repair costs of component damage states would have to be established based on engineering judgment, and possibly, on contractors' hypothetical bids for the repair of the projected damage. In any case, because the damage states are discretized descriptions of ranges of building damage and accurate estimates are difficult, the repair cost estimates should be given in terms of ranges. For example, lower and upper bound estimates of repair costs, rather than a single value, should be given. Such "low" and "high" estimates of damage factors are provided in Table 2 under the column heading "Trial Range of Central Damage Factor" for the various component damage states.

EXAMPLES

The component repair cost estimates given in Table 2 were used in Reference 9 to derive "component damage functions". These functions define the expected value of component damage factor as a function of the applicable local response parameter. An example damage function is plotted in Fig. 3. The upper and lower curves in this plot correspond to the "high" and "low" repair cost estimates for the applicable damage states; the middle curve reflects average estimates. These curves can directly be used to estimate the "percent damage" to structural frame, walls, partitions, glass, etc. based on the structural response estimate. Damage functions for additional building components are given in Reference 9.

Once the expected losses to all building components are computed the total expected building loss can be computed by summing the expected component losses.

ACKNOWLEDGEMENTS

This paper is based on a study performed by URS/John A. Blume & Associates, Engineers of San Francisco, California, for the U.S. Department of Energy, Nevada Operations Office. Contributions of Mr. David D. Miller and Mr. Steven T. Brokken to this study are gratefully acknowledged.

REFERENCES

(1) Steinbrugge, K.V., F. E. McClure, and A. J. Snow, "Studies in Seismicity and Earthquake Damage Statistics", United States Coast and Geodetic Survey, Washington, D.C., 1969.

(2) Czarnecki, R. M., "Earthquake Damage to Tall Buildings", Research Report R73-8, Department of Civil Engineering, Massachusetts Institute of Technology, Cambridge, Massachusetts, January, 1973.

(3) URS/John A. Blume & Associates, Engineers, "Effects Prediction Guidelines for Structures Subjected to Ground Motion", JAB-99-115, San Francisco, California, 1975.

(4) Culver, C. G., H. S. Lew, G. C. Hart, and C. W. Pinkham, "Natural Hazards Evaluation of Existing Buildings", Report No. BSS-61, U.S. Department of Commerce/National Bureau of Standards, Washington, D. C., 1975.

(5) Blejwas, T., and B. Bresler, "Damageability in Existing Buildings", UCB/EERC-78/12, Earthquake Engineering Research Center, University of California, Berkeley, California, August, 1979.

(6) Kustu, Onder, "A Practical Approach to Damage Mitigation in Existing Structures Exposed to Earthquakes", Proceedings of the 2nd U.S. National Conference on Earthquake Engineering, Stanford, California, August, 1979.

(7) Sauter, Franz, "Damage Prediction for Earthquake Insurance", Proceedings of the 2nd U.S. National Conference on Earthquake Engineering, Stanford, California, August, 1979.

(8) Hasselman, T. K., R. T. Eguchi, and J. H. Wiggins, "Assessment of Damageability for Existing Buildings in a Natural Hazards Environment, Volume 1: Methodology", Technical Report 80-1332-1, J. H. Wiggins Company, Redondo Beach, California, September, 1980.

(9) Kustu, O., D. D. Miller, and S. T. Brokken, "Development of Damage Functions for High-Rise Building Components", JAB-10145-2, URS/John A. Blume & Associates, Engineers, San Francisco, California, October, 1982.

(10) Scholl, Roger E., Onder Kustu, Cynthia L. Perry, and John M. Zanetti, "Seismic Damage Assessment for High-Rise Buildings", URS/JAB 8020, URS/John A. Blume & Associates, Engineers, San Francisco, California, July, 1982.

Table 1. Average Building Component Costs as Percentages of Total Construction Cost (From Reference 9)

Component	Concrete Apartment				Concrete Office				Concrete Hospital			Percentage of Total Building Cost, by Building Type				Steel Office				Steel Hospital		Steel, All Types			
	11-20 Stories	21-40 Stories	All	6-10 Stories	11-20 Stories	All	6-10 Stories	All	6-10 Stories	11-20 Stories	21-40 Stories	All	6-12 Stories	13-25 Stories	26-40 Stories	All	6-12 Stories	6-12 Stories	13-26 Stories	26-40 Stories	All
Structural																									
Substructure	3.67	3.84	3.78	3.75	5.08	4.39	4.16	4.06	3.95	4.12	4.26	4.04	4.31	2.90	3.13	3.68	4.54	4.69	2.90	3.13	4.25
Superstructure	26.73	20.83	24.17	23.58	31.60	27.52	14.69	14.30	20.01	27.12	24.34	22.86	16.56	18.80	23.91	18.29	13.68	15.43	18.80	23.91	16.68
Total	30.40	24.67	27.95	27.33	36.68	31.91	18.84	18.36	23.96	31.25	28.60	26.90	20.88	21.70	27.03	21.97	18.22	20.12	21.70	27.03	20.93
Mechanical and Electrical																									
Mechanical	16.65	16.45	17.02	16.87	10.11	14.00	23.42	23.48	21.15	15.07	15.42	18.49	20.24	16.04	18.85	18.65	26.66	22.66	16.04	18.85	21.15
Electrical	8.51	8.32	8.42	9.00	6.48	8.04	12.50	12.86	10.87	8.44	8.19	9.76	10.05	7.99	9.10	9.24	12.01	10.64	7.99	9.10	10.04
Elevator	3.71	2.87	3.57	4.41	4.03	4.75	3.07	3.21	3.88	3.85	4.57	3.94	4.32	5.61	6.40	5.03	2.87	3.49	5.61	6.40	4.10
Total	28.88	27.64	29.01	30.28	20.82	26.80	38.99	39.55	35.89	27.36	28.18	32.19	34.61	29.64	34.34	32.92	41.54	36.80	29.64	34.34	35.29
Architectural																									
Partitions	11.36	11.92	11.09	6.97	3.69	5.45	7.60	7.58	6.05	8.48	9.78	7.72	5.86	6.28	7.25	6.19	8.29	7.31	6.28	7.25	7.11
Ceilings	0.10	0.05	0.09	1.39	1.40	1.48	1.19	1.13	1.17	0.59	0.62	0.91	2.40	0.96	0.77	1.70	1.45	1.93	0.96	0.77	1.67
Exterior Walls	4.42	3.64	4.42	3.02	9.48	5.37	2.82	3.54	3.47	6.40	3.51	4.45	11.61	12.11	7.32	11.20	4.04	7.35	12.11	7.32	8.23
Glass	2.01	2.51	2.21	2.90	2.78	2.81	1.01	1.01	1.32	2.18	2.48	2.07	1.22	0.47	1.08	0.95	1.13	1.50	0.47	1.08	1.28
Ornamentation	0.49	0.81	0.53	3.25	1.28	2.24	0.68	0.63	1.97	0.74	0.61	1.81	2.02	1.19	1.82	1.72	0.63	1.55	1.19	1.82	1.80
Painting	2.00	1.97	1.98	1.19	0.63	0.91	1.21	1.21	1.31	1.48	1.57	1.40	0.54	0.67	0.58	0.82	1.21	1.18	0.67	0.58	1.04
Other	6.16	8.28	7.59	6.49	3.12	5.08	7.54	7.37	7.16	5.16	7.33	6.51	4.05	4.34	4.49	4.21	5.68	5.26	4.34	4.49	5.04
Total	26.53	29.21	27.92	25.21	22.37	23.34	22.05	22.50	23.85	25.04	25.90	24.47	28.12	26.02	23.30	26.78	22.42	26.49	26.02	23.30	26.17
Other																									
Parking, Grounds	2.03	2.04	1.99	4.85	8.38	5.79	0.94	0.85	2.09	4.00	1.80	2.70	1.58	5.26	0.51	2.66	1.02	1.25	5.26	0.51	1.94
General, Miscellaneous	12.17	16.51	13.10	14.49	11.81	13.36	19.28	18.83	14.87	12.38	15.58	14.12	14.60	17.24	14.86	15.51	16.71	15.22	17.24	14.86	15.57
Total	14.20	18.55	15.09	19.33	20.18	19.14	20.21	19.67	16.96	16.38	17.38	16.82	16.18	22.50	15.37	18.18	17.73	16.47	22.50	15.37	17.51
(Sample Size)	(7)	(3)	(11)	(6)	(4)	(11)	(6)	(7)	(20)	(12)	(4)	(36)	(8)	(5)	(1)	(15)	(9)	(20)	(5)	(2)	(27)

1,501

Table 2. Building Component Damage Threshold Statistics (From Reference 9)

Component	DS_0 ($0 < \theta < THD_1$)	Trial Range of Central Damage Factor	THD_1 Mean	THD_1 Standard Deviation	DS_1 ($THD_1 < \theta < THD_2$)	Trial Range of Central Damage Factor	THD_2 Mean	THD_2 Standard Deviation	DS_2 ($THD_2 < \theta < THD_3$)	Trial Range of Central Damage Factor	THD_3 Mean	THD_3 Standard Deviation	DS_3 ($THD_3 < \theta$)	Trial Range of Central Damage Factor	Comments
Reinforced Concrete Beams	Null	0.0	0.3511×10^{-2}	0.1503×10^{-2}	Cracked	0.02-0.10	0.8341×10^{-2}	0.3037×10^{-2}	Yielded	0.40-0.80	0.5745×10^{-1}	0.4545×10^{-1}	Maximum	1.0-1.5	
Reinforced Concrete Columns	Null	0.0	0.2569×10^{-2}	0.2175×10^{-2}	Cracked	0.02-0.10	0.9556×10^{-2}	0.4282×10^{-2}	Yielded	0.40-0.80	0.2643×10^{-1}	0.1757×10^{-1}	Maximum	1.0-2.0	
Steel Moment-Resisting Subassemblies	Null	0.0	0.1285×10^{-1}	0.4900×10^{-2}	Yielded	0.5-0.80	0.5742×10^{-1}	0.1928×10^{-1}	Maximum	1.0-1.5	--	--	--	--	
Reinforced Concrete Shear Walls	Null	0.0	0.5554×10^{-3}	0.2330×10^{-3}	Cracking	0.02-0.10	0.3694×10^{-2}	0.1928×10^{-1}	Yielded	0.4-0.8	0.1852×10^{-1}	0.1639×10^{-1}	Ultimate	1.0-1.5	
Drywall Partitions	Null	0.0	0.2828×10^{-2}	0.2422×10^{-2}	Threshold	0.02-0.10	0.6170×10^{-2}	0.2666×10^{-2}	Real	1.0-1.5	--	--	--	--	Linearly Dependent $THD_2 = 0.004 + 0.835 THD_1$
Brick Infill Walls, Reinforced	Null	0.0	0.4202×10^{-2}	0.4244×10^{-2}	Intermediate	0.30-0.60	0.3704×10^{-1}	0.2219×10^{-1}	Ultimate	1.0-1.5	--	--	--	--	
Brick Infill Walls, Unreinforced	Null	0.0	0.4202×10^{-2}	0.4244×10^{-2}	Ultimate	1.0-1.5	--	--	--	--	--	--	--	--	
Block Infill Walls, Reinforced	Null	0.0	0.8447×10^{-2}	0.5478×10^{-2}	Intermediate	0.30-0.60	0.2858×10^{-1}	0.2872×10^{-1}	Ultimate	1.0-1.5	--	--	--	--	
Block Pier Walls, Reinforced	Null	0.0	0.2637×10^{-2}	0.6935×10^{-3}	Intermediate	0.30-0.60	0.1104×10^{-1}	0.1564×10^{-2}	Ultimate	1.0-1.5	--	--	--	--	
Brick Pier Walls, Reinforced	Null	0.0	0.2995×10^{-2}	0.2258×10^{-2}	Intermediate	0.30-0.60	0.1557×10^{-1}	0.1506×10^{-1}	Ultimate	1.0-1.5	--	--	--	--	
Glass	Null	0.0	0.1405×10^{-1}	0.9852×10^{-2}	Realignment	0.05-0.15	0.2811×10^{-1}	0.1972×10^{-1}	Broken	0.30-0.50	--	--	--	--	Linearly Dependent $THD_2 = 1/2\ THD_1$

Table 3. Modal Participation Factors for Uniform Mass and Straight-Line Mode Shape

No. of Stories	1	2	3	4	5	10	20	30	40
Modal Participation Factor	1.0	1.2	1.29	1.33	1.36	1.43	1.46	1.48	1.50

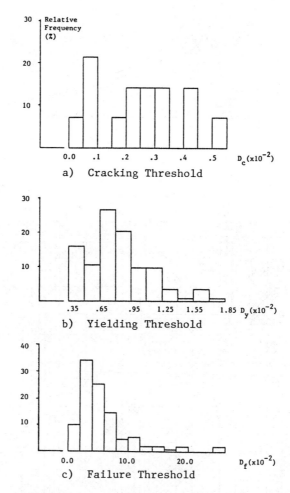

Fig.1 Relative Frequency Histograms of Damage Thresholds for Reinforced Concrete Beams

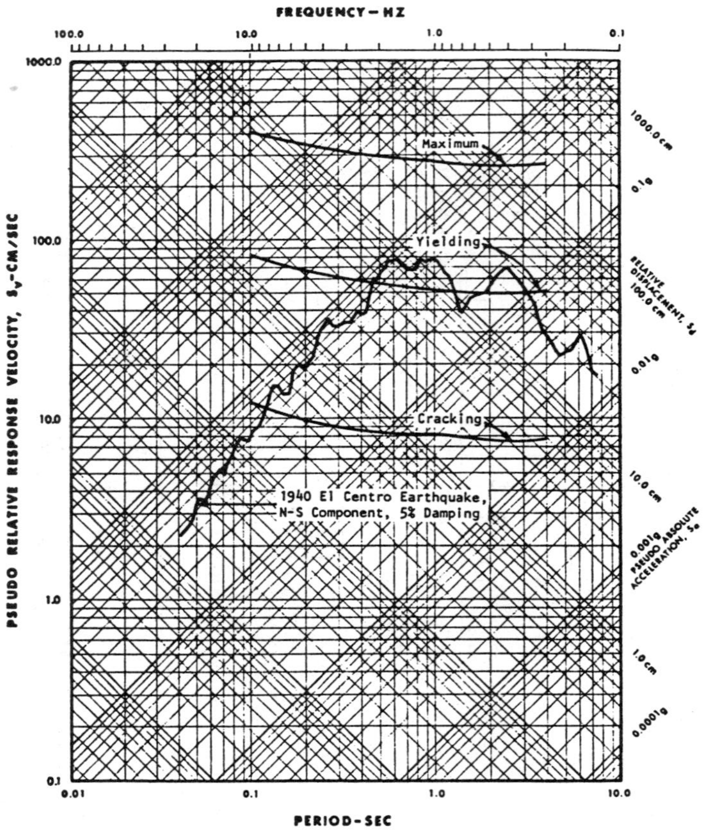

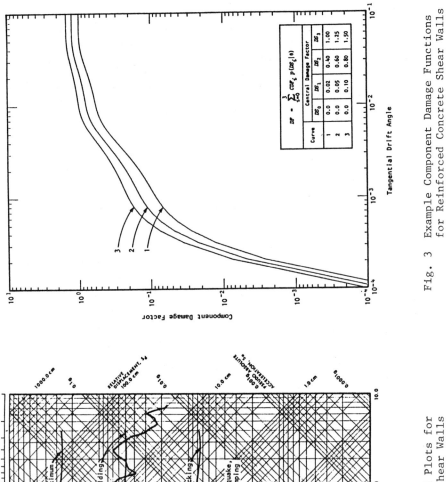

Fig. 3 Example Component Damage Functions for Reinforced Concrete Shear Walls (From Reference 9)

Fig. 2 Mean Damage Threshold Plots for Reinforced Concrete Shear Walls (From Reference 10)

SEISMIC RESPONSE OF SLAB-COLUMN FRAMES

J. P. Moehle[I]

ABSTRACT

Conventionally designed and detailed slab-column frames are generally recognized as weak seismic load resisting systems. Significant issues in design are described, and results of an experimental study of a slab-column frame are used to verify recommended procedures.

INTRODUCTION

Numerous failures of reinforced concrete flat-slab structures during the 19 September 1985 earthquake in Mexico have brought into focus the need to evaluate design methods for lateral load resistant flat-slab structures. In regions of moderate seismic risk in the United States and elsewhere, where flat-slab framing is often used as the primary lateral load resisting system, improved design methods are needed to ensure adequate strength, ductility, and stability of frames that may be subjected to occasional moderate and rare strong ground shaking. In regions of higher seismic risk, where flat-slabs may be used as the vertical load carrying system in structures braced by conventional frames or shear walls, there is a need to determine design and detail requirements to ensure that the slab will not fail as it deforms with the primary lateral load resisting system.

An experimental research program was undertaken to enable a detailed study of the design issues related to seismic resistance of slab-column frames. The research included design, construction, testing, and analyses of a two-story model flat-plate structure subjected to base motions simulating effects of earthquake ground shaking. This paper describes some of the basic issues that should be considered in seismic resistant design of slab-column frames. The experimental study and findings of the experiments in relation to these basic design issues are also described.

LATERAL LOAD DESIGN ISSUES

Three main issues in lateral-load design of slab-column frames are identified and discussed. These are (1) lateral-load stiffness, (2) lateral-load strength and ductility, and (3) resistance to progressive collapse.

Lateral-Load Stiffness

Numerous analytical studies (summarized elsewhere [1]) have established basic procedures for determining the moment-rotation characteristics of

---

I. Assistant Professor, Department of Civil Engineering, University of California, Berkeley, California.

elastic plates connected to columns. According to the analyses, when a connection is subjected to unbalanced moment, the full transverse width of the slab is not fully effective in restraining the column rotation. Two analytical models are normally used to account for this effect in design, namely, the effective beam width model and the equivalent frame model.

The effective beam width model represents the slab-column connection by an effective beam connected directly to the columns of the frame. The beam has depth equal to the slab depth and width equal to the slab transverse width multiplied by an effective width factor. Effective width factors have been determined analytically for a broad spectrum of connection configurations [1]. The other model, the equivalent frame model, represents the slab-column connection by slab-beams connected to columns through ficticious transverse torsion members. The slab-beam has properties equal to those of the slab. Properties of the transverse torsion member are determined as a function of the slab and column dimensions, as summarized elsewhere [1,2].

An essential step in lateral-load modeling is recognition that the aforementioned modeling procedures do not explicitly consider effects of cracking on stiffness. Flat-slab frames are susceptible to significant reductions in stiffness as a consequence of slab cracking near the connection. The cracking can arise from construction loads, service gravity loads, temperature and shrinkage effects, and lateral loads. Because of the broad range of loading parameters that may affect cracking, and because slab stiffness is sensitive to the degree of cracking, the stiffness reduction attributable to cracking is not easily quantifiable. Thus, simplified approaches to estimating cracking effects may be preferred. For example, Vanderbilt and Corley [1] recommend that the slab-beam effective inertia be reduced to one-third of the gross-section stiffness when using the equivalent frame method. A similar reduction can be used for the effective beam width model.

A simple numerical example can serve to illustrate the relative flexibility of slab-column frames, and, hence, the importance of conservatively estimating lateral-load stiffness. Figure 1 depicts two frames, one a beam-column frame, the other a flat-plate frame. Typical dimensions have been selected (Fig. 1). The frames were analyzed under static lateral loads with magnitudes varying from zero at the base to a maximum at the roof level. Columns in both frames were modeled using gross-section properties. To account for beam cracking in the beam-column frame, beam flexural inertias were taken equal to half the gross-section stiffnesses, a reduction typical for beams having moderate flexural reinforcement ratios. Slabs in the flat-plate frame were modeled using the effective beam width model [3]. To account for cracking effects, slab flexural inertias were reduced to one-third of the recommended effective elastic stiffnesses.

Computed lateral-load stiffnesses for the beam-column frame is approximately twice that for the slab-column frame (Fig. 1), illustrating the relatively low stiffness of the latter. Furthermore, it is noted in relation to Fig. 1 that the flat-plate frame develops computed lateral drift equal to approximately one percent of structure height under a base shear equal to ten percent of the structure weight. It is noted that a ten percent base-shear coefficient is approximately equal to the design coefficient for a moderately tall structure in a region of high seismic risk in the United States [4].

For that level of base shear, lateral drift is twice that permitted by most building codes (e.g., Reference 4). Thus, it is apparent that lateral drift is likely to be a significant design issue for slab-column frames. In many cases, lateral drift considerations, rather than strength considerations, will control frame proportions.

The relatively large flexibility of slab-column frames is cause for concern in seismic design for several reasons, including possible damage to structural and nonstructural components and overall stability of the frame under excessive drifts. For these reasons, conservative if not accurate procedures for estimating stiffness are desireable.

Lateral-Load Strength and Ductility

Shear and unbalanced moment strength of slab-column frames is well studied, and numerous procedures for determining strength have been proposed [5]. Of these, the design procedure of the ACI Building Code [2] has been shown [6] to be acceptably simple and accurate. The procedure recognizes that both shear and unbalanced moment result in slab shear stresses in the vicinity of the column and that shear stresses due to the combination of both may control strength. Under combined shear and moment transfer, a linear variation of shear stresses is presumed to develop (Fig. 2). The model predicts a linear interaction between shear and unbalanced moment transfer strengths, as depicted by the staight line in Fig. 3.

An important implication of the relationship shown in Fig. 3 is that there exists a direct relation between the gravity loads on the slab and the lateral-load strength of the building. For example, for a given design slab gravity load shear (point "a" in Fig. 3), there is a corresponding unbalanced moment capacity (point "b"). If the ultimate gravity loads have been properly selected, and the service loads at the time of an earthquake are well below design loads (point "c"), the corresponding unbalanced moment strength (point "d"), and lateral-load strength, will exceed the design requirements. Certain adverse effects may arise from the overstrength, for example, it is possible that a strong "beam"-weak column system will result.

In cases where gravity loads are estimated unconservatively, the lateral load strength of the slab-column frame is adversely affected (points "e" and "f" in Fig. 3). It is generally recognized that a reduction in lateral-load strength is likely to result in increased demand for ductility during a strong earthquake [7]. However, experimental evidence suggests that the available ductility in the connection region is reduced as the direct shear on the connection is increased (Fig. 4) [8,9].

Addition of slab shear reinforcement in the form of closed stirrups has been proposed [8] as one means of enhancing ductility of slab-column connections subjected to seismic loadings. However, such reinforcement is not favored by many engineers, as it complicates reinforcement placement. In lieu of using slab shear reinforcement, Hawkins and Mitchell [8] have suggested that the direct shear stress due to gravity loads be kept below 2 f'c if connection rotations in excess of yield are anticipated. However, even at such low shear stress levels, punching of a connection will occur eventually if connection rotations become large enough. In view of the potential for punching of slab-column connections, some provision should be made to arrest the punching and avoid progressive collapse. Appropriate provisions are summarized in the next section.

Shear and unbalanced moment strength of edge and corner slab-column connections is generally affected by the same parameters that influence behavior of interior connections, and design and analysis procedures are similar. However, for frames located in regions of high seismic risk, most codes require the use of perimeter frames comprising conventional beam-column framing. The perimeter frame can be made more resistant to shear distress and punching problems through proper detailing of the edge beam. Strengths of connections with edge beams are limited by the weaker of two mechanisms. The first involves formation of a flexural yield line across the slab column strip at the face of the transverse beam. The second (Fig. 5) involves flexural yield of the slab at the column and torsional failure of the transverse beams. Strength for either mechanism is calculated using established procedures [10].

Resistance to Progressive Collapse

Slab-column connections are susceptible to punching shear failures when subjected to inelastic load reversals. When a single connection fails, if no provision is made to suspend the slab at the punched connection, loads originally supported at the connection will be transfered to surrounding connections. These may, in turn, fail under the increased loads, possibly resulting in a progressive collapse that spreads through the building.

To avoid progressive collapse, either the original punching failure must be prevented, or progressive collapse must be arrested by suspending the slab from the columns after punching of individual connections. The first approach of avoiding the original punching failure can be achieved either by ensuring very low shear stresses under gravity loads or by providing slab stirrup reinforcement. Both of these approaches are often considered unsatisfactory because of apparent economic reasons. The alternate approach of preventing progressive collapse by suspending the slab from the columns is, in many cases, the more acceptable solution.

The model used to develop requirements for resistance to progressive collapse is illustrated in Fig. 6. The slab is envisioned as being hung from the columns by bottom longitudinal reinforcement continuous over the columns. Following the derivation of Hawkins and Mitchell [8], bottom reinforcement is assumed to develop dowel action with a shear yield force equal to half the tensile yield force. Thus, the required reinforcement is given by Eq. 1.

$$A_{sb} = \frac{w_u l_1 l_2}{f_y} \qquad (1)$$

in which A_{sb} = area of effectively continuous bottom reinforcement placed over the column in each principal direction, w_u = the ultimate uniformly distributed load, l_1 and l_2 are the span dimensions in the two orthorgonal directions, f_y = steel yield stress, and is 0.9. Alternately, Eq. 1 can be derived assuming that after punching, deformations in the connection region will result in bottom bars inclined at 30 degrees to the horizontal. Equation 1 is supported by the available experimental data [8,13].

The reinforcement required by Eq. 1 is seldom more than two or three of the main bottom slab bars provided for flexure. The bars should be placed continuously over the column with anchorage to develop the bar yield stress. This anchorage requirement is different from the minimum practice specified in the ACI Building Code [2] for nonseismic zones. However, as it is common practice to place bottom bars in a continuous mat, the requirements for continuity of a few bars over the column will not significantly influence construction economy.

Recommendations relative to postpunching resistance should be viewed with discretion. In regions of moderate seismic risk, where demands for inelastic connection rotations are limited, the aforementioned recommendations are likely to be adequate because general punching at large numbers of connections is not anticipated. In regions of high seismic risk, where demands for inelastic connection rotations are likely to be more significant, substantial loss of lateral-load resisting capacity will occur if numerous connections punch through. In such regions, alternate lateral-load resisting elements such as beam-column frames or structural walls must be used with the flat-slab system.

EXPERIMENTAL PROGRAM

To investigate the seismic performance of a flat-plate frame designed considering the effects discussed in the preceding section, an experimental research program was undertaken. The program involved design, construction, dynamic testing, and analysis of a flat-plate frame. Observed performance is used to verify the design procedure and to obtain basic data on dynamic properties. The experimental program and results are described briefly in the following sections. Further details are given elsewhere [14,15].

Test Structure

The test structure (Fig. 7) was a three-tenths scale model of a two-story, prototype structure. The prototype has three bays in one direction, and multiple bays in the transverse direction. The test structure represents the full three bays in one direction, but only a portion of the prototype in the transverse direction. As required in regions of high seismic risk [4], a spandral beam spans the perimeter of the prototype. Story heights, bay widths, and member cross sections are typical of values found in practice.

Design gravity loads comprise self weight and 2.87 kPa (60 psf) service live load, resulting in an ultimate gravity shear stress on interior critical sections equal to 2.3 f'c. The prototype is located in a region classified as seismic Zone 2 by the UBC [4], which might be expected to experience a design earthquake having Intensity VII on the Modified Mercalli Intensity Scale. Wind loads are not considered. Design concrete strength is 27.6 MPa (4000 psi). All reinforcement is Grade 60 (minimum yield stress of 414 MPa [60 ksi]). Actual mean concrete strength for the test structure was 36.5 MPa (5300 psi), and mean reinforcement yield stresses were 430 MPa (63 ksi) for the slab and 480 MPa (70 ksi) for the spandrels and columns.

The design follows requirements of the ACI Building Code [2] for gravity loads and the static lateral force method of the UBC [4] for seismic loads. Structural details satisfy requirements of the ACI Building Code, including provisions in Appendix A of that code for structures located in regions of moderate seismic risk. Typical slab details for one quadrant of the test structure are indicated in Fig. 8. The banding of slab reinforcement along column lines and continuity of top and bottom slab bars are noted.

For the experiments, the test structure was fixed to the shaking table at the Richmond Field Station of the University of California at Berkeley. Lead pigs were fastened to the top surface of slabs to simulate slab service dead load effects expected for the prototype. Live load was not simulated.

Test Description

Tests included low-amplitude free vibration tests and earthquake simulations of varying intensity. Earthquake simulations had either a single horizontal component parallel to the three-bay direction (Fig. 7) or the horizontal and corresponding vertical component. Acceleration records modeled the NS and vertical records obtained in El Centro (1940), with duration compressed by the square root of the model scale. Peak base accelerations (Table 1) were varied to obtain base motions ranging from low to high intensity. Ratios between Housner spectrum intensities [16] for test motions and for the scaled El Centro motion are presented in Table 1 [15]. It is noted that tests EQ6 and EQ7 have intensities equal to 44% of the scaled prototype motion, and may thus be gaged to be reasonable design motions for a structure in UBC Zone 2 for which the structure was designed.

TEST RESULTS

Typical Response to Earthquake Simulations

Eleven earthquake simulations were conducted, generally with increasing intensity from one test to the other. Response quantities are tabulated in Table 1. Based on observations of cracking, vibration periods, and equivalent viscous damping obtained from free-vibration decay, it was apparent that virtually no damage occurred during the first five tests. These measures indicated increasing damage during the subsequent six tests.

The sixth and seventh earthquake simulations had intensities roughly equivalent of the expected design motion. The seventh test differed from Test 6 in that Test 7 included a vertical component. Typical measured responses for Test 6 are plotted in Fig. 9. Maximum displacement corresponds to 0.003H, where H = structure height measured from the top of the footings. The peak base shear is 0.30W, where W = weight of the structure. The occurence of limited inelastic behavior is confirmed by changes in vibration periods and damping (Table 1), and the appearance of some limited slab, column, and edge beam cracking (Fig. 10a). No concrete spalling or yield of reinforcement were observed.

The most intense test (Test 11) had both horizontal and vertical input base accelerations. Peak horizontal base acceleration was 0.827 g, with corresponding spectrum intensity equal to 2.7 times that of the scaled prototype El Centro motion. Maximum top-level displacement was 0.052H, and peak base shear was 0.85W (Table 1). Extensive cracking and spalling were apparent in slabs (Fig. 10b), columns, and edge beams. Wide inclined cracks in the edge beams indicated likelihood of yield in torsion. Damage patterns were indicative of incipient punching, but no punching or collapse occurred. Extensive reinforcement yield was observed in columns and slabs.

Load-Displacement Envelope

The envelope relation between base shear and top displacement (obtained from measured responses to all tests) is plotted in Fig. 11. The relationship depicts a relatively linear portion to drifts of approximately 0.005H, followed by a gradual reduction in structure stiffness with increasing drift. Significant yield in the overall load-displacement relation occurs when top drift reaches approximately 0.015H.

The observation that significant yield in the overall load-displacement relation does not occur until lateral drifts reach 0.015H has important implications relative to seismic design. Specifically, many designers consider drift of 0.015H to be near the upper bound of acceptable lateral drifts during response to earthquakes. Thus, if drift is to be controlled to this limit, response will be essentially in the elastic range for the test structure. Consequently, design forces should be determined for elastic response, as opposed to the commonly used code design forces [4] that have been reduced on the implicit assumption that inelastic response will occur.

The large lateral drift capacity of the test structure (beyond 0.05H) is also noteworthy. This drift is well beyond the expected value for a well-designed structure. Thus, it is reasonable to conclude that suitable seismic details can be provided that will enable slab-column frames to survive the lateral deformations imposed by earthquakes. However, it is noted that for high-rise buildings, or for any frame in regions if high seismic risk, the slab-column frame by itself may not be suitably stiff to control lateral drift. More stiff lateral-load resisting elements (such as conventional frames and shear walls) may be required to keep drifts within reasonable limits.

COMPARISON BETWEEN MEASURED AND COMPUTED RESPONSE ENVELOPES

Lateral-Load Stiffness

As discussed previously in this paper, lateral-load stiffness of slab-column frames is sensitive to the cracking that results from gravity and lateral loads. For example, prior to being tested by earthquake simulations, measured stiffness was only 70 % of the calculated elastic stiffness [14]. The effect is further apparent in measured vibration periods, which increase as testing progresses (Table 1), even well before the structure has reached its lateral-load capacity.

Vanderbilt and Corley have proposed [1] that lateral-load stiffness at service loads can be estimated conservatively by reducing slab effective flexural inertias to one-third of the gross-section inertias. Stiffness computed using the equivalent frame (with the recommended stiffness reduction) is compared with the measured envelope relation between base shear and lateral drift in Fig. 11. The computed and measured stiffnesses compare reasonably well for lateral drifts in the range between 0.5 and 1.0 % of structure height. Similar results are obtained using the effective beam width model. More refined procedures of representing lateral stiffness as a function of lateral drift have been described elsewhere [14], although the practicability of such procedures for typical design office practice is questionable.

Lateral-Load Capacity

To investigate strength of the structure, two basic failure modes are investigated, one in which slab yield lines form across the transverse width of the slab at critical locations, and one in which local connection strengths limit capacity. For the test structure, the latter was found to control. For the calculations [14,15], interior connection strengths were computed using the ACI design procedure [2], with unbalanced moment limited by eccentric shear stresses. Exterior connection strengths were limited by the failure mode indicated in Fig. 5. In addition, connection strengths could were limited by flexural strengths of columns. Mean measured material properties were used for all computations. Lateral loads were assumed to be uniformly distributed over the height, as this was a typical distribution measured at times of maximum base shear [14]. The computed failure load is indicated in Fig. 11, and the mechanism is in Fig. 12. Computed strength using more refined procedures [14] is shown in Fig. 11 also. Both the failure mechanism and capacity compare well with measured quantities, suggesting that the available procedures are reasonably accurate. Sources of discrepancy have been described elsewhere [14].

SUMMARY AND CONCLUSIONS

Several important issues in design and design-oriented analysis of slab-column frames subjected to seismic loads are discussed. Included in the discussion are lateral-load stiffness, lateral-load strength, ductility, and measures to reduce the likelihood of progressive collapse. Findings from numerous previous studies are reviewed, and special implications for seismic design of slab-column frames are emphasized.

An experimental study was devised, in part, to verify these issues. The experiments included earthquake simulation tests of a two-story, multi-bay, flat-plate frame. The experiments and some of the data are described. Lateral-load stiffnesses, strengths, and ductility are analyzed and discussed. It is concluded that lateral-load stiffness is infuenced significantly by cracking at service loads. In addition, lateral-load strength can be estimated reliably using existing analysis methods. It is noted that, given proper detailing, adequate deformation capacity can be assured for flat-plate frames, but that more stiff lateral resisting elements may be needed to control lateral drift to reasonable levels.

ACKNOWLEDGEMENTS

The study was funded by the National Science Foundation under grant CEE8110050. J. Diebold is thanked for conducting the experiments.

REFERENCES

[1] Vanderbilt, M. D., and W. G. Corley, "Frame Analysis of Concrete Buildings," Concrete International: Design and Construction, Vol. 5, No. 12, December 1983, pp. 33-43.

[2] "Building Code Requirements for Reinforced Concrete (ACI 318-83)," American Concrete Institute, Detroit, Michigan, 1983, 111 pp.

[3] Pecknold, D. A., "Slab Effective Width for Equivalent Frame Analysis," ACI Journal, Vol. 72, No. 4, April 1975, pp. 135-137.

[4] Uniform Building Code, International Conference of Building Officials, Whittier, California, 1982.

[5] Park, R. and W. L. Gamble, Reinforced Concrete Slabs, John Wiley and Sons, New York, 1980, 618 pp.

[6] "The Shear Strength of Reinforced Concrete Members - Slabs," reported by ASCE-ACI Committee 426, Journal of the Structural Division, ASCE, Vol. 100, No. ST8, August 1974, pp. 1543-1591.

[7] Newmark, N. M. and E. Rosenblueth, Fundamantals of Earthquake Engineering, Prentice-Hall, Inc., Englewood Cliffs, New Jersey, 1971.

[8] Hawkins, N. M., and D. Mitchell, "Progressive Collapse of Flat-Plate Structures," ACI Journal, Vol. 76, No. 7, July 1979, pp. 775-808.

[9] Hawkins, N. M., "Seismic Response Constraints for Slab Systems," Proceedings, Workshop on Earthquake-Resistant Reinforced Concrete Building Construction, University of California, Berkeley, July 11-15, 1977, pp. 1253-1275.

[10] Park, R., and T. Paulay, Reinforced Concrete Structures, John Wiley and Sons, New York, 1975, 769 pp.

[11] Jirsa, J. O., J. L. Baumgartner, and N. C. Mogbo, "Torsional Strength and Behavior of Spandrel Beams," ACI Journal, Vol. 66, No. 11, November 1969, pp. 926-932.

[12] Rangan, B. V., and Hall, A. S., "Moment and Shear Transfer Between Slab and Edge Columns," ACI Journal, Vol. 80, No. 3, May-June 1983, pp. 183-191.

[13] Mitchell, D., and W. D. Cook, "Preventing Progressive Collapse of Slab Structures," Journal of Structural Engineering, ASCE, Vol. 110, No. 7, July 1984, pp. 1513-1532.

[14] Moehle, J. P., and J. W. Diebold, "Experimental Study of the Seismic Behavior or a Two-Story Flat-Plate Structure," Report No. UCB/EERC-84/08, Earthquake Engineering Research Center, University of California, Berkeley, California, August 1984.

[15] Moehle, J. P., and J. W. Diebold, "Lateral Load Response of Flat-Plate Frame," Journal of Structural Engineering, ASCE, Vol. 111, No. 10, October, 1985, pp. 2149-2164.

[16] Housner, G. W., "Behavior of Structures During Earthquakes," Journal of the Engineering Mechanics Division, ASCE, Vol. 85, No. EM4, October 1959, pp. 108-129.

TABLE 1.—Test and Response Summary

Test number (1)	Peak base acceleration[a] (g) (2)	Spectrum intensity ratio[b] (3)	Maximum top displacement[c] (mm) (4)	Maximum top acceleration (g) (5)	Maximum base shear (kN) (6)	Maximum response period[d] (sec) (7)	Free-vibration period[e] (sec) (8)	Damping percentage[f] (9)
EQ1	0.015	0.03	—	0.025	4.1	—	0.21	1.5
EQ2	0.012 (0.005)	0.03	—	0.024	4.1	—	0.22	1.1
EQ3	0.047	0.10	—	0.090	16.1	—	0.21	1.3
EQ4	0.048	0.10	1.1	0.090	15.5	0.23	0.21	1.5
EQ5	0.092	0.21	2.1	0.16	30.6	0.24	0.21	1.4
EQ6	0.189	0.44	5.1	0.35	61.9	0.26	0.23	2.5
EQ7	0.202 (0.042)	0.44	9.9	0.49	93.3	0.28	0.24	2.3
EQ8	0.284	0.89	20.4	0.73	137.0	0.34	0.27	2.7
EQ9	0.252 (0.106)	0.89	29.7	0.83	148.0	0.39	0.31	4.9
EQ10	0.606	1.96	61.9	1.04	165.0	0.54	0.45	4.9
EQ11	0.827 (0.197)	2.71	95.5	1.08	175.0	0.68	0.58	7.1

[a]Values given are peak horizontal base accelerations, except peak vertical accelerations are in parentheses for tests with vertical input.
[b]Ratio between 20% damped spectrum intensity of horizontal base motion and scales intensity of El Centro, NS, 1940.
[c]Values too small to read during tests EQ1, EQ2, and EQ3.
[d]Time between three successive zero crossings of top displacement during peak response cycle.
[e]First-mode value following earthquake simulation.
[f]Percent of critical damping ratio in first mode obtained by logarithmic decrement of free-vibration response.
Note: 1 mm = 0.0394 in.; 1 kN = 225 lb.

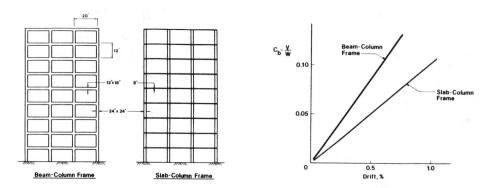

Fig. 1 Comparison of Lateral-Load Stiffness for Typical Frames.

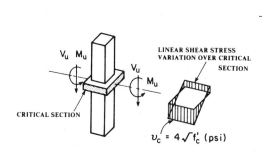

Fig. 2 Model For Interior Connection Strength.

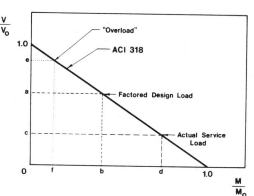

Fig. 3 Effect of Gravity Loads (V) on Lateral-Load Strength (M).

1,514

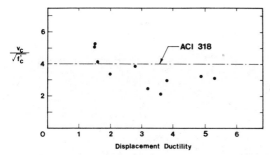

Fig. 4 Effect of Connection Shear on Connection Ductility.

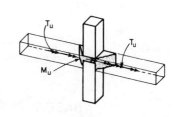

Fig. 5 Unbalanced Moment Strength of Edge Connection.

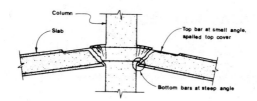

Fig. 6 Model for Resistance to Progressive Collapse.

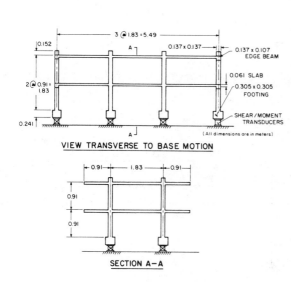

Fig. 7 Test Structure.

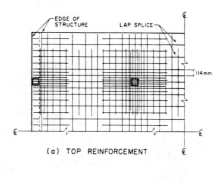

(a) TOP REINFORCEMENT

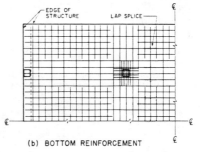

(b) BOTTOM REINFORCEMENT

Fig. 8 Slab Reinforcement.

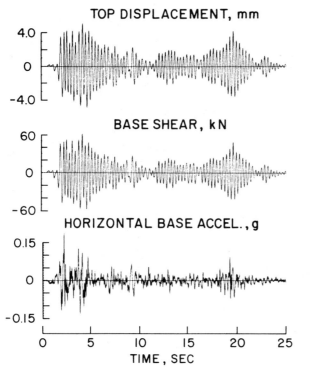

Fig. 9 Typical Responses to Test EQ6.

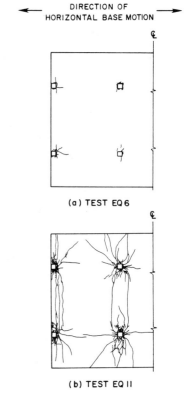

Fig. 10 Top Slab Cracks.

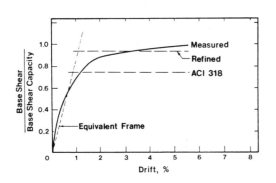

Fig. 11 Response Envelope.

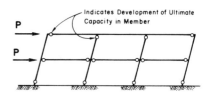

Fig. 12 Collapse Mode.

EARTHQUAKE RESPONSE OF REINFORCED CONCRETE
FRAMES WITH YIELDING COLUMNS

Arturo E. Schultz[I]

The earthquake response of r/c frame structures with columns flexurally weaker than beams (SBWC) is investigated in a combined experimental and analytical study. Two small-scale models, comprising parallel nine-story, three-bay frames, were constructed and tested on an earthquake simulator. Accelerations and displacements were measured at each level. A lumped parameter MDOF model is presented and used to interpret observed test specimen response. The model is subsequently used in a parametric study of earthquake response of SBWC frames. Variables in the study are: ground motion, frame profile, and frame strength.

INTRODUCTION

The professional community, through building codes, prohibits the use of "strong beam"-"weak column" (SBWC) r/c frames in zones of high seismic risk [1,2]. Fear of collapse due to elastoplastic story mechanisms (when all columns at a story yield), column strength decay under load reversals, joint failures, and column shear failures have motivated this code requirement. However, r/c structures exhibit post-yield stiffness rendering an elastoplastic mechanism unrealistic, and earthquake response is better understood in terms of effective stiffness (during maximum response) rather than static elastoplastic response. It may be possible to limit drift to acceptable levels and minimize strength decay. Joint failures and column shear failures are not exclusive of SBWC frames.

Architectural trends requiring open areas, which translate to long spans and for which gravity design dictates strong beams, have made SBWC frames desirable and commonly used. However, testing of SBWC structures or components has been very limited. Yet, testing is necessary to determine the feasibility of this framing system for future construction in seismic areas as well as to evaluate the safety of existing SBWC frame structures. It should be noted that while this framing configuration is prohibited in areas of high seismic risk, it is permitted and widely used in regions of moderate seismicity, including areas in the east coast and the central U.S.

This paper is a summary of an investigation carried out at the University of Illinois [3]. The purpose was to study the earthquake response of SBWC structures in a combined experimental and analytical study. Specific objectives were: (1) to design, construct, and test two small-scale specimens of SBWC frame structures, (2) to evaluate and interpret the measured response of

---

I. Assistant Professor, Department of Civil and Mechanical Engineering, Southern Methodist University, Dallas, Texas.

the test specimens, and (3) to determine the combinations of frame mass, strength, and stiffness and ground motion frequency content and intensity for which displacement response is satisfactory.

EXPERIMENT

Test Specimens

The test structures, which were models of a structural concept, were planar in configuration and comprised two parallel frames which were coupled by rigid diaphragms at each story (Fig. 1). Each frame had three bays and nine stories, where the first story was 40% taller than the others. Dimensions were selected as typical for SBWC frames assuming a length scale of 1:15 (Fig. 1). Added mass was provided by the story diaphragms at a rate of 1140 lbs (5.07 kN) per story for a total structure weight of 10 1/4 kips (45.6 kN). Fixed base conditions were simulated by rigid base girders.

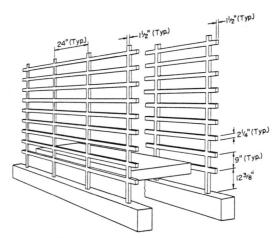

Fig. 1 Test Specimen

The specimens were designed to resist a simulated version of the NS component of El Centro 1940 along their major horizontal direction. The time scale was compressed by a factor of 2.5 to shift the frequency content of the ground motion to the frequency range of the test specimens. Base accelerations were scaled to a peak value of 0.35 g. Design forces were obtained assuming a reduced-stiffness linear model of the structures [4]. Gross-section member stiffnesses were reduced to simulate nonlinear response, and an increased equivalent damping factor was used to account for hysteretic energy dissipation with nonlinear deformation. The reduced-stiffness linear model was used in a modal response spectrum analysis to determine minimum force requirements.

The test structures were constructed of microconcrete and model reinforcement. The microconcrete, which was mixed in the laboratory from sand, gravel, and Type III Portland cement, had a compression strength of 5400 psi (37 N/mm^2). Longitudinal reinforcement for the beams was No. 7 gage wire and No. 13 gage wire was used for the columns, both which had a yield stress of 56 ksi (386 N/mm^2). Strengths were 63 and 61 ksi (435 and 420 N/mm^2) for beam and column steel, respectively.

The major variable between the test specimens was the amount of column longitudinal reinforcement. Specimen SS1 was reinforced to satisfy the minimum requirements of the design analysis, while those of SS2 were modified after consideration of the observed behavior of specimen SS1 during its design earthquake. First-story columns of specimen SS1 had a total reinforcement ratio of 2.9% while those of SS2 had 4.7% steel. The lateral strength of the columns was reduced over structure height to meet design requirements. Beams were provided enough longitudinal reinforcement so that the sum of the flexural strengths of the beams at a joint exceeded the sum of column flexural strengths. Sufficient transverse reinforcement was provided to prevent shear failures of the members. To minimize joint distress spiral reinforcement was provided at each joint and longitudinal reinforcement for the members was run continuous through the joints.

Earthquake Simulations

The structures were tested on the University of Illinois Earthquake Simulator. Response measurements during the simulations included displacements and accelerations at each level, which were measured with displacement transducers (LVDT) and accelerometers, respectively. Each structure was subjected to a series of simulated earthquakes (4 for SS1 and 6 for SS2). Low-ampltude free-vibration tests to determine dynamic properties preceded all simulations. Both specimens were subjected to a "design" earthquake, repetitions of the design motion, and motions at increased intensities. Peak base accelerations for the last run of each specimen were approximately 4 times as large as that of the "design" earthquake. Test structure SS1 collapsed after 3 sec. of the last run. This paper summarizes response during the "design" earthquake, but behavior over the entire range of simulations is discussed briefly.

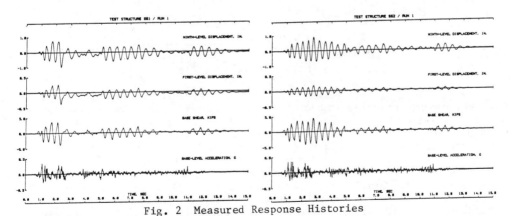

Fig. 2 Measured Response Histories

Both structures were subjected to initial simulations with peak accelerations equal to the target value of 0.35 g (Fig. 2). Specimens SS1 and SS2 responded to this motion with maximum top accelerations of 0.9 and 1.1 g, respectively. The corresponding acceleration amplification factors were 2.5 and 3.2. Base shear force histories were obtained as the sum over structure height of story inertia forces, which were computed as the product of lumped story mass and measured acceleration. Maximum values of base shear during the initial runs were 3.9 and 4.4 kips (17.3 and 19.6 kN), for specimens SS1 and SS2, respectively (Fig. 2). The corresponding base shear coefficients, assuming a total structure weight of 10 1/4 kips (45.6 kN), were 0.38 for the first specimen and 0.43 for SS2.

The structures resisted the inertia loads with maximum top displacements of 1.0 and 0.9 in. (25 and 23 mm) for specimens SS1 and SS2, respectively (Fig. 2). These displacements correspond to 1.2% and 1.1% of total frame height. In terms of overall drift, response to the ground motion was the same for both specimens. First-story displacement histories measured during the initial runs (Fig. 2) are quite similar to corresponding top displacements, indicating a dominant first mode. Maximum first-story displacement was 0.38 in. (9.7 mm) for specimen SS1, which represents 40% of maximum top displacement, and 0.18 in. (4.6 mm) for SS2, which is only 20% of maximum top displacement. Maximum first-story drift ratios were 3.1% and 1.4% for SS1 and SS2, respectively. While maximum top displacements were roughly the same for the test specimens, distribution of deformation over structure height varied considerably between the specimens.

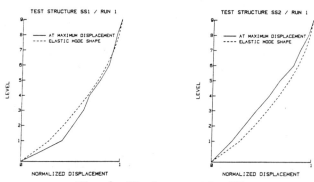

Fig. 3 Displacement Profiles

Distributions of deformation for the test structures are shown in Fig. 3. Displacements measured at each level at the time of maximum displacement were normalized by top displacement. These profiles are compared with the first mode shape for linear vibration computed for a gross-section model of the structures. During the first two low-amplitude cycles of response (Fig. 2) the profiles of both structures were very similar to the elastic mode shape. However, under large-amplitude vibrations the first-story component of the displacement profile for SS1 increased from 30% to 40%, while that for SS2 decreased from 30% to 20%. Earthquake simulation tests of r/c structures with strong vertical elements have indicated approximately constant displacement profiles regardless of the response amplitude [5].

Behavior of specimen SS1 during the "design" earthquake can be attributed to yielding of the first-story columns. First-story force-displacement response for the test structures during the first 4 sec. of the initial runs are shown in Fig. 4. The first story of specimen SS1 yielded in both directions at a displacement of 0.17 in. (4.3 mm). It was loaded to a maximum nonlinear deformation of 0.38 in. (9.7 mm), and displayed stable hysteresis loops in the nonlinear range. Yielding of the story is defined as a marked deviation from a linear force-displacement relation. The first story of specimen SS2 did not yield, thus, the observed change in displacement profile (Fig. 3) is attributed to increases in upper story components due to mechanisms other than yielding (cracking, bar slip, etc.). The hysteresis loops of specimen SS2 were of finite width (Fig. 4), indicating energy dissipation by hysteresis in the linear range.

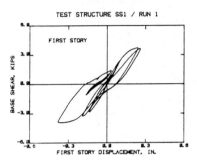 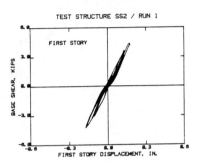

Fig. 4 Force-Displacement Hysteresis Relations

First-mode frequencies for the test specimens before the earthquake simulations, determined from free-vibration tests, were 3.9 and 4.5 Hz for specimens SS1 and SS2, respectively. First-mode frequency calculated for a gross-section linear model was 4.9 Hz. Hairline cracks produced by handling and shrinkage stresses are likely causes for this softening. During the cycle of maximum response, the frequencies associated with the effective period were 2.3 and 2.7 Hz for SS1 and SS2, respectively. During the last three sec. of the initial simulations the effective period decreased to 1.9 Hz for SS1 and 2 Hz for SS2. This phenomenon can be understood in relation to the observed force-displacement response of the test structures during low-amplitude cycles. Stiffness reductions upon unloading, at small values of force, and restiffening, as unloading proceeds to loading in the opposite direction, can be observed in the hysteresis relations (Fig. 4). This effect, also known as "pinching", results in reduced hysteresis loops and is attributed to opening and closing of cracks and reinforcing bar slip. It occurred for the first story of SS1 as well as for the first story of SS2, which did not yield. The reduced stiffness near the origin, which is roughly one-half of the cycle effective stiffness (Fig. 4), explains the low values of first-mode frequency at the end of response (Fig. 2).

The columns of the test structures were able to maintain their load-carrying capacity for a large range of deformation. Envelopes of first-story current maximum displacement, attained over the entire range of simulations, are shown in Fig. 5. Loading in both directions is lumped into a single locus. The first story of specimen SS1 yielded during the initial run. Subsequent loading produced a maximum deformation of 0.86 in. (22 mm), corresponding to a 7% drift ratio, during one of the last stable cycles before collapse (4th run). The first story of SS2 yielded during the second run at a displacement of 0.20 in. (5.1 mm), and was loaded to a maximum displacement of 0.56 in. (14 mm). The first-story response envelopes (Fig. 5) indicate positive post-yield stiffnesses for displacement ductilities of 2.5 for specimen SS1 and 3.0 for SS2. There is no significant decrease in the capacity of the first-story columns of specimen SS1 for ductilities as large as 4.5. It should be noted that column steel strength was only 9% larger than the yield stress, whereas the strength of typical deformed bars is usually more than 25% larger than the yield stress.

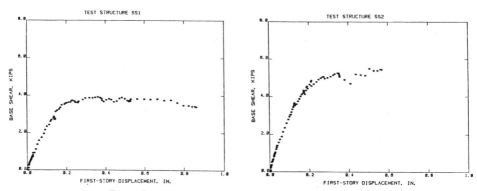

Fig. 5 Story Force-Displacement Envelopes

ANALYSIS

Model

The overall behavior of the test structures during the "design" earthquake was satisfactory. Neither structure collapsed, nor was strength deterioration, shear distress, or joint distress observed. In order to broaden the scope of the investigation, the displacement response of similar frames to various earthquake acceleration records was considered. To carry out this parametric study an analytical model of SBWC frames was required. The model had to represent observed nonlinearities of reinforced concrete, including stiffness reductions at low forces and energy dissipation before and after yielding. As a minimum each story had to be represented individually to reproduce observed changes in displacement profiles.

A lumped parameter model was selected (Fig. 6), for which a horizontal degree-of-freedom (DOF) was defined at each level. All mass between stories was lumped at each DOF, and all columns at a story were replaced by a single nonlinear spring. The response of the spring was governed by a set of hysteresis rules which controlled loading, unloading and reloading from a primary force-displacement curve. Because the beams had finite stiffness the story springs were modified to approximate beam flexibility. The Takeda hysteresis model and two modifications of it were used to represent story behavior. The best results were obtained by using a modification of the Takeda model which included stiffness reductions at low forces and energy dissipation before and after yield. Computed and measured displacements for the test structures during the initial simulation are shown in Fig. 7. Changes in effective vibration period during strong motion were reproduced well, and maximum displacements were computed within 15% of the measured vales for specimen SS1 and 10% for SS2.

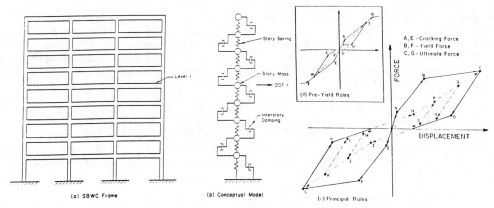

Fig. 6 Analytical Model

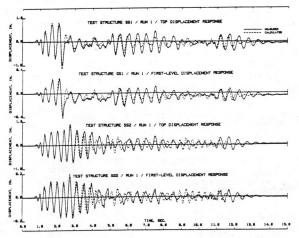

Fig. 7 Computed Response

Parametric Study

Two earthquake acceleration records, the Castaic S48E component of the 1971 San Fernando earthquake and the Santa Barbara N21E component of the 1952 Kern County earthquake, were used in addition to El Centro 1940 NS. The records were chosen for their frequency content, where Castaic had the highest and Santa Barbara the lowest. Ground period, T_g, was used as a measure of frequency content. It was defined by Shimazaki [6] as that period above which an acceleration record provides no consistent increase in intake energy for linear systems. Ground periods were computed as 0.35, 0.55, and 0.95 sec. for Castaic, El Centro, and Santa Barbara, respectively. Time and acceleration scale factors were not applied to the records, with the exception of Santa Barbara, for which accelerations were doubled so that maximum response acceleration was roughly equal for all records.

The frames analyzed had similar profiles to the test structures, but were scaled to "full-size" dimensions using a length scale of 1:15. Three-bay frames with nine, six, and three stories were included in the analyses, in addition to a SDOF oscillator for which properties were obtained assuming it to have one story and one bay. Initial member stiffness was obtained assuming uncracked sections, and stiffness at first yield was taken as one-fourth of that value. However, a single stiffness, equal to one-half the initial stiffness, was selected to characterize the frames. The multi-story frames had equal story masses of 275 kips (1220 kN), and the SDOF oscillators had 91.7 kips of weight. Characteristic periods for the frames, first-mode periods for linear models based on characteristic stiffness, were 1.4, 1.0, 0.6, and 0.24 sec. for the nine-, six-, and three-story frames and the SDOF oscillator, respectively.

For each combination of frame profile and ground motion a period ratio, TR, was defined as the ratio of characteristic period to ground period, and a modal response spectrum analysis was performed on linear models based on characteristic stiffnesses and a 2% damping factor. For each period ratio a series of frames with different strengths was considered. These strengths varied from a minimum base shear coefficient, C_b, of 0.08 to a maximum strength ratio, SR, of 100% (Table 1). Strength ratio was defined as the ratio of base shear at yield to base shear computed from modal response spectrum analysis. The minimum base shear coefficient of 0.08 was determined as a reasonable lower bound of lateral strength of yielding columns in terms of minimum code-required axial load capacity for columns [2].

Table 1 Variables in Parametric Study

Number of Stories	Record	TR	SR (%) Min.	Max.	C_b Min.	Max.
One	Castaic	0.69	6.8	100	0.08	1.20
	El Centro	0.44	6.3	100	0.08	1.30
	Santa Barbara	0.25	12	100	0.08	0.71
Three	Castaic	1.70	12	100	0.08	0.67
	El Centro	1.10	6.7	100	0.08	1.20
	Santa Barbara	0.63	8.7	100	0.08	0.90
Six	Castaic	2.90	27	100	0.08	0.29
	El Centro	1.80	12	100	0.08	0.68
	Santa Barbara	1.10	9.6	100	0.08	0.82
Nine	Castaic	4.00	49	100	0.08	0.16
	El Centro	2.60	19	100	0.08	0.41
	Santa Barbara	1.50	14	100	0.08	0.57

Top displacement response for the frames analyzed was compared with displacements computed from modal response spectrum analysis to define displacement ratio, DR. Displacement ratio is plotted against strength ratio in Fig.

8 for each of the ground motions considered. It can be observed that displacement ratio does not exceed unity if the sum of strength and period ratios is larger than or equal to 1. This observation has been made previously of nonlinear SDOF oscillator response to earthquake acceleration records [6]. The implication of this observation is that for certain combinations of frame mass, strength, and stiffness and ground motion frequency content, linear analysis is sufficient to bound maximum drifts. This range includes mid-rise r/c frames (5 to 20 stories) subjected to motions with high- and intermediate-frequency content (T_g less than 0.5 sec.).

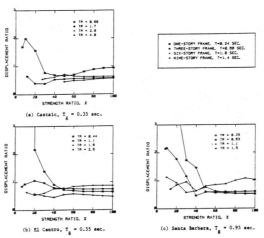

Fig. 8 Displacement Ratios

Overall drift ratio, the ratio of top displacement maxima to total frame height, was used to investigate frame distortion. Maximum tolerable drift during a "design" earthquake was selected as 1% of total height, based on consideration of a study comparing professional opinion and results of racking tests of r/c frames [7]. Castaic produced the smallest response, while drift ratios for frames subjected to Santa Barbara were the largest (Fig. 9). Drift ratio did not exceed the limit of 1% for frames with strength ratios not smaller than 30%, except for the six-story frames subjected to Santa Barbara. It should be remembered that for all frames with strength ratios of 30% or more, including the six-story frames, maximum drift could have been bounded by linear analysis (Fig. 8).

Because displacement profiles for SBWC frames can change dramatically with large-amplitude response (Fig. 3), the first-story components at maximum displacement were monitored to evaluate changes in the profiles. First-story components of the profile at maximum displacement were compared with normalized first-story components of the first vibration mode for linear models of the frames. This mode-shape ratio, MSR, is plotted in Fig. 10 against strength ratio for the frames and ground motions considered. The largest changes in displacement profile occurred for systems subjected to Castaic (up to 50%), while Santa Barbara had the least effect (MSR less than 30% for SR larger than 30%). Note that mode-shape ratio reflects higher-mode contributions, and systems with the largest period ratios exhibited the largest changes in displacement profiles. Thus, systems with the largest changes in profiles were also those with the smallest drift ratios (Fig. 9).

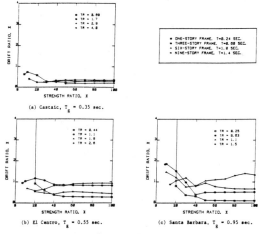

Fig. 9 Drift Ratios

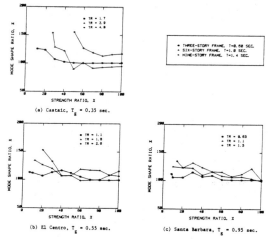

Fig. 10 Mode Shape Ratios

CONCLUSIONS

A combined experimental and analytical study of SBWC frame structures subjected to uniaxial earthquake loading was undertaken. From the experimental results it was concluded that well-detailed frames with yielding columns can be used safely in seismic zones. The overall behavior of the specimens was satisfactory during the "design" earthquake, and the columns were able to maintain their load-carrying capacity over a large range of deformation. In the course of the analytical effort, nonlinear analysis techniques were shown to provide satisfactory estimates of displacement response, and a reduced-stiffness design-analysis method was shown to provide adequate estimates of minimum strength requirements. The parametric study enabled three important observations of the response of SBWC frames to be identified. (1) For typical mid-rise r/c SBWC frames subjected to ground motions with high- and intermediate-frequency content, linear analysis is sufficient to bound maximum drift. (2) For the frames and ground motions considered in this study, a minimum strength ratio of 30% was required for a well-detailed SBWC frame to undergo inelastic deformations due to base motions and remain below acceptable drifts (1% drift ratio). (3) Large changes in the displacement profiles were observed, but for cases where drift was not critical.

REFERENCES

[1] Uniform Building Code - 1985 Edition, International Conference of Building Officials, Whittier, California.

[2] ACI 318-83, Building Code Requirements for Reinforced Concrete, American Concrete Institute, Detroit, November 1983.

[3] Schultz, A. E., "An Experimental and Analytical Study of the Earthquake Response of R/C Frames with Yielding Columns," Ph.D. Dissertation, University of Illinois, Urbana, May 1986.

[4] Shibata, A. and M. A. Sozen, "Substitute-Structure Method for Seismic Design in R/C," Journal of the Structural Division, ASCE, Vol. 102, 1-18, January 1976.

[5] Moehle, J. P. and M. A. Sozen, "Experiments to Study the Earthquake Response of R/C Structures with Stiffness Interruptions," Civil Engineering Studies, Structural Research Series No. 482, University of Illinois, Urbana, August 1980.

[6] Shimazaki, K. S. and M. A. Sozen, "Seismic Drift of Reinforced Concrete Structures," Special Research Paper, Civil Engineering Department, University of Illinois, Urbana, 1985.

[7] Algan, B. B., "Drift and Damage Considerations in Earthquake-Resistant Design of Reinforced Concrete Buildings," Ph.D. Dissertation, University of Illinois, Urbana, March 1982.

EFFECT OF CYCLIC LOADING RATE ON RESPONSE OF MODEL BEAM-COLUMN
JOINTS AND ANCHORAGE BOND

S. P. Shah[I] and L. Chung[II]

ABSTRACT

Small-scale models of reinforced concrete beam-column joints and anchorage-bond specimens were subjected to large cyclic displacements at two rates. To assess damage, free vibration tests were conducted. The reliability of the modeling techniques was established by comparison of the results for the slower rate with those obtained from the full-scale tests on prototype. The higher rate of loading caused a greater damage than that at the slower rate. This was evidenced by the measurements of the stiffness obtained from the free-vibration test. The relatively greater extent of damage appears to result from the different bond behavior at different rates of loading.

INTRODUCTION

The current practice for design of reinforced concrete structures subjected to earthquake loading is based primarily on the experimental results of structural elements and assemblages subjected to cyclic loading at the quasi-static loading rates [1, 2]. These rates are substantially lower than those corresponding to the frequencies of seismic excitation. A few studies which have been conducted to ascertain the influence of dynamic rates of loading present somewhat conflicting evidence [3, 4, 5, 6].

In this study, small-scale model reinforced concrete beam-column joints and anchorage-bond specimens were subjected to cyclic loading at two different rates. The reliability of the modeling techniques was established by comparing the results of the test at the slower rate with those of the full-scale tests conducted by previous investigators [7, 25, 32]. Several different parameters were examined to evaluate the differences due to the rate of loading.

BEAM-COLUMN JOINT SPECIMENS

Test Specimen

Three identical beam-column joints (Fig. 1) were fabricated. The first

---

I. Professor, Department of Civil Engineering, Northwestern University, Evanston, Illinois.

II. Graduate Assistant, Department of Civil Engineering, Northwestern University, Evanston, Illinois.

specimen was for the preliminary test to determine the yield displacement and to check the instrumentation. The second specimen was subjected to cyclic loading at the frequency of 2.5×10^{-3} Hz (slow rate) while the third specimen was tested at the frequency of 1.0 Hz (fast rate). The details of column design, beam design and joint design are given in Table 1.

Specimens were horizontally cast in waterproofed plywood forms. Reinforcing steel cages were assembled according to the applicable specifications of the ACI Building Code [8]. Stirrups in the joint were fabricated as per ACI-ASCE Joint Committee 352 Recommendations [9]. They were tied by 20-gage annealed wires (D = 0.0320") to the longitudinal reinforcement. A dimensional tolerance of less than 5% was achieved in the construction of the specimens.

Concrete was compacted with an electric vibrator. Specimens and cylinders were tested after 28 days of curing in a 96% relative humidity and 80° F environment.

Material Properties

The properties of the three types of steel reinforcement used for beam (D2.5 bar) and column longitudinal reinforcement (#2 bar) and for shear reinforcement (D1 bar) are given in Table 2. The reinforcement was supplied by Portland Cement Association and had special surface deformations to simulate deformed bars used in practice. Several investigators have shown the adequacy of these specially knurled small diameter bars to simulate the bond behavior of the full size bars [10, 11].

Mix proportions of concrete are also given in Table 2. Three 3" x 6" cylinders were cast and cured simultaneously with each specimen and subjected to uniaxial compressive loading on the same day the beam-column subassembly was tested. Average modulus of elasticity (secant at $0.45 f'_c$) was 3.0×10^6 psi and the average compressive strength was 4000 psi.

Loading Procedure

A reaction frame was built around the specimen-column to transfer the loads to the MTS supporting steel columns (Fig. 2). The specimen-column was subjected to a 8.0 kips applied frame load during the testing. This load was approximately 50% of the balanced design load of the column.

After a predetermined loading history, the cyclic loading was interrupted and free vibration tests were conducted. A displacement corresponding to the 10% of the yield displacement was imposed by the loading ram. This displacement was allowed to dampen out by suddenly cutting the wire connecting the beam-end at the loading ram.

The loading schedule imposed at two different rates on the two specimens is shown in Fig. 3. The displacement ductility ratio is defined here as the ratio of the displacement of the beam at the load-point at any stage in testing to the corresponding displacement at the initial yield of the beam longitudinal steel under positive shear. The yield load and corresponding displacement as measured during the preliminary test were approximately 650 lbs and 0.2 in., respectively.

Prior to the application of the cyclic loading, specimens were subjected to one cycle of positive and negative loading to one-half the yield load. This prior loading allowed the members of the test team to check the response of the data gathering equipment and to obtain the initial stiffness of the specimen. The specimens were then subjected to large scale reversed loading in four stages (Figs. 3 and 4).

The instrumentation was designed to monitor the behavior of the specimens during the test by providing a continuous time record of applied load, displacements and acceleration. The load was measured by a LEBOW Model 3157 load-cell attached to the MTS actuator (Fig. 2) while the acceleration during free vibration test was measured by an Endevco Model 7265-A-HS accelerometer attached to the tip of the beam. To measure the average rotation of beam hinging region, two linear variable differential transformers (Schaevitz Model 500 MHR LVDT) were positioned on the top and bottom of the beam near the column. They measured displacements over a gage length equal to the depth of the beam. The analog signals from the various measuring devices were conditioned, amplified, and digitized using an A/D converter and then recorded in a computer based storage system (ISAAC 2000). The real time load vs actuator displacement was also recorded on an X-Y plotter.

Test Results

The load-deflection hysteresis curves as measured by the X-Y plotter are shown in Fig. 3 for the slow rate test and in Fig. 4 for the fast rate loading test. From these two sets of curves and corresponding cyclic envelope curves shown in Figs. 3 and 4 it can be seen that the maximum load carrying capacity of the specimen is somewhat higher at the higher rate of loading. However, the damage induced by the cyclic loading seems to be higher for the faster rate of loading. This is evidenced in the sharper drop after the peak of the envelope curve for the faster rate of loading. The relatively greater extent of damage caused by the fast rate of loading was also evidenced by the results of the free vibration tests shown in Fig. 5.

During the 4th loading stage, one of the top steels of the beam fractured during the second negative cycle for the slow rate test. For the specimen tested at the faster rate of loading, all the top steel fractured during the same loading cycle.

Developments of major cracking in the model were also observed. It can be seen in Fig. 6a for the slow rate test and in Fig. 6b for the fast rate test.

Comparison with a Prototype

In order to evaluate the validity of testing small-scale models, the results of the specimen tested at the slow rate were compared with specimen No. 6 tested by Scribner and Wight [7]. The dimensions of their specimen and the design of their column, beam and beam-column joint are compared with the model specimen in Table 1.

The cyclic envelope curve obtained from the results of the specimen tested by Scribner and Wight is compared with the one obtained from the present investigation in Fig. 7. For comparison purposes the model loads and deflections were normalized according to the similitude requirements.

Development of major cracking in the model was also observed to be similar to that for the prototype (Fig. 6c). The regions of the columns away from the joint showed no cracking or degradation. Both beams showed primarily flexural cracks as can be seen in Figs. 6a and 6c.

ANCHORAGE BOND TEST

By comparing crack patterns reproduced in Fig. 6a and 6b, it was observed that flexural cracks were widely distributed for the specimen tested at the slow rate. In contrast, for the fast rate, the damage was essentially due to a single wide crack observed at the face of the column. This indicates that the observed rate effect may be related to the transfer of forces between reinforcing bars and concrete. Numerous researchers have studied the nature of deterioration in bond under monotonic loading and repeated and reversed loading conditions [12-22]. Eligehausen, Popov and Bertero observed that during cyclic loading the extent of bond deterioration depends primarily on the maximum value of the peak slip reached during prior loading cycles [21]. Fillipou, Popov and Bertero studied the effects of bond deterioration on the hysteretic behavior of reinforced concrete joints [22]. They found that the hysteretic behavior of joints is very sensitive to the bond-slip response along the anchorage length and the reinforcing bars.

Test Specimen

In this investigation, two identical cantilever type anchorage-bond specimens (Fig. 8) were subjected to cyclic loads at two rates, 2.5×10^{-3} Hz (slow rate) and 0.5 Hz (fast rate). The geometry of the specimen was selected to simulate the connection of a beam to a column. The enlarged end section represents that portion of the connection common to a column and beam framing into a point in a structure. The length of the enlarged end or anchorage zone was chosen to assure that the strength of longitudinal reinforcement was developed by anchorage of the straight portion of the bars rather than hooks.

The properties of the two types of steel reinforcement used for beam longitudinal steel (#2 bar) and for shear steel (D1 bar) are given in Table 2. Mix proportions of concrete used are the same as for the beam-column joint specimens.

Loading Procedure

A structural framework was required around the MTS supporting steel columns to act as a reaction frame. The enlarged block of the specimen was subjected to a 1 ksi applied compression stress to simulate a column during the testing. The complete sideview of this test set-up is shown in Fig. 9.

The loading schedule imposed at two different rates (2.5×10^{-3} Hz and 0.5 Hz) on the two specimens is shown in Fig. 10. At the higher rate, the expected displacements were not reached due to the mechanical limit of the testing machine. The maximum amplitude was about 80% of the expected value at each stage.

In addition to the instruments used in beam-column joint specimen tests, two more LVDTs were positioned on the end of the reinforcements inside the enlarged end block. They measured the end slip of the longitudinal reinforcements.

Test Results

Load-deflection curves of two specimens are shown in Fig. 10 for the slow rate test and in Fig. 11 for the fast rate test. The ultimate load of fast rate test was higher (about 25%) than that of slow rate test. This result is similar to the beam-column joint specimen test result.

Although the expected displacements were not reached for the fast rate test, the failure of test specimen occurred after about the same number of cycles as that for the slow-rate test. This indicates that the specimen subjected to fast rate of loading can sustain a smaller magnitude of cyclic excursion than the specimen subjected to slow rate of loading.

Failure modes compared in Fig. 12 were also observed to be similar to that for the previous beam-column joint specimens. Flexural cracks were more concentrated and developed more rapidly for the fast rate test than for the slow rate test.

No end slips were observed at the fast rate test during testing. On the other hand, 0.0005"-0.0006" of end slips were observed at the last stage (cycle 22-27) for the slow rate test specimen. This indicates that the bond stresses were transferred to the end of the embedded longitudinal reinforcement more effectively for the slow rate test than for the fast rate.

Comparison with a Prototype

The results of the specimen tested at the slow rate were compared with specimen 66-32-RV5-60 tested by Brown, Ismail and Jirsa [18, 23]. The material properties of prototype are shown in Table 2. The dimensions of prototype are four times to that of model (L_r = 4.0). The section area of longitudinal reinforcement is approximately square of length scale factor ($A_r \approx L_r^2$ = 16.0).

The cyclic curve obtained form the results of the prototype specimen is compared with the one obtained from the model slow rate test in Fig. 10. For comparison purposes the prototype loads and deflections were normalized according to the similitude requirements.

CONCLUSIONS

From the load-deflection curves reported in Figs. 3, 4, 10 and 11 it was observed that the ultimate load for the specimen subjected to fast rate was about 20-25% higher than that for the slow rate but the degradation during the cyclic loading was greater for the fast rate. This was also evidenced from the results of the free vibration tests reported in Fig. 5.

By comparing crack patterns and end slips of longitudinal reinforcement it was observed that flexural cracks were widely distributed and 0.0005" - 0.0006" end slips were obtained for the specimen tested at the slow rate. In contrast, the damage was due to a single wide crack observed at the face of the column and no end slips were obtained for the fast rate test specimen. This indicates that the observed rate effect is essentially related to the transfer of forces between reinforcing bars and concrete. It is likely that the first flexural crack in the beam developed at the face of the column for both specimens. However, because of the more efficient load-transfer occurring at the slow rate, additional cracking progressively developed at sections further away from the column face. Some support for such an argument can be obtained from the results reported by Takeda [24]. He studied the rate effect on the bond stress distribution on deformed bar during a pull-out test. Some of his results are shown in Fig. 13. The bond stress distribution at two different rates are shown. It can be seen that the bond stresses are more localized at the higher rate.

REFERENCES

1. Hawkins, N. M., Editor, "Reinforced Concrete Structures in Seismic Zones," ACI-Publication, SP-53, pp. 485, 1977.
2. Wight, J. K., Editor, "Earthquake Effects on Reinforced Concrete Structures," ACI-Publication, SP-84, pp. 428, 1985.
3. Mutsuyoshi, H. and Machida, A., "Dynamic Properties of Reinforced Concrete Piers," Transactions of the Japan Concrete Institute, Vol. 4, pp. 424-436, 1978.
4. Mutsuyoshi, H. and Machida, A., "Force Displacement Characteristics of Reinforced Concrete Piers," Transactions of the Japan Concrete Institute, Vol. 3, pp. 399-406, 1981.
5. Wilby, G. K., "Response of Concrete Structures to Seismic Motions," Ph. D. thesis, Department of Civil Engineering, University of Canterbury, New Zealand, pp. 225, 1976.
6. Arakawa, T., and Arai, Y., "Effects of the rate of cyclic loading on the inelastic behavior of R/C Columns", 8th World Conference on Earthquake Engineering, Vol. VI, San Francisco, pp. 521-528, 1984.
7. Scribner, C. F., and Wight, J. K., "Delaying Shear Strength Decay in Reinforced Concrete Load Reversals", Department of Civil Engineering, University of Michigan, pp. 220, May 1984.
8. ACI Committee 318, "Building Code Requirements for Reinforced Concrete (ACI 318-83)," American Concrete Institute, Detroit, pp. 111, 1983.
9. ACI-ASCE Joint Committee 352, "Recommendations for Design of Beam-Column Joints in Monolithic Reinforced Concrete Structures", ACI Journal, pp. 375-393, July 1976.
10. Aldridge, W. W., and Breen, J. E., "Useful Techniques in Direct Modeling of Reinforced Concrete Structures", Models for Concrete Structures, ACI-Publication, SP-24, pp. 125-140, 1970.
11. Harris, H. G., Sabnis, G. M. and White, R. N., "Reinforcement for Small Scale Direct Models of Concrete Structures", Models for Concrete Structures, ACI-Publication, SP-24, pp. 141-158, 1970.
12. Goto, Y., "Cracks Formed in Concrete Around Tension Bars," ACI Journal, Proceedings, Vol. 68, No. 4, pp. 244-251, April 1971.
13. Nilson, A. H., "Internal Measurement of Bond Slip," ACI Journal, Vol. 63, No. 7, pp. 439-441, July 1972.
14. ACI Committee 408, "Bond Stress - The State of the Art," ACI Journal, Vol. 63, No. 11, pp. 1161-1188 Nov. 1966.
15. Somayaji, S., and Shah, S. P., "Bond Stress Versus Slip Relationship and Cracking Response of Tension Members," ACI Journal, Proceedings, Vol. 78, No. 3, pp. 217-225, May-June 1981.

16. Perry, E. S., and Jundi, N., "Pullout Bond Stress Distribution Under Static and Dynamic Repeated Loadings," <u>ACI Journal</u>, Proceedings, Vol 66, No. 5, pp. 377-380, May 1969.
17. Bresler, B., and Bertero, V. V., "Behavior of Reinforced Concrete Under Repeated Loads," <u>Journal of the Structural Division</u>, ASCE, pp. 1567-1590, June 1968.
18. Ismal, M., "Bond Deterioration in Reinforced Concrete Under Cyclic Loading," Ph.D. Thesis, Rice University, Houston, Texas, pp. 164, March 1970.
19. Bertero, V. V., Popov, E. P., "Seismic Behavior of Ductile Moment Resisting Reinforced Concrete Frames," <u>ACI Special Publication</u>, SP-53, Detroit, pp. 247-292, 1977.
20. Viwathanatepa, S., Popov, E. P., and Bertero, V. V., "Seismic Behavior of Reinforced Concrete Interior Beam-column Joints," Earthquake Research Engineering Center, <u>Report No. UCB/EERC-79/14</u>, University of California, Berkeley, pp. 304, 1979.
21. Eligehausen, R., Popov, E. P., and Bertero, V. V., "Local Bond Stress-Slip Relationships of Deformed Bars under Generalized Excitations," Earthquake Engineering Research Center, <u>Report No. UCB/EERC-83/23</u>, University of California, Berkeley, pp. 169, 1983.
22. Fillipou, F. C., Popov, E. P., and Bertero, V. V., "Effects of Bond Deterioration on Hysteretic Behavior of Reinforced Concrete Joints," Earthquake Engineering Research Center, <u>Report No. UCB/EERC-83/19</u>, Univ. of California, Berkeley, pp. 183, 1983.
23. Brown, R. H., "Reinforced Concrete Cantilever Beams under Slow Cyclic Loadings," Ph.D. Thesis, Rice University, Houston, Texas, pp. 166, April 1970.
24. Takeda, J. I., "Dynamic Fracture of Concrete Structures Due to Severe Earthquakes and Some Consideration on Countermeasures", <u>8th World Conference on Earthquake Engineering</u>, Vol. VI, San Francisco, pp. 299-306. 1984.

Table 1a Comparison of column design between model and prototype

	size (in) (BxD)	column length (in)	main steel size(in^2)	ratio(%)	tie bar	shear stress (psi) $v_u = V_u / \phi bd$	allowable (psi) v_c	tie space required (in)	tie space used (in)
P	8x12	60.	4#6(1.76)	2.20	#2	239.0	155.0	6.0	2.25
M	3x3	12.	4#2(0.20)	2.50	D1	181.5	184.0	8.4	0.60

Table 1b Comparison of beam design between model and prototype

	size (in) (bxD)	beam length (in)	main steel size (in^2) top	bott.	ratio(%) top	bott.	stirrup (in^2)	max. beam shear force (kips) calculated	measured
P	8x10	31.0	2#6(0.98)	2#5(0.72)	1.23	1.0	#2(0.05)	16.9	15.2
M	2x3	13.5	2D2.5(.05)	2D2.5(.05)	0.84	0.9	D1(.0083)	0.776	0.725

	max. beam shear stress calculated (in)	measured	allowable v_c (in)	tie space required (in)	max. (d/4) (in)	tie space used (in)
P	247.0	221.0	129.0	4.13	2.0	2.0
M	169.0	158.0	126.0	7.7	0.675	0.6

(1in=25.4mm, 1kip=4448N, 1ksi=6895kPa)
P = prototype
M = model

Table 1c Comparison of joint design between model and prototype

	core size (in)	tie bar (in^2)	joint shear force (kips) $V_j = T_s - V_{u,col}$	joint shear stress (ksi) $v_j = V_j / \phi bd$	allowable (ksi) v_c	tie bar spacing required (in)	used
P	7.8x8.6x9.64	#2(.05)	67.5	1.18	0.306	2.2	1.75
M	1.4x2.4x2.4	D1(.0083)	3.38	0.74	0.371	0.91	0.4

Table 2 Comparisons of material properties between model and prototype

Reinforcement (1 ksi=6895 kPa)   * not measured ** not reported

Specimen	Type	No. of bar	f_y(ksi)	ϵ_y(x10^{-3})	E_s(ksi)	σ_{max}(ksi)
Model	Beam-column joint and anchorage-bond specimen	D2.5	85	3.2	28800	93.0
		#2	68	2.4	28000	*
		D1	40	1.7	23500	*
Prototype	Beam-column joint specimen	#6	52.7	2.0	28200	84.8
		#2	42.6	*	*	*
	Anch.-bond specimen	#6	45	1.55	29000	79.0

Concrete

Specimen	Test cylinder	f'_c(ksi)	Max. aggregate	Mix proportions(wt)		
				cement	aggregate	water
Model	3 - 3"x6"	4.0	3/16"	1.0	4.0	0.59
Beam-col. prototype	3 - 4"x8"	4.0	3/8"	1.0	6.1	0.50
Anch.-bond prototype	6"x12"	5.3	3/4"	**	**	**

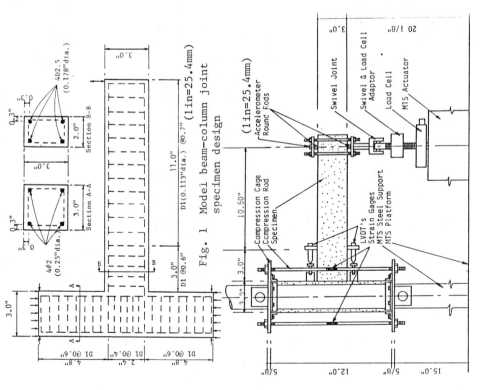

Fig. 1 Model beam-column joint specimen design

Fig. 2 Sideview of beam-column joint specimen test set-up

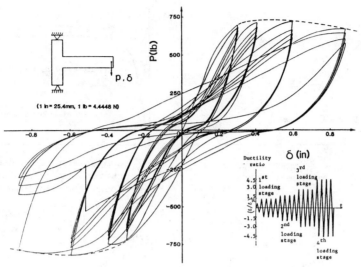

Fig. 3 Load-deflection diagram for beam-column joint specimen ($f=2.5 \times 10^{-3}$ Hz)

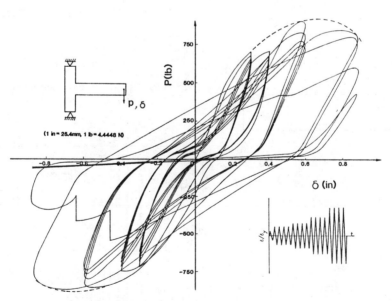

Fig. 4 Load-deflection diagram for beam-column joint specimen ($f=1.0$ Hz)

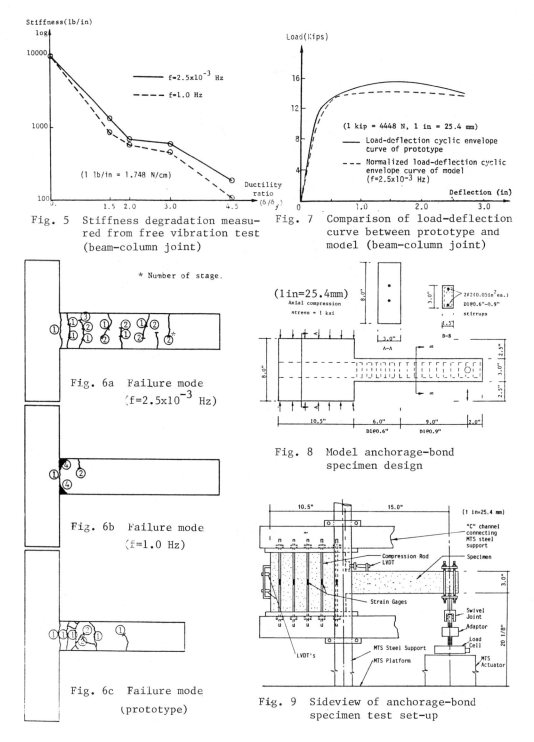

Fig. 5 Stiffness degradation measured from free vibration test (beam-column joint)

Fig. 6a Failure mode ($f=2.5 \times 10^{-3}$ Hz)

Fig. 6b Failure mode ($f=1.0$ Hz)

Fig. 6c Failure mode (prototype)

Fig. 7 Comparison of load-deflection curve between prototype and model (beam-column joint)

Fig. 8 Model anchorage-bond specimen design

Fig. 9 Sideview of anchorage-bond specimen test set-up

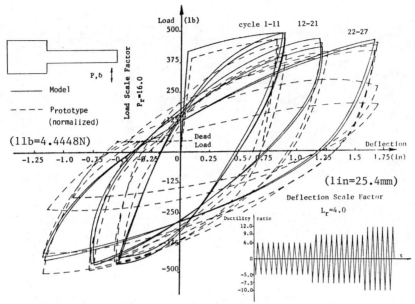

Fig. 10 Load-deflection diagram for anchorage-bond specimen(f=2.5x10^{-3} Hz)

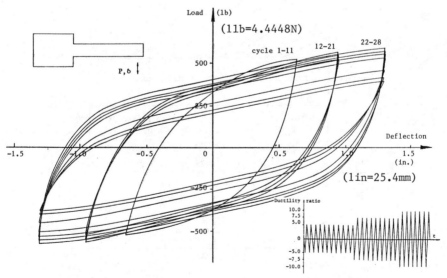

Fig. 11 Load-deflection diagram for anchorage-bond specimen(f=0.5 Hz)

* Loading stage number

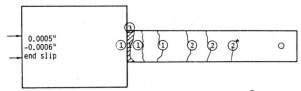

Fig. 12a Failure mode($f=2.5 \times 10^{-3}$ Hz)

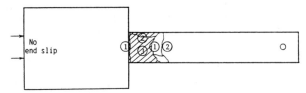

Fig. 12b Failure mode($f=0.5$ Hz)

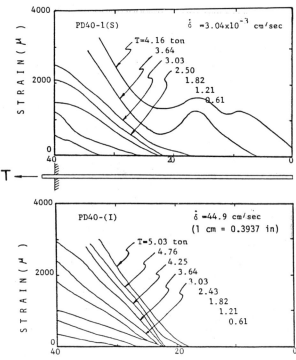

Fig. 13 Comparison of bond stress distribution
distribution on reinforcement
between static and high rate
loading test (from ref.24)

EVALUATION AND IMPROVEMENT
OF THE PSEUDODYNAMIC TEST METHOD

P. B. Shing[I]

ABSTRACT

Results of recent studies on the propagation of experimental errors in pseudodynamic tests are summarized. Criteria for determining the cumulative bounds of various experimental errors are established. It has been found that some errors of systematic nature are highly undesirable. They can severely excite the higher-mode response of a multi-degree-of-freedom structure. Methods for mitigating the error-propagation effects are presented.

INTRODUCTION

The pseudodynamic test method is recently developed [1,2,3] for evaluating the inelastic seismic performance of structures. The major advantages of this method are that it utilizes almost the same equipment as that required for conventional quasi-static testing, and that it can be applied to structural specimens that are too large, strong, or heavy to test on existing shaking tables. This method is basically a computer-controlled experimental technique in which a direct step-by-step-integration procedure is used to solve the idealized equations of motion for the structural specimen. The displacement response evaluated in each step of a test is quasi-statically imposed on the specimen; and the restoring forces developed by the deformed specimen are measured and fed back to the numerical algorithm for computing the displacement response in the next step. Hence, the most significant source of inaccuracy in this procedure is the experimental feedback error introduced into the numerical computations [4]. The resulting cumulative errors can be significant, depending on the nature of the feedback errors, the frequency characteristics of the specimen, and the numerical properties of the integration algorithm. This paper summarizes results of recent studies [4,5,6] on the error-propagation problem and presents rational criteria for estimating the effects of random and systematic errors. Numerical methods for mitigating the error-propagation effects are also proposed and verified with numerical examples.

I. Assistant Professor, Department of Civil, Environ., and Archit. Engineering, University of Colorado, Boulder, Colorado.

TEST METHOD

In a pseudodynamic test, the structural specimen is idealized as a discrete-parameter system, which has its mass concentrated at a limited number of discrete locations. The equations of motion for the idealized system can be expressed as second-order differential equations:

$$\mathbf{M}\mathbf{a} + \mathbf{C}\mathbf{v} + \mathbf{r} = \mathbf{f} \qquad (1)$$

in which $\mathbf{v}$ and $\mathbf{a}$ are the nodal velocity and acceleration vectors, which are, respectively, the first- and second-order time derivatives of the nodal displacement vector $\mathbf{d}$. The mass matrix $\mathbf{M}$ and damping matrix $\mathbf{C}$ are prescribed analytically. The displacement response of the specimen to a numerically specified external excitation $\mathbf{f}$ is evaluated by a computer during a test, based on a direct step-by-step integration procedure. The displacement response evaluated in each step is quasi-statically imposed on the specimen by means of computer-controlled electro-hydraulic actuators. The restoring forces $\mathbf{r}$ developed by the deformed specimen are measured by load transducers, and fed back to the computer for calculating the response in the next step. This procedure is repeated until the entire response history is evaluated.

The explicit type of step-by-step integration methods is generally preferred for solving the above equations in a pseudodynamic test [3]. The central difference method is one of the most accurate and widely used explicit integration techniques, and has been implemented in many early tests [1]. In this method, the velocity and acceleration responses in each step are expressed in terms of the following difference equations:

$$v_i = \frac{1}{2 \Delta t} (d_{i+1} - d_{i-1}) \qquad (2)$$

$$a_i = \frac{1}{\Delta t^2} (d_{i+1} - 2 d_i + d_{i-1}) \qquad (3)$$

in which Δt is the integration time interval and the subscript i denotes values at time equal to $i\Delta t$. Substituting the above relations into the equations of motion (Eq. 1), one can solve for the displacement response in each step as

$$d_{i+1} = (M + \frac{\Delta t}{2} C)^{-1} [\Delta t^2 (f_i - r_i) + (\frac{\Delta t}{2} C - M) d_{i-1} + 2 M d_i] \qquad (4)$$

which can be readily evaluated in a pseudodynamic test, based on the displacements and restoring forces developed in the previous steps.

The central difference method can be formulated in an alternative form (the summed form) [7,8] by introducing an additional parameter z_i that is defined as $(d_i - d_{i-1})/\Delta t$. Based on that, Eq. 4 can be replaced by the following equations:

$$z_{i+1} = (\frac{1}{\Delta t} M + \frac{1}{2} C)^{-1} [f_i - r_i + z_i (\frac{1}{\Delta t} M - \frac{1}{2} C)] \tag{5}$$

$$d_{i+1} = d_i + \Delta t \, z_{i+1} \tag{6}$$

which can be applied to pseudodynamic testing in the same fashion as the basic central difference formulation, except that an additional variable, z_{i+1}, has to be evaluated in each step of a test.

Furthermore, the central difference method is mathematically equivalent to an explicit form of the general Newmark's method [9]. The displacement and velocity responses in this explicit Newmark's formulation are expressed as

$$d_{i+1} = d_i + \Delta t \, v_i + \frac{\Delta t^2}{2} a_i \tag{7}$$

$$v_{i+1} = v_i + \frac{\Delta t}{2} (a_i + a_{i+1}) \tag{8}$$

The displacement response in each step can be evaluated with Eq. 7, based on the solution values obtained in the previous step. Physically imposing the displacement response d_{i+1} on the specimen and substituting Eq. 8 into Eq. 1, the acceleration response a_{i+1} can be evaluated from the restoring forces r_{i+1} developed by the specimen. Finally, v_{i+1} is computed with Eq. 8, and the entire procedure can be repeated to evaluate the response in the next step.

The three algorithms presented above have identical numerical properties [10]. They are stable if $\omega \Delta t \leq 2$, where ω is the highest circular frequency of the system. However, the basic central difference formulation is more sensitive to computational round-off errors than the other two algorithms [10], due to the different arithmetic operations involved.

EXPERIMENTAL ERROR PROPAGATION

In a pseudodynamic test, errors can be introduced during the control and measurement processes [4]. As shown in Fig. 1, the displacement response computed in a step may not be accurately imposed on the specimen due to displacement control errors e_i^{dc}. In addition, the imposed displacement and the restoring force actually developed by the deformed specimen may be incorrectly measured due to measurement errors e_i^{dm} and e_i^{rm}, respectively. These control and measurement errors amount to the total feedback errors introduced in each step of a test.

To study the sensitivity of numerical solutions to experimental errors, we consider an undamped, linear single-degree-of-freedom system. For such a system, the step-by-step solution of the equation of motion can be expressed in a recursive form. Introducing experimental feedback errors into the recursive formulation, one can obtain a closed-form expression for the cumulative displacement errors [4] as

$$\bar{e}_{n+1} = C \sum_{i=1}^{n} e_i^{d} \sin[\bar{\omega}\Delta t(n-i)+\theta] + D \sum_{i=1}^{n} e_i^{rd} \sin[\bar{\omega}\Delta t(n-i+1)] \quad (9)$$

in which $e_i^{d} = e_i^{dc} + e_i^{dm}$, the total displacement feedback error, and $e_i^{rd} = e_i^{r}/k$, where k is the elastic stiffness and $e_i^{r} = k\, e_i^{dc} + e_i^{rm}$, the total force feedback error. In addition, θ is a phase angle, and $\bar{\omega}$ is the apparent natural frequency of the numerical solution which can be evaluated with the following equation:

$$\bar{\omega}\Delta t = \arctan\{[4/(\omega^2\Delta t^2)-1]^{1/2} / [2/(\omega^2\Delta t^2)-1]\} \quad (10)$$

for all the three numerical algorithms considered here, where ω is the natural circular frequency of the system. The parameters C and D are error amplification factors, whose values depend on the specific numerical algorithm considered. For the basic central difference formulation (Eq. 4), we have

$$C = \{[2+1/(\omega^2\Delta t^2)] / [1-(\omega^2\Delta t^2)/4]\}^{1/2} \quad (11)$$

$$D = -2 / [4/(\omega^2\Delta t^2)-1]^{1/2} \quad (12)$$

For the summed form and the explicit Newmark's formulation, the factor D is identical to that in Eq. 12 and

$$C = -1 / [1-(\omega^2 \Delta t^2)/4]^{1/2} \tag{13}$$

Equation 9 indicates that the magnitude of cumulative errors depends strongly on the amplification factors C and D. The absolute values of these factors are plotted against $\omega \Delta t$ in Fig. 2, for the three algorithms considered. It is apparent that the basic central difference formulation is a poor scheme as far as error propagation is concerned, because the value of C approaches infinity as $\omega \Delta t$ goes to zero (Fig. 2a). Nevertheless, a small $\omega \Delta t$ value is often desirable for attaining numerical stability and accuracy. Hence, one faces a dilemma that limiting the cumulative growth of experimental errors must sacrifice the inherent accuracy of the numerical scheme. For the summed form and the explicit Newmark's formulation, |C| approaches unity and |D| goes to zero as the value of $\omega \Delta t$ diminishes (Fig. 2b). However, the same dilemma may exist for these schemes, since the value of the summation term associated with C may grow indefinitely as $\omega \Delta t$ approaches zero and n goes to infinity.

The above observations imply that test results can be substantially improved by using the computed displacement values, instead of the actual experimental feedback, for the numerical computations in a pseudodynamic test. This improvement is especially significant for the basic central difference algorithm. Under this circumstance, all the three algorithms considered have identical error-propagation characteristics. Although structural restoring forces must always be measured in a test, force feedback errors are not as significantly amplified as displacement feedback errors due to the fact that |D| is extremely small when $\omega \Delta t$ is close to zero.

SYSTEMATIC AND RANDOM ERRORS

According to Eq. 9, it is apparent that the magnitude of cumulative errors depends the nature of experimental feedback errors, as well as the error amplification factors. Force-feedback errors introduced in a pseudodynamic test can be classified into systematic and random types. In general, systematic errors are more undesirable than random errors. The worst systematic errors usually result from inherent limitations of experimental apparatus or consistently biased instrumental inaccuracies. For example, if the electro-hydraulic actuators which control the deformation of a specimen consistently over-shoot the commanded displacements, the restoring force measurements will be taken from an over- or under-deformed specimen, depending on whether the specimen is being loaded or unloaded. This will introduce an energy-dissipating hysteretic effect into the numerical computations. On the other hand, if the actuator motion tends to lag behind the commanded displacements, due to inadequate gain control or servo-capacity, then energy-adding hysteresis occurs. Owing to these hysteretic effects (see Fig. 3), systematic errors of such nature can significantly distort the response of a structure.

The cumulative error bounds due to random and systematic force-feedback errors can be established with the second term on the right-hand side of Eq. 9. Based on a statistical approximation [5], the cumulative bound due to random errors can be obtained as

$$|\bar{e}_{n+1}|_{max} \approx 2\, S_e |D| \{n/2 - [\sin \bar{\omega}\Delta t\, n \cos \bar{\omega}\Delta t(n+1)]/[2 \sin \bar{\omega}\Delta t]\}^{1/2} \quad (14)$$

in which $\bar{\omega}$ and D are defined in Eqs. 10 and 12, respectively, n is the integration time step, and S_e is the standard deviation of the feedback errors e_i^{rd}. The normalized bounds are plotted against n for various $\omega\Delta t$ values in Fig. 4. One can observe that the rate of cumulative error growth with respect to n increases rapidly with $\omega\Delta t$. Nevertheless, this rate tends to diminish as n increases. It can be shown that the cumulative errors will approach zero as Δt goes to zero and n goes to infinity. The decrease of cumulative errors with Δt can be illustrated by moving from point A to B in Fig. 4.

Energy-modifying systematic errors can be modeled as a sinusoidal function [5] which has a frequency $\bar{\omega}$. Based on this assumption, the cumulative error bound can be estimated with Eq. 9 as

$$|\bar{e}_{n+1}|_{max} \approx 0.5\, A_e |D| \{n^2 + 2n + 1/[\omega^2 \Delta t^2 (1 - \omega^2 \Delta t^2/4)]\}^{1/2} \quad (15)$$

which represents a resonance-like growth phenomenon, where A_e is the amplitude of e_i^{rd} in a sinusoidal form. The normalized bounds are plotted against n in Fig. 5, which indicates that the rate of cumulative error growth for this case is much higher than that for random errors. Furthermore, by moving from point A to B in Fig. 5, one can observe that the cumulative errors cannot be significantly mitigated by reducing Δt.

In view of the above results, one can conclude that the higher modes of a multi-degree-of-freedom system will be more sensitive to experimental errors than the lower modes. Furthermore, if systematic errors are small and of the energy-dissipating type, the higher-mode response of a system will be readily damped without significant influence on the lower modes. Since the higher modes usually have little participation in a seismic response, the resulting cumulative errors will be small. On the other hand, if the errors are energy adding, the higher-mode response can be severely excited. The resulting cumulative error can grow indefinitely and totally overwhelm the real response of the structure. This phenomenon can be illustrated with a linear two-degree-of-freedom system, of which the first- and second-mode frequencies are 11.2 and 35.7 radians/sec, respectively. Energy dissipating and adding experimental errors of equal magnitude are numerically simulated. The results evaluated by means of the explicit Newmark's algorithm with $\Delta t = 0.01$ sec. are shown in Fig. 6.

IMPROVEMENT METHODS

As discussed in the previous section, the energy-adding type of systematic errors are most undesirable in a pseudodynamic test. They tend to excite the higher-mode response of a multi-degree-of-freedom system. Hence, depending on the frequency characteristics of a structure, even small feedback errors of such nature can cause a rapid growth of cumulative errors. Furthermore, since the force response of a structure has a larger participation of higher-mode components than the displacement response, the restoring forces developed will be highly sensitive to systematic errors. Owing to these reasons and the fact that the seismic response of a structure is very often dominated by the fundamental frequency, it is always desirable to introduce an artificial damping mechanism to suppress the spurious higher-mode effects in a pseudodynamic test.

The easiest way to suppress the higher-mode response is to use an orthogonal damping matrix of the form [11]:

$$C = M \sum_{k} a_k (M^{-1}K)^k \qquad (16)$$

in which M and K are the mass and stiffness matrices, and a_k are coefficients that determine the modal damping characteristics. The values of a_k can be so selected that the higher modes are severely damped, while the lower modes are moderately damped or undamped. However, this method has certain disadvantages. First of all, to construct the damping matrix C, one needs to know the initial stiffness matrix precisely. Secondly, as non-linear deformations are developed, the change of tangent stiffness tends to alter the modal damping characteristics of C. Studies have indicated that this non-linear effect may introduce excessive damping to the fundamental mode [6].

The second approach for suppressing the higher-mode response is to use the inherent damping characteristics of a numerical integration algorithm. The explicit Newmark's method can be modified [6] to possess the desired numerical damping characteristics by introducing an additional term to the equations of motion:

$$M\,a_{i+1} + K\,d_{i+1} + \left(\alpha K + \frac{\rho}{\Delta t^2} M\right)(d_{i+1} - d_i) = f_{i+1} \qquad (17)$$

in which α and ρ are parameters controlling the numerical damping characteristics, while the expressions for d_{i+1} and v_{i+1} remain the same (Eqs. 7 and 8). The numerical damping characteristics of this algorithm can be observed from the fact that the displacement increment $(d_{i+1} - d_i)$ in Eq. 17 is approximately proportional to the velocity vector. Hence, the damping effects introduced by parameters α and ρ are very similar to those of Rayleigh damping, which consists of a mass- and stiffness-proportional damping matrix. While the magnitude of stiffness-proportional damping is proportional to the natural frequency of a system, that of mass-proportional damping is inversely proportional to the frequency. Thus, in order that damping is small for the fundamental mode and large for the higher modes, $\alpha \geq 0$ and $\rho \leq 0$. The stability of this algorithm is governed by the following condition [6]:

$$(-\rho/\alpha)^{1/2} \leq \omega \Delta t \leq \{1+[1-(1+\alpha)\rho]^{1/2}\} / (1+\alpha) \qquad (18)$$

in which the lower limit ensures positive energy dissipation. For non-linear systems, the value of ρ has to be changed continuously to prevent numerical instability, based on the following criterion:

$$-\rho \leq \Delta t^2 \, \alpha \, (\Delta d_i^T \, \Delta r_i) / (\Delta d_i^T \, M \, \Delta d_i) \qquad (19)$$

in which $\Delta d_i = d_i - d_{i-1}$ and $\Delta r_i = r_i - r_{i-1}$. The numerical damping ratio for this algorithm can be evaluated by the following equation:

$$\bar{\xi} = - \ln (1 - \alpha \omega^2 \Delta t^2 - \rho) / (2 \bar{\omega} \Delta t) \qquad (20)$$

To illustrate the efficiency of the above methods, a numerical example is presented. The structural system considered is an inelastic three-story shear frame. Experimental errors of the energy-adding type are numerically simulated. The response of the frame, with and without experimental errors, is shown in Fig. 7. We can observe that the response at the bottom story is severely contaminated by the third-mode frequency, while those at the top two stories are significantly distorted. The results with dampings are shown in Fig. 8. It is apparent that viscous and numerical dampings are equally effective in removing the spurious response. However, the use of numerical damping provides a slightly better solution than using viscous damping, because of the increase of damping effect on the fundamental mode for the latter case.

CONCLUSIONS

The effects of experimental errors on pseudodynamic test results have been examined in this paper. The error-propagation characteristics of several numerical algorithms applicable to pseudodynamic testing have been evaluated. These studies indicate that the use of experimentally measured displacements in the numerical computations can lead to significant cumulative errors, and, therefore, should be avoided in all tests. Furthermore, cumulative bounds have been established for random and systematic force-feedback errors. The results show that small errors of random nature or of the energy-dissipating type are usually acceptable in a pseusdodynamic test. However, systematic errors of the energy-adding type are very undesirable. Such errors tend to grow rapidly in a multi-degree-of-freedom system that has widely spaced frequency components. Fortunately, this type of errors can be easily identified in a test as spurious higher-mode responses, and can be effectively controlled by the numerical methods proposed in this paper.

ACKNOWLEDGMENTS

The studies presented in this paper were conducted at the University of California, Berkeley, under the U.S.-Japan Cooperative Research Program sponsored by the National Science Foundation (NSF). The author is especially grateful to his former research advisor, Professor S. A. Mahin, for his support, advice, and encouragement. However, opinions expressed in this paper do not necessarily represent those other participants in the program or of NSF.

REFERENCES

[1] Takanashi, K., et al, "Nonlinear Earthquake Response Analysis of Structures by a Computer-Actuator On-Line System," Bulletin of Earthquake Resistant Structure Research Center, No. 8, Institute of Industrial Science, University of Tokyo, 1975.

[2] Shing, P. B. and Mahin, S. A., "Pseudodynamic Test Method for Seismic Performance Evaluation: Theory and Implementation," UCB/EERC-84/01, Earthquake Engineering Research Center, University of California, Berkeley, Jan., 1984.

[3] Mahin, S. A and Shing, P. B., "Pseudodynamic Method for Seismic Testing," Journal of Structural Engineering, ASCE, Vol. 111, No. 7, pp. 1482-1503, July, 1985.

[4] Shing, P. B. and Mahin S. A., "Experimental Error Propagation in Pseudodynamic Testing," UCB/EERC-83/12, Earthquake Engineering Research Center, University of California, Berkeley, June, 1983.

[5] Shing, P. B. and Mahin, S. A., "Cumulative Experimental Errors in Pseudodynamic Tests," Earthquake Engineering and Structural Dynamics, in publication.

[6] Shing, P. B. and Mahin, S. A., "Elimination of Spurious Higher-Mode Response in Pseudodynamic Tests,", Earthquake Engineering and Structural Dynamics, in publication.

[7] Dahlquist, G., Bjorck, A., and Anderson, N., Numerical Methods, Prentice-Hall, Englewood Cliffs, N. J., 1974.

[8] McClamroch, N. H., Serakos, J., and Hanson, R. D., "Design and Analysis of the Pseudo-Dynamic Test Method," <u>UMEE 81R3</u>, University of Michigan, Ann Arbor, Sept., 1981.

[9] Newmark, N. M., "A Method of Computation for Structural Dynamics," <u>Journal of the Engineering Mechanics Division</u>, ASCE, Vol. 85, No. EM3, pp. 67-94, July, 1959.

[10] Shing, P. B. and Mahin S. A., "Computational Aspects of a Seismic Performance Test Method using On-Line Computer Control," <u>Earthquake Engineering and Structural Dynamics</u>, Vol. 13, pp. 507-526, 1985.

[11] Clough, R. W. and Penzien, J., <u>Dynamics of Structures</u>, McGraw-Hill, New York, N. Y., 1975.

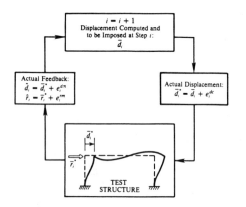

Fig. 1 Sources of Experimental Errors

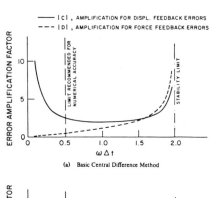

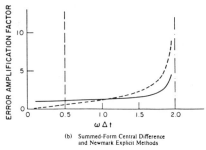

Fig. 2 Error Amplification Factors

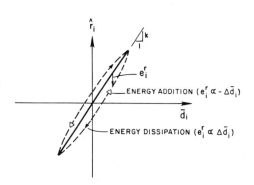

Fig. 3 Hysteretic Energy Effects due to Experimental Errors

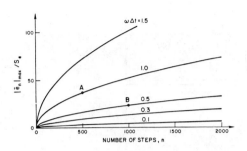

Fig. 4 Cumulative Bounds for Random Errors

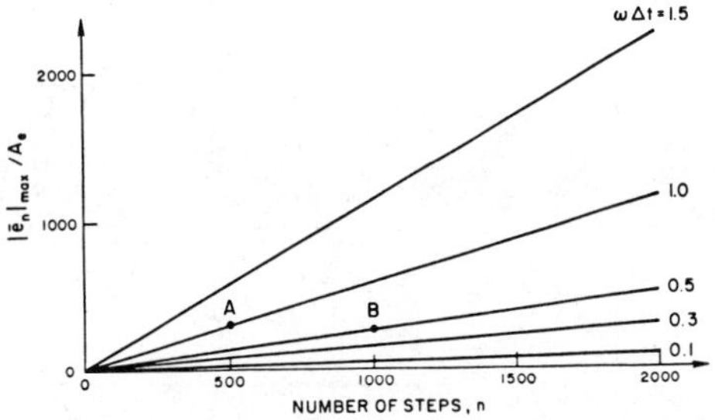

Fig. 5 Cumulative Bounds for Systematic Errors

(a) Energy-Dissipating Errors

(b) Energy-Adding Errors

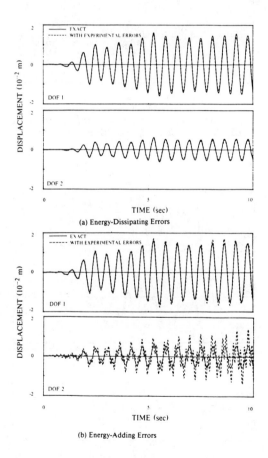

Fig. 6 Response of a Linear Two-Degree-of-Freedom System to 1940 El Centro (NS) Earthquake Ground Motion

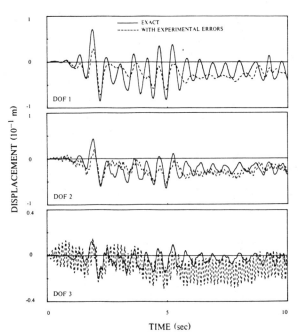

Fig. 7   Response of an Inelastic Three-Degree-of-freedom System to 1940 El Centro Earthquake Ground Motion

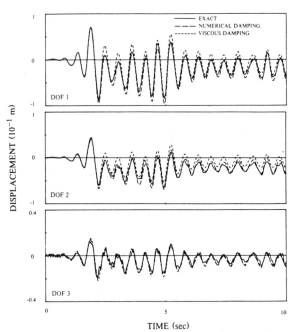

Fig. 8   Response of an Inelastic Three-Degree-of-Freedom System with Damping Effects

INTERSTORY DISPLACEMENT RECORDING

Stephen E. Pauly[I]

ABSTRACT

A displacement recorder has been developed to monitor the relative static and dynamic horizontal displacement between two points in a structure. The mechanical structure consists of two lightweight, rigid A-frames, to be mounted to the floor and ceiling with rugged leaf spring flexures. The two frames overlap so that direct recording of the relative displacements can be made with a diamond stylus and recording plate or drum mounted on the frames near the story midheight. The results of the initial testing of the recorder are presented, and they show that the basic design goals have been met. Shake table tests of the instrument are planned.

INTRODUCTION

Recording of earthquake displacements is important, because displacement directly causes damage. Velocity and acceleration are indirect measures of the forces which cause ground motion and structural damage. Historically, engineers have used accelerometers for indirect measure of displacements because accelerometers are relatively easy to build and install and because acceleration is the most direct element in dynamic equations of motion. Further, force-balance accelerometers can transduce the often large accelerations which accompany large displacements.

To obtain the displacement record from the acceleration data, the acceleration signal must be integrated twice. This double integration can introduce errors in the computed displacements, especially at the lower frequencies where the signal to noise ratio of accelerometers decreases due to lower signal output [1].

Direct measurement of dynamic and static displacement is an important research need to determine 1) how direct measurements and the integrated acceleration results compare and 2) how to compute displacement accurately from acceleration. When these questions are answered, higher confidence can be placed on the results obtained from the integration process.

The prototype frames are about 9 ft (2.7 m) in height and cover a 14 ft (4.3 m) height when overlapped, and the height can be modified for larger or smaller interstory heights.

Six recording techniques are being developed including peak displacement recorders and displacement versus time recorders using flat plates, flat film, a drum and a cartridge.

I. Project Manager, Endgahl Enterprises, Costa Mesa, California 92626.

Testing of the prototype has been performed using static and dynamic loads, and the results show that the frame stiffness keeps deflections at 1 g loading to less than the motion likely from ambient conditions.

DESIGN GOALS

The design goals for the relative displacement recorder were:

o High rigidity in the recording axis (the plane of the frame and wall). Less than 0.010 in (0.25 mm) deflection at 1g inertial load is desired.

o High vertical rigidity. Less than 0.010 in (0.25 mm) deflection at 1 g inertial load is desired.

o Moderate rigidity is tolerable in the axis perpendicular to the plane of the frame. Approximately 0.25 in (6 mm) is permissible.

o Wide base, approximately equal to half the interstory dimension, to eliminate amplification and to reduce the effects of local tilting.

o Leafspring flexures at the attachment points to accommodate cross axis motion without adding stress to the frame and to allow for irregular building dimensions.

o Both peak displacement and displacement versus time recording to be provided as options.

o The displacement time-history recording should begin when the displacement exceeds a preset level or when a signal that adjacent strong motion acceleration recorders have started.

o The recording capacity of the displacement versus time recorder should be adequate for several sequential seismic events.

o The final record should be suitable for digitization.

SENSOR DESCRIPTION

The sensing mechanism consists of two overlapping A-frames. The base of one frame is attached to the floor and the base of the other frame is attached to the underside of the next higher floor for the interstory section to be monitored. This is illustrated in Fig. 1. The attachment is made at the two outermost points along the frame base with rugged leaf spring flexures, as shown in Fig. 2. The flexures permit the system to deflect perpendicularly to the plane of the frame without adding stress to the frame. Bolts anchored at the apex of each frame slide in slots on the adjoining frame to keep the frames in close proximity. The two frames overlap each other, as shown in Fig. 3, and the displacement between them is recorded mechanically or electrically using one of several recording techniques. Each of the prototype A-frames is about 9 ft (2.7 m) in height, and when overlapped, the pair will measure the relative displacement of floors separated by about 14 ft (4.3 m). The frame dimensions can be modified for use with larger or smaller interstory heights.

Fig. 1 Overlapping A-Frames of the Relative Displacement Recorder

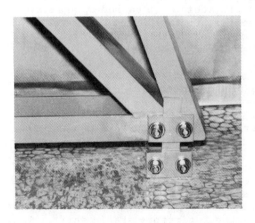

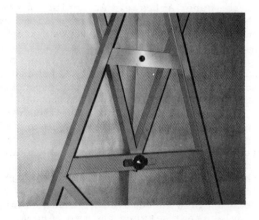

Fig. 2. Leaf Spring Attachment Detail   Fig. 3. A-Frame and Recorder Detail

RECORDING TECHNIQUES

Six recording techniques are being developed:

1. Peak horizontal displacement recording, in the plane of the A-frames, using a diamond stylus to scribe on a coated metal plate.

2. Peak horizontal and vertical displacement recording, in the plane of the A-frames, using a diamond stylus to scribe on a coated metal plate. This system is described in detail below.

3. Displacement versus time recording using either a flat coated metal plate or a flat film record (in which the developed emulsion is scratched off by the diamond stylus). This system is described in detail below.

4. Displacement versus time recording using a film wrapped around a drum for extended recording time.

5. Displacement versus time recording, as in systems 3 and 4, above, except that the flat film is wound into a roll and placed in a cartridge.

6. Peak displacement or displacement versus time recording using an extensiometer with its electrical output recorded on a strip chart, photographic film, magnetic tape or solid state recorder.

Peak Displacement Recording Using a Flat Record Plate

This recorder consists of two sections. One section contains the diamond stylus and is attached to the cross member of the lower A-frame. The other section contains the flat record plate which is attached to the cross member of the upper A-frame. An O-ring in the second section is pushed up against the flat surface of the first section, by springs, to provide a moisture seal, and desiccant is installed to help keep the interior of the recorder dry. The plate holder, O-ring, and coated record plate are shown in Fig. 4. The diamond stylus scratches a permanent record in the soft plating of the record plate. This technology has been used previously for mechanical peak acceleration recorders [2] and response spectrum recorders [3].

A zero line is created when the record plate and its holder are installed in the upper A-frame support. Since the scriber is preloaded 1/16 in (1.6 mm) against the record plate, the plate comes in contact with the scriber before it is completely inserted creating a linear scratch, as in Fig. 5. During the last 1/16 in (1.6 mm) insertion, the scriber shank is flexed backward moving the scriber upwards in contact with the record plate and creating the zero line. A window is provided to view the action of the scriber. When the record plate is removed after an event, a second zero line is scribed. The displacement between the two zero lines is the interstory static shift. A small amount of residual shift due to friction can occur. This residual was measured as about 0.001 in (0.025 mm) following the dynamic test (Fig. 6).

Fig. 4. Record Plate, Holder and O-ring Seal

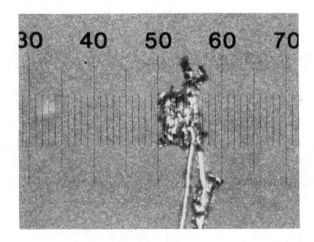

Fig. 5. Enlarged (100x) Static Force Test Record

The recorder can measure ± 3 1/2 in (± 8.9 cm) in the horizontal plane of the A-frame and ± 1/2 in (± 13 mm) in the vertical plane. The combination of the very fine scribed line (approx. 0.0005 in (0.01 mm) wide) and movement in the vertical direction, if it occurs, while recording the horizontal displacement, will produce a complete record of the biaxial displacements as well as the peak record.

Displacement Versus Time Recording Using A Flat Record Plate

This recorder is contained in a single sealed housing with a drive shaft projecting from both sides. The housing is mounted on the lower A-frame at about the center line of the overlapping A-frames. One of the projecting drive shafts is driven by the upper A-frame through a slotted drive arm, so only the horizontal motion is transmitted to the recorder. Up to $\pm$ 1/2 in (13 mm) of vertical motion can be tolerated.

The recorder contains a flat record plate which is driven upwards at a constant speed by a spring drive motor. The motor is released whenever a preset displacement is exceeded between the upper and lower A-frames. Desiccant packages are installed along with the record plate to keep the interior of the recorder dry.

When the spring drive motor is released, the record plate moves beneath the scriber and a record of displacement versus time is made for 26 seconds. The record plate can have a soft metal coating or a developed photographic film on the surface of the plate, either of which will be scratched by the diamond scriber.

Simultaneously with the start of the drive motor, an independent mechanical timer produces timing marks at one second intervals on one side of the record plate. An external signal, such as from an adjacent strong motion accelerograph, can be used to start the recorder and to control the timing marks if correlation with other recorders is desired.

The record plate is installed by removing the bottom cover of the recorder. The spring motor is attached to the record plate, and the bottom cover is replaced. The starting zero line can be seen through the window in the housing. When the record plate is removed, an exit zero line is scribed. Any shift in the zero line represents the interstory static shift.

ALARM SYSTEMS

Three alarm systems are being developed:

1. The simplest system incorporates a moveable shaft which makes contact with microswitches

2. In recorder configurations 1, 3, 4, and 5, described above, electrical contacts can be added for the alarm

3. In recorder configuration 6, the output of the extensiometer is sensed and will alarm above preset limits

In all three alarm systems, more than one alarm level can be selected, and all three systems can alarm remotely as well as locally.

DATA REDUCTION

Reduction of Peak Displacement Records

The zero-line at the beginning of the record will be found near the center of the "ambient conditions" record. This record is caused by vibrations in the structure due to daily temperature cycles, weather, construction, and passing vehicles. Also, the structure may experience slight movements due to settling of the foundation.

After an earthquake, the scriber may not locate on the original zero-line, due to structural deformations. In this case, a second zero-line will be created when the stylus and recording medium are separated.

If the recording medium is properly installed, a zero line will be inscribed in the record, because the recording stylus will deflect when it contacts the recording medium. The distance from the original zero line to each record peak is measured to determine the dynamic displacement. Usually only the largest of the peaks is recorded, as the maximum displacement. Next, the distance between the two zero lines is measured to obtain the static displacement.

Reduction of Time-History Displacement Records

Time-history records of displacement versus time can be read directly, without further processing. The timing marks and the complete record of displacements are inscribed on the same recording medium. In general, only the maximum displacements are required to be read accurately, not the duration between peaks.

If the recording medium is properly installed, a zero line should be inscribed through the length of the record.

TEST PROGRAM AND RESULTS

Static deflection tests were performed using the flat plate mechanical recorder. The overlapping A-frames were mounted securely to the foundation and roof structure of a commercial building. A weight of about 25 lbs (11 kg) was applied at the center of percussion of the A-frames, representing a force of approximately $\pm$ 1 g. The force was applied in all three orthogonal axes in each direction, making a total of six separate load applications.

The record plate was removed and examined with a microscope using 100x magnification, as shown in Fig. 5. Each mark on the reticle represents 0.0005 in (0.013 mm). At the bottom center of the record, the two parallel lines are the zero lines made during the insertion and removal of the record plate. The 0.001 in (0.025 mm) separation in these lines represents static shift due to friction between the two frames. The roughly square record made by the application of horizontal and vertical forces in the plane of the A-frames measures approximately 0.004 in (0.10 mm), horizontally, by 0.005 in (0.13 mm), vertically. The maximum zero-to-peak response of about $\pm$ 0.002 in ($\pm$ 0.05 mm) to the $\pm$ 1 g force applied meets the design goal for high rigidity in the recording axes.

Subsequently, the frame was subjected to impacts using a 1 lb (0.45 kg) rubber mallet. The A-frames were struck repeatedly with a force of approximately 25 lb (11 kg) near the center of percussion in the x, y, and z axes to drive the A-frames into vibration. The results are shown in Fig. 6. The magnification in Fig. 6 is 50, and each mark on the reticle is 0.001 in

(0.03 mm). The roughly square record measures about 0.020 in (0.51 mm), horizontally and vertically, meeting the rigidity criteria.

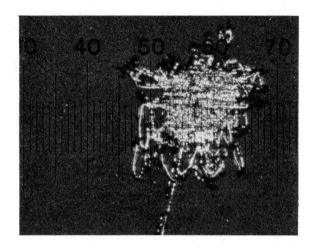

Fig. 6. Enlarged (50x) Dynamic Test Record

The accuracy of the relative displacement recorder is the maximum non-rigid response due to inertial forces divided by the full-scale displacement. For the tests described, the accuracy is:

$$\frac{0.010 \text{ in}}{3.5 \text{ in}} = 0.0028 = 0.28 \%$$

This accuracy should be sufficient for comparison of the relative displacement recorder results and the doubly integrated data from strong motion accelerographs.

Ambient conditions in structures, primarily due to temperature cycling, will cause a record to be made around the initial zero position of the scriber. The initial and final positions of the scriber can be estimated using the entry and exit scriber lines, as shown in Figs. 5 and 6, however the ambient conditions record may prevent identification of dynamic displacements from seismic events which cause very small interstory displacements. During a 24-hour period in the single story light industrial building in which the initial testing was performed, the ambient record accounted for about 0.087 in (2.21 mm) vertical and about 0.055 in (1.40 mm) horizontal displacement, peak-to-peak. The ambient conditions record in a multi-story, commercial or heavy industrial building likely would be considerably smaller.

Further tests are planned to determine the dynamic response of a full-scale or large-scale relative displacement recorder to simulated earthquake input. Initial discussions have been held with the University of California at Berkeley for use of their facilities at the Richmond Field Station.

CONCLUSIONS

A prototype interstory displacement recording system, consisting of a pair of overlapping A-frames and a mechanical, flat plate recorder has been fabricated and tested. Initial tests show that the system has adequate rigidity and accuracy so that the results can be compared to displacement data derived from accelerometers. Additional testing, using simulated seismic motion, is planned to compare the system response to direct displacement measurements on a shake table and to displacement derived from acceleration.

REFERENCES

[1] Trifunac, M.D. and Lee, V.W., "A Note on The Accuracy of Computed Ground Displacements from Strong-Motion Accelerograms," Bull. Seism. Soc. Am., Vol. 64, No. 4, 1209-1219, August 1974.

[2] Engdahl, P.D., "Operation and Maintenance Manual, Peak Acceleration Recorder, Model PAR400," Engdahl Enterprises, 1984, Revision G.

[3] Engdahl, P.D., "Operation and Maintenance Manual, Response Spectrum Recorder, Model RSR1600," Engdahl Enterprises, 1984, Revision 5.

ADVANCED ON-LINE COMPUTER CONTROL METHODS FOR SEISMIC PERFORMANCE TESTING

C.R. Thewalt[1], S.A. Mahin[2] and S.N. Dermitzakis[3]

A major concern in seismic testing is whether the loading conditions imposed on a structure adequately represent those that might occur during an earthquake. The use of shaking tables for structural testing can resolve many of these concerns for some types of structures. The recently developed pseudo-dynamic test method extends the ability to realistically simulate seismic response to a broader range of test specimens.

INTRODUCTION

The pseudo-dynamic test method is a hybrid analysis procedure used for seismic performance evaluation. Using this relatively new method, the seismic response of a test specimen can be simulated by slowly imposing displacements using generally available laboratory equipment. Software in the controlling computer is used to account for dynamic effects. This approach appears to combine the simplicity and economy of conventional quasi-static tests, the versatility and convenience of analytical studies, and the realism of shaking table testing. However, until recently, only very limited experience has been gained with the method. In order to assess the capabilities and limitations of the method, a coordinated series of investigations has been undertaken as part of the U.S. - Japan Cooperative Earthquake Research Program. These investigations have been performed in Japan at the Building Research Institute, Ministry of Construction, Tsukuba, and in the U.S. at the University of Michigan, Ann Arbor, and at the University of California, Berkeley.

Several studies at Berkeley on the pseudo-dynamic method have been reported previously by Shing and Mahin [1-4]. These relate to the stability and accuracy of the underlying numerical procedures, the effect of experimental errors on the accuracy of test results, and verification of the procedure using simple experiments and analyses. In an effort to expand the applicability of the pseudo-dynamic method and to solve some of the problems previously encountered with its use, several extensions to these studies have been made recently. In this paper, the recent work at Berkeley will be summarized.

GENERALIZED FORMULATION

The original implementation of the pseudo-dynamic test method was for complete planar test structures subjected to a single horizontal component of base excitation. Although this mode of testing is quite useful, the pseudo-dynamic method can be easily extended to permit testing of complete structures subjected to multiple component base excitation. This more general formulation would make it possible to perform complex, three dimensional tests that would not be possible using most available shaking tables.

---

1. Research Assistant, University of California, Berkeley, CA.

2. Associate Professor of Civil Engineering., University of California, Berkeley, CA.

3. Senior Engineer, Impell Corp., Walnut Creek, CA.

The time discretized equations of motion for time step i, as used in pseudo-dynamic testing, are given by:

$$\mathbf{M}\,\mathbf{a}_i + \mathbf{C}\,\mathbf{v}_i + \mathbf{r}_i - \mathbf{K}_G\,\mathbf{d}_i = \mathbf{M}\,\mathbf{B}\,\mathbf{a}_{g_i} \qquad (1)$$

where $\mathbf{M}, \mathbf{C}$: mass and damping matrices;
$\mathbf{K}_G$: geometric stiffness matrix;
$\mathbf{a}_i, \mathbf{v}_i, \mathbf{d}_i$: acceleration, velocity and displacement vectors;
$\mathbf{r}_i$: measured structural restoring force vector;
$\mathbf{B}$: ground acceleration transformation matrix;
$\mathbf{a}_{g_i}$: ground acceleration vector.

The component $\mathbf{B}_{ij}$ is the acceleration at structural degree of freedom i when the structure acts as a rigid body under a unit acceleration for ground component j. In a planar test with a single horizontal ground motion component, $\mathbf{B}$ becomes the familiar unit vector. In a test, $\mathbf{M}, \mathbf{C}, \mathbf{K}_G, \mathbf{B}$ and the ground acceleration history are defined by the user. At each step, $\mathbf{d}_{i+1}$ is computed based on the user supplied data and the measured restoring force vector, using an appropriate numerical integration strategy. The restoring force is then measured for use in the next step's calculations.

The user supplied geometric stiffness matrix is only used in tests where it is not necessary to include the full design weight on the specimen. This may be appropriate in certain pseudo-dynamic tests [2]. However, it may still be desirable to include geometric stiffness effects in the equations of motion for these tests. The improved performance of most electro-hydraulic control systems for specimens with low mass has made this technique very attractive.

Transformation of Experimental Degrees of Freedom

In performing three dimensional tests, it has been found to be convenient to allow the user to supply a test specific coordinate transformation so that the equations of motions are solved a coordinate system other than that determined by the arrangement of the hydraulic actuators. This new coordinate system can simplify calculation of the mass and damping matrices, often allowing the use of a diagonal mass matrix. The specified transformation may also need to correct for geometric nonlinearities introduced by the finite length of each actuator and its associated transducer. These geometric errors will affect both the global displacement and force vectors. The example shown in Fig. 1 shows that movement of a single actuator will give an incorrect global specimen position if the displacements are measured by monitoring the change in actuator length. Moreover, the measured actuator forces may need to be corrected for actuator angle.

The current Berkeley implementation of the pseudo-dynamic test method performs these transformations in the following order. A nonlinear transformation based on the deformed position is applied to the measured force vector to transform it it the internal coordinate system. The equations of motion are solved in the internal coordinate system, generating a new desired global displacement vector. This vector is then transformed into an actuator displacement vector using a nonlinear transformation, based on the deformed position.

Electro-Hydraulic Control

To achieve reliable results in these kinds of pseudo-dynamic tests, accurate control of a specimen's position is essential. To assess problems in achieving the required accuracy, a series of simple tests were performed. The behavior of the electro-hydraulic actuators under displacement control for both single and multiple actuator systems has been investigated [5].

The single actuator test, consisting of an actuator mounted on a wide flange steel cantilevered beam, was used to compare the performance of three different size servovalves (5, 10 and 25 gallon/minute rated flows), and also to study the effect of loop gain, force level, ramp shape and specimen stiffness on the accuracy of the control loop.

These tests demonstrated that increasing loop gains resulted in more accurate response, as long as the loop remained stable. Examples of stable and unstable response are shown in Fig's. 2(a) and 2(b). The unstable response was obtained by introducing very high loop gains. It was also found to be beneficial to use more smoothly varying ramp shapes, such as haversine curves, since this reduced the amount of overshoot at the end of the ramp. Also, as valve size or specimen mass increased, the ability to reliably track the desired signal decreased.

Considerable experience has also been gained in performing multiple actuator verification tests. These tests demonstrated the crucial importance of using accurate displacement feedback transducers and of properly adjusting the gains in the hydraulic control loops. The transducers initially used were nonlinear and introduced apparent shifts in modal response frequencies. Systematic lagging of the actuators, as a result of controller errors, numerically introduced energy into the equations of motion as previously predicted [1]. In order to reduce the lagging problem, a short wait period was introduced at the end of each displacement step to allow the actuators to converge on the specified displacements. This technique worked well in the elastic range, where the force and displacement were stable during the waiting period. However, in the inelastic range, the forces relaxed by as much as ten percent during the wait period, even though the displacements were constant, as shown in Fig. 3. The use of integral control techniques is currently being investigated, and may lead to more accurate tracking of the desired signal than is possible with conventional proportional controllers. This will hopefully lead to much shorter duration wait periods.

Verification Tests

A simple three degree of freedom structure shown in Fig. 1 is being used to test the reliability of performing pseudo-dynamic tests with multiple components of base excitation. The structure has been tested on a shaking table under a single component of lateral excitation, and is currently being tested with the pseudo-dynamic method, as well as being modeled analytically. Torsional response of the structure was induced by turning one of the corner columns 90 degrees and by mounting it on the shaking table at 45 degrees to the direction of table motion. The measured lateral table accelerations as well as the three components of rotational base acceleration, (due to imperfect table control), are used as input to the pseudo-dynamic test.

A sequence of eight earthquake records were used, varying in magnitude from low level, with linear elastic structural response, to large earthquakes resulting in considerable yielding of the structure. Previous pseudo-dynamic tests have shown experimental errors to be most significant during elastic tests, when the structure does not dissipate hysteretic energy. The tests using low level ground motion input confirmed the previous observations and allowed the investigation of a coupled electro-hydraulic system.

As suggested previously [1], Fourier amplitude spectra of the displacement error histories were used to estimate how well the hydraulic control system performed. In the initial tests, there was considerable peeking of the error spectra at the structure's natural frequencies, indicating the presence of systematic errors which in turn introduced large errors in the overall response. Fig. 4 shows spectra for two runs, one with systematic errors and one with essentially random errors. In order to reduce the systematic errors, the displacement transducers were upgraded, and a more solid reference frame was built for measurement of global displacements. The gain of the hydraulic controllers was also increased until the error histories were essentially random, with very low level amplitudes. Once the errors were reduced to acceptable levels, it can be seen from Fig. 5 that the results are quite accurate, even for low level elastic tests.

The comparison between pseudo-dynamic and shaking table results for large magnitude earthquakes is also good, as seen in Fig. 6, but slightly larger response levels can be seen in one of the degrees of freedom of the pseudo-dynamic test. This was due to the rapid relaxation of the structural restoring forces when the specimen was in the post yield region and moving in the loading direction. As mentioned previously, there was up to a ten percent force reduction during the 1.5 second wait period used in the tests, even though the displacement records show that there was no movement of the specimen. Systematic errors such as this lead to drift errors in inelastic tests, as opposed to the oscillatory error response in elastic tests, and can be seen in Fig. 6. The drifts due to such errors had been predicted by Shing and Mahin [1]. To improve reliability in the inelastic range, control procedures that assure greater accuracy with smaller wait periods are being investigated.

After the correlation between shaking table and pseudo-dynamic response has been completed, the method will be used to compare the behavior of specimens subjected to two independent horizontal ground motion inputs with the response under a single component ground motion.

PSEUDO-DYNAMIC TESTING UNDER FORCE CONTROL

In extremely stiff structures, such as those containing shear walls, the use of displacement control in pseudodynamic testing may lead to poor results. Actuator forces in such systems change rapidly with very small displacement variations. Verification tests indicate that very accurate displacement control may be possible. However, this may require the use of special high performance equipment and instrumentation, and error propagation considerations may still become critical. An alternate test method has been devised to help overcome these problems. This is based on force control. The proposed method removes problems associated with discrete time integration and essentially solves the equations of motion in analog form using transducer outputs. The rate at which an experiment is performed may be varied continuously up to the electro-hydraulic system capacity. Real inertial and viscous damping forces are explicitly accounted for in this method. This approach also provides for the possibility of performing tests in real time, which would be necessary for testing systems that exhibit rate sensitive behavior.

To implement this procedure, the test specimen must be constructed so as to satisfy all standard similitude relationships, as required for a shaking table specimen. Thus, reduced scale models would be constructed with mass added so that the modal frequencies change by a factor of the square root of the physical model scale. The method then applies forces to the structure that are based on the ground motion inputs and on the real acceleration and velocity of the structure during testing. These measured quantities are used to modify the force signal so that actual inertial and damping forces in the test specimen are accounted for in the equations of motion. A complete description of how these signals are combined in analog form is given by Thewalt [5].

SUBSTRUCTURING IN PSEUDO-DYNAMIC TESTING

All previous applications of the pseudo-dynamic method have been limited to tests of complete systems. Tests of complete structural models are not only expensive, but require special test facilities as well. Where detailed information on the local behavior of critical regions is required, it is more efficient and economical to test structural subassemblages. In addition, one may be interested in the dynamic response of components or equipment mounted in structures which are subjected to ground excitation. However, since the ground motion is not directly applied to the base of the equipment, the supporting structure has to be accounted for in such tests. This leads to a costly and inefficient test setup or significant simplifications that may reduce the accuracy of the results.

One approach to overcoming these difficulties with the pseudo-dynamic method is by application of substructuring concepts used in conventional dynamic analyses. In such analyses, different portions of a structure are grouped into substructures which are treated separately for convenience in formulating the data as well for computational economy. In a pseudo-dynamic test it may be possible to use similar methods, except that certain substructures may be analytically formulated and others are subassemblages that are physically tested. By means of substructuring techniques the displacements which are imposed on the test structure are obtained by solving the equations of motion of the complete system. The restoring force characteristics of the portion which is not subjected to experimental testing are provided by mathematical models using the computed displacements.

A study of the feasibility of applying substructuring concepts has recently been completed by Dermitzakis and Mahin [6]. The theoretical basis of the method has been formulated and some preliminary experimental and analytical studies have been performed to evaluate its reliability. A brief discussion of various applications follow.

Numerical Formulation

Using substructuring techniques, a test structure can be considered as an assembly of two distinct parts : (i) a physical subassemblage which is experimentally tested using load applying actuators ; and (ii) analytical subassemblages consisting of mathematical models of structural elements. The equations of motion of a complete system now takes the following form :

$$\mathbf{M}\,\mathbf{a}_i + \mathbf{C}\,\mathbf{v}_i + \mathbf{R}'_i + \mathbf{R}_i^* = \mathbf{F}_i \qquad (2)$$

In which $\mathbf{R}_i^*$ contains the restoring force vector of the physical specimen and $\mathbf{R}'_i$ contains the restoring force vector of the analytical subassemblage. For linear elastic systems, $\mathbf{R}'_{i+1} = \mathbf{K}\cdot\mathbf{d}_{i+1}$, whereas for nonlinear systems, $\mathbf{R}'_{i+1} = \mathbf{R}'_i + \mathbf{K}^*_{i+1}\cdot(\mathbf{d}_{i+1} - \mathbf{d}_i)$. For nonlinear systems, the tangent stiffness matrix $\mathbf{K}^*_{i+1}$ for the analytical subassemblage is continually updated.

As long as a finite mass can be associated with all active degrees of freedom, the solution of Eq. 2 can be carried out conventionally using explicit integration methods, such as the Newmark method [2,4]. This approach is identical to the general solution of the equations considered in a pseudo-dynamic test of a complete structure, except for the addition of the restoring forces resulting from the analytically modeled components. To obtain bounded solutions, the stability criteria for the particular explicit integration algorithm used must be satisfied for the complete system.

When massless degrees-of-freedom, such as rotations, must be included in the equations of motion, the explicit Newmark algorithms fail to provide a stable solution. A mixed implicit-explicit algorithm [7] can be applied to pseudo-dynamic systems having such properties. Special considerations are necessary when massless degrees of freedom lie on the boundary between the physical and analytical substructures [6].

Verification Tests

A variety of tests have been performed to evaluate the reliability of substructuring methods as applied to pseudo-dynamic testing. These relate to structural subassemblage and equipment tests using the implicit-explicit formulation as well as to nonlinear substructure tests based on an explicit formulation. Additional information on these tests is given by Dermitzakis and Mahin [6].

A single-bay, three-story steel frame building was considered to demonstrate the application of substructuring techniques to specimens with nonlinear subassemblages. For the dynamic analysis, horizontal ground excitation was considered. The prototype structure was idealized as having one lateral degree of freedom at each floor and masses were concentrated at these levels. To facilitate this verification test, beams were assumed rigid.

Analytical results for the complete structure are compared with results from a pseudo-dynamic test using substructuring. During the pseudo-dynamic test, the top two stories were modeled analytically and the bottom story was tested physically. To simplify the experimental setup, inflection points were assumed at mid-height of the first story, and overturning effects were disregarded. With these assumptions, the experimental specimen was reduced to a single degree-of-freedom cantilever column. This specimen was a 4 ft. (1.22 m) long W6×20 wide flange steel section.

Since there are no massless degrees-of-freedom in this structure, the Newmark explicit algorithm can be used as the integration method. Restoring force values in Eq. 2 are obtained from the experimental specimen and from the inelastic analytical models of the substructured stories. The inelastic behavior of the top two (substructured) stories was represented by means of a Menegotto-Pinto model. The inelastic behavior of the frame was examined considering 10 seconds of the 1940 El Centro (NS) ground motion scaled arbitrarily to 1.0g peak acceleration.

The displacement time histories for the three degrees-of-freedom are shown in Fig. 7. The experimental specimen experienced significant inelastic deformations, and its response compared well with analytical simulations. The results of this and other pseudo-dynamic tests with substructuring indicate that substructuring techniques may be successfully applied.

However, it must be recognized that the reliability of the results is limited by the accuracy of the analytical models used. Thus, care must be used in selecting the types of structures and subassemblages to be tested in this fashion. Additional research is needed to extend the applicability of the method and to more completely characterize error propagation effects.

CONCLUSIONS

The pseudo-dynamic test method has been demonstrated to be a powerful and versatile procedure for economically assessing seismic performance of structural systems. Not only can the method be used to test structures that are too large, massive or strong for testing using available shaking tables, but the generalized formulation permits tests with complex support excitations not possible with other techniques. The use of analytical substructuring will allow for more economical and realistic seismic performance evaluations of large structural subassemblages. Like all experimental procedures, the pseudo-dynamic method has inherent limitations and errors, and the limitations must be recognized in developing structural testing programs. In particular, the need for high performance control equipment and instrumentation must be taken into account.

Before the method becomes a reliable, general purpose tool, there must be continued investigation into the control of stiff, massive structures and into additional methods to mitigate the effect of experimental errors. There must be considerable software development to make the system more convenient, and to formulate a generalized substructuring methodology.

ACKNOWLEDGEMENTS

This work was performed under the U.S. - Japan Cooperative Earthquake Research Program. The authors are grateful for the advice of other participants in the program as well as the financial support of the National Science Foundation. However, opinions, findings and conclusions expressed in this report are those of the authors and do not necessarily reflect the views of the National Science Foundation.

REFERENCES

[1] Shing, P.B., and Mahin, S.A., "Experimental Error Propagation in Pseudo-Dynamic Testing," *UCB/EERC-83/12*, Earthquake Engineering Research Center, University of California, Berkeley, June 1983.

[2] Shing, P.B., and Mahin, S.A., "Pseudo-Dynamic Test Method for Seismic Performance Evaluation: Theory and Implementation," *UCB/EERC-84/01*, Earthquake Engineering Research Center, University of California, Jan. 1984.

[3] Mahin, S.A., and Shing, P.B., "Pseudo-Dynamic Method for Seismic Testing," *Journal of Structural Engineering*, ASCE, July 1985.

[4] Shing, P.B., and Mahin, S.A., "Computational Aspects of a Seismic Performance Test Method using On-Line Computer Control," *Earthquake Engineering and Structural Dynamics*, Aug. 1985.

[5] Thewalt, C.R., "Practical Implementation of the Pseudo-Dynamic Test Method," *CE 299 Report* University of California, Berkeley, April 1984.

[6] Dermitzakis, S.N., and Mahin, S.A., "Development of Substructuring Techniques for On-Line Computer Controlled Seismic Performance Testing," *UCB/EERC-85/04*, Earthquake Engineering Research Center, University of California, Feb. 1985.

[7] Hughes, T.G.R., and Liu, W.K., "Implicit-Explicit Finite Elements in Transient Analysis: Stability Theory," ASME, Journal of *ASME, Journal of Applied Mechanics*, Vol. 45, pp. 365-368, June 1978.

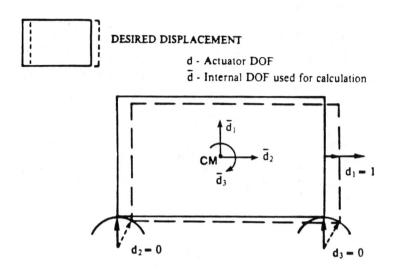

Fig. 1 Position Error due to Geometry Changes for Displacements Measured Across the Actuators

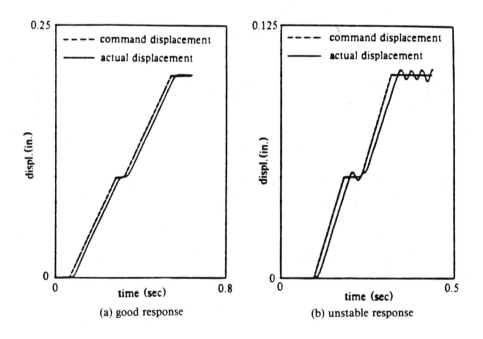

(a) good response

(b) unstable response

Fig. 2 Electrohydraulic Control Loop Response

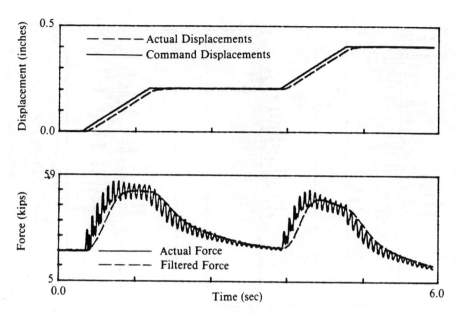

Fig. 3 Force Relaxation During Post Yield Waiting

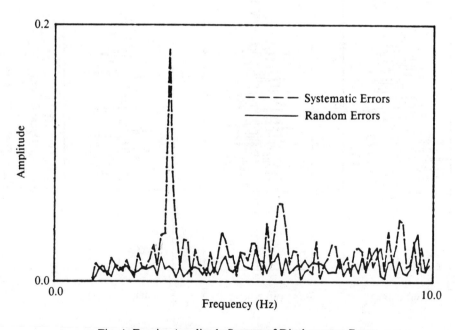

Fig. 4 Fourier Amplitude Spectra of Displacement Errors

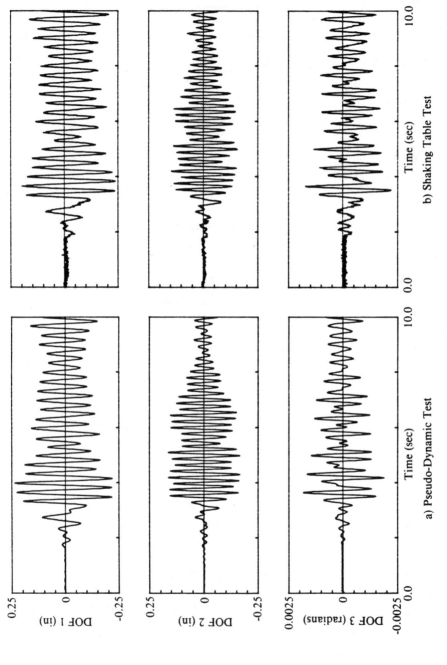

Fig. 5 Shaking Table and Pseudo-Dynamic Elastic Level Displacement Response

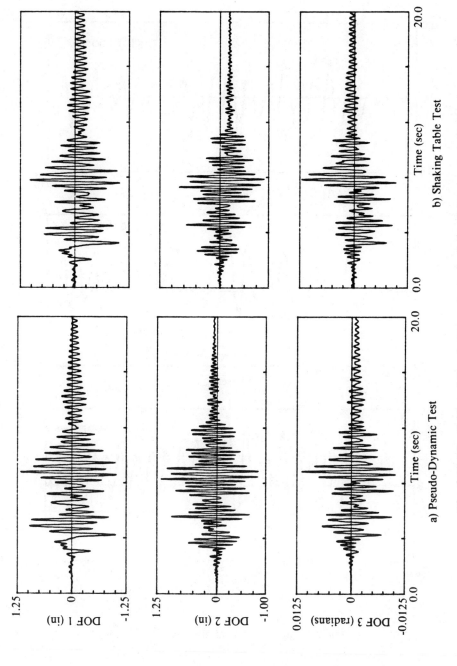

Fig. 6 Shaking Table and Pseudo-Dynamic Inelastic Level Displacement Response

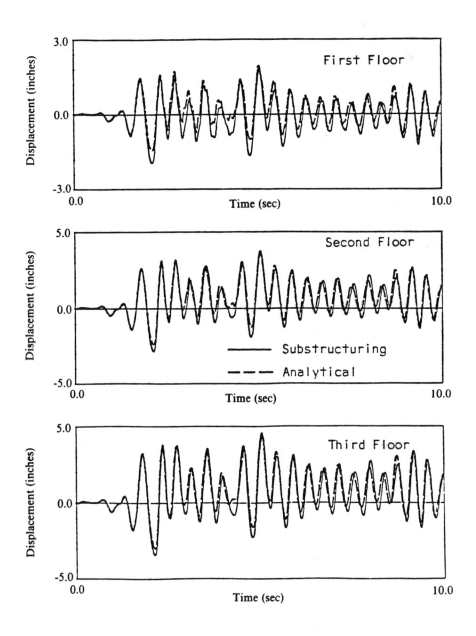

Fig. 7 Displacement Time Histories

ROOF-TOP AMBIENT VIBRATION MEASUREMENTS

J. G. Diehl[I]

ABSTRACT

Simple methods of evaluating existing structures are needed in earthquake engineering. While the ambient vibration measurement approach has been with us for many years, it is often considered only in terms of large, multi-channel research programs. However it can also be used as a simple engineering tool for many structures. This paper presents an application of ambient vibration measurements utilizing a simple two channel system placed on the roof of a multi-story building. Five modes were recovered, sufficient to verify a mathematical model. The application is illustrated with an example.

INTRODUCTION

The ambient vibration survey (AVS) testing method is becoming increasingly popular as an inexpensive and rapid means of acquiring the modal characteristics of existing structures. The AVS can be used to aid the analyst in "calibrating" his mathematical model, or it can be used to acquire modal characteristics for use in seismic risk analysis directly.

The objective of the AVS is to measure a structure's modal characteristics by utilizing the low level vibrations to which all structures are continually subjected. This is accomplished by simultaneously recording the response of the structure at several locations (the total number is dependent on the degree of definition desired in the resulting modal displacement shapes) and analyzing these data using spectral analysis techniques to calculate modal frequencies, damping and displacement shapes. [Reference 1]. These techniques have been described in the literature for nearly 20 years [Reference 2].

ROOF-TOP AMBIENT VIBRATION SURVEY (AVS)

One application of the method, which we call the "Roof-top AVS", utilizes a simple two-channel data acquisition system to measure the response of the roof of a building. By positioning the sensors at opposite ends of the roof, both translational and torsional motion can be analyzed using a portable two channel spectrum analyzer to observe the peaks in the Fourier spectra and the corresponding phases between channels at these frequencies. These measurements immediately identify the fundamental mode, and those higher modes which are translational and those which are torsional. In addition, for simple

----
I. Manager of Sales and Services, Kinemetrics/Systems, Pasadena, California.

structures where the mode ranking comes close to the 1, 3, 5 frequency ratio sequence for structures dominated by shear deformation, it is easy to select the first, second and sometimes third mode of response in each direction from the spectra. Often, verification of only 3 modes can verify a model which is accurate to ±5%.

EXAMPLE

The Roof-top AVS is illustrated using an actual structure whose plan view is shown in Figure 1. The first two channel measurement was made in the northsouth direction with one channel at Location A and the other at Location B. The output of the spectrum analyzer was plotted on-site for examination by the responsible engineer. These plots of amplitude spectra, phase and amplitude transfer functions and the coherence function are presented in Figure 2. The frequencies of the key peaks in the FFT's were identified using the cursor in the analyzer. They are noted on the channel 1 spectra in Figure 2 and presented in Table 1 along with the corresponding values of the phase, relative amplitude and coherence.

RESULTS: NORTH - SOUTH

The first peak amplitude is at 1.04 Hz. At this frequency both the east and west sides of the roof top are moving in-phase at nearly the same amplitude. The coherence is 0.99 indicating little noise contamination of the signal at this frequency. This is the characteristic motion one would expect in the first translational mode along a principal axis at the roof.

At the second frequency of 1.60 Hz, the east and west sides of the roof are moving out of phase with Location A (east side) moving at twice the amplitude of Location B (west side). The coherence of 0.80 is still quite high. This motion is characteristic of a first torsional mode. The difference in amplitude of motion between the sides of the building indicates that Locations A and B are not equidistant from the center of torsion.

At 3.36 Hz the east and west sides of the roof are again moving in phase at high coherence indicating a second translational mode. Location B is moving at about twice the amplitude of Location A indicating a torsional component in the motion.

The fourth peak at 5.23 Hz could be interpreted as a third translational mode because of the phase relationship. Also, it follows generally the 1-3-5 sequence of mode ratios. However, it is obviously not a single peak. Furthermore, the incoherence at this frequency means that either the signals are statistically independent or contaminated by noise.

RESULTS: EAST - WEST

The second two-channel measurement was made in the east-west direction with one channel at Location C and the other at Location D. The resulting plots of amplitude spectra, transfer functions and coherence are presented in Figure 3.

Analysis of the response in the east-west direction should reveal east-west translational modes not seen in the north-south plots, and confirm the torsional modes already identified. It may also give additional data on what we've called north-south modes if the response is coupled or involves torsion.

In fact, the first peak in the spectrum, at 0.92 Hz, is new and, like the 1.04 Hz peak in the north-south direction, both sides of the roof are moving in-phase at nearly the same amplitude. High coherence also indicates the dependent relationship of the signals.

The second frequency at 1.60 Hz is the torsional mode representing itself in the east-west response with the same characteristics as before.

There is a new peak at 2.80 Hz. The data indicate that this response, characterized by in-phase motion in the east-west direction with Location D moving at nearly twice the amplitude of Location C, is the second translational mode in this axis.

DAMPING

In addition to frequency and mode shape information, the damping was measured from the amplitude spectra using the halfpower bandwidth method. This is illustrated in Figure 4. The results are presented in Table 3.

DISCUSSION

It is of interest to see if the second north-south translational mode at 3.36 Hz is contained in the east-west response. This mode is observed to include some torsion evidenced by the relative amplitudes in Table 1. The east-west components of this motion measured at opposite edges of the roof are, in fact, out of phase.

If this is the case, why is there no evidence of the 2.80 Hz east-west mode in the north-south analysis? Actually, there is a peak in the Figure 2, Location B, spectrum at 2.80 Hz. It was not reported because it was not represented in the Location A spectrum and, furthermore, the signals are not coherent.

Is the third east-west translational mode visible in this analysis? There is a peak at 4.65 Hz which exhibits the necessary response characteristics. However, the coherence of 0.48 is less than desirable. Is the second torsional mode visible? In the range of 5 to 6 Hz there are several modes, probably including the second torsional mode. Additional measurements would be required to separate this response.

LIMITATIONS

What if the building had been U-shaped rather than rectangular? The Rooftop AVS allows you to recognize and identify modes of response where they are expected based on a prior understanding of what that mode looks like in plan view. The first north-south translational mode can be differentiated from the second only because they have different frequencies. Otherwise they look identical at the roof top. The same is true for more complex structures The more complex and closely spaced the modes are, the more advance knowledge is required, and more measurement-pairs are required to interpret and define results.

SUMMARY AND CONCLUSIONS

A Roof-top AVS has been performed on a typical structure. The results of these analyses are summarized in Table 3. The characteristics of the first 5 modes of this structure have been identified, based on prior understanding of the typical behaviour of rectangular, multistory buildings by interpreting the spectral analysis of roof top measurements. These results were obtained using a simple two-channel measurement system and a field program involving less than 1 day on-site.

This example illustrates the usefulness of simple ambient measurements as a diagnostic tool. This method provides a fast, inexpensive, non-destructive and effective means of acquiring modal characteristics of an existing structure. The Roof-top AVS is easiest to interpret on structures that are symmetrical about at least one axis. Complex structures can be surveyed using more measurement-pairs. These results can be used directly for seismic risk studies or inertial force analyses, or used to verify the accuracy of a mathematical model.

REFERENCES

[1] Kinemetrics/Systems Application Note No. 3 "Ambient Vibration Survey: Application, Theory and Analytical Techniques".

[2] H. S. Ward and R. Crawford <u>Bulletin of the Seismological Society of America</u> Vol 56, 1966, Pages 793-813, "Wind Induced Vibrations and Building Modes".

Table 1

Measured Spectral Values
North-South Analysis

Transfer Function (A relative to B)

Frequency	Degrees	Amplitude (dB, ratio)		Coherence
1.04 Hz.	+2°	-3.6 dB,	.66	.99
1.60	-184	+5.8,	1.95	.80
3.36	+2	-5.2,	.55	.96
5.23	-5	-10.5,	.30	.11

Table 2

Measured Spectral Values
East-West Analysis

Transfer Functions (D relative to C)

Frequency	Degrees	Amplitude (dB, ratio)		Coherence
0.92 Hz.	0°	+2.0 dB,	1.3	.96
1.60	-185	-4.2,	.62	.93
2.80	-5	+5.5,	1.9	.93
3.36	-189	-12.5,	.24	.51
4.65	-25°	-5.3,	.54	.48

Table 3

Typical Measurement Results
Rooftop AVS

Mode Description	Frequency	Damping
First E-W Translation	0.92 Hz.	2.7% of critical
First N-S Translation	1.04	3.8
First Torsion	1.60	3.2
Second E-W Translation	2.80	2.2
Second N-S Translation	3.3	1.5

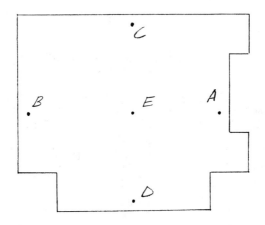

Figure 1

TYPICAL ROOFTOP
MEASUREMENT LOCATIONS

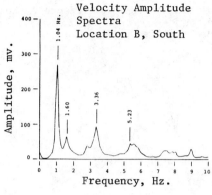

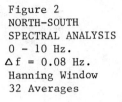

Figure 2
NORTH-SOUTH
SPECTRAL ANALYSIS
0 - 10 Hz.
$\Delta f = 0.08$ Hz.
Hanning Window
32 Averages

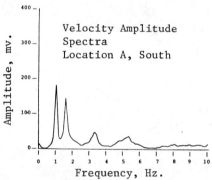

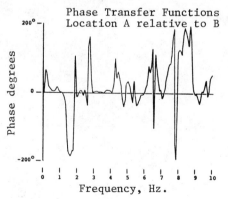

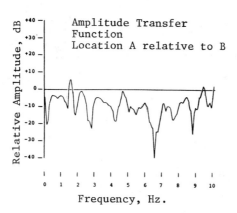
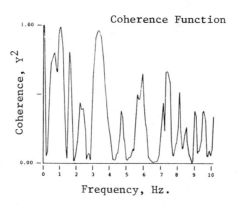

Figure 2, continued

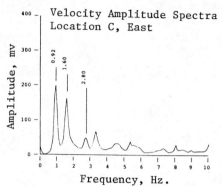
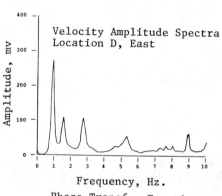
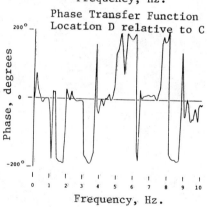
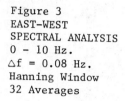

Figure 3
EAST-WEST
SPECTRAL ANALYSIS
0 - 10 Hz.
$\Delta f = 0.08$ Hz.
Hanning Window
32 Averages

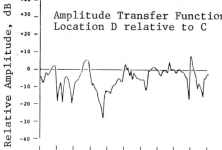

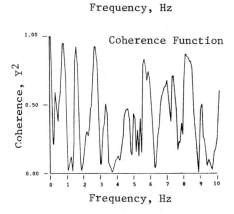

Figure 3, continued

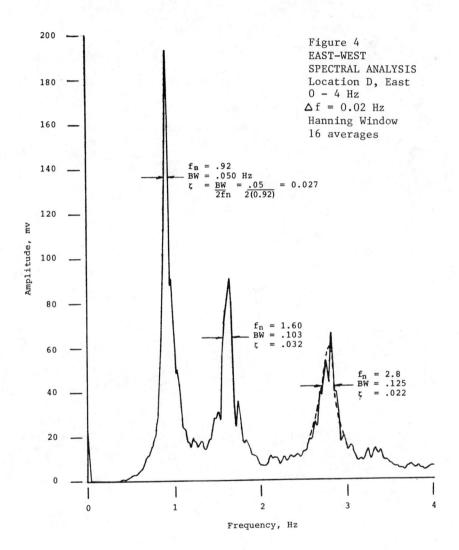

Figure 4
EAST-WEST
SPECTRAL ANALYSIS
Location D, East
0 - 4 Hz
$\Delta f = 0.02$ Hz
Hanning Window
16 averages

7. NONSTRUCTURAL SYSTEMS AND BUILDING COMPONENTS

SPECIAL ISSUES IN THE SECONDARY SYSTEM RESPONSE
BY THE FLOOR RESPONSE SPECTRUM METHOD

Jing-Wen Jaw[I] and Ajaya K. Gupta[II]

ABSTRACT

A new direct method of generating floor response spectra for multiply connected secondary systems is presented. The method is capable of accounting for the interaction between primary-secondary systems, correlation between responses from multiple connecting DOF, and the straining effect of multiple connecting DOF displacement including related correlations. The method directly uses the response spectrum at the base of primary system as seismic input without converting it into either compatible time history or power spectral density function. The method introduces certain new issues. The present paper addresses these issues which are relevant to the application of the proposed method to practice.

INTRODUCTION

In current practice, the seismic analysis of secondary systems is performed using a decoupled model of the secondary system excited by the floor response spectrum. A decoupled analysis of the secondary system ignores the effect of interaction between the primary and secondary systems, which can be significant, especially at the resonant frequencies. As a result, the secondary system response may be too conservative. To obtain floor response spectra at the support points of the secondary systems, a time history analysis of the primary system is performed. Since the primary system seismic input is usually specified in terms of a design spectrum, a time history compatible with the input response spectrum is generated. This time history is not unique and may introduce considerable uncertainty in the secondary system response values. In addition, a time history analysis to obtain the floor spectra is computationally time consuming and costly. The practice of enveloping the floor spectra for various connecting DOF may introduce unnecessary conservatism in the secondary system response calculation. The analysis implies that all the connecting DOF motions are in-phase, which may be unrealistic. To account for the effect of out-of-phase support motions, a separate static analysis is performed. This practice is irrational, and again may lead to more conservatism.

I. Graduate Student, Department of Civil Engineering, North Carolina State University, Raleigh, NC 27695-7908.

II. Professor of Civil Engineering, North Carolina State University, Raleigh, NC 27695-7908.

A review of various alternatives to the conventional approach is given by Gupta [1]. It was suggested [1] that the problems described above can be overcome if the analysis is performed on the coupled system (combined primary-secondary system). Developments leading to the coupled system analysis are presented in Ref. [2-6]. In addition, an extensive literature review is also presented in those publications.

A new direct method of generating floor response spectra for multiply supported secondary systems has been developed by us [7]. The method is capable of overcoming many problems associated with the conventional floor response spectrum method. For example, the effect of interaction is built into the IRS (Instructure Response Spectra) naturally by assigning the appropriate mass ratio for the SDOF oscillators. Also, the correlations are properly accounted for between (1) the oscillator responses at various connecting DOF; (2) the displacements at different connecting DOF; (3) the displacements at the connecting DOF and the oscillator response. Several other researchers have made attempts to improve the accuracy of the conventional floor response spectrum method also, see for example Ref. [8,9].

Though the new IRS method offers many significant advantages, it introduces certain new issues and problems. Some of the issues are summarized here. They are: (1) the amount of correlation data that is produced may be too voluminous and may discourage practical use of the new method; (2) the effect of static constraint on the primary system cannot be accounted for; (3) the algorithm for direct generation of instructure response spectra may require more computer time than that for the conventional floor response spectra. In this paper, the impact of these issues on the efficiency of the new method, and on the accuracy of the secondary system response values is addressed. Approximate expressions for the correlation coefficients are proposed. Numerical results, demonstrating the applicability of the proposed equations to several primary-secondary systems, are presented. For the purpose of completeness, a brief summary of the new IRS method is presented here.

INSTRUCTURE RESPONSE SPECTRUM (IRS)

It is shown in Ref. [7] that a majority of the attributes of the coupled analysis [5,6] can be incorporated by the IRS approach. The best way to account for interaction between primary and secondary systems is by performing a coupled analysis [1,5,6]. In the IRS method an SDOF system represents the response of the secondary system in one secondary system mode at a time. We shall denote the oscillator frequency and damping ratio by ω_s and ζ_s, respectively, and the energy mass ratio with respect to the ith primary system mode by r_i. For an SDOF system we can write

$$r_i^{1/2} = \sqrt{m_s}\ \{\phi_{ci}\} \qquad (1)$$

where $\{\phi_{ci}\}$ represents ith normalized mode shape of uncoupled primary system at the connecting DOF c. It is known that the interaction will be most significant when the frequency of the secondary system mode is close to a primary system frequency. Therefore, it can be assumed that the given mass ratio is between a secondary system mode and the primary mode 'I' whose frequency is closest to the secondary system frequency. Thus, Eq. 1 yields

$$\sqrt{\overline{m}_s} = r_I^{1/2} / \{\phi_{cI}\} \tag{2}$$

Equation 2 defines the mass of the SDOF system. The values of r_i with respect to other primary system modes now can be calculated from Eq. 1.

Coupled Frequencies and Mode Shapes

The coupled system in the IRS approach consists of the primary system and an SDOF oscillator having frequency ω_s, damping ratio ζ_s mass ratio r_I, and the corresponding mass m_s. The oscillator is connected to the primary system at the DOF at which the IRS needs to be evaluated. Let us denote the coupled eigenvalue by λ, where $\lambda = -\zeta\omega + i\omega_D$, $\omega_D = \omega\sqrt{1-\zeta^2}$; in which ω and ζ are the appropriate coupled frequency and damping ratio, respectively. The equations for the coupled system eigenvalues corresponding to Ith primary system mode, and the secondary system mode can be written as

$$\lambda_I^2 + 2\omega_{pI}\zeta_{pI}\lambda_I + \omega_{pI}^2 + \frac{\lambda_I^2(\omega_s^2 + 2\omega_s\zeta_s\lambda_I)r_I}{\omega_s^2 + 2\omega_s\zeta_s\lambda_I + \lambda_I^2} = 0$$

and

$$\lambda_s^2 + 2\omega_s\zeta_s\lambda_s + \omega_s^2 - \frac{(\omega_s^2 + 2\omega_s\zeta_s\lambda_s)^2 r_I}{\omega_{pI}^2 + 2\omega_{pI}\zeta_{pI}\lambda_s + (\omega_s^2 + 2\omega_s\zeta_s\lambda_s)r_I + \lambda_s^2} = 0 \tag{3}$$

in which ω_p and ζ_p are primary system frequency and damping ratio, respectively. Equation 3 is derived by modifying the appropriate equations in Ref. [4]. For the purpose of evaluating the two complex eigenvalues, λ_I and λ_s, Eqs. 3 are solved in an iterative manner. For the primary system modes other than the mode 'I', the changes in the eigenvalues due to coupling are usually small and are neglected here. Therefore, we can write

$$\lambda_i = -\zeta_{pi}\omega_{pi} + i\omega_{Di}, \quad i \neq I \tag{4}$$

For each uncoupled primary system mode i, including i = I, there is a coupled eigenvector which can be expressed in terms of transformed coordinates $\{X\}$ [4],

$$x_i = 1 \; ; \; x_j = 0, \; j \neq i$$

and

$$x_s = \frac{r_i^{1/2}(\omega_s^2 + 2\omega_s\zeta_s\lambda_i)}{\omega_s^2 + 2\omega_s\zeta_s\lambda_i + \lambda_i^2} \tag{5}$$

There is another eigenvector of the coupled system corresponding to uncoupled SDOF oscillator,

$$x_s = 1$$

$$x_j = \frac{r_j^{1/2}(\omega_s^2 + 2\omega_s\zeta_s\lambda_s)}{\omega_{pj}^2 + r_j\omega_s^2 + 2(\omega_{pj}\zeta_{pj} + r_j\omega_s\zeta_s)\lambda_s + \lambda_s^2} \tag{6}$$

for all values of j.

Evaluation of In-Structure Spectral Quantities

We are interested in the displacement of the connecting DOF u_c and that of the SDOF oscillator u_s. In the coupled mode corresponding to the ith uncoupled primary system mode, the complex modal values of u_c and u_s (ψ_c and ψ_s, respectively) can be written as

$$\{\psi_{ci}\} = \{\phi_{ci}\}, \quad \psi_{si} = x_s/\sqrt{m_s} \tag{7}$$

Similarly, in the coupled mode corresponding to the uncoupled oscillator mode, we can write

$$\{\psi_{cs}\} = \sum_j \{\phi_{cj}\} x_j, \quad \psi_{ss} = 1/\sqrt{m_s} \tag{8}$$

The summation in Eq. 8 is on all the primary system modes. In the response spectrum method for nonclassically damped systems [3], each complex mode shape and its conjugate give two real response vectors

$$\begin{Bmatrix} u_{ci}^d \\ u_{si}^d \end{Bmatrix} = -2\text{Re}\,\bar{\lambda}_i F_i \begin{Bmatrix} \psi_{ci} \\ \psi_{si} \end{Bmatrix} S_{Di}^d, \quad \begin{Bmatrix} u_{ci}^v \\ u_{si}^v \end{Bmatrix} = -2\text{Re}\,F_i \begin{Bmatrix} \psi_{ci} \\ \psi_{si} \end{Bmatrix} S_{Vi}^v$$

and

$$\begin{Bmatrix} u_{cs}^d \\ u_{ss}^d \end{Bmatrix} = -2\text{Re}\,\bar{\lambda}_s F_s \begin{Bmatrix} \psi_{cs} \\ \psi_{ss} \end{Bmatrix} S_{Ds}^d, \quad \begin{Bmatrix} u_{cs}^v \\ u_{ss}^v \end{Bmatrix} = -2\text{Re}\,F_s \begin{Bmatrix} \psi_{cs} \\ \psi_{ss} \end{Bmatrix} S_{Vs}^v \tag{9}$$

where S^d_{Di} and S^v_{Vi} are the spectral displacement and velocity for the ith coupled modal frequency from the displacement and velocity spectra of the input motion for the primary system; S^d_{Ds} and S^v_{Vs} are the corresponding values for the coupled modal frequency of the oscillator; $\bar{\lambda}_i$ and $\bar{\lambda}_s$ are the complex conjugate of λ_i and λ_s, respectively. The terms F_i and F_s are defined in Ref. [7].

The spectral displacement S_c is the relative displacement of the oscillator with respect to the connecting DOF.

$$S_c = u_s - u_c \tag{10}$$

The values of spectral displacements corresponding to various values of u_c and u_s in Eq. 9 can be evaluated based on Eq. 10, and denoted by S^d_{ci}, S^v_{ci}, S^d_{cs}, S^v_{cs}. To compress the notations, we shall use the subscript i to include all the coupled modes. Thus, all the instructure spectral displacements can be denoted by S^d_{ci} and S^v_{ci}.

IRS Definition and Correlations

The modal combination for a nonclassically damped system is given by [3]

$$R^2 = \sum_i \sum_j \sum_a \sum_b \varepsilon^{ab}_{ij} R^a_i R^b_j \tag{11}$$

in which both superscripts a and b can be either d or v, independently; R represents the maximum probable combined response, R^d_i is the response in the ith mode from the displacement spectrum, and R^v_i is that, due to the velocity spectrum. Further, we define ε^{ab}_{ij} as follows;

$$\varepsilon^{dd}_{ij} = \varepsilon^d_{ij}, \quad \varepsilon^{vv}_{ij} = \varepsilon^v_{ij}, \quad \varepsilon^{dv}_{ij} = \mu_{ij}, \quad \varepsilon^{vd}_{ij} = -\mu_{ji}.$$

The terms ε^d_{ij}, ε^v_{ij}, and μ_{ij} are various correlation coefficients and are defined in [3,10,11]. Based on Ref. [7], a secondary system response R can be written as

$$R^a_i = \sum_c A_c u^a_{ci} + \sum_c \sum_\alpha B_{c\alpha} S^a_{ci\alpha} \tag{12}$$

in which A_c and $B_{c\alpha}$ are known constant coefficients; u^a_{ci} (a = d or v) is the maximum displacement of the connecting DOF in the ith coupled mode; $S^a_{ci\alpha}$ represents the instructure spectral values for an oscillator with frequency $\omega_{s\alpha}$ and critical damping ratio $\zeta_{s\alpha}$ in the ith coupled mode. Let us define

$$(u_c)^2 = \sum_i \sum_j \sum_a \sum_b \varepsilon^{ab}_{ij} u^a_{ci} u^b_{cj}, \quad (S_{c\alpha})^2 = \sum_i \sum_j \sum_a \sum_b \varepsilon^{ab}_{ij} S^a_{ci\alpha} S^b_{cj\alpha}$$

$$\varepsilon_{c1c2} = \frac{\sum\limits_{i}\sum\limits_{j}\sum\limits_{a}\sum\limits_{b} \varepsilon_{ij}^{ab} u_{c1i}^{a} u_{c2j}^{b}}{u_{c1} u_{c2}}, \quad \varepsilon_{c1c2\alpha\beta} = \frac{\sum\limits_{i}\sum\limits_{j}\sum\limits_{a}\sum\limits_{b} \varepsilon_{ij}^{ab} S_{c1i\alpha}^{a} S_{c2j\beta}^{b}}{S_{c1\alpha} S_{c2\beta}}$$

and

$$\varepsilon_{c1c2\beta} = \frac{\sum\limits_{i}\sum\limits_{j}\sum\limits_{a}\sum\limits_{b} \varepsilon_{ij}^{ab} u_{c1i}^{a} S_{c2j\beta}^{b}}{u_{c1} S_{c2\beta}} \tag{13}$$

Equations 11, 12, and 13 yield

$$R^2 = \sum_{c1}\sum_{c2} \varepsilon_{c1c2} R_{c1} R_{c2} + \sum_{c1}\sum_{c2}\sum_{\alpha}\sum_{\beta} \varepsilon_{c1c2\alpha\beta} R_{c1\alpha} R_{c2\beta}$$

$$+ 2 \sum_{c1}\sum_{c2}\sum_{\beta} \varepsilon_{c1c2\beta} R_{c1} R_{c2\beta} \tag{14}$$

in which $R_{c1} = A_{c1} u_{c1}$, $R_{c1\alpha} = B_{c1\alpha} S_{c1\alpha}$

Equation 14 constitutes the modal combination equation for the proposed IRS method. The desired spectral values along with various correlations coefficients are defined in Eq. 13.

SPECIAL ISSUES

The IRS method requires the evaluation of three sets of correlation coefficients in the calculation of secondary system response. The correlation coefficients are defined in Eq. 13, and the procedure of evaluating those is straightforward. However, it involves handling of a large amount of data. It is, therefore, desirable to develop simple algorithms for the evaluation of these coefficients. First, let us consider ε_{c1c2} in Eq. 13, i.e., the correlation between the displacements at various connecting DOF. Each of the coupled analyses described in the previous section would give a set of u_{ci}^{a} values. Which set then should we use in Eq. 13, and elsewhere? It is decided to use the u_{ci}^{d} values from an uncoupled primary system analysis as a reasonable approximation. Note that the primary system, itself, is a classically damped system; therefore, the uncoupled analysis would give $u_{ci}^{v} = 0$. As the result, the evaluation of ε_{c1c2} turn out to be simple, and it does not require a great deal of storage for data transmittal. Second, consider $\varepsilon_{c1c2\alpha\beta}$, which is the correlation in the response of the oscillator at the connecting DOF c1 in the αth frequency with the response of the oscillator at the connecting DOF c2 in the βth frequency. Based on [10], we can write

$$\varepsilon_{cc\alpha\beta} = \varepsilon_{\alpha\beta} \sqrt{(1 - \alpha_{c\alpha}^{2})(1 - \alpha_{c\beta}^{2})} + \alpha_{c\alpha} \alpha_{c\beta} \tag{15}$$

in which $\varepsilon_{\alpha\beta}$ represents the correlation between the responses of two oscillators having frequencies and critical damping ratios of $(\omega_{s\alpha}, \zeta_{s\alpha})$ and $(\omega_{s\beta}, \zeta_{s\beta})$, assuming that the response is damped periodic [10,11]; $\alpha_{c\alpha}$ is a rigid response coefficient identifying the steady state content in the oscillator response. In Eq. 15 both the oscillators are subjected to the same motion at the connecting DOF c. We observed from the numerical data that the effect of the correlation between the two different connecting DOF can be approximately represented by the value of $\varepsilon_{c1c2\alpha\alpha}$ at two extreme ends of the oscillator frequency, when $\omega_{s\alpha}$ is very low (~ 0.01 Hz), and when $\omega_{s\alpha}$ is very large (~ 100 Hz). At the low frequency end, the oscillator response is primarily damped periodic and the response correlation is denoted by ε^p_{c1c2}. At the high frequency end, the oscillator response is rigid, and the support correlation is denoted by ε^r_{c1c2}. When $c1 = c2$, both ε^p_{c1c2} and ε^r_{c1c2} will become unity. The proposed equation is

$$\varepsilon_{c1c2\alpha\beta} = \varepsilon_{\alpha\beta} \, \varepsilon^p_{c1c2} \sqrt{(1 - \alpha^2_{c1\alpha})(1 - \alpha^2_{c2\beta})} + \varepsilon^r_{c1c2} \, \alpha_{c1\alpha} \, \alpha_{c2\beta} \qquad (16)$$

clearly, when $c1 = c2 = c$, Eq. 16 degenerates to Eq. 15. Finally, we shall develop the expression for $\varepsilon_{c1c2\beta}$, which is the correlation between the displacement of the connecting DOF c1 and the response of the oscillator in the βth frequency at connecting DOF c2. We note that at higher frequency, $\omega_{s\beta}$, we have $\varepsilon_{c1c2\beta} = \varepsilon_{c1c2\beta\beta} = \varepsilon^r_{c1c2}$. We propose that at lower frequencies the correlation between the motion at c1 and the oscillator response at c2 will diminish in proportion to the rigid response coefficient $\alpha_{c2\beta}$. Thus, we propose

$$\varepsilon_{c1c2\beta} = \alpha_{c2\beta} \, \varepsilon_{c1c2\beta\beta} \qquad (17)$$

It is shown in [1] that the secondary system may offer no static constraint, weak constraint, or strong constraint to the primary system. The effect of the constraint is to increase the magnitude of the coupled frequencies. To account for the effect in the evaluation of coupled frequencies, we need to know the properties of both the primary and the secondary systems at the time the IRS are being developed. It is practically impossible to do that. Furthermore, it defeats the main advantage of using the IRS. Therefore, we ignore this effect in the evaluation of the coupled frequencies. Clearly, if the static constraint offered by the secondary system significantly affects the coupled response of the primary system, the proposed IRS method--for that matter, any IRS method--cannot be used. On the other hand, we do consider the contribution of the static constraint on the response of the secondary system. In view of Eq. 12, when the secondary system does not offer any static constraint, the displacement of the connecting DOF will introduce rigid body motion in the secondary system. Therefore, in such case, for stress-related response values, the coefficient A_c will be zero. On the other hand, when the secondary system does apply constraint, the stresses in the secondary system developed by the displacement of connecting DOF can be significant, as will be shown with the numerical results.

In the conventional floor response spectrum method, r_I is assumed to be zero; as a result, the interaction between primary and secondary systems is ignored. The IRS curves are drawn for several values of ζ_s by varing ω_s. In the proposed IRS method, $r_I \neq 0$, the spectral values are drawn for several sets of (r_I, ζ_s) values by varying ω_s. Since the proposed IRS approach is an approximation of the coupled response spectrum method [5], the inherent characteristics of the coupled analysis still remain. It is obvious that in the course of obtaining a set of IRS curves, we need to solve a large number of coupled problems. However, Eqs. 3 give two coupled frequencies for a set of (r_I, ζ_s) value and a fixed ω_s. The rest of the frequencies of the coupled system remain practically the same as that of the uncoupled primary system. This procedure reduces the computational time in the evaluation of complex eigenvalues of a coupled system.

NUMERICAL RESULTS

To demonstrate application of the proposed methods presented in this paper, the primary-secondary system shown in Fig. 1 is analyzed. Ref. [2-5] provide more details about coupled and uncoupled characteristics of the system. The IRS calculated from the proposed method were compared with the IRS obtained from the time history analyses for twelve different real earthquake ground motions in Ref. [7]. The agreement between the two sets of spectra was found to be uniformly excellent. The mean floor response spectrum at the top floor of primary system for the 12 earthquake ground motions (scaled to 1g maximum ground acceleration) is given in Fig. 2, which again shows an excellent agreement between the spectra generated by the proposed method and that generated using the time history.

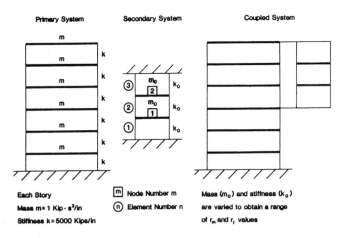

Fig. 1 Example Primary-Secondary System

To examine the accuracy of the proposed IRS method along with approximate correlation coefficients, a comparative study for secondary system response is performed. The story mass and stiffness of the secondary system, Fig. 1, are varied to give nine coupled systems, designated here as 9 cases. These are the same cases which were analyzed in [2-5]. Results from the analysis of all nine systems for the El Centro (S00E, 1940) earthquake ground motion are given in Tables 1 and 2. Nodal displacements (Table 1) and element spring forces (Table 2) were calculated using three different methods, viz, time history method, proposed IRS method and the conventional floor response spectrum method. Percent errors in the results from the proposed IRS method and the conventional floor response spectrum method are also shown in the same tables, using the time history results as the reference values. Note that the first and the third columns of the proposed IRS method in Tables 1 and 2 are the combined response values which were obtained by applying different sets of correlation coefficients. The correlation coefficients which were applied to obtain the results in the first column are given by Eq. 13. On the other hand, we applied the proposed Eqs. 16 and 17 to obtain the results in the third column. It is clear that the results from the proposed IRS method are much closer to the time history results than those from the conventional floor response spectrum method. The orders of mean percent errors and the standard deviations of the percent errors for the proposed IRS method in Tables 1 and 2 are well within the acceptable margins. Further, the use of the approximate correlation coefficients from Eqs. 16 and 17 does not appear to introduce any significant error.

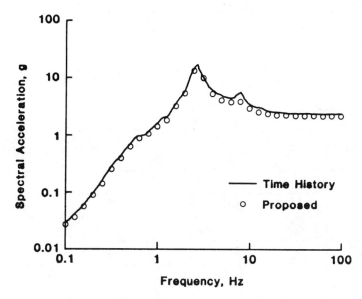

Fig. 2   Mean Floor Response Spectrum on the Top Floor, Primary System Damping: 7%, Secondary System Damping: 2%, Mass Ratio: 0

Table 1. Comparison of Nodal Displacements from the Proposed Method and the Conventional Floor Response Spectrum Method

Case	Node	Time History Displacement, Inches	Proposed IRS Method				Conventional Floor Response Spectrum Method	
			Displacement,[1] Inches	Percent Error	Displacement,[2] Inches	Percent Error	Displacement, Inches	Percent Error
1	1	1.27	1.13	-11	1.06	-17	1.07	-16
	2	1.33	1.24	-7	1.18	-11	1.15	-14
2	1	1.36	1.37	1	1.28	-6	1.91	41
	2	1.50	1.48	-1	1.40	-7	1.96	31
3	1	2.10	1.84	-12	1.70	-19	3.53	68
	2	2.24	1.95	-13	1.81	-19	3.56	59
4	1	2.59	2.32	-10	2.14	-17	4.55	75
	2	2.69	2.44	-9	2.25	-16	4.57	70
5	1	2.89	2.75	-4	2.56	-11	5.02	76
	2	2.91	2.86	-2	2.66	-9	5.04	73
6	1	4.15	3.39	-18	3.36	-19	6.30	52
	2	4.20	3.39	-19	3.36	-20	6.31	50
7	1	4.18	3.52	-16	3.48	-17	5.92	42
	2	4.38	3.52	-20	3.48	-21	5.94	36
8	1	4.90	4.02	-18	3.96	-19	6.94	42
	2	4.96	4.02	-19	3.96	-20	6.96	40
9	1	3.00	2.52	-16	2.66	-11	7.93	164
	2	3.11	2.50	-20	2.66	-14	7.94	155
Mean				-11.9		-15.2		58.0
Standard Deviation				6.8		4.7		44.1

[1] Using correlation coefficients computed numerically from Eqs. 13.
[2] Using correlation coefficients from approximate formulas, Eqs. 16 and 17.

Table 2. Comparison of Spring Forces from the Proposed IRS Method and the Conventional Floor Response Spectrum Method

Case	Element	Time History Spring Force, Kips	Proposed IRS Method				Conventional Floor Response Spectrum	
			Spring Force,[1] Kips	Percent Error	Spring Force,[2] Kips	Percent Error	Spring Force, Kips	Percent Error
1	1	167	147	-12	130	-22	278	65
	2	71.2	46.1	-35	73.8	4	164	131
	3	114	86.3	-24	83.7	-27	278	143
2	1	150	132	-12	124	-17	322	115
	2	34.9	25.9	-26	40.5	16	96.0	175
	3	130	100	-23	102	-22	322	147
3	1	152	124	-18	119	-22	345	127
	2	21.6	16.0	-26	23.2	7	54.2	151
	3	139	108	-22	110	-21	345	148
4	1	139	117	-16	113	-19	300	116
	2	12.6	11.2	-11	15.6	24	36.4	189
	3	133	106	-20	107	-20	300	126
5	1	117	106	-9	103	-12	240	106
	2	9.21	8.68	-6	11.6	26	26.3	185
	3	116	98.5	-15	99.4	-14	240	107
6	1	167	165	-1	164	-2	298	79
	2	12.9	10.2	-21	12.2	-5	26.0	102
	3	156	167	7	166	6	298	91
7	1	210	204	-3	203	-3	336	60
	2	18.2	16.2	-11	18.4	1	31.2	71
	3	200	206	3	205	3	336	68
8	1	384	343	-11	341	-11	585	52
	2	42.5	35.2	-17	38.8	-9	46.3	9
	3	384	343	-11	342	-11	585	52
9	1	88.6	83.9	-5	84.1	-5	243	174
	2	6.18	5.61	-9	7.29	18	16.8	172
	3	89.4	86.5	-3	85.4	-4	243	171
Mean				-13.2		-5.2		116.1
Standard Deviation				9.5		14.4		46.9

[1] Using correlation coefficients compiled numerically from Eqs. 13.
[2] Using correlation coefficients from approximate formulas, Eqs. 16 and 17.

As discussed earlier, when the secondary system applies static constraint on the primary system, it also develops stresses due to support displacements. The effect of this particular issue is illustrated in Table 3. The effect of the support displacement is significantly noticeable on the spring forces in element 2. Obviously, the effect of support displacement should be considered along with the response values from the IRS.

Table 3. Response Components From the Proposed IRS Method

Case	Node	Displacement, Inches			Element	Spring Force, Kips		
		Instructure Response Spectrum	Support Motion	Combined		Instructure Response Spectrum	Support Motion	Combined
1	1	0.39	0.76	1.13	1	117	32.0	147
	2	0.38	0.87	1.24	2	32.2	32.0	46.1
					3	115	32.0	86.3
2	1	0.66	0.76	1.37	1	116	18.7	132
	2	0.66	0.87	1.48	2	15.8	18.7	25.9
					3	116	18.7	100
3	1	1.17	0.76	1.84	1	115	10.6	124
	2	1.18	0.87	1.95	2	10.1	10.6	16
					3	116	10.6	108
4	1	1.67	0.76	2.32	1	111	7.09	117
	2	1.68	0.87	2.44	2	7.00	7.09	11.2
					3	112	7.09	106
5	1	2.12	0.76	2.75	1	102	5.12	106
	2	2.13	0.87	2.86	2	5.78	5.12	8.68
					3	102	5.12	98.5
6	1	3.50	0.76	3.39	1	166	5.07	165
	2	3.49	0.87	3.39	2	7.27	5.07	10.2
					3	166	5.07	167
7	1	3.61	0.76	3.52	1	205	6.07	204
	2	3.60	0.87	3.52	2	13.5	6.07	16.2
					3	205	6.07	206
8	1	4.07	0.76	4.02	1	344	9.01	343
	2	4.05	0.87	4.02	2	30.2	9.01	35.2
					3	342	9.01	343
9	1	2.78	0.76	2.52	1	85.4	3.27	83.9
	2	2.77	0.87	2.50	2	3.57	3.27	5.61
					3	85.0	3.27	86.5

CONCLUSIONS

Summary of a new instructure response spectrum (IRS) method is presented. The new method gives the IRS at various connecting DOF directly from the input primary system response spectrum without converting it into either compatible time history or a power spectral density function. It overcomes practically all the difficulties associated with the conventional floor response spectrum method.

The new method consists of essentially solving a complex coupled eigenvalue problem using an SDOF oscillator as the secondary system for each point on the IRS. This may appear to make the method computational uneconomical. It is shown that computational efficiency is achieved by (a) recalculating only two new eigenvalues and eigenvectors, one each corresponding to the oscillator frequency and that to an uncoupled primary system frequency closest to the oscillator frequency, (b) performing the recalculation of the two eigenvalues and the corresponding eigenvectors by an approximate but accurate perturbation technique, and (c) assuming that the remaining coupled eigenvalues and eigenvectors corresponding to the uncoupled primary system frequencies do not change. The IRS obtained from the new method using the above algorithms are shown to be in good agreement with those obtained from the time history analyses. Also, the secondary system response values calculated from the present IRS are in good agreement with those obtained from the time history analysis of the coupled system. The response values from the conventional floor response spectrum method have much higher error values.

The new method is capable of providing the necessary correlation data between various connecting DOF. The data so created can be voluminous, and may make the use of the method difficult. Approximate expressions for the correlations are presented to solve this problem. It is shown that the secondary system response values obtained using the correlations from these formulas are close to those calculated using the correlations computed numerically from the original definition.

When the secondary system applies static constraint on the primary system, the response of the secondary system, itself, can be significantly influenced even when the primary system response is not. It is shown that the new method is able to account for this effect quite well.

REFERENCES

[1] A.K. Gupta, "Seismic Response of Multiply Connected MDOF Primary and Secondary Systems," Nuclear Engineering and Design, Vol. 81, pp. 385-394, September 1984.
[2] A.K. Gupta and J.W. Jaw, "Seismic Response of Nonclassically Damped Systems," Nuclear Engineering and Design, Vol. 91, pp. 153-159, January 1986.
[3] A.K. Gupta and J.W. Jaw, "Response Spectrum Method for Nonclassically Damped Systems," Nuclear Engineering and Design, Vol. 91, pp. 161-169, January 1986.
[4] A.K. Gupta and J.W. Jaw, "Complex Modal Properties of Coupled Moderately Light Equipment – Structure Systems," Nuclear Engineering and Design, Vol. 91, pp. 171-178, January 1986.
[5] A.K. Gupta and J.W. Jaw, "Coupled Response Spectrum Analysis of Secondary Systems Using Uncoupled Modal Properties," Nuclear Engineering and Design, in press.
[6] A.K. Gupta and J.W. Jaw, "CREST, A Computer Program for Coupled Response Spectrum Analysis of Secondary Systems," User's Manual, Department of Civil Engineering, North Carolina State University, Raleigh, June 1985.
[7] A.K. Gupta and J.W. Jaw, "A New Instructure Response Spectrum (IRS) Method For Multiply Connected Secondary Systems With Coupling Effects," Nuclear Engineering and Design, in press.
[8] M.P. Singh, "Generation of Seismic Floor Spectra," Journal of Engineering Mechanics Division, ASCE, Vol. 101, No. EM5, Proc. Paper 11651, pp. 593-607, October 1975.
[9] A. Asfura and A. Der Kiureghian, "A New Floor Response Spectrum Method for Seismic Analysis of Multiply Supported Secondary Systems," Report No. UCB/EERC-84/04, University of California, Berkeley, June 1984.
[10] A.K. Gupta and K. Kordero, "Combination of Modal Responses," Transactions, Sixth International Conference on Structural Mechanics in Reactor Technology, Paris, Paper No. K7/5, 1981.
[11] A.K. Gupta and D.C. Chen, "Comparison of Modal Combination Methods," Nuclear Engineering and Design, Vol. 78, pp. 53-68, March 1984.

DYNAMIC RESPONSE OF NONSTRUCTURAL COMPONENTS

Alan G. Hernried[I] and Haw Jeng[II]

ABSTRACT

A method for the determination of the maximum dynamic response of nonstructural components in structures subjected to earthquake excitation is presented. The method incorporates the important effect of tuning and depends solely upon the subsystem properties, the manner the subsystems are connected, and a response spectrum associated with the excitation. Comparison between the proposed method and a Newmark integration of the combined system for several multi-degree-of-freedom primary-secondary systems subjected to the El Centro S90W earthquake is very favorable. Conclusions and recommendations for the analysis and design of nonstructural components based on the studies discussed herein are presented.

INTRODUCTION

The survivability of nonstructural components under earthquake and other excitations is an important area that has recently received significant attention [2-10]. Light equipment in structures may not only serve a vital function (such as a piping system in a nuclear power plant) but may also cost substantially more than the structure in which it is housed (sensitive scientific devices such as computers and hospital equipment are examples). It is necessary, therefore, that accurate, efficient, and reliable methods are available for the analysis and design of such systems.

In a recent study, Hernried and Sackman developed through perturbation theory "closed form" results for the maximum acceleration of a multi-degree-of-freedom equipment system multiply attached to a multi-degree-of-freedom structure [2,3]. These results depend solely on information that is already available to the structural designer - the fixed base equipment and structure properties, respectively, the manner in which they are connected, and a response spectrum associated with the ground motion. The important problem of tuning (a natural frequency of the equipment equal or close to a natural frequency of the structure) was incorporated in a rational way into the analysis. The completely detuned system (all frequencies of the fixed base subsystems well spaced) was also explored.

[I] Assistant Professor, Department of Civil Engineering, University of Utah, Salt Lake City, Utah.

[II] Research Assistant, Department of Civil Engineering, University of Utah, Salt Lake City, Utah.

In this paper, the dynamic response of several multi-degree-of- freedom equipment-structure systems is examined. The maximum acceleration of the equipment degrees-of-freedom is determined by a conventional time marching scheme of the combined system (Newmark method) as well as by the proposed method of Hernried and Sackman. The equipment is attached to the structure in several different ways to illustrate the effect of spatial coupling on the response. Several different tunings as well as a completely detuned system are considered. Conclusions and recommendations for the analysis and design of nonstructural components based on the studies discussed herein are offered.

ANALYSIS

The governing equations of motion for a primary-secondary system subjected to earthquake ground shaking are

$$\mathbf{M}\ddot{\mathbf{u}} + \mathbf{C}\dot{\mathbf{u}} + \mathbf{K}\mathbf{u} = \mathbf{C}\mathbf{R}\dot{\mathbf{u}}_g + \mathbf{K}\mathbf{R}\mathbf{u}_g \tag{1}$$

where

$$\mathbf{M} = \begin{bmatrix} \mathbf{m}^{(2)} & \mathbf{m}^{(21)} \\ \mathbf{m}^{(12)} & (\mathbf{m}^{(1)} + \tilde{\mathbf{m}}^{(1)}) \end{bmatrix} \; ; \; \mathbf{C} = \begin{bmatrix} \mathbf{c}^{(2)} & \mathbf{c}^{(21)} \\ \mathbf{c}^{(12)} & (\mathbf{c}^{(1)} + \tilde{\mathbf{c}}^{(1)}) \end{bmatrix} \; ;$$

$$\mathbf{K} = \begin{bmatrix} \mathbf{k}^{(2)} & \mathbf{k}^{(21)} \\ \mathbf{k}^{(12)} & (\mathbf{k}^{(1)} + \tilde{\mathbf{k}}^{(1)}) \end{bmatrix} \tag{2}$$

In the above matrices, the off-diagonal matrices represent coupling effects between the equipment and the structure while the matrices on the diagonals represent fixed base (fixed links) properties of the equipment (denoted by a superscript two) alone and of the structure alone without the equipment (denoted by a superscript one). We require that all elements of the fixed base equipment mass $\mathbf{m}^{(2)}$ be small (of order $\varepsilon \ll 1$) in comparison to all elements of the fixed base structure mass $\mathbf{m}^{(1)}$ (of order one) and similarly for the stiffness matrices $\mathbf{k}^{(2)}$ and $\mathbf{k}^{(1)}$. It is assumed that the subsystems are modally damped and that the modal damping ratios are of order $\sqrt{(\varepsilon)} \ll 1$. The matrices with tildes indicate contributions from the equipment to the fixed base primary subsystem property matrices. Similarly

$$\mathbf{u} = \begin{bmatrix} \mathbf{u}^{(2)} \\ \mathbf{u}^{(1)} \end{bmatrix} \; ; \; \mathbf{R} = \begin{bmatrix} \mathbf{r}^{(2)} \\ \mathbf{r}^{(1)} \end{bmatrix} \tag{3}$$

where the absolute displacements and pseudo-static influence vectors for the equipment and structure alone are given by superscripts two and one respectively. For the completely detuned system (all natural frequencies of the fixed base subsystems well-spaced) the eigenvalue problem

$$\mathbf{K}\mathbf{x} = \lambda \mathbf{M}\mathbf{x} \tag{4}$$

is solved by classical matrix perturbation theory for the eigenvalues and eigenvectors [1-3]. These expressions are then used in a classical modal solution of Eq. 1 to yield the response of the nonstructural components as a function of time. Since all modes are well-spaced a standard summation rule such as the square root of the sum of the squares procedure results in the following expression for the maximum acceleration of the secondary system degrees-of-freedom ($z = 1, 2, \ldots, n^{(2)}$)

$$|\ddot{u}_z^{(2)}|_{max} = \{ \sum_{k=1}^{n^{(2)}} [(\phi_{zk}^{(2)} P_k^{(2)} + \sum_{l=1}^{n^{(1)}} \frac{\phi_{zk}^{(2)} C_{lk}}{(\omega_k^{(2)})^2 - (\omega_l^{(1)})^2}) S_A(\omega_k^{(2)}, \beta_k^{(2)})]^2$$

$$+ \sum_{l=1}^{n^{(1)}} [\sum_{k=1}^{n^{(2)}} (\frac{\phi_{zk}^{(2)} C_{lk}}{(\omega_l^{(2)})^2 - (\omega_k^{(2)})^2}) S_A(\omega_l^{(1)}, \beta_l^{(1)})]^2 \}^{1/2} \qquad (5)$$

where

$$P_l^{(1)} = \phi_l^{(1)T} m^{(1)} r^{(1)}$$

$$P_k^{(2)} = \phi_k^{(2)T} m^{(2)} r^{(2)}$$

$$A_{lk} = \phi_l^{(1)T} k^{(12)} \phi_k^{(2)}$$

$$C_{lk} = P_l^{(1)} A_{lk} \qquad (6)$$

and $\phi_l^{(1)}$, $\omega_l^{(1)}$ are the l-th fixed base normalized (with respect to mass) structure mode shape and frequency; while $\phi_k^{(2)}$, $\omega_k^{(2)}$ are the k-th normalized equipment mode shape and frequency. These eigenproperties are determined for each of the subsystems individually. $S_A(\omega, \beta)$ is the pseudo-accleration response spectrum evaluated at frequency ω and damping factor β.

For the tuned system [the m-th natural frequency of the structure equal or close to the n-th natural frequency of the equipment, i.e. $\omega_m^{(1)^2} = (1 + \alpha_{mn}) \omega_n^{(2)^2}$ where α_{mn} is of order $\sqrt{(\epsilon)}$], the contribution to the maximum equipment acceleration from the detuned modes is found in a manner similar to that previously discussed and is given as

$$|\ddot{u}_z^{(2)}|_{max} = \{\sum_{\substack{k=1 \\ k \neq n}}^{n^{(2)}} [\phi_{zk}^{(2)}(P_k^{(2)} + \sum_{l=1}^{n^{(1)}} \frac{C_{lk}}{(\omega_k^{(2)^2} - \omega_l^{(1)^2})}) S_A(\omega_k^{(2)}, \beta_k^{(2)})]^2$$

$$+ [\phi_{zn}^{(2)}(P_n^{(2)} + \sum_{\substack{l=1 \\ l \neq m}}^{n^{(1)}} \frac{C_{ln}}{(\omega_n^{(2)^2} - \omega_l^{(1)^2})})]^2 \, S_A^2(\omega_n^{(2)}, \beta_n^{(2)})$$

$$+ \sum_{\substack{l=1 \\ l \neq m}}^{n^{(1)}} [\sum_{k=1}^{n^{(2)}} (\frac{\phi_{zk}^{(2)} C_{lk}}{(\omega_l^{(1)^2} - \omega_k^{(2)^2})}) S_A(\omega_l^{(1)}, \beta_l^{(1)})]^2$$

$$+ [\sum_{\substack{k=1 \\ k \neq n}}^{n^{(2)}} \frac{\phi_{zk}^{(2)} C_{mk}}{(\omega_m^{(1)^2} - \omega_k^{(2)^2})}]^2 \, S_A^2(\omega_m^{(1)}, \beta_m^{(1)})\}^{1/2} \qquad (7)$$

The contribution from the tuned modes to the maximum equipment acceleration is found by analyzing the equivalent two-degree-of-freedom system governing the behavior of the tuned modes and, for an excitation whose duration is less than the beat period associated with the equivalent two-degree-of-freedom system, is given as

$$|\ddot{u}_z^{(2)}|_{max} = \frac{|\phi_{zn}^{(2)} \frac{C_{mn}}{\omega_n^{(2)^2}}| e^{-K_{mn}}}{(\gamma_{mn} + \alpha_{mn}^2/4 + 4\beta_n^{(2)}\beta_m^{(1)})^{1/2}} S_A(\frac{\omega_n^{(2)} + \omega_m^{(1)}}{2}, \frac{\beta_n^{(2)} + \beta_m^{(1)}}{2})$$

$$\zeta_{mn} = [\gamma_{mn} + \alpha_{mn}^2/4 - (\beta_n^{(2)} - \beta_m^{(1)})^2]^{1/2} / (\beta_n^{(2)} + \beta_m^{(1)})$$

$$K_{mn} = (\arctan \zeta_{mn})/\zeta_{mn} \qquad (8)$$

where the effective mass ratio γ_{mn} is given by $\gamma_{mn} = A_{mn}^2 / \omega_n^{(2)^4}$.
The interested reader is referred to [2,3] for a more comprehensive treatment.

NUMERICAL STUDIES AND DISCUSSION

The systems in Fig. 1 were subjected to the El Centro 1940 S90W earthquake. The different connections were chosen to illustrate the effect of spatial coupling on the response. Several different tunings were considered. These were the first natural frequency of the equipment tuned to the first natural frequency of the structure and the first natural frequency of the equipment tuned to the fifth natural frequency of the structure. The subsystem frequencies are shown in Fig. 2. The tuned mode shapes are illustrated in Fig. 3. The completely detuned system was also examined. The subsystem frequencies are presented in Fig. 4. The properties of these systems are listed in Table 1. The structure alone was modally damped with a damping value of 4% in each mode while the equipment alone was also modally damped with 1% modal damping in each mode - values considered typical in nuclear power plant construction. The response of the light equipment items were determined by the proposed method [combining Eq. 7 with Eq. 8 by the square root of the sum of the squares procedure for the tuned systems and Eq. 5 alone for the detuned system] as well as a conventional numerical integration of the combined system (Newmark method with $\beta = 0.25$, $\gamma = 0.50$, and a time step of minimum system period divided by ten). The results are given in Table 2 where the maximum acceleration is calculated for each equipment degree-of-freedom and the largest of these given as the "maximum amplitude". The percent difference between the maximum acceleration given by the Newmark scheme and the proposed method is calculated for each equipment degree of freedom and the largest and smallest of these values entered in Table 2. A negative sign indicates that the proposed method overestimates the response in comparison to the Newmark integration scheme.

For all the systems considered, the Newmark scheme is considerably more costly than the proposed method - in most instances by as much as factor of two. As can be seen by Table 2, correlation between the proposed method and the Newmark scheme is quite good. It should also be noted that the property matrices of the combined equipment-structure system are ill-conditioned due to the large difference in the masses and stiffness of the subsystems. The reliability of a standard numerical integration of the combined system is therefore questionable.

It is evident, therefore, that the proposed method is an effective and efficient preliminary design aid. In addition, the proposed method yields insight as to where the equipment should be attached to minimize response. For example if the first modes of the subsystems are tuned, system one of Fig. 1 would give larger response than either system two or three. This is evident from the shapes of the tuned modes (Fig. 3) and substantiated by the numerical experiments (Table 2).

CONCLUSIONS AND RECOMMENDATIONS

A method is presented for use in the analysis and design of nonstructural components. The method completely bypasses the necessity of a conventional numerical integration of the combined structure-equipment system. The required data are the fixed base properties of the equipment alone, the fixed base properties of the structure alone, the manner the two subsystems are connnected, and a response spectrum associated with the ground motion.

An equipment-structure system, composed of a five-degree-of-freedom equipment item attached to a ten-degree-of-freedom structure, was subjected to the El Centro 1940 S90W earthquake. The equipment was attached to the structure in several different ways to illustrate the effect of spatial coupling on the response. Several tuned systems as well as a completely detuned system were examined. The maximum acceleration of the equipment degrees of freedom were determined by a conventional numerical integration of the combined system (Newmark scheme) as well as by the proposed method.

Although the possibility exists that spurious effects are introduced into the response when the system is analyzed by the Newmark scheme due to the ill-conditioned nature of the property matrices, agreement between the Newmark scheme and the proposed method is, in most instances, very good. In all cases the Newmark scheme is substantially more costly than the proposed method. The proposed method also yields insight as to where the equipment should be attached to minimize response if tuning, which generally results in large amplifications, is unavoidable. The method proposed herein should be an extremely effective and efficient procedure for use in the design and analysis of nonstructural components in structures subjected to earthquake excitation.

ACKNOWLEDGEMENT

The financial support of the National Science Foundation in the conduct of this investigation is gratefully acknowledged.

REFERENCES

[1] Butkov, E., Mathematical Physics, Addison Wesley, Reading, Ma., 1968.

[2] Hernried, A. G. and Sackman, J. L., "Response of Equipment in Structures Subjected to Transient Excitation," Report No. UCB/SESM-82/03, Division of Structural Engineering and Structural Mechanics, University of California, Berkeley, Ca., 1982.

[3] Hernried, A. G. and Sackman, J. L., "Response of Secondary Systems in Structures Subjected to Transient Excitation," Earthquake Engineering and Structural Dynamics, vol. 12, pp. 737-748, 1984.

[4] Hernried, A. G. and Jeng, H., "Dynamic Response of Secondary Systems," Report No. UTEC 85-043, Department of Civil Engineering, University of Utah, Salt Lake City, UT., 1985.

[5] Igusa, T. and Der Kiureghian, A., "Dynamic Analysis of Multiply Tuned and Arbitrarily Supported Secondary Systems," Report No. UCB/EERC-83/07, University of California, Berkeley, Ca., 1983.

[6] Sackman, J. L. and Kelly, J. M., "Equipment Response Spectra for Nuclear Power Plant Systems," Nuclear Engineering and Design, vol. 57, pp. 277-294, 1980.

[7] Sackman, J. L. and Kelly, J. M., "Rational Design Methods for Light Equipment in Structures Subjected to Ground Motion," Report No. UCB/EERC-78/19, Earthquake Engineering Research Center, University of California, Berkeley, Ca., 1978.

[8] Nakhata, T., Newmark, N. M. and Hall, W. J., "Approximate Dynamic Response of Light Secondary Systems," Structure Research Series Report No. 396, Civil Engineering Studies, University of Illinois, Urbana, Ill., 1973.

[9] Ruzicka, G. C. and Robinson, A. R., "Dynamic Response of Tuned Secondary Systems," Report No. UILU-ENG-80-2020, Department of Civil Engineering, University of Illinois, Urbana, Ill., Nov., 1980.

[10] Villaverde, R. and Newmark, N. M., "Seismic Response of Light Attachments to Buildings," Structure Research Series No. 469, University of Illinois, Urbana, Illinois, February, 1980.

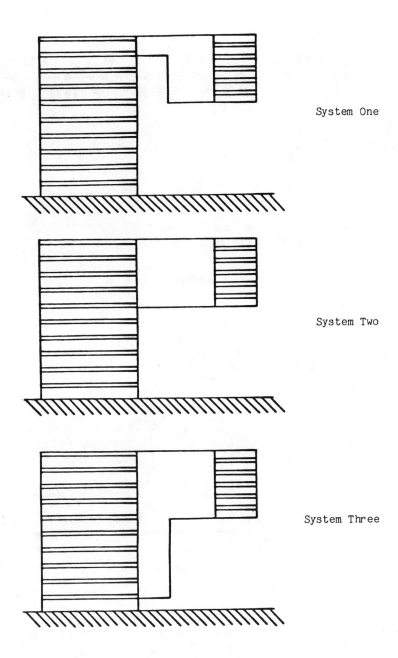

Fig. 1 Primary-Secondary Systems used in Numerical Experiments
(Floor Masses: Structure - M, Equipment - m. Inter-story
Stiffnesses: Structure - K, Equipment - k. Rigid Connection
Links).

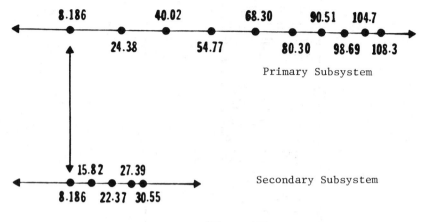
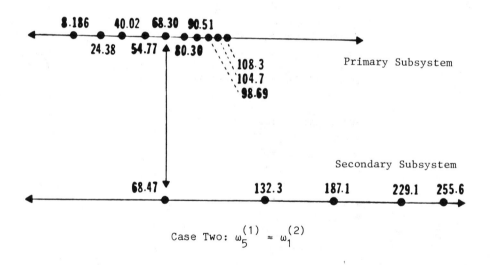

Fig. 2 Distribution of Subsystem Free Vibration Frequencies.

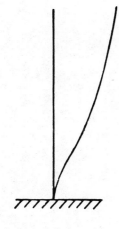
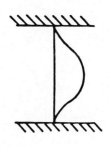

Primary Subsystem	Secondary Subsystem

Case One

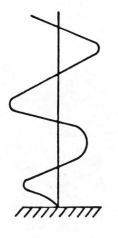

Primary Subsystem	Secondary Subsystem

Case Two

Fig. 3 Subsystem Mode Shapes of Tuned Frequencies

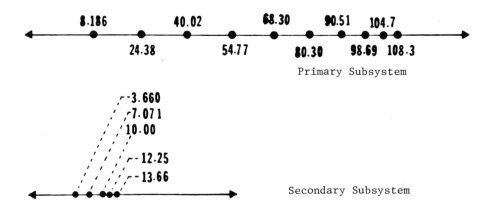

Fig. 4   Distribution of Subsystem Free Vibration Frequencies - Completely Detuned System.

Table 1

Properties of Primary-Secondary Systems used in
Numerical Experiments

Tuned Frequencies	Properties of Structure and Equipment
Case 1 $\omega_1^{(1)} = \omega_1^{(2)}$ = 8.186 (rps)	$K = 6 \times 10^{10}$, $k = 2.5 \times 10^6$ $M = 2 \times 10^7$, $m = 1 \times 10^4$
Case 2 $\omega_5^{(1)} \cong \omega_1^{(2)}$ $\cong$ 68.4 (rps)	$K = 6 \times 10^{10}$, $k = 1.75 \times 10^8$ $M = 2 \times 10^7$, $m = 1 \times 10^4$
Case 3 Completely detuned	$K = 6 \times 10^{10}$, $k = 0.5 \times 10^6$ $M = 2 \times 10^7$, $m = 1 \times 10^4$

Structure: K (kgf/m); Equipment: k (kgf/m) (1 kgf/m = 9.81 N/m)

Structure: M (kg); Equipment: m (kg)

Table 2

Results of Numerical Experiments

		γ_{mn}	MAX AMPL (g)	DIFFERENCE (%)
System One	Case 1	0.43×10^{-3}	5.81	7.0 ~ 3.9
	Case 2	0.13×10^{-4}	0.71	-2.4 ~ 10.3
	Case 3	*	0.96	-6.9 ~ -0.1
System Two	Case 1	0.35×10^{-3}	5.08	2.7 ~ 0.8
	Case 2	0.34×10^{-3}	0.65	6.5 ~ -6.1
	Case 3	*	0.86	-1.7 ~ 0.7
System Three	Case 1	0.15×10^{-3}	3.31	10.3 ~ -0.7
	Case 2	0.34×10^{-3}	0.62	8.5 ~ -2.5
	Case 3	*	0.83	20.1 ~ 4.3

SEISMIC QUALIFICATION OF INTERIOR MECHANICAL SYSTEMS

Asadul H. Chowdhury[I]

ABSTRACT

Requirements for the seismic qualification of interior mechanical systems under various conditions of building structure systems are presented. Response of mechanical systems resting on sixteen uniform shear-flexure structures representing a wide range of structural types were investigated. The importance of using consistent envelope spectra for seismic qualification of mechanical systems for a wide range of design conditions is explained. Mechanical system responses for 5%, 10% and 20% bandwidth exclusion of mechanical system-structural fundamental frequency coincidence are explored. Attention is drawn to the need for the use of bandwidth exclusion technique to reduce the degree of conservatism associated with traditional artificial broad-band spectra.

INTRODUCTION

Building structures have numerous interior mechanical systems such as heating, ventilation and air conditioning (HVAC) systems; electrical distribution equipment; scientific and hospital equipment; computer systems; etc. These systems are important to earthquake hazard mitigation. The experience gained from various earthquakes has shown that damage to internal mechanical systems can make buildings functionally useless and cause serious economic losses. But the seismic design guidelines for these mechanical systems under various conditions of the building structures are either very limited or non existent. In practice, the uncertainties and variations of site conditions, and structural and material properties are taken into account by artificially broadening the floor response spectra. The artificial broad-band spectra do not really simulate the input during any particular seismic event and give conservative results.

Thus, it becomes necessary to develop an appropriate requirement for practical ranges of parameters like site conditions, and structural and material properties which would provide for consistent seismic qualification of mechanical systems under various conditions. This may provide a method by which specifications can be written so that they are easy for mechanical system suppliers and testing laboratories to understand and employ. It may also permit universal mechanical system certification.

The analytical parametric study reported here deals with (i) the

---

I. Assistant Professor, Department of Civil Engineering, North Dakota State University, Fargo, North Dakota.

evaluation of seismic response of interior mechanical systems, and (ii) the development of requirements for the seismic qualification of these systems. The response of mechanical systems resting on sixteen uniform shear-flexure building structures representing wide range of structural types were investigated. The parameters considered are the fundamental frequencies and shear-flexure stiffness proportions of the buildings and the fundamental frequencies of the mechanical systems. The structures were subjected to the north-south component of the 1940 El Centro earthquake motions. The time history response of the floor on which the mechanical systems sit was used as the input to these mechanical systems. The response of the mechanical system mounted at the top floor of the structure was expressed in the form of response spectrum. The response spectra of the mechanical systems were used to evaluate the level of conservatism associated with traditional broad-band response spectra.

SEISMIC ANALYSIS

The north-south component of the El Centro earthquake was normalized to 1g for predominant 15 seconds (Fig. 1) and was used as the input to the structures investigated in this study.

Dynamic Properties of Structures

The structures considered in this study were idealized as uniform shear-flexure tall buildings [1]. The 16 structural configurations have been designated by ST_{ij} (i=1,4; j=1,4) in which i and j represent the fundamental frequency and shear-flexure stiffness parameters, respectively. The equation of motion of a linear elastic uniform structure with viscous damping and subjected to seismic base acceleration $\ddot{y}_g$ is given by [1].

$$\frac{\partial^4 y}{\partial x^4} - \alpha^2 \frac{\partial^2 y}{\partial x^2} + \frac{1}{a^2} \frac{\partial^2 y}{\partial t^2} + \frac{C}{EI} \frac{\partial y}{\partial t} = -\frac{1}{a^2} \ddot{y}_g \qquad (1)$$

in which $\quad \alpha^2 = \dfrac{GA}{EI} \qquad (2)$

$$a^2 = \frac{EI}{\rho A} \qquad (3)$$

and E and I = modulus of elasticity and the second moment of area, respectively, of the flexural beam; G and A = the shear modulus and cross-sectional area, respectively of the shear beam; ρ = the mass per unit volume of the uniform structure; and C = viscous damping coefficient of the structure.

The equation of motion for the undamped free vibration of the building can be obtained by eliminating the two terms involving damping C and ground acceleration $\ddot{y}_g$ from Eq. 1 and is given by

$$\frac{\partial^4 y}{\partial x^4} - \alpha^2 \frac{\partial^2 y}{\partial x^2} = -\frac{1}{a^2} \frac{\partial^2 y}{\partial t^2} \qquad (4)$$

Assuming a product solution for Eq. 4 of the form:

$$y(x,t) = \psi(x)q(t) \qquad (5)$$

in which $\psi(x)$ is the mode shape, and for free vibration $q(t)$ is the time dependent harmonic response function with frequency, ω, and substituting Eq. 5 into Eq. 4 and rearranging terms yield

$$\frac{d^4\psi(x)}{dx^4} - \alpha^2 \frac{d^2\psi(x)}{dx^2} - k^4\psi(x) = 0 \qquad (6)$$

in which $\quad k^4 = \dfrac{\omega^2}{a^2} \qquad (7)$

The solution of Eq. 6 can be written in the form [1]

$$\psi(x) = C_1 \cos\lambda_1 x + C_2 \sin\lambda_1 x + C_3 \cosh\lambda_2 x + C_4 \sinh\lambda_2 x \qquad (8)$$

in which $\quad \lambda_1^2 = \sqrt{\left(\dfrac{\alpha^2}{2}\right)^2 + k^4} - \dfrac{\alpha^2}{2} \qquad (9)$

and $\quad \lambda_2^2 = \lambda_1^2 + \alpha^2 \qquad (10)$

and C_1, C_2, C_3 and C_4 are constants of integration.

A fixed supported shear-flexure tall building has zero rotation and deflection at the base, and is free at the top. The boundary conditions for this building with foundation only under the core are [3]

Deflection: $\quad \psi(0) = 0 \qquad (11)$

Slope: $\quad \dfrac{d\psi(0)}{dx} = 0$ (12)

Moment: $\quad \dfrac{d^2\psi(H)}{dx^2} = 0$ (13)

Shear: $\quad \alpha^2 \dfrac{d\psi(H)}{dx} - \dfrac{d^3\psi(H)}{dx^3} = 0$ (14)

Substituting boundary conditions (Eqs. 11, 12, 13, 14) into Eq. 8 and expressing in matrix notation yield Eq. 15 (Table 1). The nontrivial solution of Eq. 15, that is the solution for not all C_1, C_2, C_3 and C_4 are zeros, requires that the determinant of the matrix factor of {C} be equal to zero. For a given value of shear-flexure stiffness proportion, αH, the roots $\lambda_1 H$ and $\lambda_2 H$ of Eq. 15 for each mode of vibration of the tall building can be calculated using half-interval search technique. For each pair of their values of $\lambda_1 H$ and $\lambda_2 H$, we can solve Eq. 15 for C_1, C_2, C_3 and C_4. Substituting the values of C_1, C_2, C_3 and C_4 of a given mode into Eq. 8, the mode shape for that mode of vibration of the building is obtained.

Finally, from Eqs. 3, 7 and 9 we have

$$\omega_i = \sqrt{\dfrac{EI}{\rho A} \dfrac{(\lambda_{1i}H)^2\{(\lambda_{1i}H)^2 + (\alpha H)^2\}}{H^4}}$$ (16)

in which the subscript, i, refers to the ith mode.

The structures are assumed to have the same mass per unit length, ρA, and a damping of 5% of critical damping. The stiffness parameter, αH, representing the shear-flexure stiffness proportion of the structures and the fundamental frequency variation of the 16 structures are given in Table 2. The flexural and shear stiffness of the shear-flexure structures have been expressed in terms of mass per unit of length and some power of height of structures and are given in Table 3.

The fundamental frequency, f_o of each structure is presented at 1, 3, 6 and 10 cycles per second. The second and third frequencies corresponding to the fundamental frequency of each structure have been obtained by modifying equation 16 to express higher modal frequency in terms of fundamental frequency [2,3]. The three natural frequencies for each structure are given as cycles per second in Table 2.

Table 1 Equation 15

$$\begin{bmatrix} 1 & 0 & 1 & 0 \\ 0 & \lambda_1 H & 0 & \lambda_2 H \\ (-\lambda_1^2 H^2 \cos\lambda_1 H) & (-\lambda_1^2 H^2 \sin\lambda_1 H) & (\lambda_2^2 H^2 \cosh\lambda_2 H) & (\lambda_2^2 H^2 \sinh\lambda_2 H) \\ (-\alpha \lambda_1^2 H^3 \sin\lambda_1 H - \lambda_1^3 H^3 \sin\lambda_1 H) & (\alpha \lambda_1^2 H^3 \cos\lambda_1 H + \lambda_1^3 H^3 \cos\lambda_1 H) & (\alpha \lambda_2^2 H^3 \sinh\lambda_2 H - \lambda_2^3 H^3 \sinh\lambda_2 H) & (\alpha \lambda_2^2 H^3 \cosh\lambda_2 H - \lambda_2^3 H^3 \cosh\lambda_2 H) \end{bmatrix} \begin{Bmatrix} C_1 \\ C_2 \\ C_3 \\ C_4 \end{Bmatrix} = \begin{Bmatrix} 0 \\ 0 \\ 0 \\ 0 \end{Bmatrix}$$

Table 2 Properties of Structures

Stiffness Parameters	Mode	Frequency (Hz)		Frequency (Hz)		Frequency (Hz)		Frequency (Hz)	
	Designation	ST_{11}		ST_{21}		ST_{31}		ST_{41}	
$\alpha H = 0$	First	1.00		3.00		6.00		10.00	
	Second	6.27		18.80		37.65		62.71	
	Third	17.55		52.68		105.44		175.60	
	Designation	ST_{12}		ST_{22}		ST_{32}		ST_{42}	
$\alpha H = 3$	First	1.00		3.00		6.00		10.00	
	Second	4.02		12.07		24.15		40.24	
	Third	10.34		31.04		62.09		103.49	
	Designation	ST_{13}		ST_{23}		ST_{33}		ST_{43}	
$\alpha H = 10$	First	1.00		3.00		6.00		10.00	
	Second	2.22		9.65		19.31		32.70	
	Third	5.98		17.95		35.90		60.77	
	Designation	ST_{14}		ST_{24}		ST_{34}		ST_{44}	
$\alpha H = 30$	First	1.00		3.00		6.00		10.00	
	Second	3.03		9.10		18.20		30.24	
	Third	5.32		15.98		31.97		53.27	

Table 3 Flexural[*] and Shear[**] Stiffness of Structures

Stiffness Type	Stiffness Parameter	Designation	Stiffness	Designation	Stiffness	Designation	Stiffness	Designation	Stiffness
Flexure, EI	$\alpha H = 0$	ST_{11}	3.2	ST_{21}	28.7	ST_{31}	115.0	ST_{41}	319.4
Shear, GA			0		0		0		0
Flexure, EI	$\alpha H = 3$	ST_{12}	0.82	ST_{22}	7.4	ST_{32}	29.7	ST_{42}	82.4
Shear, GA			7.4		66.7		267.0		741.6
Flexure, EI	$\alpha H = 10$	ST_{13}	0.13	ST_{23}	1.14	ST_{33}	4.6	ST_{43}	12.7
Shear, GA			12.7		114.6		458.4		1273.1
Flexure, EI	$\alpha H = 30$	ST_{14}	.016	ST_{24}	0.15	ST_{34}	0.6	ST_{44}	1.66
Shear, GA			14.9		134.2		536.7		1490.6

[*] Multiple of $\rho A H^4$
[**] Multiple of $\rho A H^2$

Floor Motion

The motion of the floor on which the mechanical system sits represents a description of the input to the equipment. This motion is the earthquake ground motion filtered through the structure and is influenced by the predominant modes of the structure. Each mode of a structure is a single degree of freedom system with the fundamental frequency equal to the natural frequency of the structure at that mode. The time history floor motion of each structure was obtained by time history seismic analysis of the structure at each of the first three modes and then superposing the modal responses. The time history response (acceleration) of structures ST_{11}, ST_{14}, ST_{41} and ST_{44} are shown in Figs. 2, 3, 4 and 5 (full results are available in Ref. [2]). These figures show that the floor motions of the structures of same fundamental frequency but of different shear-flexure stiffness proportion, αH, are of the same nature although their second and third mode frequencies vary largely. This indicates that the first mode of vibration of each structure is the dominant contributor to its floor acceleration, i.e., the contribution due to second and higher modes is very small.

Floor Response Spectrum

The floor response spectrum of a mechanical system resting on a particular location of a structure have a narrow-band characteristic. Development of envelope spectrum provides qualification of a mechanical system for a wide range of design conditions. The envelope spectrum can be developed using a number of spectra representing the response of a mechanical system siting on different environment representing variations of building and foundation properties.

The acceleration response spectra of the mechanical systems resting on the structures considered in this study are given in Figs. 6, 7, 8 and 9. These curves show that the peak response of the mechanical systems resting on the structures with fundamental frequency of 3 Hz and more is primarily at the fundamental period of those structures. But for the mechanical systems siting on the structures of fundamental frequency of 1Hz, there are also peaks at higher mode periods although the dominant peak is at the fundamental period of the structures. This may be attributed to the spacing of the frequencies of different modes of the structures. Since the ground motion has been normalized to 1g, the magnitude of the floor response spectra also represents the magnification factor of the response with respect to the ground motion. These curves also show the effect of equipment rigidity and mounting on the forces generated on the equipment.

Each of the Figs. 6 to 9 shows the envelope of the floor response spectra of mechanical systems resting on structures of a particular fundamental frequency but considering that the stiffness type of the structure can vary from pure shear to pure flexure. For clarity of drawing in each figure, the envelope curve was drawn just above the response representing the maximum of the four values at each period value of the mechanical systems. A mechanical system that can satisfy an envelope requirement as shown in these figures may be considered to be qualified for mounting on the top of any type of structures with the fundamental frequency

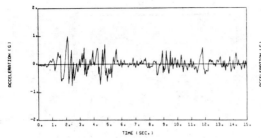

Fig. 1 N-S Component of El Centro Earthquake Normalized to 1G

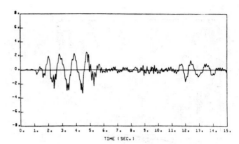

Fig. 2 Top Floor Motion of Structure ST_{11}

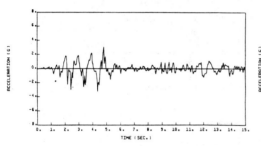

Fig. 3 Top Floor Motion of Structure ST_{14}

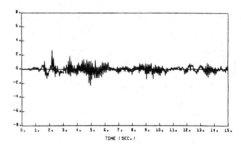

Fig. 4 Top Floor Motion of Structure ST_{41}

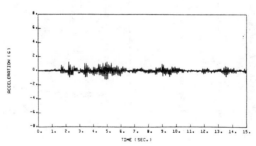

Fig. 5 Top Floor Motion of Structure ST_{44}

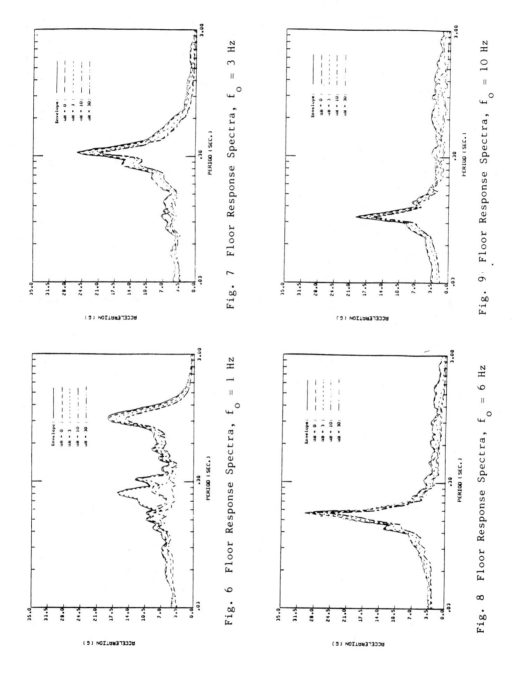

Fig. 6 Floor Response Spectra, f_o = 1 Hz

Fig. 7 Floor Response Spectra, f_o = 3 Hz

Fig. 8 Floor Response Spectra, f_o = 6 Hz

Fig. 9 Floor Response Spectra, f_o = 10 Hz

equal to that for which the envelope has been drawn when the structures are subjected to a seismic ground motion which may be as severe as up to the north-south component of the 1940 El Centro earthquake normalized to 1g.

A comparison of the four response spectra curves in each figure shows that for a given structural frequency, the stiffness parameter, αH, does not affect the nature of the equipment response significantly, except the magnitude of response. For structures with the same frequency, the peak response increases with the decrease of αH.

The envelope of the floor response spectra of mechanical systems resting on structures of a particular stiffness type but covering all possible variations of fundamental frequency are shown in Figs. 10 to 13. In each figure the peaks of the four floor response spectra curves were used to develop an envelope floor response spectrum curve. Each envelope curve is considered to represent the maximum floor response of a particular stiffness type structure at all frequencies within the frequency range considered in this analysis.

The envelope curve in each case confirms that for a given type of structural stiffness, the floor response spectra value changes with the change of the fundamental frequency of the structure. If a mechanical system satisfies an envelope requirement as shown in these figures, it may be considered to be qualified for mounting on the top of a particular type of structure for all possible variations of the fundamental frequency of that type of structure when the structure is subjected to a seismic ground motion which may be as severe as up to the N-S component of the 1940 El Centro earthquake normalized to 1g.

Determination of Level of Conservatism

The response of mechanical systems, due to floor motion developed based on broad band envelope spectrum, is likely to be somewhat larger than that when only motion compatible with the response spectrum of any one structural system is input. One of the methods of determining the level of conservatism of the various broad band spectrum is to consider the case of exclusion of mechanical system-structural fundamental frequency coincidence by a certain percentage bandwidth. Certain percentage bandwidth may be defined as a frequency band whose width is equal to that certain percentage of the mechanical system-structural frequency coincidence. In this investigation, exclusion by bandwidth of 5%, 10%, and 20% are considered.

The envelope floor response spectra for these 5%, 10%, and 20% bandwidth exclusions are shown in Figs. 10 to 13. The levels of conservatism for structures of four types of shear-flexure stiffness are shown in Figs. 14 to 16. The curves in these figures show that depending upon the type of structural stiffness, the response for 5%, 10%, and 20% bandwidth exclusion of mechanical system-structural fundamental frequency may be as low as 86.6%, 74.8%, and 52.8% of mechanical system-structural fundamental frequency coincidence, respectively.

Figs. 14 to 16 also show the variations of the level of conservatism for the four structural stiffnesses for a given percentage of bandwidth exclusion. The variation of the level of conservatism between the four

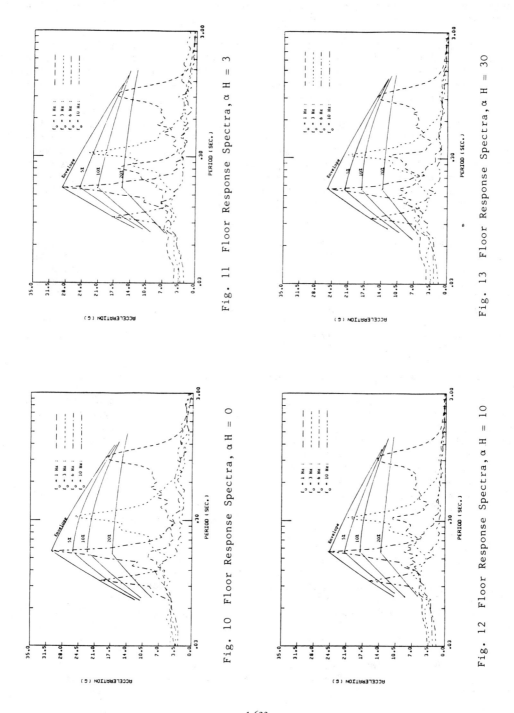

Fig. 10 Floor Response Spectra, αH = 0

Fig. 11 Floor Response Spectra, αH = 3

Fig. 12 Floor Response Spectra, αH = 10

Fig. 13 Floor Response Spectra, αH = 30

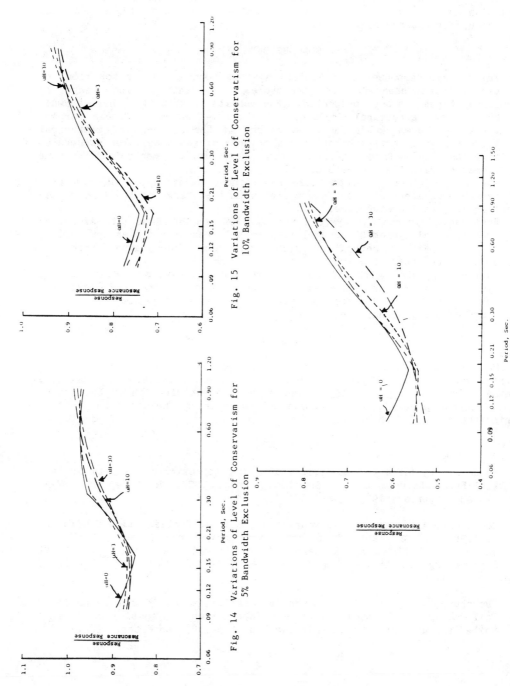

Fig. 14 Variations of Level of Conservatism for 5% Bandwidth Exclusion

Fig. 15 Variations of Level of Conservatism for 10% Bandwidth Exclusion

Fig. 16 Variations of Level of Conservatism for 20% Bandwidth Exclusion

types of structural stiffnesses is within 5% for frequency coincidence exclusion of 5% and 10%, but this variation is somewhat higher for 20% exclusion of frequency coincidence.

CONCLUSIONS

The dynamic response of mechanical systems mounted on the top floor of sixteen uniform shear-flexure structures representing wide range of structural types has been presented. The results of this study have shown:
1. For a given structural frequency, the shear-flexure stiffness parameter, αH, does not affect the nature of the response of mechanical systems significantly, except the magnitude of response. For structures with the same frequency, the peak response of mechanical systems increases with the decrease of αH.
2. For a given type of shear-flexure stiffness parameter, αH, the peak response of mechanical systems varies with the change of the fundamental frequency of the structure.
3. Response of mechanical systems for 5%, 10% and 20% bandwidth exclusion of mechanical system-structural fundamental frequency are as low as 87%, 75% and 53% of mechanical system-structural fundamental frequency coincidence, respectively.
4. Envelope spectra with 5% to 20% bandwidth exclusion of mechanical system-structural fundamental frequency coincidence give realistic seismic qualification requirements for mechanical systems for a wide range of design conditions.

ACKNOWLEDGEMENTS

Major part of this research work was carried out in the Department of Civil Engineering and Engineering Mechanics, McMaster University, Hamilton, Canada. The author gratefully acknowledges the contribution of Dr. Arthur C. Heidebrecht, Dean of the Faculty of Engineering, McMaster University, during the course of this study.

REFERENCES

1. Heidebrecht, A.C., and Stafford-Smith, B., Approximate Analysis of Tall Wall-Frame Structures, J. St. Div., Proc. ASCE, Vol. 99, No. ST-2, 199-221, February 1973.

2. Chowdhury, A.H., and Heidebrecht, A.C., Seismic Qualification Studies of Nuclear Power Plant Buildings, Equipment and Systems, McMaster Earthquake Engineering Research Group Report, Department of Civil Engineering and Engineering Mechanics, McMaster University, Hamilton, Canada, 1980.

3. Chowdhury, A.H., Effect of Foundation Flexibility on Dynamic Response of Tall Buildings, Proceedings of the International Symposium on Dynamic Soil-Structure Interaction, Minneapolis, 159-168, September, 1984.

INDEX OF PROCEEDINGS AUTHORS

Author	Page	Author	Page	Author	Page
Abdel-Ghaffar, A. M.	III-2085	Behrendt, J. C.	I-43	Chowdhury, A. H.	II-1613
Abrams, D. P.	II-1371	Berger, E.	I-313	Christian, J. T.	II-1141
Acharya, H.	I-65	Bergman, D. M.	II-1129	Christiano, P. P.	I-599
Adham, S. A.	III-1767	Bernreuter, D. L.	I-93	Chrostowski, J. D.	III-2215
Agbabian, M. S.	III-1767	Bernreuter, D. L.	I-245	Chung, L.	II-1529
Aizawa, K.	II-1107	Bertero, V. V.	II-847	Chung, L. L.	III-1971
Aktan, A. E.	II-1015	Bertero, V. V.	III-1663	Cifuentes, A. O.	II-967
Al-Haddad, M.	III-1779	Bielak, J.	I-599	Clough, G. W.	I-659
Ali, M. M.	III-1743	Bielak, J.	II-811	Clough, R. W.	II-1383
Amick, D.	I-55	Blasko, M. J.	I-477	Constantinou, M. C.	I-671
Ammerman, D. J.	III-1791	Boissonnade, A. C.	II-931	Coppersmith, K. J.	I-301
Anagnos, T.	I-281	Boore, D. M.	I-137	Corley, W. G.	II-1227
Anand, S. C.	II-1359	Boroojerdi, A.	II-1191	Coronato, J. A.	I-599
Ang, A. H. S.	III-1721	Brady, A. G.	II-823	Craig, R. J.	II-1179
Araya, R.	I-269	Brady, A. G.	III-2225	Cruz, E. F.	II-1003
Arnold, C.	III-1755	Brahimi, M.	II-919	Cundall, P. A.	I-505
Arnold, C.	III-2329	Bureau, G.	I-789	Curry, R.	III-1863
Ashour, S. A.	II-1129	Bush, Jr., T. D.	II-1203		
Aubeny, C. P.	I-405			Day, J.	III-2447
Austin, M. A.	III-1709	Caccese, V.	II-1347	Debassay, M.	I-765
Aziz, T. S.	I-427	Cadena, P. A.	II-1141	Del Valle, C. E.	III-1651
Azizinamini, A.	II-1285	Carr, J. R.	I-33	Der Kiureghian, A.	I-269
		Celebi, M.	I-125	Dermitzakis, S. N.	II-1563
Babcock, C. D.	III-2155	Celebi, M.	II-1273	Diehl, J. G.	II-1575
Baktash, P.	II-1099	Celebi, M.	III-2225	Dimitrov, N.	III-1935
Baldelli, J. A.	III-1959	Chaney, R. C.	I-517	Domer, R. G.	III-2471
Barenberg, M. E.	III-2073	Chang, C. Y.	I-369	Dong, W.	I-221
Barker, R. M.	III-2097	Chang, F. K.	I-439	Dong, W.	II-931
Bazan-Zurita, E.	I-683	Chang, K. C.	II-1451	Donovan, N. C.	II-1153
Bazan-Zurita, E.	III-1687	Chang, N. Y.	I-405	Drake, R. M.	III-2261
Beauvoir, C.	III-1851	Chang, N. Y.	I-439	Dravinski, M.	I-407
Beavers, J. E.	III-2261	Chang, T. Y. H.	II-1141	Durrani, A. J.	II-1419
Beavers, J. E.	III-2471	Chatterjee, J.	II-1087	Dusseau, R. A.	III-2049
Beck, J. L.	I-393	Chen, J. C.	I-245	Dwyer, J.	I-25
Beck, J. L.	II-1311	Cheng, F. Y.	III-1731		
Becker, A.	III-2023	Chiarito, V. P.	I-715	Easterling, W. S.	II-1251
Becker, A. M.	II-1153	Chopra, A. K.	I-741	Eder, S. J.	III-2305
Becker, D. J.	I-257	Chopra, A. K.	II-1003	Eguchi, R. T.	III-2203

Author	Page	Author	Page	Author	Page
Eguchi, R. T.	III-2215	Hamamatsu, O.	I-463	Juang, D. S.	III-1731
Ehsani, M. R.	II-1463	Hanson, C.	III-1851		
Elgohary, M. M.	I-427	Hanson, R. D.	II-1129	Kamata, M.	I-463
Elias, H. E.	II-1419	Harada, T.	III-2191	Karamchandani,	
Elton, D. J.	I-161	Harris, H. G.	II-1347	A. K.	III-1981
Elton, D. J.	I-327	Hart, G. C.	III-1839	Kareem, A.	II-1075
Elton, D. J.	I-497	Hays, W. W.	I-357	Kattell, J. R.	III-2271
Eshraghi, H.	I-417	Heidebrecht, A. C.	I-345	Kawashima, K.	II-1107
Evans, D.	III-2329	Hejazi, M.	III-2193	Kelly, J.	III-1911
		Helpingstine, M.	II-1407	Khodaveredian, M.	III-1993
Fattah, B. A.	III-1779	Hemingway, M.	I-233	Kircher, C. A.	III-2283
Fedock, J. J.	I-729	Hernried, A. G.	II-1601	Kiremidjian, A. S.	II-1039
Fedock, J. J.	I-753	Hidalgo, P.	II-1335	Kiremidjian, A. S.	I-149
Ferritto, J. M.	I-489	Hoerner, J. B.	III-1887	Kitchen, M. R.	I-83
Fletcher, J. B.	I-753	Holland, R.	III-1911	Kneifati, M. C.	I-671
Foutch, D. A.	III-2073	Hribar, J. A.	I-703	Konagai, K.	I-647
French, C. W.	II-1191	Huang, J. T.	II-835	Kreger, M. E.	II-1475
Fujino, Y.	II-895	Huang, M. J.	I-439	Kuang, J.	I-105
Fuller, G. R.	II-871	Huang, M. J.	III-1639	Kunnath, S. K.	II-1323
		Hunt, D. D.	II-1141	Kustu, O.	II-1493
Gallagher, P. J.	I-623	Hunt, R. J.	III-2471	Kyriakides, S.	III-2179
Gergely, P.	II-871	Hwang, H.	I-529		
Ghosh, S. K.	III-1803			Lamarre, M.	I-221
Ghusn, G. E.	II-1027	Idriss, I. M.	I-369	Lane, P. L.	III-2167
Glass, C. E.	I-33	Iemura, H.	II-955	Langer, W.	I-313
Gohn, G. S.	I-3	Imbsen, R. A.	III-2037	Lashkari, B.	II-1487
Gohn, G. S.	I-197	Issa, C. A.	III-2097	Lashkari, B.	III-2283
Goodman, J.	II-943	Ito, M.	II-895	Lee, G. C.	II-1451
Goodno, B. J.	II-883	Iwan, W. D.	II-967	Leon, R. T.	II-1439
Gori, P. L.	III-2341			Leon, R. T.	III-1791
Greene, M. R.	III-2341	Jacobson, R. B.	I-3	Li, X. S.	I-611
Gulkarov, B.	III-2435	Jacobson, R. B.	I-197	Lichterman, J. D.	III-2365
Gulkarov, B.	III-2447	Jaw, J. W.	II-1589	Lien, B. H.	I-439
Gupta, A. K.	II-1589	Jayakumar, P.	II-1311	Lin, A. N.	II-1407
Guthrie, L. G.	I-717	Jeng, H.	II-1601	Lin, S. T.	II-1165
		Jennings, P. C.	III-2155	Linbeck, L.	II-1475
Hadj-Hamou, T.	I-797	Jirsa, J. O.	II-1203	Liu, H. P.	I-753
Hadjian, A. H.	II-943	Johal, L. S.	II-1227	Loh, C. H.	I-381
Hadjian, A. H.	II-1165	Johnpeer, G. D.	I-233	Long, L. T.	I-25
Haldar, A.	I-575	Johnson, W. J.	III-1687	Long, L. T.	I-105
Hall, W. J.	II-1117	Jones, E. A.	II-1203	Love, D. W.	I-233
Hallbick, D. C.	I-197	Juang, C. H.	I-161	Luco, J. E.	I-553

Author	Page	Author	Page	Author	Page
Luco, J. E.	II-907	Naumoski, N.	I-345	Radziminski, J. B.	II-1285
Luders, C.	II-1335	Nazmy, A. S.	III-2085	Rahman, M. A.	II-1359
Luettich, S. M.	I-575	Nelson, G. E.	II-1015	Rao, S. J. K.	III-2435
		Noda, S.	II-955	Rao, S. J. K.	III-2447
Mahin, S. A.	II-847	Nogami, T.	I-647	Rast, B.	I-313
Mahin, S. A.	II-1427	Nordenson, G. J. P.	I-209	Razani, R.	III-1815
Mahin, S. A.	II-1563	Norris, G. M.	I-635	Redpath, B. B.	I-587
Maley, R.	II-1273	Nutt, R. V.	II-1039	Reed, J. W.	III-1981
Malik, L. E.	II-859	Nutt, R. V.	III-2037	Reich, M.	I-529
Malik, L. E.	III-2459	Nuttli, O.	II-859	Reinhorn, A. M.	II-991
Malley, J. O.	II-1215			Reinhorn, A. M.	II-1323
Malley, J. O.	III-1627	O'Hara, P. F.	I-333	Reinhorn, A. M.	III-1899
Mamoon, S. M.	I-541	O'Hara, P. F.	I-477	Reinhorn, A. M.	III-1971
Manolis, G. D.	III-1899	O'Rourke, T. D.	III-2167	Reitherman, R. K.	III-2319
Manos, G. C.	II-1383	Obermeier, S. F.	I-3	Reitherman, R. K.	III-2425
Manos, G. C.	III-2131	Obermeier, S. F.	I-197	Resheidat, M. R.	III-2143
Maragakis, E. A.	III-2237	Ohashi, H.	I-451	Richardson, G. N.	I-517
Marciano, E. A.	I-327	Ohta, Y.	I-451	Richardson, G. N.	III-2061
Mareschal, J. C.	I-83	Otani, J.	I-647	Richter, P. J.	III-2261
Mareschal, J. C.	I-105			Rihal, S. S.	II-1297
Markewich, H. W.	I-197	Pacheco, B.	II-895	Rizzo, P. C.	I-333
Marsh, C.	II-1099	Park, Y. J.	II-991	Rizzo, P. C.	I-623
Martin, II, J. R.	I-497	Park, Y. J.	III-1721	Rizzo, P. C.	I-683
Masek, J. P.	III-2203	Pauly, S. E.	II-1553	Roach, C. E.	II-1203
McCabe, S. L.	II-1117	Pauschke, J. M.	II-1087	Rogers, J. C.	I-83
McGuire, R. K.	I-293	Pavich, M. J.	I-3	Rojahn, C.	II-1039
Mensing, R. W.	I-93	Peek, R.	III-2155	Roth, W. H.	I-505
Mita, A.	II-907	Perez, V.	II-823	Roufaiel, M. S. L.	III-1827
Mittler, E.	III-2353	Petak, W. J.	III-2435	Rubin, M.	I-3
Miyamura, M.	I-463	Phipps, R. L.	I-3		
Moehle, J. P.	II-1505	Pinelli, J. P.	II-883	Saatcioglu, M.	II-1395
Monasa, F. F.	III-1827	Pires, J.	I-529	Sabina, F. J.	I-417
Morse, D. V.	II-811	Pister, K. S.	III-1709	Sabol, T. A.	III-1839
Mostaghel, N.	III-1993	Platt, C. M.	III-1947	Safak, E.	I-137
Moussa, A. E.	II-1463	Pocanschi, A.	III-1935	Safak, E.	II-823
Mozaffarian, H.	II-1407	Poland, C. D.	II-1215	Saiidi, M.	II-1027
Mozer, J. D.	I-703	Poland, C. D.	III-1627	Sauter, F. F.	II-2107
Mueller, P.	II-1239	Porter, M. L.	II-1251	Savy, J. B.	I-93
Musser, D. W.	II-1227	Powars, D. S.	I-197	Savy, J. B.	I-245
Muto, K.	I-463	Power, M. S.	I-369	Scawthorn, C.	III-1675
		Prawel, S. P.	II-1323	Scawthorn, C.	III-2023
Nau, J. M.	II-979			Schiff, A. J.	I-777

Author	Page	Author	Page	Author	Page
Scholl, R. E.	II-1039	Sweeney, B.	I-659	Wight, J.	III-1779
Scholl, R. E.	III-2005			Williams, D.	I-587
Scholl, R. E.	III-2459	Tallin, A. G.	II-919	Wilson, R. R.	II-1039
Schultz, A. E.	II-1517	Talwani, P.	I-15	Wimberly, P. M.	I-703
Scott, R. F.	I-245	Talwani, P.	I-55	Wiradinata, S.	II-1395
Scott, S.	III-2389	Tandowsky, S.	III-1851	Wisch, D. J.	I-797
SEAONC, DES	III-2377	Tarics, A. G.	III-1911	Witham, J. L.	I-477
Selvaduray, G.	III-2119	Tembulkar, J. M.	II-979	Wong, H. L.	I-553
Selvaduray, G.	III-2413	Thewalt, C. R.	II-1563	Woo, G.	I-173
Shah, H. C.	II-931	Thiel, Jr., C. C.	III-1873	Wood, R. M.	I-173
Shah, H. C.	II-1063	Thurston, H. M.	II-931	Wood, S. L.	II-1263
Shah, S. P.	II-1529	Tierney, K. J.	III-2401	Wright, W. B.	III-2061
Shakal, A. F.	III-1639	Topi, J. E.	I-257	Wyllie, L. A.	II-1215
Sharpe, R. L.	II-1039	Torkamani, A. M.	II-835		
Sharpe, R. L.	III-2017	Truman, K. Z.	III-1731	Yamada, Y.	II-955
Shen, C. K.	I-611	Turkstra, C. J.	II-919	Yang, C. Y.	I-765
Shepherd, R.	III-1947	Tyrrell, J. V.	III-1863	Yang, H. W.	I-611
Shimada, S.	II-955			Yost, R. E.	I-703
Shing, P. B.	II-1541	Usami, T.	I-463	Young, C. T.	I-83
Shinozuka, M.	III-2191			Youngs, R. R.	I-301
Siller, T. J.	I-599	Vagliente, V. N.	III-2283	Yuan, A.	I-43
Singhal, A. C.	III-2249	Vaidya, N. R.	I-623	Yun, H. D.	III-2179
Sivakumaran, K. S.	II-1051	Vaidya, N. R.	I-683		
Smietana, E. A.	III-2261	Vallenilla, C. R.	II-1463	Zaman, M. M.	I-541
Snyder, G. M.	III-1887	Van Orden, R. C.	III-1887	Zeris, C. A.	II-847
Somerville, M. R.	I-587	Von Thun, J.	I-405	Zeris, C. A.	II-1427
Somerville, P.	I-117			Zsutty, T.	III-1699
Soong, T. T.	III-1971	Wagner, M. T.	II-1215	Zullo, E. G.	I-333
Sotoudeh, V.	II-1063	Wang, L. R. L.	III-2293	Zurflueh, E. G.	I-77
Soydemir, C.	I-565	Wang, Z.	I-611	Zuroff, M. S.	III-2249
Spearman, E. L.	I-695	Way, D.	III-1911		
Statton, C. T.	I-209	Weems, R. E.	I-3		
Stepp, J. C.	I-293	Weems, R. E.	I-197		
Stiemer, S. F.	III-1923	Wen, C. Y.	III-1899		
Stone, S. D.	I-695	Wen, R. K.	III-2049		
Sudarbo, H.	II-811	Wen Y. K.	III-1721		
Sugano, T.	I-463	Werner, S. D.	I-393		
Sun, W. J.	II-1075	Werner, S. D.	III-2203		
Sunna, H.	III-2143	Wesnousky, S. G.	I-185		
Suprenant, B. A.	III-2271	White, R. N.	I-25		
Suzuki, S.	I-149	White, R. N.	II-871		
Swan, S. W.	III-2305	Whorton, R. B.	I-587		